Optical Modulation

Advanced Techniques and Applications in Transmission Systems and Networks

OPTICAL SCIENCE AND ENGINEERING

Founding Editor
Brian J. Thompson
University of Rochester
Rochester, New York

RECENTLY PUBLISHED

*Please visit our website **www.crcpress.com** for a full list of titles*

Optical Modulation
Advanced Techniques and Applications in Transmission Systems and Networks

Le Nguyen Binh

CRC Press
Taylor & Francis Group
Boca Raton London New York

CRC Press is an imprint of the
Taylor & Francis Group, an **informa** business

CRC Press
Taylor & Francis Group
6000 Broken Sound Parkway NW, Suite 300
Boca Raton, FL 33487-2742

First issued in paperback 2019

© 2018 by Taylor & Francis Group, LLC
CRC Press is an imprint of Taylor & Francis Group, an Informa business

No claim to original U.S. Government works

ISBN-13: 978-1-4987-4523-9 (hbk)
ISBN-13: 978-0-367-87504-6 (pbk)

In the memory of my father

Contents

SECTION II

SECTION III

Preface

Since the invention of the laser in 1961, the modulation of lightwaves has become the principal subjects of several lightwave systems, especially in optical guided structures for fiber optic communications systems and networks. It is even more important today with the emerging connection of the world and things: the Internet of Things and 5G mobile networks, the demand for bandwidth, ultra-high-speed minimum latency, and optical to wireless convergence. Optical networking is extremely essential to carry the enormous capacity of information in an intelligent, economical, and energy-efficient way. The modulation of lightwaves, as always, plays a critical role in the transmission of information via optical pipelines.

Integrated optics techniques and technology were created more than 30 years ago, and it describes a family of technologies where light-guided structures are integrated on planar substrates. The concept corresponds to electronic integration, where lithographical tools are used to create structures in the micron range with sub-micron precision. This term is commonly known as integrated photonics and photonic circuit technology (PIC) and thence electronic PIC (e-PIC), which is the ultimate convergent integration of electronic integrated technology and photonic integrated circuits. The modulation of guided lightwaves in these PIC structures is simple at a low voltage level, and thus within the capability of broadband microwave frequency devices.

Integrated optical modulators were then developed extensively in semiconductor heterostructures for lasers, amplifiers, electro-absorption modulators, anisotropic crystals as substrate for high-speed modulators, mainly $LiNbO_3$ by Ti diffusion or ion exchange, and electro-optic polymeric waveguides. These external modulators allow practical transmitters in which the linewidth of lasers can be preserved. Hence, both coherent and non-coherent receiver sub-systems can be realized. The external modulators have dominated the manufacturing of transmitters for a few decades due to overcoming the loss by optical amplifiers.

The baud rates of optical systems have been progressively increased from 2.5 Gbps to 10 Gbps, and then 28 GBd of 100 Gbps via polarization multiplexing and quadrature amplitude modulation (QAM) of multi-level. Currently, the bit rate can reach 400 Gbps in four wavelength lanes. Even 1.0 Tbps can be reached with optical injection locking lasers of the slave laser, which can be modulated to an effective 3 dB bandwidth to 100 GHz and higher.

Data center networking and flattening of traditional telecom networks has placed tremendous pressure on reducing the costs of the transmission equipment and reducing the latency in switching and routing. Hence, Si-based optical integrated modulators have emerged as the main contender medium for integration of electronics and photonics in CMOS and bi-CMOS platforms.

This book focuses on optical modulation techniques and external modulators, as well as direct modulation of lasers, which offer ultra-broadband signaling for the current and next generation of optical transmission systems over core, metro, or access networks. The contents can be divided into the following three sections.

Section I on direct modulation and laser generation, as well as optical injection locking master-slave direct modulation.

Section II on external modulators and modulation formats in association with optical receivers employing coherent and/or non-coherent technique, including transmission performance evaluation. This section also deals with digital signal processing technique to overcome the problems of frequency offset between carriers and local oscillator, equalization of fiber chromatic dispersion, polarization tracking of the states of multiplexed channels, non-linear impairments, etc.

Section III gives a number of basic principles of modulation, integrated devices, and fibers, which are necessary for readers who may want to strengthen the basic principles related to optical transmission.

I thank Huawei for allowing me to address the importance of optical modulation in optical transmission systems. My experiences over the last 6 years at Huawei produced insights into transmission equipment and demonstrations over installed fiber transmission lines, and have strengthened my understanding of optical modulations. I thank my colleagues—Mr Sun Chun, Dr Xie Changsong, Dr Nebojsa Stonajovic, Dr Fotini Karinou, Dr Mike Zhang Qiang, Dr Thomas Wang Tao, and Dr Mao BangNing—for technical exchanges in several experimental platforms. My sincere thanks go to Dr Thomas Lee, Vice President of SHF AG (Berlin), for several fruitful exchanges on signals and systems, and to the technical teams at Keysight Inc., Tektronix Inc., and many collegial company partners for collaborations on measurement sciences sessions and discussions on components, modules, system characterization, and performance evaluations.

Finally, this book is dedicated to my mother, Mrs Nguyen Thi Huong, who has dedicated her life to encouraging me to excel at practical and academic learning. My wife, Phuong Nguyen, and my son, Lam, have been very understanding concerning my long and quiet times of writing in my home office.

Le Nguyen Binh
Schwabing, Munich, Germany

MATLAB® and Simulink® are the registered trademark of The MathWorks, Inc. For product information, please contact:

The MathWorks, Inc.
3 Apple Hill Drive
Natick, MA 01760-2098 USA Tel: 508 647 7000
Fax: 508-647-7001
E-mail: info@mathworks.com
Web: www.mathworks.com

Author

Le Nguyen Binh earned BE (Summa Cum Laude), PhD, and Dr Eng in electronic engineering and integrated photonics, all from the University of Western Australia, Nedlands, Western Australia. He then joined the Department of Electrical Engineering of Monash University as reader after a stint three-year period with CSIRO (Commonwealth Scientific and Industrial Research Organisation) Australia (Canberra) as a research scientist conducting research on super parallel computing systems.

He has worked as principal research engineer/scientist for Siemens AG Central Research Laboratories in Munich and the Advanced Technology Centre of Nortel Networks in Harlow, UK. He was appointed as professorial fellow at the Christian Albretchs University of Kiel, Deutschland and the Tan-Chin Tuan professorial fellow of Nanyang Technological University, Singapore. He is currently acting as adjunct professor with a number of universities around the world.

Besides integrated optics and photonics, he has worked on several major advanced world-class projects, especially the Tb/s and 100 Gb/s dense wavelength division multiplexing (DWDM) non-coherent and coherent optical transmission systems for long haul, metro and access networks employing both direct and coherent detection techniques under dispersion-compensating module (DCM)-compensated and ultra-long non-DCM compensated optical transmission line. He worked at the Department of Optical Communications of Siemens AG Central Research Laboratories in Munich and the Advanced Technology Centre of Nortel Networks in Harlow, UK.

Professor Dr Eng Binh has authored and coauthored more than 300 papers in leading ISI (Institute for Scientific Information) journals and refereed conferences and 12 books in the field of photonic signal processing and digital optical communications, all published by CRC Press, Taylor & Francis Group LLC of Boca Raton, Florida, USA (www. crcpress.com). His research interests are in the fields of integrated photonics, photonic signal processing, and advanced optical transmission systems and networks.

While in the university sector, he developed and read several courses in engineering such as fundamentals of electrical engineering, physical electronics, small signal electronics, large signal electronics, signals and systems, signal processing, digital systems, micro-computer systems, electromagnetism, wireless and guided wave electromagnetics, communications theory, coding and communications, optical communications, advanced optical fiber communications, advanced photonics, integrated photonics, and fiber optics. He led several course and curriculum developments in electrical and electronic engineering (EEE) and joint courses between physics and EEE.

From January 2011 he has been appointed technical director, the European Research Center of Huawei Technologies in Munich, Deutschland. His current research responsibilities/interests include Multi-Peta-bps optical transmission systems and networks, Si-photonics and integrated nano-photonic and nano-electronics, ultra-high-speed optical modulation techniques and associate methods to mitigate non-linear impairments, convergence of wireless and optical technologies in 5G networks, novel techniques for non-linear equalization using high-order spectral methods and non-linear FFT. He has received four Gold Medal Awards by Huawei Technologies for these research advances.

Professor Dr Eng Binh chaired (1995–2005) the Commission D on electronics and photonics of the National Committee of Radio Sciences of the Australian Academy of Sciences. He currently contributes as the editor of CRC Press, *Series on Optics and Photonics*, and associate editor of *SPIE Journal of Optical Engineering*.

Le Binh is an alumnus of the Phan Chu Trinh and Phan Boi Chau high schools of Phan Thiet Vietnam.

1 Introduction

Historically, since the proposed and analytical study of dielectric waveguides by Kao and Hockham [1] more than 50 years ago, the demonstration of the guiding of lightwaves through an optical circular waveguides fiber [2,3]. Since then, advances in research and development of several aspects of optical fibers, opto-electronics components, and sub-systems, as well as transmission systems and networks based on optical fibers to-date have influenced nonstop explorations and exploitation of such guiding phenomena in the near-infrared spectral windows. The information capacity has also been reaching the Shannon's limit. The convergence of digital processing systems and optical transmission and detection by both coherence and non-coherence techniques has significantly pushed the optical systems to these limits.

1.1 EVOLUTION OF OPTICAL TRANSMISSION SYSTEMS AND NETWORKS: MODULATION, AMPLIFICATION, AND DISPERSION

In the following, we give an outline of the development phases of such optical communications systems with some emphases on the modulation of lightwaves. The developments are categorized into a number of phases according to the significant techniques that created breakthroughs in optical technology in optical fiber communications.

Dr C. Kao conducted (1966) an experiment on waveguiding in dielectric waveguide by laser beam in the former Standard Telecommunications Cables Co. Ltd. of Harlow, England, later as Advanced Technology Center of Nortel Networks.

Corning scientists Dr P. Schultz, Dr D. Keck, and Dr R. Maurer invented the first low-loss optical fiber in 1970. (Extracted from Corning website [D. Keck et al., IV Method of producing optical waveguide fibers, U.S. Patent 3,711,262, 1973-01].)

1.1.1 IN THE BEGINNING THERE WAS LIGHT

Stage 1: Invention of electronic transistor and amplification, invention of laser, and invention of optical dielectric wave guides and optical fibers.

Stage 2: Dielectric waveguide for optical frequency waves. Demonstration of optical guided waves. Multi-mode and single mode. Deployment of multi-mode optical systems—short distance. 1981 SMF fiber systems 40 km span length begins. Integrated optics starting with the whole paper by Miller and Marcatilli, and then Tien [4–6].

1.1.2   Weakly Guiding Phenomena and SMF Non-Coherent and Coherent Systems

Stage 3: Single-mode optical fibers and weakly guiding phenomena are used to explain the guiding phenomena for long-distance transmission with minimum loss, hence the design and manufacturing of single-mode optical fibers (1980s). Thence, standard single-mode fibers, dispersion shifted fibers (DSF) and nonzero dispersion shifted fibers (in 1990s), and then dispersion managed fibers.

Naturally, the multi-mode guiding waveguides or MMF (multi-mode fibers) face the problems of multi-path interference and, hence, dispersion due to delay time differences between the guided modes or rays.* Furthermore, the interferences of the modes due to the phase differences while propagating through the guided medium MMF create the fluctuation of the intensity of the receiving signals. Thus, high dynamic range opto-electronic receivers must be employed. Therefore, in the late 1970s SMF (single-mode optical fibers) were intensively researched and fabricated.

Stage 4: External modulation and SMF transmission systems: span length and loss-limited or dispersion-limited transmission. Given that the SMF had been developed using the weakly guiding phenomena, the SMF offers the possibility of the preservation of the polarization and coherence of lightwaves if the narrow linewidth of the laser is preserved. This is possible only if the laser line-width is not broadened by modulation. In the early 1980s, external modulators were thought to offer broadband property and preservation of the laser linewidth. The external modulators have thus been extensively developed, especially the uniaxial crystal-based type, such as the $LiNbO_3$ of specific orientation X- or Z-cut to use its efficient electro-optic coefficients r_{51}. However, the coupling loss between fiber and channel diffused waveguide of the $LiNbO_3$ modulator was a bit too high at about 2.2 dB, mainly due to waveguide spot-size mismatching with respect to that of fiber. These losses were overcome by the invention of the fiber amplifiers near the end of the 1989. Since then, external modulators, dominated by Ti: diffused type $LiNbO_3$, to date with 35 GHz, have been extensively exploited for long haul core transmission systems.

Stage 5: Non-coherent and coherent. Solving the attenuation problem by coherent transmission with LO boosting the signal energy so as to extend the repeater distance more than 40 km. Facing problems with coherent receivers with frequency offset, linewidth of laser, polarization rotation and matching, integrated optical components for polarization matching, and phase shift for IQ components. Optical modulation via direct modulation of the laser driving current and via external optical modulators for low voltage, high frequency, and preservation of the linewidth of the lightwaves so that coherent transmission can be implemented.

1.1.3   Optical Amplification: No-Longer Loss Limited But Dispersion, External Modulation Emerges

Stage 6: Loss-limited and dispersion-limited transmission: Optical amplification and dispersion managed transmission.

Stage 7: Optical amplifiers. Losses overcome for transmission spans and external modulators, especially the $LiNbO_3$. Dispersion-managed long-haul transmission systems with multi-Tbps.

Stage 8: Tbps dispersion managed transmission systems.

1.1.4   Modern DSP-Based Optical Transmission, Advanced Modulation Formats, Electronic—Photonic Integrated Circuit (e-PIC) Technology

Stage 9: DSP-based coherent reception. Modulation formats.

Digital signal processors have been quickly developed since the first processor, Intel Z8080, 8-bit bus and central processing unit (CPU) integrated with input–output, math processors and then

---

* Note: in MMF the number of guided modes can reach more than 600/700.

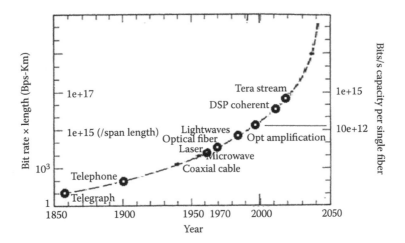

FIGURE 1.1 Transmission capacity over last 2 centuries. (Adapted from G. P. Agrawal, *Fiber-Optic Communication Systems*, 3rd ed., New York: John Wiley & Sons, 2002.)

memory capacity, since the 1970s. Development of optical fiber technology began around that same time. Over the years, these DSP chips have progressed tremendously for wireless networking. Current DSP chips can operate at a very high sampling rate and large bandwidth, which is sufficient to process sampled events of wideband electrical signals derived from the received optical signals to electrical domain. As a result, there has been an explosion of several processing algorithms to handle sampled signals of optical receivers. Coherent techniques for optical transmission systems have been made possible to overcome several serious impairments of the first coherent optical systems developed in the 1980s, such as the phase and frequency differences between the carrier and that of the local oscillator (LO).

Stage 10: Compact and ultra-high density optical transmission systems. Non-coherent and coherent transmission for metro-access and long-haul core networks. Short-distance high-capacity transmission for intra- and interdata center networking. Evolution of data center centric networking and flattening of traditional optical networks to meet challenges of DC-networking. Big data, multi-Tera-bps streaming, Exa-bps systems, and 5G networking—convergence between optical and wireless networks. DC-networking challenges (Figures 1.1 and 1.2).

1.2 TERA-STREAMS, CORE, METRO-CORE, METRO-ACCESS TECHNOLOGIES, AND MODULATION SCHEMES

Stage 11: Tbps/channel and 100 Tbps/line, Multi-Peta-bps networks: Superchannels whose total aggregate bit rate reaches Tera-bps have been demonstrated over installed multi-span optical amplified link (see Chapter 9) in which the bandwidth saving can be done by pulse shaping. Recently, we have proposed that a baud (Bd) rate of at least 112 GBd (and even 224 GBd) can be generated by direct modulation of optical injection locking laser (OILL) (see Chapter 2). Furthermore, the OILL can exhibit single-frequency and narrow linewidth as compared with free-running laser. Thence, coherent transmission and reception with polarization multiplexing can be achieved by combining two OILLs whose phases are different by pi/2. The structure of this PDM-QAM-Mary can be seen in Figure 1.3. Thus, the aggregate bit rate per wavelength channel can reach >1.0 TBps $(112\,\text{Gb} \times 2$ (polarizations) $\times 2 \times N$ bits/symbols with $M = N^2$). This technology will emerge in the near future.

Stage 12: Optical switching optical networks: Now the coherent and non-coherent transmission systems have reached the bit rates of 100–400G and Tera-Gbps. Thus, the bit or baud period becomes

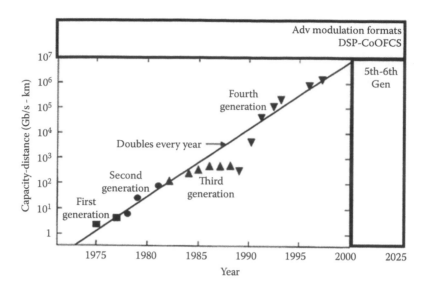

FIGURE 1.2 Technologically factors to increase capacity in optical communication systems. (Adapted from G. P. Agrawal, *Fiber-Optic Communication Systems*, 3rd ed., New York: John Wiley & Sons, 2002.)

shorter and shorter, thence any delay, the latency, longer than 1.0 ps would become very critical. The ION, particularly when high-speed channels are to be switched and routed in electronic domain, will suffer significant delay almost unacceptable to the optical domain. Thus, optical switching has now become the crucial technology for near future optical networks [8]. WDM optical switches have been demonstrated, as well as massive 384 × 384 optical spatial switching matrices.

Stage 13: The fourth revolution of information telecommunication by "networking and connecting every things." Emergence of 5G mobile networks and demands on multi-Gbps by MIMO antenna via Cloud RAN and Cloud-based data centers in distributed or centric data center networking. Internet of Things (IoT) as shown in Figure 1.4, require massive capacity interconnections of users and networked things, including sensor networks. Thus, the flattening of traditional carrier telecom networks to future cloud-based networking. Optical networking must evolve to meet new demands and challenges toward the year 2030. Si integrated photonics can be the platform for high-speed optical transponder and optical interconnects to facilitate low latency and ultra-high-speed modulation in very compact packaged modules CFP, CFP2, and CFP4.

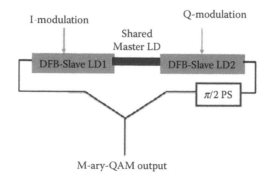

FIGURE 1.3 Schematic representation of a M-QAM transmitter employing Master-slave-slave distributed feedback lasers and a mater laser in integrated form.

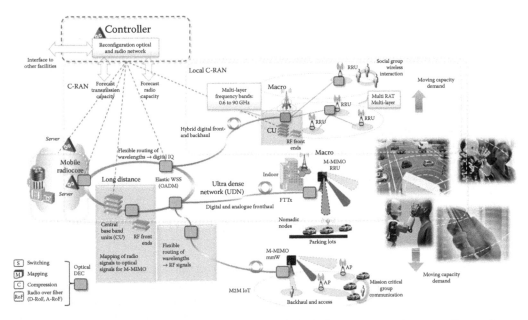

FIGURE 1.4 Global information optical-wireless networks including core network, metropolitan and access networks for 5G, ultra-high-speed enterprise networks and Internet of Things.

1.3   INTEGRATED OPTICS AND INTEGRATED PHOTONICS

The term "integrated optics" was created more than 30 years ago and describes a family of technologies where light-guiding structures are integrated on planar substrates. The concept corresponds to electronic integration, where lithographical tools are used to create structures in the micron range with submicron precision.

Integrated optics is a technology aimed at constructing the so-called integrated optical devices, photonic integrated circuits, or planar lightwave circuits (PLC), and integrating several optical components which are combined to fulfill a number of complex functions. Such functional components can be multiplexing and demultiplexing, modulators, switching, amplifiers, lasers and photodetectors, etc. They can, for example, be fabricated on the surface of some polycrystalline material (such as silicon, silica, or $LiNbO_3$) or anisotropic material platform. These functional devices are implemented in waveguiding structures where the width can possibly be a few micrometers or less than 1 micrometer, thus the term nano-photonics, such as in silicon over insulator (SOI).

The original inspiration for integrated optics came from the technology of electronic integrated circuits, which has shown rapid development over several decades and has led to amazing achievements, such as complex and powerful microprocessors containing many millions of transistors, specialized signal processors, and computer memory chips with huge data storage capacity. Unfortunately, integrated optics has not been able to match the progress of microelectronics in terms of the complexity of possible devices. This results from a number of technical limitations.

While electronic circuits can contain extremely small wires, optical components need to be interconnected via waveguides, the dimensions of which usually cannot be much smaller than the wavelength, and which often cannot tolerate very sharp bends. (This limitation might be eliminated by using waveguides with very high index contrast, e.g., nanofibers or photonic bandgap waveguides.) Optical connections, for example, between waveguides, and couplers are significantly more critical than electrical connections unless density of the devices is high, then the limitation will be placed on electrical connections. Waveguides, device connections, and passive optical components exhibit optical losses, which often need to be compensated with optical amplifiers. These are larger and more complex than electronic amplifiers based on transistors.

Some types of optical components can hardly be miniaturized, for example, lithium niobate-based devices [9–11]. In contrast, SOI devices can be highly compact and integrable with CMOS linear and digital VLSI integrated electronic systems.*

For these reasons, integrated optical circuits have not reached the complexity of electronic integrated circuits. However, devices of moderate complexity can still be useful, for example for optical fiber communications, where they can host multiple data transmitters and/or receivers, consisting of distributed feedback lasers, optical modulators, photodiodes, and optical filters (e.g., in the form of arrayed waveguide gratings). Recently, new hope for a powerful and cost-effective integrated optical technology has arisen from developments in silicon photonics [12–15].

Recently, demonstrations and commercial e-PIC modules have been announced and available for 100 G over 2 and 10 and 80 km [16,17].

1.4 DATA CENTER ECCENTRIC NETWORKING AND FLATTENED TRADITIONAL TELECOM NETWORKS

Currently, optical communications offer a significant improvement in energy performance and data transfer speed within and outside data centers (DC), and specialized data centers that are driven by security or speed of data access need components that will support those characteristics. This is particularly important to the financial sector, especially for high-speed trading. Data centers are the next big market opportunity for optical communications companies and firms, and the European Union offers the key elements necessary to be a key global player in the DC equipment market—major end users, a strong research base, a billion-pound specialist local market, core supplier capability, and capacity for substantial supplier growth. However, the market has failed to generate the necessary links between these core elements to create a market-driven development and uptake of EU technology.

"One of the key problems the report has highlighted is the disconnection of the supply chain: the end-user is unaware of the technological advances, and the technology providers having no connections to the end-user," as commented by Knowledge Transfer Network. "KTN can help those sectors together across the supply chain, leading to collaborations in R&D and providing a competitive advantage for UK companies across the supply chain." Currently, EU data centers use as much electricity as 3.5 million homes combined, indeed a huge amount of energy demand. If the growth of DC continues at the current pace, DC would consume the entire electricity supply by 2030, a bigger problem than electric cars. Working with photonics technology across global industries to find energy- and cost-efficient solutions is vital if data-oriented information societies want to sustain this growth. Thus, the challenges are not on the available capacity but on the energy and safety of such information networks.

Optical modulation techniques are well established for transmission systems and efficient for highly integral long-distance and metro- and access-networking systems. It is expected that the optical switching and routing are critical in DC and DC networking to save energy and reduce their demands on energy. Such optical or photonic switching matrices for very large-scale DC interconnects are not yet available. The switching is also considered as a part of the modulation must be highly efficient with appropriate materials so that lowered switching voltage levels can be employed.

Si photonic technology is expected to offer large switching matrices in which the required switching voltages are sufficiently low, as well as the high-speed switching. However, the losses in propagation and coupling in Si waveguides may not be low enough for the demands on the OSNR. It is thus expected that optical amplification by Er:doped waveguide amplifiers will contribute to these integrated photonic switches [18–21].

---

* CMOS = Complementary metal-oxide-semiconductor; VLSI = Very-large-scale integration.

Alternatively, it is also highly possible that build-free space-switching matrices can offer such optical interconnections in very dense scenarios [22–25].

1.5 DIGITAL OPTICAL COMMUNICATIONS AND TRANSMISSION SYSTEMS: CHALLENGING ISSUES

Starting from the proposed dielectric waveguides by Kao and Hockham [26] in 1966, the first research phase attracted intensive interests around the early 1970s in demonstration of fiber optics. Optical communications have tremendously progressed over the last three decades. The first-generation lightwave systems were commercially deployed in 1983 and operated in the first wavelength window of 800 nm over MMF at transmission bit rates of up to 45 Mb/s [27,28]. After the introduction of ITU-G652 Standard Single Mode Fibers (SSMF) in the late 1970s [29], the second generation of lightwave transmission systems became available in the early 1980s [30,31]. The operating wavelengths were shifted to the second window of 1300 nm, which offers much lower attenuation for silica-based optical fibers compared to the previous 800 nm region. The chromatic dispersion (CD) factor is almost zero. These second-generation systems could operate at bit rates of up to 1.7 Gb/s and have a repeater-less transmission distance of about 50 km [7]. Further research and engineering efforts were also pushed for the improvement of the receiver sensitivity by coherent detection techniques and the repeaterless distance has reached 60 km in installed systems with a bit rate of 2.5 Gb/s. Optical fiber communications then evolved to third-generation transmission systems, which utilized the lowest attenuation 1550 nm wavelength window and operated up to 2.5 Gb/s bit rate. These systems were commercially available in 1990 with a repeater spacing of around 60–70 km [32]. At this stage, the generation of optical signals was mainly based on direct modulation of the semiconductor laser source and either direct detection. Since the invention of erbium-doped fiber amplifiers (EDFA) in the early 1990s [33–35], lightwave systems have rapidly evolved to wavelength division multiplexing (WDM) and shortly after that, dense WDM (DWDM) optically amplified transmission systems that are capable of transmitting multiple 10 Gb/s channels. This is due to the fact that the loss is no longer a major issue for external optical modulators that normally suffer an insertion loss of at least 3 dB. These modulators allow the preservation of the narrow linewidth of distributed feedback lasers (DFB). These high-speed and high-capacity systems extensively exploited the external modulation in their optical transmitters. The present optical transmission systems are considered to be the fifth generation, having a transmission capacity of a few terra bytes per second.

Coherent detection, homodyne or heterodyne, was the focus of extensive research and developments during the 1980s and early 1990s [36–41], and was the main detection technique in the first three generations of lightwave transmission systems. At that time, the main motivation for the development of coherent optical systems was to improve the receiver sensitivity, commonly by 3–6 dB [36,40]. The repeaterless transmission distance was able to be extended to more than 60 km of SSMF (with 0.2 dB/km attenuation factor). However, coherent optical systems suffer severe performance degradation due to fiber dispersion impairments. In addition, the phase coherence for lightwave carriers of the laser source and the local laser oscillator was very difficult to maintain. On the contrary, incoherent detection technique minimizes the linewidth obstacles of the laser source, as well as the local laser oscillator, and thus, relaxes the requirement of the phase coherence. Moreover, incoherent detection mitigates the problem of polarization control in the mixing of transmitted lightwaves and the local laser oscillator at multi-THz optical frequency range. The invention of EDFAs that are capable of producing optical gains of 20 dB and above, has also greatly contributed to the progress of incoherent digital photonic transmission systems to-date.

Recent years have shown a huge increase in demand for broadband communications driven mainly by the rapid growth of multi-media services, peer-to-peer networks, and IP streaming services, in particular the IP TV. It is most likely that such tremendous growth will continue in the coming years. This is the main driving force for local and global telecommunications service carriers to develop high-performance and high-capacity next-generation optical networks. The overall capacity of

WDM or DWDM optical systems can be boosted either by increasing the base transmission bit rate of each optical channel, multiplexing more channels in a DWDM system, or, preferably, by combining both of these schemes. However, while implementing these schemes, optical transmission systems encounter a number of challenging issues that are outlined in the following paragraphs.

Current 100 Gb/s transmission systems employ intensity modulation (IM), also known as on–off keying (OOK) and utilize non-return-to-zero (NRZ) pulse shapes by EML (in-line external modulator laser) or VSCEL (vertical cavity surface emitting laser) of four lanes, each carrying 28 Gb/s. The term OOK can also be used interchangeably with amplitude shift keying (ASK).* Moving toward high bit rate transmission such as 40 Gb/s, the performance of OOK photonic transmission systems is severely degraded due to fiber impairments, including fiber dispersion and fiber non-linearities.

The issues of dispersion-limited or attenuation-limited transmission distance were first raised in the early 1980s when SMF was commercially available, and the dispersion-shifted or dispersion-zero fibers were implemented. However, SSMF spectral windows of the O-band (1310 nm) and (C + L)-bands 1520–1650 nm are now seriously taken into consideration of the design of the transmission systems for data-center interconnect or access networks.

The fiber dispersion is classified into chromatic dispersion (CD) and polarization mode dispersion (PMD), causing an intersymbol interference (ISI) problem. On the other hand, severe deteriorations on the system performance due to fiber non-linearities result from high-power spectral components at the carrier and signal frequencies of OOK-modulated optical signals. It is also of concern that existing transmission networks are comprised of millions of kilometers of SSMF which have been installed for approximately two decades. These fibers do not have as advanced properties as state-of-the-art fibers used in recent laboratory "hero" experiments, and they have degraded after many years of use.

The total transmission capacity can be enhanced by increasing the number of multiplexed DWDM optical channels. This can be carried out by reducing the frequency spacing between these optical channels, for example, from 100 GHz down to 50 GHz, or even 25 and 12.5 GHz [14,15]. The reduction of the channel spacing also results in narrower bandwidths for the optical multiplexers (mux) and demultiplexers (demux). Passing through these narrowband optical filters, signal waveforms are distorted and optical channels suffer the problem of interchannel cross-talks. The narrowband filtering problems are getting more severe at high data bit rate, for example, 40 Gb/s, thus degrading the system performance significantly.

Together with the demand of boosting the total system capacity, another challenge for the service carriers is to find cost-effective solutions for the upgrading process. These cost-effective solutions should require minimum renovation to the existing photonic and electronic sub-systems, that is, the upgrading should only take place at the transmitting and receiving ends of an optical transmission link. Another possible cost-effective solution is to significantly extend the uncompensated reach of optical transmission links, that is, without using dispersion compensation fibers (DCF), thus reducing a considerable number of required inline EDFAs. This network configuration has recently received much interest from the photonic research community as well as service carriers.

Therefore, the principal motivations of this chapter are to describe the employment of digital communications in modern optical transmission technology. The fundamental principles of digital communications, both coherent and incoherent transmission and detection techniques, are described with focus on the technological development and limitations in the optical domain. The enabling technologies, research results, and demonstration in laboratory experimental platforms for the development of high-performance and high-capacity next-generation optical transmission systems impose significant challenges to the engineering of optical transmission systems in the near future and techniques for network monitoring.

---

* The OOK format simply implies the on–off states of the lightwaves where only the optical intensity is considered. On the other hand, the ASK format is a digital modulation technique representing the signals in the constellation diagram by both the amplitude and phase components.

1.6 MODULATION FORMATS AND OPTICAL SIGNALS GENERATION

Modulation is a process of facilitating the transfer of information over a medium. In optical communications, the process of converting information so that it can be successfully sent through the optical fiber is called optical modulation.

There are three basic types of digital modulation techniques. These are ASK, frequency shift keying (FSK), and phase shift keying (PSK), in which the parameter, the carrier whose amplitude, frequency, or phase are varied to represent the information that is to be sent. Digital modulation is a process of mapping such that the digital data of "1" and "0" or symbols of "1" and "0" that convert it into some aspect of the carrier, the amplitude and phase, and then transmit the carrier, the lightwave. The carrier is then remapped back to a near copy of the information data.

1.6.1 BINARY LEVEL

Modulation is a process that facilitates the transport of information over the medium. In this book, our medium is the optical-guided fiber and associated photonic components. In digital communications, there are three basic types of digital modulation techniques: ASK, PSK, and FSK. All these techniques vary a parameter of a sinusoidal carrier to represent the information "1" and "0."

In ASK, the amplitude of the lightwave carrier normally generated by a narrow linewidth laser source, is changed in response to the digital data and all else is kept fixed. That is, bit "1" is transmitted by the lightwave carrier of a particular amplitude. To transmit "0," the amplitude is changed, keeping the frequency unchanged as shown in Figure 1.5. NRZ or return-to-zero (RZ) can be assigned depending on the occupation of the state "1" during the timelength of a bit period. For RZ, normally only half of the bit period is occupied by the digital data.

In addition to NRZ and RZ formats, in optical communications the carrier can be suppressed under these formats so as to achieve non-return-to-zero carrier suppression (NRZ-CS) and return-to-zero carrier suppression (RZ-CS). This is normally generated by biasing the optical modulator in such a way that the carriers passing through the two parallel paths of an interferometric modulator are pi phase shift difference with each other. Thus, the carrier at the center frequency is suppressed by the sidebands of the modulated signals.

In PSK, the phase of the lightwave carrier is changed to represent the information. The phase in this context is the shift of angle at which the sinusoidal carrier starts. To transmit a "0," the phase would be shifted by pi and a "1" with no change of phase. The phase angle can be changed and take a value of a set of phases corresponding to the mapping of the symbols as shown in Figure 1.6.

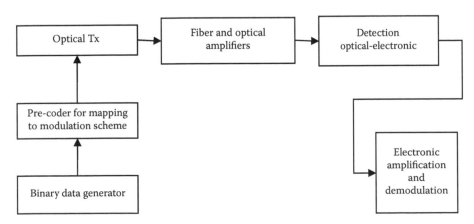

FIGURE 1.5 Schematic diagram of the modulation and the electronic detection and demodulation of an advanced modulation format optical communications system.

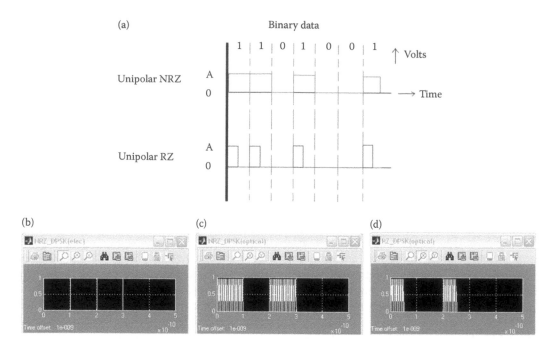

FIGURE 1.6 (a) NRZ and RZ pulse amplitude modulated formats for a sequence of {1 0 1 0 1 0 1 0 1 0 1 0} and (b) Generated ASK signals with carrier (not to scale)-data (c) and carrier modulated NRZ and NZ formats (d).

In FSK, the frequency of the carrier represents the digital information. One particular frequency is assigned to a "1" and another frequency is assigned to the "0," as shown in Figure 1.7. An FSK can be considered a continuous phase modulation, for example, the continuous phase modulation MSK (minimum phase shift keying) whose frequency separation f_d is selected such that the signals carried by these frequencies are orthogonal.

ASK can be combined with PSK to create a hybrid modulation scheme such as quadrature amplitude modulation (QAM) where the phase and amplitude of the carrier are changed at the same time. The carriers are expected to follow a similar pattern to that of the DPSK in Figure 1.6d, but with

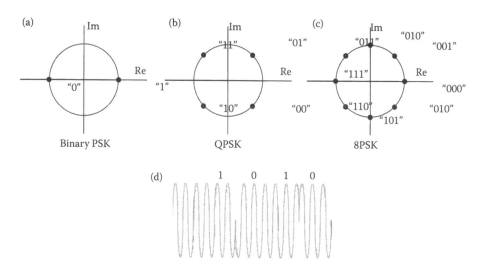

FIGURE 1.7 Signal-space constellation of discrete PM (a) binary PSK, (b) quadrature binary PSK, (c) 8-PSK, and (d) phase of the carrier under modulation with π phase shift of the BPSK at the edge of the pulse period.

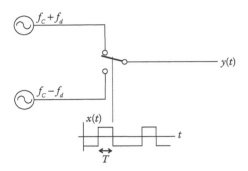

FIGURE 1.8 Schematic diagram of a FSK transmitter. f_d is the deviation frequency.

different frequencies of the carrier under the envelope of the bit "0" and "1." For MSK signals the carrier frequency is chirped up or down depending on the "0" or "1." That is, the phase of the carrier is continuously varied during the bit period, and the carrier frequencies of the bits are such that there is an orthogonality of the carriers and the signal envelope.

1.6.2 BINARY AND MULTI-LEVEL

Additional degrees of freedom for detection can be used to effectively enhance the capacity due to effective equivalence of the multi-level and symbol rate, thence the detection of the received optical signals. A widely used and matured detection scheme for optical signals is the direct detection in which the optical power $P = [E]^2$, the square of a complex optical field amplitude. The photodetector would not be able to distinguish between a "0" or "pi" phase shift of the carrier lightwave embedded within the pulse. The carrier phase can only be possibly extracted if, and only if, photonic processing was used to extract the phase at the front end of the receiver. Thus, a $+$ or $-$ field complexity would be seen as identical in the photodetector.

This ambiguity of the phase detection process would allow one to shape the optical spectra of optical signals to induce a modulation format more resilient to the distortion effects accumulated during the transmission process.

Formats making use of the tri-level are illustrated in Figures 1.8 and 1.9 and can be termed as pseudo-multi-level or tri-level or poly binary signals. These tri-level signals can be represented in terms of the phase or frequency of the lightwave carrier.

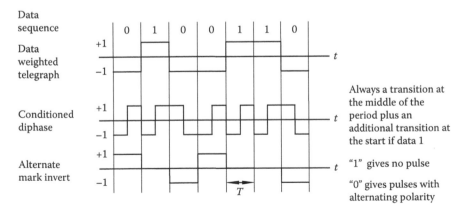

FIGURE 1.9 Illustration of pseudo-multi-level or polybinary baseband signals. Binary data sequence, weighted signals, diphase RZ, and alternate mark inversion formats.

The use of more than two symbols to encode a single bit of information, but transmit at the symbol rate B_s. Under optical transmission, the -1 and $+1$ can be coded in terms of the phase of the carrier, as there is no negative intensity representation unless the field of the lightwaves is used. The tri-level uses $\{+IEI, -IEI, $ and $0\}$ with its equivalent phase representation mapped to $\{0, IEI^2\}$ at the optical receiver, for example, the duo-binary format that would be described in a later chapter. A phase difference of "pi" and "0" between the three levels would minimize the pulse dispersion as they are propagating along the fiber due to the relative phase difference of pi, hence destructive interference of any pulse spreading due to dispersion.

This tri-level must not be mixed up with the truly multi-level signaling in which $\log_2 M$ bits are encoded on N-symbols, and then transmitted at a reduced rate $B_s/\log_2 N$. Both multi-level amplitude or amplitude-phase shift keying and DQPSK are multi-level optical modulation techniques.

1.6.3 IN-PHASE AND QUADRATURE PHASE CHANNELS

Another form of modulation that would enhance the capacity of the transmission is the use of the orthogonal channels in which the information can be coded into the in-phase and quadrature as shown in its polar or Cartesian coordinates as shown in Figure 1.6.

Quadrature phase shift keying (QPSK) is most commonly used in the differential phase modulation in which I and Q components are used extensively. It is an extension of the binary PSK signals, but with the phase change of only $\pi/2$ instead of π. Mathematically, the signal $s(t)$ can be expressed as

$$s(t) = A_c p_s(t) \cos\left(2\pi f_c t + \frac{2\pi i}{M}\right) \qquad (1.1)$$

where $p_s(t)$ is the pulse shaping of the data, M is the quantized level, or the total number of phase states of the modulation and I is the phase modulation index. The QPSK can be combined with ASK to generate quadrature amplitude modulation (QAM) where the phase and amplitude can be used to map a symbol of data information into one of the point on the signal space.

1.6.4 EXTERNAL OPTICAL MODULATION

External modulation is the essential technique for modulating the lightwaves so that its linewidth preserves its narrowness and only the sidebands of the modulation scheme dominate the spectral property of the generated passband characteristics. Figures 1.10 through 1.12 show the typical structure of optical transmitters for generation of NRZ and RZ optical signals. The laser is always switched on and its lightwaves are modulated via the electro-optic modulator using the principles of interferometric constructive and destructive interference to represent the ON and OFF states of the lightwaves.

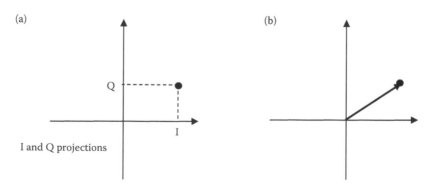

FIGURE 1.10   Signal vectors plotted in signal space (a) Cartesian coordinate and (b) polar form.

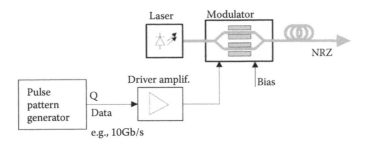

FIGURE 1.11 Generation of optical signals of format NRZ using an external modulator.

The RZ can be generated similarly, but with an additional optical modulator that would generate periodic optical pulses in which width is half of that of the bit period. The phase and frequency modulation can also be generated using these electro-optic modulators by biasing conditions and control of the amplitudes of the electrical pulses. These optical transmitters will be described in Chapter 2. We note here that the fiber that connects the two modulators of Figure 1.11 must be of polarization maintaining (PM) type, otherwise there would be polarization fluctuation and, hence, reduction of the coupling of the lightwave power one to the other.

The laser source would normally be a narrow linewidth laser that is turned on at all times to preserve its narrowness characteristics. The lightwaves are generated and coupled to the optical modulator via the pigtails of both devices. The modulator would be driven by data pulse sequence output of a bit pattern generator conditioned to appropriate driving level required by the V_π and the phase variation of the carrier if phase modulation is needed. When a modulation format is necessary, an electronic pre-coder is required to code the serial sequence to the appropriate coding. The pre-coder can be a differential coding, multi-level coding, or IFFT to generate orthogonal data sub-channels in case of the orthogonal frequency division multiplexing (OFDM).

The pulses of all modulation formats can take the form of NRZ or RZ. For RZ format, an additional optical modulator is required to generate or condition the "1" NRZ to RZ as shown in Figure 1.11. The second modulator can exchange its position with that of the other modulator without affecting the generation of the modulation formats. This second modulator is usually called the *pulse carver*.

It is noted that if the RZ modulator is biased such that the phase difference at the biasing condition is pi phase difference, then we would have a carrier suppression of the carriers at the central location of the generate spectra but the sidebands of the optical signals. The bandwidth of the modulator determines the rise time and fall time of edges of the pulse sequence shown in Figures 1.10 and 1.11. The details of the optical transmitters for different direct modulation formats are given in Chapters 3 and 4.

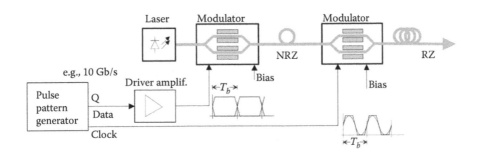

FIGURE 1.12 Generation of optical signals of format RZ.

1.6.5 ADVANCED MODULATION FORMATS

The above-described problems facing contemporary optical fiber communications can be effectively overcome by utilizing spectrally efficient transmission schemes via the implementation of advanced modulation formats. A number of modulation formats have recently been reported as alternatives for the OOK format, including RZ pulses in OOK/ASK systems, differential phase shift keying (DPSK) [42–44] and, more recently, minimum-shift keying (MSK) [22–29]. These formats are adopted into photonic communications from the knowledge of wire-line and wireless communications.

DPSK has received much attention over the last few years, particularly when it is combined with RZ pulses. The main advantages of RZ DPSK are (i) a 3-dB improvement in the receiver sensitivity over OOK format by using an optical balanced receiver [43,45] and (ii) high resilience to fiber non-linearities [46,47] such as intrachannel self-phase modulation (SPM) and interchannel cross phase modulation (XPM). Several experimental demonstrations of DPSK long-haul DWDM transmission systems for 10, 40, 100 Gb/s and higher bit rates have been reported recently [48–51]. However, there are few practical experiments addressing the performance of cost-effective 40 Gb/s DPSK—10 Gb/s OOK hybrid systems for gradually upgrading the existing installed SSMF transmission infrastructure [52,53]. In addition, the performance of 40 Gb/s DPSK for use in this hybrid transmission scheme has not been thoroughly studied. Therefore, one of the main contributions of this research is to prove the feasibility of overlaying 40 Gb/s DPSK channels on the existing 10 Gb/s network infrastructure for implementing the hybrid systems.

MSK format offers a spectrally efficient modulation scheme compared to the DPSK counterpart at the same bit rate. As a sub-set of continuous phase frequency shift keying (CPFSK), MSK possesses spectrally efficient attributes of the CPFSK family. The frequency deviation of MSK is equal to a quarter of the bit rate and this frequency deviation is also the minimum spacing to maintain the orthogonality between two FSK-modulated frequencies. On the other hand, MSK can also be considered a particular case of offset differential quadrature phase shift keying (ODQPSK) [54–57], which enables MSK to be represented by I and Q components on the signal constellation. The advantages of optical MSK format can be summarized as follows: (i) Compact spectrum, which is of particular interest for spectrally efficient and high-speed transmission systems. This also provides robustness to tight optical filtering; (ii) high suppression of spectral side lobes in the optical power spectrum compared to DPSK. The roll-off factor follows f^{-4}, rather than f^{-2}, as in the case of DPSK. This also reduces effects of interchannel cross-talks; (iii) no high power spectral spikes in the power spectrum, thus reducing fiber non-linear effects compared to OOK; (iv) as a sub-set of either CPFSK or ODQPSK, MSK can be detected incoherently based on the phase or frequency of the lightwave carrier, or coherently based on the popular I-Q detection structure; and (v) constant envelop property, which eases the measure of the average optical power.

Several studies have been conducted recently on the generation and direct detection of externally modulated optical MSK signals [58–61]. However, there is little research investigating the performance of externally modulated MSK formats for digital photonic transmission systems, particularly at high bit rates such as 40 Gb/s. Furthermore, if MSK can be combined with a multi-level modulation scheme, the transmission baud rate would be reduced in addition to the spectral efficiency of the MSK formats. This is of great interest for long-haul and metropolitan optical networks, and thus provides the main motivation for proposing the dual-level MSK modulation format in this research. In addition, the potential of optical dual-level MSK format transmission has yet to be explored. Therefore, another main contribution of this research is to provide comprehensive studies on the performance of MSK and dual-level MSK modulation formats for long-haul and metropolitan optical transmission systems.

1.6.6 INCOHERENT OPTICAL RECEIVERS

The modulation formats studied in this research, optical DPSK and MSK-based formats, can be demodulated incoherently by using an optical balanced receiver that employs a Mach–Zehnder delay interferometer (MZDI). In the case of optical DPSK format, MZDI is used to detect differentially coded phase information between every two consecutive symbols [45,62]. This detection is carried out in photonic domain, as it is beyond the speed of electrical domain, especially at very high bit rates of 40 Gb/s or above. The MZDI balanced receiver is also used for incoherent detection of optical MSK signals by also detecting the differential phase of MSK-modulated optical pulses. However, using the MZDI-based detection scheme, it is found that optical MSK provides a slight improvement for the CD tolerance over DPSK and OOK counterparts.

As a sub-set of the CPFSK family, MSK-modulated lightwaves can also be incoherently detected based on the principles of optical frequency discrimination. Thus, an optical frequency discrimination receiver (OFDR) employing dual narrowband optical filters and an optical delay line (ODL) is proposed in this research. This receiver scheme effectively mitigates CD-induced ISI effects and enables breakthrough CD tolerances for optical MSK transmission. In addition, the feasibility of this novel receiver is based on recent advances in the design of optical filters, in particular, the micro-ring resonator filters. Such optical filters have very narrow bandwidths, for example, less than 2 GHz (3-dB bandwidth), and they have been realized commercially by Little Optics. This research thus provides a comprehensive study on this OFDR scheme, from the operational principles to the analysis of the receiver design, and onto the performance of OFDR-based MSK optical transmission systems.

1.6.7 COHERENT OPTICAL RECEIVERS

Coherent detection and transmission techniques have been extensively exploited in the mid-1980s to extend the repeaterless distance a further 20–40 km of SSMF with an expected improvement of the receiver sensitivity of 10–20 dB, depending on the modulation format and receiver structure using phase or polarization diversity.

In general, a coherent receiver would operate on the beating of received optical signals and that of the field of a local laser oscillator. The beating optical signals are then detected by the photodetector with the phase of the carrier preserved that permits the detection of the phase of the carriers. Hence, the phase modulation and continuous phase or frequency modulation signals can be processed in the electronic domain. With the advancement of digital electronic processors, the processing of the received signals either in the IF or base band of the heterodyne and homodyne detection, respectively, can be processed to determine the phase of the modulated and transmitted signals. Coherent receivers for different modulation formats are described at appropriate sections of the chapters of this book, especially in Chapters 6 through 8.

Linear and non-linear equalization are possible, as they are well known in the field of signal processing. The principles of equalization with equalizers placed at the transmitter, or the receiver or sharing between the transmitter and receiver are critical for practical networks. Equalizers in hardware form are placed after the TIA and linear amplifier or automatic gain control state with its own clocking and decision circuits. However, for the soft digital equalizer, the equalization is implemented in the DSP (digital signal processors), which follows the ADC (analog-to-digital converter). The ADC follows an opto-electronic receiver.

The three types of coherent receivers, namely homodyne, heterodyne, and intradyne detection techniques are possible and dependent on the frequency difference of zero, intermediate frequency greater or smaller than the passband of the signals between the local oscillator laser and that of the signal carrier. With modern optically amplified fiber communications, broadband ASE noises always exist and under coherent detection, the beating between local laser source and ASE dominate the electronic noise of the receiver at the front end.

1.6.8 ELECTRONIC EQUALIZATION

Electronic equalizers have recently become one of the potential solutions for future high-performance optical transmission systems. The Si–Ge technological development has enhanced the electron mobility and, hence, shortened rise and fall time of the pulse propagation, thus increasing the processing speed. Sampling rate can now reach several Giga-samples/s which enables the processing of 10 Gb/s bit rate data channel without any difficulty. Hence, it is very probable that the electronic processing and equalization can be implemented in real systems in the very near future.

The channel is single mode optical fibers of dispersive in the negative or positive factors and with some residual dispersion. Thus, the distortion of the pulse is purely phase distortion prior to the detection by the photodiode which follows the square law rule for direct detection case. On the other hand, for coherent detection, the beating between the local oscillator and the signal in the photodetection (PD) device would lead to the preservation of the phase and one could be considered a pure phase distortion. Again, in the case of direct detection after the square law detection, the phase distortion is then transferred to the amplitude distortion. In order to conduct the equalization process, it is important for us to know the impulse and step responses of the fiber channel $h_F(t)$ and $s_F(t)$. We will then give the fundamental aspects of equalization using feed forward equalization, decision feedback equalization with maximum mean square error (MMSE), or maximum likelihood sequence estimation of Viterbi algorithm (MLSE).

1.6.8.1 Feed-Forward Equalization

Feed-forward equalization (FFE) is a linear equalizer which has been the most widely studied and a transversal filter structure would offer a linear processing of the signal prior to decision. The structure of a FFE transversal filter consists of a cascade delay of the input sample and at each delay the signal is tapped and multiplied with a coefficient. These tapped signals whose delay tap time is the bit period are then summed to give the output sample. The coefficients of the transversal filter must take values that matched with the channel so the convolution of the channel impulse response and that of the filter result to a unity so as to achieve complete equalized pulse sequence at the output. Figure 1.13a and b shows the linear equalization scheme that uses either feed-forward or feedback equalization, with the difference between these two schemes being that the tapped signals are either at the output of the transversal filter or at the output of the feedback that minimize the input sequence.

On the other hand, a decision feedback equalizer (DFE) differs from that of the linear equalizer with a decision detector that would determine the signal amplitude required for feedback to the difference error at the input as shown in Figure 1.14.

1.6.8.2 Decision Feedback Equalization

The feed-forward equalization (FFE) method is based on the use of a linear filter with adjustable coefficients. The equalization method that exploits the use of previously detected symbols to suppress the ISI in the present symbol being detected, is termed as decision-feedback equalization.

The DFE is illustrated in Figure 1.14. It consists of m coefficients and m delay taps. Each tap spacing equals the bit duration T. From Figure 1.14 we can see that the received signal sequence goes through the forward filter first. After making decisions on previously detected symbols, the feedback filter provides information from the previously detected symbols for present estimating.

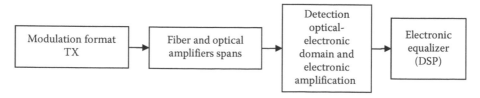

FIGURE 1.13 Schematic diagram of the location of the electronic equalizer at the receiver of an advanced optical communications system. DSP = digital signal processor.

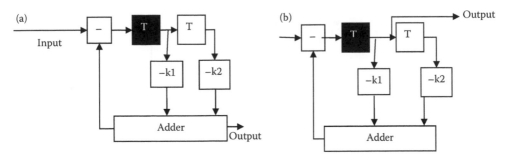

FIGURE 1.14   Linear (a) feed forward equalization (transversal equalization) and (b) feedback equalization scheme.

1.6.8.3   Minimum Mean Square Error Equalization and Maximum Likelihood Sequence Estimation

Consider next the general case where the linear equalizer is adjusted to minimize the mean square error due to both ISI and noise. This is called the MMSE (minimum mean square error equalization).

Amongst the electronic equalization techniques is maximum likelihood sequence estimation (MLSE) which can be implemented effectively with the Viterbi algorithm has attracted considerable research interest [45–50]. However, most of the studies on MLSE equalizers have focused on either ASK or DPSK formats [47,63]. Apart from a coauthored paper reported recently [28], there has not been any study on the performance of MLSE equalizers for optical MSK transmission, especially when OFDR is used as the detection scheme. The performance of OFDR-based MSK optical transmission systems is significantly enhanced with the incorporation of post-detection MLSE electronic equalizers as the ISI problem caused by either fiber dispersion impairments or tight optical filtering effects are effectively mitigated. Therefore, this decision process considered as an equalizer is included as a case study in a chapter of this book to comprehensively investigate the performance of MLSE equalizers for OFDR-based MSK optical transmission systems.

1.6.8.4   Volterra Filtering

In this sub-section, the Volterra digital filters are described. Aside from the back propagation to reverse the non-linear propagation impairments in the digital domain, the Volterra filter (VF) is the new and emerging effective technique for non-linear compensations. The mathematical processes of Volterra series have been extensively presented [64,65].

For the DML-based IM/DD transmission operating at O-band, the effect of CD can be considered to be negligible. Therefore, the transmission impairments mainly come from (a) limited bandwidth of the optoelectronic devices, (b) signal-to-signal beating noise (SSBN), (c) fiber non-linearities, and (d) fiber attenuation. The input–output relationship of third-order VF is expressed as [66]

$$y(n) = \sum_{l_1=0}^{L_1=1} h_1(l_1)x(n - l_1) + \sum_{l_1=0}^{L_2=1}\sum_{l_2=0}^{l_1=1} h_2(l_1, l_2) \prod_{m=1}^{2} x(n - l_m)$$

$$+ \sum_{l_1=0}^{L_2=1}\sum_{l_2=0}^{l_1=1}\sum_{l_2=0}^{l_2} h_3(l_1, l_2, l_3) \prod_{m=1}^{3} x(n - l_m) + e(n)$$

(1.2)

where $x(n)$ is the nth sample of *the received signal, $y(n)$ is the nth sample of the output signal after VF equalizer, $h_i(.)$ is* the ith order Volterra kernel, L_i is the ith order memory length, and $e(n)$ is VF error. The number of three kernels are L_1, $L_2(L_2 + 1)$, $L_3(L_3 + 1)(L_3 + 2)/6$, respectively. Although *VF* with higher order or longer memory length can model the non-linear system more precisely, the

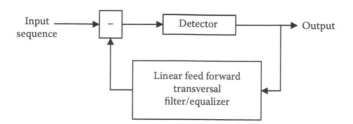

FIGURE 1.15   Schematic diagram of receiver using non-linear equalization by decision directed cancellation of ISI.

increment of memory length or system order will increase the number of kernels exponentially, leading to complicated calculations.

The first-order kernels of Volterra filter can be used to mitigate the linear impairments, because it is a FIR filter. The SSBN resulting from the square-law detection can be compensated by the second-order kernels of Volterra filter. Generally, the SSBN can be described as

$$v_D(n) = |E_{\lambda_0} + E_0(n)|^2 + 2\Re e(E_{\lambda_0} \cdot E_0) + |E_0(n)|^2 \tag{1.3}$$

where $v_D(n)$ is the nth sample of detected signal, E_{λ_0} and E_0 are the electric field component of the optical waves of the optical carrier and the SSBN, respectively. The filter can thus include the non-linear effects and filter out those higher-order kernels as shown in Figures 1.15 and 1.16, hence leaving the linear components of the samples of the received signals.

It might be possible to compensate of the optical non-linear impairment effects by using integrated optical device by non-linear optical waveguides [67,68] that not demonstrated so far.

1.7   ORGANIZATION OF THE CHAPTERS

The chapters are organized as follows:

> Chapter 1 introduces the challenges of the network evolution and demands on capacity leading to novel multi-level modulations in both non-coherent and coherent transmission techniques. The rest of the chapters can be classified into three sections:
>
> *Section I*: Chapters 2 and 3 give the basic principles of optical direct and the most basic external modulation. By direct modulation, the injection current to the gain cavity of the laser is manipulated and the maximum modulation bandwidth can be expected to about 10 GHz. Until very recently, the use of the external optical injection to extend the bandwidth to several tens of GHz [69–71]. External modulation requires an independent optical modulator coupled by the lightwaves in order to preserve the frequency and linewidth of the laser.

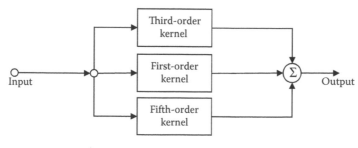

FIGURE 1.16   Sampled Volterra filter.

Manipulation of the lightwaves in the external modulator can be via acousto-optic (AO), electro-optic (EO), or magneto-optic (MO) effects in anisotropic crystals or electro-absorption phenomena in semiconductor. However, EO effects always offer much wider modulation bandwidth than the others. With the high-speed DSP of several tens of sampling rate, pre-distortion of pre-emphasis techniques can be applicable. The modulation can thus be implemented in phase, intensity and amplitude modulation in both time and frequency domains (OFDM, DMT—discrete multi-tone), etc. Furthermore, the uses of phasors are illustrated to describe the states of modulations in complex plane. Thus, one can observe the effects of noises on modulated signals before and after transmission. Hence, the term error vector magnitudes (EVM) in digital sampling oscilloscope or real-time sampling oscilloscope is used to measure the magnitude of the noise added phasor states of a sampled constellation.

Section II: Chapters 4 through 9 describes the implementation of such modulators in coherent and non-coherent optical transmission systems. More emphasis in the modulation effects on transmission performances is given. Optical reception techniques under direct and coherent techniques are briefly outlined due to their common use for long haul (core network or line-side) and metro or access (client side) transmission networks. Coherent photonics are also emerging for use in access and metro networks.

Section III: Chapters 11 through 15 gives an overview of fabrication techniques of integrated guided-wave photonic modulators and basic principles of modulation schemes and transmission medium, the single mode or few mode fibers. In particular, the Si integrated photonic circuits (Si-PIC) currently attract significant interest due to its intergrability with CMOS micro-electronics. This technology may emerge as the critical technique to integrate both electronics and photonics on the same chip e-PIC (electro-photonic integrated circuits), hence leading to significantly reduction of the total product cost using automatic CMOS technology production line. Readers who think that basic principles are necessary can go directly to these chapters to review.

Chapter 11 gives a brief account of planar optical waveguide, the guiding principles and method for estimating the propagation vector and the effective refractive index. This is essential to understand the wave propagation phenomena so that a 3D guided structure can be followed in Chapter 12. These 3D waveguides are fabricated by diffusion of impurities into crystalized structures, thus to create buried diffused waveguides, and thence optical modulation by applied voltage signals onto electrodes deposited across them.

Chapter 13 discusses 3D waveguide, the ridge type which is commonly employed in the design of lasers, optical amplifiers.

Chapter 14 gives a brief account of optical fibers, circular 3D optical waveguide, which are commonly employed to establish optical transmission lines. Optical amplifiers are not outlined and readers are expected to understand the optical gain (in dB) and noise figure (in dB). These amplifiers are required to compensate for the losses of the fibers and optical components.

Chapter 15 gives a brief account of design of optical transmission systems, especially in bi-directional optical fiber transmission links. A number of examples are given for bidirectional optical transmission. Readers who need to review to understand some basic design of optical fiber transmission may find this chapter useful.

REFERENCES

1. C. K. Kao and G. A. Hockham, Dielectric-fibre surface waveguides for optical frequencies, *Proceedings IEE*, vol. 113, no. 7, S.1151–1158, 1966.
2. D. Keck et al., IV Method of producing optical waveguide fibers, U.S. Patent 3,711,262, 1973-01.
3. R. Maurer et al., Fused silica optical waveguide, U.S. Patent 3,659,915, 1972-05.
4. S. Miller, Integrated optics: An introduction, *Bell Systems Tech. J.*, vol. 48, no. 7, pp. 2059–2069, 1969.

5. E. A. J. Marcatili, Dielectric rectangular waveguide and directional coupler for integrated optics, *Bell Syst. Tech. J.*, vol. 48, pp. 2071–2102, 1969.

6. P. K. Tien, Light waves in thin films and integrated optics, *Appl. Opt.*, vol. 10, no. 11, p. 2395, 1971.

7. G. P. Agrawal, *Fiber-Optic Communication Systems*, 3rd ed., New York: John Wiley & Sons, 2002.

8. N. Calabratta, A new scheme of developing all-optical 4×4 cross-connect switch for WDM network, *Journal of Optics*, vol. 42, no. 4, pp. 376–381, 2013.

9. L. Thylen, Integrated optics in $LiNbO_3$: Recent developments in devices for telecommunications, *J. Lightwave Technol.*, vol. 6, no. 6, p. 847, 1988.

10. I. Baumann et al., Er-doped integrated optical devices in $LiNbO_3$, *IEEE J. Sel. Top. Quantum Electron.*, vol. 2, no. 2, p. 355, 1996.

11. K Okamoto, Recent progress of integrated optics planar lightwave circuits, *Opt. Quantum Electron.*, vol. 31, no. 2, p. 107, 1999.

12. B. Jalali et al., Advances in silicon-on-insulator optoelectronics, *IEEE J. Sel. Top. Quantum Electron.*, vol. 4, no. 6, p. 938, 1998.

13. I. García López, P. Rito, D. Micusik, J. Borngräber, L. Zimmermann, A. C. Ulusoy, and D. Kissinger, A 2.5 Vppd broadband 32 GHz BiCMOS linear driver with tunable delay line for InP segmented Mach-Zehnder modulators, *2015 IEEE MTT-S International Microwave Symposium*, May 17–22, 2015, Phoenix, Arizona, USA, 2015, pp. 1–4.

14. R. Soref, The past, present, and future of silicon photonics, *IEEE J. Selected Topics in Quantum Electronics*, vol. 12, no. 6, pp. 1678–1687, 2016.

15. R. Soref and B. Bennett, Electro-optical effects in silicon, *IEEE J. Quantum Electronics*, vol. 23, no. 1, pp. 123–129, 1987.

16. D. Knoll et al., High-performance BiCMOS Si photonics platform, *2015 IEEE Bipolar/BiCMOS Circuits and Technology Meeting*, October 26–28, 2015, Boston, MA, USA, pp. 88–96.

17. Inphi Inc., Inphi Debuts 100 G DWDM Solution for 80 km Data Center Interconnects, https://www.inphi.com/media-center/press-room/press-releases-and-media-alerts/inphi-debuts-100g-dwdm-solution-for-80km-data-center-interconnects.php

18. L. Agazzi, J. D. B. Bradley, M. Dijkstra, F. Ay, G. Roelkens, R. Baets, K. Wörhoff, and M. Pollnau, Monolithic integration of erbium-doped amplifiers with silicon-on-insulator waveguides, *Optics Express*, vol. 18, no. 26, pp. 27703–27711, 2010.

19. J. D. B. Bradley, M. C. E. Silva, M. Gay, L. Bramerie, A. Driessen, K. Wörhoff, J.-C. Simon, and M. Pollnau, 170 Gbit/s transmission in an erbium-doped waveguide amplifier on silicon, *Optics Express*, vol. 17, no. 24, pp. 22201–22208, 2009.

20. S. Keyvaninia, G. Roelkens, D. Van Thourhout, J.-M. Fedeli, S. Messaoudene, G.-H. Duan, M. Lamponi, F. Lelarge, E. J. Geluk, and B. Smalbrugge, A highly efficient electrically pumped optical amplifier integrated on a SOI waveguide circuit, Paper Th-D6, *Proc. IEEE OFC 2012*, Anaheim, CA, 2012.

21. R. Stabile, A. Albores-Mejia, A. Rohit, and K. A. Williams, Integrated optical switch matrices for packet data networks, *Microsystems & Nanoengineering*, vol. 2, Article number: 15042, 2016. doi:10.1038/micronano.2015.42.

22. D. S. Greywall, P. A. Busch, F. Pardo, D. W. Carr, G. Bogart, and H. T. Soh, Crystalline silicon tilting mirrors for optical cross-connect switches, *J. Microelectromechanical Systems*, vol. 12, no. 5, pp. 708–712, 2003.

23. L. Wosinska, L. Thylen, and R. P. Holmstrom, Large-capacity strictly non-blocking optical cross-connects based on micro electro opto-mechanical systems (MEOMS) switch matrices: Reliability performance analysis, *IEEE Journal of Lightwave Technology*, vol. 19, no. 8, p. 1065, 2001.

24. A. Ejnioui, Routing on field-programmable switch matrices, *IEEE Transactions on Very Large Scale Integration (VLSI) Systems*, vol. 11, no. 2, pp. 283–287, 2003.

25. Huber and Suhner, Data Centre Networks—All Optical Switches, http://www.polatis.com/data-center-colocation-network-optical-switch-solutions-cloud-computing-datacenter-low-loss-switches-cross-connect.asp.

26. Kao, C. and G. Hockham, Dielectric fiber surface waveguides for optical frequency, *IEE Proceedings IEE*, vol. 113, no. 7, 1966.

27. I. P. Kaminow and T. Li, *Optical Fiber Communications*, vol. IVB, USA: Elsevier Science, 2002.

28. R. S. Sanferrare, Terrestrial lightwave systems, *AT&T Technology Journal*, vol. 66, pp. 95–107, 1987.

29. C. Lin, H. Kogelnik, and L. G. Cohen, Optical pulse equalization and low dispersion transmission in single-mode fibers in the 1.3–1.7 mm spectral region, *Optics Letters*, vol. 5, pp. 476–478, 1980.

30. A. H. Gnauck, S. K. Korotky, B. L. Kasper, J. C. Campbell, J. R. Talman, J. J. Veselka, and A. R. McCormick, Information bandwidth limited transmission at 8 Gb/s over 68.3 km of single mode optical fiber, in *Proceedings of OFC'86*, Paper PDP6, Atlanta, GA, 1986.

31. H. Kogelnik, High-speed lightwave transmission in optical fibers, *Science*, vol. 228, pp. 1043–1048, 1985.

32. A. R. Chraplyvy, A. H. Gnauck, R. W. Tkach, and R. M. Derosier, 8×10 Gb/s transmission through 280 km of dispersion-managed fiber, *IEEE Photonics Technology Letters*, vol. 5, pp. 1233–1235, 1993.

33. C. R. Giles and E. Desurvire, Propagation of signal and noise in concatenated erbium-doped fiber amplifiers, *IEEE Journal of Lightwave Technology*, vol. 9, no. 2, pp. 147–154, 1991.

34. P. C. Becker, N. A. Olsson, and J. R. Simpson, *Erbium-Doped Fiber Amplifiers, Fundamentals and Technology*, San Diego: Academic Press, 1999.

35. M. C. Farries, P. R. Morkel, R. I. Laming, T. A. Birks, D. N. Payne, and E. J. Tarbox, Operation of erbium-doped fiber amplifiers and lasers pumped with frequency-doubled Nd:YAG lasers, *IEEE Journal of Lightwave Technology*, vol. 7, no. 10, pp. 1473–1477, 1989.

36. T. Okoshi, Heterodyne and coherent optical fiber communications: Recent progress, *IEEE Transactions on Microwave Theory and Techniques*, vol. 82, no. 8, pp. 1138–1149, 1982.

37. T. Okoshi, Recent advances in coherent optical fiber communication systems, *IEEE Journal of Lightwave Technology*, vol. 5, no. 1, pp. 44–52, 1987.

38. J. Salz, Modulation and detection for coherent lightwave communications, *IEEE Communications Magazine*, vol. 24, no. 6, pp. 38–49, 1986.

39. T. Okoshi, Ultimate performance of heterodyne/coherent optical fiber communications, *IEEE Journal of Lightwave Technology*, vol. 4, no. 10, pp. 1556–1562, 1986.

40. P. S. Henry, *Coherent Lightwave Communications*, New York: IEEE Press, 1990.

41. A. F. Elrefaie, R. E. Wagner, D. A. Atlas, and A. D. Daut, Chromatic dispersion limitation in coherent lightwave systems, *IEEE Journal of Lightwave Technology*, vol. 6, no. 5, pp. 704–710, 1988.

42. G. Charlet et al., Comparison of system performance at 50, 62.5 and 100 GHz channel spacing over transoceanic distances at 40 Gbit/s channel rate using RZ-DPSK, *Electronics Letters*, vol. 41, no. 3, pp. 145–146, 2005.

43. W. A. Atia and R. S. Bondurant, Demonstration of return-to-zero signaling in both OOK and DPSK formats to improve receiver sensitivity in an optically preamplified receiver, in *Proceedings of IEEE LEOS '99*, San Francisco, CA, USA, vol. 1, pp. 226–227, 1999.

44. G. Bosco, A. Carena, V. Curri, R. Gaudino, and P. Poggiolini, On the use of NRZ, RZ, and CSRZ modulation at 40 Gb/s with narrow DWDM channel spacing, *IEEE Journal of Lightwave Technology*, vol. 20, no. 9, pp. 1694–1704, 2002.

45. J. A. Lazaro, W. Idler, R. Dischler, and A. Klekamp, BER depending tolerances of DPSK balanced receiver at 43 Gb/s, in *Proceedings of IEEE/LEOS Workshop on Advanced Modulation Formats 2004*, San Francisco, CA, USA, pp. 15–16, 2004.

46. H. Kim and A. H. Gnauck, Experimental investigation of the performance limitation of DPSK systems due to nonlinear phase noise, *IEEE Photonics Technology Letters*, vol. 15, no. 2, pp. 320–322, 2003.

47. A. H. Gnauck and P. J. Winzer, Optical phase-shift-keyed transmission, *IEEE Journal of Lightwave Technology*, vol. 23, no. 1, pp. 115–130, 2005.

48. S. Bhandare, D. Sandel, A. F. Abas, B. Milivojevic, A. Hidayat, R. Noe, M. Guy, and M. Lapointe, 2/spl times/40 Gbit/s RZ-DQPSK transmission with tunable chromatic dispersion compensation in 263 km fibre link, *Electronics Letters*, vol. 40, no. 13, pp. 821–822, 2004.

49. T. Mizuochi, K. Ishida, T. Kobayashi, J. Abe, K. Kinjo, K. Motoshima, and K. Kasahara, A comparative study of DPSK and OOK WDM transmission over transoceanic distances and their performance degradations due to nonlinear phase noise, *IEEE Journal of Lightwave Technology*, vol. 21, no. 9, pp. 1933–1943, 2003.

50. C. Xu, X. Liu, L. F. Mollenauer, and X. Wei, Comparison of return-to-zero differential phase-shift keying and ON-OFF keying in long-haul dispersion managed transmission, *IEEE Photonics Technology Letters*, vol. 15, no. 4, pp. 617–619, 2003.

51. L. N. Binh, *Digital Processing: Optical Transmission and Coherent Receiving Techniques*, Boca Raton, FL: CRC Press, 2015.

52. L. N. Binh and T. L. Huynh, Phase-modulated hybrid 40 and 10 Gb/s DPSK DWDM long-haul optical transmission, in *Proceedings of OFC '07*, paper JWA94, Anaheim, CA, 2007.

53. T. Ito, K. Sekiya, and T. Ono, Study of 10 G/40 G hybrid ultra-long haul transmission systems with reconfigurable OADM's for efficient wavelength usage, in *Proceedings of ECOC'02*, paper 1.1.4, Copenhagen, Denmark, 2002.

54. K. Iwashita and N. Takachio, Experimental evaluation of chromatic dispersion distortion in optical CPFSK transmission systems, *IEEE J. Lightwave Tech.*, vol. 7, no. 10, pp. 1484–1487, 1989.

55. J. G. Proakis, *Digital Communications*, 4th ed., New York: McGraw-Hill, 2001.

56. J. G. Proakis and M. Salehi, *Communication Systems Engineering*, 2nd ed., New Jersey: Prentice Hall, Inc., pp. 522–524, 2002.

57. K. K. Pang, *Digital Transmission*, Melbourne, Australia: Mi-Tec Media Pty. Ltd., 2005.

58. L. N. Binh and T. L. Huynh, Linear and nonlinear distortion effects in direct detection 40 Gb/s MSK modulation formats multi-span optically amplified transmission, *Optics Communications*, vol. 237, no. 2, pp. 352–361, 2007.

59. J. Mo, D. Yi, Y. Wen, S. Takahashi, Y. Wang, and C. Lu, Optical minimum-shift keying modulator for high spectral efficiency WDM systems, in *Proceedings of ECOC'05*, Glasgow, Scotland, vol. 4, pp. 781–782, 2005.

60. J. Mo, Y. J. Wen, Y. Dong, Y. Wang, and C. Lu, Optical minimum-shift keying format and its dispersion tolerance, in *Proceedings of OFC'05*, paper JThB12, Anaheim, CA, 2005.

61. M. Ohm and J. Speidel, Optical minimum-shift keying with direct detection (MSK/DD), in *Proceedings of SPIE on Optical Transmission, Switching and Systems*, San Francisco, USA, vol. 5281, pp. 150–161, 2004.

62. A. H. Gnauck and P. J. Winzer, Optical phase-shift-keyed transmission, *IEEE Journal of Lightwave Technology*, vol. 23, no. 1, pp. 115–130, 2005.

63. N. Alic, G. C. Papen, R. E. Saperstein, L. B. Milstein, and Y. Fainman, Signal statistics and maximum likelihood sequence estimation in intensity modulated fiber optic links containing a single optical preamplifier, *Optics Express*, vol. 13, no. 12, pp. 4568–4579, 2005.

64. L. N. Binh, *Advanced Digital Optical Communications*, 3rd ed., Boca Raton, FL: CRC Press, 2015, Series: Optics and Photonics.

65. L. N. Binh, *Digital Processing: Optical Transmission and Coherent Receiving Techniques*, Boca Raton, FL: CRC Press, October 22, 2013, Series: Optics and Photonics.

66. F. Gao, S. Zhou, X. Li, S.N. Fu, L. Deng, M. Tang, D. Liu, and Qi Y., 2 × 64 Gb/s PAM-4 transmission over 70 km SSMF using O-band 18 G-class directly modulated lasers (DMLs), *Optics Exp.*, vol. 25, no. 7, p. 7230, 2017.

67. G. I. Stegeman and C. T. Seaton, Nonlinear integrated optics, *J. Appl. Phys.*, vol. 58, no. 12, p. R57, 1985.

68. G. I. Stegeman et al., Third order nonlinear integrated optics, *J. Lightwave Technol.*, vol. 6, no. 6, p. 953, 1988.

69. Z. Liu, J. Kakande, B. Kelly, J. O'Carroll, R. Phelan, D. J. Richardson, and R. Slavík, Modulator-free quadrature amplitude modulation signal synthesis, *Nature Communications*, doi:10.1038/ ncomms6911.

70. E. Lamothe, C. M. Long, A. Caliman, V. Iakovlev, A. Mereuta, A. Sirbu, G. Suruceanu, and E. Kapon, Optical injection locking of polarization modes in VCSELs emitting at 1.3 μm wavelength, *IEEE J. Quant. Elect.*, vol. 49, no. 11, pp. 939–943, 2013.

71. J. P. Toomey, C. Nichkawde, D. M. Kane, K. Schires, I. D. Henning, A. Hurtado, and M. J. Adams, Stability of the nonlinear dynamics of an optically injected VCSEL, *Opt. Express*, vol. 20, no. 9, pp. 10256–10270, 2012.

Section I

2 Direct Modulation of Laser and Optical Injection Locking Sources

This chapter deals with the modulation of lightwave sources by either direct or external modulation, as a brief introduction. For direct modulation, the generation of the lightwave within the cavity of the laser is manipulated by "injection" or "withdrawing" of electrons, and hence the excitation or turning-off of stimulated emission process of the output lightwaves. On the other hand, the laser is "turned-on" at all times for continuous waves (CW) operation. The generated CW light is then modulated in frequency, phase, or amplitude through an optical modulator, the external modulator. The properties of modulated optical signals are described. The uses of these transmitters in optical communication transmission systems are given, especially those for long haul transmission at very high bit rate. MATLAB® Simulink models are briefly introduced. Furthermore, an introduction to external modulation, phase, and interferometric types is briefly introduced so that a comparison can be made with complex modulation by optical injection cavity, thus extending the limited passband of the direct modulated short-cavity laser to a bandwidth wider than 100 GHz, allowing the possibility of bit rate of 1.0 Tera-bps per wavelength by complex modulation.

2.1 INTRODUCTION

A photonic transmitter would consist of single or multiple lightwave sources that can either be modulated directly by manipulating the driving current of the laser diode or externally via an integrated optical modulator. These are called direct and external modulation techniques.

This chapter presents the techniques for generation of lightwaves and modulation techniques of lightwaves, either by direct or external means. Direct modulation is the technique that directly manipulates the stimulated emission from inside the laser cavity, via the use of electro-optic effects. In external modulation, the laser is turned on at all times and the generated lightwaves are coupled to an integrated optic modulator through which the electro-optic effect is used with the electrical traveling waves (Signals), the amplitude or phase of the lightwaves are modulated. Advanced modulation formats have recently attracted much attention for enhancement of the transmission efficiency since the mid-80s for coherent optical communications. Hence, the preservation of the narrow linewidth of the laser source is critical for operation bit rates in the range of several tens of Gb/s. Thus, external modulation is essential.

For direct modulation, fundamentally, electrical signals are injecting electronic carriers into the lasing cavity which are then converted to optical modulated ligtwave signals that would then be transmitted through the optical fiber transmission line, the optical fibers.

In this module, a number of practical lasers such as special lasers, constricted mesa lasers, distributed-feedback (DFB) lasers, and Fabry–Perot (FP) lasers are modeled for analysis and simulation of the ultra-long high-speed optical fiber communication system.

A complete analysis and study of the effect of physical parameters of the laser rate equation to its dynamic behavior are described. The simulated results of the laser output are confirmed with the experimental and analytical results published in various literatures.

For external modulation, three typical types of optical modulators are briefly presented in this chapter, including the modulation of lightwaves using lithium niobate (LiNbO$_3$) electro-optic modulators. Their operating principles, device physical structures, device parameters, and their applications and driving condition for generation of different modulation formats, as well as their impacts on system performance. Further details of modulation of lightwaves to generate different modulation formats such as phase and frequency shift keying schemes will be treated in the next chapter.

2.2 DIRECT MODULATION

2.2.1 GENERAL INTRODUCTION

The principal optical component of an optical fiber communications transmitter is the optical source. The crucial role of the optical transmitter is to convert an electrical information input signal into its corresponding optical domain which would then be launched into optical communication channel, the single mode optical fiber (SMF). Most long-haul optical communication systems use semiconductor lasers as optical sources by direct modulation or incorporated with an external modulator for modulation without switching the laser on and off. With no exception, our analysis for ultra-long high-speed optical communication systems examined in this chapter, employs semiconductor lasers with direct-intensity modulation.

Laser is a coherent source generated by stimulated emission process and therefore produces a relatively narrow spectral width of emitted light that allows operation for ultra-long high-speed optical fiber communication systems. Lightwaves emitted from the semiconductor laser can be either modulated directly by controlling the diode injected current, or externally coupled to an optical modulator, normally an electro-optic integrated device is used, through which guided lightwaves are electro-optically modulated by applying a traveling electric wave. The phase modulated optical lightwaves are combined with non-modulated phase path giving amplitude-modulated lightwaves output [1–9].

Optical systems operating at Gigabit-per-second range (up to 100 Gbit/s) can be directly modulated without using an external modulator [10], while external modulators such as LiNbO$_3$ Mach Zehnder modulator and electro-absorption (EA) modulators can operate up to several tens of GHz.

Several designs of semiconductor laser have been produced to cater to ultra-high speed optical transmission systems, for example, Fabry–Perot (*FP*) laser, distributed feedback (*DFB*) laser, and the constricted mesa laser. They are capable of handling approximately 200 Mbps, 600 Mbps, and 15 Gbps system bit rate, respectively [1]. These laser sources are used for modeling the optical transmitter except for the "special" laser source that is modeled in a separate section.

Further understanding of the semiconductor laser acting as an optical transmitter is presented in Section 2.3. Three laser rate equations, namely, the photon rate equation, the carrier rate equation, and the optical phase rate equation are described for generating temporally coherent light waves. The effects of each physical parameter of the rate equation to the laser response are studied and analyzed in Section 2.2.

For ultra-long high-speed optical transmission systems, dynamic response of laser source is critical for the output characteristic of an optical source. Relaxation oscillation and switch-on delay are two examples of dynamic response as covered in Section 2.3.1. These effects are simulated and discussed in Section 2.2.2.8.

In order to model a real practical system, Langevin force (a noise term) is introduced into each rate equation and noise considerations are given in Section 2.3.3. In Section 2.6, eye diagrams of laser output light waves are generated and compared to those obtained in practical systems. Further research and development in optical semiconductor laser done by various research centers and institutions is summarized in Section 2.7.

2.2.2 Physics of Semiconductor Lasers

Under normal conditions, all semiconductor materials absorb light rather than emitting it. The absorption process can be understood by referring to Figure 2.1, where the energy levels E_1 and E_2 correspond to the ground state and the excited state of atoms. If the photon energy $h\upsilon$ of the incident light of frequency υ is about the same as the energy difference $E_g = E_2 - E_1$, the photon is absorbed by the atom, which ends up in the excited state.

Excited atoms eventually return to their normal "ground" state and emit light in the process. Light emission can occur through two fundamental processes known as spontaneous emission and stimulated emission. Both are shown schematically in Figure 2.1. In the spontaneous emission process, photons are emitted in random directions with no phase relationship. Whereas in stimulated emission, the process is initiated by an existing photon and the emitted photons match the original photon in energy (or frequency), phase, and the direction of propagation. All lasers, including semiconductor lasers, emit light through the process of stimulated emission and are said to emit coherent light. Therefore, over a large number of stimulated emission events, the laser source acts like a linear amplifier [10].

At room temperature, spontaneous emission always dominates stimulated emission in thermal equilibrium, and thus, it can never emit coherent light. Therefore, lasers should necessarily operate away from thermal equilibrium by pumping lasers with an external energy source. Even for an atomic system pumped externally, stimulated emission may not be the dominant process since it has to compete with the absorption process. Thus, population inversion is a pre-requisite for laser operation where the atomic density in excited states must be relatively greater than the ground level [11].

2.2.2.1 The Semiconductor *p-n* Junction for Lasing Lightwaves

The *p-n* junction as shown in Figure 2.2 is a homojunction type because the same type of semiconductor material is fabricated on both sides of the junction. The disadvantage of a homojunction is that electron-hole recombination can occur over a relatively wide region (10 μm) that is determined by the diffusion length of charged carriers (electron and holes). However, the homojunction structure leads to spatial dispersion of charged carriers and non-confinement to the immediate vicinity of the junction, and thus it is difficult to realize high carrier densities.

Shown in Figure 2.3 is the heterojunction in which the carrier confinement occurs as a result of bandgap discontinuity at the junction between two different layers of semiconductors that have the same crystalline structure but different bandgaps. The carrier confinement problem for homojunction can thus be resolved by sandwiching this semiconductor material between the *p*-type and *n*-type layers. This is shown as the lasing active region. The bandgap of the active region is smaller than the

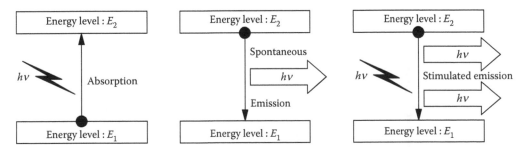

FIGURE 2.1 Three fundamental processes in a semiconductor laser—absorption, spontaneous emission, and stimulated emission—occurring between the two energy state of an atom where E_1 and E_2 are valance band and conduction band, respectively and $E_2 > E_1$.

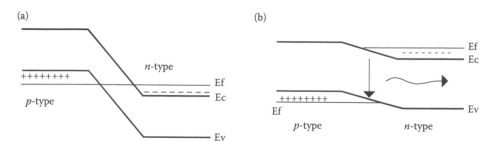

FIGURE 2.2 Energy-band diagram of a *p-n* homo-junction of a semiconductor laser: (a) thermal equilibrium and (b) under forward bias.

layers surrounding it. Therefore, the electron-hole recombination process only occurs in a relatively narrow region (0.1 μm) and high carrier densities can be realized at a given injection current, which is far better than that for the case of homogeneous junction types.

In addition, the refractive index of the active layer is slightly larger than the surrounding layers. As a result of the refractive-index difference, the active layer acts as a dielectric waveguide. The main function of this waveguide and active layer is to confine the generated optical energy in the active layer so that the resonance can occur to effectively generate lightwaves in the cavity. Thus, the heterostructure *p-n* junction semiconductor laser is a very efficient device where the optical waves are mostly confined to the active region, and similarly for the carriers, thus maximizing stimulated emission and hence the laser optical gain.

2.2.2.2 Optical Gain Spectrum

Optical gain occurs in the active region when the injected carrier density in the active layer exceeds a certain limit, known as the transparency value where population inversion is realized. Spectra distribution of emitted light affects the performance of optical communication systems through fiber dispersion. The optical gain spectrum is found by considering all possible transitions from conduction to valance band shown in Figure 2.4. The width of the gain spectrum ranges from 30 to 100 nm. It can be further reduced by the laser resonance cavity that consists of two partially reflected mirrors placed at the ends of a very short lasing length (in the order of 10–40 ppm).

2.2.2.3 Types of Semiconductor Lasers

Semiconductor lasers are mainly classified by their structure. To achieve high bit-rate transmission, one has to narrow the laser spectral width and improve the optical pulse response by minimizing the

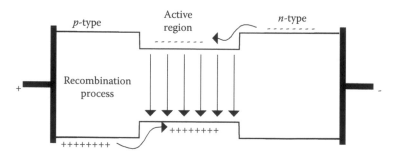

FIGURE 2.3 Energy-band diagram of a heterojunction.

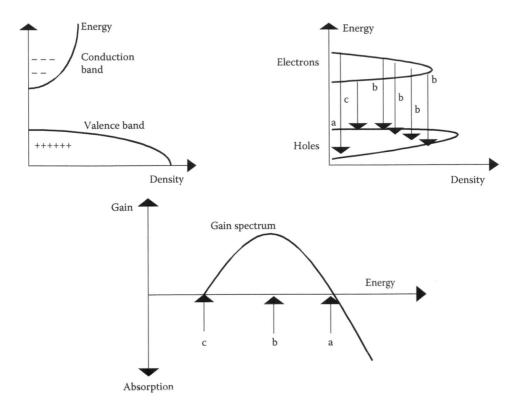

FIGURE 2.4 Optical gain and absorption as a function of bandgap energy.

rise-time constant. Some common types of semiconductor lasers commercially available in the production are Fabry–Perot (*FP*) lasers, distributed feedback (*DFB*) lasers, distributed Bragg reflector (*DBR*) lasers, single quantum well (*SQW*) lasers, multi-quantum well (*MQW*) lasers, constricted mesa lasers, and buried heterostructure (*BH*) lasers.

2.2.2.4 Fabry–Perot (*FP*) Heterojunction Semiconductor Laser

The optical gain alone is essential, but not sufficient for laser operation. The other necessary ingredients are the optical feedback mechanism and the optical guiding in the active layer.

Similar to the resonance in second- and third-order electrical circuitry, the optical energy feedback phenomenon from the two end mirrors located at the end of the guiding cavity as indicated in Figure 2.5a, is a positive feedback. This would thus stimulate the optical resonance and oscillation. This cavity is commonly known as the Fabry–Perot (*FP*) cavity. Each modal resonance frequency in and along the cavity is called a longitudinal mode. Since a standing wave is formed along the cavity, it is bound to have several longitudinal modes (due to multiple resonance modes) to exist within the gain spectrum of the device (Figure 2.5). However, improved longitudinal mode selectivity can be achieved using structures that give adequate loss discrimination between the desired mode and all unwanted modes of the laser resonator, for example, by using short cavity resonator, coupled cavity resonators, and distributed feedback. In Figure 2.5, the broad spectral width ($\approx$5 nm) limits the *FP* lasers transmission rate to about 565 Mb/s for *NRZ* line coding.

The lightwaves are reflected at the boundaries between the active core and cladding layers. They are then traveling back and forth and tightly confined in the active core layer. That means that the active core layer is designed such that it acts as an optical waveguiding structure whose refractive index must be higher than that of cladding regions. This confinement would give better power conservation and field matching to the fiber pigtail.

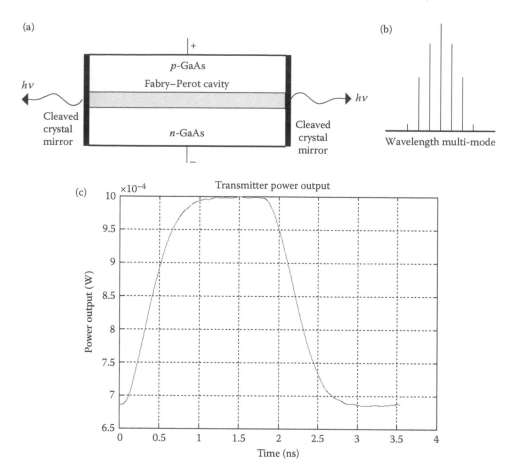

FIGURE 2.5 (a) Schematic diagram of Fabry–Perot laser, (b) *FP* laser output spectral, and (c) *FP* laser output response (565 MHz) obtained by simulation.

2.2.2.5 Distributed-Feedback (*DFB*) Semiconductor Laser

In a *DFB* laser, as the name implies, optical energy feedback mechanism is not localized at the facets, but is distributed throughout the cavity length. This is achieved through an internal built-in corrugated grating that leads to a periodic perturbation of the refractive index (Figure 2.6a). Feedback occurs by means of Bragg diffraction that couples the waves propagating in the forward and backward directions. We can now refer to Figure 2.6b, in which each impedance boundary causes a reflection and the grating period (Λ) must satisfy the Bragg condition

$$\Lambda = m\left(\frac{\lambda_B}{2n_{eq}}\right) \tag{2.1}$$

where n_{eq} denotes equivalent (or effective, normally derived from the egeinvalue of the wave equation) refractive index of the optical guided waves in the active waveguide cavity, λ_B denotes operating wavelength, and the integer m represents the order of Bragg diffraction. The region of periodically varying refractive index serves to couple two counter-propagating traveling waves. The coupling is a maximum for wavelengths close to λ_B. First-order gratings provide the strongest coupling, but second-order gratings are sometimes used because they are easier to fabricate, with their larger spatial period (Λ). Thus, *DFB* lasers by means of Bragg diffraction could produce single frequency mode with a very narrow spectral width ($\approx a$ few hundred MHz). The modulation rate of *DFB* lasers can be about 600 Mb/s to 2 Gb/s for *NRZ* line coding without broadening its line width (Figure 2.7).

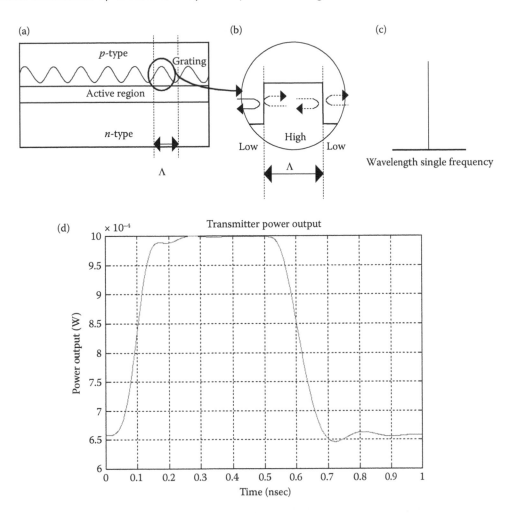

FIGURE 2.6 (a) Space schematic diagram of DFB laser, (b) each impedance boundary causes a reflection (c) DFB laser output spectral, and (d) DFB laser output response (2 GHz) simulated by MOCSS©.

2.2.2.6 Constricted-Mesa Semiconductor Laser

The constricted mesa laser is one component of the family of index-guided lasers. Index-guided laser structure is modified from the gain-guided laser [1–6] that eliminates the problems of having "kinks," astigmatism, unstable far-field patterns, and self-pulsations [1]. In other words, this is to linearize the optical power-driven current (P-I) characteristic of a laser, and hence reduce the distortion of the emitted laser pulse.

Real refractive index variation is introduced into the lateral structure of the laser to achieve this purpose. In high-speed modulation laser, parasitic capacitance could be the limiting factor in achieving excellent high-speed transmission. In a laser with significant parasitic capacitance, the electrical pulse reaching the laser is broadened, and the capacitance provides a source of current during the time when the photon density is high. Consequently, the laser output may consist of two or more pulses as the electron density is repetitively built up and extinguished (relaxation oscillations in Section 2.2.2.2). Thus, constricted mesa lasers are fabricated with current-blocking layers by the regrowth of semi-insulating material, SiO_2 or $SiNx$, shown in Figure 2.2. Hence, modulation speed of 20 GHz for *NRZ* line coding format is achieved in modeling the simulation package.

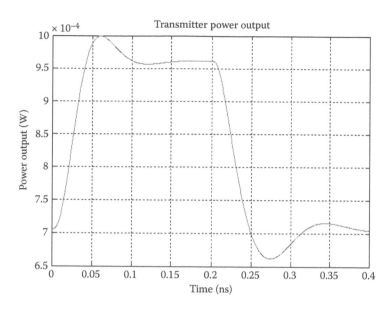

FIGURE 2.7   Constricted mesa laser output response (5 GHz) from MOCSS*© simulation.

2.2.2.7   Special Semiconductor Laser Source

In addition to the above lasers available for MOCSS© simulation, a "special" semiconductor laser has been modeled to tailor for the purpose of analysis of the high-speed ultra-long optical fiber communication systems. With optical confinement factor of 0.9 (which is remarkably higher than normal laser) and narrow spectral line-width of 0.05 nm, these enable it to modulate very high speed signal up to 20 GHz without any observable defects. At around 2.5 GHz, it provides a smooth-running response with slight relaxation oscillation (in Section 2.3.1) and overshooting (Figure 2.8). Thus, it produces a reasonable optical pulse shape for us to analyze, especially at the fiber output and receiver terminal.

2.2.2.8   Single Mode Optical Laser Rate Equations

The operating characteristics of semiconductor lasers are well described by a set of rate equations that govern the interaction of photons and electrons inside the active region. A rigorous derivative of these rate equations generally starts from Maxwell's equations together with a quantum-mechanical approach for the induced polarization [10,12]. However, the rate equations can also be obtained heuristically by considering the physical phenomena through which the number of photons S and the number of electrons N change with time inside the active region as illustrated in Figure 2.9.

For a single-mode laser, three rate equations are given in Equations 2.2 through 2.4. These rate equations can be used for computer simulation of the frequency chirp and output power waveform.

$$\frac{dN(t)}{dt} = \frac{I(t)}{qV_a} - \frac{N(t)}{\tau_n} - v_g a_o \frac{N(t) - N_o}{1 + \varepsilon S(t)} S(t) \tag{2.2}$$

$$\frac{dS(t)}{dt} = \left(\Gamma a_o v_g \frac{N(t) - N_o}{1 + \varepsilon S(t)} - \frac{1}{\tau_p} \right) S(t) + \frac{\beta \Gamma N(t)}{\tau_n} \tag{2.3}$$

$$\frac{d\phi_m(t)}{dt} = \frac{\alpha}{2} \left(\Gamma v_g a_o (N(t) - N_o) - \frac{1}{\tau_p} \right) \tag{2.4}$$

where the parameters of the above rate equations are Γ—optical confinement factor, v_g—guided lightwave group velocity, a_0—gain coefficient, N_0—carrier density at transparency, ε—gain compression

---

* MOCSS = Monash Optical Communication Systems Simulator.

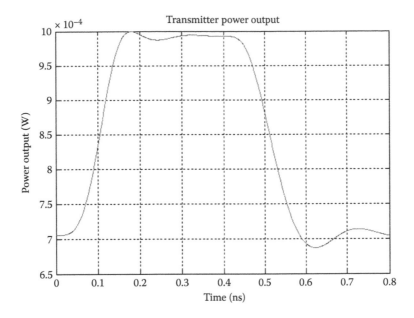

FIGURE 2.8 Special laser output response (2.5 GHz) by simulation.

factor, τ_p—photon lifetime, β—fraction of spontaneous emission coupled into the lasing mode, τ_n—electronic carrier lifetime, q—electronic charge, V_a—active (lasing) layer volume, α—optical linewidth enhancement factor, and ϕ_m—optical phase.

In Equation 2.2, the electron density $N(t)$, increases due to the injection of a current $I(t)$ into the active layer volume V_a, and decreases due to stimulate and spontaneous emission of photon density $S(t)$. Similarly in Equation 2.3, the photon density $S(t)$, is increased by stimulated and spontaneous emission $S(t)$ and decreased by internal and mirror losses with a photon lifetime τ_p given by

$$\tau_p = \frac{1}{v_g(\alpha_i + \alpha_m)} \tag{2.5}$$

where α_i and α_m denotes waveguide loss and α_m the mirror loss, respectively. The carrier lifetime, τ_n is related to the loss of electrons due to spontaneous emission and non-radiative recombination [13] given by

$$\tau_n = \frac{N(t)}{R_{sp} + R_{nr}} \tag{2.6}$$

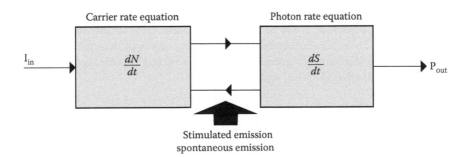

FIGURE 2.9 Single mode laser model as expressed by the rate Equations 2.2 through 2.4.

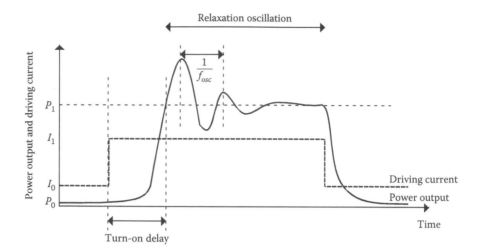

FIGURE 2.10   Typical response to a drive rectangular pulse.

where R_{sp} denotes rate of spontaneous emission and R_{nr} denotes non-radiative recombination. The time variations of the optical power output [14] is given by

$$m(t) = \frac{S(t)V_a \eta_o h\nu}{2\Gamma \tau_p}$$
(2.7)

where $h\nu$ denotes the photon energy and η_0 denotes the total differential quantum efficiency.

2.2.2.9 Dynamic Response of Laser Source

The dynamic behavior of the injection laser is critical, especially when it is used in ultra-long high-speed optical fiber communication systems. The application of a step current to the device results in a switch-on delay, often followed by high frequency damped oscillations ($\approx 10\,\text{GHz}$) known as the relaxation oscillations (RO). Relaxation oscillation occurs when the electron (as the carrier) and photon populations within the structure come into equilibrium and are illustrated in Figure 2.9. The input electrical pulse causes the electron density to rise to a maximum, which is maintained during a turn-on delay until a large photon density builds up and depletes the carriers. This behavior is easily seen from the carrier rate in Equations 2.2 through 2.4, and will be explained in Section 2.3 (Figure 2.10).

In addition, when a current pulse reaches a laser that has significant parasitic capacitance after the initial delay time, the pulse will be broadened because the capacitance provides a source of current over the period that the photon density is high. The turn-on delay is caused by the initial build-up of photon density resulting from stimulated emission. It is related to the minority carrier lifetime and the current through the device. It can be reduced by biasing the laser current near threshold current level (pre-biasing). However, further increase in laser current will decrease the extinction ratio (P_1/P_0). The resonant or oscillating frequency f_{osc} can be increased by increasing I_1. The overshooting can be reduced by shaping electrical input pulse [15,16].

2.2.2.10 Frequency Chirp

The direct current modulation of a single longitudinal mode semiconductor laser can cause a dynamic shift of the peak wavelength emitted from the device (Figure 2.11). This phenomenon, which results in dynamic line-width broadening under direct modulation of injection current, is referred to as frequency chirping. It arises from gain-induced variations in the laser refractive index due to the strong

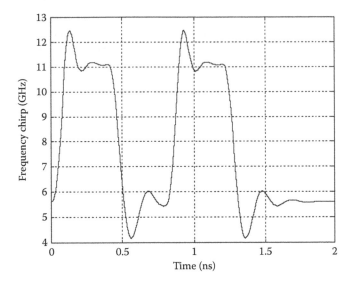

FIGURE 2.11 Frequency chirp effects on direct modulated optical pulses.

coupling between the free carrier density and the index of refraction which is present in any semiconductor structure [17]. Hence, even small changes in carrier density will result in a phase shift of the optical field, giving an associated change in the resonance frequency within both *FP* and *DFB* laser structures.

$$\Delta\nu(t) = \frac{1}{2\pi}\left(\frac{d\phi_m(t)}{dt}\right) = \frac{1}{2\pi}\left[\frac{\alpha}{2}\left(\Gamma v_g a_o(N(t) - N_o) - \frac{1}{\tau_p}\right)\right] \quad (2.8)$$

where $\Delta\nu$ denotes frequency chirp, $N(t)$ denotes the carrier density and $d\phi_m/dt$ denotes optical phase rate Equation 2.4.

A number of techniques can be employed to reduce the frequency chirp effects by

1. Biasing the laser sufficiently above threshold so that the modulation current does not drive the device below the threshold where the rate of change of optical output power varies rapidly with time. However this method gives an extinction ratio penalty of the order of several decibels at the receiver.
2. Damping the relaxation oscillations by shaping the electrical drive pulses that would result in small fluctuations of optical output power [18].
3. Using quantum well lasers, Bragg wavelength detuned *DFB* lasers and multi-electrode *DFB* lasers which provides an improvement in frequency chirp. However, it requires complex fabrication process.

Using an external modulator such as LiNbO$_3$ Mach–Zehnder interferometric amplitude type with the laser source emits continuous lightwaves of appropriate wavelengths or frequencies [19].

2.2.2.11 Laser Noises

The output of a semiconductor laser exhibits fluctuations in its intensity, phase, and frequency chirp (as described in Section 2.2.2.9) even when the laser is biased at a constant current with negligible current fluctuations. This is due to two fundamental noise mechanisms, the spontaneous emission and the electron-hole recombination (shot noises). Usually, noises generated in semiconductor lasers are dominated by photons randomly generated by spontaneous emission. The intrinsic intensity fluctuation in a semiconductor laser is a noise source in direct modulated optical communication systems and in some specific applications may significantly reduce the signal-to-noise ratio (SNR), whereas phase fluctuation leads to a limited spectral line-width when semiconductor lasers are operated continuously at a constant current. The rate equations can be used to study laser noises by adding an extra noise term, known as the Langevin force, to each of them. Then the rate equations are obtained as

$$\frac{dS(t)}{dt} = \left(\Gamma a_o v_g \frac{N(t) - N_o}{1 + \varepsilon S(t)} - \frac{1}{\tau_p}\right) S(t) + \frac{\beta \Gamma N(t)}{\tau_n} + F_S(t) \tag{2.9}$$

$$\frac{dN(t)}{dt} = \frac{I(t)}{q V_a} - \frac{N(t)}{\tau_n} - v_g a_o \frac{N(t) - N_o}{1 + \varepsilon S(t)} S(t) + F_N(t) \tag{2.10}$$

$$\frac{d\phi_m(t)}{dt} = \frac{\alpha}{2}\left(\Gamma v_g a_o(N(t) - N_o) - \frac{1}{\tau_p}\right) + F_\phi(t) \tag{2.11}$$

where $F_S(t)$, $F_N(t)$, and $F_\phi(t)$ denote Langevin noise sources due to the spontaneous emission, the carrier generation recombination process in photon number and generated phase respectively [16]. They are assumed to be Gaussian random processes with zero mean values. Under the Markovian assumption (system has no memory) and the correlation function of the form by Markovian approximation is

$$\langle F_i(t) F_j(t') \rangle = 2 D_{ij} \delta(t - t') \tag{2.12}$$

where $i, j = S$, N, or ϕ, angle brackets denote the ensemble average, and D_{ij} is called the diffusion coefficient [20] and are listed as

$$D_{SS} = \frac{\beta V_a N_{sd}(V_a S_{sd} + 1)^3}{\tau_n} \tag{2.13}$$

$$D_{NN} = \frac{V_a N_{sd}}{\tau_n}[\beta V_a S_{sd} + 1] \tag{2.14}$$

$$D_{\phi\phi} = \frac{R_{sp}}{4S} \tag{2.15}$$

where N_{sd} and S_{sd} represent the steady-state average values of the carrier and photon populations, respectively (Equations 2.16 and 2.17), and R_{sp} denotes rate of spontaneous emission (2.18).

$$N_{sd} = \frac{1}{\Gamma a_o v_g \tau_p} + N_o \tag{2.16}$$

$$S_{sd} = \frac{\tau_p}{\tau_n} N_{sd}\left(\frac{I}{I_b} - 1\right) \tag{2.17}$$

$$R_{sp} = 2\Gamma v_g \sigma_g(N(t) - N_o) \tag{2.18}$$

where I_b denotes bias current and σ_g, the gain cross section normally is $2 \cdot 10^{-20}\,\text{m}^2$ By using the Wiener–Kinchen theorem which states that the Fourier transform of the auto-correlation function of a process is equal to the power spectral density of that process, we take the Fourier transform of a delta function that is simply white Gaussian noise. For ergodic random process and since

$\overline{Fi}^2 = 0$, variance of F_i or $VAR(F_i) = 2D_{ij}$, (refer to (2.19)). Thus, the Langevin force is white Gaussian with a mean of zero and standard deviation of $\sqrt{2D_{ij}}$ given as

$$F_S(t) = N(0,1)\sqrt{2D_{SS}} \qquad (2.19)$$

$$F_N(t) = N(0,1)\sqrt{2D_{NN}} \qquad (2.20)$$

$$F_\phi(t) = N(0,1)\sqrt{2D_{\phi\phi}} \qquad (2.21)$$

where $N(0,1)$ denotes Gaussian distributed random process with mean 0 and standard deviation of 1 and these three Langevin forces are to be fitted in (Equations 2.19 through 2.21). The noise power spectrum $s(f)$ of the photon density as a function of the Fourier frequency, f, is written using the rate equation as

$$s(f) = \frac{\tau_p f_r^4 \langle F_N^2 \rangle + \Psi^2 \dfrac{\langle F_s^2 \rangle}{4\pi^2} + \tau_p f_r^4 \Psi \dfrac{\langle F_s F_N \rangle}{\pi} + \dfrac{\langle F_s^2 \rangle}{4\pi^2} f^2}{\left(f^2 - f_r^2\right) + \Psi^2 f^2} \qquad (2.22)$$

and Relative Intensity Noise (*RIN*) of laser source is given by

$$RIN = 10\mathrm{Log}_{10}\frac{s(f)^2}{S_{sd}^2} \qquad (2.23)$$

where f_r denotes the resonant frequency given by

$$fr = \frac{1}{2\pi}\sqrt{\frac{1}{\tau_p \tau_n}\left(1 + \Gamma a_o v_g N_o \tau_p\right)\left(\frac{I}{I_b} - 1\right)} \qquad (2.24)$$

and $\langle F_S F_N \rangle$ denotes cross-correlation given by

$$\langle F_s F_N \rangle = -\frac{\beta V_a N_{sd}(V_a S_{sd} + 1)}{\tau_N} + \frac{V_a S_{sd}}{\tau_p} \qquad (2.25)$$

and constant Ψ denotes (Figures 2.12 through 2.15)

$$\Psi = \frac{1}{2\pi}\left(\Gamma a_o v_g S_{sd} + \frac{1}{\tau_n}\right) \qquad (2.26)$$

2.3 EXTERNAL CAVITY LASERS

Having discussed the generation of lightwaves for lasing and direct detection, external cavity can be employed to form a single frequency light source whose linewidth can be as narrow as a few KHz, which are in high demand for modern digital coherent communication systems.

An external-cavity diode laser is a diode laser based on a laser diode chip that typically has one end anti-reflection coated, and the laser resonator is completed with, for example, a collimating lens and an external mirror as shown in Figure 2.1. Another type of external-cavity laser uses a resonator based on an optical fiber rather than on free-space optics. Narrowband optical feedback can then come from a fiber Bragg grating (FBG).

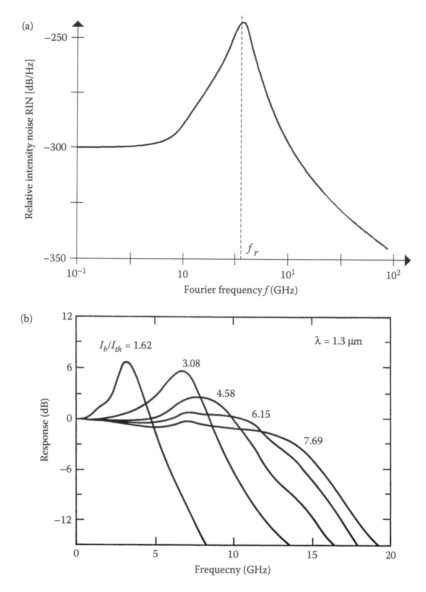

FIGURE 2.12   Relative intensity noise for intensity modulation (a) modeling using rate equations (b) measured frequency response.

The external laser resonator introduces various new features and options:

- The longer resonator increases the damping time of the intracavity light and thus allows for lower phase noise and a smaller emission linewidth (in single-frequency operation). An intra-cavity filter such as the diffraction grating can further reduce the linewidth. Typical linewidths of external-cavity diode lasers are below 100 KHz, comparing with 5 MHz for DFB laser (Figure 2.16).
- Wavelength tuning is possible by including some adjustable optical filter as a tuning element. Most often, a diffraction grating or, recently, a micro-ring-resonator (MRR) is used for this purpose. For details, see below.
- The external resonator also adds important features for mode locking (see Section 2.3.1).

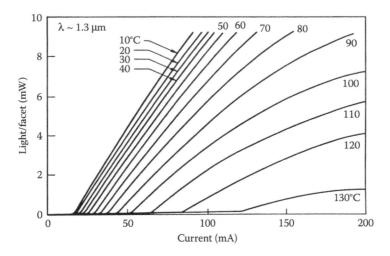

FIGURE 2.13 Laser P-I characteristics with temperature as a parameter.

Note that there are external-cavity semiconductor lasers that are not diode lasers, but optically pumped vertical external-cavity surface-emitting lasers (VECSELs).

2.3.1 TUNABLE EXTERNAL-CAVITY DIODE LASERS

Tunable external-cavity diode lasers (→*tunable lasers*) usually use a diffraction grating as the wavelength-selective element in the external resonator. They are also called *grating-stabilized diode lasers*.

The common Littrow configuration (see Figure 2.17a) contains a collimating lens and a diffraction grating as the end mirror. The first-order diffracted beam provides optical feedback to the laser diode chip, which has an anti-reflection coating on the right-hand side. The emission wavelength can be tuned by rotating the diffraction grating. A disadvantage is that this also changes the direction of the output beam, which is inconvenient for many applications.

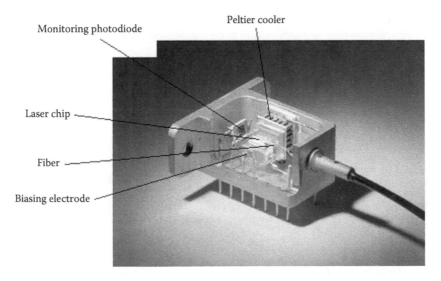

FIGURE 2.14 Packaging of a semiconductor laser.

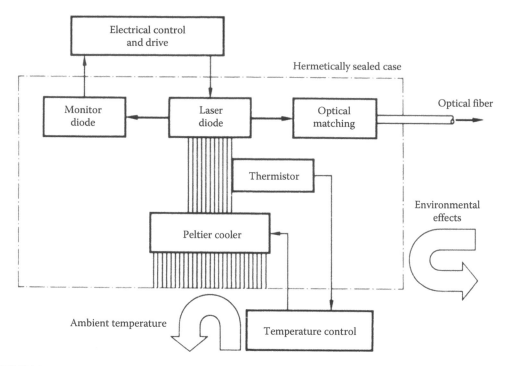

FIGURE 2.15 Block diagram of driving circuitry for a semiconductor laser.

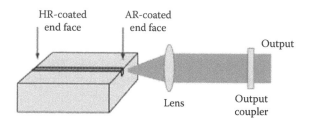

FIGURE 2.16 Simple setup of a diode laser with external cavity. The semiconductor chip is anti-reflection coated on one side, and the laser resonator extends to the output coupler mirror on the right-hand side.

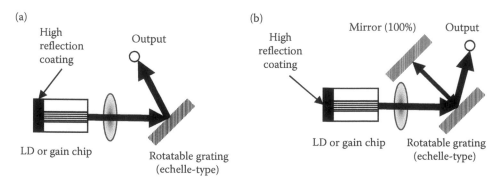

FIGURE 2.17 Tunable external-cavity diode lasers in Littrow and Littman–Metcalf configuration (a) rotatable grating—echelle type and (b) fixed grating with an additional mirror to create resonant cavity.

In the Littman–Metcalf configuration, the grating orientation can be fixed, and an additional mirror is used to reflect the first-order beam back to the laser diode cavity. The wavelength can be tuned by rotating that mirror. This configuration offers a fixed direction of the output beam, and also tends to exhibit a smaller linewidth, as the wavelength selectivity is stronger. (The wavelength-dependent diffraction occurs twice instead of once per resonator round trip.) A disadvantage is that the zero-order reflection of the beam reflected by the tuning mirror is lost, so that the output power is lower than that for a Littrow laser.

Competing types of tunable lasers are DBR laser diodes and small fiber lasers.

2.3.2 Mode-Locked External-Cavity Diode Lasers

In the context of mode locking [18] (→*mode-locked diode lasers*), external-cavity diode lasers have various interesting properties:

- Additional optical elements, such as a saturable absorber for passive mode locking or an optical filter, can be inserted in the laser resonator.
- The longer laser resonator allows for lower pulse repetition rates (although still usually above 1 GHz), and also for tuning the repetition rate by changing the resonator length.
- Even for high repetition rates of tens of gigahertz, external-cavity lasers, then operated with harmonic mode locking, can be interesting, because they exhibit lower laser noise, for example, in the form of timing jitter.

More details are found in the article on mode-locked diode lasers.

Mode-locked external-cavity diode lasers sometimes compete with mode-locked fiber lasers. They do not reach their potential for clean pulses and high output power, but are much more compact and cheaper to manufacturer.

A design circle for integrated laser structure based on quantum dot instead of multi-quantum well mode locked laser can be shown in a design chart in Figure 2.18.

2.4 MEASUREMENTS OF LASER LINEWIDTH AND PHASE NOISES

2.4.1 Generic Aspects of Intensity and Phase Noise

Lasers can exhibit various kinds of "noises," with manifold influences on applications. This subsection discusses where such noises can come from, how they can be quantified, and how to mitigate/minimize their influences in optical transmission systems, especially the DSP-based coherent optical reception-sub-systems.

"Noise" of lasers [19] is a short term for random fluctuations of various output parameters. This is a frequently encountered phenomenon that has a profound impact on many applications in photonics, particularly in the area of precision measurements. We can consider, for example, interferometric position measurements, which can be directly affected by fluctuations of the optical phase, or spectroscopic measurements of transmission, where intensity fluctuations limit the possible sensitivity. Similarly, the date rate and the transmission distance for fiber-optic links are at least partly limited by noise issues.

Many system engineers and physicists feel more or less uneasy about the laser noise. One reason is that the causes are often hard to evaluate and eliminate. Further technical and mathematical difficulties are related to measuring and quantifying noise; specifications found in many data sheets reveal less about the laser than about the competence of the person who made the specification. Furthermore, it is desirable but not always easy to estimate the influence of different kinds of noise in some application. For such reasons, trial and error approaches without a decent understanding are often used, for example, for minimizing noise influences, but this often turns out to be ineffective or inefficient.

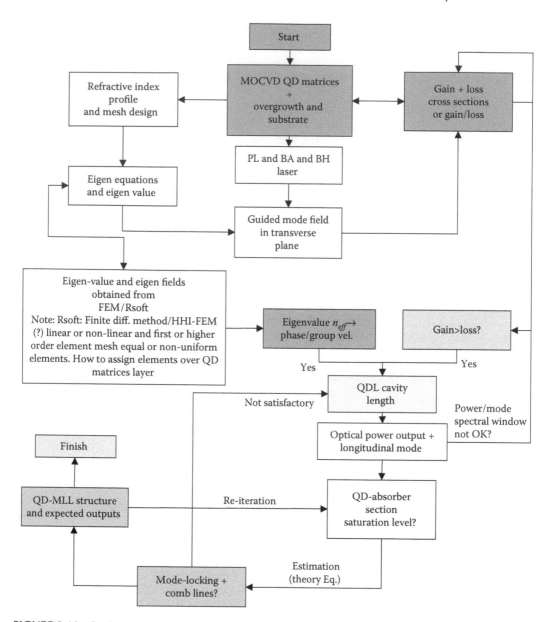

FIGURE 2.18   Design circle for integrated laser structure: QD-ML Laser: Design flowchart and circle.

This section is intended to treat the basic understandings of the intensity and phase noises with a demonstration of experimental measurement of the linewidth and phase noises of semiconductor lasers aiming at the coherent communication systems.

2.4.2   Intensity Noise

Intensity noise is usually understood to quantify fluctuations of the laser output power (not actually an optical intensity), and, in most cases, normalized to the average power. The measurement is based on recording the temporally varying output power, using a photodiode, for example. The normalization is the simplest aspect; other aspects, to be discussed in the following sections, are subtler.

Frequently encountered specifications like "$\pm 1\%$" appear simple, but are quite meaningless. They suggest that the power always stays within 1% of its average value, while in reality, there is usually a smooth probability distribution without sharp edges. A good way to avoid this problem is to specify root mean squared (RMS) values, meaning the square root of the average of the squared power fluctuations:

$$\partial P_{RMS} = \sqrt{\langle (P(t) - P_{av})^2 \rangle} \tag{2.27}$$

where P_{av} is the average power. This is usually applied to *relative* power fluctuations, thus specifying relative intensity noise (RIN). The RMS RIN is of course dimensionless.

Another problem is that the registered fluctuations can strongly depend on the measurement bandwidth. For example, a laser may exhibit fast fluctuations of the output power, which are seen by a fast photodetector, while being averaged out and thus not registered by a slower detector. Furthermore, a limited measurement time may not be sufficient to detect slow fluctuations (drifts); essentially, one does not know how far the average power in the chosen time interval deviates from the average over longer times. For these reasons, a simple number without specification of a measurement bandwidth (lower and upper noise frequency) is actually meaningless and should never occur in a data sheet. The measurement bandwidth may extend from some very small frequency, limited by the inverse measurement time, to some maximum frequency, determined by the speed of the detector.

Note also that any noise specification is actually a statistical measure, which is only estimated by one data trace for a given time interval, or (better) with an average over several traces. Further, the laser noise may depend on ambient conditions. Therefore, one should know whether a specification applies for constant room temperature, after a long warm-up time, and in a vibration-free environment.

If it is of interest to which extent different noise frequencies contribute to the overall noise, a power spectral density (PSD) $S_I(f)$ is most useful. Conceptually, one can imagine that many noise measurements are done for different (small) noise frequency intervals. This is how many electronic spectrum analyzers record noise spectra: the detector is subsequently tuned to different noise frequencies. However, there are also techniques based on Fourier transforms, which allow the estimation of whole noise spectra from single (or a few) data traces recorded in the time domain. One may, for example, use a sampling card to record a signal proportional to the laser power over a certain time and with a certain temporal resolution. Before doing a Fourier transform, a so-called "window function" must normally be applied to avoid certain artifacts. The lowest noise frequency in such a measurement is the inverse total duration, while the maximum frequency is, at most, half the sampling rate. The latter limit results from the Nyquist theorem.

Once we have the power spectral density $S_I(f)$ of the relative intensity, we can calculate the RMS relative intensity noise (RIN) according to

$$\left. \frac{\partial P}{\partial P} \right|_{RMS} = \sqrt{\int_{f_1}^{f_2} S_I(f) df} \tag{2.28}$$

where f_1 and f_2 are the lower and upper noise frequency, respectively. (This is assuming a one-sided PSD; physicists often use two-sided PSDs, where the integration also has to include negative frequencies.) The units of $S_I(f)$ are Hz^{-1}, indicating that $S_I(f)$ is the noise contribution per frequency interval. It is also common to specify $10 \cdot Log_{10}(S_I(f))$, arriving at units of dBc/Hz, meaning dB relative to the carrier in a 1-Hz noise bandwidth.

Figure 2.19 shows the simulated relative intensity noise spectrum (i.e., $S_I(f)$ versus frequency) of a diode-pumped single-frequency Nd:YAG solid state laser at 1080 nm, using logarithmic scales for

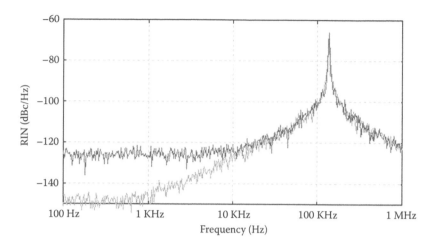

FIGURE 2.19 Simulated intensity noise spectrum of a diode-pumped miniature Nd:YAG laser. Grey curve: with quantum-limited pump source. Dark grey curve: with 30 dB excess noise of the pump source.

both axes. The RIN exhibits a characteristic peak around 140 kHz, related to so-called relaxation oscillations, which result from the dynamic interaction of the intracavity power and the laser gain. Obviously, a detector with 20 kHz bandwidth would not be able to detect such fast oscillations and would thus record much weaker noise than a 200-KHz detector. Laser diodes exhibit relaxation oscillations with much higher frequencies (multiple GHz) and stronger damping due to their short carrier lifetime and short resonator. Generally, different laser types can exhibit very different noise properties, as characteristic parameters may be totally different.

Around the relaxation oscillation frequency f_{ro}, a laser is particularly sensitive to external noise influences, for example, from the pump diode, and even to quantum noise. Well above f_{ro}, noise of any kind has little impact. Well below f_{ro}, pump noise may directly affect the output. For the dark light (light grey) curve in Figure 2.19, some frequency-independent excess noise of the pump diode has been assumed; in reality, there can be an increase toward lower frequencies, often determined by the quality of the diode driver. Even for a perfect shot-noise-limited pump source (grey curve), noise around f_{ro} is relatively strong. The impact of quantum noise depends on various parameters; in particular, it is stronger in cases with low intracavity power, high resonator losses, and a short round-trip time of the resonator. Additional contributions to intensity noise of a laser can result from acoustic influences. For example, a cooling system may cause vibrations of the laser resonator, which translate into intensity and phase noise. In many cases, such noise contributions appear in the form of sharp peaks in the spectrum.

An entirely different phenomenon is noise from mode beating in lasers where multiple resonator modes are oscillating. The occurring beat frequencies are differences of resonator mode frequencies, which may be rather high (multiple GHz) for laser diodes and very low (few kHz) for long fiber lasers. The pattern of beat frequencies can reveal whether or not the laser operates on axial modes only.

Minimization of laser noise can be done on entirely different routes:

- A first approach is the minimization of external noise influences, for example, by using single-frequency pump diodes operated with a carefully stabilized current source, and making a mechanical stable laser setup.
- A second possibility is the optimization of laser parameters such that the impact of quantum noise and/or external noise influences is minimized. For example, one may minimize quantum noise influences by using a long low-loss laser resonator, or move the relaxation oscillation frequency into a region where noise is less strongly disturbing the application.

- Finally, one may reduce laser noise with a feedback system, automatically adjusting, for example, the pump power based on measured output power fluctuations. The characteristics of the feedback loop need to be optimized based on the knowledge of laser parameters.

Obviously, the potential of these approaches depends very much on the circumstances, which should be analyzed carefully beforehand. Without a proper understanding of the sources and types of noise, such exercises are prone to fail or at least to be inefficient.

The intensity noise of nearly all light sources is limited by shot noise. In an intuitive (and somewhat simplified) picture, this can be understood as the random occurrence of photons (packets of light energy). Even if the probability of detecting a photon within some short time interval is constant, the actually recorded photon absorption events exhibit some randomness if there are no correlations between photons. This leads to a PSD of the relative intensity noise of $2h\nu/P_{av}$, where $h\nu$ is the photon energy. The relative intensity noise of a laser is often well above the shot noise level, but the latter rises if the output is more and more attenuated (e.g., with some linear absorber). With sufficiently stronger power attenuation, the intensity noise will be at the shot noise limit.

Squeezed states of light can exhibit intensity noise below the shot noise limit, but require special methods to be generated, and tend to lose that special property, for example, when the light is attenuated.

2.4.3 Phase Noise and Linewidth

Phase noise is related to fluctuations of the optical phase of the output. As simple as this sounds, the optical phase may not even be defined for a laser oscillating on multiple resonator modes. We thus assume to be dealing with a single-frequency laser, where essentially all power is in a single resonator mode. (For multi-mode lasers, one may consider phase noise for different modes separately.)

Due to various influences, even a single-frequency laser will not exhibit a perfect sinusoidal oscillation of the electric field at its output. There are fluctuations of the power (see above) and the optical phase φ. The latter can be quantified with a phase noise PSD $S_{\varphi}(f)$, having units of rad^2/Hz or Hz^{-1} as rad. is dimensionless.

Initially, we may consider only quantum noise, in a simplified picture described as the random phase of photons added by spontaneous emission. This leads to a "random walk" of the optical phase, which is related to a phase noise PSD $S_{\varphi}(f) \sim f^{-2}$. The divergence at $f = 0$ is related to the unbounded drift of the phase, which, unlike the output power, lacks a "restoring force." It also causes a finite value of the linewidth, which is the width of the main peak in the power spectral density of the optical field. A famous paper from 1958 by Schawlow and Townes presents a simple formula to calculate that linewidth, and shows that the linewidth decreases for increasing intracavity power, decreasing resonator losses, and increasing resonator length. However, the quantum-limited linewidth is often difficult to reach, as other (technical) noise influences are dominating.

In many cases, the linewidth, being a single value, is of primary interest for applications. However, the complete phase noise spectrum may be required in other cases, and may reveal important noise contributions even if these are not relevant for the linewidth. Essentially, the linewidth is determined by low-frequency noise.

The measurement of phase noise or the linewidth is substantially more difficult than that of intensity noise, partly because the phase evolution must be compared with some reference. A conceptually simple but often impractical method is based on measuring a beat note of the laser with a second laser, exhibiting a similar optical frequency (keeping the beat frequency low enough) and much lower phase noise. Alternatively, two similar lasers may be used, and the relative phase noise provides an estimate for the noise of a single laser.

An often convenient variant is the delayed self-heterodyne method, using a setup as shown in Figure 2.20. Here, one shifts the optical frequency with an acousto-optic modulator and derives a reference from the laser itself by introducing a substantial time delay with a long optical fiber. For a very long delay (which is often inconvenient to provide, however), the noise of the delayed

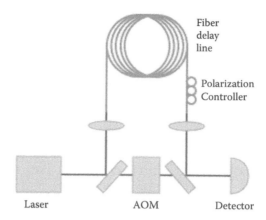

FIGURE 2.20 Setup for delayed self-heterodyne measurements. The mirrors next to the acousto-optic modulator (AOM) have a reflectivity of, for example, 50%. The detector records the beat note between the frequency-shifted part and the delayed part of the laser light. This contains information on phase noise and linewidth.

reference would be uncorrelated to the laser noise, allowing a simple analysis. Noise measurements are still possible with shorter delays, but require a more sophisticated mathematical analysis. This should also take into account the possibility that the phase noise PSD does not exhibit an f^{-2} law.

In principle, the phase noise spectrum can also be obtained from the optical spectrum, as recorded, for example, with a scanning high-finesse Fabry–Pérot interferometer. Here, the optical resonator provides a phase reference. This can work for a semiconductor laser, whereas it is hard to get the resonator linewidth below that of a typical solid-state laser.

2.4.4 Remarks

Lasers exhibit intensity and phase noise resulting from various internal and external influences, the impact of which is strongly influenced by the internal dynamics. Noise can be minimized in various ways; the best way strongly depends on the circumstances.

Intensity noise is, in principle, easy to measure, but proper specifications also need a decent understanding of certain mathematical aspects, and, in particular, of the influence of measurement time and bandwidth, as well as of power spectral densities.

Phase noise measurements are subject to additional difficulties, partly related to the need for some phase reference and the mathematical complication of a diverging phase noise PSD. The laser linewidth is directly related to phase noise, but of course contains much less information than the whole phase noise spectrum. Quantum noise alone in a laser leads to the Schawlow–Townes linewidth, which is often not reached because of additional noise influences.

An experimental measurement set up to determine these parameters of a single mode semiconductor laser is described in the next Section 2.48.

2.4.5 Experimental Phase Noise Measurement of a Semiconductor Narrow Linewidth CW Laser by Self-Heterodyne Detection

2.4.5.1 Experimental Platform

Two different laser phase noise measurement techniques are compared. One of these two techniques is based on a conventional and low-cost delay line system, which is usually set up for the linewidth measurement of semiconductor lasers. The results obtained with both techniques on a high-spectral-purity laser agree well and confirm the interest of the low-cost technique. Moreover, an extraction of the laser linewidth using computer-aided design tools is performed.

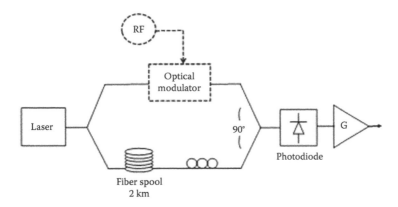

FIGURE 2.21 Measurement bench for both approaches; the AO modulator is used in the self-heterodyne technique and removed for the homodyne technique, G, amplifier gain.

The delay line technique is a low-cost and efficient approach for the measurement of the frequency fluctuations of frequency sources. Long optical delay lines (a few kilometers) are generally used in optics to reach a complete signal de-correlation and measure the laser linewidth by simply mixing this signal to the original laser signal on a photodiode [21]. In the microwave range, the same long optical delay lines are used in frequency discriminators, thus featuring an ultra-high sensitivity [20].

When a high-spectral-purity laser is measured on this type of bench, the measurement conditions are similar to the ones observed in microwave frequency discriminators: the coherence length is no longer than the delay line length, and the data measured at the system output are the frequency fluctuation spectrum of the laser [16,22,23]. However, compared to the frequency discriminator used for microwave source measurement, the use of an optical high-sensitivity frequency discriminator faces different problems related to the frequency stability of the laser during the measurement time.

In this section, two techniques are compared for the measurement of laser short-term frequency fluctuations using an optical delay line. One of these techniques is based on a homodyne delay line technique, and uses a fast Fourier transform (FFT) signal analyzer; meanwhile, the other technique is based on a self-heterodyne approach and on the spectrum analysis close to an intermediate frequency of 80 MHz using a costly piece of equipment: an radio frequency (RF) phase noise measurement bench. The measurement bench is depicted in Figure 2.21. The signal is delayed on one path using a 2 km fiber spool, followed by a polarization controller (optional). In the homodyne approach, this signal is directly recombined with the reference path, and the output of the coupler feeds a photodiode, which performs the quadratic detection. When the laser being tested is a high-quality laser, such as a fiber laser or an externally stabilized semiconductor laser, both of which typically feature a linewidth smaller than 10 kHz, the signal on both arms remains correlated and the output is proportional to the frequency fluctuations of the laser at low frequency offsets (frequency discriminator).

2.4.5.2 Linewidth and Phase Noises: Analytical Derivations and Measurements

A classical approach to describe this system is to replace the frequency noise by a deterministic sinusoidal modulation [17]. In this case, the signal in both arms before the coupler can be written as

$$A_I(t) = \frac{A}{2}\cos\left(2\pi f_0(t - \tau) + \frac{\Delta f}{f_m}\cos 2\pi f_m(t - \tau)\right)$$

$$A_I(t) = \frac{A}{2}\cos\left(2\pi f_0(t) + \frac{\Delta f}{f_m}\cos 2\pi f_m(t)\right)$$

(2.29)

where τ is the delay between the two arms of the interferometer, f_0 the optical frequency, f_m the noise frequency, and Δf the modulation amplitude (frequency noise). The quadratic detection leads to (mixing term only)

$$I(t) = S\frac{A^2}{4}\cos\left(-2\pi f_0(\tau) + 2\frac{\Delta f}{f_m}\sin \pi f_m(t)\sin 2\pi f_m(t - \tau/2)\right) \tag{2.30}$$

S being the photodiode sensitivity. Equation 2.29 may be written as

$$I(t) = S\frac{A^2}{4}\cos(\varphi + x) \tag{2.31}$$

φ being a constant phase and x a small amplitude component (noise). The derivative of this function versus x near $x = 0$ allows us to linearize the output current and to determine the ratio of the output current amplitude and Δf, which is the sensitivity factor of the frequency discriminator, K_m

$$K_m = S\frac{P^2}{4}\sin\varphi 2\pi\frac{\sin(\pi f_m\tau)}{\pi f_m\tau} \tag{2.32}$$

P being the optical power received on the photodiode. While varying the phase φ, the output current changes from its minimum, which should be zero in a perfect interferometer, and its maximum SP. In practice, the minimum is never equal to 0, and SP is generally approximated by the peak-to-peak excursion of the output current. $SP = 2$ is thus the peak excursion of this current, which we note ΔI peak [21].

Ideally, the phase φ between the two arms should be adjusted to $\pi = 2$ in order to maximize the frequency fluctuation detection. This is what is classically performed in a microwave frequency discriminator. However, in an optical experiment, such a precise phase control after a 2 km delay is impossible because of temperature variations. The phase φ rotations due to the slow temperature drift of the fiber spool can be observed at the system output. In our case, 360° rotations were reached every 200 or 300 ms, meaning that, on a 15 min acquisition time, the complete phase rotations will reach 4000. Because of such a large number of phase rotations and a large number of averaged spectrum (about 200 spectra), we expect the phase states to be equally spaced on 360° during the measurement. Such an approach is very close to the setting given in Reference 23, in which a phase modulator had been introduced in the system. However, with kilometer-long delay lines, the phase modulator is no longer necessary, as a natural phase shift occurs due to the slow temperature drift of the fiber spool. When a large set of phase states are considered, the effective K_m can be calculated from Equation 2.31 by integrating on a complete phase rotation the quadratic value of K_m (it is a power spectrum that is detected):

$$K_{m\_eff} = \frac{K_m}{\sqrt{2}} = \frac{\Delta I_{peak}}{2}2\pi\tau\frac{\sin(\pi f_m\tau)}{\pi f_m\tau} \tag{2.33}$$

In the experiment, the photodiode is loaded on a resistance R, and the voltage signal is measured on a 100 kHz FFT spectrum analyzer (Advantest R9211B). The system calibration is performed by measuring ΔI peak using the oscilloscope mode of the FFT analyzer, and then K_{meff} is computed for a set of frequency fm. In our case, the delay resulting from the optical line is 9.5 μs. The laser frequency fluctuation spectrum can then be calculated,

$$S_{\Delta f} = \frac{S_{V_{output}}}{R_2 K_{m\_eff}^2} \tag{2.34}$$

$$K_{m\_eff} = \frac{K_m}{\sqrt{2}} = \frac{\Delta I_{peak}}{2} 2\pi\tau \frac{\sin(\pi f_m \tau)}{\pi f_m \tau} \tag{2.35}$$

The system sensitivity is determined by K_{m_eff}, and the output noise due to any source of noise different from

$$S_{\Delta f} = \frac{S_{V_{output}}}{R_2 K_{m\_eff}^2} \tag{2.36}$$

$$L(f_m)_{dBc/Hz} = 10\text{Log}_{10} \frac{S_{V_{output}}}{R_2 K_{m\_eff}^2} \tag{2.37}$$

$S_{\Delta f}$ laser amplitude modulation (AM) noise, photodiode noise, Brillouin scattering in the fiber, etc. Generally, laser AM noise and photodiode noise are negligible in a long-line interferometer. However, with a delay in the kilometer range, the Brillouin threshold is lowered, and this noise may become the system noise floor. A compromise has thus to be found between long delay (high K_{meff}), high optical power, and the setting up of scattering phenomena in the fiber.

The self-heterodyne technique applied to high-quality lasers was presented in Reference 25 and thus is described briefly. The measurement set-up schematic is depicted in Figure 2.21, incorporating an optical AO (acousto-optic) modulator in order to shift the output signal around the 80 MHz RF (TeO$_2$ AO crystal). Then the signal is analyzed using a phase noise measurement bench at 80 MHz. Using the same approach developed for the homodyne case, it can be demonstrated [24] that the phase fluctuations of this 80 MHz RF signal are proportional to the laser frequency fluctuations $S_{\Delta f}$ at low frequency offsets. More precisely,

$$L_{RF}(f_m)_{dBc/Hz} = 20\log\left(2\pi\tau \frac{\sin(\pi f_m \tau)}{\pi f_m \tau}\right) + 10\log S_{\Delta f} \tag{2.38}$$

From Equations 2.34 and 2.35, we may calculate the laser phase noise from the measured RF signal phase noise as

$$L_{laser}(f_m) = L_{RF}(f_m)_{dBc/Hz} - 20\log(2\sin(\pi f_m \tau)) \tag{2.39}$$

The calibration in this case is performed by replacing the 80 MHz output signal with a frequency-modulated source of the same frequency and amplitude than the source being tested, which is automatically performed in such a modern phase noise measurement platform. Compared to the first approach, this approach is quite easy, but requires much more sophisticated and costly equipment, particularly an RF phase noise measurement bench. To compare the two approaches, the phase noise characterization of a commercial 1:55 µm fiber laser (Koheras Adjustik) has been performed. Using the homodyne approach, 200 spectra are averaged on three different and overlapping frequency bandwidths (600 spectra for the whole measurement).

The acquisition time (10–15 min) allows the random phase repartition, and the spectrum becomes remarkably well defined and stable after about 50 averages. The FFT analyzer is a relatively low-cost piece of equipment, and, moreover, no RF synthesizer or optical modulator is required. The self-heterodyne approach has been performed using a high-power 80 MHz source to drive the acousto-optic modulator, and the signal is measured at the output using an Agilent E5052B signal source analyzer, which is a high-performance phase noise test set that includes a self-calibration procedure.

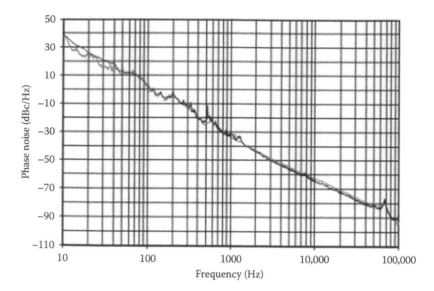

FIGURE 2.22 (Color online) Phase noise measurement of a highquality fiber laser delivering 30 mW optical power at $\lambda = 1:55\,\mu m$ using the two techniques (homodyne technique in blue and self-heterodyne technique in dark grey).

The result of both measurements is depicted in Figure 2.22. A good agreement is observed, which validates our approach for the homodyne technique.

From the frequency noise or the phase noise spectrum, it is possible to evaluate the laser linewidth. It is common to specify a laser through its linewidth, although for very-high-spectral-purity sources this parameter cannot be measured directly and thus has a very limited practical interest. What is measured effectively is the frequency noise or the phase noise spectrum, and the linewidth evaluation requires a computation process from these data. Such a link between phase noise and power spectrum is a complex problem that has been the purpose of many papers [20,25–29], namely only a few. The only case that can be computed analytically is the one of white frequency noise [8], which leads to a Lorentzian line shape. However, for the type of laser concerned here, almost all the laser power is determined by the $1 = f$ part of the spectrum, and the Lorentzian formula is useless. The computation of the linewidth in the case of pure low frequency problem for an arbitrary phase noise spectrum with a combination of different slopes [27].

To get an estimate of the laser linewidth, we can use an original approach based on the simulation of the power spectrum using commercially available microwave computer-aided design software, Agilent Advanced Design System (ADS). This software allows the description of a noisy frequency source through its phase noise spectrum. Moreover, it includes a simulation module based on the envelope technique [29], which allows the computation of slow perturbations of the carrier, such as modulations by deterministic or noisy signals. The result of the simulated spectrum for our laser is represented in Figure 2.23. As the simulated spectrum changes with the window used for both the phase noise description and the baseband temporal analysis window, we have considered for the simulation the phase noise data between 100 Hz and 100 kHz, and we have set a maximum computation time of 0:5 s. The simulated result is a 3 dB full linewidth of about 5.0 kHz. This result has been compared to the simplified approach of [28]. The frequency noise spectral density of our laser can be roughly described by

$$S_{\Delta f}(f) = \frac{k}{f} with\_k = 10^6 (Hz^2) \tag{2.40}$$

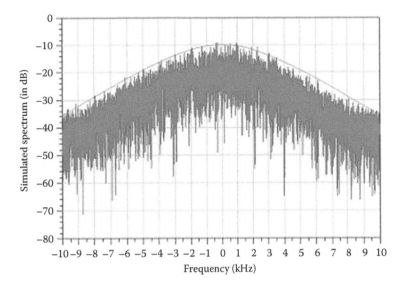

FIGURE 2.23 (Dark grey online) Laser power spectrum versus the offset from the carrier frequency simulated from the phase noise data using Agilent Advanced Design System and the envelope simulation technique.

Using Equation 2.17 of* and taking into account an integration time $T_o = 0{:}5$ s, a linewidth of 6 kHz is obtained for this laser, which is in relatively good agreement with the result of the envelope simulation approach.

2.5 INTRODUCTION TO EXTERNAL MODULATION

The modulation of lightwaves via an external optical modulator can be classified into three types depending on the special effects that alter the lightwaves property, especially the intensity or the phase of the lightwave carrier. In an external modulator, the intensity is normally manipulated by manipulating the phase of the carrier lightwaves guided in one path of an interferometer. Mach–Zehnder interferometric structure is the most common type [30,31].

Electro-absorption (EA) modulator employs the Franz and Keldysh effect, which is observed as lengthening the wavelength of the absorption edge of a semiconductor medium under the influence of an electric field [30,32,33]. In quantum structure such as the multi-quantum well structure, this effect is called the Stark effect, or the electro-absorption (EA) effect. The EA modulator can be integrated with a laser structure on the same integrated circuit chip. For LiNbO$_3$ modulator, the device is externally connected to a laser source via an optical fiber.

The total insertion loss of semiconductor intensity modulator is about 8–10 dB, including fiber-waveguide coupling loss, which is rather high. However, this loss can be compensated by a semiconductor optical amplifier (SOA) that can be integrated on the same circuit. Comparing with LiNbO$_3$, its total insertion loss is about 3–4 dB, which can be affordable as Er-doped fiber amplifier, is now readily available and matured in several installed optical systems.

The driving voltage for EA modulator is usually lower than that required for the LiNbO$_3$. However, the extension ratio is not as high as that of LiNbO$_3$ type, which is about 25 dB as compared to 10 dB for EA modulator. This feature contrasts the operating characteristics of the LiNbO$_3$ and EA modulators. Although the driving voltage for the EA modulator is about 4–3 V and 5–7 volts for LiNbO$_3$, the former type would be preferred for intensity or phase modulation formats due to this extinction ratio that offers much lower "zero" noise level and hence high quality factor.

2.5.1 PHASE MODULATORS

The phase modulator is a device that manipulates the "phase" of optical carrier signals under the influence of an electric field created by an applied voltage. That means that the applied voltage across an optical waveguide will, in turn, change the refractive index of the guided medium, change its refractive index, and change of the wave phase velocity of the guided mode. Thus, a change of the phase velocity in a fast action way of the guided wave can be generated, thence the modulation of the phase of the guided wave passing through the waveguide. When voltage is not applied to the RF-electrode, the number of periods of the lightwaves, n, exists in a certain path length. When voltage is applied to the RF-electrode, one or a fraction of one period of the wave is added, which means $(n + 1)$ waves exist in the same length. In this case, the phase has been changed by 2π and the half voltage of this is called the driving voltage. In case of long distance optical transmission, waveform is susceptible to degradation due to non-linear effect such as self-phase modulation, etc. A phase modulator can be used to alter the phase of the carrier to compensate for this degradation. The magnitude of the change of the phase depends on the change of the refractive index created via the electro-optic effect that in turn depends on the orientation of the crystal axis with respect to the direction of the established electric field by the applied signal voltage.

An integrated optic phase modulator operates in a similar manner except that the lightwave carrier is guided via an optical waveguide which a diffused or ion-exchanged confined regions for $LiNbO_3$, and rib-waveguide structures for semiconductor type. Two electrodes are deposited so an electric field can be established across the waveguiding cross section so that a change of the refractive index via the electro-optic or EA effect as shown in Figure 2.24. For ultra-fast operation, one of the electrodes is a traveling wave type or hot electrode and the other is a ground electrode. The traveling wave electrode must be terminated with matching impedance at the end so as to avoid wave reflection. Usually a quarter wavelength impedance is used to match the impedance of the traveling wave electrode to that of the 50 Ω transmission line.

A phasor representation of a phase-modulated lightwave can be by the circular rotation at a radial speed of ω_c. Thus, the vector with an angle ϕ represents the magnitude and phase of the lightwave.

2.5.2 INTENSITY MODULATORS

A basic structured LN modulator is comprised of (i) two waveguides, (ii) two Y-junctions, and (iii) RF/DC travelling wave electrodes (Figure 2.25). Optical signals coming from the lightwave source are launched into the $LiNbO_3$ (LN) modulator through the polarization maintaining fiber, it are then equally split into two branches at the first Y-junction on the substrate. When no voltage is applied to the RF electrodes, the two signals are recombined constructively at the second Y-junction and coupled into a single output. In this case, output signals from the LN modulator are recognized as "ONE." When voltage is applied to the RF electrode, due to the electro-optic effects of LN crystal substrate, the waveguide refractive index is changed, and hence the carrier phase in one arm is advanced though retarded in the other arm. Thence, the two signals are recombined destructively at the second Y-junction and are transformed into a higher order mode and radiated at the junction. If the phase retarding is in multiple odd factor of π, the two signals are completely out of phase, the combined signals are radiated into the substrate, and the output signal from the LN modulator is recognized as a "ZERO." The voltage difference that induces this "ZERO" and "ONE" is called the driving voltage of the modulator, and is one of the important parameters in deciding a modulator's performance.

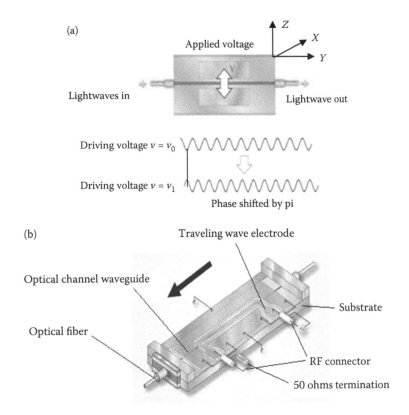

FIGURE 2.24 Electro-optic phase modulation in an integrated modulator using $LiNbO_3$. Electrode impedance matching is not shown (a) schematic diagram and (b) integrated optic structure.

2.5.3 PHASOR REPRESENTATION AND MODULATION TRANSFER CHARACTERISTICS

A phasor can be considered as a vector representing the mean amplitude of the wave and its phase angle rotating around the polar coordinate at the angular speed of the waves. For optical waves, the angular rotation speed is very fast and thus can be considered to be stationary for the human eye observation. Thus, we can represent an optical wave by its mean amplitude and its phase angle with respect to the reference direction, commonly defined as zero phase horizontal direction.

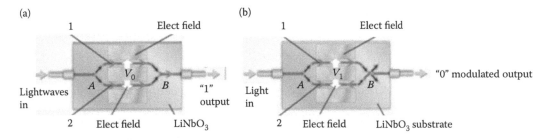

FIGURE 2.25 Intensity modulation using interferometric principles in guide wave structures in $LiNbO_3$ (a) ON—constructive interference mode and (b) destructive interference mode—OFF. Optical guided wave paths 1 and 2. Electric field is established across the optical waveguide of either path 1 or 2 or both (differential mode).

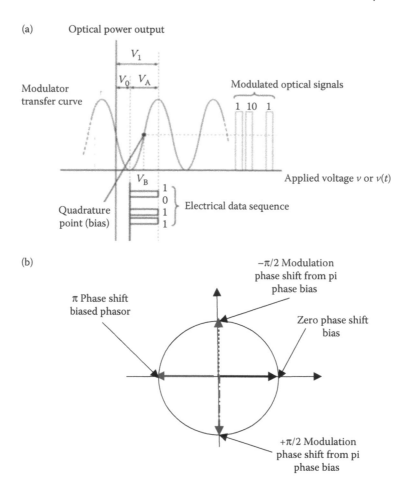

FIGURE 2.26 (a) Electrical to optical transfer curve of an interferometric intensity modulator. (b) Phasor representation of a pi bias and modulation by $+\pi/2$ ("1") and $-\pi/2$ ("0").

Now, considering an interferometic intensity modulator consists of an input waveguide split into two branches and then recombined to a single output waveguide. If the two electrodes are initially biased with voltages V_{b1} and V_{b2}, then the initial phases exerted on the lightwaves would be $\phi_1 = \pi V_{b1}/V_\pi = -\phi_2$ which are indicated by the bias vectors shown in Figure 2.26b. From these positions, the phasors are swinging according to the magnitude and sign of the pulse voltages applied to the electrodes. They can be switched to the two positions that can be constructive or destructive. The output field of the lightwave carrier can be represented by

$$E_0 = \frac{1}{2} E_{iRMS} e^{j\omega_c t} \left(e^{j\phi_1(t)} + e^{j\phi_2(t)} \right) \tag{2.41}$$

where ω_c is the carrier radial frequency, E_{iRMS} is the root mean square value of the magnitude of the carrier, and $\phi_1(t)$ and $\phi_2(t)$ are the temporal phase generated by the two time-dependent pulse sequences applied to the two electrodes. With the voltage levels varying according to the magnitude of the pulse sequence, one can obtain the transfer curve as shown in Figure 2.26a. This phasor representation can be used to determine exactly the biasing conditions and magnitude of the RF or digital signals required for driving the optical modulators to achieve 50%, 33%, or 67% bit period pulse shapes.

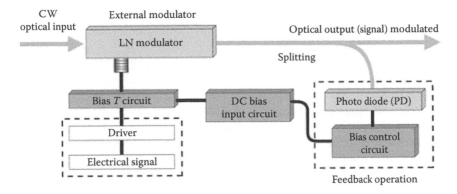

FIGURE 2.27   Arrangement of bias control of integrated optical modulators.

The power transfer function of Mach–Zehnder modulator is expressed as*

$$P_0(t) = \alpha P_i \cos^2 \frac{\pi V(t)}{V_\pi} \tag{2.42}$$

where $P_0(t)$ is the output transmitted power, α is the modulator total insertion loss, P_t is the input power (usually from the laser diode), $V(t)$ is the time-dependent signal applied voltage, V_π is the driving voltage so that a π phase shift is exerted on the lightwave carrier. It is necessary to set the static bias on the transmission curve through Bias electrode. It is common practice to set bias point at 50% transmission point or a $\pi/2$ phase difference between the two optical waveguide branches, the quadrature bias point, as shown in Figure 2.26.

2.5.4   BIAS CONTROL

One factor that affects the modulator performance is the drift of the bias voltage. For the Mach–Zehnder interferometric modulator (MZIM) type, it is very critical when it is required to bias at the quadrature point (+pi/2), or at minimum or maximum locations on the transfer curve. DC drift is the phenomena occurring in LiNbO$_3$ due to the built up of charges on the surface of the crystal substrate. Under this drift, the transmission curve gradually shifts in the long term [34,35]. In the case of the LiNbO$_3$ modulator, the bias point control is vital, as the bias point will shift long term. To compensate for the drift, it is necessary to monitor the output signals and feed it back into the bias control circuits to adjust the DC voltage so that operating points stay at the same point as shown in Figure 2.27, for example, the quadrature point. It is the manufacturer's responsibility to reduce DC drift so that DC voltage is not beyond the limit throughout the life time of the device.

2.5.5   CHIRP FREE OPTICAL MODULATORS

Due to the symmetry of the crystal refractive index of the uniaxial anisotropy of the class m of LiNbO$_3$, the crystal cut and the propagation direction of the electric field affect both modulator efficiency, denoted as driving voltage and modulator chirp. The uniaxial property of LiNbO$_3$ affect the modulation property of the lightwaves due to the electro-optic coefficients of the crystal orientation. This can be explained later in Figure 2.3 of Chapter 4. The chirping mechanism in different cut of the crystal substrate in which the guided wave interferometer is formed can be explained in Chapter 4.

---

* Note this equation is representing for single drive MZIM—it is the same for dual drive MZIM provided that the bias voltages applied to the two electrodes are equal and opposite in signs. The transfer curve of the field representation would have half the periodic frequency of the transmission curve shown in Figure 2.26.

2.5.5.1 Structures of Photonic Modulators

Figure 2.28a and b shows the structure of a MZ intensity modulator using single and dual electrode configurations, respectively. The thin line electrode is called the "hot" electrode, or traveling wave electrode. Radio frequency (RF) connectors are required for launching the RF data signals to establish the electric field required for electro-optic effects. Impedance termination is also required. Optical fibers pig tails are also attached to the end faces of the diffused waveguide. The mode spot size of the diffused waveguide is not symmetric, so some diffusion parameters are controlled so that maximizing the coupling between the fiber and the diffused or rib waveguide can be achieved. Due to this mismatching between the mode spot sizes of the circular and diffused optical waveguides, coupling loss occurs. Furthermore, the difference between the refractive indices of fiber and $LiNbO_3$ is quite substantial, and thus Fresnel reflection loss would also incur.

Figure 2.28c shows the structure of a polarization modulator, which is essential for the multiplexing of two polarized data sequences so as to double the transmission capacity, for example, 40 G to 80 Gb/s. Furthermore, this type of polarization modulator can be used as a polarization rotator in a polarization dispersion compensating sub-system [15].

2.5.5.2 Typical Operational Parameters of Optical Intensity Modulators

Table 2.1 shows the operational parameters of external $LiNbO_3$ optical intensity modulators.

2.5.6 MATLAB® SIMULINK MODELS OF EXTERNAL OPTICAL MODULATORS

2.5.6.1 Phase Modulation Model and Intensity Modulation

In MATLAB® Simulink, there are a number of blocksets such as Simulink Common blockset, Communications Blockset, Control System Blockset, Signal Processing Blockset. A phase modulation section of optical waveguide can be implemented by using the phase shift block given in "Utility Blocks" of the Communication Blockset, as shown in Figure 2.29.

The integration of phase shift blocks to form an intensity modulator is shown in Figure 2.30.

This data modulator can be integrated with another optical modulator, the pulse carver to generate the RZ pulse shaping with carrier suppression or not. This is implemented by setting an appropriate biasing of the pulse carver as shown in Figure 2.31. Note that the sinusoidal signal generators fed into the two phase shifters of the modulator are complement to each other and the voltage peak-to-peak magnitude must be two V_{pi} in order to create periodic sequence with suppression of carrier. The output pulse period is half of that of the input sinusoidal wave, thus a frequency doubling effect due to the non-linear property of the voltage-output intensity transfer characteristics of the intensity modulator. If a dual drive modulator is used, then the biasing voltage must be at $pi/2$ so that there is always a pi phase difference between the two arms of the Mach–Zehnder interferometer. The overall transmitter

(a)   (b)   (c)

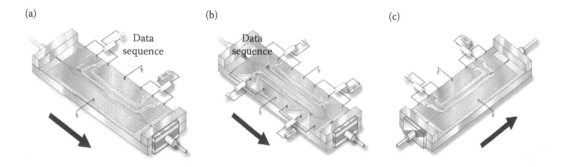

FIGURE 2.28 Intensity modulators using $LiNbO_3$ (a) single drive electrode, (b) dual electrode structure, and (c) electro-optic polarization scrambler using $LiNbO_3$. Similar structures are also given in Chapter 3 in more details.

TABLE 2.1
Typical Operational Parameters of Optical Intensity Modulators

Parameters	Typical Values	Definition/Comments
Modulation speed	10 Gb/s, 40 Gb/s	Capability to transmit digital signals
Insertion loss	Max 5 dB	Defined as the optical power loss within the modulator including coupling loss between fiber and channel waveguides and propagation in the structure.
Driving voltage	Max 4 V	The RF voltage required to have a full modulation
V_π		Voltage required for generating a phase difference of pi between the two arms of the interferometer. That is the minimum transmission point.
Optical bandwidth	Min 8 GHz	3 dB roll-off in efficiency at the highest frequency in the modulated signal spectrum
ON/OFF extinction ratio	Min 20 dB	The ratio of maximum optical power (ON) and minimum optical power (OFF)
Polarization extinction ratio	Min 20 dB	The ratio of two polarization states (TM and TE guided modes) at the output

is shown in Figure 2.32. In addition to the optical modulation, for DPSK there must be a differential coder shown in Figure 2.33 for coding the different feature of the bit sequence to map two bits into the port representing the in-phase and quadrature components to be fed into the optical modulator.

2.5.6.2 DWDM Optical Multiplexers and Modulators

In DWDM transmission systems, the optical transmitters are multiplexed before conduction of the transmission via optical transmission spans. The Simulink model is shown in Figure 2.34. Different modulation formats, the MSK and DNPSK transmitters, are employed in this model. It is noted that the optical spectrum analyzer shown in this figure should be set sufficiently wide to accommodate the spectrum of all multiplexed channels. The $1/z$ block is required for sampling the waveform for appropriate setting for observation in the spectrum analyzer.

2.6 OPTICAL IQ MODULATORS FOR COHERENT 100G AND BEYOND

IQ modulation is critical for generation of optical signals of M-QAM employed in modern optical coherent communication systems. Over the years of development of integrated optics since 1974, lithium-niobate-($LiNbO_3$)-based optical modulators of Mach–Zehnder interferometric type have dominated the optical modulation because of its low insertion loss, which includes the waveguide propagation loss, fiber-waveguide coupling, and its ease of fabrication by diffusion. Dual-polarization IQ (In-Phase Quadrature-Phase) optical modulators (DP-IQOM) have been dominated in the engineering market by Fujitsu for a number of years [36]. However, in the small form-factor, the foot print of the DP-IQOM is bulky and the operating speed may not be higher than the newly emerging InP-based IQOM. In such IQ, modulation two orthogonal in the optical domain

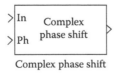

Complex phase shift

FIGURE 2.29 Phase shift block for phase modulation in optical phase modulators or intensity modulators.

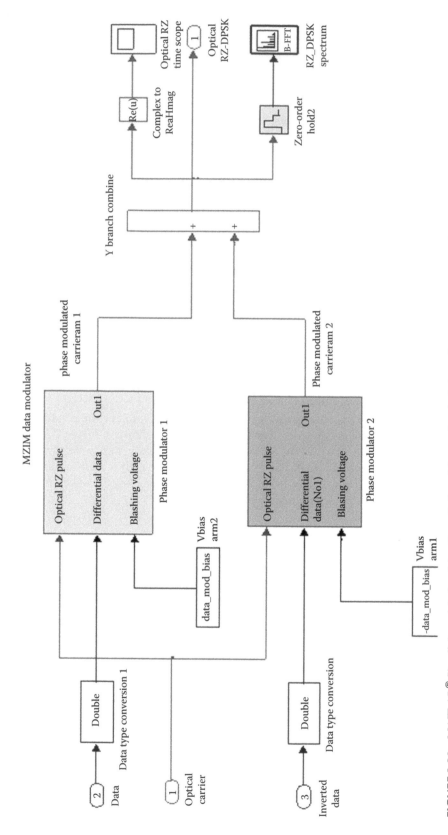

FIGURE 2.30 MATLAB® Simulink model of an intensity optical modulator using dual electrode structures represented as two phase shift blocks with phase bias, laser lightwave input port and electrical data modulation port.

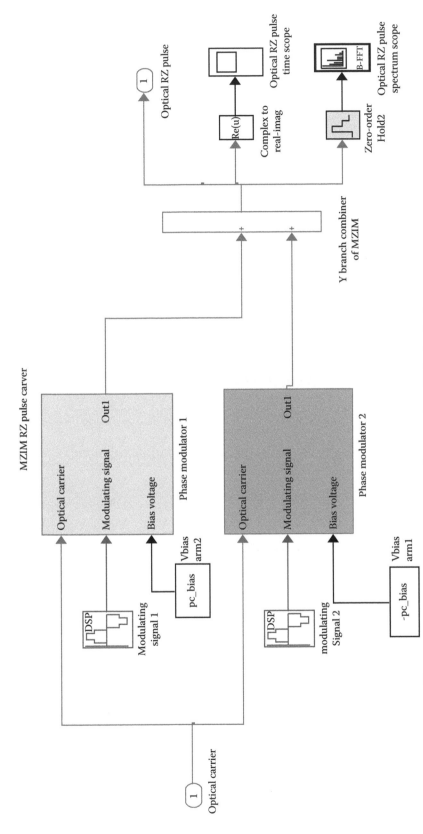

FIGURE 2.31 MATLAB® Simulink model of an optical pulse carver for generation of RZ pulse shaping.

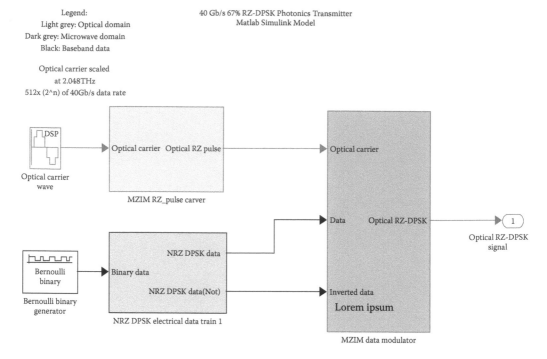

FIGURE 2.32 Photonic transmitters for carrier suppressed RZ and data modulation using DPSK modulation scheme. (Adapted from L. N. Binh, *J. Crystal Growth*, vol. 288, no. 1, pp. 180–187, 2006.)

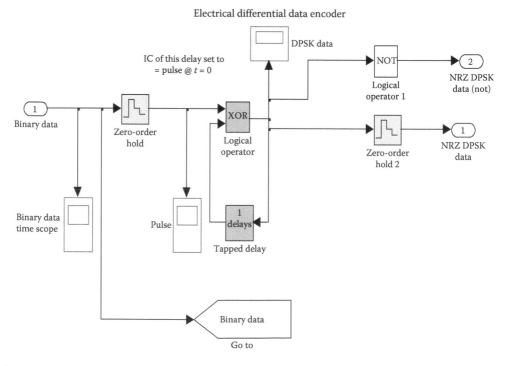

FIGURE 2.33 Electrical differential encoder for DPSK modulation scheme.

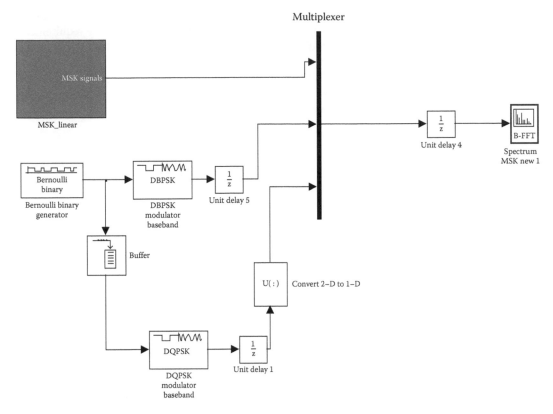

FIGURE 2.34 Simulink model for multiplexing of different modulation format.

intensity-field modulation is conducted in each direction, thus a complex signal in optical domain can be obtained. Furthermore, a new modulation technique by optical injection (OIJ) can offer an extension of bandwidth to tens of GHz from a narrow passband direct modulate devices, such as vertical cavity semi-conductor laser (VCSEL).

Therefore, in this section we address two types of IQOM, the InP IQ modulator and the OIJ IQ OM. One (InP type) is to offer a small footprint for compact module size, but the performances are nearly comparable with LiNbO$_3$ type. The other is a laser injection type with an external master laser running at continuous wave mode (CW) and narrow linewidth injecting its lasing lines into a band-limited laser, such as VCSEL, to extend it bandwidth of modulation and forming complex optical signals serving as in-phase and quadrature (IQ) phase components for coherent optical transmission systems. Thus, this chapter is organized as follows: in the next section, the structures and operational performances of InP dual polarization IQ integrated optical modulators are described. Then, the modulation extension of bandwidth by injection laser is described together with an example of system transmission performance.

2.6.1 InP-Based Direct Modulation Sources and Optical Injection

The continuing increase in fiber capacity demands are driving advances in coherent optical-communication systems. The first generation 100G coherent systems have already been deployed in major central offices for a few years. However, the need to address bandwidth requirements, port density, and system power consumption continue to influence development of technology for 200G, 400G, and beyond. The IQOM is a critical platform used in transmitter architectures designed to address these problems. Hence, an exploration in modulator requirements for next generation coherent communication and their system impacts related to key modulator parameters are discussed. In particular, the benefits of indium phosphide (InP) modulator technology for these requirements will be clarified. Recent InP modulators indicate that low drive voltage and high bandwidth performance can be certainly achieved [38].

Thus we can appreciate a strong possibility that InP-based modulators can overcome the limitations of LiNbO$_3$, opening the door to improving the performance of next generation coherent transmission systems in long-haul as well as metropolitan networking.

2.6.2 LIMITS OF LITHIUM NIOBATE

The development of electro-optic (EO) Mach–Zehnder (MZ) modulators using the linear EO effects of lithium niobate (LiNbO$_3$) crystals was critical for the early advances of optical-fiber networks. While transmitter designs using directly modulated high speed laser or electro-absorption modulator (EAM) technologies may offer advantages in size and cost [39], their low extinction ratio (ER) always limits the modulator performance. In contrast, high-amplitude ER can be achieved with ease with an MZ modulator design. Efficient high speed conversion of electrical signals to modulated light using an external LiNbO$_3$ MZ modulator has enabled ultra-long haul optical-fiber links.

Although LiNbO$_3$ IQ modulators are widely used in today's 100G deployments, there are still significant technology limitations for next generation coherent systems in order to reduce the effective system costs. As the port density and data rate of coherent systems increase, optical components must shrink while offering improved performance. A 100G CFP [40] digital coherent optics (DCO) modules require modulators with a smaller form factor (or footprint) than the existing optical internetworking forum (OIF) [41] standard based on LiNbO$_3$. A new modulator standard with a smaller form factor based on InP is can be defined by the OIF as shown in Figure 2.35. For a CFP2 analog coherent optics (ACO) module, a compact integrated modulator and tunable-laser package may be necessary to reduce the footprint of component even further.

As the cooling capacity of systems remains at the maximum limit, an increase in the component density has to be offset by a lower modulator drive voltage to reduce the total system power consumption. With the incumbent LiNbO$_3$ technology, a lower drive-voltage is difficult to achieve without an increase in the modulator length, that is, the length-bandwidth (BL) product term, and negative impacts to other key parameters critical to next regeneration coherent systems.

Next generation coherent systems thus require modulators with low drive voltage, small size, and proven reliability while keeping the insertion loss to an acceptable value. Polymer- or semiconductor-based modulator technologies might offer such small size and low drive voltage. While research on polymer modulators has shown promising results [42], the stability of the polymer material over the system's life is an important concern that limits broad deployment. Meanwhile, recent interest in silicon photonics [43] has led to many silicon-based modulator developments. However, ER and

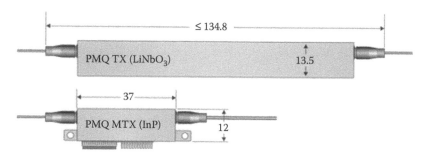

FIGURE 2.35 Form factors/footprint of current OIF modulator standards: PMQ TX (134.8 mm × 13.5 mm) and PMQ MTX (37, 12 mm) of LiNbO$_3$ type and InP type.

insertion loss could be limiting factors for long haul systems. Although an optical amplifier can be used to overcome such insertion loss, the increased power consumption and added noise are undesirable.

Much more detailed structures and operations of LiNbO$_3$ modulators are described in Chapter 3.

2.6.3 InP Traveling-Wave MZ Modulator

InP has paved the way for major advances in high speed optical-fiber communications. The ability to epitaxially tailor the material properties in III–V semiconductors has benefited tunable lasers and high-speed receivers while maintaining the proven reliability of InP devices [44]. Wafer-scale fabrication with precise process controls combined with low-cost packaging has dramatically reduced the cost of components, enabling a lower cost per transmitted bit. These benefits make InP material an attractive candidate for creating a modulator for next gen coherent systems.

A high-speed and small-sized MZ modulator with a low drive voltage requires a material with a large optical phase shift per unit length. Ternary and quaternary alloy materials epitaxially grown on InP can be bandgap engineered to alter the characteristics of the material to suit a particular device application. Using quantum confined stark effect (QCSE) in an InGaAsP alloy multiple-quantum well (MQW) structure lattice matched to InP can create a substantial phase shift per unit length [3] Furthermore, modulators with a high bandwidth can be achieved with a traveling-wave electrode design, where broadband matching of the RF and optical wave group velocities can be achieved.

Figure 2.36 illustrates the basic device concept for a dual-polarization traveling-wave IQ modulator. Recent advances have produced commercially available InP IQ modulators with low drive voltage and high bandwidth [44]. The devices are inherently small in size and ideally suited for

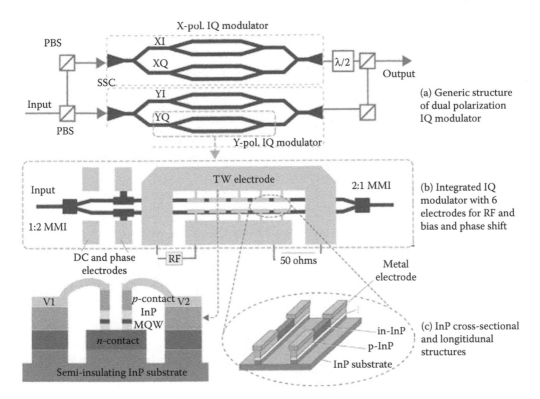

FIGURE 2.36 Basic schematic representation of an InP MQW dual-polarization traveling-wave IQ modulator. IQ modulators structured with integrated optical waveguides and electrodes using either (a) generic structure of DP-IQOM and (b) LiNbO$_3$ or InP-based material systems.

integration with other InP-based devices such as tunable lasers and high speed receivers. This size advantage will be critical to enable compact coherent optics modules like CFP and CFP2.

2.6.4 MODULATOR REQUIREMENTS

The key modulator parameters for next-gen coherent systems are the drive voltage required to induce a pi phase shift (V_π), linearity, ER, and modulation bandwidth.

The drive voltage directly affects the power consumption of the module or line card being integrated into the coherent system. Modulators with large drive voltages will require high-power drivers, and their applications in 100G modules such as CFP and CFP2 will be limited.

CFP-DCO specifications allow 24 W maximum power dissipation for a class 3 module, while for CFP2-ACO, only 12 W is allowable for a class 2 module [5] The modulator driver power must be limited to enable applications like CFP2-ACO. Modulators with a V_π of 1.5 V or less are highly desirable for such applications. Additionally, low-V_π modulators enable the use of lower-voltage drivers, decreasing the complexity of the amplifier design and reducing the number of gain chips required in a package, thus leading to a potential cost benefit.

Linearity is a key requirement for 200 and 400G applications, where more advanced modulation formats will be needed. To provide a linear output, driver amplifier design requires an increased voltage supply level to compensate for the distortion at higher output voltages. A smaller modulator V_π naturally reduces this requirement, enabling a more efficient amplifier design with a lower supply voltage.

The ER of each child and parent MZ is defined as the ratio between the maximum and minimum optical intensities measured at the same port. Poor ERs and any imbalance between the two MZ arms will induce chirp in the optical signal. Chirp is the optical phase variation due to relative variation of optical intensity.

The presence of chirp in a transmitted signal will distort the transitions between constellation points and increase the minimum required OSNR for the system. With closely spaced constellation points, higher order modulation formats such as 16QAM will require better ERs than the values defined in current 100G standards.

Although the DP-QPSK modulation format is now common for 100G, there are many approaches for future 400G systems. Regardless, modulators with higher bandwidth can provide better linearity and spectral efficiency in such next generation coherent systems. As shown in Table 2.2, some recent advanced development and typical operational parameters of InP-based traveling-wave (TW) electrode MZ modulators have shown improved bandwidth that can lead to significant system benefits.

2.6.5 APPLICATION REQUIREMENTS OF InP IQ MODULATORS

New and improved technologies often bring different requirements to system applications. The operation of a LiNbO$_3$ modulator is based on a linear electro-optic effect. The modulation bias point is set by a control voltage on each MZ arm, either via a bias-tee through the RF port, or a separated phase

---

TABLE 2.2

Key Parameters for Integrated InP IQ Modulators

Operational Parameters (Typical)	IQ Modulator	Advantages for Transmission Systems
V_π	~1.5 Volts	Low power dissipation, less impact on electronic driver
Device footprint	~40 mm	Small module size
Extension ratio	<25 dB	Improved OSNR
Modulation bandwidth	<35 GHz	Higher order baud rate → 56 GBaud

electrode. InP modulator phase control is accomplished via either reverse or forward biased phase electrodes to adjust the operating points. As with all InP-based lasers or photodiodes, proper attention is required for the voltage and current limits of the control circuits.

It is well known that the strong thermal drift of $LiNbO_3$ material requires a very fast bias control to stabilize the operation point in a system. The fast phase change can be compensated by applying a fast control signal to a phase electrode.

For InP material, this fast thermal drift is absent, leading to a lower speed, simpler control loop. For InP devices, the material characteristics still need to be stabilized using a thermo-electric cooler (TEC) to ensure constant operation over the environmental temperature range. To maintain the suppressed carrier at null bias point over the operational lifetime, a slow control loop will be needed to compensate for the device's aging.

2.6.6 Low-V_π, High Bandwidth InP IQ Modulator

An InP modulator based on QCSE requires a DC bias to provide the necessary pn-junction electric field. To maintain a constant drive voltage across the wavelength, this DC bias needs to be adjusted across the C-band. An example for wavelength dependence of DC bias is shown in Figure 2.3 using a commercially available InP IQ modulator. A −5.0 V DC bias is required at 1528 nm to set V_π at 1.5 V, while a DC bias of −9 V is necessary @ 1567 nm. This 3 dB corner frequency of the modulator can also be observed to be >30-GHz and very high extinction ration (ER) that is very close to the $LiNbO_3$ type. The low V_π and high bandwidth S_{21} shown here are important characteristics for the next generation coherent technology (Figure 2.37).

Although $LiNbO_3$ modulators offer excellent performance for today's 100G networks, next-gen large-capacity coherent systems with high port density will require small-form-factor modulators with low drive voltages and high bandwidth. Intrinsic material limits bound the performance of today's $LiNbO_3$ technology. A new modulator technology is required to satisfy future advances in coherent systems.

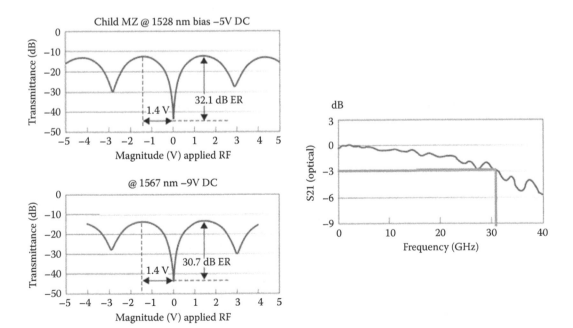

FIGURE 2.37 Typical measured transfer function of transmittance and S_{21} of a wideband InP IQ modulator with $V_\pi \sim 1.5$ V and >30-GHz 3 dB modulation bandwidth.

Combined with an advanced material engineering capability and reliable device technology, the InP platform opens new opportunities for advanced modulator developments. Low-V_π and high-bandwidth InP modulator technology is available today and will prove a key enabler for next-generation high port density coherent systems that require compact modules.

2.7 LASER INJECTION FOR BANDWIDTH EXTENSION TO DIRECT MODULATION

This section describes the use of optical injection locking of directly modulated semiconductor lasers to extend the modulation passband without the uses of an external modulator. Furthermore, the method of the integrated external injection laser can be employed to generate complex modulation format signals showing distinct advantages over other external modulation solutions.

2.7.1 INTRODUCTORY REMARKS

Direct modulation of a laser has been used in optical telecommunications since the days of multimode fiber employed as the guided medium for optical communications. Although distance–capacity product capabilities using this approach is limited, it is a very cost-effective solution, especially in access networks and possibility metropolitan networks, making it a prime choice, especially when transmission distance is limited, for example, within a data center, supercomputer (hundreds of thousands within a single supercomputer), or in access and metro telecommunications networking. For long-distance or high-capacity transmission, however, digital optical signals are generated using a continuous-wave (CW) laser source and an external Mach–Zehnder modulator in order to preserve the linewidth of the optical carrier sources.

To respond to the ever-increasing requirements on data transfer capacity, on–off-keyed (OOK) signaling is gradually being replaced by quadrature amplitude modulation (QAM) signaling in association digital signal processing (DSP) to mitigate impairments that enables increasing the capacity without the need to scale up the bandwidth of the electrical parts of the system, which is both costly and power hungry. However, the complex QAM signaling requires in-phase-quadrature (IQ) modulator structure which is more complicated than OOK and more stringent requirements are imposed on the modulator performance.

Modulators based on the electro-optic (EO) effect (such as the industry-standard LiNbO$_3$-based devices) offer superior performance relative to modulators exploiting charge-carrier effects in semiconductors, which, although capable of generating larger refractive index changes, also inherently impart undesirable amplitude modulation. Although LiNbO$_3$-based modulators offer unparalleled performance, they are long (centimeter-scale), expensive, and difficult to integrate with the laser, making the dense integration of tens to hundreds of transmitters as now required challenging. Consequently, major effort is being directed to obtain equivalent performance using other materials for the modulator, including InP, GaAs, and Silicon on insulator silica (SOI). However, besides parasitic residual amplitude modulation, these modulators suffer from high propagation loss, allowing only for short devices requiring large radio frequency (RF) drive voltages. Although intensive research, devices generating QAM signals have recently appeared enabled both by improved modulator design and the use of digital signal processing (DSP) at both transmitter and receiver to compensate for the non-ideal phase and amplitude transfer functions. The current state of the art includes: a 40-GBaud InP modulator delivering 160 Gbps via external polarization multiplexing of two QPSK (quadruple phase shift keying) signals, an InP-based photonic integrated circuit (laser and modulator integration as well as photodetectors and electronics) capable of 256-Gbit/s operation using 32-GBaud polarization-multiplexed 16QAM GaAs-based modulator at 150 Gbps using 25-GBaud QAM signals; and finally a silicon-based modulator capable of 224 Gbps 116QAM. Although the progress has been impressive, several drawbacks still remain, specifically, that the number of high-speed RF connections to the modulator is twice that ideally is needed due to the requirement for push-pull operation (increasing packaging cost and power requirements); the required RF power is high; computationally heavy DSP is needed, leading to a requirement for high effective number of bits and high bandwidth converters.

An IQ optical modulator requires all together six electrodes, two RF high frequency electrodes, and four DC bias and phase shift electrodes with the complex pin arrangement as shown in Figure 2.38. The detailed arrangement of optical waveguide of a dual polarization Mach–Zehnder Interferometric modulator (MZIM) and electrodes is shown in Figure 2.36.

The new capability of the laser injection cavity modulator demonstrates a significant impact within the many scientific and engineering communities that are directly concerned with or exploit laser radiation. For example, the coherent combination of high-power lasers could greatly benefit from the control of amplitude and phase that we have demonstrated from low-cost semiconductor devices and may open the way to significantly higher powers and useful functionalities (e.g., beam steering), which require precise control of the phase properties of the combined beams.

2.7.2 MODULATION BY OPTICAL INJECTION LASER (OIL)

The schematic arrangement of an optical injection modulator is shown in Figure 2.39. A direct modulated laser is injected with a narrow linewidth laser so that a composite cavity is formed via an optical circulator. The composite resonance cavity enhances the passband of the OIL to several decades wider than that of the direct modulated laser.

An IQ modulation scheme can be constructed using both the in-phase and quadrature-phase components by direct modulation shown in Figures 2.40 and 2.41, and can be generated by combing two OILs with a phase shift of pi/2 between them. Depending on either level (OOK) or 4-level (PAM4 in one direction) direct modulation QPSK or 16 QAM can be generated.

In the following sections, we describe the implementation of the OIL using modulators in InP material platform for C-band.

A brief experimental demonstration of an OIL can be outlined as follows. The slave lasers could be a pair of 250-mm long, 1550-nm AlGaInAs Fabry–Perot (FP) laser designed for operation at 2.5 Gbps, packaged in a commercial high-speed butterfly module with thermo-electric control and an electrical 3 dB bandwidth of 18 GHz. The laser output is directly coupled into a polarization-maintaining (PM) single-mode fiber without any in-built isolator (as needed to allow the formation

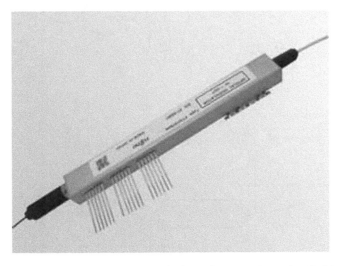

FIGURE 2.38 IQ modulators in package module with electrode connection points for bias and phase shifting pi/2 for IQ; RF electrode connectors placed on other side of the module (hidden).

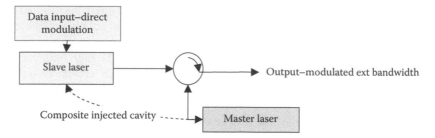

FIGURE 2.39 Schematic representation of optical injection direct modulation.

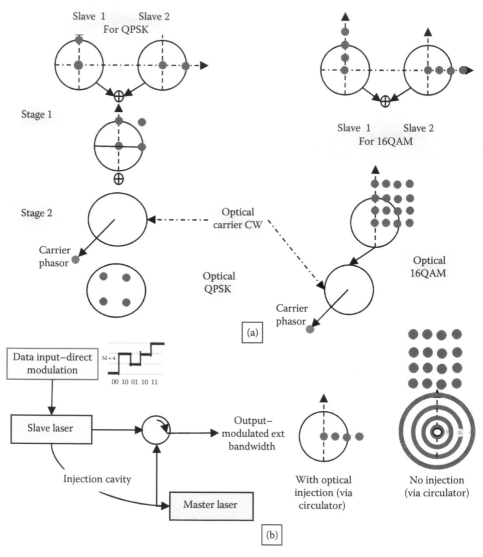

FIGURE 2.40 (a) Complex QAM modulation constellation of 4QAM and 16QAM and (b) structure of master-slave modulation by laser injection and illustrated constellation under with and without optical injection.

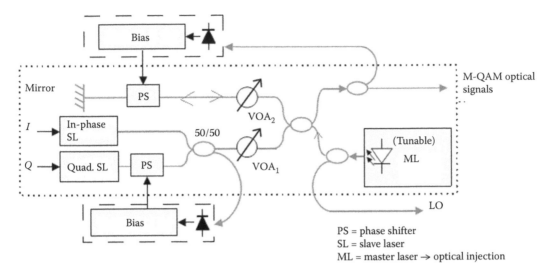

FIGURE 2.41 Schematic diagram of an injection laser master-slave arrangement.

of an OIL cavity). The master is either a 40-mW CW telecom-grade C-band tunable laser. Alternatively, a 10-kHz linewidth fiber laser emitting a CW power of 13–16 dBm at 1,555.7 nm could be used. A variable optical attenuator (VOA) can be integrated before launching into the fiber so that the signal can be generated with desired signal to generate the constellation for required optical signal-to-noise-ratio (OSNR) at the receiver.

The VOA_1 attenuates the master laser signal by 3 dB to obtain the optimal fiber injection power of 2 dBm. In an optimized design, VOA_1 could be eliminated, thereby requiring less power from the master laser and, at the same time permitting higher output signal power, as the slave laser output would also not need to pass through this attenuator. The free-running slave laser wavelengths can be set via the temperature control, to be slightly higher than the master laser wavelength. The generated wavelengths can be fine-tuned to obtain optimum linearity of the modulation scheme and minimum residual modulation chirp. The outputs from the two slave lasers are orthogonally combined by setting and maintaining a relative phase shift between their carriers in quadrature. The relative phase shift between the two slave laser carriers can be controlled via a piezoelectric (PZT)-driven fiber stretcher (6-cm long) placed at the output of one of the slave lasers (SL). Alternatively, a low-speed optical phase modulator can be used. The feedback to keep the two slave lasers in quadrature consists of a slow photodiode followed by a 1-kHz BPF (bandpass filter) and a proportional integral (PI) loop controller. Constructive/destructive interference of the two slaves (corresponding to a relative phase of pi radians, respectively) produced maximum/minimum signal at the photodiode. Hence, in quadrature (pi/2), the signal at the output of the photodiode changes monotonically with the relative phase between the two slave lasers. It could be used directly as the error signal at the PI controller input. Once the two slave lasers are combined in quadrature, the carrier part of their combination is removed by destructively interfering it with a component of the master laser signal obtained via reflection from a mirror.

To keep the destructive interference, another feedback loop is necessary. Again, a slow PD (photodiode/photodetector) followed by a 1-kHz filter, PI loop filter, and a PZT stretcher can be used. However, unlike in the previous feedback loop, it is needed to obtain the phase lock at the minimum of the signal coming out of the slow PD as operating at the minimum interference. We introduced a small dither signal at the PZT fiber stretcher (15 kHz, introducing phase dither), which, in conjunction with lock-in detection, allowed for a suitable error signal. In the experiments, the performance of the operating and locking depends on the followings: the power of the optical injection laser (OIL) and the master laser (ML), the frequency, the biasing points of both injection and master lasers, the variable attenuation factor and the coupling coefficients, the tunability of the local oscillator, the modulation scheme M-QAM on the optical signals. We also denote, in Figure 2.14, the followings: PS = phase shifter; SL = slave laser; ML = master laser and IL as optical injection. The frequency detuning between the free-running ML and SL can be control and adjusted to obtain the optimum locking. In practice, these could be maintained by implementing further simple, low-bandwidth phase-lock loops, as was done previously for slave lasers under CW operation. The transmitter set-up is made entirely of PM fibers and PM-coupled components to avoid any polarization drift. Further, to maintain mutual coherence between the three interfering signals (two slaves and a portion of the master), the difference in paths travelled by these three signals was kept to 20 cm, well below the coherence length of the master laser. Finally, to minimize phase drifts due to environmental factors (temperature, acoustic pick-up), we kept the fibers carrying these signals as physically close together as possible. As a result, feedback with only sub-kHz loop bandwidth was needed to stabilize the path difference as applied via two PZT fiber stretchers.

The slave lasers can be biased in the linear range, for example, at 43 mA and modulated by electrical signals generated either using a dual-channel arbitrary wave generator (AWG), Agilent M8040A with a bandwidth of 30 GHz and sampling rate of 100 GS/s or a dual-output PRBS generator (SHF10002B/12003) capable of operating up to 56 GBaud in association with a 6-bit digital to analog converter (ADC). The two RF data streams (two AWG outputs [used as I and Q] or two complementary outputs of a PRBS generator properly delayed to be decorrelated) were amplified by two RF amplifiers to 0.8Vp-p and used to directly drive the two slave lasers. The RF signals sent to the two slave lasers were synchronized using a manual RF phase shifter delay line when the PRBS generator was used, or directly using the AWG. The AWG was used for the generation of the OFDM signals (QPSK and 16QAM), 10-GBaud QPSK and 16QAM as shown in Figure 2.40. When baud rates beyond the AWG capabilities are necessary, the SHF PRBS generator can be employed, which allows only for the generation of QPSK or M-QAM signals.

Recently, intensive attraction on optical injection techniques have been reported on such techniques for high speed optical transmission [45–47].

The injection lightwaves generate a perturbation of the phasor component superimposing on the electric field phasor as shown in Figure 2.42. The resultant phasor magnitude can then be self-adjusted in such a way that the magnitude remains unchanged over a frequency range, thus extension of the bandwidth of the response of the directly modulated laser. The frequency response represented

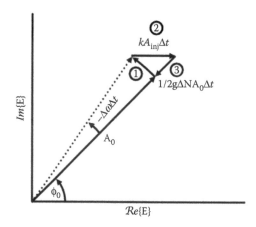

FIGURE 2.42 Phasor model for injection locking, showing phasor perturbation in a time interval, Δt. Vector 1 (circled number 1) corresponds to the free-running slave angular rotation, with respect to the frame-of-reference of the master laser frequency. Vector 2 (Circled 2) is the vector addition of the injected master light at phase ϕ. Vector 3 (circled no. 3) is the reduction in amplitude due to the reduced gain. [Extracted from E. K. Lau et al., *IEEE Journal of Selected Topics in Quantum Electronics*, vol. 15, no. 3, pp. 618–633, May–June 2009 with permission).

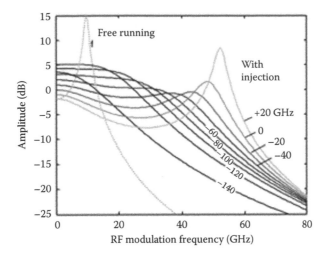

FIGURE 2.43 Frequency response with square of the magnitude of the output over the input as a function of excitation frequency curves for various detuning frequencies (labeled in GHz), at an injection ratio of $R_{inj} = 4$ dB. The dotted curve is the free-running response.

by the amplitude of the output optical waves with respect to the modulation frequency under free running and OIJ condition with an injection ratio of 5 dB, is shown in Figure 2.43 demonstrating the effective passband of the modulation by OIJ technique. Furthermore, the extension of such OIJ direct modulation can be seen in References 48 and 49. The phase response as a function of the excitation frequency of the direct modulation section will allow an amplification of the phase dependent signal and thus phase sensitive amplifiers.

A recent demonstration of FB-FB injection locking [ ] in which 20-Gbit/s PAM4 signal was generated by directly modulating two injection-locked Fabry–Perot lasers. The control can be implemented on the full field of the optical signal and achieved error-free transmission over 300-km SMF-28 (standard single mode fiber) and with digital signal processing for pre-emphasis. The eye diagram is shown in Figure 2.44. The small signal frequency response is seen in Figure 2.45, which shows a significant bandwidth extension under pre-emphasis. Figure 2.46 shows the transmission performance BER versus OSNR with distance as parameter.

These transmission results confirm the feasibility of long distance transmission of directly modulated laser under optical injection with phase locking.

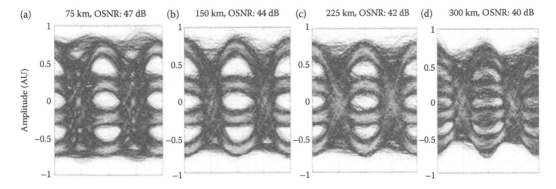

FIGURE 2.44 Eye diagrams of the direct detected PAM4 signal and the OSNRs after transmission over different length of SMF-28 of different lengths of 75–300 km (a), (b), (c), and (d) respectively with required optical amplifiers to compensate for fiber losses.

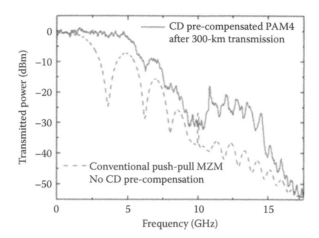

FIGURE 2.45   (a) RF spectrum of the detected signal after 300-km transmission. Solid line: Our transmitter with CD pre-compensation; Dashed line: Simulated push-pull MZM transmitter without CD compensation. (From Z. Liu et al., 300-km transmission of dispersion pre-compensated PAM4 using direct modulation and direct detection, in *Proc. OFC, Optical Fiber Conf.*, OFC2017, L.A. Annaheim, March 2017.)

2.7.3 INTEGRATED MUTUAL COUPLED CAVITY FOR OPTICAL INJECTION LOCKING

A laser cavity can be realized with the monolithic integration of two distributed Bragg reflector (DBR) lasers allowing one to extend the modulation bandwidth. Such an extension is obtained by introducing a photon–photon resonance (PPR) into the dynamic response at a frequency higher than the modulation bandwidth of the corresponding single-section laser. Design guidelines are proposed in Reference 51 and dynamic small and large signal simulations results are calculated using a finite difference traveling wave (FDTW) numerical simulator. The coupled cavity for optical

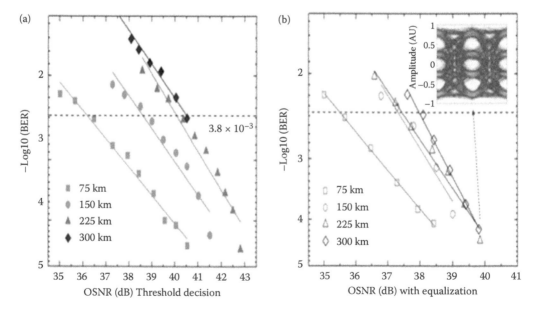

FIGURE 2.46   Measured BER after transmission over different lengths of SSMF under decision using (a) threshold level and (b) forward error correction-decision + decision feedback equalization (FFE-DFE) equalization. Markers in both figures: Square: 75 km; Circle: 150 km; Triangle: 225 km; Diamond: 300 km.

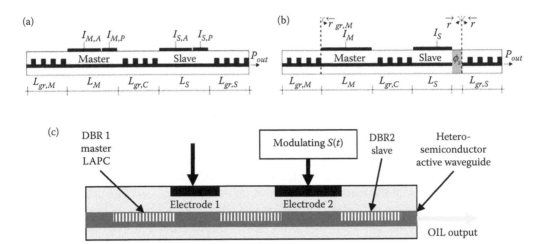

FIGURE 2.47 (a) Schematic representation of a coupled-cavity DBR laser including active and phase-control sections of the two laser cavities. The currents $I_{M,A}$ ($I_{M,P}$) and $I_{S,A}$ ($I_{S,P}$) are injected in the active (phase-control) sections of the master (left) and slave (right) laser cavities, respectively. (b) Analysis model of the mutual cavity laser. The phase tuning sections are replaced by a lumped phase term φ_S. The right dashed line indicates the reference plane at which the equivalent reflectivities $\vec{r}$ and $\overleftarrow{r}$ can be estimated by using the transfer matrix method (c) A simplified schematic diagram of mutual coupled DFB-DFB laser.

injection and phase control is shown in Figure 2.47, in which two DBRs reflect the guided waves back and forth in the guided region, hence forming the cavity. One laser section (the slave laser SL) is thus directly modulated with high frequency broadband driver and the other (the master laser ML) is operating in continuous wave (CW) mode. A phase section is employed to control to the phase of the locking.

Similarly, two VCSELs can be coupled laterally to generate optical injection resonant modes of operation as shown in Figure 2.48 to obtain the band width extension by lateral optical injection.

2.8 CONCLUDING REMARKS

Direct modulation of laser sources is described with modeling and practical considerations. These lightwave sources dominate the optical communications in the bit rate up to 28 Gb/s, but are no longer capable to operate in the higher bit rates, especially at greater than 56 Gb/s. In this case, external modulators are employed. These modulators have been described and coupled with advanced

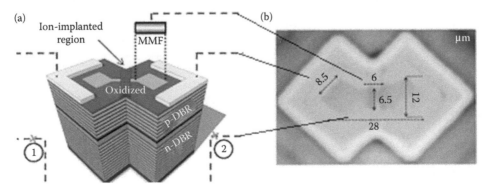

FIGURE 2.48 (a) Schematic view and (b) oxidation formation of transverse coupled cavity VCSEL.

modulation techniques as given in the rest of this chapter. However, recent experimental results have shown that if optical injection can be employed with phase locking the 3 dB bandwidth can be extending to more than 100 GHz allowing more than 200 GBaud rate. Thence, under polarization multiplexing and quadrature phase modulation, the aggregate can go up to more than 1.0 Tbps/carrier/100 GHz. This has proven the direct modulation transmitter more superior than an external modulation laser transmitter.

Since the proposal of dielectric waveguide and the advent of optical circular waveguide, the employment of modulation techniques has only been extensively exploited since the availability of optical amplifiers. The modulation formats allow the transmission efficiency, and hence the economy of ultra-high capacity information telecommunications. Optical communications have evolved significantly through several phases from single mode systems to coherent detection and modulation, which was developed with the main aim to improve on the optical power. The optical amplifiers defeated that main objective of modulation formats and allow the possibility of incoherent and all possible formats employing the modulation of the amplitude, the phase and frequency of the light-wave carrier.

Currently, photonic transmitters play a principal part in the extension of the modulation speed into several GHz range and make possible the modulation of the amplitude, the phase and frequency of the optical carriers and their multiplexing. Photonic transmitters using LiNbO$_3$ have been proven in laboratory and installed systems. The principal optical modulator is the MZIM, which can be a single or a combined set of these modulators whereby to form binary or multi-level amplitude or phase modulation and even more effective for discrete or continuous phase shift keying techniques.

Optical modulators employed for the generation of optical signals using advanced modulation techniques at ultra-high speed will be presented in the next chapter. The effects of the modulation on transmission performance will be given. MATLAB Simulink assists the modeling of the photonic transmitters. Optical modulation can be implemented using the phase shift blocks. These techniques will be further developed in Chapter 6 for the implementation of models for advanced modulation formats and pulse shaping.

Furthermore, optical injection technique is introduced as a new emerging source that can be under direct modulation in association with optical injection locking to generate IQ modulation. This IQ modulation subject will be described in details in later chapters. An introduction of the IQ modulation is briefly introduced here with optical phase modulation and intensity or interferometric structures for amplitude modulation are to complete the pictures and concept of modulation using non-optical modulators. This technique has now attracted significant attention from industries and academic communities as external modulator-free transmission systems using both coherent and non-coherent receptions.

APPENDIX 2A: MATLAB® PROGRAM FOR SOLVING THE LASER RATE EQUATION

```
%%%%%%%%%%%%%%%%%%%%%%%%%%%%%%%%%%%%%%%%%%%%%%%%%%%%%%%%%%%%%%%%%%
%                D F B L A S E R D I O D E P A R A M E T E R S
%                1.55micron; InGaAsP/InP; 2 Gbit/s; 40 Km
%                          Le Nguyen Binh  2017
%%%%%%%%%%%%%%%%%%%%%%%%%%%%%%%%%%%%%%%%%%%%%%%%%%%%%%%%%%%%%%%%%%
global q
global Vg
global Beta
global Imax
global Mune
global Gamma
global Anull
```

```
global Nnull
global Epsil
global Tphot
global Tcarr
global Vactv
global Alpha
global Trise
global Ibias
global MidLamda
global SpecWidth

Gamma = 0.8      ;    % OPTICAL CONFINEMENT FACTOR 0.8 %
Anull = 3.2e-20 ;    % GAIN COEFFICIENT                       %
Vg    = 7.5e+7  ;    % GROUP VELOCITY                         %
Nnull = 1.0e+12;     % CARRIER DENSITY AT TRANSPARENCY       %
Epsil = 2.5e-23 ;    % GAIN COMPRESSION FACTOR                %
Tphot = 3.0e-12 ;    % PHOTON LIFE-TIME                       %
Beta  = 3.0e-5  ;    % FRACTION OF SPONTANEOUS EMISSION     %
Tcarr = 0.3e-9  ;    % CARRIER LIFE-TIME                      %
q     = 1.6e-19 ;    % ELECTRON CHARGE                        %
Vactv = 1.5e-16 ;    % ACTIVE VOLUME LAYER                    %
Alpha = 5        ;    % LINE WIDTH ENHANCEMENT FACTOR        %
Trise = 1e-12    ;    % RISE TIME—1 ps version oc2000_1

Ith      = q/Tcarr*Vactv*(Nnull+1/Gamma/Anull/Vg/Tphot);  %  Threshold
current
Ibias    = 7.1*Ith ; % BIAS CURRENT
Imax     = 9.0*Ith ; % Maximum Input Current
Mune     = 0.042 ; % TOTAL DIFFERENTIAL QUANTUM EFFICIENCY %
% if Channel ==1
% MidLamda = 1553.3*1e-9; % Operating Wavelength of Laser Source = 1553.8
nm to match with centre of AWG see mulplex %
%change from 1553.8 nm to 1553.3 nm to match with ITU-grid AC and GA
% else
    MidLamda= OpLamda(Channel);
% end

SpecWidth= 40e-15;      % Laser Source Spectral Line Width       = 10 MHz
linewidth set %
%%%%%%%%%%%%%%%%%%%%%%%%%%%%%%%%%%%%%%%%%%%%%%%%%%%%%%%%%%%%%%%%%%%%%
Solving The Rate equations
%%%%%%%%%%%%%%%%%%%%%%%%%%%%%%%%%%%%%%%%%%%%%%%%%%%%%%%%%%%%%%%%%%%%
function rate=rate_equ(t,y)
global q
global Vg
global Nsd
global Ssd
global Beta
global Gamma
global Anull
global Nnull
global Epsil
global Tphot
global Tcarr
global Vactv
```

```
global Alpha
global Trise
global Ibias
global Sigma

global Imax
global dummy
global Initial
global InPulse
global BitPeriod
Nsd = 1/(Gamma*Anull*Vg*Tphot)+Nnull ;
Ssd = (Tphot/Tcarr)*Nsd*(Imax/Ibias-1);
Sigma = 2e-20;

%---- Computing Rate Equations ------------------------------
------------
rate = zeros(3,1);

rate(1) = (Gamma*Anull*Vg*((y(2)-Nnull)/...
          (1+Epsil*y(1)))-1/Tphot)*y(1)+Beta*Gamma*y(2)/Tcarr +...
          randn*sqrt(2*Beta*Vactv*Nsd*(Vactv*Ssd+1)^3/Tcarr) ;

rate(2) = dgcoding(t,InPulse)/(q*Vactv)-y(2)/...
          Tcarr-Vg*Anull*y(1)*(y(2)-Nnull)/(1+Epsil*y(1)) +...
          randn*sqrt(2*Vactv*Nsd/Tcarr*(Beta*Vactv*Ssd+1));

rate(3) = Alpha*(Gamma*Vg*Anull*(y(2)-Nnull)-1/Tphot)/2 +...
          randn*sqrt(Gamma*Vg*Sigma*(y(2)-Nnull)/y(1)) ;

  if t<0
    rate=Initial;
  end

  dummy=dummy+1;
  if dummy>=50
%    waitbar(t/(BitPeriod*size(InPulse,2))) ;
    dummy=0 ;
end
```

REFERENCES

1. R. C. Alferness, Optical guided-wave devices, *Science*, vol. 234, no. 4778, pp. 825–829, 1986.
2. M. Rizzi and B. Castagnolo, Electro-optic intensity modulator for broadband optical communications, *Fiber and Integrated Optics*, vol. 21, pp. 243–251, 2002.
3. J. Noda, Electro-optic modulation method and device using the low-energy oblique transition of a highly coupled super-grid, *IEEE J. Lightw. Tech.*, vol. LT-4, pp. 1445–1453, 1986.
4. H. Takara, High-speed optical time-division-multiplexed signal generation, *Optical and Quantum Electronics*, vol. 33, no. 7–10, pp. 795–810, 10 July 2001.
5. E. L. Wooten et al., A review of lithium niobate modulators for fiber-optic communications systems, *IEEE J. Sel. Topics Quant. Elect.*, vol. 6, no. 1, pp. 69–80, Jan./Feb 2000.
6. K. Noguchi, O. Mitomi, H. Miyazawa, and S. Seki, A broadband Ti: LiNb03 optical modulator with a ridge structure, *J. Lightwave Tech.*, vol. 13, no. 6, pp. 1164–1168, June 1996.
7. M. Suzuki, Y. Noda, H. Tanaka, S. Akiba, Y. Kuahiro, and H. Isshiki, Monolithic integration of InGaAsP/InP distributed feedback laser and electroabsorption modulator by vapor phase epitaxy, *IEEE J. Lightwave LT-Tech.*, vol. 5, no. 9, p. 12, Sept. 1987.

8. P. K. Tien, Integrated optics and new wave phenomena in optical waveguides, *Rev. Mod. Phys.*, vol. 49, pp. 361–420, 1977.

9. A. Yariv, C. V. Mead, and J. V. Parker, 5C3-GaAs as an electrooptic modulator at 10.6 microns, *IEEE J Quantum Electron.*, vol. QE-2, no. 8, pp. 243–245, 1966.

10. C. Langton, Basic concept of modulation, www.complextoreal.com. Access date: December 2007.

11. T. Kawanishi, S. Shinada, T. Sakamoto, S. Oikawa, K. Yoshiara, and M. Izutsu, Reciprocating optical modulator with resonant modulating electrode, *Electronics Letters*, vol. 41, no. 5, pp. 271–272, 2009.

12. R. E. Epworth, K. S. Farley, and D. Watley, Polarization mode dispersion compensation, US Patent 398/152, 398/202, 398/6.

13. G. P. Agrawal, *Fiber-Optic Communication Systems*, 3rd ed., New York: Wiley, 2002.

14. P. Kaminow and T. Li, *Optical Fiber Communications*, vol. IVA, New York: Elsevier Science, Chapter 16, 2002.

15. F. Amoroso, Pulse and spectrum manipulation in the minimum frequency shift keying (MSK) format, *IEEE Trans. Commun.*, vol. 24, pp. 381–384, March 1976.

16. H. Ludvigsen, M. Tossavainen, and M. Kaivola, Laser linewidth measurements using self-homodyne, detection with short delay, *Opt. Commun.*, vol. 155, p. 180, 1998.

17. H. Packard, *Phase Noise Characterization of Microwave Oscillators*, product note 11729C-2, Hewlett-Packard, San Jose, CA, USA, 1985.

18. L. N. Binh and N. Q. Nam, *Ultra-Short Pulse Fiber Lasers*, Boca Raton, FL USA: CRC Press, 2009.

19. R. Paschotta, H. R. Telle, and U. Keller, eds., Encyclopedia of laser physics and technology: Articles on laser noise, intensity noise, phase noise and others, in *Noise of Solid State Lasers, in Solid-State Lasers and Applications* (ed. A. Sennaroglu), Boca Raton, FL: CRC Press, Chapter 12, pp. 473–510, 2007.

20. O. Llopis, P. H. Merrer, H. Brahimi, K. Saleh, and P. Lacroix, Phase noise measurement of a narrow linewidth CW laser using delay line approaches, *Optics Letters*, vol. 36, no. 14, pp. 2713, July 15, 2011.

21. D. Derickson, *Fiber Optic Test and Measurement*, Upper Saddle River, NJ: Prentice-Hall, 1998.

22. H. Ludvigsen and E. Bodtker, New method for self-homodyne laser linewidth measurements with a short delay fiber, *Opt. Commun.*, vol. 110, p 595, 1994.

23. W. V. Sorin, K. W. Chang, G. A. Conrad, and P. R. Hernday, Frequency domain analysis of an optical FM discriminator, *J. Lightwave Technol.*, vol. 10, p. 787, 1992.

24. S. Camatel and V. Ferrero, Narrow linewidth CW laser phase noise characterization methods for coherent transmission system applications, *J. Lightwave Technol.*, vol. 26, p. 3048, 2008.

25. P. Gallion and G. Debarge, Analysis of the phase-amplitude coupling factor and spectral linewidth of distributed feedback and composite-cavity semiconductor lasers, *IEEE J. Quantum Electron.*, vol. 20, p. 343, 1984.

26. L. B. Mercer, 1/f frequency noise effects on self-heterodyne linewidth measurements, *J. Lightwave Technol.*, vol. 9, p. 485, 1991.

27. A. Godone, S. Micalizio, and F. Levi, Metrological characterization of the pulsed Rb clock with optical detection, *Metrologia*, vol. 49, no. 4, p. 313, 2012.

28. G. Di Domenico, S. Schilt, and P. Thomann, Simple approach to the relation between laser frequency noise and laser line shape, *Appl. Opt.*, vol. 49, no. 25, pp. 4801–4807, 1 Sep 2010.

29. E. Ngoya and R. Larcheveque, Envelop transient analysis: A new method for the transient and steady state analysis of microwave communication circuits and systems, in *Proc. IEEE MTT-S International Microwave Symposium, IEEE,* San Francisco, CA, USA, vol. 3, p. 1365, 1996.

30. Y. Yamada, H. Taga, and K. Goto, Comparison between VSB, CS-RZ and NRZ format in a conventional DSF based long haul DWDM system, in *Proceedings of ECOC'02*, Amsterdam, The Netherlands, vol. 4, pp. 1–2, 2002.

31. A. H. Gnauck, X. Liu, X. Wei, D. M. Gill, and E. C. Burrows, Comparison of modulation formats for 42.7-gb/s single-channel transmission through 1980 km of SSMF, *IEEE. Photonics Technology Letters*, vol. 16, no. 3, pp. 909–911, 2004.

32. G. P. Agrawal, *Fiber-Optic Communication Systems*, 3rd ed., New York: Wiley, 2002.

33. Hirano, Y. Miyamoto, and S. Kuwahara, Performances of CSRZ-DPSK and RZ-DPSK in 43-Gbit/s/ch DWDM G.652 single-mode-fiber transmission, in *Proc. Optical Fiber Conference OFC'03*, vol. 2, pp. 454–456, 2003; A. H. Gnauck, G. Raybon, P. G. Bernasconi, J. Leuthold, C. R. Doerr, and L. W. Stulz, 1-Tb/s (6/spl times/170.6 Gb/s) transmission over 2000-km NZDF using OTDM and RZ-DPSK format, *IEEE Photonics Technology Letters*, vol. 15, no. 11, pp. 1618–1620, 2000.

34. I. B. Djordjevic and B. Vasic, 100-Gb/s transmission using orthogonal frequency-division multiplexing, *IEEE Photonics Tech. Lett.*, vol. 18, no. 15, pp. 1576–1578, Aug 1, 2006.

35. A. J. Lowery, L. Du, and J. Armstrong, *Orthogonal Frequency Division Multiplexing for Adaptive Dispersion Compensation in Long Haul WDM Systems*, OFC 2006, Anaheim, CA, USA: IEEE, Paper PDP39, March 9, 2006.

36. Fujitsu, 100G IQ modulator, Japan 2010. http://www.fujitsu.com/jp/group/foc/en/products/optical-devices/100gln/.

37. L. N. Binh, LiNbO$_3$ optical modulators: Devices and applications, *J. Crystal Growth*, vol. 288, no. 1, pp. 180–187, 2006.

38. S. Chandrasekhar, X. Liu, P. J. Winzer, J. E. Simsarian, and R. A. Griffin, Compact all-InP laser-vector-modulator for generation and transmission of 100-Gb/s PDM-QPSK and 200-Gb/s PDM-16-QAM, *J. Lightwave Technol.*, vol. 32, pp. 736–742, 2014.

39. D. A. B. Miller, D. S. Chemla, T. C. Damen, A. C. Gossard, W. Wiegmann, T. H. Wood, and C. A. Burrus, Band-edge electroabsorption in quantum well structures: The quantum-confined stark effect, *Physical Review Letters*, vol. 53, no. 22, pp. 2173–2176, 1984.

40. CFP MSA Specifications, www.cfp-msa.org.

41. https://www.opennetworking.org/about/liaisons/1196-oif.

42. Hiromu Sato, Hiroki Miura, Feng Qiu, Andrew M. Spring, Tsubasa Kashino, Takamasa Kikuchi, Masaaki Ozawa, Hideyuki Nawata, Keisuke Odoi, and Shiyoshi Yokoyama, Low driving voltage Mach-Zehnder interference modulator constructed from an electro-optic polymer on ultra-thin silicon with a broadband operation, *Optics Express*, vol. 25, no. 2, pp. 768–775, 2017.

43. P. Dong, X. Chongjin, L. L. Buhl, Chen Young-Kai, J. H. Sinsky, and G. Raybon, Silicon in-phase/quadrature modulator with on-chip optical equalizer, in *Proc. ECOC 2014, European Optical Communications Conference,* Cannes France, Paper We.1.4.5, September 14–16, 2014.

44. G. Letal et al., Low loss InP C-band IQ modulator with 40 GHz bandwidth and 1.5 V V_π, in *Proc. of OFC 2015, Optical Fiber Conference,* Annaheim, CA, USA, Paper Th4E.3, March 2015.

45. B. Zhang, X. Zhao, L. Christen, D. Parekh, W. Hofmann, M. C. Wu, M. C. Amann, C. J. Chang-Hasnain, and A. E. Willner, Adjustable chirp injection-locked 1.55-μm VCSELs for enhanced chromatic dispersion compensation at 10-Gbit/s, in *Proc. Opt. Fiber Conference 2008, paper OWT7,* OFC 2008, Anaheim, CA USA, Paper OWT7, March 2008.

46. C. J. Chang-Hasnain and X. Zhao, Ultrahigh-speed laser modulation by injection locking, in *Optical Fiber Telecommunications Vol.A: Components and Subsystems*, Tingye Li, Alan E. Willner, and Ivan Kaminow (eds.), 3rd ed., London: Academic Press, pp. 145–182, 2008.

47. E. K., Lau, Member, Liang Jie Wong, and M. C. Wu, Enhanced modulation characteristics of optical injection-locked lasers: A tutorial, *IEEE Journal of Selected Topics in Quantum Electronics*, vol. 15, no. 3, pp. 618–633, May–June 2009.

48. N. K. Fontaine et al., Chirp-Free Modulator using Injection Locked VCSEL Phase Array, paper PDP Th.3.A.3, in *Proc. Europ. Conf. Opt. Comm. 2016,* Duesseldorf, Germany, Sept 2016.

49. R. Slavík, J. Kakande, R. Phelan, J. O'Carroll, B. Kelly, and D. J. Richardson Richards, 24 Gbit/s Synthesis of BPSK signals via Direct Modulation of Fabry–Perot Lasers under Injection Locking, online http://www.optics.rochester.edu/workgroups/knox/myweb/ISUPT2013/archive/Slavik.pdf. Accessed on January 15, 12017.

50. Z. Liu, G. Hesketh, B. Kelly, J. O'Caroll, R. Phelan, D. J. Richardson, and R. Slavík, 300-km transmission of dispersion pre-compensated PAM4 using direct modulation and direct detection, in *Proc. OFC, Optical Fiber Conf.*, OFC2017, L.A. Annaheim, March 2017.

51. P. Bardella, W. W. Chow, and I. Montrosset, Design and analysis of enhanced modulation response in integrated coupled cavities DBR lasers using photon-photon resonance, *Photonics*, vol. 3, no. 4, 2016. doi:10.3390/photonics3010004.

3 Binary Digital Optical Modulation

Since the invention of optical fibers, the modulation of lightwaves for transmission in optical communication systems has been mainly based on on–off switching of the intensity of the waves, rather than manipulating the phase or frequency of the field of the waves. In the 1980s, interests in extending the repeaterless distance from 4 km to longer 6 km coherent optical communications attract the uses of external modulation of the carrier wave. This external modulation concept was briefly introduced in Chapter 2.

However, there was no sufficiently narrow line width laser or integrated optical components that would facilitate the development of such techniques. The invention of optical amplification devices diverted the attention from coherent communications, until recently when the bit rate was extended to 40 Gb/s, 100G, 400G, and even 1.0 Tera-bps. The necessity of narrowing the signal bandwidth to minimize the dispersion effects proves the exploitation of modulation technique for direct detection systems has been successfully demonstrated. Hence, several modulation techniques have been studied. Coherent communications have also been revived in this trend and are associated with electronic digital processing and processors at ultra-high sampling rate for estimation of the phase of the demodulated signals, compensation of frequency offset, compensation of dispersion, and non-linear impairments. This chapter gives an account of optical modulation of binary signals in pulse shaping of return zero (RZ) or non-return to zero (NRZ), and the manipulation of either the amplitude or phases (discrete or continuous) of the lightwave carrier. Furthermore, the single sideband (SSB) modulation technique is also illustrated. We must note that pulse shaping of the modulated optical channels such as Nyquist filtering is important for spectral efficient modulation. This technique is not given here but in Chapter 9.

3.1 ADVANCED PHOTONIC COMMUNICATIONS AND CHALLENGING ISSUES

3.1.1 BACKGROUND

Beginning from the first research phase around the early 1970s, fiber-optic communications have progressed tremendously over the last three decades. The first-generation lightwave systems were commercially deployed in 1983 and operated in the first wavelength window of 800 nm at transmission bit rates of up to 45 Mb/s [1,2]. After the introduction of ITU-G652 Standard Single Mode Fibers (SSMF) in the 1970s [2,3], the second-generation of lightwave transmission systems became available in the early 1980s [4,5]. The operating wavelengths were shifted to the second window of 1300 Nm, and this window offers much lower attenuation for the optical fiber compared to the previous 800 nm region. These systems could operate at transmission rates of up to 1.7 Gb/s and have a repeaterless transmission distance of about 50 km [6]. Optical communications then evolved to third-generation transmission systems that utilized the lowest attenuation 1550 nm wavelength window and operated at 2.5 Gb/s bit rate [7]. These systems were commercially available in 1990 [6,8] with a repeater spacing of around 60–70 km. At this stage, the generation of optical signals was mainly based on the direct modulation of the semiconductor laser source. Since the invention of the Erbium-Doped Fiber Amplifiers (EDFA) in the early 1990s [9,10], lightwave systems have rapidly evolved to Wavelength Division Multiplexing (WDM) and then to Dense WDM (DWDM) optically amplified transmission systems, which are capable of transmitting multiple 10 Gb/s channels. These high-speed and long-haul systems employ external modulation in their optical transmitters. The present fiber-optic transmission systems are considered to be at the fifth-generation with the transmission capacity of a few Tb/s [6].

Coherent detection was the focus of extensive research and developments during the 1980s and early 1990s [11–16], and was the main detection technique in the first three generations of lightwave transmission systems. At that time, the main motivation for the development of coherent optical systems was to improve the receiver sensitivity, commonly by 3–6 dB [12,15]. The repeaterless transmission distance was thus able to be extended to more than 60 km of SSMF (with 0.2 dB/km attenuation factor). However, coherent optical systems suffer severe performance degradation due to fiber dispersion impairments. In addition, the phase coherence for the lightwave carriers of the laser source and the local laser oscillator was very difficult to maintain. On the contrary, incoherent detection technique minimizes the linewidth obstacles of the laser source as well as the local laser oscillator, and relaxes the requirement of the phase coherence. Moreover, incoherent detection mitigates the problem of polarization control for the mixing of the transmitted lightwaves and the local laser oscillator at multi-THz optical frequency range. The invention of Erbium-doped optical amplifier (EDFAs) [17] capable of producing optical gains of 25 dB and above, has also greatly contributed to further advances of incoherent optical transmission systems. Since the mid-90s, incoherent detection has become the preferred choice in digital photonic transmission systems.

3.1.2 CHALLENGING ISSUES

In recent years, enormous increases in demand for broadband communications have mainly been driven by the rapid growth of multi-media services, peer-to-peer networks, and IP streaming services, in particular, IP TV, the Internet of Things (IoT), and 5G mobile networking, as well as cloud serves. It is most likely that such tremendous growth will continue in the coming years. This is the main driving force for local and global telecommunications service carriers to develop high-performance and high-capacity next-generation optical networks. The overall capacity of WDM and DWDM optical systems can be boosted by increasing the base transmission bit rate of each optical channel, multiplexing more channels in a DWDM system, or, preferably, by combining both of these schemes. While implementing these schemes, optical transmission systems encounter a number of challenging issues, which are outlined in the following paragraphs.

Current 10 Gb/s transmission systems employ intensity modulation (IM), also known as on-off keying (OOK), and utilize the non-return-to-zero (NRZ) pulse shapes. The term OOK can be used interchangeably with amplitude shift keying (ASK). It is noted here that the OOK format simply implies the on–off states of the optical lightwaves where only the manipulation of the intensity is considered. On the other hand, the ASK format takes into account not only the amplitude, but also the phase in the constellation diagram. Moving towards high bit rate transmission, the performance of OOK photonic systems is severely degraded due to fiber impairments, including fiber dispersion and fiber non-linearities. The fiber dispersion is classified into chromatic dispersion (CD) and polarization mode dispersion (PMD), causing the intersymbol interference (ISI) problem. On the other hand, severe deterioration on the performance due to fiber non-linearities are the result of the high-power spectral components at the carrier and signal frequencies of OOK-modulated optical signals. It is also of concern that existing transmission networks are comprised of millions of kilometers of SSMF, which have been installed for approximately two decades. The properties of these fibers are not only low-performance compared to the state-of-the-art fibers used in recent laboratory "hero" experiments, but have also degraded after many years of use.

The total transmission capacity can also be enlarged by increasing the number of multiplexed DWDM optical channels. This can be carried out by reducing the spacing between these channels, for instance, from 100 GHz down to 50 GHz, or even 25 GHz have been reported [18,19]. The reduction of the channel spacing also results in narrower bandwidths for the optical multiplexers (mux) and demultiplexers (demux). Passing through these narrowband optical filters, signal waveforms are distorted and optical channels suffer from interchannel cross-talks. These narrowband filtering problems are becoming more severe, especially at a high transmission bit rate, for example, 40 Gb/s, thus degrading the system performance significantly.

Together with the demand of boosting the total system capacity, another challenge for the service carriers is to find cost effective solutions for the upgrading process. These cost-effective solutions should require minimum renovation to the existing photonic and electronic sub-systems, that is, the upgrading should only take place at the transmitter and receiver ends of an optical transmission link. In short, next-generation optical networks are required to meet the desired high performance, while the cost of the system upgrading should also be considered.

All of the above-described problems facing contemporary fiber-optic communications can be overcome with spectrally efficient transmission schemes. The main motivation of this research is to find effective solutions to upgrade the aggregate transmission capacity, while at the same time minimizing the effects of the fiber shortfalls, particularly the dispersion impairments. The enabling technologies presented in this research aim to overcome the challenging issues addressed above.

3.1.3 Enabling Technologies

3.1.3.1 Digital Modulation Formats

Spectrally efficient transmission schemes can be achieved by utilizing advanced modulation formats. A number of modulation formats have recently been proposed as alternatives for the OOK format. These include return-to-zero (RZ) pulses in OOK/ASK systems [20–22], differential phase shift keying (DPSK) [18,23–25] and, more recently, minimum-shift keying (MSK) [26–33]. These formats are adopted into photonic communications from the knowledge of wire-line and wireless communications.

DPSK has received much attention over the last few years, particularly when combined with RZ pulses. The main advantages of this format is (i) 3-dB improvement over OOK on the receiver sensitivity by using an optical balanced receiver [21,34,35] and (ii) high tolerance to fiber non-linearities [6,20,34,36] such as intrachannel self phase modulation (SPM) and interchannel cross phase modulation (XPM) effects. Several experimental demonstrations of DPSK long-haul DWDM transmission systems for 10 Gb/ps, 40 Gb/s and higher bit rates have been reported recently [18,23,37–39]. However, there are few practical experiments addressing the performance of cost-effective 40 G DPSK-10 G OOK hybrid systems on the existing old-vintage SSMF infrastructure [40,41]. In addition, the performance of 40 Gb/s DPSK for use in this hybrid transmission scheme has not been thoroughly studied. Therefore, one of the main contributions of this chapter and others in this book is to present the feasibility of overlaying 40 Gb/s DPSK channels on the existing 10 Gb/s OOK channels for implementing the hybrid systems.

The MSK format offers a spectrally efficient modulation scheme compared to the DPSK counterpart at the same bit rate. As a sub-set of continuous phase frequency shift keying (CPFSK), MSK possesses spectrally efficient attributes of the CPFSK family. The frequency deviation of MSK is equal to a quarter of the bit rate and this frequency deviation is also the minimum spacing to maintain the orthogonality between two FSK-modulated frequencies. On the other hand, it can also be considered as a particular case of offset differential quadrature phase shift keying (ODQPSK) [42–46], which enables MSK to be represented by I and Q components on the constellation. The advantageous characteristics of the optical MSK format can be summarized as follows:

- Compact spectrum, which is of particular interest for the spectrally efficient high-speed transmission systems. In addition, this provides robustness to the problem of tight optical filtering.
- High suppression of the side lobes in the optical power spectrum compared to the DPSK. The roll-off factor follows f^{-4} rather than f^{-2} as in the case of DPSK.
- Constant envelop property, which eases the measure of the average optical power.
- No high power spikes in the power spectrum and high energy concentration in the main lobe offers the advantage of lower peak input powers into the fiber, thus reducing the fiber nonlinear effects.
- Being a sub-set of either CPFSK or ODQPSK, MSK can be detected either incoherently based on the phase or the frequency of the lightwave carrier or coherently based on the popular I-Q detection structure.

In addition, if MSK can be combined with multi-level modulation schemes, the baud rate would be reduced and the spectral efficiency can be even more effective. This is of great interest for long-haul and metropolitan optical networks and provides the main motivation for the investigation on the performance of the proposed dual-level MSK modulation format in this research.

Several studies have been conducted [26–29] focusing on the generation and direct detection of externally modulated optical MSK, as well as the format's CD tolerance. However, there is little research investigating the potential of externally modulated MSK format for digital photonic transmission systems, particularly at high bit rates such as 40 Gb/s [26,47,48]. Therefore, another key contribution of this research is to provide comprehensive studies on the performance of MSK-based modulation formats for long-haul and metropolitan optical transmission systems.

3.1.3.2 Self-Coherent Optical Receivers

The modulation formats described in this section, the optical DPSK and MSK-based formats, can be demodulated incoherently by using an optical balanced receiver that employs a Mach–Zehnder delay interferometer (MZDI). In the incoherent detection of optical DPSK, MZDI is used to detect the differentially coded phase information between every two consecutive symbols [6,34,35,49]. This detection is carried out in the photonic domain, as it is beyond the speed of the electrical domain, especially at very high bit rates of 40 Gb/s or above. The MZDI balanced receiver is also used for the incoherent detection of optical MSK signals [26,28,29]. However, using the MZDI-based detection scheme, optical MSK does not improve the CD tolerance significantly over to the DPSK counterpart [28,30].

In order to fully exploit the advantages of optical MSK spectral characteristics, MSK-modulated lightwaves can be incoherently detected based on the principles of optical frequency discrimination. Thus, an optical frequency discrimination receiver (OFDR) is proposed in this research, which employs dual narrowband optical filters and an optical delay line (ODL). This receiver scheme effectively mitigates the CD-induced ISI effects and enables breakthrough CD tolerances for optical MSK transmission systems. These results have been reported in References 31 and 32. The feasibility of this novel optical receiver is based on recent advances in the design of optical filters, the microring resonator filters, in particular. Such optical filters have very narrow bandwidths, for example, less than 2 Hz (3-dB bandwidth), and they have been realized commercially by Little Optics [47–50]. This research provides a comprehensive study, not only on this OFDR scheme from the receiver operational principles to the analysis of the receiver design, but also on the performance of OFDR-based optical MSK systems.

In phase modulation techniques, the phase transition between two consecutive bit slots can be either non-continuous or continuous. As a non-continuous format, DPSK phase, in particular RZ DPSK, has received considerable research attention, as RZ DPSK displays advantages such as

3-dB improvement on the receiver sensitivity with the use of balanced receivers and robustness to fiber non-linear effects.

Spectral efficiency is a critical requirement for modern high-performance digital photonic transmission systems. The fact that non-continuous phase modulation schemes, such as DPSK, are not spectrally efficient motivates the exploration of continuous phase modulation (CPM), especially continuous phase frequency shift keying (CPFSK) formats, which are popularly known for its compact power spectrum [10,11]. Among CPFSK formats, MSK is most widely used in bandwidth-limited digital wireless and satellite communications. However, the potential of MSK using external modulation techniques in non-coherent optical communications has been approached. In MSK, the frequency deviation from the lightwave carrier is equal to a quarter of the bit rate and this is also the minimum spacing to maintain the orthogonality attribute between the two modulated frequencies. Additionally, MSK is a special case of ODQPSK in which I and Q components are interleaved by one bit period [10,18,19]. The two approaches from CPFSK and ODQPSK enable two possible schemes for generating optical MSK signals.

If a multi-level scheme can be incorporated with the MSK format, the transmission baud rate would be reduced in addition to the spectral efficiency. This effectively mitigates the fiber impairments for optically amplified long-haul and ultra-long-haul transmission systems, as well as relax the requirement of high-speed processing electronics. The above demand is the principal motivation for proposing the dual-level MSK modulation scheme in fiber-optic communications, which utilizes two binary bits for each modulated symbol. The transmission symbol rate is thus halved compared to the data rate from a bit pattern generator.

The advanced modulation formats described so far can be detected coherently or incoherently. The incoherent detection is preferred, as it overcomes the laser's linewidth obstacles facing homodyne or heterodyne coherent detection systems. Furthermore, the incoherent detection scheme can be easily implemented by using a Mach–Zehnder delay interferometer (MZDI). This structure reflects the main advantages of the optical domain over the electrical domain for systems operating at ultrahigh bit rates, such as 40 Gb/s. This MZDI is normally utilized in balanced receivers for incoherent detection of both DPSK and MSK optical signals. Additionally, optical MSK signals can also be incoherently detected by using an optical frequency discrimination receiver (OFDR) employing dual optical narrowband filters and an optical delay line (ODL).

This chapter is organized as follows: Section 3.3.1 presents principles for generating several types of RZ optical pulses, with and without the suppression of the lightwave carrier. The modulation process of these RZ pulses can be clearly explained by using a phasor representation. Section 3.3.3 then revises optical transmitter and receiver configurations of DPSK format. In Section 3.3.4, two MSK optical transmitter schemes are presented. These two schemes are derived from the knowledge that MSK is either a sub-set of CPFSK or a special case of ODQPSK. Section 3.3.4 also describes incoherent detection techniques for optical MSK systems, which are either a MZDI balanced receiver or an OFDR scheme. In Section 3.3.5, the novel dual-level MSK modulation format is proposed for optical communications together with its generation and incoherent detection schemes. Finally, Section 3.3.6 compares key spectral characteristics of all modulation formats under the study before a summary of this chapter is provided.

3.2 RETURN-TO-ZERO OPTICAL PULSES

3.2.1 GENERATION PRINCIPLES

A conventional RZ OOK optical transmitter comprises of two external LiNbO$_3$ optical Mach-Zehnder Interferometric Modulators (MZIM), as shown in Figure 3.1.

The first MZIM serves as a pulse carver to generate a periodic pulse train of a specific RZ pulse shape format and it is usually implemented with a chirp-free X-cut single-drive MZIM. Different

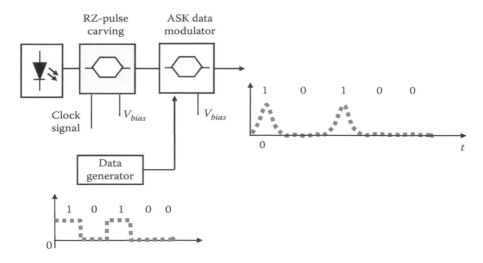

FIGURE 3.1   Conventional structure of an OOK/ASK optical transmitter utilizing two MZIMs.

types of RZ pulses can be generated depending on the amplitude of sinusoidal electrical driving voltages and the biasing schemes of MZIM. Suppression of the lightwave carrier can be carried out at this stage to form Carrier Suppressed—RZ (CS-RZ) pulses. Compared to other RZ types, CS-RZ is found to have interesting attributes for long-haul transmission, including the π phase difference between adjacently modulated bits, the suppression of high power carrier, and the narrower spectral width. Another type of RZ pulses, RZ33, also has several advantages because of its soliton-like properties. The RZ33 pulse has full-width half maximum (FWHM) positions occupying 33% of the duty circle, whereas a FWHM of 67% is the case of CS-RZ. These optical RZ pulses are represented in the optical field $E(t)$ as follows [20]:

$$E(t) = \begin{cases} \sqrt{\dfrac{E_b}{T}}\,\sin\left[\dfrac{\pi}{2}\cos\left(\dfrac{\pi t}{T}\right)\right] & 67\% \text{ duty-ratio RZ pulses or CS-RZ} \\[4mm] \sqrt{\dfrac{E_b}{T}}\,\sin\left[\dfrac{\pi}{2}\left(1 + \sin\left(\dfrac{\pi t}{T}\right)\right)\right] & 33\% \text{ duty-ratio RZ pulses or RZ33} \end{cases} \tag{3.1}$$

where E_b is the pulse energy per a transmitted bit and T is the bit period.

The art in generating these two RZ pulse types relies on the biasing position on the transmittance characteristic curve of the MZIM pulse carver and the amplitude of the sinusoidal driving signal. The bias point is situated at minimum or maximum position of the characteristic curve for the case of CS-RZ or RZ, respectively. The peak-to-peak amplitude of the RF driving signal for these two RZ pulses is similar and equal to $2V_\pi$ where V_π is the required driving voltage to obtain a π phase shift on the lightwave carrier at the output of MZIM. It should be emphasized that that the electrical driving signal is operating at only half of the bit rate and hence, the pulse carver is actually implementing the frequency doubling. Figure 3.2a and b demonstrates all the above-described settings as well as the pulse shapes of generated CS-RZ and RZ33 optical pulse trains.

The pulse carver for generating RZ optical pulses can also utilize a dual-drive MZIM that is driven by two complementary sinusoidal RF signals. The settings of the biasing scheme and driving signal as well as the modulation process in these dual-drive MZIMs are explained in the next section by using the phasor representation.

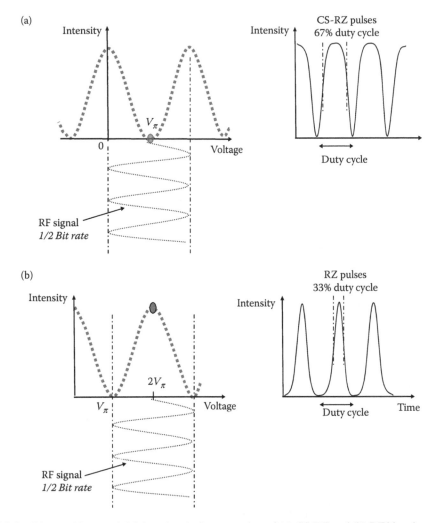

FIGURE 3.2 Bias positions and driving signals for generation of (a) CS-RZ and (b) RZ33 pulses.

3.2.2 PHASOR REPRESENTATION

The output optical field of a dual-drive MZIM is given as

$$E_o = \frac{E_i}{2}\left[e^{j\varphi_1(t)} + e^{j\varphi_2(t)}\right] = \frac{E_i}{2}\left[e^{j\pi V_1(t)/V_\pi} + e^{j\pi V_2(t)/V_\pi}\right] = E_1 + E_2 \tag{3.2}$$

This equation suggests that a phasor diagram (see Figure 3.3) that involves vector addition and simple trigonometric calculus can give a clear view on the superposition of optical fields at the coupling output of dual-drive MZIM. The RF driving signals follow a push-pull configuration, that is, the data $V_1(t)$ and the inverse data ($\overline{\text{data}}$) $V_2(t)$ ($V_2(t) = -V_1(t)$) are applied into each arm of the MZIM, respectively.

It should be noted that the widths of RZ33 and CS-RZ pulses are measured at FWHM positions that are defined for intensity pulses, whereas the phasor diagram considers optical field vectors. Thus, the normalized output optical field (E_o) has a magnitude value of $1/\sqrt{2}$ at FWHM positions. In addition, the time interval between these FWHM points defines the time ratios such as 33% or 67% and they are normalized with respect to the duty cycle.

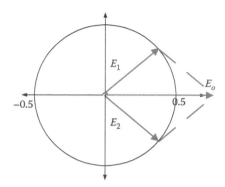

FIGURE 3.3   Phasor representation for generation of optical signals using a dual-drive MZIM.

3.2.2.1   Phasor Representation for CS-RZ Pulses

Figure 3.4a shows key parameters in the generation scheme of CS-RZ pulses, including the biasing voltage V_{bias} and the amplitude of RF driving signals. Accordingly, its initialized phasor representation is demonstrated in Figure 3.4b where the red dot indicates the sum of two initialized field vectors of two arms.

Values of these key parameters are outlined as follows:

$$V_{bias} = \pm \frac{V_\pi}{2}$$

RF driving signal on each arm has the amplitude of $V_\pi/2$ (i.e., $V_{p-p} = V_\pi$ swings in opposite directions and operates at half of the bit rate $(R/2)$).

Considering the generation of 40 Gb/s CS-RZ pulses $(R = 40 \text{ Gb/s})$, the modulating frequency is $f_m = R/2 = 20 \text{ GHz}$.

At FWHM positions of the optical intensity pulse, $E_o = \pm 1/\sqrt{2}$ and the field vectors E_1 and E_2 form with the vertical axis a phase of $\pi/4$, as shown in Figure 3.5.

The conditions shown in Figure 3.5 lead to:

$$\frac{\pi}{2}\sin(2\pi f_m) = \frac{\pi}{4} \Rightarrow \sin 2\pi f_m = \frac{1}{2} \Rightarrow 2\pi f_m = \left(\frac{\pi}{6}, \frac{5\pi}{6}\right) + 2n\pi \tag{3.3}$$

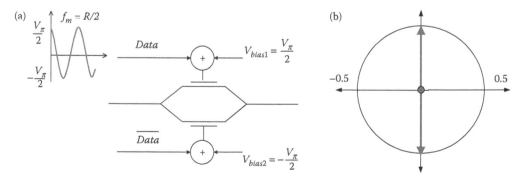

FIGURE 3.4   Initialized stage for generation of CS-RZ optical pulses: (a) RF driving signal and biasing voltages and (b) initial phasor representation.

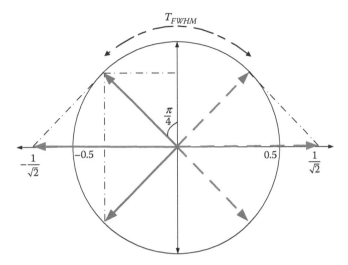

FIGURE 3.5 Phasor representation of CS-RZ generation at FWHM positions using dual-drive MZIM.

Thus, the normalized FWHM (with reference to the duty cycle) of a CS-RZ optical pulse can be obtained as follows:

$$T_{FWHM} = \left(\frac{5\pi}{6} - \frac{\pi}{6}\right)\frac{1}{2\pi R} = \frac{1}{3}\pi \times \frac{1}{R} \Rightarrow \frac{T_{FWHM}}{T} = \frac{1.66 \times 10^{-4}}{2.5 \times 10^{-11}} = 66.67\% \qquad (3.4)$$

Thus, the generation of CS-RZ pulses is clearly explained via the phasor representation.

3.2.2.2 Phasor Representation for RZ33 Pulses

Key parameters and initial phasor representation for generation of RZ33 pulses are shown in Figure 3.6a and b. The long light grey vector is the sum vector of two initialized field vectors.

Values of the key parameters are $V_{bias} = V_\pi$ for both arms; RF driving signal on each arm has the amplitude of $V_\pi/2$ (i.e., $V_{p-p} = V_\pi$), swings in opposite directions and operates at half of the bit rate ($R/2$); and modulating frequency: $f_m = 20\,\text{GHz}$.

At FWHM positions of the optical intensity pulse, $E_o = \pm 1/\sqrt{2}$ and the field component vectors E_1 and E_2 form with the horizontal axis a phase of $\pi/4$, as displayed in Figure 3.7.

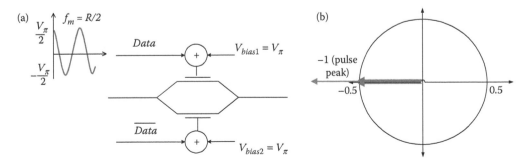

FIGURE 3.6 Initialized stage for generation of RZ33 pulses: (a) RF driving signals and the biasing voltages and (b) initial phasor representation.

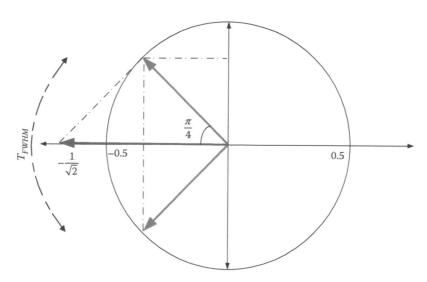

FIGURE 3.7 Phasor representation of RZ33 generation at FWHM positions using dual-drive MZIM.

The conditions in Figure 2.7 result in:

$$\frac{\pi}{2}\cos\left(2\pi f_m t\right) = \frac{\pi}{4} \Rightarrow t_1 = \frac{1}{6f_m} \tag{3.5}$$

$$\frac{\pi}{2}\cos\left(2\pi f_m t\right) = -\frac{\pi}{4} \Rightarrow t_2 = \frac{1}{3f_m} \tag{3.6}$$

Thus, the normalized FWHM of RZ33 optical pulses can be calculated:

$$T_{FWHM} = \frac{1}{3f_m} - \frac{1}{6f_m} = \frac{1}{6f_m} \therefore \frac{T_{FWHM}}{T} = \frac{1/6f_m}{1/2f_m} = 33\% \tag{3.7}$$

Equation 3.7 clearly verifies the generation of RZ33 optical pulses from the phasor representation.

3.3 DIFFERENTIAL PHASE SHIFT KEYING (DPSK)

3.3.1 BACKGROUND

Digital encoding of data information by modulating the phase of lightwave carrier is referred as optical phase shift keying (PSK). Optical PSK was extensively researched for coherent photonic transmission systems [21–24]. These systems strictly require the coherence between the phase of the lightwave carrier and the phase of the laser local oscillator. This phase coherence was difficult to be maintained because of problems of the phase noise, broad linewidth and chirping effects of the laser source. In contrast, the DPSK format minimizes the above problems. The detection of DPSK modulated optical signals is conducted incoherently by comparing differentially coded phases between every two consecutive optical pulses. The existing incoherent receiver scheme for optical DPSK format utilizes the MZDI balanced receiver.

In DPSK, a binary "1" is encoded if the present and the past encoded bits are opposite logics, whereas a binary "0" is encoded if the logics are similar. This operation is equivalent to an XOR logic operation. Thus, a XOR gate is employed as a differential pre-coder. In addition, the XOR gate can be replaced by a NOR gate (see Figure 3.8a). In DPSK, data "1" indicates a π phase change on the optical

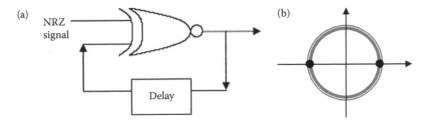

FIGURE 3.8   (a) DPSK pre-coder and (b) DPSK signal constellation diagram.

carrier between consecutive data bits, while binary "0" implies no phase change between them. Thus, two discrete points, which are located exactly at π phase difference with respect to each other in a signal constellation diagram, are utilized to represent this encoding scheme (refer to Figure 3.8b). In the case of a continuous phase shift keying format, for example, MSK, signal phase evolves continuously between one state to the other as indicated by the inner bold circle in Figure 3.8b.

3.3.2   OPTICAL DPSK TRANSMITTER

The conventional structure of a DPSK optical transmitter employs two external LiNbO$_3$ optical MZIMs (see Figure 3.9). The first MZIM is an optical pulse carver that was described in the previous section. Generated RZ optical pulses are then fed into the second MZIM, and these RZ pulses are modulated with pre-coded binary data to generate RZ DPSK optical signals.

Binary electrical data is differentially encoded in a pre-coder using a XOR or a NOR operation. Without the pulse carver, the structure shown in Figure 3.9 becomes a NRZ DPSK optical transmitter. In the DPSK data modulation, the second MZIM is biased at the minimum transmission point. The pre-coded RF driving signal has a peak-to-peak amplitude equal to $2V_\pi$ and operates at the transmission bit rate. The data modulation of NRZ DPSK optical signals are demonstrated in Figure 3.10.

In addition, an EOPM might also be used for the generation of NRZ DPSK optical signals. However, EOPM produces chirps whereas MZIM, in particular the X-cut type, generates chirp-free optical signals. In practice, a small amount of chirp might be desirable in transmission [25].

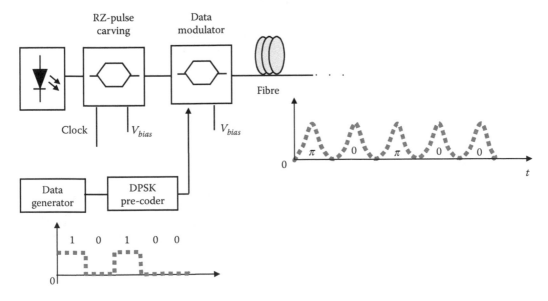

FIGURE 3.9   DPSK optical transmitter with RZ pulse carver.

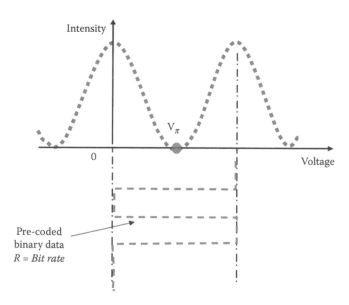

FIGURE 3.10 Bias point and RF driving signal for generation of optical NRZ DPSK signals.

3.3.3 SELF-COHERENT DETECTION OF OPTICAL DPSK

An optically delayed balanced receiver is well known for the self-coherent detection of DPSK-modulated lightwaves. It consists of a photonic MZDI followed by dual photo-detectors connected in a balanced structure, as shown in Figure 3.11a [6,34]. At the input of MZDI, the lightwaves are split into two guided paths and one path is delayed with respect to the other by a physical length that equals to a propagation time delay of one bit interval.

MZDI can be effectively seen as an optical self-heterodyne structure and it performs the differential phase demodulation in the photonic domain. This photonic processing overcomes the bandwidth and speed limitations of the electrical domain. MZDI has constructive and destructive ports that are input to the balanced-structure photo-detectors. Hence, a push-pull eye diagram is produced (see Figure 3.11b). This detection scheme offers a 3-dB improvement over the direct detection using a single photo-detector [34]. In addition, the DPSK MZDI balanced receiver is inherently insensitive to the absolute phase of the lightwave carrier. The received signal current at the output of the

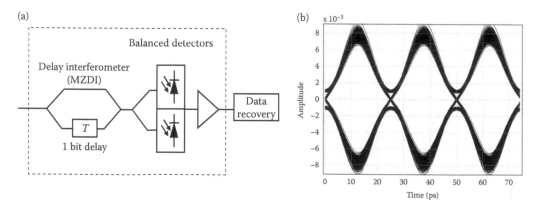

FIGURE 3.11 (a) MZDI balanced receiver for incoherent detection of optical DPSK signals and (b) detected electrical eye diagrams.

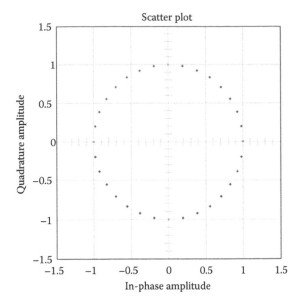

FIGURE 3.12   Signal trajectory of optical MSK signals.

push-pull photo-detectors can be expressed as [51]:

$$i(t) = |E(t) + E(t-T)|^2 - |E(t) - E(t-T)|^2$$
$$= 4\Re\{E(t)E^*(t-T)\} = 4|E(t)|^2 \cos(\Delta\phi + \varsigma) \tag{3.8}$$

where $E(t)$ and $E(t-T)$ are the current and the one-bit-delay version of the optical field, respectively; $\Delta\phi$ represents the differential phase and ς indicates the phase noise caused by either the non-linear phase noise or the imperfection of waveguide path lengths in MZDI. This MZDI imperfection can be overcome by thermally fine-tuning the thin-film heater integrated on the waveguide of MZDI [34,52].

3.3.4   Minimum Shift Keying (MSK)

MSK can be formulated from either continuous phase frequency shift keying (CPFSK) or offset DQPSK (OQPSK) format [43,53]. The signal trajectory of optical MSK signals shown in Figure 3.12 is continuously evolving. This is unlike the discrete DPSK format where encoded bits are located exactly at the "0" and "π" phase positions. Two novel schemes are proposed for generating optical MSK signals.

3.3.5   CPFSK Approach

3.3.5.1   Theoretical Background

In CPFSK modulation, binary data is modulated to either upper side band (USB) frequency f_1 or lower side band (LSB) frequency f_2 such that $f_1 = f_c + f_d$ and $f_2 = f_c + f_d$ where f_c is the lightwave carrier frequency and f_d is the frequency deviation from f_c. In the case of MSK format, $f_d = 1/4T$ and the frequency modulation index, defined as $h = 2f_dT$, is equal to $1/2$ ($h = 1/2$). It is emphasized that the phase of the optical MSK carrier is continuous at symbol transitions as MSK meets the condition of the frequency spacing between USB and LSB to be a multiple of $1/2T$ to obtain the phase

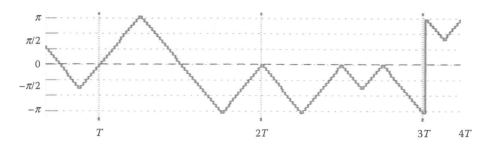

FIGURE 3.13 Phase trellis of MSK signals.

continuity [19,28,43,53]. In addition, this frequency spacing ($h = 1/2$) is also the minimum value to maintain the orthogonality between the two frequency modulated signals [43,53]. The base-band representation of an optical MSK pulse train $x(t)$ at the nth symbol interval is expressed as [43]:

$$x(t) = \sqrt{\frac{E_b}{T}}e^{j\Phi_n(t,a)}nT - T/2 \leq t \leq nT + T/2 \tag{3.9}$$

where E_b is the bit energy. The phase term, $\Phi_n(t;a)$, is given by

$$\Phi_n(t,a) = \left[\pi/2\sum_{k=-\infty}^{n-1}a_k\right] + 2\pi f_d(t + T/2 - nT)a_n \tag{3.10}$$

$$= \theta_{n-1}(t,a) + 2\pi f_d(t + T/2 - nT)a_n$$

where $a_n \in \{+1,-1\}$ is the data logics transmitted at the nth interval.

Equation 3.10 shows that the MSK phase evolution follows a linear phase trellis (see Figure 3.13). It increases linearly up to $\pi/2$ within a bit slot with a transmitted data "1," and linearly decreases by $\pi/2$ with a transmitted data "0."

3.3.5.2 Proposed Generation Scheme

MSK optical transmitter employing two external EOPMs can be arranged in a cascaded configuration, as shown in Figure 3.14. The first EOPM enables the frequency modulation of data logics into either USB or LSB frequencies. The RF driving voltage $V_d(t)$ requires a periodic ramp signal source with a duty cycle of $4T$. It should be noted that the differential phase pre-coder and bias voltages are not required in this configuration. This reduces the complexity of this MSK optical transmitter scheme.

The second external EOPM enforces the phase continuity of the lightwave carrier at transitions between every two consecutive symbols. The driving voltage into the second EOPM (V_{prep}) is pre-computed to fully compensate transitional phase discrepancy at the output $E_{01}(t)$ of the first EOPM. An electrical phase shifter is usually used to control the synchronization between the EOPMs. The calculation of V_{prep} at the nth instance is given by the following algorithm:

$$V_{prep_n}(t,a) = \frac{V_\pi}{2}\left(\sum_{k=0}^{n-1}a_k - a_nI_n\sum_{k=0}^{n-1}I_k\right) \tag{3.11}$$

where $a_n = \pm 1$ are the logic levels; $I_n = \pm 1$ represents two stages of a clock pulse $V_c(t)$ whose duty cycle is equal to $4T$. It is emphasized that the values of V_{prep} are either 0 or V_π.

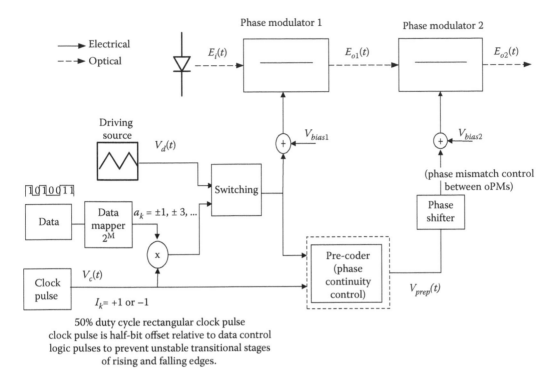

FIGURE 3.14 Block diagram of MSK optical transmitter employing two cascaded external optical phase modulators.

The time-domain signal phase evolution during the modulation process for an optical MSK sequence [−1 1 1 −1 1 −1 −1 1] is demonstrated step by step in Figure 3.14. The key points of this figure are

- Figure 3.15a shows the periodic ramp signal $V_d(t)$ with a duty cycle of $4T$.
- Figure 3.15b illustrates the clock pulse $V_c(t)$ with a duty cycle of $4T$. In order to mitigate the overshooting effect at rising and falling edges of electronic circuits, the clock pulse $V_c(t)$ is delayed, for example, by $2T$, compared to the driving signal $V_d(t)$.
- Figure 3.14c outlines the phase of the carrier at the output of the first EOPM $E_{o1}(t)$. It can be observed that optical signals at this stage are of FSK type and thus, the phase at transitions of consecutive symbols is not continuity.
- In Figure 3.14d, the transitional phase discrepancy is fully compensated by the pre-computed driving signal $V_{prep}(t)$ which is calculated from Equation 3.11.
- Figure 3.15e demonstrates the continuous phase trellis of generated optical MSK sequence [−1 1 1 −1 1 −1 −1 1] at the output of the second EOPM.

The proposed generation scheme is realizable with the advance in the design of EOPM. EOPMs operating up to 40 Gb/s using resonant-type electrodes have been reported in some previous research [54,55]. In addition, high-speed electronic driving circuits evolved with the ASIC technology using 0.1 μm GaAs P-HEMT or InP HEMTs have recently been realized [56]. Furthermore, a simplified version of the proposed optical MSK transmitter can be implemented by alternating the ramp driving signals $V_d(t)$ with a sinusoidal driving source. Finally, this transmitter scheme is also capable of generating M-ary FSK optical signals. The next section describes optical transmitter configuration for generating optical MSK signals based on the I-Q approach of ODQPSK format.

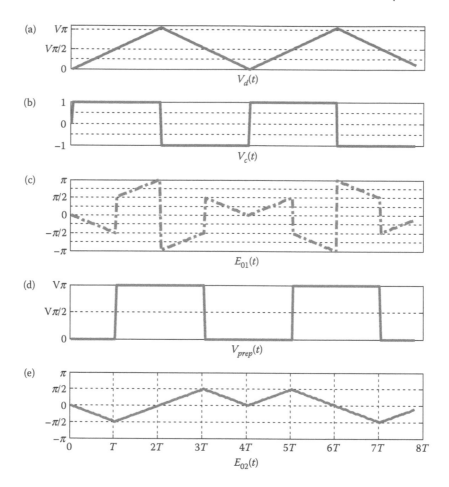

FIGURE 3.15 Evolution of time-domain phase trellis of an optical MSK sequence [−1 1 1 −1 1 −1 −1 1] at different stages of the MSK optical transmitter. (a) periodic ramp signal $V_d(t)$ with a duty cycle of 4T; (b) Clock pulse $V_c(t)$ with a duty cycle of 4T; (c) phase of the carrier at the output of the first EOPME01(t); (d) transitional phase discrepancy as fully compensated by the pre-computed driving signal; and (e) continuous phase trellis of generated optical MSK sequence at the output of the second EOPM.

3.3.6 ODQPSK Approach

3.3.6.1 Theoretical Background

MSK format can be formulated in the form of DQPSK by using in-phase I and quadrature-phase Q components [43,54]:

$$s(t) = \pm \sqrt{\frac{E_b}{T}} \cos\left(\frac{\pi t}{2T}\right) \cos\left(2\pi f_c t\right) \pm \sqrt{\frac{E_b}{T}} \sin\left(\frac{\pi t}{2T}\right) \sin\left(2\pi f_c t\right) \qquad (3.12)$$

The I and the Q components consist of a half-cycle cosine and sine waveforms, respectively, expressed as

$$s_I(t) = \pm \sqrt{\frac{E_b}{T}} \cos\left(\frac{\pi t}{2T}\right) - T < t < T \qquad (3.13)$$

$$s_Q(t) = \pm \sqrt{\frac{E_b}{T}} \sin\left(\frac{\pi t}{2T}\right) 0 < t < 2T \qquad (3.14)$$

During even bit intervals, the I component consists of positive cosine waveforms for the zero phase and negative cosine waveforms for the π phase. On the other hand, during odd bit intervals, the Q component consists of positive sine waveforms for the phase of $\pi/2$, and negative sine waveforms for the phase of $-\pi/2$. Hence, any of the four states "0," "$\pi/2$," "$-\pi/2$," "π" can arise. However, only state 0 or π can occur during any even bit intervals and only state $\pi/2$ or $-\pi/2$ can occur during any odd bit intervals. The output signal is the superposition of I and Q signal components.

Two important characteristics in generating optical MSK signals using the I-Q approach of ODQPSK format are the I and Q driving signals operate at half of the bit rate and they are interleaved by one bit period. Therefore, only either I or Q component can change at a time, that is, when one is at zero-crossing, the other is at maximum peak.

3.3.6.2 Generation Scheme

The MSK optical transmitter shown in Figure 3.16 consists of two dual-drive MZIMs in parallel, reflecting I-Q structure. The binary logic data is pre-coded and de-interleaved into even and odd symbol streams that are staggered with each other by one bit duration. These pre-coded I and Q data are operating at half of the data bit rate. In this generation scheme, phase shaping filters can also be used for reducing the spectral width of driving signals before modulating the dual-drive MZIMs. In addition, two arms of dual-drive MZIMs are biased at $V_\pi/2$ and $-V_\pi/2$ and driven by push-pull driving voltages, that is, data and $\overline{data}$.

The phase-shaping driving source can be a periodic ramp source in the case of linear MSK or, preferably, a sinusoidal source for non-linear optical MSK. By using sinusoidal waveforms, the hardware complexity for implementing this I-Q MSK optical transmitter scheme is significantly reduced. This is the main motivation for generating weakly non-linear and strongly non-linear types of optical MSK signals. The waveform amplitude of non-linear MSK optical signals are fluctuating and this ripple level depends on the magnitude of the electrical sinusoidal driving source. The main properties in the generation of linear and non-linear optical MSK signals are described below:

3.3.6.2.1 Linear MSK

The phase-shaping signal to drive the dual-drive MZIMs is a periodic ramp signal with a duty cycle of $4T$. The Q driving path is delayed by one bit period with respect to the I path. Optical MSK signals are obtained at the output of two MZIMs as the superposition of I and Q modulated signals and generated MSK waveforms clearly display the constant amplitude characteristic of the CPM format. In additions, its signal trajectory is a perfect circle. These characteristics are illustrated in Figure 3.17a.

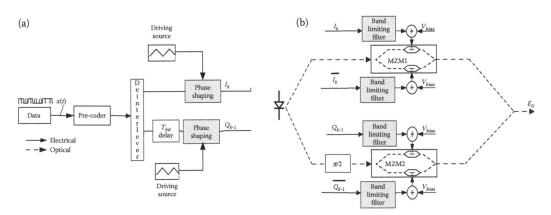

FIGURE 3.16   Block diagram of the I-Q MSK optical transmitter employing two dual-drive MZIMs (a) general block structure and (b) device and signals.

3.3.6.2.2   Weakly Non-linear MSK

The generation of weakly non-linear MSK format requires a sinusoidal driving signal having the amplitude of about $V_\pi/4$ and operating at half of the data bit rate. The inphase pulse shaper is a cosine waveform, while the quadrature counterpart is a sine waveform. The time-domain variation of the carrier phase is thus non-linear. As a result, this causes the mismatch between two MZIMs when I- and Q-waveforms are superimposed, resulting in the ripples on generated optical MSK waveforms. The signal trajectory is thus not a perfect circle. The magnitude ripples of weakly non-linear optical MSK waveforms are less than 10% of the push–pull signal levels. All of the above characteristics are illustrated in Figure 3.17b.

3.3.6.2.3   Strongly Non-linear MSK

Sinusoidal phase shaping waveforms are also used for generating strongly non-linear MSK optical signals. However, the magnitude of this driving signal is close to V_π compared to $V_{\pi}/4$, as in the case of weakly non-linear MSK. Strongly non-linear optical MSK signals have the amplitude ripples of above 10% of the push pull signal levels. This fluctuation is also caused by the mismatch of the waveforms on two arms at the output of the MZIMs. Figure 3.17c illustrates the signal trajectory and the ripples of strongly non-linear optical MSK signals.

3.3.6.2.4   Pre-Coder

The logic operations of the pre-coder are constructed based on a state diagram as shown in Figure 3.18a.

It can be observed that the current state of the signal is dependent on the previous state. Therefore, memory is needed to store the previous state. The state diagram in Figure 3.18a is then developed into a logic state diagram demonstrated in Figure 3.18b. In this case, S_0, S_1 and S_0'', S_1'' represent current states and previous states, respectively. The pre-coding logic block computes the current state and the output based on the feedback state (previous state) and binary data (b_n) from the bit pattern generator. Hence, a Karnaugh map is constructed to derive the logic gates

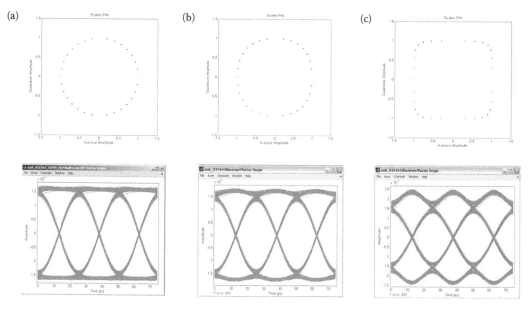

FIGURE 3.17   Signal constellation and eye diagrams of MSK optical signals: (a) linear, (b) weakly non-linear, and (c) strongly non-linear types.

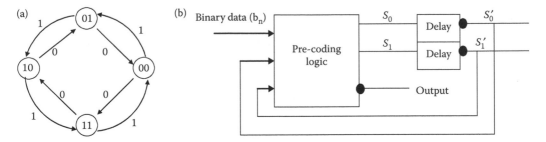

FIGURE 3.18 (a) State diagram for MSK (the arrows indicate continuous increment or decrement of the phase of the carrier) and (b) combinational logic for constructing the truth table of the pre-coder.

within the pre-coding logic block:

$$S_0 = \overline{b_n S_0' S_1'} + b_n \overline{S_0'} S_1' + \overline{b_n} S_0' S_1' + b_n S_0' \overline{S_1'}$$

$$S_1 = \overline{S_1'} = \overline{b_n S_1'} + b_n \overline{S_1'} \qquad (3.15)$$

$$\text{Output} = \overline{S_0}$$

As a result, the truth table (Table 3.1) is constructed. For a positive half cycle cosine wave and a positive half cycle sine wave, the output is "1," whereas for a negative half cycle cosine wave and a negative half cycle sine wave, the output is "0."

3.3.6.3 Incoherent Detection of Optical MSK

3.3.6.3.1 MZDI Balanced Receiver

Similar to optical DPSK, a MZDI balanced receiver is used for the direct detection of MSK-modulated optical signals (see Figure 3.19). However, the main difference in this MZDI configuration compared to that of the DPSK counterpart is an additional $\pi/2$ phase shift inserted on one arm of the MZDI [27,29,33,57]. This $\pi/2$ phase shift can be located on either arm of the MZDI.

3.3.6.3.2 Optical Frequency Discrimination Receiver

The detection of optical MSK signals using frequency discrimination principles was studied in the early 1990s, however, mainly for heterodyne coherent receivers [23,34–37]. The frequency discrimination was conducted in the electrical domain, at the intermediate frequency (IF) range. The non-coherent frequency discrimination in photonic domain for the direct detection of MSK

TABLE 3.1

Truth Table Based on MSK State Diagram

$b_n S_0' S_1'$	$S_0 S_1$	Output
100	01	1
001	00	1
010	01	1
101	10	0
110	11	0
111	00	1
000	11	0
011	10	0

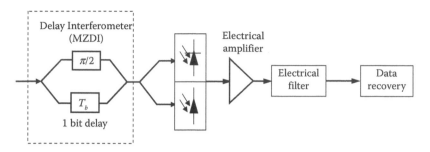

FIGURE 3.19 MZDI balanced receiver for incoherent detection of optical MSK signals.

lightwaves could not be effectively implemented at that time due to the unavailability of optical filters with very narrow bandwidths (as a fraction of signal bandwidth).

However, with recent advances in the design and fabrication of optical filters, in particular using micro-ring resonator type, the 3-dB bandwidth of optical filters can be narrowed down to around 2 GHz [47,48,58]. The availability of such narrowband optical filters thus enables the feasibility of the non-coherent optical frequency discrimination for detecting modulated frequencies of optical MSK signals. An implementable receiver scheme employing dual narrowband optical filters and an optical delay line is proposed in this research for optical MSK transmission systems. Detailed descriptions on the design and performance characteristics of this receiver are provided in Chapter 6.

3.4 CONTINUOUS PHASE DUAL-LEVEL MSK

3.4.1 THEORETICAL BACKGROUND

Dual-level MSK is a multi-level scheme encoding two data bits into a modulated symbol. One of the information bits is for coding amplitude levels, whereas the other is for the linear phase trellis coding of $\pm\pi/2$ phase change within one symbol period. Therefore, the transmission baud rate is halved compared to the bit rate from the data pattern generator. This clearly offers the transmission bandwidth efficiency, and significantly reduces the requirement of high speed electronic processing. Figure 3.20 illustrates the signal constellation of dual-level MSK format.

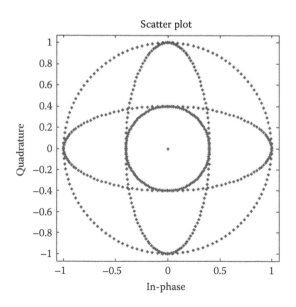

FIGURE 3.20 Signal constellation of dual-level MSK.

The mapping scheme of data information a and b in dual-level MSK is shown in Table 3.2.

The dual-level MSK modulation format can be generated from the superposition of two MSK signals having different amplitudes, given by [43]:

$$s(t) = A \cos\left(2\pi f_c t + \phi_n(t,a)\right) + B \cos\left(2\pi f_c t + \phi_n(t,b)\right) \tag{3.16}$$

where f_c is the frequency carrier, A and B are signal amplitude levels. At the transition of the nth symbol interval, these two amplitudes are either in phase or π phase shift with each other. The phase terms $\phi_n(t,a)$ and $\phi_n(t,b)$ are defined as

$$\phi_n(t,a) = \pi h \frac{a_n(t - nT)}{T} + \pi h \sum_{k=-\infty}^{n-1} a_k \quad nT \le t \le (n+1)T \tag{3.17}$$

$$\phi_n(t,b) = \pi h \frac{b_n(t - nT)}{T} + \pi h \sum_{k=-\infty}^{n-1} b_k \quad nT \le t \le (n+1)T \tag{3.18}$$

The logic values a_n and b_n are statistically independent and taken from the set of $\{\pm 1\}$. The frequency modulation index h of the dual-level MSK is equal to $1/2$.

3.4.1.1 Pre-Coder

A pre-coder is compulsory in the generation of dual-level MSK signals. At the nth instance, the logics of a_n and b_n taken from $\{\pm 1\}$ are pre-coded from the binary logics d and d'' as follows:

$$a_n = 2d_n - 1; \quad \text{and} \quad b_n = a_n\left(1 - \frac{d'_n}{h}\right) \tag{3.19}$$

where the logics d and d'' are statistically independent and taken from $\{0,1\}$.

3.4.2 GENERATION SCHEME

Any configuration of MSK optical transmitters which were either presented in Section 3.3.5 or reported in some previous research [28,29,33,55], can be utilized for the following proposed generation scheme of dual-level MSK optical signals. Figure 3.21 shows the block diagram of a dual-level MSK optical transmitter in which two MSK optical transmitters are arranged in a parallel configuration.

It can be seen from Figure 3.21, as well as from Equations 3.16, that the amplitude levels A and B which are input into the two MSK optical transmitters play significant roles in the proposed generation scheme as they determine the amplitude levels of transmitted dual-level MSK optical signals. In practice, optical components are normally specified by the optical intensity rather than magnitude of

TABLE 3.2

Mapping of Data Information in Dual-Level MSK Format

a	b	Remarks on Signal Constellation
1	1	Amplitude unchanged, phase increased
1	−1	Amplitude unchanged, phase decreased
−1	1	Amplitude changed, phase increased
−1	−1	Amplitude changed, phase decreased

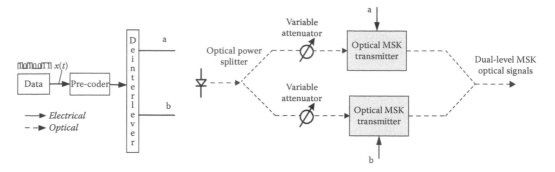

FIGURE 3.21 Block diagram of the proposed dual-level MSK optical transmitter.

the optical field. Thus, we consider the optical powers input into the two MSK optical transmitters rather than the amplitude levels *A* and *B*. Moreover, the splitting ratio between these two optical powers is more important than their individual values. This intensity-splitting ratio can be obtained by using a high-precision optical power splitter or simply by using a 3-dB or a 50:50 optical power splitter followed by a variable optical attenuator on each path.

3.4.3 Non- and Self-Coherent Detection of Optical Dual-Level MSK

As a combination between optical MSK and multi-level modulation formats, the demodulation of dual-level MSK optical signals requires the incoherent detection of both amplitude and phase components, as shown in Figure 3.22.

Similar to optical MSK, a MZDI-balanced receiver is implemented for the phase detection and an additional $\pi/2$ phase shift is also inserted on one arm of MZDI. On the other hand, the lightwave amplitude is detected by a single photo detector. It should be noted that a differential decoder is required to detect the amplitude changes between two consecutive bits. Detected eye diagrams for the amplitude and phase detections of dual-level MSK received signals are demonstrated in Figure 3.23a and b. In these figures, the dotted line represents the decision threshold voltage.

It can be observed that the phase detection of dual-level MSK gives a push-pull eye diagram having multiple levels. However, the interest is only on either the positive or the negative attribute of these levels, which corresponds to either zero phase or π phase change, respectively. Thus, the threshold level for the phase detection is set at zero level, whereas the amplitude threshold is a non-zero value to distinguish two modulated signal levels.

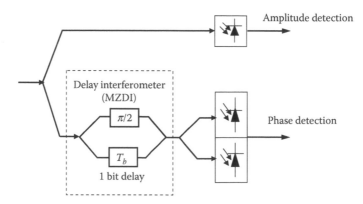

FIGURE 3.22 Incoherent and self-coherent detection for both amplitude and phase of dual-level MSK optical signals.

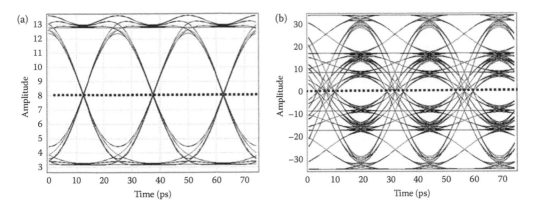

FIGURE 3.23 Eye diagrams for (a) amplitude and (b) phase detection of dual-level MSK received signals. The decision threshold is plotted in broken-line style.

3.4.4 Spectral Characteristics of Advanced Modulation Formats

Optical power spectra of 40 Gb/s NRZ/RZ33/CS-RZ DPSK and OOK optical signals are shown in Figures 3.24 and 3.25, respectively.

Several main points observed in Figures 3.24 and 3.25 are

- The optical power spectra of OOK formats have high power components at the carrier and signal frequencies, thus causing fiber non-linear effects. On the contrary, DPSK spectra do not contain high power frequency components. Therefore, optical DPSK formats are more resilient to fiber non-linear effects.
- RZ pulses are more susceptible to fiber CD due to their broader spectra compared to NRZ pulses. In particular, RZ33 type has the broadest spectrum (at 20-dB position down from the peak) and thus, RZ33 is least tolerant to fiber CD.

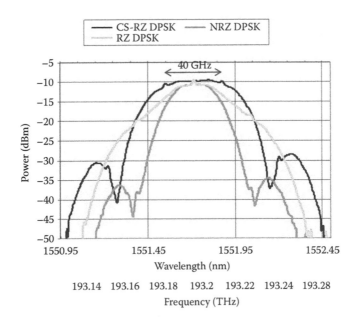

FIGURE 3.24 Spectra of 40 Gb/s CS-RZ/RZ33/NRZ DPSK optical signals.

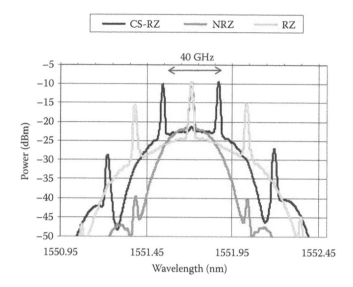

FIGURE 3.25   Spectra of 40 Gb/s CS-RZ/RZ33/NRZ OOK optical signals.

- However, the susceptibility of RZ33 and CS-RZ pulses to fiber CD provides a tradeoff in the improvement of the resilience to fiber non-linearities. This is because time-domain peaks of these fast spreading optical pulses decrease rapidly when propagating along the fiber, hence lowering effects of fiber non-linearities.

Optical power spectra of 40 Gb/s linear, weakly non-linear and strongly non-linear MSK formats are compared in Figure 3.26. It can be observed that the power spectra of these formats are similar,

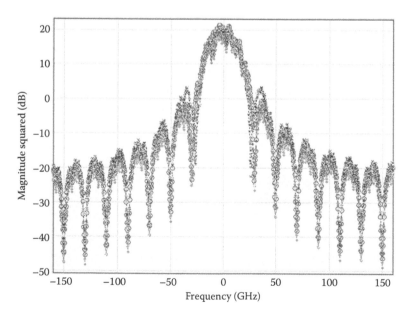

FIGURE 3.26   Optical power spectra of three 40 Gb/s optical MSK formats: linear (mid-grey), weakly non-linear (light grey) and strongly non-linear (black).

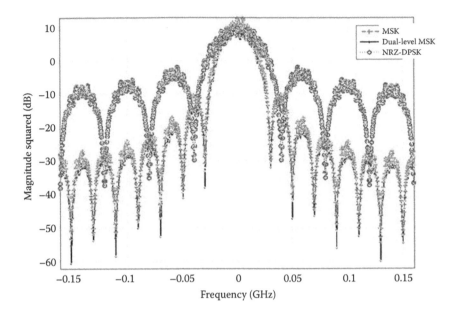

FIGURE 3.27 Spectral properties of three modulation formats: 40 Gb/s MSK (red, dash), 80 Gb/s dual-level MSK (black, solid) and 40 Gb/s NRZ DPSK (blue, dot).

with the exception point that sidelobes in the power spectrum of strongly non-linear MSK are not highly suppressed as in the case of linear and weakly non-linear MSK. All three formats offer better spectral efficiency compared to the DPSK counterpart.

Figure 3.27 compares the power spectra of three modulation formats: 80 Gb/s dual-level MSK, 40 Gb/s linear MSK, and 40 Gb/s NRZ DPSK. The intensity-splitting ratio for the optical dual-level MSK format is set at "0.8/0.2."

A number of key points are observed:

- The signal power spectrum of 80 Gb/s optical dual-level MSK has identical characteristics to that of the 40 Gb/s MSK format. The main lobe spectral widths of these two formats are narrower than that of NRZ DPSK. More specifically, the base-width takes a value of approximately ± 32 GHz on either side compared to ± 40 GHz in the case of DPSK. Hence, it is expected that these MSK-based formats are more tolerant to fiber CD and more robust to tight optical filtering than NRZ DPSK.
- High suppression of side lobes and the confinement of signal energy in the main spectral lobe of 80 Gb/s dual-level MSK and 40 Gb/s MSK optical power spectra significantly mitigate inter-channel cross-talks between DWDM channels.

3.4.5 REMARKS

This chapter presents the principles for generating RZ33 and CS-RZ optical pulses. The phasor representation providing a clear explanation of the modulation process was verified mathematically. The generation and detection of optical DPSK signals were also described.

This chapter section describes two MSK optical transmitters' schemes. The first scheme is based on the CPFSK approach and implements two cascaded electro-optic phase modulators. The second scheme consists of two dual-drive MZIMs, reflecting the I-Q configuration of the ODQPSK format. These I and Q driving signals are staggered by one bit period and operate at half of the bit rate. This scheme can generate linear or non-linear types of optical MSK pulses. Non-linear optical MSK signals are distinguished based on the magnitude ripple levels of transmitted waveforms. Optical MSK signals with a small ripple of less than 10% of the push–pull amplitude levels are referred as weakly non-linear type, whereas ripples of more than 10% of the signal levels are observed on strongly non-linear optical MSK waveforms. Key differences in the implementations of these MSK optical transmitters are the waveforms and the magnitudes of the phase shaping electrical driving signals. In the case of linear MSK, periodic ramp waveforms are utilized, whereas sinusoidal waveforms are used for generation of weakly non-linear and strongly non-linear MSK. In order to generate weakly non-linear MSK optical pulses, sinusoidal waveforms have an amplitude of approximately $V_\pi/4$ compared to a value of close to V_π in the case of strongly non-linear optical MSK. The pre-coder for I-Q optical MSK structure was also derived. The incoherent detection of MSK-modulated optical signals is based on the MZDI balanced receiver. The main difference in this MSK receiver compared to the DPSK is the additional $\pi/2$ phase shift inserted onto one arm of the MZDI.

The dual-level MSK format has proposed for fiber-optic communications. The optical transmitter of this format is constructed from two MSK optical transmitters arranged in a parallel structure, while the demodulation requires the non-coherent detection of both amplitude and phase components of the lightwaves. The phase detection for dual-level MSK format is similar to that of optical MSK whereas the amplitude detection is simply implemented with a single photo-detector.

Finally, spectral properties of optical 80 Gb/s dual-level MSK, 40 Gb/s MSK, and 40 Gb/s NRZ/ RZ33/CS-RZ DPSK and OOK formats were discussed. There is a tradeoff between RZ33/CS-RZ and NRZ pulse types. Compared to NRZ pulses, RZ33 and CS-RZ pulses are more resilient to fiber non-linear effects but less tolerant to fiber CD. The spectral properties of 40 Gb/s MSK and 80 Gb/s dual-level MSK are similar to each other and more advantageous than 40 Gb/s optical DPSK and OOK. These advantages include (i) the spectral efficiency and filtering robustness due to the narrow spectral main lobe and (ii) the mitigation of interchannel cross-talks because of high suppression of side lobes. Therefore, optical MSK and dual-level MSK formats offer promising potential for high-speed and high-capacity optical long-haul and metropolitan transmission systems.

3.5 PARTIAL RESPONSES AND SINGLE-SIDEBAND MODULATION FORMATS

Optical fiber communication systems have continuously evolved over the years. The increasing demand for a higher transmission capacity has driven the development of communication systems at ultra-high capacity ultra-high bit rates. The fact that 40 Gb/s optical fiber communication systems have an extended reach and improved capacity has made them an attractive alternative to the 10 Gb/s optical fiber communication systems. System performance is further enhanced by employing various advanced modulation formats, such as DB, RZ-DPSK, NRZ-DPSK. Research and investigations have been carried out to determine the most appropriate and efficient formats that meets the current, as well as future demand.

Duo-binary format, in photonic domain, offers three-level coding, but unlike electronic or wireless communication systems, the "-1" and "$+1$" are coded using the phase of the lightwave carrier, that is either "0" or "π." This coding would overcome the dispersion because of its single sideband property in which the detection scheme is much simpler, as direct detection technique can be employed. The single sideband can also be implemented using vestigial sideband modulation technique; that is, an optical filter can be inserted after the optical modulator to filter half of the band of the spectrum. Alternatively, the modulators can be conditioned with two Hilbert transform signals and hence the suppression of half of the band.

The first part of this chapter presents a comprehensive modeling platform for the duo-binary modulation format optical fiber transmission system. The modeling of the system is developed on the MATLAB® Simulink 7.0 or higher. Simulink has been chosen because of the availability of blocks, such as the communication block sets and signal processing block sets, which eased the process of implementation. Further the transmission of 40 Gb/s alternating phase 0 and π radians duo-binary modulation format with 33% and 50% pulse width is demonstrated with 850 ps/nm dispersion tolerance. At least 2 dB receiver sensitivity improvement is achieved as compared with carrier suppressed DPSK.

The second part of this chapter presents the transmission of optical multiplexed channels of 40 Gb/s using vestigial single side band modulation format over a long reach optical fiber transmission system. Thus, it is essential that an optical filter is designed to follow the optical modulator to filter half of the signal band. The effects on the Q-factor of fibers dispersion, the pass band, and roll off frequency of the optical filters and the channel spacing are described. The performance of the optical transmission using low and non-zero dispersion fibers and/or dispersion compensation is discussed. It has been demonstrated that BER of 10^{-12} or better can be achieved across all channels and minimum degradation of the channels can be obtained under this modulation format. Optical filters are designed with asymmetric roll-off bands. Simulations of the transmission system are also given and compared for channel spacing of 20, 30, and 40 GHz. It is shown that the passband of 28 GHz and 20 dB cut-off band performs best for 40 GHz channel spacing.

3.5.1 PARTIAL RESPONSES: DUO-BINARY MODULATION

3.5.1.1 Introductory Remarks

The demand for high-capacity long-haul telecommunication systems has increased over recent years. To achieve high throughput of signals with minimum errors, different advanced modulation formats, such as ASK, PSK (coherent and differential in-coherent), and FSK have been proposed and comparisons are made to determine which modulation format would offer the best transmission performance. In countering performance degradation, modulation formats aim to narrow the optical spectrum to enable close channel spacing in the network. They increase symbol duration so that more uncompensated dispersion accumulates before ISI becomes significant. Furthermore, this format is more resilient to fiber non-linearities and optical signal distortion.

Modulation format is important in determining the performance of 40 Gb/s optical fiber communication systems. Duobinary (DB) and continuous phase modulation (CPM) DB are shown to offer high spectral efficiency [59,60]. DB modulation formats minimize intersymbol interference (ISI) impairments in a controlled way instead of trying to eliminate it. It is possible to achieve a signaling rate of equal to the Nyquist rate of 2 W symbols per second in a channel of bandwidth W Hz. Optical DB technique has received much attention due to its high dispersion tolerance and high frequency-utilization efficiency by means of spectral narrowing. DBM format is similar to the non-return-to-zero (NRZ) format, with inclusion of phase coding. The phase characteristics of DBM signals compensate for the group velocity dispersion by reducing the spectral component in conventional NRZ modulation. ISI is reduced since bit patterns such as 101 are transmitted with the ones carrying opposite phase. Therefore, if pulses spread out into the zero time slot, due to dispersion, they tend to cancel each other out. The recovering of signals at the receiver is relatively simple. Furthermore, conventional direct detection receiver is applicable, hence simple receiver structure. There are two types of DBM schemes, which are constant phase and alternating phase in blocks of logics "1s."

This chapter presents the models for photonic transmission with optical channels operating under duo-binary (DB) modulation format. This includes the development and implementation of the photonic transmitter, the optical fiber propagation, and the opto-electronic receiver. DBM encodes two-level electrical signal to three-level electrical signal before modulating the lightwave carrier. The transmitter of the SIMULINK model will consist of a DB encoder and a dual drive MZIM. A baseband modulation is first implemented in the DB encoder, which encodes the binary signal into three levels signals of "1," "0" and "−1." MZIM is an electro-optic modulator that converts the electrical signal to optical signal.

The DB or phase-shape binary modulation formats can be generated by modulating a dual drive MZIM. Recent works have shown that the driving voltages for the modulator can be reduced to generate variable pulse width DB optical signals. However, the pulse width of the DB DPSK has not been thoroughly investigated under the alternating phase of the "1" coded bits, meaning the "0" "π" "0" "π" phases of consecutive "1" in contrary of conventional duo-binary formats. We also present modeling performance of alternating phase DB modulation with a full width half mark (FWHM) ratio with respect to the bit period of 100%, 50%, and 33%, and compared with experimental transmission of carrier-suppressed DPSK over 50 km of SSMF and dispersion compensation. For the DB case, the transmission without dispersion compensation over the same SSMF length offers better performance for 50% FWHM DB modulation and slightly worse for 33%. The transmission performance, the bit-error-rate versus receiver sensitivity, of these DB modulation formats are compared with the carrier-suppressed DPSK experimental transmission.

Section 3.5.1.2 describes the fundamental aspects of DB modulation format. It is followed by the description of each component in the 40 Gb/s DBM photonic transmission systems in Section 3.5.1.3. Section 3.5.1.7 describes the implementation of Simulink model of the communication system. Lastly, Sections 3.5.1.7 and 3.5.1.8 briefly outlines some simulated results, then an alternative structure for another alternative implementation using phases in optical domain.

3.5.1.2 The DBM Formatter

Modulation format aims to modulate one or more field properties to suit system needs. There are four types of field properties: Intensity, phase, polarization, and frequency. Symbols are constelled in one or more dimensions in order to carry more information and to travel a further distant. Data modulation format is the information-carrying property of the optical field.

DBM format has become an attractive modulation format over recent years, compared to other formats, such as non-return-to-zero (NRZ) and return-zero (RZ). This is due to the fact that it can overcome the fiber chromatic dispersion in high-capacity transmissions. It is the characteristics of the DBM format that makes it a preferred format.

DBM schemes can be described as correlative-level coding or a partial-response signaling schemes. Correlative-level coding schemes involves adding ISI to transmitted signal in a controlled manner, by which a signaling rate equal to the Nyquist rate of 2 W symbols per second in a channel of bandwidth $W\,Hz$ can be achieved. "Duo" in the word DB indicates the doubling of transmission capacity of a conventional binary system. DBM format is, in fact, NRZ modulation with an inclusion of phase coding. The one bits in the data input are phase-modulated. For instance, for a bit pattern of "101," this data will be transmitted with the ones carrying opposite phase, 0 and π. If the pulse of the one bits spread out to the zero time-slot in between, they can cancel each other. This effect increases the dispersion tolerance and allows the signal to be transmitted over a longer distance.

DB coding converts a two-level binary signal of 0s and 1s into a three-level signal of "-1," "0," and "$+1$." This is done by first applying the binary sequence to a pulse-amplitude modulator to produce two-level short pulses of amplitude of -1 and $+1$, with -1 corresponding to 0 and $+1$ corresponding to 1. This sequence is then applied to DB encoder to produce a 3-level output of "-2," "0," and "2."

As shown in Figure 3.28 input sequence, $\{a_k\}$ of uncorrelated two-level pulses is transformed into $\{c_k\}$, which is a sequence of correlated three-level pulses. The correlation between adjacent pulses is equivalent to introducing ISI into transmitted signal in an artificial manner. The DB encoder is simply a filter involving a single delay element and summer, as shown in Figure 3.29. However, once errors are made, they tend to propagate through the output. This is because a decision made on the current input a_k depends on the decision made on the previous input a_{k-1}. Therefore, precoding is needed to avoid this error propagation phenomenon. Binary sequence, $\{b_k\}$ is converted into another binary

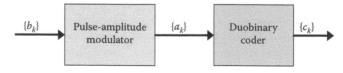

FIGURE 3.28   Brief overview of DB signaling.

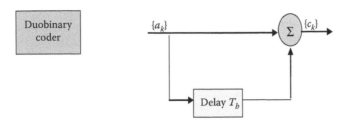

FIGURE 3.29 DB Encoder—the block at the left is represented by the signal flow diagram shown on the right.

sequence, $\{d_k\}$ by modulo-two addition, exclusive OR of b_k and d_{k-1}, as shown in Figure 3.30

$$c_k = a_k + a_{k-1} \tag{3.20}$$

$$d_k = b_k \oplus d_{k-1} \tag{3.21}$$

The three-level DB output, $\{c_k\}$ is, then modulated into a two-level optical signal by an external modulator. The most commonly used external modulator is the MZIM. The optical DB signal has two intensity levels, "on" and "off." The "on" state signal can have one of two optical phases, 0 and π. The two "on" states correspond to the logic states "1" and "-1" of the DB encoded signal, $\{a_k\}$ and the "off" state correspond to the logic state "0." Figure 3.31a shows an example of the original binary signal, the DB encoded signal, optical DB signal, and then a summary of coding rule Figure 3.31b.

3.5.1.3 40 Gb/s DB Optical Fiber Transmission Systems

Ultra-long terrestrial networks transmitting signal at a bit rate of 40 Gb/s has matured over recent years. Various advanced modulation schemes, such as RZ, NRZ, NRZ-DPSK, and RZ-DPSK, have been proposed to achieve an extended reach and an improved capacity of the communication system.

Figure 3.32 shows the typical DWDM optical fiber communication system. Signals are modulated at the transmitters and are multiplexed together at the wavelength multiplexer before transmitting them into the fiber. The fiber link is divided into a number of spans. Each span consists of a SSMF and DCF. The EDFA is used to compensate for the optical power loss of the transmission span. At the end of the fiber, the signals are demultiplexed and detected at the receivers.

DBM format has become popular compared to other modulation formats because it extends the transmission distance as limited by fiber loss, without regenerative repeaters. It extends the dispersion limit without additional optical components, such as dispersion compensating fiber and DCF. Chromatic dispersion has become a main effect that limits the transmission distance. The optical three-level transmission can overcome this limitation since narrowband signal has higher tolerance to chromatic dispersion compared to broadband signal. Furthermore, DB optical fiber communication system can suppress stimulated Brillouin scattering (SBS).

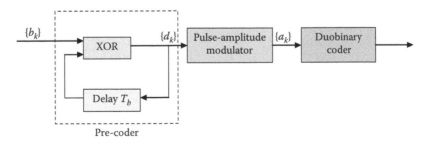

FIGURE 3.30 DBM scheme with pre-coder.

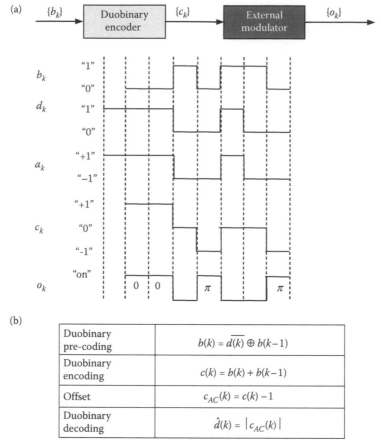

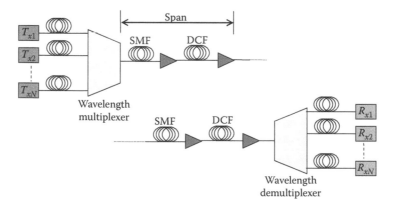

FIGURE 3.31 (a) Example of original binary signal (b_k), pre-coded signal (d_k), DB encoded signal (c_k) and optical DB signal (o_k). (b) Summary of coding rule.

The main modules of the communication system are the transmitter, optical fiber, and the opto-electronic receiver, as shown in Figure 3.33. The transmitter consists of the DB encoder and the MZIM. A series of 0s and 1s is modulated under DBM scheme. This three-level electrical data is then used to modulate the laser source, producing a two-level optical signal. This modulated signal is transmitted along an optical fiber transmission link toward the electro-optic receiver. The signal

FIGURE 3.32 Ultra-long-haul fiber transmission.

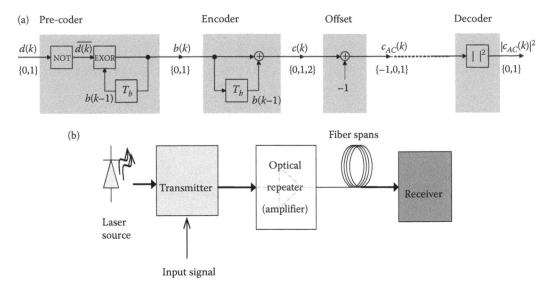

FIGURE 3.33   Main modules of a DBM optical communication system (a) coder and decoder and (b) generic transmission system.

will be detected using photodetector, which converts the two-level optical signal back to electrical signal. Optical amplification can be done at some points along this transmission link to minimize the effect of fiber loss.

3.5.1.4   Electro-Optic Duobinary Transmitter

Transmitter modulates and converts incoming electrical signal into optical domain. Depending on the nature of the signal, the resulting modulated light may be turned on and off, or may vary linearly in intensity between two levels. The output of the DB transmitter is the modulated lightwaves switched on and off at transitional instances of the input electrical signal. In general, a DBM transmitter is shown in Figure 3.34, consisting of a monochromatic laser source, a coder, and a photonic modulator.

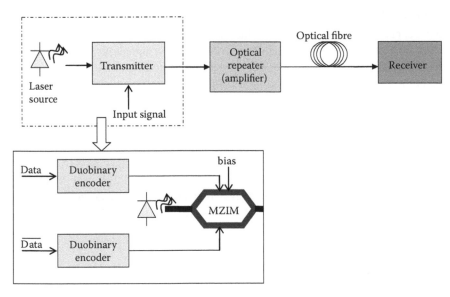

FIGURE 3.34   The transmitter of the 40 Gb/s DBM photonic transmission system.

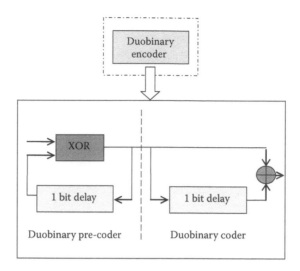

FIGURE 3.35 The DBM encoder of the 40 Gbps DBM photonic transmission system.

Binary data is encoded by a DB encoder (see Figure 3.35). This resulting three-level electrical signal is converted into two-level optical signal using the folding characteristic of an optical MZIM. It is then transmitted into the optical fiber.

There are two types of DB transmitters. The conventional DB transmitter, as previously mentioned, includes a dual-drive MZIM driven by three-level electrical signals in a push–pull configuration. The fact that MZIM is normally driven by two-level signal, the effect of driving it with a three-level signal has its uncertainties. It is proposed that three-level signals may experience significant distortion in electrical amplifiers operating in saturation leading to penalties for long word lengths. It may also cause the degradation of receiver sensitivity. For these reasons, the second type of DB transmitter has been proposed. This type of transmitter has the MZ modulator driven by only two-level electrical signals. The optical DB signal generated is the same as DB transmitter type one, that is, constant phase in blocks of 1s.

3.5.1.5 The Duobinary Encoder

The DB encoder encodes the binary signals, which is a sequence of 0s to a three-level electrical signal. DB signal is a fundamental correlative coding in partial response signaling. A DB encoder consists of a pre-coder and a DB coder. A pre-coder is used before DB coding to allow for easier recovery of binary data at the receiver, and to avoid error propagation. The pre-coder is a simple binary digital circuit that consists of an exclusive OR (XOR) and a 1 bit delay feedback. The DB coder is a filter consisting of a single delay element and a summer.

Binary data input is pre-coded with initialization of the 1-bit delay to 0. The output of the DB pre-coder is, then, modulated by a pulse-amplitude modulator, to produce a two-level electrical signal with amplitude of −1 and 1. The DB signal is produced by adding data delayed by 1 bit period to the present data. This DB signal is a three-level electrical signal of −2, 0, and 2. Finally, it is converted to a level of −1, 0, and 1. The three-level is mapped into optical domain by modulating both amplitude and phase. The "+1" and "−1" levels have the same optical intensity but opposite optical phase.

3.5.1.6 The External Modulator

Although the electro-optic modulator has been given in Chapter 2, it is essential to revisit the operation of the MZIM for DB operation. In a MZIM, the input optical carrier is split into two paths via a Y

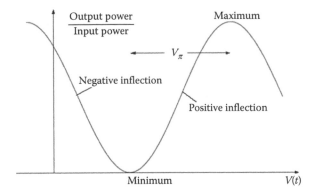

FIGURE 3.36 Input-output transfer characteristics of the MZIM.

junction. This Y junction splits the input signal into $E_i/\sqrt{2}$ each. The resultant signal is

$$E_o = \frac{E_i}{2}\left[1 + \exp\left(j\pi\frac{V(t)}{V\pi}\right)\right]$$ (3.22)

where V_π is the voltage to provide a π phase shift of each phase modulator and $V(t)$ is the driving volt-age. The input and output relationship of this MZIM is as shown in Figure 3.36. It is accompanied by a phase modulation of $\exp(j\varphi(t))$ with $\varphi(t) = \pi V(t)$. For $V(t)$ from 0 to V_π, E_o and E_i have the same phase, and as for $V(t)$ from V_π to $2V_\pi$, E_o and E_i have different phase.

MZIM can be single-drive or dual-drive. Single-drive x-cut LiNbO$_3$ MZM has no phase modula-tion along with the amplitude modulation. It follows the transfer characteristics of Figure 3.36. Dual-drive X-cut LiNbO$_3$ MZIM, on the other hand, has two paths phase-modulated with opposite phase shifts in a push-pull operation. The V_π in Figure 3.9 is reduced by half in this case. For dual-drive y-cut LiNbO3 MZIM, two paths are driven by complementary signal with V_1 equals to $-V_2$. The out-put electric of a dual-drive MZIM is

$$E_o = \frac{E_i}{2}\left[\exp\left(j\pi\frac{V_1}{V_\pi} + \exp\left(j\pi\frac{V_2}{V_\pi}\right)\right)\right]$$ (3.23)

DB optical signal is generated by driving a dual-drive MZIM with push-pull operation, as shown in Figure 3.37. One arm is driven by the DB signal and the second arm is driven by the inverted DB signal. Figure 3.38 shows the operation of the MZIM. The output electric field $E_o(t)$ can be expressed as

$$E_o(t) = E_i \cos\frac{\Delta\phi(t)}{2}\cdot\exp\left(-j\cdot\frac{\phi_0}{2}\right)$$ (3.24)

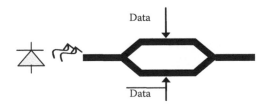

FIGURE 3.37 Dual-drive MZIM.

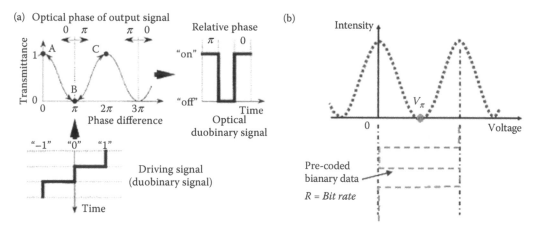

FIGURE 3.38 Driving operation of dual-drive MZM [61] (a) DB signaling and (b) biasing to generate optical DB signals with phase states.

where E_i is the input electric field, $\Delta\phi(t)$ is the phase difference between the lightwaves propagating in two optical waveguides, and ϕ_0 is a constant when the MZIM is driven in a push-pull operation. At point B of Figure 3.38, the phase of the output optical signal is inverted. The optical DB signal is dependent on the biasing point of the driving signal, which is the electrical DB signal. By biasing at point B in Figure 3.11, "−1" and "+1" level of the electrical DB signal will correspond to the "on" state of the optical signal, while the "0" level will correspond to the "off" state. To achieve the effect of carrier suppressed, there must be a π phase different between the two arms.

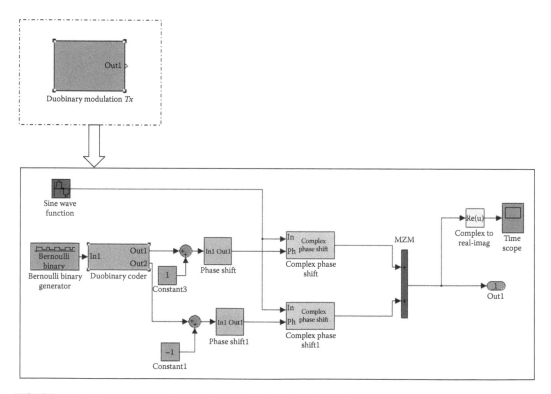

FIGURE 3.39 The conventional transmitter model of the 40 Gb/s DB optical fiber communication system.

3.5.1.7 Duobinary Transmitters and Pre-Coder

The transmitter model generally consists of the DB coder and the MZIM. The DB coder encodes the incoming binary sequence of 0s and 1s to DB electrical signal. This signal is then used to drive the arms of the dual-drive MZIM. One arm is driven by the DB signal, and the other arm is driven by the inverted DB signal. The Bernoulli binary generator generates a random sequence of binary electrical signal. It is set to generate the data at a rate of 40 Gb/s as seen in Figure 3.39 in Simulink model. This signal is encoded by the DB encoder, which consists of a DB pre-coder and a DB coder. The first output of the encoder is shifted up by 1 to produced levels of "0," "1," and "2." This electrical DB signal is sent to the phase shift block, as shown in Figure 3.40, to represent these levels with a certain phase. This, in fact, represents the biasing point on the transmittance curve. For dual-drive MZIM, the driving signal is biased at $V\pi/2$. The second output of the DB encoder is the inversion of output 1. This output is used to modulate the second arm of MZIM. The output 2 signal is shifted down by -1, to bias at the point $-V\pi/2$ of the transmittance curve. MZIM is an amplitude modulator, accompanied by a shift of phase. This modulation is also called AM-PSK. This lightwave, which is the sine wave produced by the sine wave function, is modulated by the DB signal through the complex phase shift block. The input sine wave is shifted by the amount specified at the input.

The DB coder consists of a pre-coder and a coder, as shown in Figure 3.41. The pre-coder is a differential coder, with an exclusive OR (XOR) gate and a 1-bit delay feedback path. The addition of -0.5 and division by two function as the amplitude modulator, which shifts levels of the signal from "0" and "1" to "-1" and "$+1$." The signal is then added to its one bit delay to produce a three-level DB signal of level "-2," "0," and "$+2$," followed by a conversion to a level of "-1," "0," and "$+1$." The summation of the signal with its one bit delay is the DB coder. For the second output, the output of the differential coder is inverted, before going through the same operation as Out1. Zero-order hold is placed before the output of the DB encoder functions to discretize the signals to have a fast-to-slow transition of signals. It holds and samples the signal before transmitting it out. If the signals are transmitted out without the zero-order hold, the transition to "0" level will be overseen. The signal will only have two levels, which are "-1" and "$+1$."

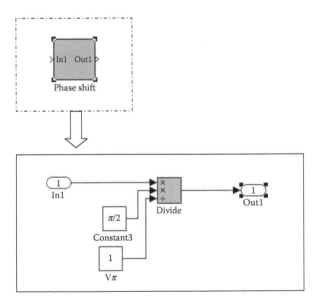

FIGURE 3.40 Phase shift block of the transmitter model.

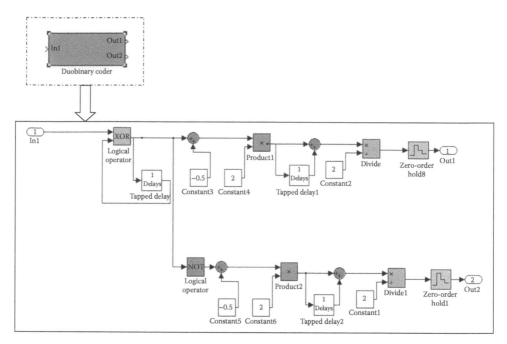

FIGURE 3.41 DB coder of the transmitter model.

3.5.1.8 Alternative Phase DB Transmitter

Two types of DB transmitter models are proposed. The conventional DB transmitter, as mentioned previously, uses a dual-drive Mach–Zehnder modulator driven by three-level DB electrical signals. The MZIM shown in Figure 3.42 is usually driven by a two-level electrical signal. In some cases, it occurs that this three-level driving signal may experience significant distortion in electrical amplifiers operating at saturation. This may lead to penalties for long word lengths. It may also occur that there will be a degradation of receiver sensitivity. Due to these uncertainties, an alternative DB transmitter, as shown in Figure 3.43 is proposed. This second type of DB transmitter has the MZIM driven by two-level electrical signals. It consists of a differential encoder, a one-bit-period electrical time delay and a MZ modulator. One arm of the dual-drive MZ modulator is driven with the signal from the electronic driver output stage, whereas the other arm is driven by the inverted signal, delayed by one bit period. Both DB transmitters produce the same result, which is constant phase in blocks of 1s.

3.5.2 SINGLE SIDE BAND MODULATION

3.5.2.1 Hilbert Transform SSB MZ Modulator Simulation

The Hilbert transform of signal $m(t)$ is defined to be the RF signal whose frequency components are all phase shifted by $\pi/2$ radians [62]. Thus we have

$$\hat{m}(t) = H\{m(t)\} \rightarrow m(t) = A\cos 2\pi f_o t \quad \text{then} \quad \hat{m}(t) = A\cos\left(2\pi f_o t - \frac{\pi}{2}\right) \quad (3.25)$$

Taking the Fourier transform, we have

$$M(f) = -j\mathrm{sgn}(f)\frac{A}{2}[\delta(f+f_o)+\delta(f-f_o)] = \frac{A}{j2}[-\delta(f+f_o)+\delta(f-f_o)] \quad (3.26)$$

Thus,

$$\hat{m}(t) = A\cos\left(2\pi f_o t - \frac{\pi}{2}\right) = A\sin 2\pi f_o t$$

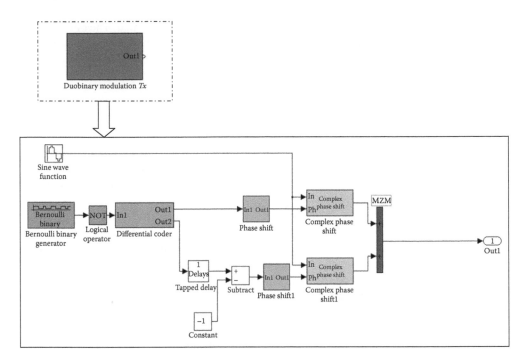

FIGURE 3.42 Simulink model of MZIM.

3.5.2.2 SSB Demodulator Simulation

The modulated signal is

$$u(t) = A_c/2\{\cos(\omega_c t + \gamma\pi + \alpha\pi\cos\omega_{rf}t) + cos(\omega_c t + \alpha\pi\cos(\omega_{rf}t + \theta)\} \qquad (3.28)$$

The demodulated signal is

$$s(t) = A_c \cos 2\pi f_c t * u(t) \qquad (3.29)$$

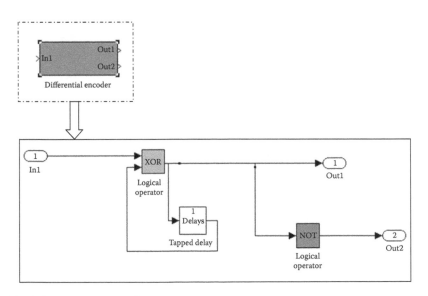

FIGURE 3.43 DB transmitter (type 2).

The low-pass filter is used to filter high-frequency components in the signal. The only component left is the modulating signal (10 Gbps [binary*cosine signal]) after demodulation. It is required to multiply the demodulate signal by the in-phase cosine signal. There is also a high-frequency signal, so that LPF is required to filter the high-frequency components in the signal.

3.5.2.3 SSB in Comb-Generation Techniques

As an example for such SSB modulation, we illustrate an application of SSB technique to create a multi-carrier generator (comb-generator) as follows (see Figure 3.44):

1. A continuous-wave (CW) external cavity laser (ECL) or multi-laser bank of ECLs whose line width is sufficiently narrow, possible in order of less than 100 KHz, can be employed as the original lightwave carrier/carriers.
2. An MZM for pulse shaping (CSRZ), which will make the output spectrum of the super-channel flat.
3. An MZM used as phase modulator, the phase control signal coming from a RF signal generator.
4. RF generator, its frequency is the spacing of the sub-channels.
5. Sinusoidal signal generator whose output is split into two branches, one branch applied to the amplitude modulator and the other is phase shifted by $\pi/2$ radians and applied to other branch of the optical modulator.
6. 90° phase shifter is used as a Hilbert transformer to generate SSB carrier, hence the spectrum to be only single carrier shifted in frequency domain, the single-side band carrier.

Figure 3.45 shows a laboratory bench set up of the SSB modulation of a laser carrier and then recirculating to continually shift the carriers to several multiple carrier output. The spectra of the multiple-carrier laser are shown in Figure 3.46 showing the shift right and shift left by positive or negative phase shift of $\pi/2$ rad, respectively, which is applied to the optical modulator for single sideband frequency shifting.

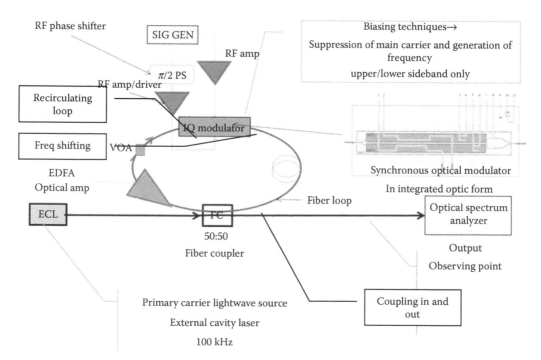

FIGURE 3.44 Principle of the experimental setup of the comb-generator using recirculating SSB frequency shifting. Insert: plane view structure of a dual drive optical modulator.

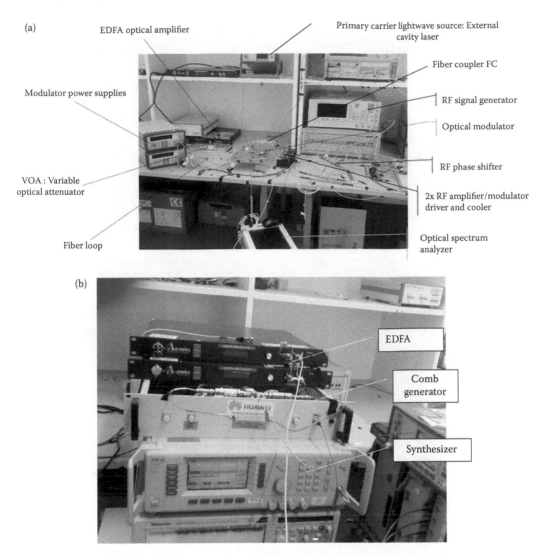

FIGURE 3.45 Laboratory setup of a comb-generator prototype employing recirculating frequency shifting technique (a) comb generator under development; (b) packaged comb generator with external EDFA and signal synthesizer.

3.6 SI OPTICAL MODULATOR

Optical modulators formed in silicon are the keystone to many low-cost optical applications. Increasing the data rate of the modulator benefits the efficiency of channel usage and decreases power consumption per bit of data, especially in data center environment. SOI based optical modulators are considered to be the most important and economical devices for the near-future optical networks, especially in access networking. The main driving force is due to the CMOS compatible process in production of such modulators in integration with electronic integrated circuits.

3.6.1 OPTICAL PHASE MODULATOR

A typical phase modulator based on SOI integrated photonic technology is shown in Figure 3.47. The device is formed in rib waveguides of height 220 nm and width 400 nm. A 100 nm slab height is used, as this provides lower access resistance to the junction and lower sidewall roughness interaction than (a)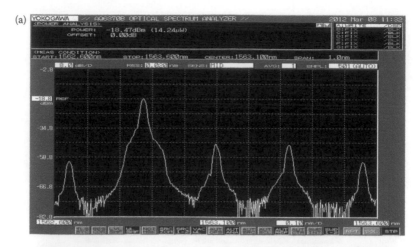

(b)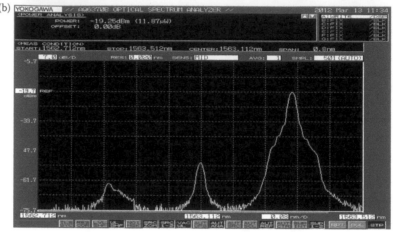

FIGURE 3.46 Spectrum of the SSB lightwave at output of the optical modulator (a) shifting right ($+\pi/2$ phase shift applied) in frequency scale and (b) shifting right in frequency ($-\pi/2$ phase shift applied).

a 50 nm slab height. The rib section of the device and the slab to one side is p-type doped. The slab on the other side of the rib is n-type doped. The target doping concentration of the p- type region (3×10^{17} cm^{-3}) is made lower than the n-type region (1.5×10^{18} cm^{-3}) such that the depletion extends mainly into the rib region during reverse bias conditions. It is noted that the width of the rib is about 1/3 of the guided lightwave wavelength. This is reasonable, as the wavelength is equal to the free

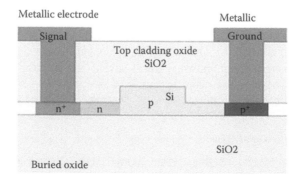

FIGURE 3.47 Cross-sectional view of a Si-on-insulator (SOI) optical phase modulator.

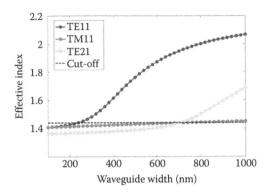

FIGURE 3.48 Effective index of the guided modes TE and TM in the SOI ridge waveguide. n_{eff} versus waveguide modes at wavelength 1550 nm of three guided modes TE_{11}, TM_{11}, and TE_{21} simulation by Rsoft.

space wavelength divided by the effective refractive index of the guided wave in silicon which is about 3.4.

This configuration also allows for reduced optical loss. The p- and n-type regions contact to highly doped p- and n-type regions, respectively, which, in turn, connect to coplanar waveguide (CPW) electrodes that are used to drive the device at high speed. The phase modulators are incorporated into asymmetric MZI structures with an arm length mismatch of 180 μm to convert between phase and intensity modulation. A key advantage of the device design is the simplicity of the fabrication process required for its formation.

The effective index of the Si on SOI ridge waveguide for TE and TM modes cab be estimated as shown in Figure 3.48. Figure 3.49 [63] shows the guided mode field distribution with the height of the waveguide as a parameter. This shows that the mode still holds the single mode conditions but with different group velocity, and hence, the dispersion of the lightwave channel would vary as shown in Figure 3.50. The dispersion is important if one wants to use the waveguide as a wideband phase matching for non-linear interaction in four wave mixing for phase conjugation.

3.6.2 SI MZI Modulator

Figure 3.51 shows the structure in plain view of a MZI modulator in which the incoming lightwaves are split into two branches via a multi-mode interference (MMI) coupler. The phases of guided

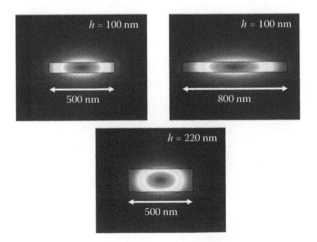

FIGURE 3.49 Structure and near field of guided modes Different thickness h of waveguide 100–220 nm.

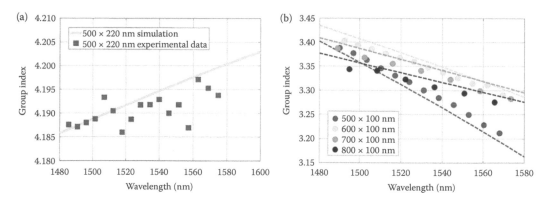

FIGURE 3.50 (a) Experimentally derived (grey squares) group index values for standard 500×220 nm waveguides, and values predicted from simulations (light grey line). (b) Experimentally derived (dots) group index values for 100 nm high waveguides, and values predicted from simulations (dashed lines). Note the y-axis scale is 10 times smaller in (a) with respect to (b).

lightwaves are modulated by applying signal voltages on the CPW electrodes. A lateral PN junction is embedded into a 450 nm-wide waveguide of a MZI modulator by the standard 180 nm CMOS process. The PN junction's interface is accurately offset by 100 ± 20 nm to the waveguide center by improving the locating accuracy of the doping masks. The p- and n-doped regions have a concentration of 2×10^{17} cm^{-3}. The fabricated MZI modulator with the ground-signal-ground (GSG) traveling electrodes can be seen in Figure 3.51a. The optical transmission power of the MZI is shown in Figure 3.51b under different biasing voltages. A 56 Gbit/s NRZ PRBS sequence is applied through the MZI modulator and the observed eye diagram is shown in Figure 3.52 with BER of 1e-7. After transmission through 2 km, the BER is about 1e-3 that can be error free under FEC.

3.6.3 MICRO-RING MODULATOR (MRM)

Micro-ring resonator modulator in integrated waveguide structure offers much lower modulation voltage level and very short-length electrodes as shown in Figure 3.53. The resonant condition is such that the total phase distributed around the ring equals to a multiple number of half wavelength. Thus, any applied voltage that departs from this resonant condition, the intensity circulating in the ring is reduced, hence modulation amplitude. The MRM frequency passband can be very wide up to few tens of GHz.

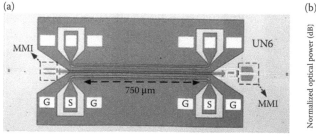
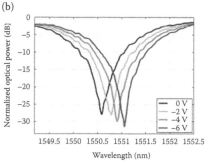

FIGURE 3.51 MZI modulator. (a) Top-view microscope figure of the 750 micro-m-long MZI modulator with the GSG electrodes. (b) Optical transmission of the asymmetric MZI modulator under different reversed bias voltages.

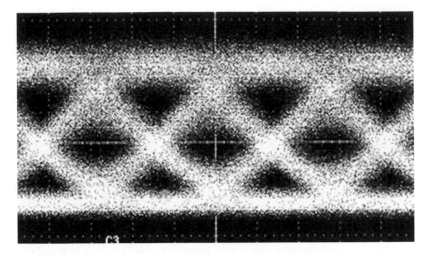

FIGURE 3.52 NRZ eye diagram of the 56 Gb/s modulation bit sequence.

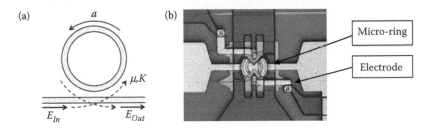

FIGURE 3.53 Micro-ring modulator (MRM) (a) sketch of the MRM with field propagation and coupling directions and (b) plane view of the fabricated MRM.

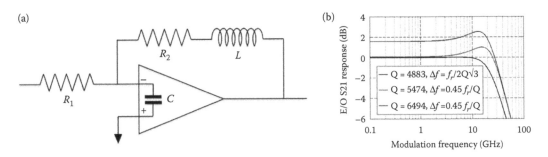

FIGURE 3.54 MRM peaking technique for equalization of optical transfer responses. (a) Circuit diagram of an idealized peaking amplifier RLC with an extended transfer function and (b) ideal transfer function of equalized modulator by peaking to a 3 dB bandwidth to 40 GHz. Note: peaking of the transfer function.

The investigated MRM has a radius of 10.3 mm and is constituted out of a 400 nm wide ridge waveguide partially etched into the 220 nm thick silicon device layer of a silicon-on-insulator wafer with a remaining silicon slab height of 90 nm around the waveguide on an insulator SiO_2 thin layer. A reversed biased PIN junction with a nominal implant concentration of P- of $3 \times 10^{17}\,\text{cm}^{-3}$ and N of $1\text{e}17\,\text{cm}^{-3}$ overlapping by 50 nm serves as phase shifter. Electrical connectivity is obtained via highly doped regions of a doping concentration about one thousand times higher than the lightly

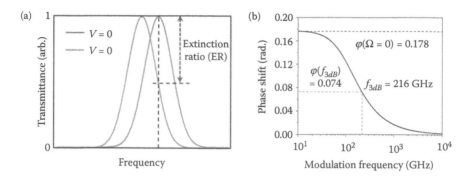

FIGURE 3.55 Transfer function of MRM and its variation with applied voltage across the electrode along the ring perimeter. (a) Intensity amplitude response and (b) phase response.

doped regions. The total junction capacitance is 17 femto-Farad (fF) at 21 V bias and the series resistance 215 Ohms, resulting in an electrically limited RC-bandwidth of 43 GHz [64]. It is noted that the electrode for MRM is not a CPW type but more or less a lumped type. The limited bandwidth of this type of electrode can be equalized by using a peaking technique which is commonly used in electrical circuit theory [65]. This extends the 3 dB bandwidth to about 40 GHz permitting the modulation of the modulator to 56 Gbps or 64 Gbps without any problem (see Figure 3.54 for bandwidth peaking circuitry). The main issue of the MRM is the accurate positioning of the wavelength of the optical carrier due to the resonant shifting when applying the electrical signals as shown in Figure 3.55a and b. Thus, a biasing voltage is required in association with the RF signals for tuning the resonant wavelength and locating he applied signals to ensure linearity in phase and amplitude.

The frequency shift of the resonant peak is given by

$$\Delta\omega = \frac{n_{eff}^3 r\omega^2}{2\sqrt{3}\pi c_0} V_{pp} \tag{3.30}$$

where ω is the angular frequency of the optical waves, n_{eff} is the effective refractive index, c_0 is the velocity of light in vacuum, r is the electro-optic coefficient, and V_{pp} is the peak-to-peak voltage applied to the ring. The driving voltage level can be reduced further by constructing a dual electrode structure so that differential push–pull mode of operation can be implemented, hence reducing the complexity if RF signaling.

The resonant characteristics of the micro-ring can be modified by combining with other rings or interferometers, such as the one shown in Figure 3.56. In addition, the free spectral range of the MRM must be considered so that for multiple wavelength channel systems no cross talk would happen [66].

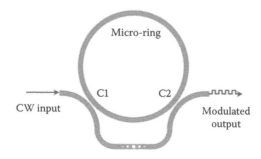

FIGURE 3.56 Schematic design of the dual photonic crystal/microring cavity modulator (top view); C1 and C2: integrated couplers to couple lightwaves into and out of the ring.

3.6.4 SI CMOS MODULATOR

Silicon has long been the optimal material for electronics, but it is only relatively recently that it has been considered as a material option for photonics due to its propagation loss as compared with other crystallized electro-optic materials, such as lithium niobate. Furthermore, one of the key limitations for using silicon as a photonic material has been the relatively low speed of silicon optical modulators compared to those fabricated from III–V semiconductor compounds. To date, the fastest silicon-waveguide-based optical modulator that has been demonstrated experimentally has a modulation frequency of only 20 MHz, although it has been predicted, theoretically, that a 1-GHz modulation frequency might be achievable in some device structures.

This section describes an approach based on a metal–oxide–semiconductor (MOS) capacitor structure shown in Figure 3.57, embedded in a silicon waveguide that can produce high-speed optical phase modulation. It is demonstrated that such an all-silicon optical modulator whose 3dB bandwidth exceeds more than 40 GHz. As this technology is compatible with conventional complementary MOS (CMOS) processing, monolithic integration of the silicon modulator with advanced electronics on a single silicon substrate becomes possible. Modulation of the refractive index in silicon can be achieved via the free-carrier plasma dispersion effect. In contrast to the conventional way of inducing a free carrier density change in silicon (through injection or depletion of electrons and holes in the intrinsic region of a forward biased silicon p-i-n diode or a three-terminal device), we demonstrate the use of an MOS capacitor phase shifter to realize the charge density modulation. Under accumulation conditions, the majority carriers in the silicon waveguide modify the refractive index so that phase shift is induced in the optical mode. The advantage of using a MOS capacitor phase shifter is the high achievable modulation speed, as there are no slow carrier generation and/or recombination processes involved in the accumulation operation. The silicon-waveguide-based MOS capacitor phase shifter comprises an n-type doped crystalline silicon slab (the silicon layer of the "silicon-on-insulator" wafer) and a p-type doped polysilicon rib with a gate oxide sandwiched between them. The polysilicon is formed by first depositing amorphous silicon using low pressure chemical vapor deposition (CVD) at 5508C. This amorphous layer is then crystallized into polysilicon by high temperature annealing. The annealing process is designed to minimize polysilicon grain boundaries, which can cause optical loss and limit the electric activation of dopants.

The frequency response of such a typical modulator is shown in Figure 3.58, showing that up to 1.0 GHz bandwidth can be achieved. The modulator is structure in a MZM with the phase change by

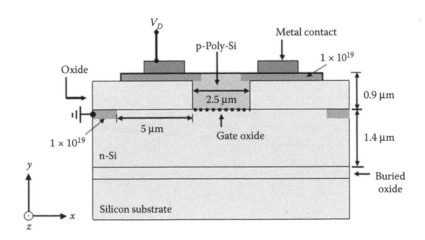

FIGURE 3.57 Cross section schematic representation of a CMOS modulator.

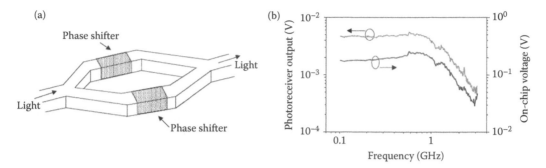

(a)

Phase shifter

Light

Light

Phase shifter

(b)

FIGURE 3.58 Frequency response of the CMOS modulator (b) of the MZIM employing the phase modulation in one of the branch of the structure (a).

an applied voltage is given by

$$\Delta\Phi = \frac{2\pi}{\lambda}\Delta n_{eff}L \tag{3.31}$$

where the change of the effective refractive index due to the applied voltage to the CMOS is given by the change of the impurity of the gate carrier density. This change in phase can be given as shown in Figure 3.59. The voltage-induced charge density change ΔN_e (for electrons) and ΔN_h (for holes) is related to the drive voltage applied to the drain and the gate forward bias and given as

$$\Delta N_e = \Delta N_h = \frac{\varepsilon_0\varepsilon_r}{qt_{ox}t}[V_D - V_{FB}] \tag{3.32}$$

$$\Delta n_e = -8.8 \times 10^{-22}\Delta N_e$$
$$\Delta n_h = -8.5 \times 10^{-18}(\Delta N_h)^{0.8} \tag{3.33}$$

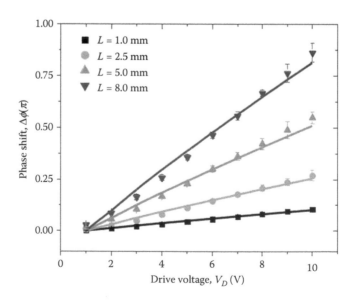

FIGURE 3.59 Phase shift $\Delta\Phi$ versus drive voltage V_D of the MOS capacitor phase shifter in Figure 3.1 at a wavelength of l¼1.55 mm for different phase shifter lengths as a parameter. The symbols represent the measured phase shifts, and the solid lines are the simulated phase.

REFERENCES

1. I.P. Kaminow and T. Li, *Optical Fiber Communications*, vol IVB, USA: Elsevier Science, 2002.
2. R.S. Sanferrare, Terrestrial lightwave systems, *AT&T Technology Journal*, vol. 66, pp. 95–107, 1987.
3. C. Lin, H. Kogelnik, and L. G. Cohen, Optical pulse equalization and low dispersion transmission in single-mode fibers in the 1.3–1.7 mm spectral region, *Optics Letters*, vol. 5, pp. 476–478, 1980.
4. A. H. Gnauck, S. K. Korotky, B. L. Kasper, J. C. Campbell, J. R. Talman, J. J. Veselka, and A. R. McCormick, Information bandwidth limited transmission at 8 Gb/s over 68.3 km of single mode optical fiber, in *Proceedings of OFC'86, Paper PDP6*, Atlanta, GA, 1986.
5. H. Kogelnik, High-speed lightwave transmission in optical fibers, *Science*, vol. 228, pp. 1043–1048, 1985.
6. G. P. Agrawal, *Fiber-Optic Communication Systems*, 3rd ed., New York: Wiley, 2002.
7. A. R. Chraplyvy, A. H. Gnauck, R. W. Tkach, and R. M. Derosier, 8×10 Gb/s transmission through 280 km of dispersion-managed fiber, *IEEE Photonics Technology Letters*, vol. 5, pp. 1233–1235, 1993.
8. H. Kogelnik, High-capacity optical communications: Personal recollections, *IEEE Journal on Selected Topics in Quantum Electronics* vol. 6, no. 6, pp. 1279–1286, November/December 2000.
9. C. R. Giles and E. Desurvire, Propagation of signal and noise in concatenated erbium-doped fiber amplifiers, *IEEE Journal of Lightwave Technology*, vol. 9, no. 2, 1991.
10. P. C. Becker, N. A. Olsson, and J. R. Simpson, *Ebium-Doped Fiber Amplifiers, Fundamentals and Technology*, San Diego: Academic Press, 1999.
11. T. Okoshi, Heterodyne and coherent optical fiber communications: Recent progress, *IEEE Transactions on Microwave Theory and Techniques*, vol. 82, no. 8, pp. 1138–1149, 1982.
12. T. Okoshi, Recent advances in coherent optical fiber communication systems, *IEEE Journal of Lightwave Technology*, vol. 5, no. 1, pp. 44–52, 1987.
13. J. Salz, Modulation and detection for coherent lightwave communications, *IEEE Communications Magazine*, vol. 24, no. 6, 1986.
14. T. Okoshi, Ultimate performance of heterodyne/coherent optical fiber communications, *IEEE Journal of Lightwave Technology*, vol. 4, no. 10, pp. 1556–1562, 1986.
15. P. S. Henry, *Coherent Lightwave Communications*, New York: IEEE Press, 1990.
16. A. F. Elrefaie, R. E. Wagner, D. A. Atlas, and A. D. Daut, Chromatic dispersion limitation in coherent lightwave systems, *IEEE Journal of Lightwave Technology*, vol. 6, no. 5, pp. 704–710, 1988.
17. M. C. Farries, P. R. Morkel, R. I. Laming, T. A. Birks, D. N. Payne, and E. J. Tarbox, Operation of erbium-doped fiber amplifiers and lasers pumped with frequency-doubled Nd:YAG lasers, *IEEE Journal of Lightwave Technology*, vol. 7, no. 10, pp. 1473–1477, 1989.
18. G. Charlet, E. Corbel, J. Lazaro, A. Klekamp, W. Idler, R. Dischler, S. Bigo, P. Tran, T. Lopez, H. Mardoyan, A. Konczykowska, and J.-P. Thierry, Comparison of system performance at 50, 62.5 and 100 GHz channel spacing over transoceanic distances at 40 Gbit/s channel rate using RZ-DPSK, *Electronics Letters*, vol. 41, no. 3, pp. 145–146, 2005.
19. P. S. Cho, V. S. Grigoryan, Y. A. Godin, A. Salamon, and Y. Achiam, Transmission of 25-Gb/s RZ-DQPSK signals with 25-GHz channel spacing over 1000 km of SMF-28 fiber, *IEEE Photonics Technology Letters*, vol. 15, no. 3, pp. 473–475, 2003.
20. K. Ishida, T. Kobayashi, J. Abe, K. Kinjo, S. Kuroda, and T. Mizuochi, A comparative study of 10 Gb/s RZ-DPSK and RZ-ASK WDM transmission over transoceanic distances, in *Proceedings of OFC'03*, vol. 2, pp. 451–453, 2003.
21. W. A. Atia and R. S. Bondurant, Demonstration of return-to-zero signaling in both OOK and DPSK formats to improve receiver sensitivity in an optically preamplified receiver, in *Proceedings of IEEE LEOS '99*, vol. 1, pp. 226–227, 1999.
22. G. Bosco, A. Carena, V. Curri, R. Gaudino, and P. Poggiolini, On the use of NRZ, RZ, and CSRZ modulation at 40 Gb/s with narrow DWDM channel spacing, *IEEE Journal of Lightwave Technology*, vol. 20, no. 9, pp. 1694–1704, 2002.
23. A. H. Gnauck, G. Raybon, P. G. Bernasconi, J. Leuthold, C. R. Doerr, and L. W. Stulz, 1-Tb/s (6/spl times/170.6 Gb/s) transmission over 2000-km NZDF using OTDM and RZ-DPSK format, *IEEE Photonics Technology Letters*, vol. 15, no. 11, pp. 1618–1620, 2003.
24. B. Zhu, L. E. Nelson, S. Stulz, A. H. Gnauck, C. Doerr, J. Leuthold, L. Gruner-Nielsen, M. O. Pedersen, J. Kim, and R. L., Lingle, Jr., High spectral density long-haul 40-Gb/s transmission using CSRZ-DPSK format, *IEEE Journal of Lightwave Technology*, vol. 22, no. 1, pp. 208–214, 2004.
25. A. Hirano, Y. Miyamoto, and S. Kuwahara, Performances of CSRZ-DPSK and RZ-DPSK in 43-Gbit/s/ch DWDM G.652 single-mode-fiber transmission, in *Proceedings of OFC'03*, vol. 2, Anaheim, CA, pp. 454–456, 2003.

26. L. N. Binh and T. L. Huynh, Linear and nonlinear distortion effects in direct detection 40 Gb/s MSK modulation formats multi-span optically amplified transmission, *Optics Communications*, vol. 237, no. 2, pp. 352–361, May 2007.

27. J. Mo, D. Yi, Y. Wen, S. Takahashi, Y. Wang, and C. Lu, Optical minimum-shift keying modulator for high spectral efficiency WDM systems, in *Proceedings of ECOC'05*, vol. 4, pp. 781–782, 2005.

28. J. Mo, Y. J. Wen, Y. Dong, Y. Wang, and C. Lu, Optical minimum-shift keying format and its dispersion tolerance, in *Proceedings of OFC'05, Paper JThB12*, Anaheim, CA, 2005.

29. M. Ohm and J. Speidel, Optical Minimum-shift keying with direct detection (MSK/DD), in *Proceedings of SPIE on Optical Transmission, Switching and Systems*, vol. 5281, pp. 150–161, 2004.

30. T. L. Huynh, T. Sivahumaran, L. N. Binh, and K. K. Pang, Narrowband frequency discrimination receiver for high dispersion tolerance optical MSK systems, in *Proceedings of Coin-Acoft'07, Paper TuA1-3*, Melbourne, Australia, June 2007.

31. T. L. Huynh, T. Sivahumaran, L. N. Binh, and K. K. Pang, Sensitivity improvement with offset filtering in optical MSK narrowband frequency discrimination receiver, in *Proceedings of Coin-Acoft '07, Paper TuA1-5*, Melbourne, June 2007.

32. T. Sivahumaran, T. L. Huynh, K. K. Pang, and L. N. Binh, Non-linear equalizers in narrowband filter receiver achieving 950 ps/nm residual dispersion tolerance for 40Gb/s optical MSK transmission systems, in *Proceedings of OFC'07, Paper OThK3*, Anaheim, CA, 2007.

33. T. Sakamoto, T. Kawanishi, and M. Izutsu, Optical minimum-shift keying with external modulation scheme, *Optics Express*, vol. 13, pp. 7741–7747, 2005.

34. A. H. Gnauck and P. J. Winzer, Optical phase-shift-keyed transmission, *IEEE Journal of Lightwave Technology*, vol. 23, no. 1, pp. 115–130, 2005.

35. J. A. Lazaro, W. Idler, R. Dischler, and A. Klekamp, BER depending tolerances of DPSK balanced receiver at 43Gb/s, in *Proceedings of IEEE/LEOS Workshop on Advanced Modulation Formats 2004*, pp. 15–16, 2004.

36. H. Kim and A. H. Gnauck, Experimental investigation of the performance limitation of DPSK systems due to nonlinear phase noise, *IEEE Photonics Technology Letters*, vol. 15, no. 2, pp. 320–322, 2003.

37. S. Bhandare, D. Sandel, A. F. Abas, B. Milivojevic, A. Hidayat, R. Noe, M. Guy, and M. Lapointe, 2/spl times/40 Gbit/s RZ-DQPSK transmission with tunable chromatic dispersion compensation in 263 km fibre link, *Electronics Letters*, vol. 40, no. 13, pp. 821–822, 2004.

38. T. Mizuochi, K. Ishida, T. Kobayashi, J. Abe, K. Kinjo, K. Motoshima, and K. Kasahara, A comparative study of DPSK and OOK WDM transmission over transoceanic distances and their performance degradations due to nonlinear phase noise, *Lightwave Technology, Journal of*, vol. 21, no. 9, pp. 1933–1943, 2003.

39. C. Xu, X. Liu, L. F. Mollenauer, and X. Wei, Comparison of return-to-zero differential phase-shift keying and ON-OFF keying in long-haul dispersion managed transmission, *Photonics Technology Letters, IEEE*, vol. 15, no. 4, pp. 617–619, 2003.

40. L. N. Binh and T. L. Huynh, Phase-modulated hybrid 40Gb/s and 10Gb/s DPSK DWDM long-haul optical transmission, in *Proceedings of OFC '07, Paper JWA94*, Anaheim, CA, 2007.

41. T. Ito, K. Sekiya, and T. Ono, Study of 10 G/40 G hybrid ultra long haul transmission systems with reconfigurable OADM's for efficient wavelength usage, in *Proceedings of ECOC'02, Paper 1.1.4*, Copenhagen, Denmark, 2002.

42. K. Iwashita and N. Takachio, Experimental evaluation of chromatic dispersion distortion in optical CPFSK transmission systems, *IEEE Journal of Lightwave Technology*, vol. 7, no. 10, pp. 1484–1487, 1989.

43. J. G. Proakis, *Digital Communications*, 4th ed., New York McGraw-Hill, 2001.

44. J. G. Proakis and M. Salehi, *Communication Systems Engineering*, 2nd ed., New Jersey: Prentice Hall, Inc, pp. 522–524, 2002.

45. K. K. Pang, *Digital Transmission*, Melbourne, Australia: Mi-Tec Media Pty. Ltd., 2005.

46. K. Iwashita and T. Matsumoto, Modulation and detection characteristics of optical continuous phase FSK transmission system, *IEEE Journal of Lightwave Technology*, vol. 5, no. 4, pp. 452–460, 1989.

47. B. E. Little, Advances in micro-ring resonator, *Integrated Photonics Research Conference 2003*, Washington, DC, USA, Invited Paper, June 16–20, 2003.

48. V. Van, B. E. Little, S. T. Chu, and J. V. Hryniewicz, Micro-ring resonator filters, in *Proceedings of LEOS'04*, vol. 2, pp. 571–572, 2004.

49. P. P. Absil, S. T. Chu, D. Gill, J. V. Hryniewicz, F. Johnson, O. King, B. E. Little, F. Seiferth, and V. Van, Very high order integrated optical filters, in *Proceedings of OFC'04*, vol. 1, Anaheim, CA, 2004.

50. L. Brent et al., Advanced ring resonator based PLCs, *IEEE Lasers & Electro-Optics Society*, pp. 751–752, 2006.
51. U.-V. Koc and X. Wei, Combined effect of polarization-mode dispersion and chromatic dispersion on strongly filtered /spl pi//2-DPSK and conventional DPSK, *IEEE Photonics Technology Letters*, vol. 16, no. 6, pp. 1588–1590, 2004.
52. H. Kim and P. J. Winzer, Robustness to laser frequency offset in direct-detection DPSK and DQPSK systems, *IEEE Journal of Lightwave Technology*, vol. 21, no. 9, 2003.
53. F. Xiong, *Digital Modulation Techniques*, Boston, London: Artech House Inc, 2000.
54. T. Kawanishi, S. Shinada, T. Sakamoto, S. Oikawa, K. Yoshiara, and M. Izutsu, Reciprocating optical modulator with resonant modulating electrode, *Electronics Letters*, vol. 41, no. 5, pp. 271–272, 2005.
55. R. Krahenbuhl, J. H. Cole, R. P. Moeller, and M. M. Howerton, High-speed optical modulator in LiNbO3 with cascaded resonant-type electrodes, *IEEE Journal of Lightwave Technology*, vol. 24, no. 5, pp. 2184–2189, 2006.
56. I.P. Kaminow and T. Li, *Optical Fiber Communications*, vol IVA, San Diego: Elsevier Science, Chapter 16, 2002.
57. T.L. Huynh, L.N. Binh, K. K. Pang, and L. Chan, Photonic MSK transmitter models using linear and non-linear phase shaping for non-coherent long-haul optical transmission, in *Proceedings of SPIE APOC'06*, paper 6353–85, Gwangju, Korea, Sep 2006.
58. L. Brent, C. Sai, C. Wei, C. Wenlu, H. John, G. Dave, K. Oliver, J. Fred, D. Roy, D. Kevin, and G. John, Advanced ring resonator based PLCs, *IEEE Lasers & Electro-Optics Society*, pp. 751–752, 2006.
59. D. Bolvin et al., *IEEE Photon. Tech. Lett.*, vol. 14, No. 6, pp. 843, 2002.
60. Y. Kim et al., *Opt. Fib. Tech.*, vol. 10, pp. 312–324, 2004.
61. T. Frank, P. B. Hansen, T. N. Nielsem, L. Eskildson, DB transmitter with lower inter-symbol interference, *IEEE Photonics Tech. Lett.*, vol. 10, no. 4, April 1998.
62. Mathworks, Single sideband modulation via the Hilbert transform, in Signal Processing Tool Box, Cambridge, U.K. https://uk.mathworks.com/help/signal/examples/single-sideband-modulation-via-the-hilbert-transform.html?requestedDomain=www.mathworks.com.
63. Riccardo Marchetti et al., Group-velocity dispersion in SOI-based channel waveguides with reduced-height, *Optics Express*, vol. 25, no. 9, 9761, 1 May 2017.
64. J. Müller, F. Merget, S. Sharif Azadeh, J. Hauck, S. Romero Garcıa, B. Shen & J. Witzens, *Optical Peaking Enhancement in High-Speed Ring Modulators*, Scientific Reports, vol. 4, doi:10.1038/srep06310, 2014, pp. 1–9.
65. E.M. Cherry and D. Hooper, *Amplifying Devices And Low Pass Amplifier Design*, John Wiley, New York, 1968.
66. Iraj Sadegh Amiri, Abdolkarim Afroozeh, Harith Ahmad, *Integrated Micro-Ring Photonics: Principles and Applications as Slow Light Devices, Soliton Generation and Optical Transmission*, Boca Raton, FL, USA: CRC Press, 2017. ISBN.

Section II

4 Guided-Wave Photonic Transmitters

External Modulation and Formats

A direct modulated photonic transmitter can consist of a single or multiple lightwave sources that can be modulated directly by manipulating the driving current of the laser diode. On the other hand, the laser source can be turned on at all times, thus operating in the continuous wave (CW) mode. The output CW lightwaves are externally modulated via an integrated optical modulator. The combination of the laser and this modulator is commonly known as external modulated optical transmitter. This externally modulated transmitter preserves the line width of the laser and, hence, its coherence.

The idea of external modulation was first proposed by P.K. Tien in 1969 in an article on integrated optics and reviewed in 1977 [1]. This opens several versions of guided-wave integrated optical modulation structures and phenomena such as electro-optics, maneto-optics, acousto-optics, electro-absorptions, pn junction depletion modulation in Si ion-implanted structures, etc. Electro-optic effect has been demonstrated as the most effective technique in high-speed and low-driving voltage level, hence high extension ratio. This chapter presents the techniques for modulation of lightwaves externally, not directly manipulating the stimulated emission from inside the laser cavity, through the use of electro-optic effects. Advanced modulation formats have attracted much attention for enhancement of the transmission efficiency since the mid-80s for coherent optical communications. Hence, the preservation of the narrow linewidth of the laser source is critical for operation bit rates in the range of several tens of Gb/s. Thus, external modulation is essential.

Three typical types of optical modulators are presented in this chapter, including the lithium niobate (LiNbO$_3$) electro-optic modulators, the electro-absorption modulators, and polymeric integrated modulators. Their operating principles, devices physical structures, device parameters, and their applications and driving condition for generation of different modulation formats, as well as their impacts on system performance.

4.1 INTEGRATED OPTICAL MODULATORS

The modulation of lightwaves via an external optical modulator can be classified into three types depending on the special effects that alter the lightwaves property, especially the intensity or the phase of the optical carrier. In an external modulator, the intensity is normally manipulated by manipulating the phase of the carrier waves guided in one path of an interferometer. Mach–Zehnder interferometric structure is the most common type, especially the lithium niobate type [2–5].

On the other hand, electro-absorption (EA) type modulator employs the Franz and Keldysh effect, which is observed as lengthening the wavelength of the absorption edge of a semiconductor medium under the influence of an electric field [6,7]. In quantum structure such as the multi-quantum well structure, this effect is called the Stark effect, or the EA effect. The EA modulator can be integrated in-line with a laser structure on the same integrated circuit chip, now commonly called EML (external modulation laser [8]). For the LiNbO$_3$ modulator, the device is externally connected to a laser source via an optical fiber.

The total insertion loss of semiconductor intensity modulator is about 8–10 dB, including fiber-waveguide coupling loss which is rather high. However, this loss can be compensated by a semiconductor optical amplifier (SOA) that can be integrated on the same circuit. Compared with $LiNbO_3$, its total insertion loss is about 3–4 dB, which can be affordable, as erbium-doped fiber amplifier (EDFA) is now readily available.

The driving voltage for EA modulator is usually lower than that required for $LiNbO_3$. However, the extension ratio is not as high as that of $LiNbO_3$ type, which is about 25 dB, as compared to 10 dB for EA modulator. This feature contrasts the operating characteristics of the $LiNbO_3$ and EA modulators. Although the driving voltage for EA modulator is about 2–3 V and 5–7 volts for $LiNbO_3$, the former type would be preferred for intensity or phase modulation formats due to this extinction ratio that offers much lower "zero" noise level and, hence, high quality factor.

4.1.1 Phase Modulators

The phase modulator is a device that manipulates the "phase" of optical carrier signals under the influence of an electric field created by an applied voltage. When voltage is not applied to the RF-electrode, the number of periods of the lightwaves, n, exists in a certain path length. When voltage is applied to the RF-electrode, one or a fraction of one period of the wave is added, which means $(n + 1)$ waves exist in the same length. In this case, the phase has been changed by 2π and the half voltage of this is called the driving voltage. In case of long distance optical transmission, waveform is susceptible to degradation due to non-linear effect such as self-phase modulation etc. A phase modulator can be used to alter the phase of the carrier to compensate for this degradation. The magnitude of the change of the phase depends on the change of the refractive index created via the electro-optic effect that, in turn, depends on the orientation of the crystal axis with respect to the direction of the established electric field by the applied signal voltage.

An integrated optic phase modulator operates in a similar manner except that the lightwave carrier is guided via an optical waveguide with a diffused or ion-exchanged confined regions for $LiNbO_3$, and rib-waveguide structures for semiconductor type. Two electrodes are deposited so an electric field can be established across the waveguiding cross section so that a change of the refractive index via the electro-optic or EA effect as shown in Figure 4.1. For ultra-fast operation, one of the electrodes is a travelling wave type or hot electrode and the other is a ground electrode. The travelling wave electrode must be terminated with matching impedance at the end so as to avoid wave reflection. Usually, a quarter wavelength impedance is used to match the impedance of the travelling wave electrode to that of the 50 Ω transmission line.

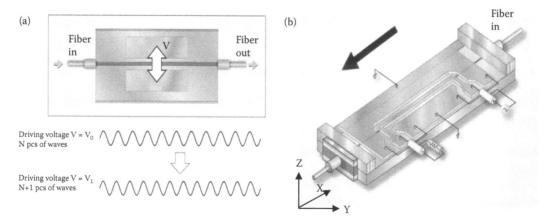

FIGURE 4.1 Electro-optic phase modulation in an integrated modulator using LiNbO3. Electrode impedance matching is not shown (a) schematic diagram and (b) integrated optic structure.

A phasor representation of a phase-modulated lightwave can be by the circular rotation at a radial speed of ω_c. Thus, the vector with an angle ϕ represents the magnitude and phase of the lightwave.

4.1.2 INTENSITY MODULATORS

Basic structured LN modulator comprises of (i) two waveguides, (ii) two Y-junctions, and (iii) RF/DC traveling wave electrodes. Optical signals coming from the lightwave source are launched into the LN modulator through the polarization maintaining fiber, it is then equally split into two branches at the first Y-junction on the substrate. When no voltage is applied to the RF electrodes, the two signals are recombined constructively at the second Y-junction and coupled into a single output. In this case, output signals from the LN modulator are recognized as "ONE." When voltage is applied to the RF electrode, due to the electro-optic effects of LN crystal substrate, the waveguide refractive index is changed, and hence the carrier phase in one arm is advanced, though retarded, in the other arm. Thence, the two signals are recombined destructively at the second Y-junction, they are transformed into higher order mode and radiated at the junction. If the phase retarding is in multiple odd factor of π, the two signals are completely out of phase, the combined signals are radiated into the substrate and the output signal from the LN modulator is recognized as a "ZERO." The voltage difference that induces this "ZERO" and "ONE" is called the driving voltage of the modulator, and is one of the important parameters in deciding modulator's performance.

4.1.2.1 Phasor Representation and Transfer Characteristics

Consider that an interferometic intensity modulator consists of an input waveguide that is split into two branches and then recombined to a single output waveguide, this forms an optical inteferometric structure. If the phase of the guided wave in one branch is modulated with an electric field via the electrodes across the waveguide, then this field alters the refractive index of the waveguide, slowing down or speeding up the group velocity of the optical guided waves, hence their phases. Depletion (destructive) or constructive interference are resulted at the output of the interferometers, so that is stage "0" or "1" of the lightwave output.

We must distinguish the single electrode or single drive MZIM and dual drive MZIM types. The difference is that a dual drive can be used to double the effects on the phases of the two branches of the interferometers by a push–pull operation. Thus, there is reduction of the amplitude of the applied microwave signals at very broadband (35 GHz) region, hence the complexity of the electrical driving circuitry.

The control and level applied to the electrodes at DC level set the initial phase states of the lightwaves passing through the two branches of the interferometers. Therefore, if the two electrodes are initially biased with voltages V_{b1} and V_{b2} then the initial phases exerted on the lightwaves would be $\phi_1 = \pi V_{b1}/V_\pi = -\phi_2$, which are indicated by the bias vectors shown in Figure 26 of (Chapter 2). From these biased phasor positions, the signal phasors are swinging according to the magnitude and the signs of the pulse voltages applied to the electrodes, the electric field and thence the optical phase swinging. They can be switched to the two biased positions which can reach constructive or destructive interferometric states at the merged and combined ouput of the guided-wave interferometer. Mathematically, the output field of the lightwave carrier can thus be represented by

$$E_0 = \frac{1}{2} E_{iRMS} e^{j\omega_c t} \left(e^{j\phi_1(t)} + e^{j\phi_2(t)} \right) \tag{4.1}$$

where ω_c is the carrier radial frequency, E_{iRMS} is the root mean square value of the magnitude of the carrier and $\phi_1(t)$ and $\phi_2(t)$ are the temporal phase generated by the two time-dependent pulse sequences applied to the two electrodes. With the voltage levels vary according to the magnitude of the pulse sequence one can obtain the transfer curve as shown in Figure 26 of Chapter 2. This phasor representation can be used to determine exactly the biasing conditions and magnitude of the RF or digital signals required for driving the optical modulators to achieve 50%, 33%, or 67% bit period pulse shapes.

The power transfer function of Mach–Zehnder modulator is expressed as[*]

$$P_0(t) = \alpha P_i \cos^2 \frac{\pi V(t)}{V_\pi} \tag{4.2}$$

where $P_0(t)$ is the output transmitted power, α is the modulator total insertion loss, P_i is the input power (usually from the laser diode), $V(t)$ is the time-dependent signal applied voltage, V_π is the driving voltage so that a π phase shift is exerted on the lightwave carrier. It is necessary to set the static bias on the transmission curve through bias electrode. It is common practice to set the bias point at 50% transmission point or a $\pi/2$ phase difference between the two optical waveguide branches, the quadrature bias point. As shown in Figure 26 (Chapter 2), electrical digital signals are transformed into optical digital signal by switching voltage to both ends of quadrature points on the positive and negative.

4.1.2.2 Bias Control

One factor that affects the modulator performance is the drift of the bias voltage, hence the fluctuation of the modulated amplitude or phases of the output lightwaves. For the Mach–Zehnder interferometric modulator (MZIM) type, it is very critical when it is required to bias at the quadrature point or at minimum or maximum locations on the transfer curve. The DC drift is the phenomena occurred in LiNbO₃ due to the build up of charges on the surface of the crystal substrate. Under this drift, the transmission curve gradually shift in the long term [9,10]. In the case of the LiNbO₃ modulator, the bias point control is as vital as the bias point shift in the long term. To compensate for the drift, it is necessary to monitor the output signals and feed it back into the bias control circuits to adjust the DC voltage so that operating points stay at the same point as shown in Figure 27 of Chapter 2. Thus, an optical coupler to tap the optical signal for monitoring is normally implemented from the ouput waveguide of the modulator and fed to a photodetector for electrical signel output to feedback to the electronic control circuitry. For example, the quadrature point and the output level are known, and a control algorithm can be generated to stabilize the biasing point at quadrature through a look-up table. It is the manufacturer's responsibility to reduce DC drift so that DC voltage is not beyond the limit throughout the lifetime of device.

4.1.2.3 Chirp-Free and Chirped Modulation

Chirping of the lightwaves indicates the varying of the frequency of the optical waves. This chirping can be in opposite direction of the chirping of the lightwaves via the propagation of the optical modulated signals through the optical fiber transmission line. Thus, the chirping can be considered an optical compensating technique to counter the distortion effects in transmission over the fiber [11]. This section gives an account of the chirpping of optical modulation via different structures and applied signal conditions. Due to the symmetry of the crystal refractive index of the uniaxial anisotropy of the class m of LiNbO₃, the crystal cut and the propagation direction of the electric field affect modulator efficiency, denoted as driving voltage and modulator chirp. The uniaxial property of LiNbO₃ is shown in Figures 4.2 and 4.4. As shown in Figure 4.4, in the case of the Z-cut structure, as a hot electrode is placed on top of the waveguide, RF field flux is more

---

[*] Note this equation is representing for single drive MZIM–it is the same for dual drive MZIM, provided that the bias voltages applied to the two electrodes are equal and opposite in signs. The transfer curve of the field representation would have half the periodic frequency of the transmission curve shown in Figure 4.7.

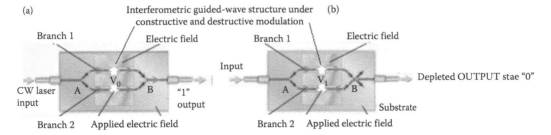

FIGURE 4.2 Intensity modulation using interferometric principles in guide wave structures in LiNbO$_3$ (a) ON—constructive interference mode and (b) destructive interference mode—OFF. Optical guided wave paths 1 and 2. Electric field is established across the optical waveguide.

concentrated, and this results in the improvement of overlap between RF and optical field. However, overlap between RF in ground electrode and waveguide is reduced in the Z-cut structure so that overall improvement of driving voltage for Z-cut structure compared to X-cut is approximately 20% (see Figure 4.3). The different overlapping area for the Z-cut structure results in a chirp parameter of 0.7, whereas X-cut and Z-propagation has almost zero-chirp due to its symmetric structure. A number of commonly arranged electrode and waveguide structures are shown in Figure 4.4 to maximize the interaction between the travelling electric field and the optical guided waves. Furthermore, a buffer layer, normally SiO$_2$, is used in order to match the travelling velocities between the electric microve field and that of the optical guided-waves so as to optimize to modulation bandwidth.

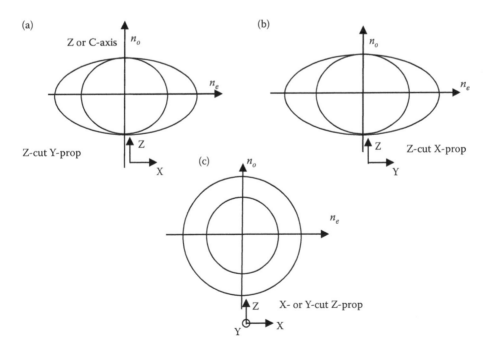

FIGURE 4.3 Refractive index contours of LiNbO$_3$ uniaxial crystal with Z- or C- denoting as the principal axis (a) lightwaves propagation in the Z-axis polarised along Y-cut LiNbO$_3$, (b) prop-direction Z-axis and X-cut crystal, and (c) Y-prop and Y-cut crystal.

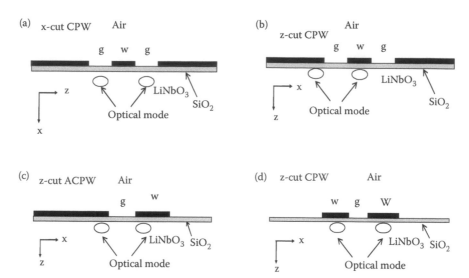

FIGURE 4.4 Cross section of the optical modulator and crystal orientation for inteferometric modulation to maximize the use of the overlap integral between the optical guided mode and the electric field distribution for largest electro-optic coefficients. (a) X-cut LiNbO$_3$ with co-planar travelling waves (CPW) type electrode; (b) Z-cut LiNbO$_3$ with CPW electrode; (c) Z-cut LiNbO3 with asymmetric co-planar wave (ACPW) electrode; and (d) Z-cut with co-planar wave electrode.

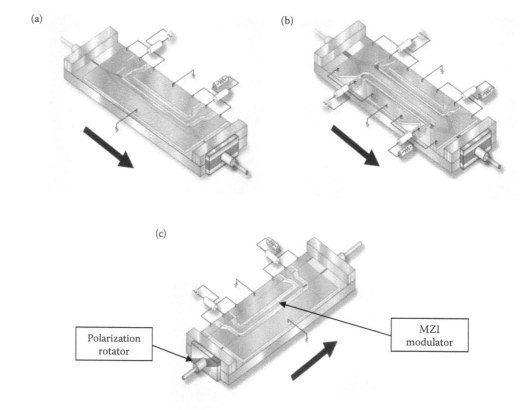

FIGURE 4.5 Intensity modulators using LiNbO$_3$ (a) single drive electrode, (b) dual electrode structure, and (c) electro-optic polarization scrambler using LiNbO$_3$.

TABLE 4.1

Typical Operational Parameters of LiNbO$_3$ Optical Intensity Modulators

Parameters	Typical Values	Definition/Comments
Modulation speed	56 Gbps or 64 Gbps	Capability to transmit fast or slow of optical digital signals
Insertion loss	Max 5 dB	Defined as the optical power loss within the modulator
V_π	3–4 Volts	
Driving voltage	Max 4V p-p	The RF voltage required to have a full modulation
Optical bandwidth	Min 35 GHz	3dB roll-off in efficiency at the highest frequency in the modulated signal spectrum
ON/OFF extinction ratio	Min 20 dB	The ratio of maximum optical power (ON) and minimum optical power (OFF)
Polarization extinction ratio	Min 20 dB	The ratio of two polarization states (TM and TE guided modes) at the output

4.1.2.4 Structures of Photonic Modulators

Figure 4.5a and b shows the structure of a MZ intensity modulator using single and dual electrode configurations, respectively. The thin line electrode is called the "hot" electrode, or traveling wave electrode (TWE). RF connectors are required for launching the RF data signals to establish the electric field required for electro-optic effects. Impedance termination is also required. Optical fibers pigtails are also attached to the end faces of the diffused waveguide. The mode spot size of the diffused waveguide is not symmetric and hence, some diffusion parameters are controlled so that maximizing the coupling between the fiber and the diffused or rib waveguide can be achieved. Due to this mismatching between the mode spot sizes of the circular and diffused optical waveguides, coupling loss occurs. Furthermore, the difference between the refractive indices of fiber and LiNbO$_3$ is quite substantial, and thus, Fresnel reflection loss would also incur.

Figure 4.5c shows the structure of a polarization modulator, which is essential for multiplexing of two polarized data sequences so as to double the transmission capacity, for example, 40 G to 80 Gb/s. Furthermore, this type of polarization modulator can be used as a polarization rotator in a polarization dispersion compensating sub-system [12,13]. Currently, these types of modulators with IQ modulation have been reported [14] (Table 4.1).

4.1.3 RETURN-TO-ZERO OPTICAL PULSES

4.1.3.1 Generation

Figure 4.6 shows the conventional structure of an RZ-ASK transmitter in which two external LiNbO$_3$ MZIMs can be used. The MZIM shown in this transmitter can be a single- or dual-drive (push–pull) type. Operational principles of the MZIM were presented in Section 4.2.2 of the previous chapter. The optical OOK transmitter would normally consist of a narrow-linewidth laser source to generate lightwaves whose wavelength satisfies the ITU grid standard.

The first MZIM, commonly known as the pulse carver, is used to generate the periodic pulse trains with a required return-to-zero format (RZ) format. The suppression of the lightwave carrier can also be carried out at this stage if necessary, which is commonly known as the carrier-suppressed RZ (CSRZ). Compared to other RZ types, CSRZ pulse shape is found to have attractive attributes for long haul WDM transmissions, including the π phase difference of adjacent modulated bits, suppression of the optical carrier component in optical spectrum, and narrower spectral width.

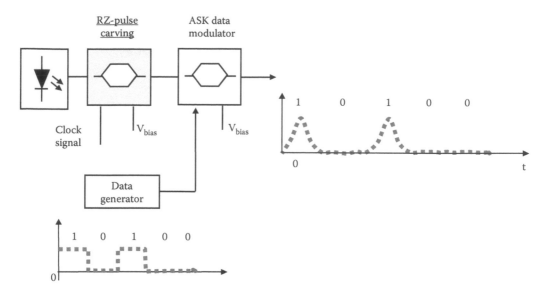

FIGURE 4.6 Conventional structure of an OOK optical transmitter utilizing two MZIMs.

Different types of RZ pulses can be generated depending on the driving amplitude of the RF voltage and the biasing schemes of the MZIM. The equations governing the RZ pulses electric field waveforms are

$$
E(t) = \begin{cases} \sqrt{\dfrac{E_b}{T}} \sin\left[\dfrac{\pi}{2}\cos\left(\dfrac{\pi t}{T}\right)\right] & \text{67\% duty-ratio RZ pulses or CSRZ} \\[3ex] \sqrt{\dfrac{E_b}{T}} \sin\left[\dfrac{\pi}{2}\left(1 + \sin\left(\dfrac{\pi t}{T}\right)\right)\right] & \text{33\% duty-ratio RZ pulses or RZ33} \end{cases}
\tag{4.3}
$$

where E_b is the pulse energy per a transmitted bit and T is one bit period. The 33% duty-ratio RZ pulse are denoted as RZ33 pulse, whereas the 67% duty cycle RZ pulse is known as the CSRZ type. The art in generation of these two RZ pulse types stays at the difference of biasing point on the transfer curve of an MZIM.

The bias voltage conditions and the pulse shape of these two RZ types, the carrier suppression and non-suppression of maximum carrier can be implemented with the biasing points at the minimum and maximum transmission point of the transmittance characteristics of the MZIM, respectively. The peak-to-peak amplitude of the RF driving voltage is $2V_\pi$ where V_π is the required driving voltage to obtain a π phase shift of the lightwave carrier. Another important point is that the RF signal is operating at only a half of the transmission bit rate. Hence, pulse carving is actually implementing the frequency doubling. The generations of RZ33 and CSRZ pulse train are demonstrated in Figure 4.7a and b.

The pulse carver can also utilize a dual drive MZIM that is driven by two complementary sinusoidal RF signals. This pulse carver is biased at $-V_{\pi/2}$ and $+V_{\pi/2}$ with the peak to peak amplitude of $V_{\pi/2}$. Thus, a π phase shift is created between the state "1"and "0" of the pulse sequence and hence the RZ with alternating phase 0 and π. If the carrier suppression is required, then the two electrodes are applied with voltages V_π and swing voltage amplitude of V_π.

Although RZ modulation offers improved performance, RZ optical systems usually require more complex transmitters than those in the NRZ ones. Compared to only one stage for modulating data on the NRZ optical signals, two modulation stages are required for generation of RZ optical pulses [15–18].

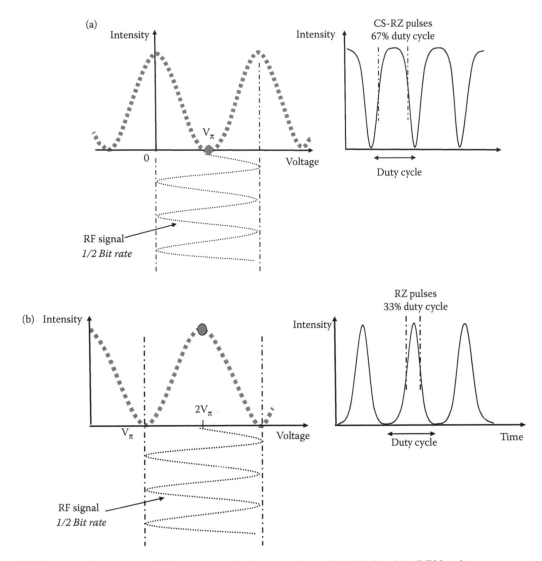

FIGURE 4.7 Bias point and RF driving signals for generation of (a) CSRZ and (b) RZ33 pulses.

4.1.3.2 Phasor Representation

Recalling (1) we have

$$E_o = \frac{E_i}{2}\left[e^{j\varphi_1(t)} + e^{j\varphi_2(t)}\right] = \frac{E_i}{2}\left[e^{j\pi v_1(t)/V_\pi} + e^{j\pi v_2(t)/V_\pi}\right] \tag{4.4}$$

The modulating process for generation of RZ pulses can be represented by a phasor diagram, as shown in Figure 4.8, where the optical frequency component $e^{j\omega t}$ has been removed to indicate that the wave vector is rotating at this angular frequency and considered to be stationary. This technique gives a clear understanding of the superposition of the fields at the coupling output of two arms of the MZIM. Here, a dual-drive MZIM is used, that is, the data driving signals $[V_1(t)]$ and the inverse data signals ($\overline{\text{data}}$: $V_2(t) = -V_1(t)$) are applied into each arm of the MZIM, respectively, and the RF voltages swing in inverse directions. Applying the phasor representation, vector addition, and simple trigonometric calculus, the process of generation RZ33 and CSRZ is explained in detail and verified.

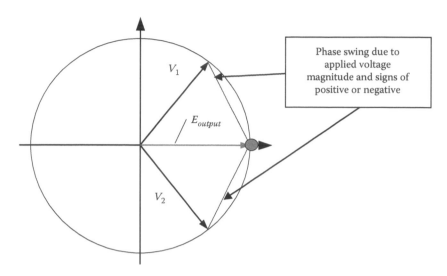

FIGURE 4.8   Phasor representation for generation of output field in dual-drive MZIM.

The width of these pulses are commonly measured at the position of full-width half maximum (FWHM). It is noted that the measured pulses are intensity pulses, whereas we are considering the addition of the fields in the MZIM. Thus, the normalized E_o field vector has the value of $\pm 1/\sqrt{2}$ at the FWHM intensity pulse positions and the time interval between these points gives the FWHM values.

4.1.3.3   Phasor Representation of CSRZ Pulses in Dual-Drive MZIM

Key parameters including the V_{bias}, the amplitude of the RF driving signal are shown in Figure 4.9a. Accordingly, its initialized phasor representation is demonstrated in Figure 4.9b.

The values of the key parameters are outlined as follows: (i) V_{bias} is $\pm V_\pi/2$. (ii) Swing voltage of driving RF signal on each arm has the amplitude of $V_\pi/2$ (i.e., $V_{p\text{-}p} = V_\pi$). (iii) RF signal operates at half of bit rate $(B_R/2)$. (iv) At the FWHM position of the optical pulse, the $E_{out} = \pm 1/\sqrt{2}$ and the component vectors V_1 and V_2 form with vertical axis a phase of $\pi/4$ as shown in Figure 4.10.

Considering the scenario for generation of 40 Gb/s CSRZ optical signal, the modulating frequency is f_m $(f_m = 20\,\text{GHz}. = B_R/2)$. At the FWHM positions of the optical pulse, the phase is given

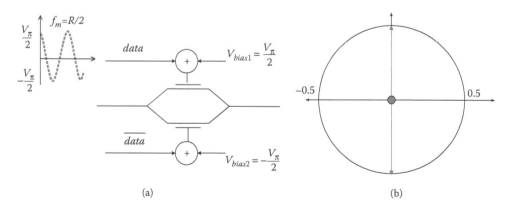

(a)   (b)

FIGURE 4.9   Initialized stage for generation of CSRZ pulse: (a) RF driving signal and the bias voltages. (b) Initial phasor representation.

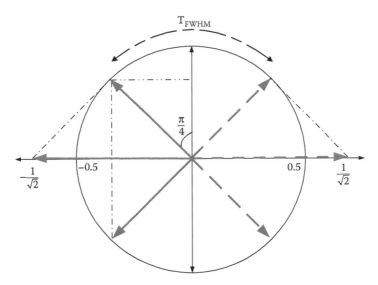

FIGURE 4.10   Phasor representation of CSRZ pulse generation using dual-drive MZIM.

by the following expressions:

$$\frac{\pi}{2}\sin(2\pi f_m) = \frac{\pi}{4} \Rightarrow \sin 2\pi f_m = \frac{1}{2} \Rightarrow 2\pi f_m = \left(\frac{\pi}{6}, \frac{5\pi}{6}\right) + 2n\pi \quad (4.5)$$

Thus, the calculation of TFWHM can be carried out and hence, the duty cycle of the RZ optical pulse can be obtained as given in the following expressions:

$$T_{FWHM} = \left(\frac{5\pi}{6} - \frac{\pi}{6}\right)\frac{1}{R2\pi} = \frac{1}{3}\pi \times \frac{1}{R} \Rightarrow \frac{T_{FWHM}}{T_{BIT}} = \frac{1.66 \times 10^{-4}}{2.5 \times 10^{-11}} = 66.67\% \quad (4.6)$$

The result obtained in Equation 4.6 clearly verifies the generation of CSRZ optical pulses from the phasor representation.

4.1.3.4   Phasor Representation of RZ33 Pulses

Key parameters including the V_{bias}, the amplitude of driving voltage and its correspondent initialized phasor representation are shown in Figure 4.11a and b, respectively.

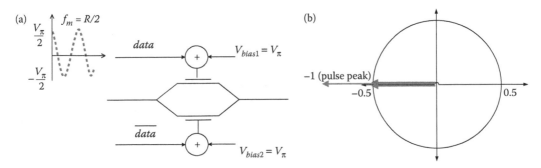

FIGURE 4.11   Initialized stage for generation of RZ33 pulse: (a) RF driving signal and the bias voltage. (b) Initial phasor representation.

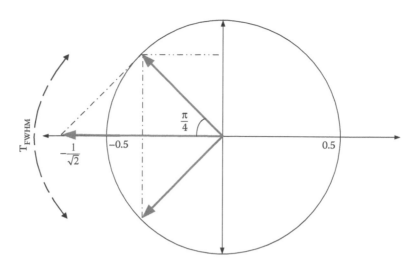

FIGURE 4.12   Phasor representation of RZ33 pulse generation using dual-drive MZIM.

The values of the key parameters are (i) V_{bias} is V_π for both arms; (ii) Swing voltage of driving RF signal on each arm has the amplitude of $V_\pi/2$ (i.e., $V_{p-p} = V_\pi$); and (iii) RF signal operates at half of bit rate ($B_R/2$).

At the FWHM position of the optical pulse, the $E_{out} = \pm 1/\sqrt{2}$ and the component vectors V_1 and V_2 form with a phase of $\pi/4$ with horizontal axis, as shown in Figure 4.12.

Considering the scenario for generation of 40 Gb/s CSRZ optical signal, the modulating frequency is f_m ($f_m = 20$ GHz. $= B_R/2$). At the FWHM positions of the optical pulse, the phase is given by the following expressions:

$$\frac{\pi}{2}\cos(2\pi f_m t) = \frac{\pi}{4} \Rightarrow t_1 = \frac{1}{6f_m} \tag{4.7}$$

$$\frac{\pi}{2}\cos(2\pi f_m t) = -\frac{\pi}{4} \Rightarrow t_2 = \frac{1}{3f_m} \tag{4.8}$$

Thus, the calculation of TFWHM can be carried out and the duty cycle of the RZ optical pulse can be obtained as given in the following expressions:

$$T_{FWHM} = \frac{1}{3f_m} - \frac{1}{6f_m} = \frac{1}{6f_m} \therefore \frac{T_{FWHM}}{T_b} = \frac{1/6f_m}{1/2f_m} = 33\% \tag{4.9}$$

The result obtained in Equation 4.10 clearly verifies the generation of RZ33 optical pulses from the phasor representation.

4.2   DIFFERENTIAL PHASE SHIFT KEYING MODULATION

4.2.1   BACKGROUND

Digital encoding of data information by modulating the phase of the lightwave carrier is referred to as optical phase shift keying (PSK). In early days, optical PSK was studied extensively for coherent photonic transmission systems. This technique requires the manipulation of the absolute phase of the lightwave carrier. Thus, precise alignment of the transmitter and demodulator center frequencies for the coherent detection is required. These coherent optical PSK systems phase severe obstacles,

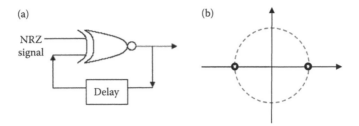

FIGURE 4.13   (a) DPSK pre-coder and (b) signal constellation diagram of DPSK. (From P. K. Tien, *Rev. Mod. Phys.*, Vol. 49, pp. 361–420, 1977)

such as broad line-width and chirping problems of the laser source. Meanwhile, the differential phase shift keying (DPSK) scheme overcomes those problems, since the DPSK optically modulated signals can be detected incoherently. This technique only requires the coherence of the lightwave carriers over one bit period for the comparison of the differentially coded phases of the consecutive optical pulses.

A binary "1" is encoded if the present input bit and the past encoded bit are of opposite logic, whereas a binary 0 is encoded if the logics are similar. This operation is equivalent to an XOR logic operation. Hence, an XOR gate is employed as a differential encoder. NOR can also be used to replace an XOR operation in differential encoding as shown in Figure 4.13a. In DPSK, the electrical data "1" indicates a π phase change between the consecutive data bits in the optical carrier, while the binary "0" is encoded if there is no phase change between the consecutive data bits. Hence, this encoding scheme gives rise to two points located exactly at π phase difference with respect to each other in signal constellation diagram. For continuous phase shift keying such as the minimum shift keying, the phase evolves continuously over a quarter of the section, thus a phase change of $\pi/2$ between one phase state to the other. This is indicated by the inner bold circle as shown in Figure 4.13b.

4.2.2   Optical DPSK Transmitter

Figure 4.14 shows the structure of a 40 Gb/s DPSK transmitter in which two external LiNbO$_3$ MZIM are used. Operational principles of a MZIM were presented above. The MZIMs shown in Figure 4.14 can be of single or dual drive type. The optical DPSK transmitter also consists of a narrow linewidth laser to generate a lightwave whose wavelength conforms to the ITU grid.

The RZ optical pulses are then fed into the second MZIM through which the RZ pulses are modulated by the pre-coded binary data to generate RZ-DPSK optical signals. Electrical data pulses are differentially pre-coded in a pre-coder using the XOR coding scheme. Without pulse carver, the structure shown in Figure 4.14 is an optical NRZ-DPSK transmitter. In data modulation for DPSK format, the second MZIM is biased at the minimum transmission point. The pre-coded electrical data has a peak-to-peak amplitude equal to $2V_\pi$ and operates at the transmission bit rate. The modulation principles for generation of optical DPSK signals are demonstrated in Figure 4.6.

The electro-optic phase modulator might also be used for generation of DPSK signals instead of MZIM. Using optical phase modulator, the transmitted optical signal is chirped whereas using MZIM, especially the X-cut type with Z-propagation, chirp-free signals can be produced. However, in practice, a small amount of chirp might be useful for transmission [19].

4.3   GENERATION OF MODULATION FORMATS

Modulation is the process facilitating the transfer of information over a medium, for example, a wireless or optical environment. Three basic types of modulation techniques are based on the

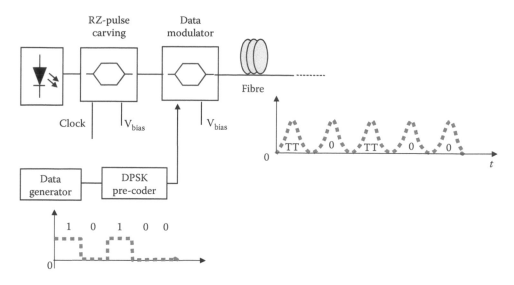

FIGURE 4.14 DPSK optical transmitter with RZ pulse carver.

manipulation of a parameter of the optical carrier to represent the information digital data. These are amplitude shift keying (ASK), phase shift keying (PSK), and frequency shift keying (FSK). In addition to the manipulation of the carrier, the occupation of the data pulse over a single period would also determine the amount of energy concentrates and the speed of the system required for transmission. The pulse can remain constant over a bit period or return to zero level within a portion of the period. These formats would be named non-return-to-zero (NRZ) or return-to-zero (RZ) (see Figure 4.15). They are combined with the modulation of the carrier to form various modulations formats which are presented in this section. Figure 4.16 shows the base band signals of the NRZ and RZ formats and its corresponding block diagram of a photonic transmitter.

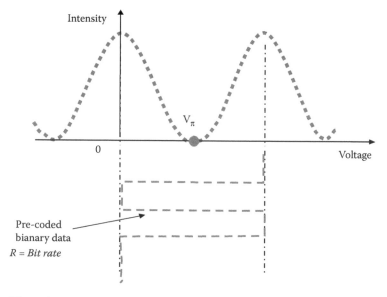

FIGURE 4.15 Bias point and RF driving signals for generation of optical DPSK format.

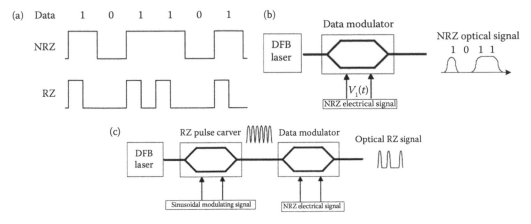

FIGURE 4.16 (a) Baseband NRZ and RZ line coding for 101101 data sequence. (b) Block diagram of NRZ photonics transmitter. (c) RZ photonics transmitter incorporating a pulse carver.

4.3.1 Amplitude–Modulation ASK-NRZ and ASK-RZ

4.3.2 Amplitude–Modulation OOK-RZ Formats

There are a number of advanced formats used in advanced optical communications. Based on the intensity of the pulse, these may include NRZ, RZ, and duo-binary. These amplitude shift keying (ASK) formats can also be integrated with the phase modulation to generate discrete or continuous phase NRZ or RZ formats. Currently, the majority of 10 Gb/s installed optical communications systems have been developed with NRZ because of its simple transmitter design and bandwidth efficient characteristic. However, RZ format has higher robustness to fiber non-linearity and polarization mode dispersion (PMD). In this section, the RZ pulse is generated by MZIM commonly known as pulse carver, as arranged in Figure 4.17. There are a number of variations in RZ format based on the biasing point in transmission curve shown in Table 4.2. The phasor representation of the biasing and driving signals can be observed in Table 4.2.

CSRZ has been found to have more attractive attributes in long haul WDM transmissions compared to conventional RZ format because of the possibility of reducing the upper level of the power contained in the carrier that serves no purpose in the transmission, but increases the total energy level and approaches the non-linear threshold level faster. CSRZ pulse has an optical phase difference of π in adjacent bits, removing the optical carrier component in optical spectrum and reducing the spectral width. This offers an advantage in compact WDM channel spacing.

4.3.3 Amplitude–Modulation Carrier-Suppressed RZ (CSRZ) Formats

The suppression of the carrier can be implemented by biasing the MZ interferometer in such a way so that there is a pi phase shift between the two arms of the interferometer. The magnitude of the

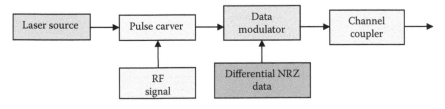

FIGURE 4.17 Block diagrams of RZ-DPSK transmitter.

TABLE 4.2

Summary of RZ Format Generation and Characteristics of Single Drive MZIM Based on Biasing Point, Drive Signal Amplitude, and Frequency

Biasing point in the cos² transmission transfer curve	RZ generation characteristics	Phasor representation
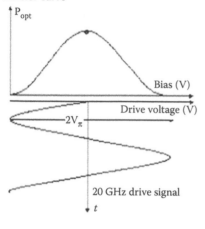	a. Biasing point: Maximum, $V_{2\pi}$ b. Drive signal amplitude: $2\,V_\pi$ c. Drive sinusoidal signal frequency: 20 GHz d. Pulse width: 9–10 ps about 70% of half the data rate e. RZ pulse frequency: 40 GHz frequency doubling effect	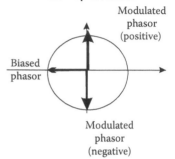
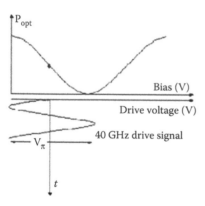	a. Biasing point: Linear region, $V_{\pi/2}$ b. Drive signal amplitude: V_π c. Drive sinusoidal signal frequency: 20 GHz d. Pulse width: 9–10 ps or about 70% of half the data rate e. RZ pulse frequency: 20 GHz	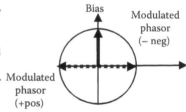
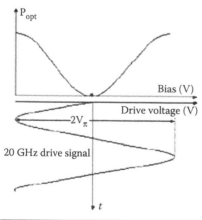	a. Biasing point: Minimum, V_π b. Drive signal amplitude: $2\,V_\pi$ c. Drive sinusoidal signal frequency: 20 GHz d. Pulse width: 9–10 ps about 70% of half the data rate e. RZ pulse frequency: 40 GHz frequency doubling effect. (Carrier suppressed RZ is generated using this scheme)	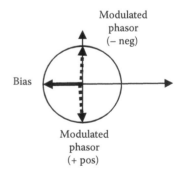

sinusoidal signals applied to an arm or both arms would determine the width of the optical output pulse sequence. The driving conditions and phasor representation are shown in Table 4.2.

4.3.4 DISCRETE PHASE–MODULATION NRZ FORMATS

The term discrete phase modulation is referred to as the differential phase shift keying whether DPSK or quadrature DPSK (DQPSK) to indicate that the states of the phases of the lightwave carrier are switch from one distinct location on the phasor diagram to the other state, for example, from 0 to π or $-\pi/2$ to $-\pi/2$ for binary PSK (BPSK), or even more evenly spaced phase shift keying levels as in the case of M-ary PSK.

4.3.4.1 Differential Phase Shift Keying (DPSK)

Information encoded in the phase of an optical carrier is commonly referred to as optical phase shift keying. In early days, PSK required precise alignment of the transmitter and demodulator center frequencies [15]. Hence, PSK systems are not widely deployed. With the DPSK scheme introduced, coherent detection is not critical since DPSK detection only requires source coherence over one bit period by comparison of two consecutive pulses.

A binary "1" is encoded if the present input bit and the past encoded bit are of opposite logic and a binary 0 is encoded if the logic is similar. This operation is equivalent to XOR logic operation. Hence, an XOR gate is usually employed in differential encoder. NOR can also be used to replace XOR operation in differential encoding as shown in Figure 4.18.

In optical application, electrical data "1" is represented by a π phase change between the consecutive data bits in the optical carrier, while state "0" is encoded with no phase change between the consecutive data bits. Hence, this encoding scheme gives rise to two points located exactly at π phase difference with respect to each other in signal constellation diagram as indicated in Figure 4.18b.

A RZ-DPSK transmitter consists of an optical source, pulse carver, data modulator, differential data encoder, and a channel coupler. The channel coupler model is not developed in simulation by assuming no coupling losses when optical RZ-DPSK modulated signal is launched into the optical fiber. This modulation scheme has combined the functionality of dual drive MZIM modulator of pulse carving and phase modulation.

The pulse carver, usually a MZ interferometric intensity modulator, is driven by a sinusoidal RF signal for single-drive MZIM and two complementary electrical RF signals for dual drive MZIM, to carve pulses out from optical carrier signal forming RZ pulses. These optical RZ pulses are fed into second MZ intensity modulator where RZ pulses are modulated by differential NRZ electrical data to generate RZ-DPSK. This data phase modulation can be performed using a straight-line phase modulator, but the MZ waveguide structure has several advantages over phase modulator because of its chirpless property. Electrical data pulses are differentially pre-coded in a differential pre-coder as shown in Figure 4.18a. Without pulse carver and sinusoidal RF signal, the output pulse sequence

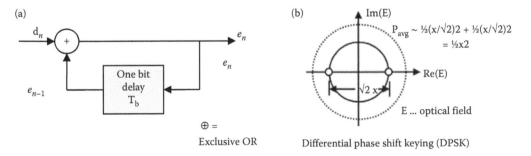

FIGURE 4.18   (a) The encoded differential data are generated by $e_n = d_n \oplus e_{n-1}$. (b) Signal constellation diagram of DPSK.

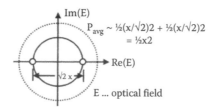

Differential phase shift keying (DPSK)

FIGURE 4.19 Signal constellation diagram of DQPSK. Two bold dots are orthogonal to the DPSK constellation.

follows NRZ-DPSK format; that is, the pulse occupies 100% of the pulse period and there is no transition between the consecutive "1s."

4.3.4.2 Differential Quadrature Phase Shift Keying (DQPSK)

This differential coding is similar to DPSK except that each symbol consists of two bits that are represented by the two orthogonal axial discrete phases at $(0, \pi)$ and $(-\pi/2, +\pi/2)$, as shown in Figure 4.19. Two additional orthogonal phase positions are located on the imaginary axis of Figure 4.18b.

4.3.4.2.1 NRZ-DPSK

Figure 4.20 shows the block diagram of a typical NRZ-DPSK transmitter. Differential pre-coder of electrical data is implemented using the logic is explained in the previous section. In phase modulating of an optical carrier, the MZ modulator, known as the data phase modulator, is biased at minimum point and driven by a data swing of $2V_\pi$. The modulator shows an excellent behaviour that the phase of the optical carrier will be altered by π exactly when the signal transiting the minimum point of the transfer characteristic.

4.3.4.2.2 RZ-DPSK

The arrangement of RZ-DPSK transmitter is, essentially, similar to RZ-ASK, as shown in Figure 4.21 with the data intensity modulator replaced with data phase modulator. The difference between them is the biasing point and the electrical voltage swing. Different RZ formats can also be generated.

4.3.4.3 M-ary Amplitude DPSK (M-ary ADPSK)

As an example, a 16-ary-MADPSK signal can be represented by the constellation shown in Figure 4.22. It is a combination of a 4-ary ASK and a DQPSK scheme. At the transmitting end, each group of four bits $[D_3D_2D_1D_0]$ of user data are encoded into a symbol. Among them, the two least significant bits $[D_1D_0]$ are encoded into four phase states $[0, \pi/2, \pi, 3\pi/2]$ and the other two

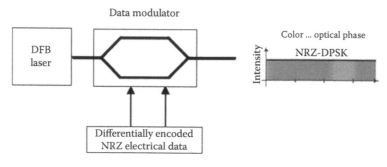

FIGURE 4.20 Block diagram of NRZ-DPSK photonics transmitter.

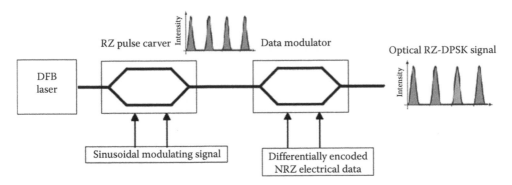

FIGURE 4.21   Block diagram of RZ-DPSK photonics transmitter.

most significant bits, $[D_3D_2]$, are encoded into four amplitude levels. At the receiving end, as MZ delay interferometers (DI) are used for phase comparison and detection, a diagonal arrangement of the signal constellation shown in Figure 4.22a is preferred. This simplifies the design of transmitter and receiver and minimizes the number of phase detectors, leading to high receiver sensitivity.

In order to balance the bit error rate (BER) between ASK and DQPSK components, the signal levels corresponding to four circles of the signal space should be adjusted to a reasonable ratio that depends on the noise power at the receiver. As an example, if this ratio is set to $[I_0/I_1/I_2/I_3] = [1/1.4/2/2.5]$, where I_0, I_1, I_2, and I_3 are the intensity of the optical signals corresponding to circle 0, circle 1, circle 2, and circle 3, respectively, then by selecting E_i equal to the amplitude of the circle 3 and V_π equal to 1, the driving voltages should have the values given in Table 4.3. Inversely speaking, one can set the outer most level such that its peak value is below the non-linear SPM threshold, the voltage level of the outer most circle would be determined. The innermost circle is limited to the condition that the largest signal-to-noise ratio (SNR) should be achieved. This is related to the optical SNR required for a certain BER. Thus, we can design the other points of the constellation from the largest and smallest amplitude levels.

Furthermore, to minimize the effect of intersymbol interference, 66%-RZ and 50%-RZ pulse formats are also used as alternatives to the traditional NRZ counterpart. These RZ pulse formats can be created by a pulse carver that precedes or follows the dual-drive MZIM modulator. Mathematically,

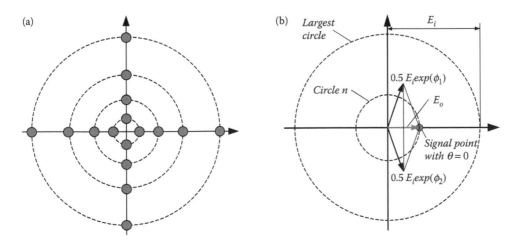

FIGURE 4.22   Signal constellation of 4-ary ADPSK format and phasor representation of a point on the constellation point for driving voltages applied to dual dive MZIM.

TABLE 4.3

Driving Voltages for 16-ary MADPSK Signal Constellation

Circle 0			Circle 1			Circle 2			Circle 3		
Phase	$V_t(t)$ volt	$V_2(t)$ volt	Phase	$V_t(t)$ volt	$V_2(t)$ volt	Phase	$V_t(t)$ volt	$V_2(t)$ volt	Phase	$V_t(t)$ volt	$V_2(t)$ volt
0	0.38	−0.38	0	0.30	−0.30	0	0.21	−0.21	0	0.0	0.0
$\pi/2$	0.88	0.12	$\pi/2$	0.80	0.20	$\pi/2$	0.71	0.29	$\pi/2$	0.5	0.5
π	−0.62	0.62	π	−0.7	0.70	π	−0.79	0.79	π	−1.0	1.0
$3\pi/2$	−0.12	−0.88	$3\pi/2$	−0.20	−0.8	$3\pi/2$	−0.29	−0.71	$3\pi/2$	−0.5	−0.5

waveforms of NRZ and RZ pulses can be represented by the following equations, where E_{on}, $n = 0, 1, 2, 3$ are peak amplitude of the signals in the circle 0, circle 1, circle 2, and circle 3 of the constellation, respectively:

$$p(t) = \begin{cases} E_{on} & \text{for NRZ} \\ E_{on} \cos\left(\dfrac{\pi}{2}\cos^2\left(\dfrac{1.5\pi t}{T_s}\right)\right) & \text{for 66\%-RZ} \\ E_{on} \cos\left(\dfrac{\pi}{2}\cos^2\left(\dfrac{2\pi t}{T_s}\right)\right) & \text{for 50\%-RZ} \end{cases} \quad (4.10)$$

A typical arrangement of the signals of the pre-coder and driving signals for the MZIM is shown in Figure 4.23.

4.3.5 Continuous Phase–Modulation

In the previous section, the optical transmitters for discrete PSK modulation formats have been described. Obviously, the phase of the carrier has been used to indicate the digital states of the bits or symbols. These phases are allocated in a non-continuous manner around a circle

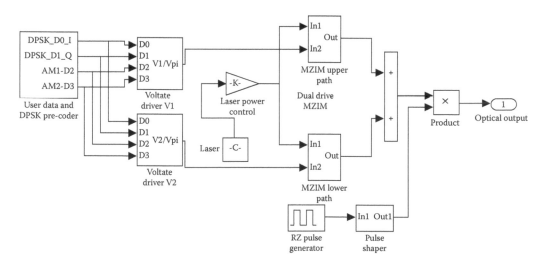

FIGURE 4.23 Matlab Simulink model of a MADPSK photonic transmitter. The MZIM is represented by two phase shifter blocks.

corresponding to the magnitude of the wave. Alternatively, the phase of the carrier can be continuously modulated and the total phase changes at the transition instants, usually at the end of the bit period, would be the same as those for discrete cases. Since the phase of the carrier continuously varies during the bit period, this can be considered a FSK modulation technique, except that the transition of the phase at the end of one bit and the beginning of the next bit would be continuous. The continuity of the carrier phase at these transitions would reduce the signal bandwidth, making it more tolerable of dispersion effects and a higher energy concentration for effective transmission over the optical guided medium. One of the examples of the reduction of the phase at the transition is the offset DQPSK, which is a minor but important variation on the QPSK or DQPSK. In OQPSK, the Q-channel is shifted by half a symbol period so that the transition instants of I and Q channel signals do not happen at the same time. The result of this simple change is that the phase shifts at any one time are limited and hence the offset QPSK is a more constant envelope than the normal QPSK.

The enhancement of the efficiency of the bandwidth of the signals can be further improved if the phase changes and these transitions are continuous. In this case, the change of the phase during the symbol period is continuously changed by using a half-cycle sinusoidal driving signal with the total phase transition over a symbol period being a fraction of π, depending on the levels of this phase shift keying modulation. If the change is $\pi/2$, then we have a minimum shift keying (MSK) scheme. The orthogonality of the I- and Q-channels will also reduce the bandwidth of the carrier-modulated signals, the continuous phase modulation (CPM), even further. In this section, we describe the basic principles of optical MSK and the photonic transmitters for these modulation formats. Ideally, the driving signal of the phase modulator should be a triangular wave so that a linear phase variation of the carrier in the symbol period is linear. However, when a sinusoidal function is used, there are some non-linear wave variations. We call this type of MSK a non-linear MSK format. This non-linearity contributes to some penalty in the optical signal-to-noise ratio (OSNR), which will be explained in a later chapter. Furthermore, the MSK as a special form of ODQPSK is also described for optical systems.

4.3.5.1   Linear and Non-Linear MSK

MSK is a special form of continuous phase FSK (CPFSK) signal in which the two frequencies are spaced in such way that they are orthogonal and have minimum spacing between them. They are defined by

$$s(t) = \sqrt{\frac{2E_b}{T_b}} \cos\left[2\pi f_1 t + \theta(0)\right] \quad \text{for symbol "1"} \tag{4.11}$$

$$s(t) = \sqrt{\frac{2E_b}{T_b}} \cos\left[2\pi f_2 t + \theta(0)\right] \quad \text{for symbol "0"} \tag{4.12}$$

As shown by the equations above, the signal frequency change corresponds to higher frequency for data and lower frequency for data-0. Both frequencies, f_1 and f_2, are defined by

$$f_1 = f_c + \frac{1}{4T_b} \tag{4.13}$$

$$f_2 = f_c - \frac{1}{4T_b} \tag{4.14}$$

Depending on the binary data and the phase of signal changes, data-1 increases the phase by $\pi/2$, while data-0 decreases the phase by $\pi/2$. The variation of phase follows paths of sequence of straight lines in phase trellis (Figure 4.24), in which the slopes represent frequency changes. The change in

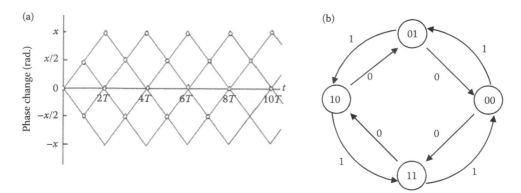

FIGURE 4.24   (a) Phase trellis for MSK and (b) state diagram for MSK. (From E. Ngoya and R. Larcheveque, Envelop Transient analysis: A new method for the transient and steady state analysis of microwave communication circuits and systems, in *Proc. IEEE MTT-S International Microwave Symposium, IEEE*, vol. 3, p. 1365, 1996.)

carrier frequency from data-0 to data-1, or vice versa, is equal to half the bit rate of incoming data [5]. This is the minimum frequency spacing that allows the two FSK signals representing symbols 1 and 0, to be coherently orthogonal in the sense that they do not interfere with one another in the process of detection.

A MSK signal consists of both I and Q components that can be written as

$$s(t) = \sqrt{\frac{2E_b}{T_b}} \cos\left[\theta(t)\right] \cos\left[2\pi f_c t\right] - \sqrt{\frac{2E_b}{T_b}} \sin\left[\theta(t)\right] \sin\left[2\pi f_c t\right] \qquad (4.15)$$

The in-phase component consists of half-cycle cosine pulse defined by

$$s_I(t) = \pm\sqrt{\frac{2E_b}{T_b}} \cos\left(\frac{\pi t}{2T_b}\right), \qquad -T_b \leq t \leq T_b \qquad (4.16)$$

While the quadrature component would take the form

$$s_Q(t) = \pm\sqrt{\frac{2E_b}{T_b}} \sin\left(\frac{\pi t}{2T_b}\right), \qquad 0 \leq t \leq 2T_b \qquad (4.17)$$

During even bit interval, the I-component consists of a positive cosine waveform for phase of 0, while a negative cosine waveform exists for phase of π; during odd bit interval, the Q-component consists of a positive sine waveform for phase of $\pi/2$, while a negative sine waveform exists for phase of $-\pi/2$ (as shown in Figure 4.24). Any of these four states can arise: $0, \pi/2, -\pi/2, \pi$. However, only state 0 or π can occur during any even bit interval, and only $\pi/2$ or $-\pi/2$ can occur during any odd bit interval. The transmitted signal is the sum of I- and Q-components and its phase is continuous with time.

Two important characteristics of MSK are that each data bit is held for two-bit period, meaning the symbol period is equal two-bit period ($h = 1/2$) and the I- and Q-components are interleaved. I- and Q-components are delayed by one bit period with respect to each other. Therefore, only the I- or Q-component can change at a time (when one is at zero-crossing, the other is at maximum peak). The pre-coder can be a combinational logic, as shown in Figure 4.25.

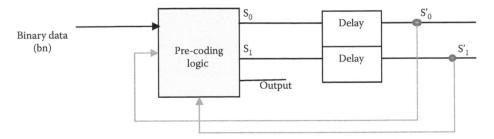

FIGURE 4.25 Combinational logic, the basis of the logic for constructing the pre-coder.

A truth table can be constructed based on the logic state diagram and combinational logic diagram above. For positive half cycle cosine wave and positive half cycle sine wave, the output is 1; for negative half cycle cosine wave and negative half cycle sine wave, the output is 0. A K-map can be constructed to derive the logic gates of the pre-coder, based on the truth table. The following three pre-coding logic equations are derived:

$$S_0 = \overline{b_n S_0' S_1'} + b_n \overline{S_0'} S_1' + \overline{b_n} S_0' S_1' + b_n S_0' \overline{S_1'} \tag{4.18}$$

$$S_1 = \overline{S_1'} = \overline{b_n S_1'} + b_n \overline{S_1'} \tag{4.19}$$

$$Output = \overline{S_0} \tag{4.20}$$

The resultant logic gates construction for the pre-coder is as shown in Figure 4.25.

4.3.5.2 MSK as a Special Case of Continuous Phase FSK (CPFSK)

CPFSK signals are modulated in upper and lower side band frequency carriers f_1 and f_2 as expressed in Equations 4.18 and 4.19.

$$s(t) = \sqrt{\frac{2E_b}{T_b}} \cos\left[2\pi f_1 t + \theta(0)\right] \quad \text{for "1"} \tag{4.21}$$

$$s(t) = \sqrt{\frac{2E_b}{T_b}} \cos\left[2\pi f_2 t + \theta(0)\right] \quad \text{for "0"} \tag{4.22}$$

where $f_1 = f_c + 1/4T_b$ and $f_2 = f_c + 1/4T_b$ with T_b is the bit period.

The phase slope of lightwave carrier changes linearly or non-linearly with the modulating binary data. In case of linear MSK, the carrier phase linearly increases by $\pi/2$ at the end of the bit slot with data "1," while it linearly decreases by $\pi/2$ with data "0." The variation of phase follows paths of well-defined phase trellis in which the slopes represent frequency changes. The change in carrier frequency from data-0 to data-1, or vice versa, equals half the bit rate of incoming data [12]. This is the minimum frequency spacing that allows the two FSK signals representing symbols 1 and 0, to be coherently orthogonal in the sense that they do not interfere with one another in the process of detection. MSK carrier phase is always continuous at bit transitions. The MSK signal in Equations 4.21 and 4.22 can be simplified as

$$s(t) = \sqrt{\frac{2E_b}{T_b}} \cos\left[2\pi f_c t + d_k \frac{\pi t}{2T_b} + \Phi_k\right], \quad kT_b \le t \le (k+1)T_b \tag{4.23}$$

and the base-band equivalent optical MSK signal is represented as

$$\tilde{s}(t) = \sqrt{\frac{2E_b}{T_b}} \exp\left\{j\left[d_k \frac{\pi t}{2T} + \Phi(t, k)\right]\right\}, \quad kT \leq t \leq (k+1)T$$

$$= \sqrt{\frac{2E_b}{T_b}} \exp\{j[d_k 2\pi f_d t + \Phi(t, k)]\} \tag{4.24}$$

where $d_k = \pm 1$ are the logic levels; f_d is the frequency deviation from the optical carrier frequency and $h = 2f_d T$ is defined as the frequency modulation index. In case of optical MSK, $h = 1/2$ or $f_d = 1/(4T_b)$.

4.3.5.3   MSK as Offset Differential Quadrature Phase Shift Keying (ODQPSK)

Equation 4.18 can be rewritten to express MSK signals in form of I-Q components as

$$s_I(t) = \pm\sqrt{\frac{2E_b}{T_b}}\cos\left(\frac{\pi t}{2T_b}\right)\cos[2\pi f_c t] \pm \sqrt{\frac{2E_b}{T_b}}\sin\left(\frac{\pi t}{2T_b}\right)\sin[2\pi f_c t] \tag{4.25}$$

The I- and Q-components consist of half-cycle sine and cosine pulses defined by

$$s_I(t) = \pm\sqrt{\frac{2E_b}{T_b}}\cos\left(\frac{\pi t}{2T_b}\right) \quad -T_b < t < T_b \tag{4.26}$$

$$s_Q(t) = \pm\sqrt{\frac{2E_b}{T_b}}\sin\left(\frac{\pi t}{2T_b}\right) \quad 0 < t < 2T_b \tag{4.27}$$

During even bit intervals, the in-phase component consists of a positive cosine waveform for phase of 0, and a negative cosine waveform for phase of π; during odd bit interval, the Q-component consists of a positive sine waveform for phase of $\pi/2$, and a negative sine waveform for phase of $-\pi/2$. Any of these four states can arise: 0, $\pi/2$, $-\pi/2$, π. However, only state 0 or π can occur during any even bit interval, and $\pi/2$ or $-\pi/2$ can only occur during any odd bit interval. The transmitted signal is the sum of I- and Q-components and its phase is continuous with time.

Two important characteristics of MSK are that each data bit is held for a two-bit period, meaning that the symbol period is equal to a two-bit period ($h = 1/2$), and the I- and Q-component are interleaved. I- and Q-components are delayed by one bit period with respect to each other. Therefore, only the I- or Q-component can change at a time (when one is at zero-crossing, the other is at maximum peak).

4.3.5.4   Configuration of Photonic MSK Transmitter Using Two Cascaded Electro-Optic Phase Modulators

Electro-optic phase modulators and interferometers operating at a high frequency using resonant-type electrodes have been studied and proposed in References 21–23. In addition, high-speed electronic driving circuits evolved with the ASIC technology using 0.1 μm GaAs P-HEMT or InP HEMTs [24] enables the feasibility in realization of the optical MSK transmitter structure. The base-band equivalent optical MSK signal is represented in Equation 4.25.

The first electro-optic phase modulator (E-OPM) enables the frequency modulation of data logics into upper side bands (USB) and lower side bands (LSB) of the optical carrier with frequency deviation of f_d. Differential phase pre-coding is not necessary in this configuration because of the nature of the continuity of the differential phase trellis. By alternating the driving sources $V_d(t)$ to sinusoidal waveforms for simple implementation or combination of sinusoidal and periodic ramp signals, which was first proposed by Amoroso in 1976 [15], different schemes of linear and non-linear phase shaping MSK transmitted sequences can be generated whose spectra are shown later in Figure 4.35.

The second E-OPM enforces the phase continuity of the lightwave carrier at every bit transition. The delay control between the E-OPMs is usually implemented by the phase shifter shown in Figure 4.26. The driving voltage of the second E-OPM is pre-coded to fully compensate the transitional phase jump at the output $E_{01}(t)$ of the first E-OPM. The phase continuity characteristic of the optical MSK signals is determined by the algorithm given in Equation 4.25.

$$\Phi(t, k) = \frac{\pi}{2} \left(\sum_{j=0}^{k-1} a_j - a_k I_k \sum_{j=0}^{k-1} I_j \right) \qquad (4.28)$$

where $a_k = \pm 1$ are the logic levels; $I_k = \pm 1$ is a clock pulse whose duty cycle is equal to the period of the driving signal, $V_d(t)$ f_d is the frequency deviation from the optical carrier frequency, and $h = 2f_d T$ is previously defined as the frequency modulation index, in case of optical MSK, $h = 1/2$ or $f_d = 1/(4T)$. The phase evolution of the continuous phase optical MSK signals are explained in Figure 4.26. In order to mitigate the effects of unstable stages of rising and falling edges of the electronic circuits, the clock pulse $V_c(t)$ is offset with the driving voltages $V_d(t)$ (see Figure 4.27).

4.3.5.5 Configuration of Optical MSK Transmitter Using Mach-Zehnder Intensity Modulators: I-Q Approach

The conceptual block diagram of optical MSK transmitter is shown in Figure 4.28. The transmitter consists of two dual-drive electro-optic MZM modulators generating chirpless I- and Q-components of MSK modulated signals, which is considered a special case of staggered or offset QPSK. The binary logic data is pre-coded and de-interleaved into even and odd bit streams that are interleaved with each other by one bit duration offset.

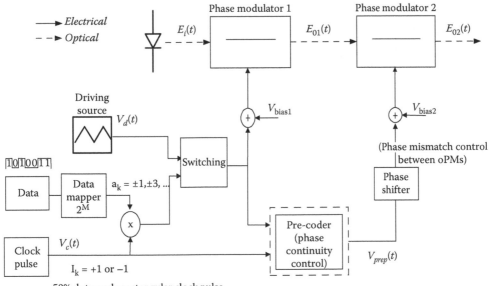

FIGURE 4.26 Block diagram of optical MSK transmitter employing two cascaded optical phase modulators.

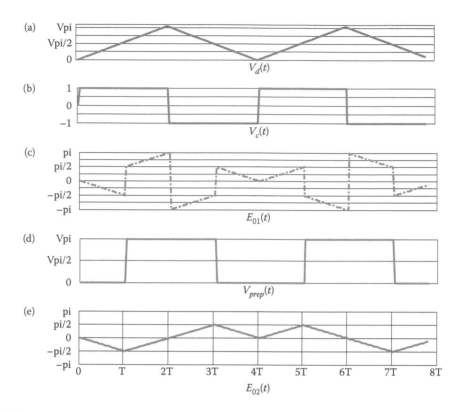

FIGURE 4.27 Evolution of time-domain phase trellis of optical MSK signal sequence $[-1\ 1\ 1\ -1\ 1\ -1\ 1\ 1]$ as inputs and outputs at different stages of the optical MSK transmitter. The notation is denoted in Figure 4.26 accordingly; (a) $V_d(t)$: periodic triangular driving signal for optical MSK signals with duty cycle of 4 bit period, (b) $V_c(t)$: the clock pulse with duty cycle of $4T$, (c) $E_{01}(t)$: phase output of oPM1 (d) $V_{prep}(t)$: pre-computed phase compensation driving voltage of oPM2, and (e) $E_{02}(t)$: phase trellis of a transmitted optical MSK sequences at output of oPM2.

Two arms of the dual-drive MZM modulator are biased at $V_\pi/2$ and $-V_\pi/2$ and driven with *data* and $\overline{data}$. Phase-shaping driving sources can be a periodic triangular voltage source in case of linear MSK generation or simply a sinusoidal source for generating a non-linear MSK-like signal which also obtains a linear phase trellis property but with small ripples introduced in the magnitude. The

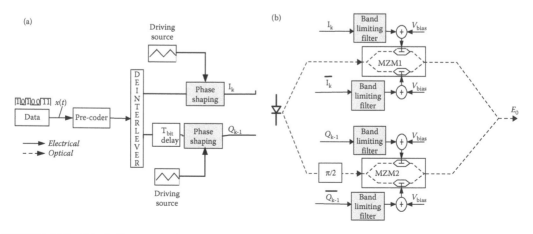

FIGURE 4.28 Block diagram configuration of a band limited phase shaped optical MSK.

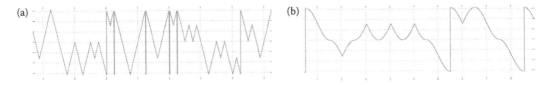

FIGURE 4.29   Phase trellis of linear and non-linear MSK transmitted signals (a) linear and (b) non-linear.

magnitude fluctuation level depends on the magnitude of the phase shaping driving source. High spectral efficiency can be achieved with tight filtering of the driving signals before modulating the electro-optic MZMs. Three types of pulse shaping filters are investigated including Gaussian, raised cosine and squared-root raised cosine filters. The optical carrier phase trellis of linear and non-linear optical MSK signals are shown in Figure 4.29.

4.3.5.6   Single Side Band (SSB) Optical Modulators

4.3.5.6.1   Operating Principles

A SSB modulator can be formed using a primary interferometer with two secondary MZM structures, the optical Ti-diffused waveguide paths that form a nested primary MZ structure as shown in Figure 4.30 [25]. Each of the two primary arms contains MZ structures. Two RF ports are for RF modulation and three DC ports are for biasing the two secondary MZMs and one primary MZM. The modulator consists of X-cut Y-propagation LiNbO$_3$ crystal, where you can produce a SSB modulation just by driving each MZ. DC voltage is supplied to produce the π phase shift between upper and lower arms. DC bias voltages are also supplied from DC$_B$ to produce the phase shift between third and fourth arms. A DC bias voltage is supplied from DC$_C$ to achieve a $\pi/2$ phase shift between MZIM$_A$ and MZIM$_B$. The RF voltage applied $\Phi_1(t) = \Phi \cos\omega_m t$ and $\Phi_2(t) = \Phi \sin\omega_m t$ are inserted from RF$_A$ and RF$_B$, respectively, by using a wideband $\pi/2$ phase shifter. Modulation level is Φ and ω_m is the RF angular frequency.

4.3.5.6.2   Optical RZ-MSK

The RZ format of the optical MSK modulation scheme can also be generated. A bit is used to generate the ASK-like feature of the bit. A Simulink structure of such a transmitter is shown in Figure 4.21. The encoder, as shown in the far left of the model, provides two outputs, one for MSK signal generation and the other for amplitude modulation for generation of the RZ or NRZ format. The amplitude and phase of the RZ signals at the receivers are shown in Figure 4.31 after one 100 km span

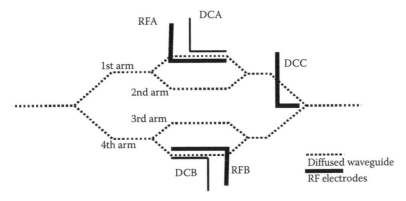

FIGURE 4.30   Schematic diagram (not to scale) of a single sideband (SSB) FSK optical modulator formed by nested MZ modulators.

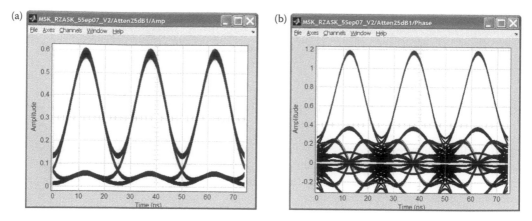

FIGURE 4.31 RZ eye diagram at output of the amplitude receiver (a) and phase detection (b).

transmission and fully compensated. The phase of the RZ MSK must be non-zero so as to satisfy the continuity of the phase between one state to the other (see Figure 4.32). The coding from non return to zero format to MSK can be referred to Table 4.4.

4.3.5.7 Multi-Carrier Multiplexing (MCM) Optical Modulators

Another modulation format that can offer much higher single-channel capacity and flexibility in dispersion and non-linear impairment mitigation is the employment of multi-carrier multiplexing. When these sub-carrier channels are orthogonal, the term orthogonal frequency division multiplexing

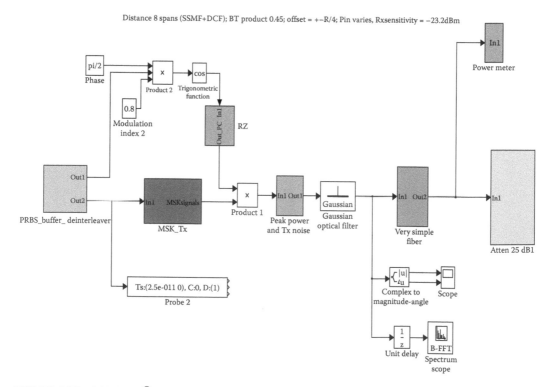

FIGURE 4.32 MATLAB® Simulink model of a RZ-MSK optical transmission system.

TABLE 4.4
Truth Table Based on MSK State Diagram

$b_n S_0' S_1'$	$S_0' S_1'$	Output
100	01	1
001	00	1
100	01	1
101	10	0
010	01	1
101	10	0
110	11	0
111	00	1
000	11	0
011	10	0

(OFDM) can be used [19,17,26]. Chapter 8 will give more details of this type of multi-orthogonal carrier, OFDM modulation.

Our motivation in the introduction of OFDM is because of its potential as a ultra-high-capacity channel for the next-generation ethernet, the optical internet. The network interface cards for 1- and 10-Gb/s ethernet are readily commercial available. Traditionally, the ethernet data rates have grown in 10-Gb/s increments, so the data rate of 100 Gb/s can be expected as the speed of the next generation of ethernet. The 100-Gb/s all-electrically time-division-multiplexed (ETDM) transponders are becoming increasingly important because they are viewed as a promising technology that may be able to meet speed requirements of the new generation of Ethernet. Despite the recent progress in high-speed electronics, ETDM [27] modulators and photodetectors are still not widely available, so that alternative approaches to achieving a 100-Gb/s transmission using commercially available components and QPSK are very attractive. However, the use of polarization division multiplexing (PDM) and 25 GBaud QPSK would offer 100 Gb/s transmission with superiority under coherent reception incorporating digital signal processing (DSP) [28]. This PDM-QPSK technique will be described in later chapters.

OFDM is a combination of multiplexing and modulation. The data signal is first split into independent sub-sets and then modulated with independent sub-carries. These sub-channels are then multiplexed to for OFDM signals. OFDM is a special case of FDM, but instead of one stream, it is a combination of several small streams into one bundle.

A schematic signal flow diagram of a MCM is shown in Figure 4.33. The basic OFDM transmitter and receiver configurations are given in Figure 4.34a and b, respectively [29]. Data streams (e.g., 1Gb/s) are mapped into a 2D signal point from a point signal constellation such as QAM. The complex-valued signal points from all sub-channels are considered as the values of the discrete Fourier transform (DFT) of a multi-carrier OFDM signal. The serial to parallel converter arranges the sequences into equivalent discrete frequency domain. By selecting, the number of sufficiently-large sub-channels, the OFDM symbol interval can be made much larger than the dispersed pulse width in a single-carrier system, resulting in an arbitrary small intersymbol interference (ISI). The OFDM symbol, shown in Figure 4.34c and d, is generated under software processing (for example using an arbitrary waveform generator), as follows: input QAM symbols are zero-padded to obtain input samples for inverse fast Fourier transform (IFFT), the samples are inserted to create the guard band , and the OFDM symbol is multiplied by the window function which can be represented by a raised cosine function. The purpose of cyclic extension is to preserve the orthogonality among sub-carriers even when the neighboring OFDM symbols partially overlap due to dispersion.

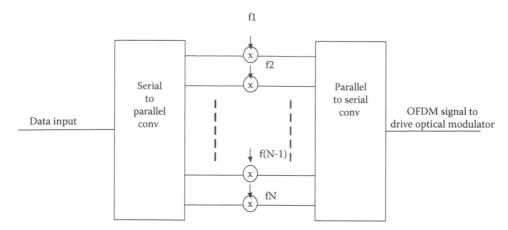

FIGURE 4.33 Multi-carrier FDM signal arrangement. The middle section is the RF converter as shown in Figure 4.26.

4.3.5.8 Spectra of Modulation Formats

Utilizing this double phase modulation configuration, different types of linear and non-linear CPM phase shaping signals including MSK, weakly non-linear MSK, and linear-sinusoidal MSK can be generated. The third scheme was introduced by Amoroso [30] and its side lobe decays with a factor of f-8 compared to f-4 of MSK. The simulated optical spectra of DBPSK and MSK schemes at 40 Gb/s are contrasted in Figure 4.35. Table 4.5 outlines the characteristics and spectra of all different modulation schemes.

4.3.6 SPECTRA OF MODULATION FORMATS

Figure 4.36 shows the power spectra of the DPSK modulated optical signals with various pulse shapes including NRZ, RZ33, and CSRZ types.

For the convenience of the comparison, the optical power spectra of the return-to-zero OOK counterparts are also shown in Figure 4.35.

Several key notes observed from Figures 4.36 and 4.37 are outlined as follows: (i) The optical power spectrum of the OOK format has high power spikes at the carrier frequency or at signal modulation frequencies that contribute significantly to the severe penalties caused by the non-linear effects. Meanwhile, the DPSK optical power spectra do not contain these high power frequency components; (ii) RZ pulses are more sensitive to the fiber dispersion because of their broader spectra. In particular, RZ33 pulse type has the broadest spectrum at the point of −20 dB down from the peak. This property of the RZ pulses thus leads to faster spreading of the pulse when propagating along the fiber. Thus, the peak values of the optical power of these CSRZ or RZ33 pulses decrease much more quickly than the NRZ counterparts. As a result, the peak optical power quickly turns to be lower than the non-linear threshold of the fiber, which means that the effects of fiber non-linearity are significantly reduced; and (iii) NRZ optical pulses have the narrowest spectrum. Thus, they are expected to be most robust to the fiber dispersion. As a result, there is a tradeoff between RZ and NRZ pulse types. RZ pulses are much more robust to non-linearity, but less tolerant to the fiber dispersion. The RZ33/CSRZ—DPSK optical pulses are proven to be more robust against impairments especially self-phase modulation and polarization mode dispersion compared to the NRZ-DPSK and the CSRZ/RZ33-OOK counterparts.

Optical power spectra of three I-Q optical MSK modulation formats that are linear, weakly non-linear, and strongly non-linear types are shown in Figure 4.38. It can be observed that there are no significant distinctions of the spectral characteristics between these three schemes. However, the

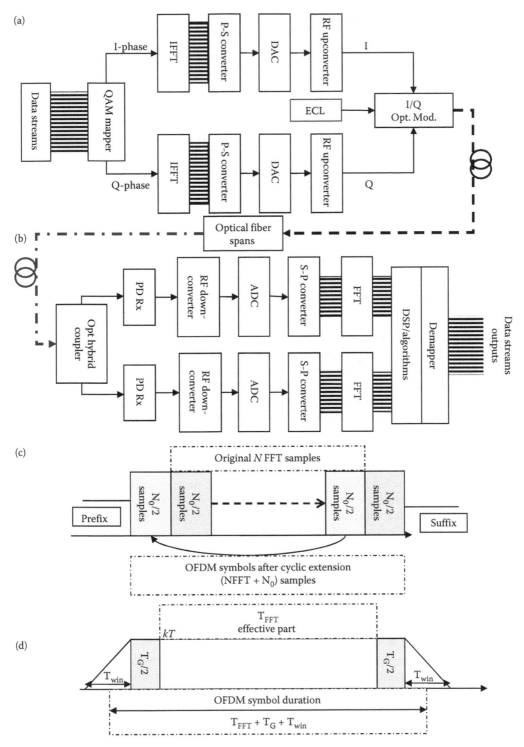

FIGURE 4.34 Schematic diagram of the principles of generation and recovery of OFDM signals. (a) Electronic processing and optical transmitter; (b) opto-electronic and DSP-based receiver configurations; (c) OFDM symbol cyclic extension; and (d) OFDM symbol after windowing (T_{win}).

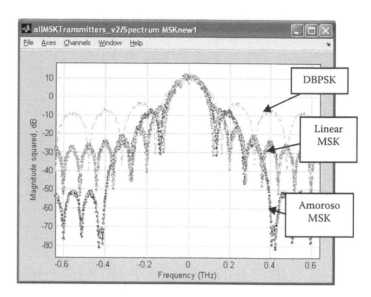

FIGURE 4.35   Spectra of 40Gbps DBPSK, and linear and non-linear MSK.

strongly non-linear optical MSK format does not highly suppress the side lobe as compared to the linear MSK type. All three formats offer better spectral efficiency compared to the DPSK counterpart as shown in Figure 4.39. This figure compares the power spectra of three modulation formats: 80 Gb/s dual-level MSK, 40 Gb/s MSK, and NRZ-DPSK optical signals. The normalized amplitude levels into the two optical MSK transmitters comply with the ratio of 0.75/0.25.

Several key notes can be observed from Figure 4.39 are outlined as follows: (i) The power spectrum of the optical dual-level MSK format has identical characteristics to that of the MSK format. The spectral width of the main lobe is narrower than that of the DPSK. The base-width takes a value of approximately ± 32 GHz on either side, compared to ± 40 GHz in the case of the DPSK format. Hence, the tolerance to the fiber dispersion effects is improved; (ii) high suppression of the side lobes with a value of approximately greater than 20 dB in the case of 80 Gb/s dual-level MSK and 40 Gb/s optical MSK power spectra; thus, more robustness to interchannel crosstalk between DWDM channels; and (iii) the confinement of signal energy in the main lobe of spectrum leads to a better signal to noise ratio. Thus, the sensitivity to optical filtering can be significantly reduced [2–6]. A summary of the spectra of different modulation formats is given in Table 4.5.

4.3.7   INTEGRATED COMPLEX MODULATORS

4.3.7.1   Inphase and Quadrature Phase Optical Modulators

We have described in the above sections the modulation of QPSK and M-ary-QAM schemes using single drive or dual drive MZIM devices. However, we have also witnessed another alternative technique to generate the states of constellation of QAM, using I-Q modulators. I-Q modulators are devices in which the amplitude of the inphase and the quadrature components are modulated in synchronization as illustrated in Figure 4.40b. These components are $\pi/2$ out of phase, thus, we can achieve the inphase and quadrature of quadrature amplitude modulation. In optics, this phase quadrature at optical frequency can be implemented by a low-frequency electrode with an appropriate applied voltage as observed by the $\pi/2$ block in the lower optical path of the structure given in Figure 4.40a. This type of modulation can offer significant advantages when multi-level QAM schemes are employed, for example, 16-QAM (see Figure 4.40c) or 64-QAM. Integrated optical modulators have been developed in recent years, especially in electro-optic structures, such as LiNbO$_3$ for coherent QPSK and even with

TABLE 4.5

Typical Parameters of Optical Intensity Modulators for Generation of Modulation Formats

Modulation Techniques	Spectra	Formats	Definition/Comments
Amplitude modulation – ASK-NRZ	DSB + carrier	ASK – NRZ	Biased at quadrature point or offset for pre-chirp
AM – ASK-RZ	DSB + carrier	ASK-RZ	Two MZIM s required—one for RZ pulse sequence and other for data switching
ASK –RZ-Carrier suppressed	DSB - CSRZ	ASK-RZCS	Carrier suppressed, biased at π phase difference for the two electrodes
Single sideband	SSB + carrier	SSB NRZ	Signals applied to MZIM are in phase quadrature to suppress one side band. Alternatively, an optical filter can be used
CSRZ DSB	DSB – carrier	CSRZ-ASK	RZ pulse carver is biased such that a pi phase shift between the two arms of the MZM to suppress the carrier and then switch on and off or phase modulation via a data modulator
DPSK-NRZ DPSK-RZ, CSRZ-DPSK		Differential BPSK RZ or NRZ / RZ-carrier suppressed	
DQPSK		DQPSK – RZ or NRZ	Two bits per symbol
MSK	SSB equivalent	Continuous phase modulation with orthogonality	Two bits per symbol and efficient bandwidth with high side-lobe suppression
Offset-DQPSK			
MCM (Multi-Carrier Multiplexing- e.g., OFDM)	Multiplexed bandwidth – base rate per sub-carrier		
Duo-binary	Effective SSB		Electrical low-pass filter required at the driving signal to the MZM
FSK			
Continuous Phase FSK			
Phase Modulation (PM)	Chirped carrier phase		Chirpless MZM should be used to avoid inherent crystal effects, hence carrier chirp

polarization division multiplexed optical channels. In summary, the I-Q modulator consists of two MZIM performing ASK modulation incorporating a quadrature phase shift.

Multi-level or multi-carrier modulation formats such as quadrature amplitude modulation (QAM) and orthogonal frequency division multiplexing (OFDM) have been demonstrated as the most promising technology that supports high capacity and high spectral efficiency in ultra-high speed optical transmission systems. Several QAM transmitter schemes have been experimentally demonstrated using commercial modulators [24,31,32] and integrated optical modules [33,34] with binary or multi-level driving electronics. The integration techniques could offer a stable performance, and effectively reduce the complexity of the electrics in QAM transmitter with binary driving electronics. The integration schemes based on parallel structures usually require hybrid integration between LiNbO3 modulators and silica-based planar lightwave circuits (PLCs). Except for the DC electrodes for the bias control of each sub-MZM (child MZM), several additional electrodes are required to adjust the relative phase offsets among embedded sub-MZMs. Shown in Figure 4.40a is a

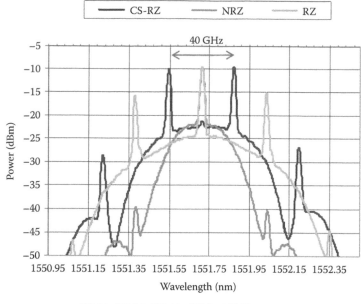

FIGURE 4.36 Spectra of CSRZ/RZ33/NRZ—DPSK modulated optical signals.

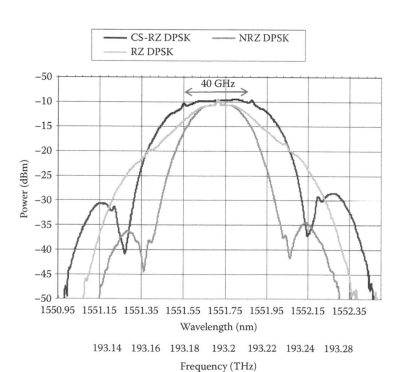

FIGURE 4.37 Spectra of CSRZ/RZ/NRZ—OOK modulated optical signals.

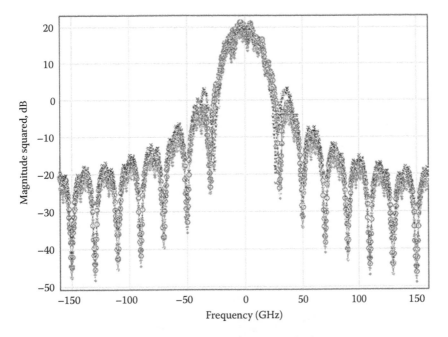

FIGURE 4.38 Optical power spectra of three types of I-Q optical MSK formats: linear (light grey), weakly non-linear (dark grey) and strongly non-linear (black).

16-QAM transmitter using a monolithically integrated quad Mach-Zehnder in-phase/quadrature (QMZ-IQ) modulator. As distinguishable from the parallel integration, four sub-MZMs are integrated and arranged in a single IQ superstructure, where two of them are cascaded in each of the arms (I and Q arms). These two pairs of child MZMs are combined to form a master or parent MZ interferometer with a pi/2 phase shift incorporated in one arm to generate the quadrature phase shift between them,

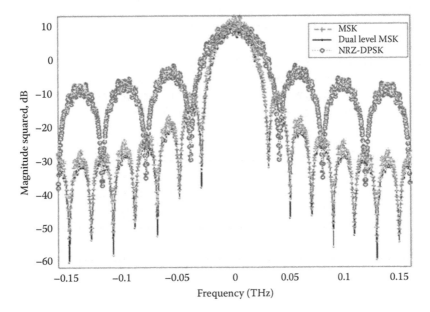

FIGURE 4.39 Spectral properties of three modulation formats: MSK (grey, dash), dual-level MSK (black, solid) and DPSK (dark grey, dot).

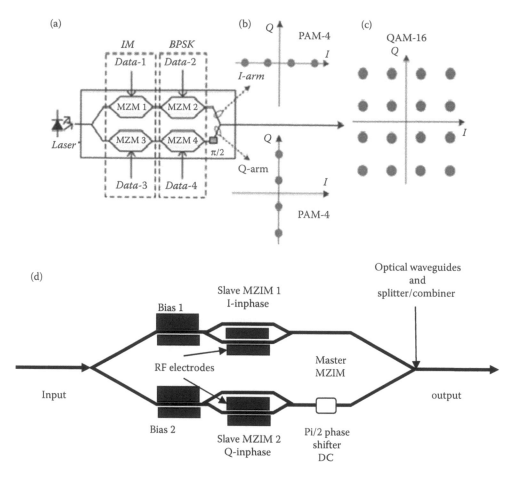

FIGURE 4.40 (a) Schematic representation of integrated IQ optical modulator, (b) amplitude modulation of lightwave path in the inphase and quadrature components, (c) Constellation of a 16 QAM modulation scheme, and (d) alternate structure of an I-Q modulator using two salve MZIM and one master MZIM.

and thus we have the inphase arm and the quadrature optical components. In principle, only one electrode is required to obtain orthogonal phase offset in these IQ superstructures, which makes the bias-control much easier to handle, and thus leads to stable performance. A 16-QAM signal can be generated using the monolithically-integrated quadrature Mach–Zehnder inphase and quadraturee (QMZ-IQ) modulator with binary driving electronics, shown in Figure 4.40c, by modulating the amplitude of the lightwaves guided through both I (inphase) and Q (quadrature) paths, as indicated in Figure 4.40b. We can see that the QAM modulation can be implemented by modulating the amplitude of these two orthogonal I- and Q-components so that any constellation of Mary-QAM., for example, 16QAm or 256QAM, can be generated. Alternatively, the structure of Figure 4.40d gives an arrangement of two slave MZIM and one master MZIM for the I-Q modulator, which is commonly used in Fujitsu type.

In addition, a number of electrodes would be incorporated so that biases can be applied to ensure that the amplitude modulation operating at the right point of the transfer curve is characteristic of the output optical field versus the applied voltage of the MZIM in each interferometric branch of the master MZ interferometer. This can be commonly observed and simplified as in the IQ modulator manufactured by Fujitsu [35] shown in Figure 4.41 (top view only).

The arrangement of the high-speed IQ Mach–Zehnder modulator using Ti-diffused Lithium Niobate (LiNbO$_3$) waveguide technology in which the DC bias can be adjusted at the separate DC

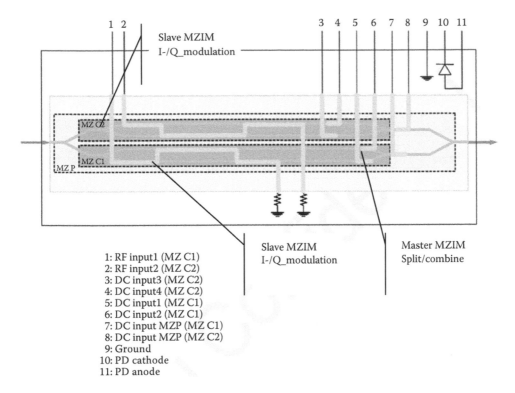

1: RF input1 (MZ C1)
2: RF input2 (MZ C2)
3: DC input3 (MZ C2)
4: DC input4 (MZ C2)
5: DC input1 (MZ C1)
6: DC input2 (MZ C1)
7: DC input MZP (MZ C1)
8: DC input MZP (MZ C2)
9: Ground
10: PD cathode
11: PD anode

FIGURE 4.41 Schematic diagram of Fujitsu PDM-IQ modulator with assigned electrodes.

port of the modulator. This type of modulator can be employed for various modulation formats such as NRZ, DPSK, Optical Duo-Binary, DQPSK, DP-BPSK, DP-QPSK, M-ary QAM, etc. There are built-in PD monitors and coupler functions for auto bias control. 100 Gb/s optical transmission equipment (NRZ, DPSK, Optical Duo-Binary, DQPSK, DP-BPSK, PDM-QPSK) can be generated by this IQ modulator in which four wavelength carriers are used with 28–32 Gb/s per channel to form 100 Gb/s including extra error coding bits and payload 25 Gb/s for each channel.

4.3.7.2 IQ Modulator and Electronic Digital Multiplexing

Recently, we [21–23,36,37] have seen reports on the development of electrical multiplexer whose speed can reach 165 Gb/s. This multiplexer will allow the interleaving of high speed sequence so that we can generate a higher bit rate with the symbol rate as shown in Figure 4.42. This type of multiplexing in the electrical domain has been employed for generations of superchannel in Reference 38. Thus, we can see that the data sequence can be generated from the DACs, and then analog signals at the output of DACs can be conditioned to the right digital level of the digital multiplexer with assistance from a clock generator so the multiplexing occurs. Note that the multiplexer operates in digital mode so any pre-distortion of the sequence for dispersion pre-compensation will not be possible unless some pre-distortion can be done at the output of the digital multiplexer.

This time domain interleaving can be combined with the I-Q optical modulator to generate M-level QAM signal and further increase the bit rate of the channels. Several of these high bit rate channels can be combined with pulse shaping, for example, the Nyquist shape, to generate superchannels, which will be illustrated in Chapter 7. With a digital multiplexer operating higher than 165 Gb/s, a 128 Gb/s bit rate can be generated with a 32 GBd/s (Bd = baud) data sequence. Thus, with polarization moultiplexing and 16QAM, we can generate 8 × 128 Gb/s for one channel or 1.32 Tb/s per channel. If eight of these 1.32 Gb/s form a superchannel, then the bit rate capacity reaches higher than

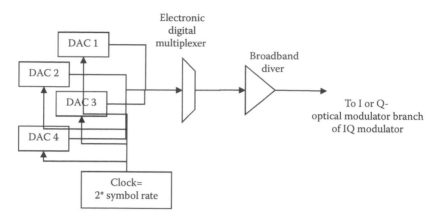

FIGURE 4.42 Time division multiplexing using high speed sequences from DACs.

10 Tb/s. This is all possible with a symbol rate of 32 GBd (Bd = baud) within the bandwidth available electronic components, as well as the photonic devices at the transmitter and receiver.

4.3.8 Digital-to-Analog Converter for DSP-Based Modulation and Transmitter

Recently, we have witnessed the development of ultra-high sampling rate DAC and ADC by Fujitsu and NTT of Japan. Figure 4.43a shows an IC layout of the InP-based DAC produced by NTT [39].

4.3.8.1 Digital-to-Analog Converter (DAC)

A differential input stage incorporating with D-type flip flops, then sums up circuits to produce analog signals representing the digital samples, as shown in Figure 4.43a and b. These DAC allow the generation and programmable sampling and digitalized signals to form analog signals to modulate the I-Q modulator, as described in Section 4.5. These DACs allow shaping of the pulse sequence for pre-distortion to combat dispersion effects when necessary to add another dimension of compensation in combination with that function implemented at the receiver. Test signals used are ramp waveform for assurance of the linearity of the DAC at 27 GBd as shown in Figure 4.44a, and eye diagrams of generated 16QAM signals after modulation at 22 and 27 GBd are shown in Figure 4.44b.

4.3.8.2 Hardware Structure

A DSP-based optical transmitter can incorporate digital-to-analogue converter (DAC) for pulse shaping, pre-equalization, and pattern generation, as well as digitally multiplexing for higher symbol rates. A schematic structure of the DAC and functional blocks fabricated by Si-Ge technology is shown in Figure 4.45a and b, respectively. An external sinusoidal signal is fed into the DAC so that multiple clock source can be generated for sampling at 56–64 GSa/s. Thus, the noises and clock accuracy depends on the stability and noise of this synthesizer/signal generator. Four DACs sub-modules are integrated in one IC with four pairs of eight outputs of $(V_I^+, V_Q^+)(H_I^+, H_Q^+)$_ and _$(V_I^-, V_Q^-)(H_I^-, H_Q^-)$.

4.3.8.3 Generation of Nyquist Pulse-Shaped I- and Q-Components

The electrical outputs from the quad DACs are a mutually complementary pair of positive and negative signs. Thus, it would be able to form two sets of four output ports from the DAC development board. Each output can be independently generated with offline uploading of pattern scripts. The arrangement of the DAC and PDM-IQ optical modulator is similar to others with the optical modulator driven by channels generated from the AWG of the data channels and the encoding section. Note that we require two PDM IQ modulators for generation of odd and even optical channels.

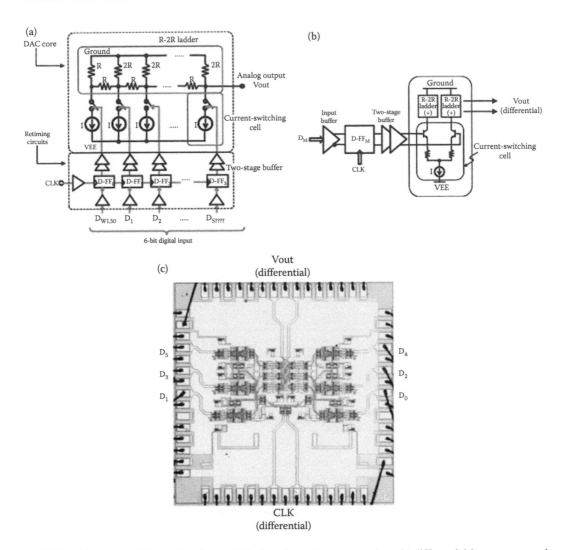

FIGURE 4.43 NTT InP based DAC with 6-bit (a) schematic representation; (b) differential input stage; and (c) IC layout (Extracted with permission from P. J. Winzer et al., *J. Lightwave Technol.*, 28, pp. 547–556, 2010.)

As the Nyquist pulse shaped sequences are required, a number of pressing steps can be implemented:

- Characterization of the DAC transfer functions
- Pre-equalization in the RF domain to achieve equalized spectrum in the optical domain, that is, at the output of the PDM IQ modulator.

The characterization of the DAC is conducted by launching to the DAC sinusoidal wave at different frequencies and measured the waveforms at all eight output ports. The electrical spectrum of the DAC is quite flat, provided that pre-equalization is done in the digital domain launching to the DAC. The spectrum of the DAC output without equalization is shown in Figure 4.46a and b. The amplitude spectrum is not flat because of the transfer function of the DAC as given in Figure 4.47, which is obtained by driving the DAC with sinusoidal waves of different frequencies. This shows that the DAC acts as a low-pass filter with the amplitude of its pass-band gradually decreasing when the frequency is increased. This effect can occur when the number of samples is reduced and the frequency is increased, as the sampling rate can only be set in the range of 56–64 GSa/s. The equalized RF

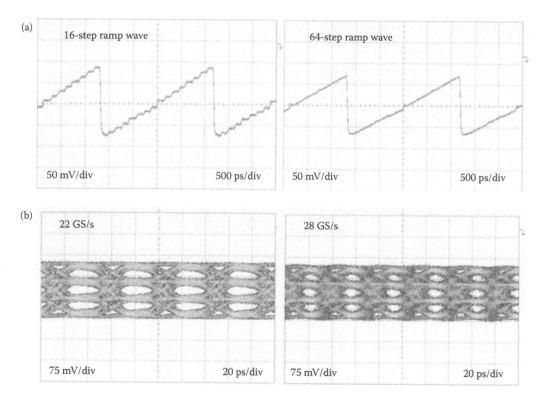

FIGURE 4.44 (a) Test signals and converted analog signals for linearity performance of the DAC and (b) 16QAM 4-level generated signals at 22 and 27GSy/s.

spectra are depicted in Figure 4.46c and d. The time domain waveforms corresponding to the RF spectra are shown in Figure 4.46e and f and thence 4.46g and h for the coherent detection after the conversion back to electrical domain from the optical modulator via the real time sampling oscilloscope Tektronix DPO 73304A or DSA 720004B. Furthermore, the noise distribution of the DAC is shown in Figure 4.47b, indicating that the sideband spectra of Figures 4.46 and 4.48 come from these noise sources.

It notes that the driving conditions for the DAC are very sensitive to the supply current and voltage levels which are given in Appendix 2 with resolution of as little as to 1 mV. With this sensitivity, care must be taken when new patterns are fed to DAC for driving the optical modulator. Optimal procedures must be conducted with the evaluation of the constellation diagram and BER derived from such constellation. However, we believe that the new version of the DAC supplied from Fujitsu Semiconductor Pty Ltd. of England have somehow overcome these sensitive problems. We still recommend that care should be taken, and the constellation should be inspected after the coherent receiver is paid to ensure that the B2B connection is error-free. Various time-domain signal patterns obtained in the electrical time domain generated by DAC at the output ports can be observed. Obviously, the variations of the inphase and quadrature signals give raise to the noise, hence blurry constellations.

TOSA—Transmitter Optical Sub-assembly: Fiber optic transceivers are key components of fiber optic transmission network, especially in access networks and data center interconnections. They are designed in a small form-factor with some integrated optical sub-assemblies that can be suitable for high-density network. The major cost components of a transceiver module is the transmitter optical sub-assembly (TOSA), which converts an electrical and the receiver optical sub-assembly (ROSA). However, inside a BiDi (Bi-Directional) transceiver, there are components called

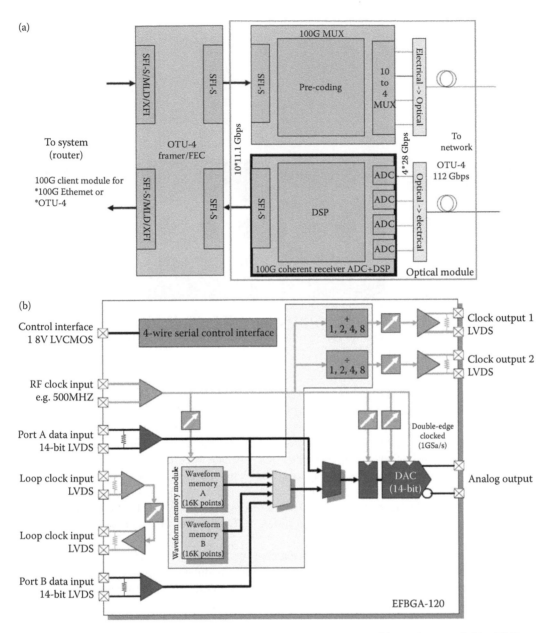

FIGURE 4.45   (a) Structure of the Fujitsu DAC, note 4 DACs are structured in one integrated chip. (b) Functional block diagram.

bidirectional optical sub-assemblies (BOSA), which act in the roles of TOSA and ROSA, but with different principle.

Miniature optical transmitter (MTx) and transceiver (MTRx) for the high luminosity LHC (HL-LHC) have attracted much attention from equipment manufacturers for deployment of high bit rates, for example, 25Gbpsx4 (100G) module. These assemblies are commonly known as TOSA and ROSA (T = transmitter; R = Receiver). There have been two major developments: the vertical cavity surface emitting laser (VCSEL) driver ASIC and the external modulator in-line with a laser in assembly. The external modulators are now offering performance to enable access networking or data-center interconnect.

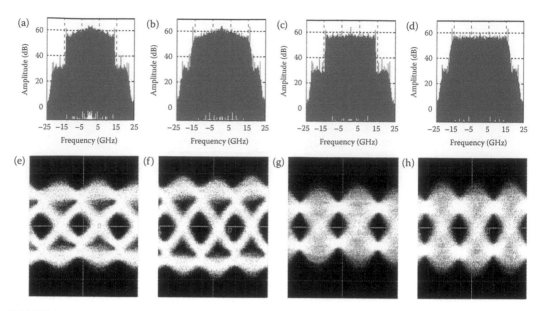

FIGURE 4.46 Spectrum (a) and eye diagram (e) of 28Gbuad RF signals after DAC without pre-equalization, (b) and (f) for 32Gbaud; Spectrum (c) and eye diagram (g) of 28Gbuad RF signals after DAC with pre-equalization, (d) and (h) for 32Gbaud.

TOSA and ROSA optical packages are assemblies comprising optical (lenses, prisms, apertures, filters, etc.) and electronic components (LD, PD, amplifiers, controllers, etc.). Applications can be found in communication technologies where optical signals are transformed into electrical signals and vice versa. To have a package working properly, it is mandatory that the optical and electronic parts are aligned to each other with highest precision. The task becomes even more complex as, by design, these packages are typically actively tempered by thermo-electric coolers (TEC), which need to be integrated into the assembly, and are shown in Figure 4.39. The TOSA can be operating in direction modulation (DML) or EML with external modulation. For VCSEL TOSA, then, the laser can be directly modulated. Figure 4.39b also shows a ROSA (receiver optical sub-assembly) as a reverse operation of TOSA (Figure 4.49).

The VCSEL can also be associated with an external optical injection, as described in Chapter 2, to obtain an extended passband, allowing much higher bit rate modulation. However, it is now common that TOSAs are designed for operating in the 1310 nm so that they can suffer minimum dispersion in SSMF over which the fiber dispersion factor crosses the null point. Thus, without the availability of optical amplifiers in this spectral window, the transmission links are limited by attenuation factor.

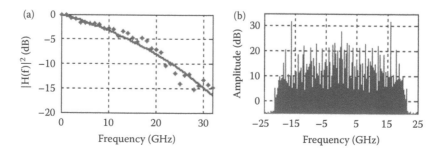

FIGURE 4.47 (a) Frequency transfer characteristics of the DAC. Note the near linear variation of the magnitude as a function of the frequency. (b) Noise spectral characteristics of the DAC.

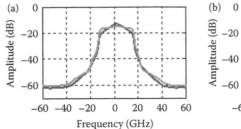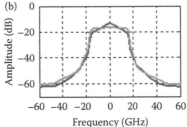

FIGURE 4.48 Optical spectrum after PDM IQM, dark grey line for without pre-equalization, light grey line for with pre-equalization (a) 28 GSym/s and (b) 32 GSym/s.

The received eyes of the transmitted signals can be compensated by using digital signal processors and associated algorithm.

Integrated Modulators and Modulation Formats: Since the proposed of dielectric waveguide and then the advent of optical circular waveguide, the employment of modulation techniques has only been extensively exploited recently because of the availability of optical amplifiers. The modulation formats allow the transmission efficiency, and hence, the economy of ultra-high capacity information telecommunications. Optical communications have evolved significantly through several phases from single mode systems to coherent detection and modulation, which was developed with the main aim to improve on the optical power. The optical amplifiers defeated that main objective of modulation formats and allow the possibility of incoherent and all possible formats employing the modulation of the amplitude, the phase, and frequency of the lightwave carrier.

Currently, photonic transmitters play a principal part in the extension of the modulation speed into the several GHz range and make possible the modulation of the amplitude, the phase, and frequency of the optical carriers and their multiplexing. Photonic transmitters using $LiNbO_3$ have been proven in laboratory and installed systems. The principal optical modulator is the MZIM, which can be a single or a combined set of these modulators whereby to form binary, multilevel amplitude, or phase modulation and even more effective for discrete or continuous phase shift keying techniques. The effects of the modulation on transmission performance will be given in the coming chapters.

Spectral properties of the optical 80 Gb/s dual-level MSK, 40 Gb/s MSK, and 40 Gb/s DPSK with various RZ pulse shapes are compared. The spectral properties of the first two formats are similar.

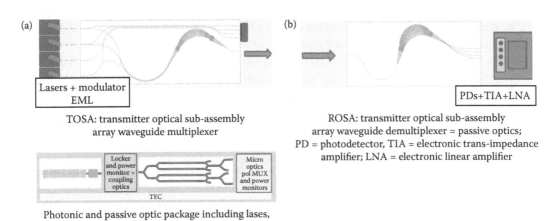

FIGURE 4.49 Integrated module for 100G (a) TOSA and (b) ROSA: integrated solutions in 100G applications necessary to enable CFP, CFP2, CFP4, QSFP28 with 4 channels of 28GBd per channel. CFP = C form-factor pluggable.

Compared to the optical DPSK, the power spectra of optical MSK and dual-level MSK modulation formats have more attractive characteristics. These include the high spectral efficiency for transmission, higher energy concentration in the main spectral lobe, and more robustness to inter-channel crosstalk in DWDM due to greater suppression of the side lobes. In addition, the optical MSK offers the orthogonal property, which may offer a great potential in coherent detection, in which the phase information is reserved via I and Q components of the transmitted optical signals. In addition, the multi-level formats would permit the lowering of the bit rate and lead to substantial reduction of the signal effective bandwidth and the possibility of reaching the highest speed limit of the electronic signal processing, the digital signal processing, for equalization and compensation of distortion effects. The demonstration of ETDM receiver at 80 G and higher speed would allow the applications of these modulation formatted scheme very potential in ultra-high speed transmission.

High-speed operation can only be possible if the guided lightwave can be modulated by a high-speed broadband electrode. Thus, the next section describes the modeling and fundamental characteristics of the microwave travelling wave electrodes for such integration with optical channel waveguide to maximize the bandwidth of optical modulators.

4.3.9 REMARKS

Currently, photonic transmitters play a principal part in the extension of the modulation speed into several GHz range and make possible the modulation of the amplitude, the phase and frequency of the optical carriers and their multiplexing. Photonic transmitters using LiNbO$_3$ have been proven in laboratory and installed systems. The principal optical modulator is the MZIM which can be a single or a combined set of these modulators whereby to form binary or multi-level amplitude or phase modulation and even more effective for discrete or continuous phase shift keying techniques. The effects of the modulation on transmission performance will be given in the coming chapters.

Spectral properties of the optical 80 Gb/s dual-level MSK, 40 Gb/s MSK and 40 Gb/s DPSK with various RZ pulse shapes are compared. The spectral properties of the first two formats are similar. Compared to the optical DPSK, the power spectra of optical MSK and dual-level MSK modulation formats have more attractive characteristics. These include the high spectral efficiency for transmission, higher energy concentration in the main spectral lobe, and more robustness to interchannel crosstalk in DWDM because of greater suppression of the side lobes. In addition, the optical MSK offers the orthogonal property, which may offer a great potential in coherent detection, in which the phase information is reserved via I and Q components of the transmitted optical signals. In addition the multi-level formats would permit the lowering of the bit rate and hence substantial reduction of the signal effective bandwidth and the possibility of reaching the highest speed limit of the electronic signal processing, the digital signal processing, for equalization and compensation of distortion effects. The demonstration of ETDM receiver at 80 G and higher speed would allow the applications of these modulation formatted scheme very potential in ultra-high speed transmission.

In recent years, research for new types of optical modulators using silicon waveguides has attracted several groups. In particular, graphene thin layer deposited [40] on silicon waveguides enables the improvement of the electro-absorption effects and enable the modulator structure to be incredibly compact and potentially performing at speeds up to ten times faster than current technology allows, reaching higher than 100 GHz and even to 500 GHz. This new technology will significantly enhance our capabilities in ultrafast optical communication and computing. This may be the world's smallest optical modulator, and the modulator in data communications is the heart of speed control. Furthermore, these grapheme Silicon modulators can be integrated with Si- or SiGe-based microelectronic circuits, such as DAC, ADC, and DSP so that the operating speed of the electronic circuits can reach much higher than the 25–32 GGSy/s of today's technology.

The linewidth of the laser employed in the coherent reception and transmission systems are critical, as phase noises result from the fluctuation of the laser frequency. Thus, a further explanation on this laser intensity and phase noises is given in the Appendix, for reference. We will see that the DSP in association with the coherent receiver can minimize these effects, as described in more detail in Chapter 10.

APPENDIX 4A: LASER INTENSITY AND PHASE NOISES

Lasers can exhibit various kinds of "noises," with manifold influences on applications. This appendix discusses where such noises can come from, how they can be quantified, and how to mitigate/minimize their influences in optical transmission systems, especially the DSP-based coherent optical reception-sub-systems.

4A.1   Generic Aspects

"Noise" of lasers [41] is a short term for random fluctuations of various output parameters. This is a frequently encountered phenomenon that has a profound impact on many applications in photonics, particularly in the area of precision measurements. Consider, for example, the interferometric position measurements that can be directly affected by fluctuations of the optical phase, or spectroscopic measurements of transmission, where intensity fluctuations limit the possible sensitivity. Similarly, the date rate and the transmission distance for fiber-optic links are at least partly limited by noise issues.

Scientific communities, especially the optical communications engineering community, feel more or less uneasy about laser noise. One reason is that the causes are often hard to evaluate and eliminate. Further technical and mathematical difficulties are related to the measurements and quantifying of the noises; specifications found in several commercial data sheets reveal less about the laser. Furthermore, it is desirable but not always easy to estimate the influence of different kinds of noise in some application. For such reasons, trial and error approaches without a decent understanding are often used, for example, for minimizing noise influences, but this often turns out to be ineffective or inefficient. This appendix explains some basics of such noises of lasers. It discusses intensity and phase noise; many aspects, of course, can be applied to other types of noises.

4A.2   Intensity Noise

Intensity noise is usually understood to quantify fluctuations of the laser output power (not actually an optical intensity), and, in most cases, normalized to the average power. The measurement is based on recording the temporally varying output power, using a photodiode, for example. The normalization is the simplest aspect; other aspects, to be discussed in the following, are more subtle. Frequently encountered specifications like "$\pm 1\%$" appear simple, but are quite meaningless. They suggest that the power always stays within 1% of its average value, while in reality there is usually a smooth probability distribution without sharp edges. A good way to avoid this problem is to specify the RMS (root mean square) values, meaning the square root of the average of the squared power fluctuations, or mathematically

$$\delta P_{RMS} = \sqrt{\langle (P(t) - P_{av})^2 \rangle} \tag{4A.1}$$

where P_{av} is the average power. This is usually applied to *relative* power fluctuations, thus specifying relative intensity noise (RIN). The RMS RIN is of course dimensionless.

Another problem is that the registered fluctuations can strongly depend on the measurement bandwidth. For example, a laser may exhibit fast fluctuations of the output power, which can be detected by a fast photodetector, while being averaged out and thus not registered by a slower detector. Furthermore, a limited measurement time may not be sufficient to detect slow fluctuations (drifts); essentially, one does not know how far the average power in the chosen time interval deviates from the average over longer times. For these reasons, a simple number without specification of a measurement bandwidth (lower and upper noise frequency) is actually meaningless and should never occur in a data sheet. The measurement bandwidth may extend from some very small frequency, limited by the inverse measurement time, to some maximum frequency, determined by the speed of the detector. It is also noted that any noise specification is actually a statistical measure, which is only estimated by one data trace for a given time interval, or (better) with an average over several traces. Furthermore, the laser noise may depend on the ambient conditions. Therefore, one should know whether a specification applies for constant room temperature, after a long warm-up time, and in a vibration-free environment.

If it is of interest to which extent different noise frequencies contribute to the overall noise, a power spectral density (PSD) $S_I(f)$ is most useful. Conceptually, one can imagine that many noise measurements are done for different (small) noise frequency intervals. This is how many electronic spectrum analyzers record noise spectra: the detector is subsequently tuned to different noise frequencies. However, there are also techniques based on Fourier transforms that allow to estimate whole noise spectra from single (or a few) data traces recorded in the time domain. One may, for example, use a sampling card to record a signal proportional to the laser power over a certain time and with a certain temporal resolution. Before doing a Fourier transform, a so-called "window function" must normally be applied to avoid certain artifacts. The lowest noise frequency in such a measurement is the inverse total duration, while the maximum frequency is at most half the sampling rate. The latter limit results from the Nyquist theorem.

Once we have the power spectral density $S_I(f)$ of the relative intensity, we can calculate the RMS relative intensity noise according to

$$\frac{\delta P}{\overline{P}} = \sqrt{\int_{f_1}^{f_2} S_I(f)df} \qquad (4A.2)$$

where f_1 and f_2 are the lower and upper noise frequency, respectively. (This is assuming a one-sided PSD; physicists often use two-sided PSDs, where the integration also has to include negative frequencies.) The units of $S_I(f)$ are $1/Hz$, indicating that $S_I(f)$ is the noise contribution per frequency interval. It is also common to specify $10 \cdot Log_{10}(S_I(f))$, arriving at units of dBc/Hz, meaning dB relative to the carrier in a 1-Hz noise bandwidth.

Figure 4A.1 shows the simulated relative intensity noise spectrum (i.e., $S_I(f)$ versus frequency) of a diode-pumped single-frequency Nd:YAG solid state laser at 1080 nm , using logarithmic scales for both axes. The RIN exhibits a characteristic peak around 140 kHz, related to so-called relaxation oscillations, which result from the dynamic interaction of the intracavity power and the laser gain. Obviously, a detector with 20 kHz bandwidth would not be able to detect such fast oscillations and would thus record much weaker noise than a 200-kHz detector. Laser diodes exhibit relaxation oscillations with much higher frequencies (multiple GHz) and stronger damping due to their short carrier lifetime and short resonator. Generally, different laser types can exhibit very different noise properties, as characteristic parameters may be totally different.

Around the relaxation oscillation frequency f_{ro}, a laser is particularly sensitive to external noise influences, for example, from the pump diode, and even to quantum noise. Well above f_{ro}, noise of any kind has little impact. Well below f_{ro}, pump noise may directly affect the output. For the red curve in Figure 4A.1, some frequency-independent excess noise of the pump diode has been assumed; in reality, there can be an increase toward lower frequencies, often determined by the quality of the diode driver. Even for a perfect shot-noise-limited pump source (grey curve), noise around f_{ro} is relatively strong. The impact of quantum noise depends on various parameters; in particular, it is stronger in cases with low intracavity power, high resonator losses and a short round-trip time of the resonator. Additional contributions to intensity noise of a laser can result from acoustic influences.

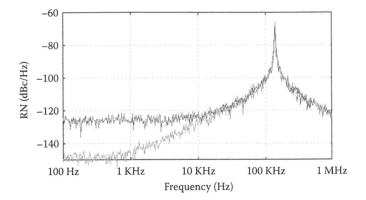

FIGURE 4A.1   Simulated intensity noise spectrum of a diode-pumped miniature Nd:YAG laser. Grey curve: with quantum-limited pump source. Dark grey curve: with 30 dB excess noise of the pump source.

For example, a cooling system may cause vibrations of the laser resonator, which translate into intensity and phase noise. In many cases, such noise contributions appear in the form of sharp peaks in the spectrum.

An entirely different phenomenon is noise from mode-beating in lasers where multiple resonator modes are oscillating. The occurring beat frequencies are differences of resonator mode frequencies, which may be rather high (multiple GHz) for laser diodes and very low (few kHz) for long fiber lasers. The pattern of beat frequencies can reveal whether or not the laser operates on longitudinal modes only.

Minimization of laser noise can be done on entirely different routes:

- A first approach is the minimization of external noise influences, for example, by using singly-frequency pump diodes operated with a carefully stabilized current source, and making a mechanical stable laser setup.
- A second possibility is the optimization of laser parameters such that the impact of quantum noise and/or external noise influences is minimized. For example, one may minimize quantum noise influences by using a long low-loss laser resonator, or move the relaxation oscillation frequency into a region where noise is less strongly disturbing the application.
- Finally, one may reduce laser noise with a feedback system, automatically adjusting, for example, the pump power based on measured output power fluctuations. The characteristics of the feedback loop need to be optimized based on the knowledge of laser parameters.

Obviously, the potential of these approaches depends very much on the circumstances, which should thus be analyzed carefully beforehand. Without a proper understanding of the sources and types of noise, such exercises are prone to fail or at least to be inefficient.

The intensity noise of nearly all light sources is limited by shot noise. In an intuitive (and somewhat simplified) picture, this can be understood as the random occurrence of photons (packets of light energy). Even if the probability of detecting a photon within some short time interval is constant, the actually recorded photon absorption events exhibit some randomness if there are no correlations between photons. This leads to a PSD of the relative intensity noise of $2\,h\nu/P_{av}$, where $h\nu$ is the photon energy. The relative intensity noise of a laser is often well above the shot noise level, but the latter rises if the output is more and more attenuated (for example with some linear absorber). With sufficiently stronger power attenuation, the intensity noise will be at the shot noise limit. The squeezed states of light can exhibit intensity noise below the shot noise limit, but require special methods to be generated, and tend to lose that special property, for example when the light is attenuated.

4A.3 Phase Noise and Linewidth

Phase noise is related to fluctuations of the optical phase of the output. As simple as this sounds, the optical phase may not be defined for a laser oscillating on multiple resonator modes. We thus assume to be dealing with a single-frequency laser, where essentially all power is in a single resonator mode. (For multi-mode lasers, one may consider phase noise for different modes separately).

Because of various influences, even a single-frequency laser will not exhibit a perfect sinusoidal oscillation of the electric field at its output. There are fluctuations of the power (see above) and the optical phase φ. The latter can be quantified with a phase noise PSD $S_\varphi(f)$, having units of rad^2/Hz or Hz^{-1}, as rad is dimensionless. Initially, we may consider only quantum noise, in a simplified picture described as the random phase of photons added by spontaneous emission. This leads to a "random walk" of the optical phase, which is related to a phase noise PSD $S_\varphi(f) \sim f^{-2}$. The divergence at $f = 0$ is related to the unbounded drift of the phase, which, unlike the output power, lacks a "restoring force." It also causes a finite value of the linewidth, which is the width of the main peak in the power spectral density of the optical field. A famous paper from 1958 by Schawlow and Townes presents a simple formula to calculate that linewidth, and shows that the linewidth decreases for increasing intracavity power, decreasing resonator losses, and increasing resonator length. However, the quantum-limited linewidth is often difficult to reach, as other noises may dominate. In many cases, the linewidth being a single value is of primary interest for applications. However, the complete phase noise spectrum may be required in other cases, and may reveal important noise contributions even if these are not relevant for the linewidth. Essentially, the linewidth is determined by low-frequency noise.

The measurement of phase noise or the linewidth is substantially more difficult than that of intensity noise, partly because the phase evolution must be compared with some reference. A conceptually simple but often impractical method is based on measuring a beat note of the laser with a second laser, exhibiting a similar optical frequency (keeping the beat frequency low enough) and much lower phase noise. Alternatively, two similar lasers may be used, and the relative phase noise provides an estimate for the noise of a single laser.

An often convenient variant is the delayed self-heterodyne method, using a setup as shown in Figure 4A.2. Here, one shifts the frequency of the lightwave beam by an acousto-optic modulator and derives a reference from the laser itself by introducing a substantial time delay with along an optical fiber. For a very long delay (say, about 1.0 Km, depending on the expected linewidth), the

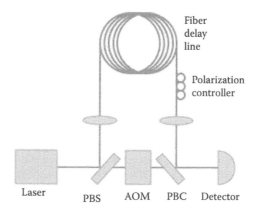

FIGURE 4A.2 Setup for delayed self-heterodyne measurements. The mirrors next to the acousto-optic modulator (AOM) have a reflectivity of, for example, 50%. The detector records the beat note between the frequency-shifted part and the delayed part of the laser light. This contains information on phase noise and linewidth. Legend: PBS = polarization beam splitter; OPC = polarization beam combiner.

noise of the delayed reference would be uncorrelated to the laser noise, allowing a simple analysis. Noise measurements are still possible with shorter delays, but require a more sophisticated mathematical analysis. This should also take into account the possibility that the phase noise PSD does not exhibit an f^{-2} law.

In principle, the phase noise spectrum can also be obtained from the optical spectrum, as recorded, for example, with a scanning high-finesse Fabry–Pérot interferometer. Here, the optical resonator provides a phase reference. This can work for a semiconductor laser, whereas it is hard to get the resonator linewidth below that of a typical solid-state laser.

4A.4   Phase Noise Measurement of a Narrow Linewidth CW Laser

Two different laser phase noise measurement techniques are compared. One of these two techniques is based on a conventional and low-cost delay line system, which is usually set up for the linewidth measurement of semiconductor lasers. The results obtained with both techniques on a high-spectral-purity laser agree well and confirm the interest of the low-cost technique. Moreover, an extraction of the laser linewidth using computer-aided design tools is performed.

The delay line technique is a low-cost and efficient approach for the measurement of the frequency fluctuations of frequency sources. Long optical delay lines (a few kilometers) are generally used in optics to reach a complete signal decorrelation and measure the laser linewidth by simply mixing this signal to the original laser signal on a photodiode [42]. In the microwave range, the same long optical delay lines are used in frequency discriminators, thus featuring an ultrahigh sensitivity [43].

When a high-spectral-purity laser is measured on this type of bench, the measurement conditions are similar to the ones observed in microwave frequency discriminators: the coherence length is no longer shorter than the delay line length, and the data measured at the system output are the frequency fluctuation spectrum of the laser [44–46]. However, compared to the frequency discriminator used for microwave source measurement, the use of an optical high sensitivity frequency discriminator faces different problems related to the frequency stability of the laser during the measurement time.

In this section, two techniques are compared for the measurement of laser short-term frequency fluctuations using an optical delay line. One of these techniques is based on a homodyne delay line technique, and uses a fast Fourier transform (FFT) signal analyzer; meanwhile, the other technique is based on a self-heterodyne approach and on the spectrum analysis close to an intermediate frequency of 80 MHz using a costly piece of equipment: an RF phase noise measurement bench. The measurement bench is depicted in Figure 4A.3. The signal is delayed on one path using a 2 km fiber

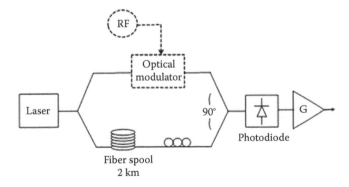

FIGURE 4A.3   Measurement bench for both approaches; the acousto-optic modulator is used in the self-heterodyne technique and removed for the homodyne technique, G, amplifier gain.

spool, followed by a polarization controller (optional). In the homodyne approach, this signal is directly recombined with the reference path, and the output of the coupler feeds a photodiode, which performs the quadratic detection. When the laser being tested is a high-quality laser, such as a fiber laser or an externally stabilized semiconductor laser, which typically feature a linewidth smaller than 10 kHz, the signal on both arms remains correlated and the output is proportional to the frequency fluctuations of the laser at low frequency offsets (frequency discriminator).

4A.5 ANALYTICAL DERIVATIONS

A classical approach to describe this system is to replace the frequency noise by a deterministic sinusoidal modulation [47]. In this case, the signal in both arms before the coupler can be written as

$$A_1(t) = \frac{A}{2}\cos\left(2\pi f_0(t - \tau) + \frac{\Delta f}{f_m}\cos\left(2\pi f_m(t - \tau)\right)\right)$$

$$A_2(t) = \frac{A}{2}\cos\left(2\pi f_0(t - \tau) + \frac{\Delta f}{f_m}\cos\left(2\pi f_m t\right)\right)$$
(4A.3)

where τ is the delay between the two arms, f_0 the optical frequency, f_m the noise frequency, and Δf the modulation amplitude (frequency noise). The quadratic detection leads to (mixing term only)

$$I(t) = S\frac{A^2}{4}\cos\left(-2\pi f_0\tau + \frac{2\Delta f}{f_m}\sin\left(2\pi f_m(t - \tau/2)\sin(\pi f_m\tau)\right)\right)$$
(4A.4)

With S being the photodiode sensitivity, Equation 4A.4 may be written as

$$I(t) = S\frac{A^2}{4}\cos(\varphi + x)$$
(4A.5)

with φ being a constant phase and x a small amplitude component (noise). The derivative of this function versus x near $x = 0$ allows us to linearize the output current and to determine the ratio of the output current amplitude and Δf, which is the sensitivity factor of the frequency discriminator, K_m

$$K_m = S\frac{P}{4}\sin(\varphi)2\pi\tau\frac{\sin(\pi f_m t)}{(\pi f_m t)}$$
(4A.6)

with P being the optical power received on the photodiode. While varying the phase φ, the output current changes from its minimum, which should be zero in a perfect interferometer, and its maximum SP. In practice, the minimum is never equal to 0, and SP is generally approximated by the peak-to-peak excursion of the output current. $SP = 2$ is thus the peak excursion of this current, which we note ΔI peak [43].

Ideally, the phase φ between the two arms should be adjusted to $\pi = 2$ in order to maximize the frequency fluctuation detection. This is what is classically performed in a microwave frequency discriminator. However, in an optical experiment, such a precise phase control after a 2 km delay is impossible because of temperature variations. The phase φ rotations due to the slow temperature drift of the fiber spool can be observed at the system output. In our case, 360° rotations were reached every 200 or 300 ms. This means that, on a 15 min acquisition time, the complete phase rotations will reach 4000. Because of such a large number of phase rotations, and using also a large number of averaged spectrum (about 200 spectra), we expect the phase states to be equally spaced on 360° during the measurement. Such an approach is very close to the setting given in Reference 46, in which a phase modulator had been introduced in the system. However, with kilometer-long delay lines, the phase modulator is no longer necessary, as a natural phase shift occurs due to the slow temperature drift of the fiber spool. When a large set of phase state is considered, the effective kilometers can be calculated from Equation 32 by integrating on a complete phase rotation the quadratic value of Km (it is a power spectrum that is detected):

$$K_{m-eff} = \frac{K_m}{\sqrt{2}} = \frac{\Delta I_{peak}}{\sqrt{2}} 2\pi\tau \frac{\sin(\pi f_m t)}{(\pi f_m t)} \tag{4A.7}$$

In our experiment, the photodiode is loaded on a resistance R, and the voltage signal is measured on a 100 kHz FFT spectrum analyzer (Advantest R9211B). The system calibration is performed by measuring ΔI peak using the oscilloscope mode of the FFT analyzer, and then K_{meff} is computed for a set of frequency fm. In our case, the delay resulting from the optical line is 9:5 µs. The laser frequency fluctuation spectrum is then calculated,

$$S_{\Delta f} = \frac{S_{V_{output}}}{R^2 K_{m-eff}^2} \tag{4A.8}$$

$$K_{m-eff} = \frac{K_m}{\sqrt{2}} = \frac{\Delta I_{peak}}{\sqrt{2}} 2\pi\tau \frac{\sin(\pi f_m t)}{(\pi f_m t)} \tag{4A.9}$$

The system sensitivity is determined by K_{meff} and the output noise due to any source of noise different from $S_{\Delta f}$

$$L(f_m)_{dBc/Hz} = 10Log_{10}\left(\frac{S_{\Delta f}(f_m)}{(2f_m^2)}\right) \tag{4A.10}$$

With $S_{\Delta f}$ laser amplitude modulation (AM) noise, photodiode noise, Brillouin scattering in the fiber, etc. Generally, laser AM noise and photodiode noise are negligible in a long-line interferometer. However, with a delay in the kilometer range, the Brillouin threshold is lowered, and this noise may become the system noise floor. A compromise has thus to be found between long delay (high K_{meff}), high optical power, and the setting up of scattering phenomena in the fiber.

The self-heterodyne technique applied to high-quality lasers was presented in Reference 48 and thus is described briefly. The measurement set-up schematic is depicted in Figure 4A.3, incorporating an optical AO (acoustooptic) modulator in order to shift the output signal around the 80 MHz RF (TeO2 AO crystal). Then the signal is analyzed using a phase noise measurement bench at 80 MHz. Using the same approach developed for the homodyne case, it can be demonstrated [49] that the phase fluctuations of this 80 MHz RF signal are proportional to the laser frequency fluctuations $S_{\Delta f}$ at low frequency offsets. More precisely,

$$L_{RF}(f_m) = 20Log_{10}\left(\sqrt{2}\pi\tau\frac{\sin(\pi f_m\tau)}{(\pi f_m\tau)}\right) + 10Log_{10}(S_{\Delta f}) \tag{4A.11}$$

From Equations 34 and 35, we may calculate the laser phase noise from the measured RF signal phase noise as

$$L_{laser}(f_m) = L_{RF}(f_m) - 20Log_{10}(2\sin(\pi f_m\tau)) \tag{4A.12}$$

The calibration in this case is performed by replacing the 80 MHz output signal with a frequency-modulated source of the same frequency and amplitude than the source under test, which is automatically performed in such modern phase noise measurement platform. Compared to the first approach, this approach is quite easy but requires much more sophisticated and costly equipment, particularly an

RF phase noise measurement bench. To compare the two approaches, the phase noise characterization of a commercial 1:55 μm fiber laser is performed. Using the homodyne approach, 200 spectra are
averaged on three different and overlapping frequency bandwidths (600 spectra for the whole
measurement).

The long acquisition time (10–15 min) allows the random phase repartition, and the spectrum
becomes remarkably well defined and stable after about 50 averages. The FFT analyzer is a low-
cost processor, and, moreover, no RF synthesizer or optical modulator is required. The self-heterodyne approach can be performed with a 200 mW power source at 80 MHz sinusoidal waves to drive
the AO modulator, and the signal is measured at the output by an signal source analyzer, which is a
high-performance phase noise test set incorporating a self-calibration. The results of both measurements are shown in Figure 4A.4. A good agreement is observed, which validates our approach for
the homodyne technique.

From the frequency noise or the phase noise spectrum, it is possible to evaluate the laser linewidth.
It is common to specify a laser through its linewidth, although for very-high-spectral-purity sources
this parameter cannot be measured directly and has thus a very limited practical interest. What is measured effectively is the frequency noise or the phase noise spectrum, and the linewidth evaluation
requires a computation process from these data. Such a link between phase noise and power spectrum
is a complex problem that has been the purpose of many papers, and [49–54] only represent a short
selection of these papers. The only case that can be computed analytically is the one of white frequency noise [8], which leads to a Lorentzian line shape. However, for the type of laser concerned
in this Letter, almost all the laser power is determined by the 1 = f part of the spectrum and the Lorentzian formula is useless. The computation of the linewidth in the case of pure 1 = f noise is more
complex [51–54], and only a computer-based approach may solve the problem for an arbitrary phase
noise spectrum with a combination of different slopes [53].

To get an estimate of the laser linewidth, we can use an original approach based on the simulation
of the power spectrum using commercially available microwave computer-aided design software,
Agilent Advanced Design System (ADS). This software allows the description of a noisy frequency
source through its phase noise spectrum. Moreover, it includes a simulation module based on the
envelope technique [20], which allows the computation of slow perturbations of the carrier, such
as modulations by deterministic or noisy signals. The result of the simulated spectrum for our laser
is represented in Figure 4A.5. As the simulated spectrum changes with the window used for both the
phase noise description and the baseband temporal analysis window, we have considered for the

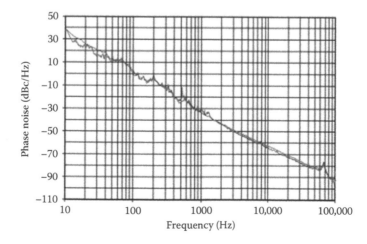

FIGURE 4A.4   Phase noise measurement of a highquality fiber laser delivering 30 mW optical power at λ = 1:55
μm using the two techniques (homodyne technique in dark grey and self-heterodyne technique in light grey).

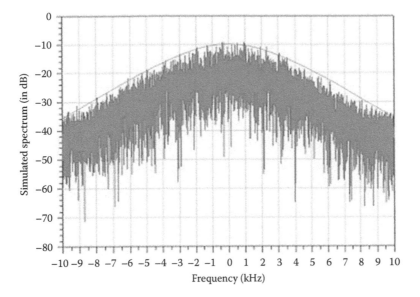

FIGURE 4A.5   Laser power spectrum versus the offset from the carrier frequency simulated from the phase noise data using Agilent Advanced Design System and the envelope simulation technique.

simulation the phase noise data between 100 Hz and 100 kHz, and we have set a maximum computation time of 0:5 s. The simulated result is a 3 dB full linewidth of about 5.0 kHz. This result has been compared to the simplified approach of Reference 54. The frequency noise spectral density of our laser can be roughly described by the following equation:

$$S_{\Delta f} = \frac{k}{f}; \ k = 10^6 Hz \tag{4A.13}$$

taking into account an integration time T_o 1/4 0:5 s, a linewidth of 6 kHz is computed for this laser, which is in relatively good agreement with the result of the envelope simulation approach.

4A.6   REMARKS

The laser exhibits an intensity and phase noise resulting from various internal and external influences, the impact of which is strongly influenced by the internal dynamics. Noise can be minimized in various ways; the best way strongly depends on the circumstances. Intensity noise is in principle easy to measure, but proper specifications need a decent understanding also of certain mathematical aspects, in particular of the influence of measurement time and bandwidth, and of power spectral densities. Phase noise measurements are subject to additional difficulties, partly related to the need for some phase reference and the mathematical complication of a diverging phase noise PSD. The laser linewidth is directly related to phase noise, but of course contains much less information than the whole phase noise spectrum. Quantum noise alone in a laser leads to the Schawlow–Townes linewidth, which is often not reached because of additional noise influences. An experimental measurement set up to determine these parameters of a single mode semiconductor laser is also given.

REFERENCES

1. P. K. Tien, Integrated optics and new wave phenomena in optical waveguides, *Rev. Mod. Phys.*, Vol. 49, pp. 361–420, 1977.

2. R. C. Alferness, Optical guided-wave devices, *Science*, vol. 234, no. 4778, pp. 825–829, 14 November 1986.

3. M. Rizzi and B. Castagnolo, Electro-optic intensity modulator for broadband optical communications, *Fiber and Integrated Optics*, vol. 21, pp. 243–251, 2002.

4. H. Takara, High-speed optical time-division-multiplexed signal generation, *Optical and Quantum Electronics*, vol. 33, no. 7–10, pp. 795–810, 10 July 2001, E. L. Wooten, K. M. Kissa,, A. Yi-Yan, E. J. Murphy, D. A. Lafaw, P. F. Hallemeier, D. Maack, D. V. Attanasio, D. J. Fritz, G. J. McBrien, and D. E. Bossi, A review of lithium niobate modulators for fiber-optic communications systems, *IEEE J. Sel. Topics Quant. Elect.*, vol. 6, no. 1, pp. 69–80, Jan./Feb 2000.

5. K. Noguchi, O. Mitomi, H. Miyazawa, and S. Seki, A broadband Ti: LiNbO3 optical modulator with a ridge structure, *J. Lightwave Tech.*, vol. 13, no. 6, pp. 1164–1168, June 1995.

6. J. Noda, Electro-optic modulation method and device using the low-energy oblique transition of a highly coupled super-grid, *IEEE/Journal of Lightwave Technology*, vol. LT-4, pp. 1445–1453, 1986.

7. M. Suzuki, Y. Noda, H. Tanaka, S. Akiba, Y. Kuahiro, and H. Isshiki, Monolithic integration of InGaAsP /InP distributed feedback laser and electroabsorption modulator by vapor phase epitaxy, *IEEE J. Lightwave Tech.*, Vol. LT-5, No. 9, p. 127, Sept. 1987.

8. http://www.acronymeo.com/10gbemlla.html

9. H. Nagata, Y. Li, W. R. Bosenberg, and G. L. Reiff, DC drift of X-Cut LiNbO3 modulators, *IEE Photonics Tech. Lett.*, vol. 16, No. 10, pp. 2233–2335, Oct. 2004.

10. H. Nagata, DC drift failure rate estimation on 10 Gb/s X-Cut lithium niobate modulators, *IEEE Photonics Tech Lett.*, Vol. 12, No. 11, pp. 1477–1479, Nov. 2000.

11. L. N. Binh, On the linear and nonlinear transfer functions of single mode fiber for optical transmission systems, *Journal of Optical Society of America A*, vol. 26, no. 7, pp. 1564–1575, 2009.

12. R. Krahenbuhl, J. H. Cole, R. P. Moeller, and M. M. Howerton, High-speed optical modulator in LiNbO3 with cascaded resonant-type electrodes, *J. Light. Technol.*, vol 24, no. 5, pp. 2184–2189, 2006.

13. A. Sano, T. Kobayashi, A. Matsuura, S. Yamamoto, S. Yamanaka, Z. Yoshida, Y. Miyamoto, M. Matsui, M. Mizoguchi, and T. Mizuno, 100 × 120-Gb/s PDM 64-QAM transmission over 160 km using linewidth-tolerant pilotless digital coherent detection, Proc. OFC 2012, L.A., 2010.

14. P. Dong, C. Xie, L. Chen, L. L. Buhl, and Y.-K. Chen, 12-Gb/s monolithic PDM-QPSK modulator in silicon, *Optics Express*, vol. 20, no. 26, pp. B624–B629, 2012.

15. A. Hirano, Y. Miyamoto, and S. Kuwahara, Performances of CSRZ-DPSK and RZ-DPSK in 43-Gbit/s/ch DWDM G.652 single-mode-fiber transmission, in *Proceedings of OFC'03*, Annaheim, LA, USA, vol. 2, pp. 454–456, 2003.

16. A. H. Gnauck, G. Raybon, P. G. Bernasconi, J. Leuthold, C. R. Doerr, and L. W. Stulz, 1-Tb/s (6/spl times/170.6 Gb/s) transmission over 2000-km NZDF using OTDM and RZ-DPSK format, *IEEE Photonics Technology Letters*, vol. 15, no. 11, pp. 1618–1620, 2003.

17. Y. Yamada, H. Taga, and K. Goto, Comparison between VSB, CS-RZ and NRZ format in a conventional DSF based long haul DWDM system, in *Proceedings of ECOC'02*, Annaheim, LA, USA, vol. 4, pp. 1–2, 2002.

18. A. H. Gnauck, X. Liu, X. Wei, D. M. Gill, and E. C. Burrows, Comparison of modulation formats for 42.7-Gb/s single-channel transmission through 1980 km of SSMF, *IEEE Photonics Technology Letters*, vol. 16, no. 3, pp. 909–911, 2004.

19. W. Shieh, Yi, X, Tang Y., Transmission experiment of multi-gigabit coherent optical OFDM systems over 1000 km SSMF fiber, *Elect Lett.*, vol. 43, no. 3, 1st February, 2007.

20. E. Ngoya and R. Larcheveque, Envelop Transient analysis: A new method for the transient and steady state analysis of microwave communication circuits and systems, in *Proc. IEEE MTT-S International Microwave Symposium, IEEE*, Palo Alto, CA, USA. vol. 3, p. 1365, May 16–19, 1996.

21. J. Hallin, T. Kjellberg, and T. Swahn, A 165-Gb/s 4:1 multiplexer in InP DHBT technology, *IEEE J. Solid-State Circ.*, Vol. 41, No. 10, pp. 2209–2214, Oct 2006.

22. K. Murata, K. Sano, H. Kitabayashi, S. Sugitani, H. Sugahara, and T. Enoki, 100-Gb/s multiplexing and demultiplexing IC operations in InP HEMT technology, *IEEE J. Solid-State Circuits*, vol. 39, no. 1, pp. 207–213, Jan. 2004.

23. M. Meghelli, 132-Gb/s 4:1 multiplexer in 0.13-μm SiGe-bipolar technology, *IEEE J. Solid-State Circuits*, vol. 39, no. 12, pp. 2403–2407, Dec. 2004.

24. M. Nakazawa, M. Yoshida, K. Kasai, and J. Hongou, 20 Msymbol/s, 64 and 128 QAM coherent optical transmission over 525 km using heterodyne detection with frequency-stabilized laser, *Electron. Lett.*, vol. 43, pp. 710–712, 2006.

25. K. K. Pang, *Digital Transmission*, Melbourne, Australia: Mi-Tec Publishing, 2002, pp. 58.
26. I. B. Djordjevic, and B. Vasic, 100-Gb/s Transmission using orthogonal frequency-division multiplexing, *IEEE Photonics Tech. Lett.*, vol. 18, no. 15, pp. 1576–1578, 2006.
27. W. Shieh, Q. Yang, and Y. Ma, High-speed and high spectral efficiency coherent optical OFDM, *Digest of IEEE/LEOS Summer Topical Meetings*, Acapulco, Mexico, July 21–23, 2008, paper No. TuC2.3 (Invited).
28. C. Schubert, R. H. Derksen, M. Möller, R. Ludwig, C-J. Weiske, J. Lutz, S. Ferber, A. Kirstädter, G. Lehmann, and C. Schmidt-Langhorst, Integrated 100-Gb/s ETDM receiver, *IEEE J. Lightw. Tech.*, vol. 25, no. 1, Jan. 2007.
29. Le Nguyen Binh, *Digitl Processing: Optical Transmission and Coherent Reception Techniques*, New York: CRC Press, Taylor and Francis Grp, 2013.
30. A. Ali, Investigations of OFDM transmission for direct detection optical systems, *Dr. Ing. Dissertaion*, Albretchs Christian Universitaet zu K[iel, 2012.
31. F. Amoroso, Pulse and spectrum manipulation in the minimum frequency shift keying (MSK) format, *IEEE Trans. Commun.*, vol. 24, pp. 381–384, 1976.
32. P. J. Winzer, A. H. Gnauck, C. R. Doerr, M. Magarini, and L. L. Buhl, Spectrally efficient long-haul optical networking using 112-Gb/s polarization-multiplexed 16-QAM, *J. Lightwave Technol.*, 28, pp. 547–556, 2010.
33. X. Zhou and J. Yu, 200-Gb/s PDM-16QAM generation using a new synthesizing method, ECOC 2009, paper 10.3.5, 2009.
34. T. Sakamoto, A. Chiba, and T. Kawanishi, 50-Gb/s 16-QAM by a quad-parallel Mach-Zehnder modulator, ECOC 2007, paper PDP2.8, 2007.
35. H. Yamazaki, T. Yamada, T. Goh, Y. Sakamaki, and A. Kaneko, 64QAM modulator with a hybrid configuration of silica PLCs and LiNbO3 phase modulators for 100-Gb/s applications, ECOC 2009, paper 2.2.1, 2009.
36. Fujitsu Optical Components Ltd., Japan, www.fujitsu.com
37. Y. Suzuki, Z. Yamazaki, Y. Amamiya, S. Wada, H. Uchida, C. Kurioka, S. Tanaka, and H. Hida, 120-Gb/s multiplexing and 110-Gb/s demultiplexing ICs, *IEEE J. Solid-State Circuits*, vol. 39, no. 12, pp. 2397–2402, Dec. 2004.
38. T. Suzuki, Y. Nakasha, T. Takahashi, K. Makiyama, T. Hirose, and M. Takikawa, 144-Gbit/s selector and 100-Gbit/s 4:1 multiplexer using InP HEMTs, *in IEEE MTT-S Int. Microwave Symp. Dig.*, New York, NY, pp. 117–120, Jun. 2004.
39. X. Liu, S. Chandrasekhar, P. J. Winzer, T. Lotz, J. Carlson, J. Yang, G. Cheren, and S. Zederbaum, 1.5-Tb/s Guard-Banded Superchannel Transmission over 56 GSymbols/s 16QAM Signals with 5.75-b/s/Hz Net Spectral Efficiency, Paper Th.3.C.5.pdf ECOC Postdeadline Papers, ECOC 2012, Netherland.
40. M. Nagatani[†] and H. Nosaka, High-speed low-power digital-to-analog converter using, In P. Heterojunction *Bipolar Transistor Technology for Next-generation Optical Transmission Systems*, NTT Technical Reviewl, Tokyo, Japan, vol. 9, no. 4, 2011.
41. M. Liu, X. Yin, E. Ulin-Avilal, B. Geng, T. Zentgraf, L. Ju, F. Wang and X. Zhang, A graphene-based broadband optical modulator, Nature, letter section, *Nature*, vol. 474, p. 64, 2 June, 2011.
42. R. Paschotta, H. R. Telle, and U. Keller, eds. *Encyclopedia of Laser Physics and Technology: Articles on Laser Noise, Intensity Noise, Phase Noise and Others, Noise of Solid State Lasers*, in Solid-State Lasers and Applications (ed. A. Sennaroglu), Boca Raton, FL: CRC Press, 2007, Chapter 12, pp. 473–510.
43. D. Derickson, *Fiber Optic Test and Measurement*, Upper Saddle River, NJ: Prentice-Hall, 1998.
44. B. Onillon, S. Constant, and O. Llopis, in *Proceedings of the 2005 IEEE International Frequency Control Symposium and Exposition* (IEEE, 2005), p. 545.
45. H. Ludvigsen, M. Tossavainen, and M. Kaivola, *Opt. Commun.* vol. 155, p. 180, 1998.
46. H. Ludvigsen and E. Bodtker, *Opt. Commun.* vol. 110, p. 595, 1994.
47. W. V. Sorin, K. W. Chang, G. A. Conrad, and P. R. Hernday, *J. Lightwave Technol.* vol. 10, p. 787, 1992.
48. Hewlett Packard, *Phase Noise Characterization of Microwave Oscillators*, product note 11729C-2, Hewlett Packard, Palo Alto, USA, 1985.
49. S. Camatel and V. Ferrero, Narrow linewidth CW laser phase noise characterization methods for coherent transmission system applications, *J. Lightwave Technol.* vol. 26, p. 3048, 2008.
50. P. Gallion and G. Debarge, Analysis of the phase-amplitude coupling factor and spectral linewidth of distributed feedback and composite-cavity semiconductor lasers, *IEEE J. Quantum Electron.*, vol. 20, p. 343, 1984.

51. L. B. Mercer, 1/f frequency noise effects on self-heterodyne linewidth measurements, *IEEE J. Lightwave Technol*, vol. 9, pp. 485, 1991.

52. O. Llopis, P. H. Merrer, H. Brahimi, K. Saleh, and P. Lacroix, Phase noise measurement of a narrow linewidth CW laser using delay line approaches, *Optics Letters*, vol. 36, no. 14, p. 2713, July 15, 2011.

53. A. Godone, S. Micalizio, and F. Levi, Metrological characterization of the pulsed Rb clock with optical detection, *Metrologia*, vol. 49, no. 4, p. 313, 2012.

54. G. Di Domenico, S. Schilt, and P. Thomann, Simple approach to the relation between laser frequency noise and laser line shape, *Appl Opt.*, vol. 49, no. 25, pp. 4801–4807, Sep 1, 2010.

5 Binary and Multi-Level Complex Signal Optical Modulation

This chapter presents the modulation formats to generate complex optical signaling that combine the amplitude modulation and differential phase modulation schemes, the multi-level amplitude, and phase shift keying. Comparisons between multi-level and binary modulation are made. Critical issues of transmission performance for multi-level are identified. A simulation platform based on MATLAB Simulink [1] is briefly described. Furthermore, for reaching the 100 Gb/s ethernet, a number of multi-level modulation, such as PSK and OFDM are proposed and given in the last section of this chapter.

5.1 INTRODUCTION

Under the conventional on–off keying (ASK) modulation format, the transmission bit rate beyond 40 Gb/s per optical channel is very costly because the electronic signal processing technology may have reached its fundamental speed limit. It is expected that advanced photonic modulation formats such as M-ary amplitude and differential phase shift keying will replace ASK soon. These advanced formats would offer efficient spectral properties and be able to increase transmission rate without placing stringent requirements on high-speed electronics and to use the same existing photonic communication infrastructure.

Coherent communications developed in the mid-1980s have extensively exploited different modulation techniques to improve the optical signal-to-noise ratio (OSNR). However, the coherence detection has faced considerable difficulties because of the stability of the source spectrum and the laser linewidth for a mere gain in the receiver sensitivity of 3dB for heterodyne detection and 6 dB for homodyne detection to extend the repeaterless distance 60–80 km of standard single mode fiber (SSMF).

The invention of the optical amplifiers (OA) in the early 1990s has overcome the fiber attenuation limit and offers a significant improvement in optical transmission technology. As a result, ultra-long haul and ultra-high capacity optical transmission systems have been deployed widely around the world in the last decade. The technology has been matured with ASK modulation reaching 10 Gb/s per optical channel, total channel count of hundreds, and with 100/50 GHz channel spacing.

Based on the proven efficient spectra and transmission technology, especially the controllable total dispersion of the transmission and compensating fibers, it is much more advantageous that these spectral regions be efficiently used. Therefore, the contribution of advanced modulation techniques and formats would offer higher spectral efficiency for photonic transmission.

Furthermore, digital modulation techniques have been well established over the last half century with amplitude, frequency, or phase modulations. These techniques, especially phase modulation, rely principally on the detection schemes, that is, on whether it is coherent or pseudo-coherence differential detection, and have been heavily exploited in wireless communication networks. In the photonic domain, the technological difficulties associated with manufacturing narrow linewidth lasers have prevented the use of coherent and differential phase modulation for a long time. Only over the last several years, due to the maturity of the laser technology, particularly the successful development of distributed feedback (DFB) laser, has laser linewidth reached a level that is much smaller than the modulation bandwidth. The coherence of the sources is now sufficient for differential phase modulating and detecting applications which requires the phase of the sources remain stable over at least two consecutive symbol periods.

Recently, advanced modulation techniques have attracted significant interest from the photonic transmission research and system engineering communities. Several modulation schemes and formats such as binary differential phase shift keying (BDPSK), differential-quadrature phase shift keying (DQPSK), duo-binary ASK associated with non-return-to-zero (NRZ), return-to-zero (RZ), and carrier-suppress return-to-zero (CSRZ) formats have been widely studied. However, what have not been widely explored are optical multi-level modulation schemes. Although multi-level schemes have been intensively exploited in wireless communications, there are minimum works to date that incorporate the in-coherent multi-level optical amplitude-phase shift keying modulation schemes [3] that offer the following advantages: (a) Lower symbol rate, hence for the same available spectral region a multi-level modulation scheme would offer a transmission capacity higher than binary modulation counterparts. (b) Efficient bandwidth utilization, photonic transmission of these multi-level signals could be implemented over the existing optical fiber communications infrastructure without significant alteration of the system architecture, thus saving the cost of capital investment and easing the system management. (c) The complexity of the coder and demodulation sub-systems falls within the technological capabilities of current microwave and photonic technologies.

The principal objectives of this chapter are as follows: (i) To evaluate different modulation and coding techniques and signal pulse formats for long haul ultra-high capacity transmission; then, determine novel modulation schemes, the multi-level amplitude-phase shift keying, and others to be determined, for the research studies. (ii) To develop analytical, simulation and experimental test-beds to demonstrate the uniqueness and superiority of our novel schemes. (iii) A comparative study of the modulations formats to unveil the principal directions for photonic modulation and transmission technologies for the next transmission generation. (iv) A novel photonic communication system based on advanced multi-level optical modulation format and implementation of the system on Simulink platform to demonstrate its effectiveness and superiority to its other counterparts can be demonstrated to be useful platform for desk top computer simulation.

A conceptual photonic transmission system is proposed based on a hybrid technique that combines the phase and amplitude modulation, the multi-level amplitude-differential phase shift keying (MADPSK) format. This technique combines two modulation formats: the well-known M-ary ASK and the M-ary DPSK to take the advantages of high receiver sensitivity and dispersion tolerance (DPSK), and the enhancement of total transmission capacity (M-ASK) as compared to the traditional ASK format.

The models of MADPSK transmitter and receiver have been structured for MADPSK signaling. A simulation model based on MATLAB Simulink platform has been developed for the proof-of-concept. The system performance is evaluated for back-to-back and long-haul transmission. Analytical and simulation results of the transmission configurations are demonstrated. The followings are presented: (i) noise mechanisms, for example, quantum shot noises, quantum phase noises, optically amplified noises, noise statistics, non-linear phase noises; hence design of an optimum detection and decision-level schemes for MADPSK; (ii) linear and non-linear and polarization dispersion impairments and their impacts on MADPSK system performance; (iii) matched filter design for optimum MADPSK signal detection; (iv) offset MADPSK (O-MADPSK) modulation schemes; (v) multi-level amplitude-minimum shift keying (MAMSK) modulation, (vi) MADPSK modulation for applications in sub-carrier transmission systems, especially for metropolitan wide area multi-add/drop networks. (vii) Other issues or additional modulation formats suitable for MADPSK.

This chapter is organized as follows: Section 5.1 gives a brief review of a number of advanced photonic modulation formats. Section 5.2 reviews and compares different modulator structures used for generating advanced photonic modulation signals and emphasizes the advantages of dual-drive MZIM as modulator for generating MADPSK signal, the main object of the research.

In Section 5.4, a novel photonic transmission system with MADPSK modulation format is proposed. Section 5.5 summarizes the preliminary works and results. Section 5.3 identifies a number of critical issues and alternative multi-level signaling for optical systems.

5.2 AMPLITUDE AND DIFFERENTIAL PHASE MODULATION

5.2.1 ASK MODULATION

5.2.1.1 NRZ-ASK Modulation

ASK has been the dominant modulation technique from the early days of optical communications. The main advantage of this modulation is that ASK signal is not sensitive to the phase noise. ASK modulation can take two principal formats: the first is called NRZ-ASK, in which the "1" optical bit occupies the whole bit period; the second format, RZ-ASK, has the "1" bit presented in only the first half of the bit period.

Figure 5.1 shows the spectrum of a 40 Gb/s NRZ-ASK signal, with the carrier seen at the highest peak and the 3 dB bandwidth reaches the bit rate. The main advantage of NRZ-ASK signal, is that its spectrum is generally the most compact compared with that of other formats such as RZ-ASK, CSRZ-ASK. On the other hand, however, NRZ-ASK signal is affected by fiber chromatic dispersion and more sensitive to fiber non-linear effects as compared to its RZ- and CSRZ-ASK counterparts.

5.2.1.1.1 RZ-ASK Modulation

An RZ-ASK signal, shown Figure 5.2, is similar to the NRZ-ASK one, except for the "1" bit occupying only the first half of the bit period. This signal can be generated by a transmitter shown in the same figure in which a NRZ-ASK transmitter is followed by a pulse caver driven by a pulse train synchronized with the data source. The pulse train has frequency equal to the data rate. The RZ-ASK pulse width can take the form of 33%, 50%, or 66% duty-ratio. Because of its narrower pulse width, the spectrum of RZ-ASK signal, shown in Figure 5.3, is larger than that of NRZ-ASK signal, leading to less spectrum efficiency. In this spectrum, the carrier is seen as highest peak; the two side peaks are RF modulating signals positioned at 80 GHz apart.

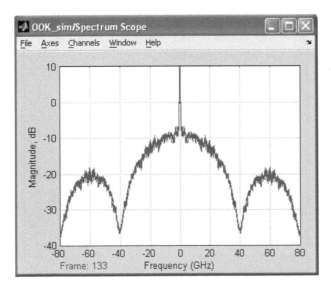

FIGURE 5.1 Spectrum of 40 Gb/s NRZ-ASK signal.

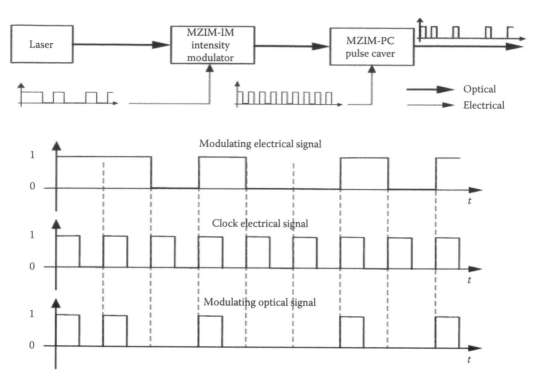

FIGURE 5.2 RZ-ASK transmitter and electrical for generation of optical signal.

5.2.1.1.2 CSRZ-ASK Modulation

The CSRZ-ASK modulation format is similar to standard the RZ-ASK one, except that the neighboring optical pulses have π-phase difference. The carrier in neighboring time slots is thus cancelled out and effectively excluded from the signal spectrum. CSRZ-ASK signal can be generated by a transmitter with the scheme shown in Figure 5.4. In this scheme, the first MZIM modulates the intensity of optical signal coming from a laser source, while the second MZIM, driven by a clock signal at the

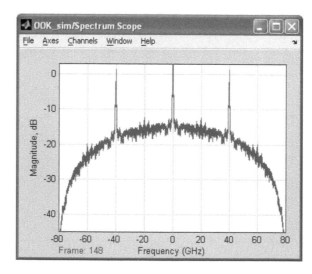

FIGURE 5.3 Spectrum of 40 Gb/s with 50% RZ-ASK signal.

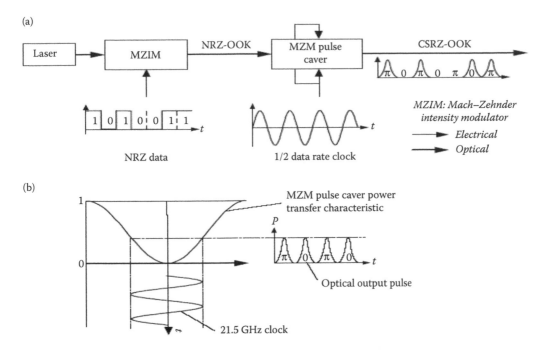

FIGURE 5.4 (a) Block diagrams of CSRZ-ASK transmitter and (b) generation of optical pulse with alternative phase using biasing control and amplitude.

haft data rate, caves the NRZ pulses into RZ ones. Because the second MZIM is biased at the minimum-intensity point, it provides a RZ pulse train at the data rate with alternating phase 0 and π for neighboring time slots. CSRZ-ASK signal can be also detected by a direct-detection receiver as it would not be phase sensitive.

The main advantages of CSRZ-ASK include narrower spectrum, higher tolerance to dispersion, and stronger robustness against fiber non-linear effects as compared with standard RZ-ASK. Because its peak optical power is much lower than that of other formats, it is less affected by self-phase modulation (SPM) and cross-phase modulation (XPM). Figure 5.5 shows the spectrum of 40 Gb/s CSRZ-ASK signal with very low carrier power level.

Amplitude shift keying (ASK) is a modulation technique that generates a signal $s(t)$ by multiplying a digital signal $m(t)$ by a carrier f_c:

$$s(t) = A \cdot m(t) \cos 2\pi f_c t; \quad \text{for } 0 < t < T, \tag{5.1}$$

where A is amplitude envelop; digital signal $m(t)$ may take one of M levels $[b_0, b_1, \ldots, b_M]$. When $M = 2$, $s(t)$ is a binary ASK signal with ASK is a special case. ASK is also implemented in NRZ, RZ, and CSRZ formats, whose spectra are shown in Figure 5.6 in the same graph for comparison. Like their ASK analogues, NRZ-ASK has the most compact spectrum, while RZ-ASK has the broadest. In terms of energy, CSRZ-ASK has the lowest peak power because the carrier signal has been effectively removed.

5.2.1.2 Differential Phase Modulation

Under ASK/ASK modulation schemes with the associated NRZ, RZ, and CSRZ formats, the amplitude of optical carrier varies accordingly. Phase modulation, on the other hand, modulates carrier phase and thus facilitates the use of bi-polar signals "± 1." This distinguished feature means that phase modulation offers significant improvement in receiver sensitivity, as compared with ASK modulation. With the recent advancement in photonic lightwave technology, especially the integrated

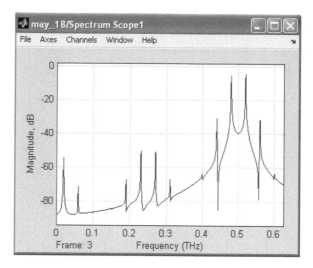

FIGURE 5.5   Spectrum of 40 Gb/s with CSRZ-ASK modulation format.

optic delay interferometer, differential phase modulation and demodulation and balanced receiver have become realizable. This section gives a brief overview on the differential modulation techniques and their implementations in photonic domain, especially the MADPSK.

The term NRZ-BPSK, or traditionally NRZ-DPSK, is commonly used for denoting a modulation technique in which optical carrier is always present with a constant power, only its phase is alternated between 0 and π. The modulation rule is as follows: (i) At the transmitter, initially a reference "0" bit is entered as the present encoded bit. Then the next data bit is compared with the present encoded bit. If they are different then the next encoded bit is "1" for which a phase change of π occurs, else the next encoded bit is "0" which cause no (or 0) phase change. (ii) At the receiver, the phase of the carrier at the present bit slot is compared with that of the previous one. If the phase difference is π, then the data is decoded as "1," otherwise the data is "0" when phase difference is 0.

One of the NRZ-DPSK transmitter structures is shown in Figure 5.7. User data are first encoded by a differential encoder into the driving voltage that then alternates phase of the carrier signal between 0

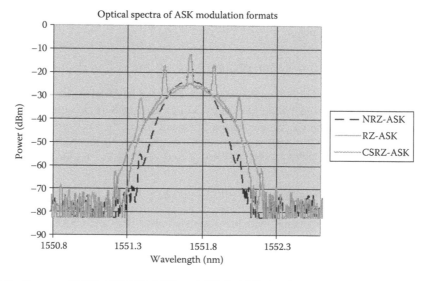

FIGURE 5.6   Spectra of NRZ-ASK, RZ-ASK, and CSRZ-ASK signals.

and π. In detecting a NRZ-DPSK signal, a delay Mach–Zehnder Interferometer (MZI) in combination with balanced optoelectronic receiver can be used. The interferometer acts as the phase comparator with constructively and destructively interfered outputs. As shown in Figure 5.7, the received optical signal is split into two arms of a MZI, one of which has a one-bit optical delay. The MZI compares the phase of each bit with the phase of the previous bit and the photodetector converts the phase difference to intensity. When there is no phase shift between two bits, they are added constructively and give maximum rise to the output signal, otherwise they cancel out when the phase shift equal to π. If the differential phase shift is $\Delta\phi$, then the differential current at the output of the balanced photodetector can be written as

$$i = A^2 \cos \Delta\phi \tag{5.2}$$

Because the balance receiver uses both constructive and destructive ports of the MZI, the detected signal level can swing from "1" to "−1." Compared with ASK or with the use of unbalanced receiver where signal amplitude is limited between "1" and "0." DPSK can offer a 3dB improvement in receiver sensitivity.

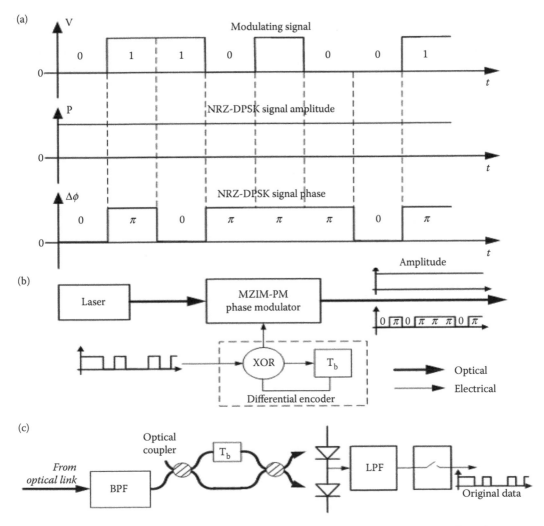

FIGURE 5.7 NRZ-DPSK modulation format system (a) signal generation and corresponding phase states; (b) Schematic of an optical transmitter; and (c) Schematic of the optical receiver.

Because of its constant envelope, NRZ-DPSK signal is less sensitive to power modulation-related non-linear effects, such as SPM and XPM, than its NRZ-ASK counterpart. On the other hand, however, long-haul DPSK systems, including both NRZ and RZ, with optical amplifier are affected by non-linear phase noise. Amplified spontaneous emission (ASE) noise of optical amplifiers is converted into phase noise leading to the waveform distortion and, consequently, signal degradation. The spectrum of NRZ-DPSK signal is shown in Figure 5.8, together with other DPSK formats. NRZ-DPSK signal has the most compact spectrum compared with that of other DPSK formats. This can be explained by the fact that the NRZ-DPSK signal amplitude remains constant, regardless whether bit "1" or bit "0" is transmitted, and thus the energy is distributed more equally when comparing with RZ- and CSRZ-DPSK signals.

RZ-DPSK format is like the NRZ-DPSK one, with the only difference is that instead of constant optical power, pulse narrower than bit period appears in each bit slot as shown in Figure 5.9. RZ-DPSK transmitter, however, resembles a RZ-ASK transmitter with the phase modulator (PM) replaces the intensity modulator (IM). RZ-DPSK signal can also be detected by the same receiver used for NRZ-DPSK signal. Because of its narrow pulse, RZ-DPSK format is expected to minimize the effects of intersymbol interference and thus capable of achieving a longer transmission distance. Narrow pulse, however, spreads spectrum of RZ-DPSK signal wider than that of NRZ-DPSK, making RZ-DPSK systems more susceptible to chromatic dispersion (CD). To reduce the effect of this impairment, CD compensation devices are used.

RZ-DPSK signal energy does not distribute equally as in the case of NRZ-DPSK. Most of it concentrates in only a fraction of bit duration while reducing nearly to zero for the rest of time. This large energy fluctuation makes the signal more susceptible to fiber non-linearity and signal detection more difficult.

The carrier suppression technique can also be used in conjunction with RZ-DPSK modulation to produce CSRZ-DPSK signal, which has been demonstrated as one of the most attractive modulation formats in high spectral efficiency wavelength division multiplexing (WDM) and dense WDM (DWDM) systems.

Due to the energy contained in the CSRZ-DPSK signal spectrum, this signal modulation format offers better tolerance to fiber non-linearity induced impairments as well as the CD and polarization-mode dispersion (PMD) in comparison to its RZ-DPSK counterpart. The spectra of CSRZ-DPSK, RZ-DPSK, and NRZ-DPSK are shown together in Figure 5.8 to illustrate this difference.

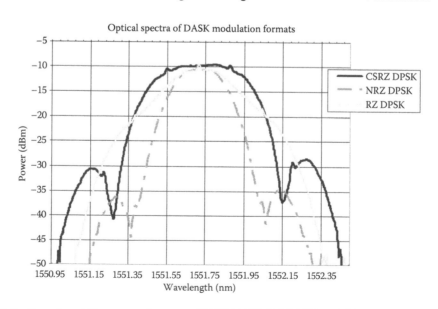

FIGURE 5.8 Experimentally measured spectra of NRZ-DPSK, RZ-DPSK, and CSRZ DPSK signals.

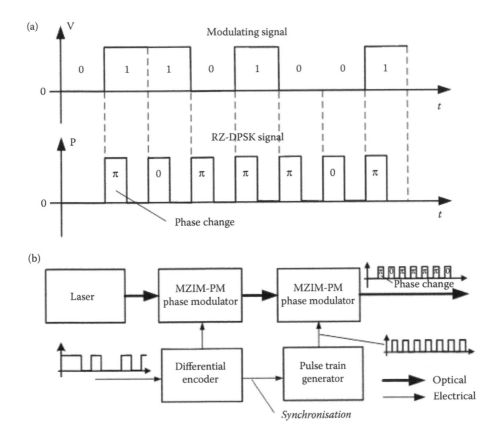

FIGURE 5.9   (a) RZ-DPSK signal and (b) transmitter structure.

CSRZ-DPSK signal can be generated by a transmitter whose scheme is shown in Figure 5.10. The ASK parts are similar to that of CSRZ-ASK. The main difference is for the CSRZ-DPSK transmitter, a PM is replacing the IM in the CSRZ-ASK transmitter. The receiver for CSRZ-DPSK has the same structure as that of the NRZ-DPSK scheme.

To increase transmission bit rate without suffering bandwidth requirement, one can code more than one bit into a data symbol. DQPSK modulation is the first step in the realization of this idea.

A signal constellation or signal space (see Figure 5.11) is the best way to represent a DQPSK signal, in which the points representing phase modulated signals are in two orthogonal axes called I and Q (for in-phase and quadrature components, respectively). Each two data bits $[D_0D_1]$ are first pre-encoded into a symbol, then the symbol is encoded into phase shift that may take one of four values $[0, \pi/2, \pi, 3\pi/2]$ depending on the bit combination it represents. DQPSK symbol rate is thus equal to only half of bit rate. Intuitively, one can say that with the same bandwidth available, DQPSK can offers twice the transmission capacity compared with ASK and binary DPSK counterpart.

DQPSK signal can be generated by a transmitter shown in Figure 5.12. This structure consists of two MZIMs connected in parallel. A $+\pi/2$ phase shift is introduced in one of these MZIM making optical signals in two paths orthogonal to each other. A pre-coder encodes user data in accordance with the differential rule to generate I and Q driving voltages which then modulate carrier's phase in two optical paths. Modulated carrier components are then combined at the output of the MZI. If the two normalized driving signals are denoted by I and Q, respectively, then the output signal is

$$E_{output} = I \cos 2\pi f_c t + Q \sin 2\pi f_c t, \qquad (5.3)$$

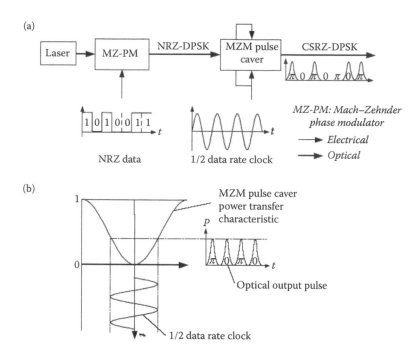

FIGURE 5.10 (a) Block diagrams of CSRZ-DPSK transmitter and (b) generation of optical pulse with alternative phase by driving the dual drive MZIM with a 2 V_π voltage swing.

where f_c is frequency of optical carrier. The coding and mapping bits $[D_1, D_0]$ into I and Q and signal constellation points follow the rule:

DQPSK receiver uses two set of MZ delay interferometers (DI) and balance receivers to detect in-phase (I) and quadrature-phase (Q) components of the received signal, each set is like the one used in NRZ-DPSK receiver [12]. There are, however, two main differences: First, the delay introduced in the first branches of interferometers is now replaced by the symbol duration T_s and second, the phases

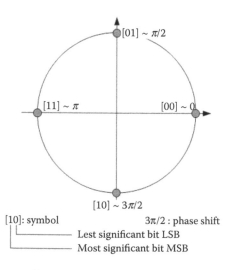

FIGURE 5.11 DQPSK signal constellation.

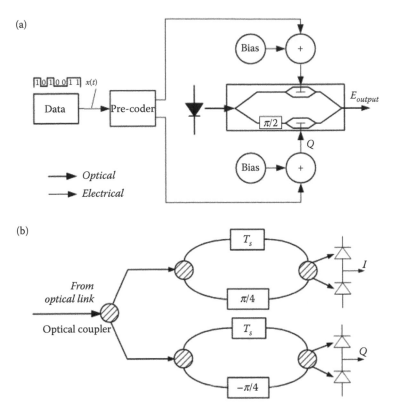

FIGURE 5.12   Parallel structure of DQPSK transmitter, T_s = symbol duration.

of signal in the second branches are shifted by $+\pi/4$ and $-\pi/4$ for I and Q components, respectively. These additional phase shifts are needed to separate two orthogonal phase components I and Q.

Figure 5.13a shows the spectrum of 40 Gb/s NRZ-DQPSK signal with the single-sided bandwidth of the main lobe equal 20 GHz, which is only half of the transmitted bit rate. The spectra of RZ-DQPSK signal [12], Figure 5.13b, is much broader with strong harmonics beside the main lobe.

Despite numerous advancements in optical modulation techniques, the number of levels encoded in a signal symbol falls far behind the 256 or 1024 achieved in microwave modulation schemes. The phase noise associated with optical sources and OAs has hindered the use of phase-related modulation schemes to current fluctuations in the photodetection (PD), leading to the degradation of the BER. Differential phase demodulation process based on the phase comparison of two consecutive symbols requires that the phase should remain stable over two symbol periods. Thus, narrow linewidth lasers are critical for phase modulated systems. It has been shown that to achieve a power penalty less than 1 dB, $\Delta v/B < 1\%$ with Δv and B, the laser linewidth and system bit rate, respectively. In optical transmission systems where OAs are used, the ASE noises intermingle with the fiber non-linear phase effect, thus enhance the non-linear phase noise. While SPM-induced non-linear phase noise is dominant phase noise in single optical channel systems, XPM-induced phase noise is main phase noise for multi-channel (WDM) systems.

Significant phase noises caused by optical sources and OA have prevented optical DPSK schemes from having many levels in each symbol. To increase the number of levels in the signal space, and thus the number of bits per symbol, higher than four one of the most preferred solution is a combination of DQPSK and ASK.

Recently, Hayase et al. [3] have demonstrated experimentally a 30 Gb/s 8-states per symbol optical modulation system using a combined ASK and DQPSK modulation scheme as shown in Figure 5.14,

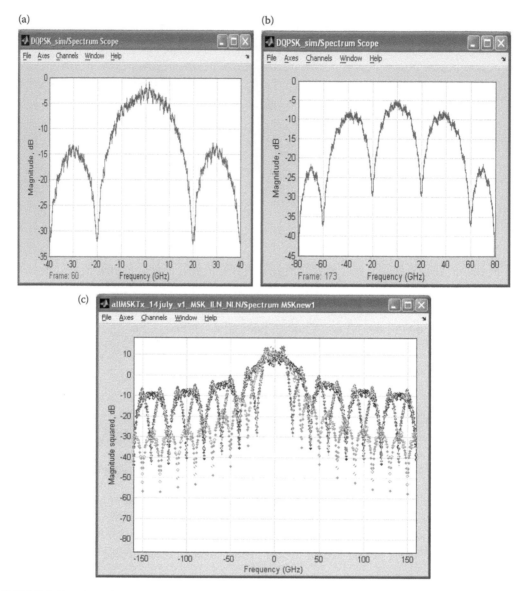

FIGURE 5.13 Optical spectra of 40 Gb/s (a) NRZ-DQPSK, (b) 50% RZ-DQPSK, and (c) DPSK as compared with MSK (light color curve).

maps three bits into a symbol, and creates a transmission bit rate three times higher than symbol rate. The transmitter consists of two cascaded PMs and an amplitude modulator (AM). The first PM, driven by data bit D_0, creates 0 and π phase shifts, while the second, driven by D_1, forces two further phase shifts 0 and $\pi/2$, the quadrature phase, to generate four distinct phases of DQPSK signal. The AM, driven by D_2 bit, shifts the four phases between two amplitudes to create eight total signal points.

At the receiver side, optical signals are detected in both amplitude and differential phase. An ASK demodulator detects the D_2 bit. The other is a DQPSK demodulator and is detected to recover D_1 and D_0 bits. Sekine et al. [4] reported an experimentally similar scheme, but with 4-bits $[D_3, D_2, D_1, D_0]$ mapped into a symbol: $[D_1, D_0]$ bits are used to generate a "normalized" DQPSK signal, while $[D_3, D_2]$ bits manipulate the amplitude of this DQPSK signal between four concentric circles. Thus, a

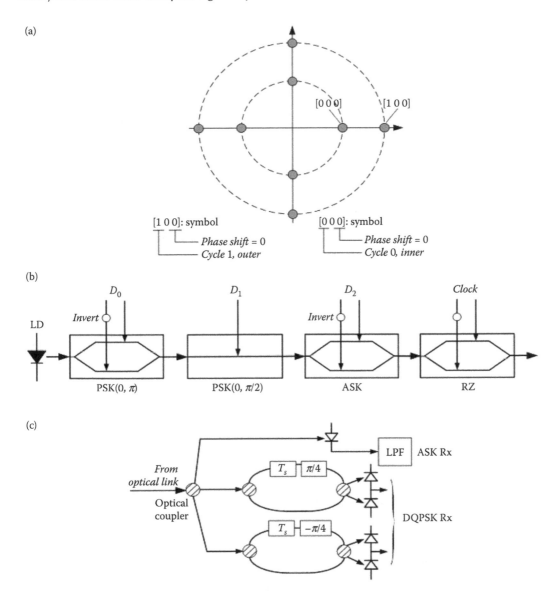

FIGURE 5.14 8-ary APSK modulation experimental configuration (extracted from **0**), 10 GHz clock assign synchronization of symbol rate, data modulator and qudrature phase shift in optical domain using the PM. Two balanced receivers for differential phase shift detection and direct detection for amplitude detection. (a) 8-ary ASK-DPSK signal, (b) transmitter configuration, and (c) receiver configuration.

16-ary MADPSK signal can be generated. This would offer 40 Gb/s bit rate with the symbol rate of only 10 Gbauds.

5.2.2 Comparison of Different Optical Modulation Formats

Different amplitude and phase optical modulation formats are summarized in Table 5.1. In most cases, NRZ-ASK parameters are used as references. From the comparison, it can be concluded that MADPSK takes advantage over other modulation formats in term of spectral efficiency and ability to significantly increase transmission bit rate which are very, if not the most, important parameters

TABLE 5.1

Comparison of Different Optical Modulation Formats

Mod. Format	Spectral Width	Receiver Sensitivity	Resilience to Dispersion	Resilience to SPM	Resilience to XPM	Current Transmission Bit Rate Limits
NRZ-ASK	Narrowest	Lowest	Worst	High	High	40 Gb/s
RZ-ASK	2xRZ-ASK (at 50% duty ratio)	Higher NRZ-ASK	Higher NRZ-ASK	High	High	40 Gb/s
CSRZ-ASK	Same as RZ-ASK	Higher NRZ-ASK	Higher NRZ-ASK	High	High	40 Gb/s
NRZ-DPSK	Same as NRZ-ASK	3dB better than NRZ-ASK	Higher NRZ-ASK	Worse than ASK	Worse than ASK	40 Gb/s
RZ-DPSK	Same as RZ-ASK	3dB better than RZ-ASK	Higher NRZ-ASK	Worse than ASK	Worse than ASK	40 Gb/s
CSRZ-DPSK	Same as CSRZ-ASK	3dB better than CSRZ-ASK	Higher CSRZ-ASK	Higher than ASK	Higher than ASK	40 Gb/s
DQPSK	1/2 DPSK	1.5dB better than ASK	UR*	Worse than ASK	Worse than ASK	2xDPSK
MADPSK	1/M DPSK					$(\log_2 M)$xDPSK - expected

for an optical transmission system. It is also expected that MADPSK inherits good properties (and of course the bad ones—if there are any) from two basic ASK and DPSK modulation formats. The spectra of different modulation formats such as NRZ-DQPSK, 67% CSRZ-DQPSK, and 100 Gb/s CSRZ 16-ADPSK are shown in Figure 5.15, indicating their spectral efficiency.

5.2.3 MULTI-LEVEL OPTICAL TRANSMITTER

In this section, several optical transmitter structures used for generating DQPSK signal are described. It is necessary because a novel optical transmission system will be developed based on DQPSK modulation format. All these structures have MZIM as their base component, which can be single- or dual-electrode structure.

Unlike single-drive MZIM, a dual-drive electrode structure with two traveling wave RF electrodes can modulate the phase of optical signals in both of its branches, hence push–pull operation. Interference at the output of dual-drive MZIM will produce phase modulated signal. However, when the effects of phase modulation in the two branches are exactly equal but opposite in sign, the output signal becomes intensity modulated. In this manner, dual-drive can be used for phase and intensity modulation. The relationship between input and output signals of a dual-drive MZIM can be described by

$$E_{output} = \frac{E_{input}}{2}\left[\exp\left(j\pi\frac{V_1(t)}{V_\pi}\right) + \exp\left(j\pi\frac{V_2(t)}{V_\pi}\right)\right] \tag{5.4}$$

where $V_1(t)$ and $V_2(t)$ are driving voltages applied to modulator, V_π is voltage required to provide a π phase shift of the carrier in each branch of MZIM. It is also noted here that unlike in single-drive MZIMs, chirp effect does not exist in dual-drive MZIMs.

The transmitter structure shown in Figure 5.12 is called parallel type. It is only one of the several structures that can be used specifically for generating DQPSK signal, namely parallel

(a)   (b)

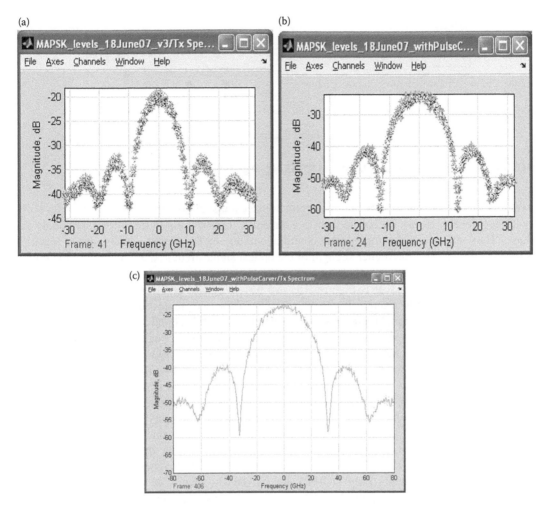

FIGURE 5.15   Optical spectra of 40 Gb/s (a) NRZ-DQPSK, (b) 67% CSRZ-DQPSK, and (c) 100 Gb/s CSRZ 16-ADPSK.

structure, serial structure, single PM structure, and dual-drive MZIM structure. These terms are used to indicate the structuring of MZIMs whether they are connected in tandem, parallel, or just a pure PM with a single electrical drive port. Another possible modulation structure for generating optical multi-level amplitude and phase can be shown in Figure 5.16. It is an electro-optic transmitter of the serial type, a MZIM generating in-phase component and a PM generating quadrature component are connected in tandem. Pre-encoded data generate two signals: one is used for driving the MZIM, and the other for driving the PM. Usually, the square shape of the pre-encoded waveforms are replaced by the raised cosine one before being fed to the modulators. Furthermore, the biasing conditions and the amplitude of the modulators can be used to generate 33% to 67% pulse-width RZ formats. It is also noted that the pulse shape would follow a $\cos^2$ profile because of the property of the intensity modulator. This transmitter would suffer the chirping effects because of the rise time of the electrical driving signals, and hence would contribute to the distortion of the lightwave signals, particularly when switching between the lowest level and the highest level.

A typical multi-level eye diagram of a multi-level modulation scheme is shown in Figure 5.17. The accuracy and noises levels are critical to ensure the degree of eye opening and the distance of the eye

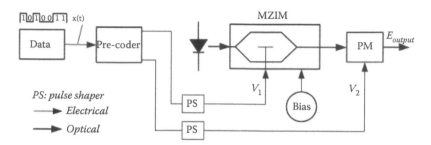

FIGURE 5.16 Cascade PM and MZIM for DQPSK signal generation.

levels. Due to the threshold of the non-linear effects in the transmission fibers, the maximum level of a multi-level level must not exceed so there is a maximum eye opening in multi-level amplitude and phase. Thus, minimum electronic noise contributions due to the output signals, hence requiring higher OSNR and then the higher strength of the optical signals required at the input of the optical receiver. This would thus requires higher-power signals at the transmitter.

The single PM structure seen in Figure 5.18 uses only one MZIM as phase modulator. Pre-encoded data are added up to create a single driving voltage. One of the two pre-encoded data is amplified, and together with the other signal represents four positions of DQPSK signal.

The dual-drive MZIM structure, Figure 5.19, uses two driving voltages for modulating optical carrier phase in two branches of a MZIM. Data are first pre-encoded following the differential rule and then used to create driving voltages $V_1(t)$ and $V_2(t)$ corresponding to the signal constellation points.

In the fours transmitter structures described above the parallel and serial structures are the most complex and difficult to implement because they have discrete devices connected together. The dual-drive MZIM and single PM structures, on the other hand, are much simpler because of the fact that they require less discrete devices. Furthermore, as it will be shown in the next section, dual-drive MZIM can be configured to work as phase and amplitude modulators at the same time, so it can easily generate not only DQPSK, but MADPSK signals as well. Thus, a dual-drive MZIM is the principal part of the MAPSK transmission system. Table 5.1 gives a comparison

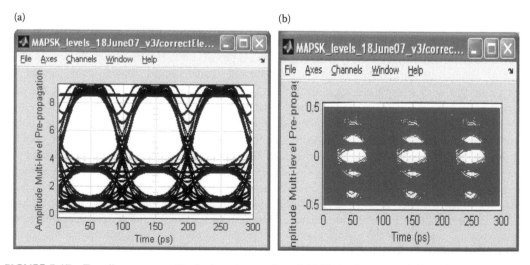

FIGURE 5.17 Eye diagram—amplitude detection section of 40 Gb/s after transmission of (a) 5 km SSMF transmission under direct detection and (b) quadrature phase—coherent detection without digital signal processing.

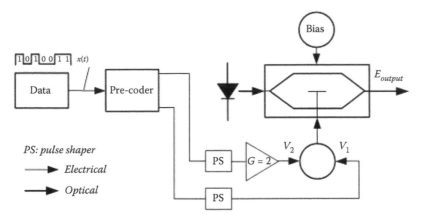

FIGURE 5.18   Single-drive PM structure of MZIM.

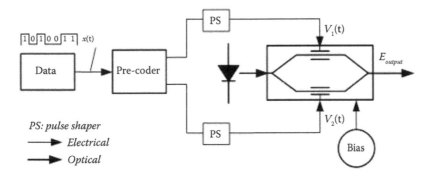

FIGURE 5.19   Dual-drive MZIM structure.

between different transmitter structures. The dual drive modulator is outstanding as a combined amplitude and for phase switching between the states of a multi-circular constellation.

5.3   MADPSK OPTICAL TRANSMISSION

In general, the structures of the MADPSK can be given as shown in Figure 5.20. A model has been constructed for investigating the performance of systems based on MADPSK modulation format. It consists of signal coding model, transmitter model, receiver model, and transmission and dispersion compensation fiber models.

The 16-ary MADPSK signal model described in Section 0 will be used. To balance the ASK and DQPSK sensitivities, ASK signal levels are preliminary adjusted to the ratio $I_3/I_2/I_1/I_0 = 3/2/1.5/1$ as shown in Figure 5.21. These level ratios can be determined from the signal-to-noise ratio at each separation distance of the eye diagram or q-factor. The noise is assumed dominated by the beat noise between the signal level and that of the ASE noise.

The transmitter model shown in Figure 5.20 is used to produce the 16-ary MADPSK signal. It consists of a DFB laser source generating CW light (carrier) which is then modulated in both phase and amplitude by a dual-drive MZIM. Each four bits of user data $[D_3D_2D_1D_0]$ are first grouped into a symbol and then encoded to generate two electrical driving signals $V_1(t)$ and $V_2(t)$ under which amplitude and phase of the carrier in two optical paths of the dual-drive MZIM will be modulated to produce NRZ 16-ary MADPSK signal. The following RZM-PC caves NRZ pulse train into RZ one for minimizing the effects of intersymbol interference.

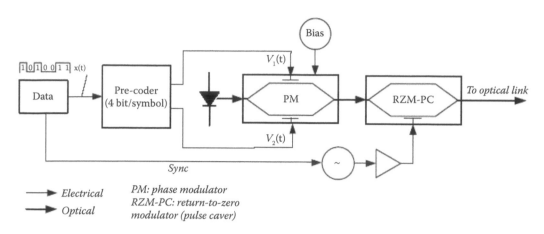

FIGURE 5.20 MADPSK transmitter.

The receiver model shown in Figure 5.22 consists of two phase demodulators, an amplitude demodulator and a data multiplexer MUX. Two phase demodulators are used for extracting [D_1, D_0] bits and they work exactly in the same way as ones in DQPSK receiver described in the above section. The amplitude demodulator (AD) is used for detecting four amplitude levels of the MADPSK signal. It is a well-known direct-detection scheme consisting of a photodiode followed by an electronic receiver. The amplitude modulated signal is then threshold detected in association with a clock recovery circuit to recover two bits [D_3, D_2]. Two bits [D_3, D_2] are interleaved with two bits [D_1,D_0] by the MUX to reconstruct the original binary data stream.

5.3.1 SINGLE DUAL-DRIVE MZIM TRANSMITTER FOR MADPSK

The main reason explaining why the dual-drive MZIM structure has attracted our attention in this chapter is that it can play the role of both amplitude and phase modulators simultaneously, which is impossible with other transmitter structures. This means that to generate a MADPSK signal, there is no need to employ separate phase and amplitude modulators, as it has been implemented in the works of Seikine et al. [4] and Hayase et al. [3]. This section describes a method for generating 16-ary MADPSK signal using this dual-drive MZIM structure.

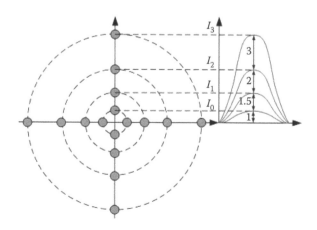

FIGURE 5.21 ASK interlevel spacing.

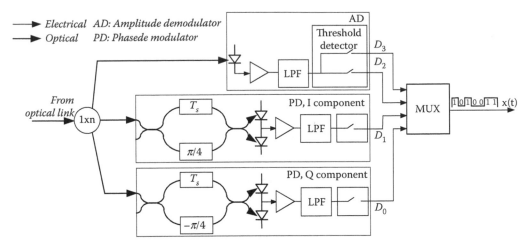

FIGURE 5.22 Amplitude direct detection and photonic phase comparator with balance receiver for MADPSK demodulation.

The 16-ary MADPSK signal constellation of interest is shown in Figure 5.23. It is actually a combination of a 4-ary ASK and a DQPSK signal, with four bits $[D_3, D_2, D_1, D_0]$ mapped into a symbol. Among them, two bits $[D_1, D_0]$ are coded into four phases $[0, \pi/2, \pi, 3\pi/2]$ and two bits $[D_3, D_2]$ are coded into four amplitude levels $[I_3, I_2, I_1, I_1, I_0]$. With the use of balanced receiver and DI, which is a solely available today practical optical phase demodulator, the MADPSK signal produces clear DQPSK eye patterns centered at a zero-voltage decision level only when constellation points are positioned in a radial pattern.

Recall that the signal at the output of the dual-drive MZIM can be represented as

$$E_o = \frac{E_i}{2} e^{j\phi_1} + \frac{E_i}{2} e^{j\phi_2}, \qquad (5.5)$$

with $\phi_1 = \pi(V_1(t)/V\pi)$, $\phi_2 = \pi(V_2(t)/V\pi)$ where E_i and E_o are electrical fields of the input and output optical signal, respectively; $V_1(t)$, $V_2(t)$ are driving voltages applied to the modulator, and V_π is voltage required to provide a π phase shift of for the carrier in each MZIM branch.

Equation (5.7) suggests that with properly chosen input signal E_{input} and driving voltages $V_1(t)$, $V_2(t)$, all signal points of the constellation in Figure 5.23 can be constructed from two phasor signals $E_i/2\, e^{j\phi_1}$ and $E_i/2\, e^{j\phi_2}$. Indeed, if E_i is chosen to be equal the electrical field corresponding to signal points in the largest circle of the constellation, then a constellation signal point E_{output} with the phase shift θ_i in the circle n can be found as a sum of two vectors $E_i/2e^{j\phi_{ni1}}$ and $E_i/2e^{j\phi_{ni2}}$, where $\phi_{ni1} = \theta_i + \arccos(E_n/E_{input})$, $\phi_{ni2} = \theta_i - \cos^{-1}(E_n/E_i)$. The subscriptions i and n denote the phase position and the order of the circle of interest. Figure 5.24 illustrates the relationship between E_i, E_0, ϕ_{ni1} and ϕ_{ni2}. For simplicity, the signal point is chosen with $\theta_i = 0$.

By substituting ϕ_1 and ϕ_2 into (7) the driving voltages, this point can be obtained:

$$V_{ni1}(t) = \frac{V_\pi}{\pi} [\theta_i + \cos^{-1}(E_n/E_i)], \quad V_{ni2}(t) = \frac{V_\pi}{\pi} [\theta_i - \cos^{-1}(E_n/E_i)] \qquad (5.6)$$

5.3.2 PERFORMANCE EVALUATION

Figure 5.25 shows a schematic diagram of the photonic transmitters and receivers for the 16ADQPSK transmission system in which the receiver consists of branches for detection of the amplitude, and the in-phase and quadrature-phase components. Under performance evaluation, the following main

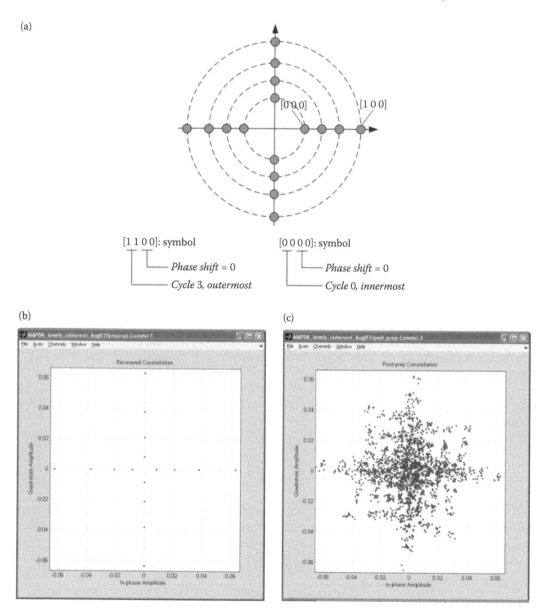

FIGURE 5.23 16-ary MADPSK signal bit-phase mapping (a) design, (b) Simulink scattering plot before transmission, and (c) after 200 km transmission with 2 km mismatch of fiber dispersion factors.

parameters are investigated: (i) system bit error rate BER versus SNR: a solution for system BER will be found analytically and BER will be computed against different SNR values and bit rates. System BER versus SNR will also be obtained by system simulation and cross-checked with BER versus obtained analytically. Graphs of BER versus SNR can be obtained; (ii) system BER versus receiver sensitivity: BER versus receiver sensitivity will be obtained analytically and by simulation, and the results will be cross-checked. Graphs of BER versus receiver sensitivity will be plotted; (iii) dispersion tolerance: transmission over fibers of types ITU-G652, ITU-G.655, and LEAF with corresponding dispersion factors will be considered. Graphs of the power penalty due to the dispersion as compared with back-to-back will be plotted against the dispersion factor in ps/nm; and (iv) tolerance to other system impairments: like with dispersion tolerance, power penalty due to other impairments such as laser and OA phase noise, receiver phase error will be investigated.

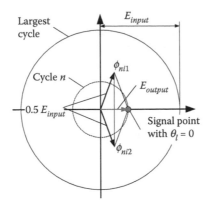

FIGURE 5.24 Relationship between E_i, E_O, ϕ_{ni1}, and ϕ_{ni2} using phasor representation.

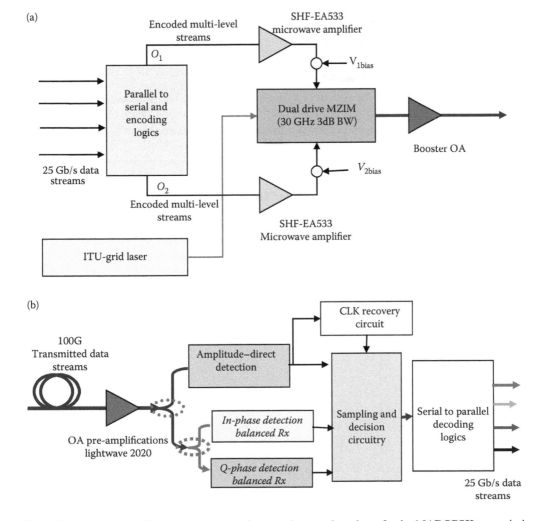

FIGURE 5.25 Schematic diagram of the photonic transmitters and receivers for the 16ADQPSK transmission scheme (a) transmitter and (b) receivers with branches for detection of amplitude, inphase and quadrature phase components.

Performance evaluation is conducted under the effects of the following conditions or contexts: (i) different pulse shapes: raised cosine, rectangular, Gaussian, (ii) modulation formats: NRZ, RZ, and CSRZ, (iii) ASE noise of optical amplifiers, (iv) transmitter impairments: laser noise, (v) receiver impairments: phase error of DI-based phase demodulators, (vi) change of ASK interlevel spacing, (vii) optical and electrical filtering, and (viii) multi-channel environment: system performance in combination with DWDM technology would be reported in future.

5.3.3 IMPLEMENTATION OF MADPSK TRANSMISSION MODELS

5.3.3.1 System Modeling

The following simulation models have been built on the MATLAB Simulink platform for proving the working principles and for investigating the performance of systems using optical MADPSK modulation. A transmitter is simulated to generate 16-ary MADPSK signal, and a receiver is to reconstruct the original binary signal. These models run over a simulated single mode optical fiber. Laser chirp, OA phase noise, non-linearities, CD, PMD, and other impairments will be involved in later stages to evaluate different performance characteristics of the modulation format: system BER, receiver sensitivity, power penalties due to different impairments.

The phases and the driving voltages for creating signal points of the 16-ary MADPSK constellation can be computed and tabulated in Table 5.2.

5.3.3.2 Simulink Transmitter Model

MATLAB Simulink model of the system is shown in Figure 5.26. The transmitter model using the dual-drive MZIM structure is shown in Figure 5.27. The purpose of blocks are as follows: (i) "User data and ADPSK pre-coder" block generates a pseudo-random data sequence to simulate user data

TABLE 5.2
Phase and Driving Voltages for 16-ary MADPSK Constellation

Positions	θ_i	V_{i1}	V_{i2}
Circle 3			
1100	0^0	$0.0\ V_\pi$	$0.0\ V_\pi$
1101	90^0	$0.5\ V_\pi$	$0.5\ V_\pi$
1111	180^0	$1.0\ V_\pi$	$1.0\ V_\pi$
1110	270^0	$1.5\ V_\pi$	$1.5\ V_\pi$
Circle 2			
1000	0^0	$0.2952\ V_\pi$	$-0.2952\ V_\pi$
1001	90^0	$0.7949\ V_\pi$	$0.2046\ V_\pi$
1011	180^0	$1.2947\ V_\pi$	$0.7043\ V_\pi$
1010	270^0	$1.7944\ V_\pi$	$1.2041\ V_\pi$
Circle 1			
0100	0^0	$0.3919\ V_\pi$	$-0.3919\ V_\pi$
0101	90^0	$0.8917\ V_\pi$	$0.1078\ V_\pi$
0111	180^0	$1.3914\ V_\pi$	$0.6076\ V_\pi$
0110	270^0	$1.8912\ V_\pi$	$1.1073\ V_\pi$
Circle 0			
0000	0^0	$0.4575\ V_\pi$	$-0.4575\ V_\pi$
0001	90^0	$0.9573\ V_\pi$	$0.0422\ V_\pi$
0011	180^0	$1.4570\ V_\pi$	$0.5420\ V_\pi$
0010	270^0	$1.9568\ V_\pi$	$1.0417\ V_\pi$

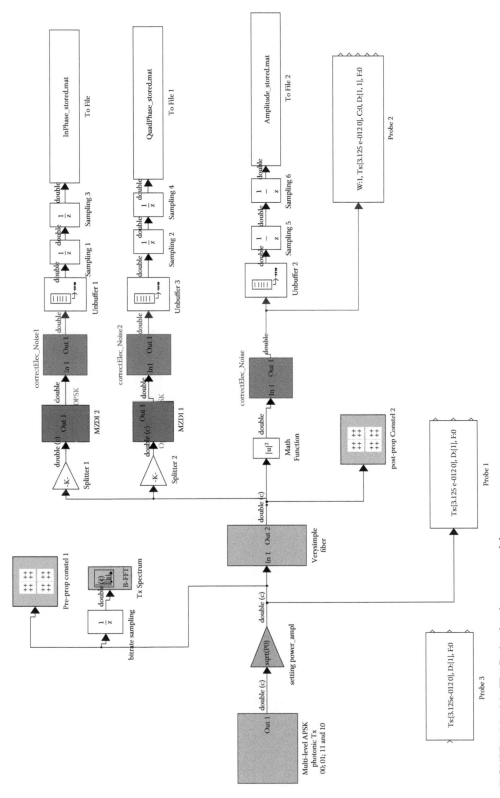

FIGURE 5.26   MATLAB simulated system model.

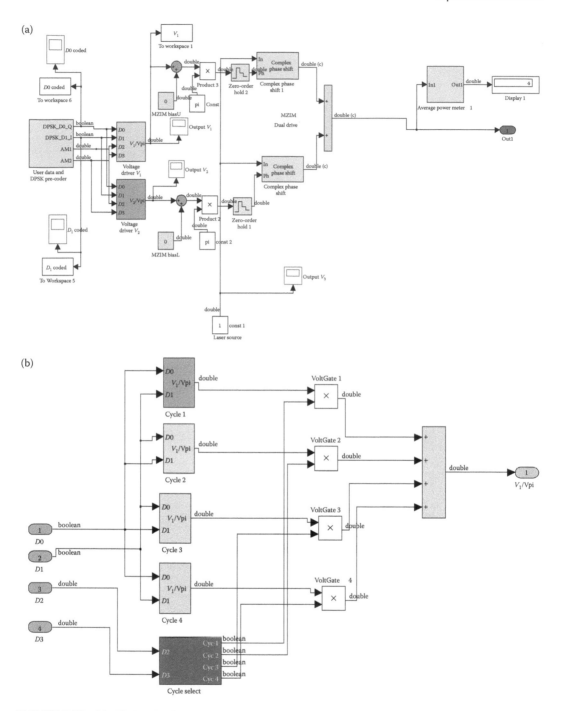

FIGURE 5.27 MATLAB simulated MADPSK (a) transmitter and (b) logic pre-coder.

stream and encodes each group of 4 data bits into a symbol. (ii) "Voltage driver 1" and "Voltage driver 2" blocks map pre-encoded data into driving voltages for modulating amplitude and phase of the carrier in the dual-drive MZIM. (iii) Two "complex phase shift" blocks simulate two optical paths of the dual-drive MZIM. (iv) "Sum block" simulates the combiner at the output of MZIM. (v) "Gaussian noise generator" block simulates noise source. (vi) "Amplifier" block simulates optical amplifier.

5.3.3.3 Simulink Receiver Model

The receiver structure is shown in Figure 5.28. The functions of the blocks are as follows: (i) Each DI is simulated by a set of two "Magnitude-Angle" blocks, a "Delay block," and a "Sum" block. The "Delay block" stores the phase of the previous symbol and the "Magnitude-Angle" blocks extract the phase and amplitude of present and previous symbols, which will be used in the followed different phase demodulation and detection operations. (ii) "Constant pi/4" and "Constant -pi/4" and the following two "Sum" blocks simulate extra phase delay in each branch of DI. (iii) Two "Cos" blocks and two "Product" blocks simulate two balanced receivers. (iv) The "Amplitude Detector D2 and D3" block simulates ASK detector for D_2 and D_3 bits. (v) Three "Analog Filter Design" blocks simulate electrical low pass filters. (vi) "Phase Detector D0_I" and "Phase Detector D1_Q" blocks simulate the threshold detectors for D_0 and D_1 bits (I and Q component of a DQPSK signal), respectively.

(a)

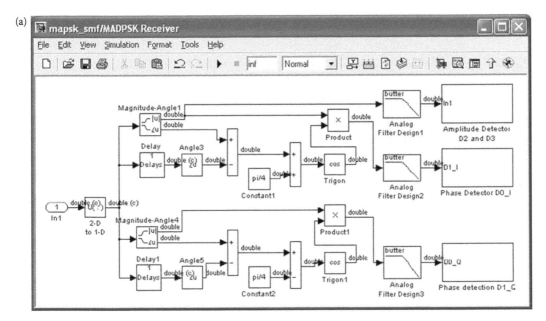

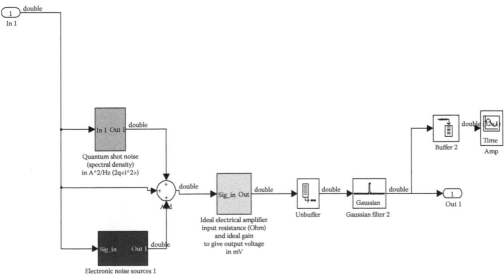

FIGURE 5.28   MATLAB simulated MADPSK receiver (a) amplitude direct detection.   (*Continued*)

(b)

Balanced receiver

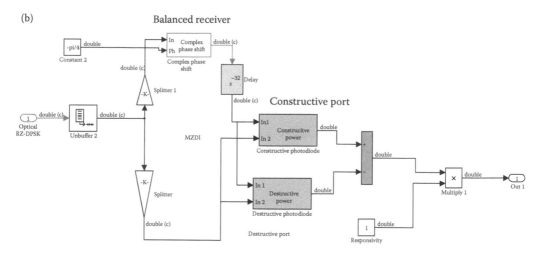

FIGURE 5.28 (Continued)   (b) Balanced receiver detection—inphase and quadrature.

5.3.3.4 Dispersion-Compensating Optical Transmission Line Model

The propagation of optical signal in a fiber medium that is dispersive and non-linear is best described by the non-linear Schrödinger equation (NLSE) [1], as described in Chapter 13. Other parameters are explained below. The transmission fiber model shown in Figure 5.29 is used to simulate the propagation of optical signal. This fiber model simulates the impairments that have impacts on the system performance.

All characteristic parameters of the fiber medium together with optical input signal are taken by the "Matrix Concatenation" block and then processed by a MATLAB function that solves the NLSE

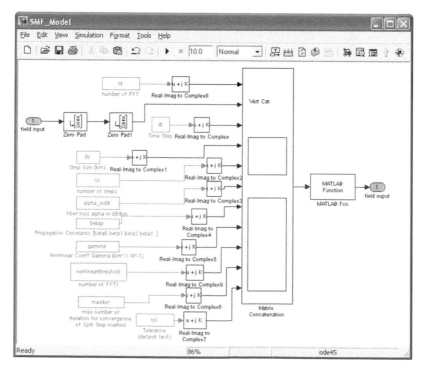

FIGURE 5.29   Input parameters for single mode fiber SSMF model.

using the split-step Fourier method [4]. The dispersion compensation fiber model has the same structure of the propagation fiber model, except that the sign of the propagation constant β_2 in the two models are opposite.

5.3.3.5 Transmission Performance

5.3.3.5.1 Signal Spectrum, Signal Constellation, and Eye Diagram

The spectrum of 40 Gb/s 16-ary MDAPSK signal obtained by running transmitter model is given in Figure 5.30. As it has been seen clearly in the graph, the single-sided bandwidth of the main lobe equals 10 GHz. Numerically, it amounts to only one-fourth of the transmission bit rate, and from that it can be concluded that MADPSK is a high bandwidth efficient modulation format.

Figure 5.31 shows the signal constellation recovered at the receiving end. Noise and non-linear property of fiber cause amplitude and phase fluctuations and scatter signal points around some mean value. The MADPSK eye diagram is shown in Figure 5.32 for the I component (the Q component should have a similar diagram). This eye diagram clearly shows four amplitude levels associated with two phase shifts 0 and π.

5.3.3.5.2 BER Evaluation

The MADPSK system can be considered as consisting of two sub-systems, ASK and DQPSK, and its error probability can be evaluated as a join error probability of the two:

$$P_{ADPSK} = \left[\frac{1}{2}P_{ASK} + \frac{1}{2}P_{DPSK} - \frac{1}{2}P_{ASK} \cdot \frac{1}{2}P_{DPSK}\right] = \frac{1}{2}[P_{ASK} + P_{DPSK} - P_{ASK} \cdot P_{DPSK}] \quad (5.7)$$

where P_{ASK} and P_{DPSK} are error probabilities of ASK and DQPSK sub-systems, respectively.

5.3.3.5.2.1 ASK Sub-System Error Probability

Figure 5.33 shows four ASK signal levels b_0, b_1, b_2, b_3, three decision levels d_1, d_2, d_3, and standard deviation of noise at different signal levels σ_0, σ_1, σ_2, σ_3. The error probability of the ASK sub-system can be evaluated by

$$P_{ASK} = \frac{2}{M+1}\sum_{1}^{M} Q\left(\frac{b_i - d_i}{\sigma_i}\right) = \frac{2}{3+1}\left[\begin{array}{l} Q\left(\dfrac{b_1 - d_1}{\sigma_1}\right) \\ +Q\left(\dfrac{b_2 - d_2}{\sigma_2}\right) \\ +Q\left(\dfrac{b_3 - d_3}{\sigma_3}\right) \end{array}\right] \quad (5.8)$$

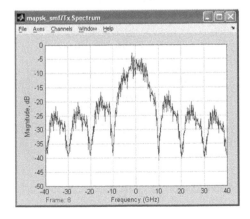
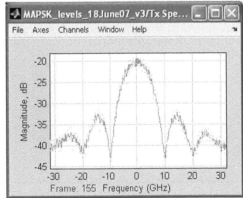

FIGURE 5.30 40 Gb/s MADPSK spectrum.

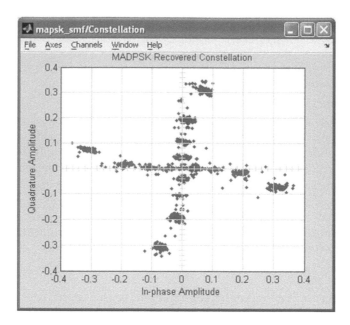

FIGURE 5.31   40 Gb/s MADPSK constellation recovered at the receiver.

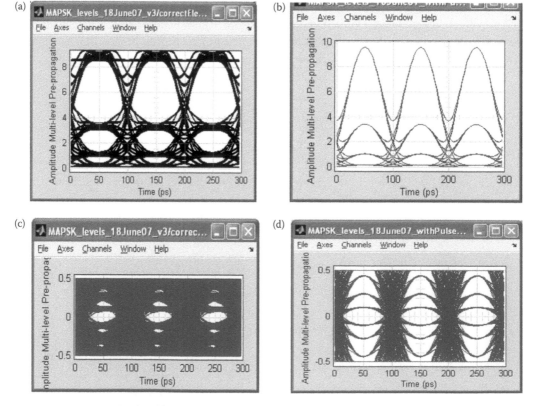

FIGURE 5.32   40 Gb/s MADPSK eye diagram at OSNR = 20 dB (a) NRZ amplitude, (b) CSRZ amplitude, (c) NRZ in phase and (d) CSRZ in phase.

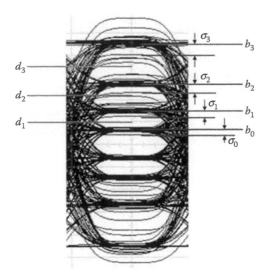

FIGURE 5.33 MADPSK eye diagram: signal levels, decision levels, and standard deviation of noise.

For example, in our system: (i) $b_1 = 8.08e - 2$, $b_2 = 1.45e - 1$, $b_3 = 2.42e - 1$; (ii) $d_1 = 5.11e - 2$, $d_2 = 1.08e - 1$, $d_3 = 1.88e - 1$; and (iii) $\sigma_1 = 5.00e - 3$, $\sigma_2 = 6.70e - 3$, $\sigma_2 = 8.65e - 3$ at an OSNR $= 20$ dB

The error probability of the ASK sub-system thus equals:

$$P_{ASK} = \frac{1}{2}\left[\begin{array}{c} Q\left(\dfrac{(8.08e - 2) - (5.11e - 2)}{5.00e - 3}\right) + Q\left(\dfrac{(1.45e - 1) - (1.08e - 1)}{6.70e - 3}\right) \\ + Q\left(\dfrac{(2.42e - 1) - (1.88e - 1)}{8.65e - 3}\right) \end{array}\right] \rightarrow$$

$$P_{ASK} = \frac{1}{2}[Q(5.94) + Q(5.52) + Q(6.24)]$$

$$= \frac{1}{2}[(1.47e - 9) + (1.73e - 8) + Q(2.26e - 10)]$$

$$= 9.49e - 9$$

The error probability of ASK sub-system over a range of OSNR from 6 dB to 24 dB is evaluated and shown in Figure 5.34.

5.3.3.5.2.2 DQPSK Sub-System Error Probability Evaluation   In terms of differential phase shift keying modulation, the system can be broken up into four independent DQPSK sub-systems corresponding to circle 0, circle 1, circle 2, and circle 3 of the signal constellation. The error probability of each sub-system is evaluated first and then they are averaged to obtain the error probability of the DQPSK sub-system.

Each DQPSK sub-system in turn can be thought of as made from two 2-ary DPSK sub-systems. The error probability of each 2-ary DPSK sub-system is evaluated and then they are averaged to get the error probability of DQPSK sub-system:

$$P_{DQPSK} = 1 - (1 - P_{DPSK\_I})(1 - P_{DPSK\_Q}) = P_{DPSK\_I} + P_{DPSK\_Q} - P_{DPSK\_I}.P_{DPSK\_Q} \quad (5.9)$$

where P_{DPSK_I} and P_{DPSK_Q} are error probability of in-phase (I) and quadrature-phase (Q) components of each DPSK sub-system (circle). Because I is coded by bit D_0, Q is coded by bit D_1, I and

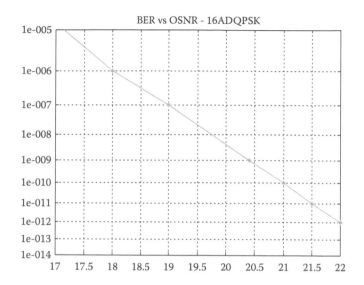

FIGURE 5.34   Error probability of ASK sub-system versus OSNR.

Q are detected in the same way, and D_0 and D_1 are supposed to be equally probable, then (20) becomes:

$$P_{DQPSK} = 2P_{DPSK\_I} - P_{DPSK\_I}^2 = 2P_{DPSK\_Q} - P_{DPSK\_Q}^2 \qquad (5.10)$$

P_{DPSK_I} is evaluated based on the δ-factor [4] as

$$P_{DPSK\_I} = \frac{1}{2}\left(\frac{\delta}{\sqrt{2}}\right) \approx \frac{\exp\left(-\delta^2/2\right)}{\delta\sqrt{2\pi}} \qquad (5.11)$$

where $Q = i_H - i_L/\sigma_H + \sigma_L$, i_H, i_L and σ_H, σ_L are mean value and standard deviation of signal currents at high and low levels at the input of the receiver, respectively. For example, the transmission parameters can be set as follows: $i_H = 3.23e - 02$, $i_L = (-3.23e - 02)$, at OSNR = 20 dB $\sigma_H = \sigma_H = 3.16e - 3$. The δ-factor for a single DQPSK sub-system of circle 0 thus corresponds to an error probability is $P_{DPSK\_I\_CYCLE0} = 1/2erfc(10/\sqrt{2}) \approx 7.7e - 24$.

The error probability of circle 0 (inner most circle) is $P_{DQPSK\_CYCLE\ 0} = 2*(7.7e - 24) - (7.7e - 24)^2 = 1.54*10e - 23$. Thus the error probability of all four circles is

$$P_{DQPSK} = \frac{1}{4}\left[P_{DQPSK\_CYCLE\ 0} + P_{DQPSK\_CYCLE\ 1} + P_{DQPSK\_CYCLE\ 2} + P_{DQPSK\_CYCLE\ 3}\right] \qquad (5.12)$$

P_{DQPSK} over a range of OSNR from 6 dB to 24 dB is evaluated and shown in Figure 5.35.

5.3.3.5.2.3   MADPSK System BER Evaluation   The MADPSK system error probability is evaluated based on the Equation 5.17. Figure 5.36 shows the graphs of error probability for the ASK sub-system, DQPSK sub-system, and MADPSK system in the same coordinates for comparison purpose. As can be observed from Figure 5.36, at OSNR = 24 dB the MADPSK it is also clear that for the same value of OSNR, especially when it is high, DQPSK sub-system outperforms its ASK counterpart, and the overall performance of MADPSK system is dominated by the ASK sub-system performance. Thus, the spaces between ASK levels could be adjusted for a better balance between BER

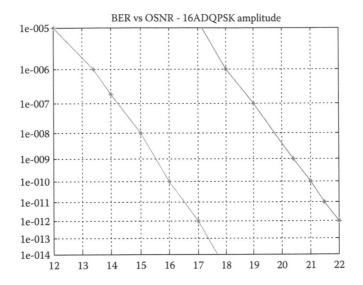

FIGURE 5.35 Error probability of DQPSK sub-system versus OSNR.

ASK and BER DQPSK to achieve a better overall MADPSK BER performance. This is probably mainly caused by the intersymbol interference during the transition of different levels.

Figure 5.36 shows the simulation results of 16ADPSK at 100 Gb/s transmission (extreme left graph) in comparison with other modulation formats such as duobinary 50 and duobinary 67 and experimental results of CSRZ DPSK. The bit rates of these other transmission results are at 40 Gb/s. It is observed that for 16MADPSK, the receiver sensitivity is close to the −28 dBm performance standard used in 10 Gb/s NRZ transmission, and performs better at 100 Gb/s than the other modulations operating at the lower rate of 40 Gb/s. However, this superior performance at 100 Gb/s

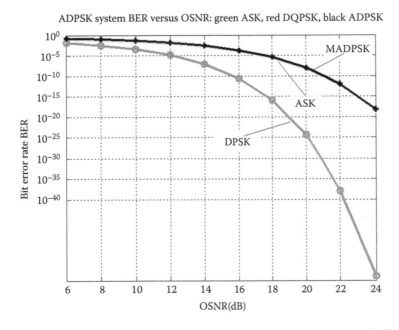

FIGURE 5.36 Error probability of MADPSK (black) system versus OSNR, logarithm scale. Error probability of ASK (grey) and MADPSK are nearly coincided.

is still at penalty of approximately 3 dB compared with 10 Gb/s transmission systems. Fortunately, this penalty can easily be compensated for using a low-noise optical pre-amplifier at the receiver end. For example, a 15 dB gain optical pre-amplifier with a 3 dB noise figure would adequately resolve the issue. The BER versus the receiver sensitivity of 16ADPSK and duobinary formats and ASK is shown in Figure 5.37. It indicates a 2–3 dB improvement of the MADPSK.

The detection of the lowest level may have been affected by the noise level of the optical pre-amp when only the amplitude information is used. This can be improved significantly if the phase detection and amplitude detection are used, as we see in Figure 5.32 (c) and (d).

5.3.3.5.2.4 Chromatic Dispersion Tolerance The residual chromatic dispersion of the optical link is characterized by the *DL* product defined as product of the dispersion coefficient *D* and the total fiber length *L*. Figure 5.38 shows the signal phase evolution under the effect of chromatic dispersion. With a predetermined $DL = 50$ ps/nm, all signal points are rotated around the [0,0] origin by the same angle of approximately 0.125 radians. This confirms the parabolic phase shift because of the chromatic dispersion. This phenomenon is called linear phase distortion in contrary to the non-linear phase distortion caused by the fiber non-linearity.

Figure 5.39 shows the BER penalty versus different values of *DL* product. It can be seen very clearly that the BER performance of NRZ format is severely affected by the fiber dispersion. When *DL* increases from 0 ps/nm (fully CD compensated) to 35 ps/nm, its BER performance is improved by 1.5 dB, but sharply degraded by 28 dB penalty at $DL = 50$ ps/nm, and should be worse for a higher value of dispersion. This leads to the conclusion that it is undesirable to use NRZ format in MADPSK systems because the optical link residual dispersion usually cannot be compensated to a small amount, and ineffective dispersion management and control plan could lead to a very high BER.

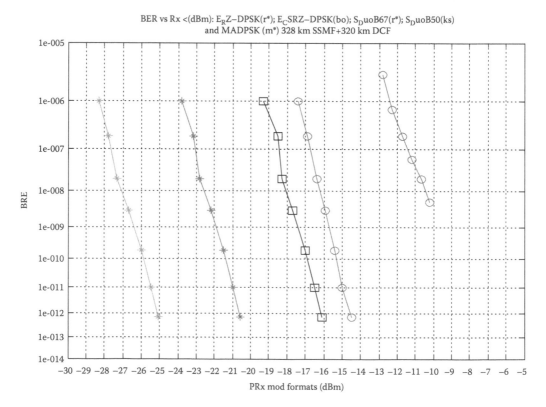

FIGURE 5.37 BER versus receiver sensitivity for MADPSK format and other Duo-binary and ASK (simulation) and CSZ and CSRZ DPSK (experimental). Legend: lightest * is the MDAPSK.

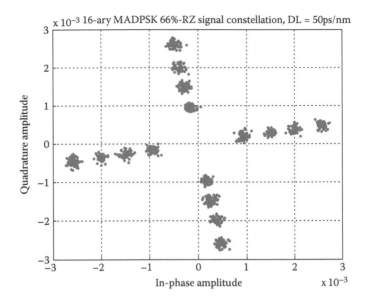

FIGURE 5.38   Evolution of the phase scattering of the MADPSK signal constellation under chromatic dispersion effects.

The 66%-RZ format, on the other hand, can tolerate a much higher degree of chromatic dispersion. Its BER performance is even slightly improved at DL ≈ 50 ps/nm, and the BER penalty is just less than 1 dB at DL = 100 ps/nm. That is equivalent to the transmission over a 6kms of uncompensated standard SMF fiber without significantly giving up the BER performance.

The MAPSK offer a lower symbol rate and hence higher channel capacity that would allow the upgrading of higher rate merging in a low bit rate optical fiber transmission system without modifying the photonic infrastructure of the optical networks.

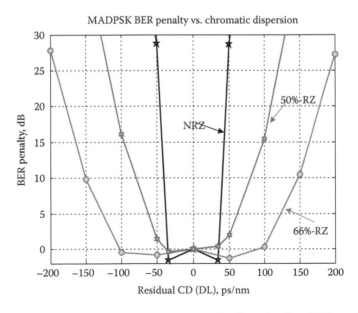

FIGURE 5.39   Error probability of MADPSK system versus the dispersion-length DL product.

5.3.3.6 Critical Issues

This section outlines the critical issues to evaluate the performance of MADPSK systems.

5.3.3.6.1 *Noise Mechanism and Noise Effect on MADPSK*

Although receiver noises in multi-level amplitude modulation were investigated intensively in the 1980s, little has been reported for multi-level phase and differential phase modulation. One of the principal goals in the system design, especially for long-haul transmission systems, is to achieve high receiver sensitivity. At a given optical power, the error probability depends on the noise power, and hence the receiver sensitivity.

Quantum shot noise is the fundamental noise mechanism in photodiode, which leads to the fluctuation in the detected electrical current even when the incident optical signal has a constant or varied power. Thus, it is signal dependent. Furthermore, the beating of the currents of the signal and the optical phase noise would generate an amplitude-dependent noise at different level signals of the MADPSK. It is caused by random generation of electrons contributing to the photoelectric current, which is a random variable. All photodiodes generate some current even in the absence of optical signal because of the stray light and/or thermal generation of electron-hole pairs, the dark current.

In MADPSK, the amplitude of the signal of the outer most circle of the constellation would be affected by the quantum shot noises which are strongly signal amplitude dependent, especially when there is an optical pre-amplifier. On the other hand, it is desirable that the inner most constellation would have the largest magnitude to maximize the optical signal energy for long haul transmission. Therefore, an optimum receiving scheme must be developed by both analytically and by modeling and eventually by experimental demonstration.

However, the amplitude of the outer most constellations is limited by the non-linear self-phase modulation effects that would be further explained in the next few sections. Thus, the lower and upper limits of the amplitude of the MADPSK would be extensively investigated in the next phases of the research.

The electronic equivalent noise as seen from the input of the electronic pre-amplifier following the PD (photodetctor) can be measured and considered as part of the total noise process caused by thermal noises of the input impedance, the biasing current shot noise, and the noises at the output of the electronic pre-amplifier. These noises are combined with the signal dependent quantum shot noises to gauge their contribution to the MADPSK receiver. Thus, we may consider new structures of electronic amplifiers or matched filter at the input of the receiver to achieve optimum MADPSK receiver structure.

For long-haul transmission systems, ASE of optical amplifier is probably the most important noise mechanism. In optical amplifiers, even in the absence of input optical signal, spontaneous emission always occurs stochastically when electron-hole pairs recombine and release energy in the form of light. This spontaneous emission is noise and it is amplified by the optical amplifiers together with the useful optical signal and accumulated along optical transmission link.

Noises reduce the SNR, as well as system BER and receiver sensitivity. Noise models also affect the design of optimum detection schemes, such as decision thresholds. To the best of my knowledge, a thoroughly investigation of the noise mechanism and their impacts on multi-level signaling has never been reported except some preliminary results for 10 Gb/s 4-ary ASK schemes. Thus, all noise sources and mechanism by which they affect the system performance must be thoroughly investigated. These noises are used to estimate the optimum decision level of the detection of the amplitude of the multi-level eye diagram.

5.3.3.6.2 *Transmission Fiber Impairments*

For optical signals, the transmission medium is an optical fiber with associated OAs and dispersion compensation devices, or a leased wavelength running on top of a DWDM system. Impairments are always part of the transmission medium, among them CD, PMD, and non-linearity are critical.

When an optical pulse propagates along a fiber, its spectral components disperse due to the differential group delay (DGD), and the output pulse will be broadened. CD is proportional to the fiber length and the laser linewidth, especially the spectrum of the lightwave modulate signals. CD may cause optical pulses to overlap each other, thus leading to intersymbol interference and increase system BER, especially for ASK systems. DPSK systems, on the other hand, are more CD tolerant. For MADPSK systems, the phase constellation, as shown in Figure 5.40, is rotating when the MADPSK is under the linear chromatic dispersion effect. It is also well known and developed in our model that this CD can be compensated by dispersion compensating fiber modules. However, the mismatching of the dispersion slopes of the transmission and compensating fibers is very critical for multi-channel multi-level modulation schemes.

Optical pulse is also broadened by PMD which is the time mismatching between two orthogonal polarizations of the optical pulse when they traverse along a fiber. In the ideal optical fiber having truly homogeneous glass and truly coaxial geometry of the core, the two optical polarizations would propagate with the same velocity. However, it is not the case for a real fiber, so the two polarizations have different speeds and will reach the fiber end at different times.

Like the CD effects, PMD can cause pulse overlapping and thus increase system BER. However, unlike CD, which is practically constant over time and can be in a large scale compensated, PMD is a stochastic process and cannot be managed easily. It has been well known that PMD has the Maxwellian probability density function with a mean value $\langle PMD \rangle = K_{PMD}.\sqrt{L}$ where K_{PMD} is defined as PMD coefficient whose measured values vary from fiber to fiber in the range $[0.01 - 1\,ps/\sqrt{km}]$, and L is fiber length. Under the MADPSK, the signal-space of the constellations would be affected either in the magnitude or phases by PMD, but expected to be dominated by the phase distortion. It is well known that the PMD first and second effects are critical for ASK modulation. For DPSK, it is expected that the principal axes of the polarization modes propagating through the fiber would be minimally affected. Thus, under hybrid amplitude-phase modulation scheme, several issues remain to be resolved. Under the MADPSK scheme, the delay of the polarization modes would generate the phase difference or phase distortion on I and Q components, hence enhancement of the distortion effects of the ISI. The amplitude distortion would then be increased but considered to be secondary effect.

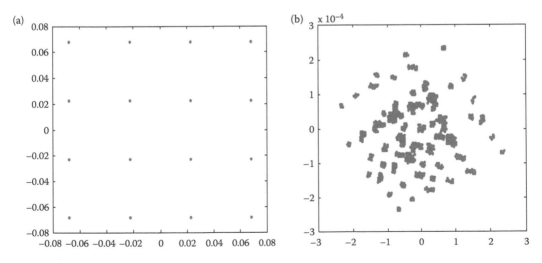

FIGURE 5.40 Constellation of 16-square QAM after two optically amplified spans and 2 km SSMF dispersion mismatch (a) pre-propagation and (b) post-propagation.

5.3.3.6.3   Non-linear Effects on MADPSK

Non-linear effects occur because of the non-linear response of the fiber glass to the applied optical power. Fiber non-linearity can be classified into stimulated scattering and Kerr effect. Among several stimulated scattering effects, stimulated Raman scattering, caused by interaction between light and acoustical vibration modes in the fiber glass, is the most critical. Under this mechanism, optical signal is reflected back to the transmitter, and in WDM systems its power is also transferred from shorter to longer wavelengths, thus attenuating signal and causing crosstalk. Kerr effect is the root of intensity-dependent phase shift of the optical field. It is shown in three forms: self-phase modulation (SPM), cross-phase modulation (XPM), and four-wave mixing (FWM) provided the phase matching is satisfied.

SPM is usually the dominant effect in a single-channel DPSK systems. The changes in instantaneous power of optical pulses, together with the ASE from associated OAs, lead to intensity-dependent changes, as well as the Kerr effect in the guided medium refractive index, hence the effective index of the guided mode. These changes are converted to the phase shifts or phase noise of the lightwave carriers. At the receiver, the phase noise is transferred back to intensity noise that degrades BER. As mentioned above, the contribution of noises into different levels of the MADPSK scheme is very critical to determine optimum decision thresholds. This is further complicated by these additional non-linear effects, especially the non-linear phase noises (NLPN) usually contributed by the SPM because of the outer most constellation. These NLPN effects from the outermost constellation to other inner circle signal spaces have never been investigated.

XPM becomes the most critical non-linearity in WDM systems where the phase shifts (noise) in one channel comes from refractive index fluctuations caused by power changes in other channels. XPM becomes more pronounced when neighboring channels have equal bit rate. Four-wave mixing (FWM) is basically a crosstalk phenomenon in WDM systems. When three wavelengths with frequencies, ω_1, ω_2, and ω_3 propagate in a non-linear fiber medium at which the dispersion is zero. They combine and create a degenerate fourth wavelength which would fall on the location of an active wavelength channel. If these parametric wavelengths fall into other channels, they cause crosstalk and degrade the performance of the system. FWM is expected to reduce the receiver sensitivity in the proposed system. To minimize the effects of fiber non-linearity, the maximum power of optical signal should not be set higher a certain threshold. This maximum power dictates the amplitude of signal points in the outermost circle (circle 3), and hence other circles, of the signal space. Thus, optimization of the signal amplitude levels for MADPSK is very critical.

5.3.3.7   Offset Detection

The 16-ary MADPSK signal model described in section 0 can be modified. To balance the ASK and DQPSK sensitivities, ASK signal levels are preliminarily adjusted to the ratio $I_3/I_2/I_1/I_0 = 3/2/1.5/1$ and rotated by $\pi/4$, as shown in Figure 5.41. These level ratios can be determined from the signal-to-noise ratio at each separation distance of the eye diagram or q-factor. The noise is assumed to be dominated by the beat noise between the signal level and that of the ASE noise. The eye opening is expected to improve significantly as shown in Figure 5.42.

5.4   STAR 16-QAM OPTICAL TRANSMISSION

This section gives a briefing on the simulation of the transmission performances of optical transmission systems over 10 spans of dispersion compensated and optically amplified fiber transmission systems. The modulation format is focused on the Star 16-QAM with two-level and eight-phase state constellations. Optical transmitters and coherent receivers are the main transmission terminal equipment. Other constellations of the 16- QAM are also given very briefly. Simulated results have shown that it is possible to transmit and detect the data symbols for 43 Gb/s, with the possibility of scaling to

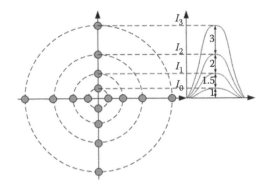

FIGURE 5.41 ASK interlevel spacing and offset modulation and detection line (A) 0 kilometer mismatch SSMF.

107 Gb/s without much difficulty. The OSNR with 0.1 nm optical filters is achieved with 18 dB and 23 dB for back to back and long haul transmission cases with a dispersion tolerance of 300 ps/nm.

5.4.1 Introductory Remarks

To increase the channel capacity and bandwidth-efficiency in optical transmission, the multi-level modulation formats like QAM formats are of interest. In the digital transmission with multi-level (*M*-levels) modulation, m bits are collected and mapped onto a complex symbol from an alphabet with $M = 2$ m possibilities at the transmitter side.

The symbol duration is $T_s = mT_B$ with T_B as the bit duration and the symbol rate is $fs = f_B/m$ with $f_B = 1/T_B$ as the bit rate. This shows that, for a given bit rate, the symbol rate decreases if the modulation level increases. That means higher bandwidth efficiency can be achieved by a higher order modulation format. For 16-QAM format, $m = 4$ bits are collected and mapped to

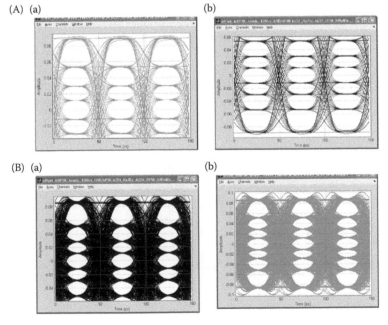

FIGURE 5.42 40Gb/s MADPSK eye diagrams of the *I* (a) and *Q* (b) components (A) 0 kilometer – back to back and (B) 2 km SSMF mismatch over three 100 km SSMF transmission spans (dispersion compensated).

one symbol from an alphabet with $M = 16$ possibilities. In comparison with the case of binary modulation format, only $m = 1$ bit is mapped to one symbol from an alphabet with $M = 2$ possibilities. With 16-QAM format and a data source with a bit rate of $f_B = 40$ Gb/s, only a symbol rate of f_s of 10 Gbaud/s is necessary. In commercial point of view, it means a 40 Gb/s data rate can be transmitted with 10 Gb/s transmission devices. In the case of binary transmission, the transmitter needs a symbol rate of f_s of 40 Gbaud/s. It means 16-QAM transmission requires four times slower transmission devices than that for the binary transmission. It is noted here that a 10.7 GSy is used as the symbol rate to compare the simulation results with the well-known 10.7 Gb/s modulation schemes such as DPSK, CSRZ DPSK, etc. For 10.7 Gb/s bit rate, the transmission performance, that is, the sensitivity and OSNR, can be scaled accordingly without any difficulty.

This section gives a general approach regarding the design and simulation of Star 16-QAM with two amplitude levels and eight phase states forming two star circles. We term this Star 16-QAM as 2A-8P Star 16-QAM, two-amplitude–level and eight-phase states. The transmission format is discussed with theoretical estimates and simulated results to determine the transmission performance. The optimum Euclidean distance is defined for the design of star 16-QAM. Then, in the second section, the two detection schemes, namely the direct detection and the coherent detection for star-QAM constellations are discussed.

5.4.2 Design of 16-QAM Signal Constellation

There are many possibilities when designing a 16-QAM signal constellation. Three of the most popular constellations can be introduced. For 16-QAM, modulation schemes are (i) Star 16-QAM, (ii) Square 16-QAM, and (iii) Shifted-square 16-QAM. The first two of these constellations are implemented. However, only the Start 16-QAM with two amplitudes and eight phases per amplitude level are employed in this section.

5.4.3 Signal Constellation

The signal constellation for star 16-QAM with Grey coding is shown in Figure 5.43. The binary presentation of the symbols in the figure is shown in mapping Table 5.4.3 of Table 5.4. As can be seen from this figure, the symbols are evenly distributed on two rings and the phase difference between the neighboring symbols on the same ring are equal (pi/4). To detect a received symbol, its phase and amplitude must be determined. In other words, between two amplitude levels of the rings and among 8 phase possibilities there are a number of ways to form this constellation.

The ring ratio (RR) for this constellation is defined as: $RR = b/a$ where a and b are the ring radii, as shown in Figure 5.1. The RR can be set to different values to optimize the transmission performance.

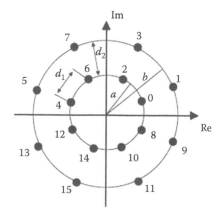

FIGURE 5.43 Theoretical arrangement of the Modulation constellation for Start 16-QAM.

TABLE 5.3
DQPSK Signal Bit-Phase Mapping

D_1	D_0	I	Q	Shifted Phase (rad.)
0	0	0	0	0
0	1	0	1	$\pi/2$
1	1	1	1	π
1	0	1	0	$3\pi/2$

There is a possibility of using the same coding scheme as described above, but the codes are randomly distributed so that the adjacent codes that give the worst error rate are avoided, thus improvement of the BER can be achieved. Improvement in the OSNR for the same BER can be 1 to 2 dB. This is quite substantial.

5.4.4 OPTIMUM RING RATIO FOR STAR CONSTELLATION

From Figure 5.43 it can be seen that there are many possibilities to choose the RR for the star 16-QAM constellation. Here the theoretical best RR is defined to minimize the error probability in an AWGN channel by maximizing the minimum distance d_{min} between the neighboring symbols.

TABLE 5.4

Symbol Mapping and Coding for Star 16-QAM

Table 5.4.1 symbol→→serial bits

0→0000	4→0100	8→1000	12→1100
1→0001	5→0101	9→1001	13→1101
2→0010	6→0110	10→1010	14→1110
3→0011	7→0111	11→1011	15→ 1111

Table 5.4.2 Grey code→Star 16-QAM

0→1	4→7	8→15	12→9
1→0	5→6	9→12	13→8
2→3	6→5	10→13	14→11
3→2	7→4	11→12	15→ 10

Table 5.4.3 mapping for Star QAM

0→1	4→9	8→13	12→5
1→0	5→8	9→12	13→4
2→3	6→11	10→15	14→7
3→2	7→10	11→14	15→ 6

Table 5.4.4 Grey Code for Square QAM

0→12	4→11	8→13	12→4
1→10	5→2	9→5	13→3
2→15	6→8	10→14	14→7
3→9	7→1	11→6	15→ 0

Table 5.4.5 mapping for Square QAM

0→10	4→11	8→4	12→15
1→6	5→1	9→14	13→3
2→5	6→2	10→13	14→0
3→9	7→8	11→7	15→ 12

The results for AWGN channel can be used approximately for optical transmission. For Star 16-QAM the minimum distance d_{min} is maximized, when

$$d_1 = d_2 = b - a = d_{\min} \tag{5.13}$$

With some geometrical calculations, it can be obtained that

$$d_{\min} = 2a \cdot \sin(22.5^0) \tag{5.14}$$

which leads to the optimal ring ratio of

$$RR_{opt} = b/a = (d_{\min} + a)/a = (2 \cdot a \cdot \sin 22.5 + a)/a \approx 1.77 \tag{5.15}$$

The average power of the star 16-QAM constellation can be determined as

$$P_0 = (8a^2 + 8b^2)/16 = (a^2 + b^2)/2 \tag{5.16}$$

Thus, we have the relationship between the average optical power and the minimum distance between the two rings of the two amplitude levels as

$$d_{\min} \approx 0.53(P_0)^{1/2} \tag{5.17}$$

The obtained $RR_{opt} = 1.77$ does not depends on P_0 and is constant for each P_0 value. For an average power of 5 dBm (3.16 mW), $d_{\min} = 2.98 \cdot 10 - 2\sqrt{W}$, $a = 3.89 \cdot 10 - 2\sqrt{W}$ and $b = 6.87 \cdot 10 - 2\sqrt{W}$ are obtained.

5.4.4.1 Square 16-QAM

The signal constellation of the square 16-QAM with Grey coding is shown in Figure 5.44. The square 16QAM can be generated by a combination of two rings and a QPSK in the outer most that can also be a ring. Thus, the circular phasor generation scheme can be used to form signals amplitude and phases to be applied to the optical modulators. The binary presentation of the symbols in the figure is shown in mapping Table 5.4.1 in Table 5.4. In the constellation of the square 16-QAM, the 16 symbols have equal distance with direct neighbors and totally 12 different phases, that is, three phases per quarter, distributed on three rings. The phase differences between neighboring symbols on the inner and outer rings are equal ($\pi/2$), but the phase differences between neighboring symbols on the middle ring are

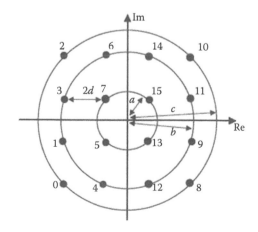

FIGURE 5.44 Square 16-QAM signal constellation represented in three circular levels.

different ($37°$ or $53°$). If the distance between direct neighbors in the square 16-QAM be rotated as 2d, the average symbol power (P_0) of the constellation is:

$$P_0 = 10 \cdot d_2 \tag{5.18}$$

For an average power of 5 dBm (3.16 mW), it can be computed that $d = 1.77 \cdot 10^{-2}\sqrt{W}$ and from it: $a = 2.5 \cdot 10^{-2}\sqrt{W}$, $b = 5.6 \cdot 10^{-2}\sqrt{W}$ and $c = 7.5 \cdot 10^{-2}\sqrt{W}$. In comparison with star 16-QAM, here the distances between the middle ring and the outer ring are much smaller. It means, to achieve the same BER, square 16-QAM needs a higher average power than star 16-QAM. The decision method for the square 16-QAM is more complicated than that for Star 16-QAM. First, the decision between three amplitude possibilities of each ring should be made then, depending on the ring level, the decision is made between four or eight phase possibilities.

5.4.4.2   Offset-Square 16-QAM

To optimize the phase detection of the middle ring, it is envisaged that the phase differences between neighboring symbols on the middle ring in square 16-QAM should be equal. Thus, the shifted-square 16-QAM is introduced by shifting (rotation) of symbols on the middle ring to obtain equal phase differences between all neighboring symbols as shown in Figure 5.7. After shifting the symbols on the middle ring, the distances between all direct neighbors are not necessarily equal. In comparison to square 16-QAM, this constellation may offer more robust detection against phase distortions according to our amplitude and phase detection method introduced in later section.

5.4.5   Detection Methods

In the case of differential encoding for the 16-QAM format, there are two different detection methods to demodulate and recover the data in the receiver (i) direct detection and (ii) coherent detection.

In this section, direct detection means detection with MZDI or (2×4) $90°$ hybrid and coherent is similar except that a local oscillator, a very narrow linewidth laser, is used to mix the signal and its lightwaves to generate the IF or base band signals with preservation of the modulated phase states. Each of these two receiving methods have different implementations that can be introduced as follows.

5.4.5.1   Direct Detection

Contrary to the coherent detection, the differential decoding is done for direct detection in the optical domain. This is equivalent to a self-homodyne coherent detection. This has the disadvantage that the transmitted absolute phase is lost after differential decoding. However, the relative phase (the phase of differential decoded signal) remains in electrical domain and it makes the electrical equalization still possible. The equalization with relative phases is more difficult and the results are worse than equalization with absolute phases. The advantage of direct detection, compared to coherent detection, is that the synchronization of local laser with that of the signal light wave is omitted. There are two methods to implement the direct detection. One is with MZDI and the other is with a (2×4) $90°$ hybrid coupler.

5.4.5.2   Coherent Detection

In a coherent receiver, a local oscillator (LO) is used to mix its signal with the incoming signal lightwave for demodulation. As a result, it makes possible to preserve the phase in electrical domain. This makes the electrical equalizing very effective in the coherent detectors. For coherent detectors, the differential decoding is done in electrical domain. Dependent on the intermediate frequency (f_{IF}) defined as $f_{IF} = f_S - f_{LO}$, three different coherent methods can be distinguished (i) homodyne receiver, (ii) heterodyne receiver, and (iii) Intradyne receiver. Only homodyne receiver is included in this section and the other two are only briefly mentioned.

5.4.5.2.1 Homodyne Receiver

A receiver is called homodyne when the carrier frequency (f_s) and the local oscillator frequency (f_{LO}) are the same. It means

$$f_{IF} = f_s - f_{LO} = 0 \qquad (5.19)$$

In practice, because of the laser linewidth, carrier synchronization must be implemented to set the center frequency and the phase of LO to the same values as those in the incoming signal. For homodyne receivers, carrier synchronization can be implemented in the optical domain via an optical phase locked loop (OPLL). Carrier synchronization failure causes degradation in the receiver's performance, but in this document, this effect is not considered and a perfect synchronization in the receiver (a perfect single spectrum line) is assumed. Alternatively, as mentioned later, a heterodyne receiver using only one 90° hybrid coupler with associated electronic demodulation circuitry can be used to simplify the receiver configuration for coherent detection. Polarization control is another critical difficulty in all coherent receivers, which is also not included in this chapter. The implementation of homodyne receiver for the star 16-QAM is described in several text books.

5.4.5.2.2 Heterodyne Receiver

For this kind of receiver, it applies that:

$$f_{IF} = f_s - f_{LO} \neq 0 > B_{opt} \qquad (5.20)$$

B_{opt} is the optical bandwidth of the transmitted signal. The IF will be mixed in electrical domain with a synchronous or asynchronous method in the low pass domain. In the case of the synchronous demodulation, the phase synchronization can be done in the electrical domain. The implementation complexity of heterodyne receivers in optical domain is less than that of homodyne receivers.

5.4.5.2.3 Intradyne Receiver

The intradyne receiver requires

$$f_{IF} = f_s - f_{LO} \neq 0 < B_{opt} \qquad (5.21)$$

The phase synchronization in the intradyne receiver can be done in the digital domain. That makes the intradyne receiver less complex in optical domain than the homodyne receiver. The intradyne receiver, compared to heterodyne receiver, has the advantage of the processing bandwidth being smaller. The disadvantage of the intradyne receiver is the higher requirement on laser linewidth than that of heterodyne receiver.

5.4.6 STAR 16-QAM FORMAT

In this section, the transmission performance of star 16-QAM are evaluated by simulations and compared. The implementation of the transmitter for star 16-QAM is introduced in the first section. Both direct and coherent homodyne receivers for star 16-QAM are described in the second section. In the case of the homodyne receiver (without laser phase noise and ideal frequency locking), two different receiver models, one without phase estimation and the other one with phase estimation, are realized. The simulation results of all three receivers are shown in the third section and the comparison of them is in the fourth section.

5.4.6.1 Transmitter Design

There are many possibilities for implementing the transmitter for star 16-QAM described in the previous section. For the simulations in this work, the parallel transmitter shown in Figure 5.45 is implemented. The bit stream enters the differential encoder module after serial-to-parallel converting.

The differential encoder provides the following processes: (i) the four bits that have parallel arrived at the module are mapped (Grey coding) into symbols according to the mapping Table 5.4.2

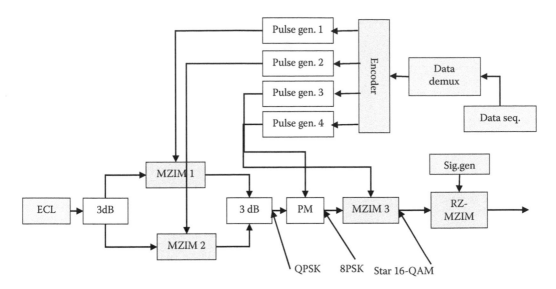

FIGURE 5.45 Schematic diagram of the optical transmitter for Start 16-QAM. Legend: Sig gen. = signal generator pulse gen = pulse generator; ECL= external cavity laser; MZIM = opt. MZ intensity modulators; PM = opt. phase modulator.

in Table 5.4; (ii) the pre-coded symbols are differentially encoded (differential coding); and (iii) the differentially encoded symbols are mapped again to other symbols to drive the MZMs according to the mapping Table 5.4.3 of Table 5.4.

Each symbol at the output of differential encoder module is represented by four bits. The bits are sent to pulse formers. The first two bits drive the first two Mach–Zehnder modulators (MZM), with lightwaves generated from the Continues Wave (CW) laser. If the input bit is equal to "1," then the output of MZM is "−1," and the output of MZM is a "1" (after sampling). After combining the output signals of these two MZMs and considering the 90° phase delay in one arm, we obtain the QPSK signal shown in Figure 5.45.

The third bit from the differential encoder output drives a Phase Modulator (PM) to obtain the 8-PSK signal constellation from the QPSK signal. If this bit is equal to "1," then the QPSK symbol will rotate by pi/4. The 8-PSK signal constellation is shown in Figure 5.46. To achieve the two-level star

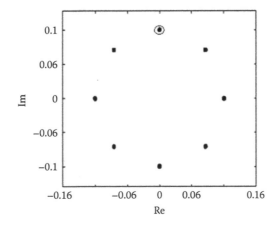

FIGURE 5.46 Constellation of the first amplitude level generated from the optical transmitter for Star 16-QAM.

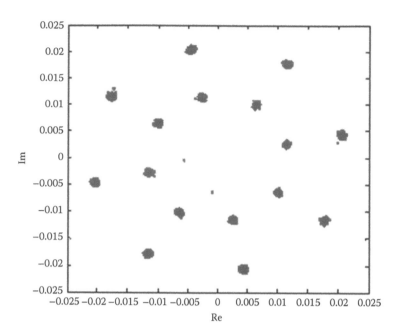

FIGURE 5.47 Constellation of the first and second amplitude level at the receiver of Start 16-QAM after 10-spans of dispersion compensated standard SMF links.

16-QAM signal constellation, another MZM is used to generate the second amplitude. If the fourth bit of differential encoder output is "1," then this output symbol is set on the outer ring of the constellation, otherwise on the inner ring. This MZM sets the RR of constellation. The signal constellation after MZM3 is shown in Figure 5.47.

The signal constellation in Figures 5.46 and 5.47 can be constructed from that the whole constellation in Figure 5.43 with a rotation of pi/6. The advantage of this rotation is, that on the real and imaginary axis of the constellation, only eight different amplitude levels exist instead of nine. Another PM can be used between MZM3 and MZM-RZ to rotate the constellation by pi/6°. This additional PM is not shown in Figure 5.45. To increase the receiver sensitivity and reduce the signal chirp, a return to zero (RZ) pulse curving with a duty cycle of 50% should be implemented at the end of the transmitter with a MZM driven by a sinus signal generator (SG). In our simulations, the MZMs in Figure 5.45 work in push-pull operation and the PMs are MZMs working as phase modulators. The received constellation of this Star QAM after transmission through 10 spans of 100km and optically amplified is shown in Figure 5.47, using coherent reception and DSP to recover the original constellation.

5.4.6.2 Receiver for 16-Star QAM

In this section, the implementation of direct detection receiver and coherent receiver for star 16-QAM are explained. For coherent detection, there are many possibilities in the digital domain of receiver to recover the data. The two methods implemented in this study detect the symbol before realizing the differential decoding. The difference is one detects the symbol directly using the method described in Section 4.2.1, while the other employs a phase estimation algorithm, described in Section 4.2.2, before the symbol detection in order to cancel the phase distortions (phase synchronization between LO and the received signal). Another possibility, which is not implemented in this work, is doing at first the differential decoding of the incoming signal and then symbol detection.

5.4.6.3 Coherent Detection Receiver without Phase Estimation

The structure of the coherent detection receiver is shown in Figure 5.48a. After transmission over fiber, the signal is amplified by an EDFA. The input power of EDFA can be changed via an attenuator to set the OSNR. The output signal of EDFA is sent to a band pass (BP) filter to reduce the noise bandwidth.

An attenuator used to set the OSNR is required. The output signal of EDFA is sent to a band pass (BP) filter to reduce the noise bandwidth. The signal from LO and the output of th BP filter are sent to a (2×4) $\pi/2$-hybrid and after that, to two balanced detectors. The (2×4) $90°$ hybrid and the balanced detectors demodulate and separate the received signal into inphase (I) and quadrature (Q) components. The structure of the (2×4) $\pi/2$-hybrid coupler, the balanced detectors and their mathematical description can be found in may published works for coherent optical communication technology. This coherent detector can be simplified further if heterodyne detection is used as shown in Figure 5.48b and commercially available by Discovery Semiconductor [5]. However, the electrical signals should be demodulated into I and Q components as shown above, rather than the balanced receiver of Discovery Semiconductor.

Furthermore, an amplitude direct detection of in the electronic digital processor must be able to process the magnitude of the vector formed by I and Q components to decide the amplitude and phase of the received signals, and hence their corresponding position on the constellation and hence the decoding of the data symbols.

Following the low-pass (LP) filtering and sampling of I and Q components, the samples are sent to the symbol detection and differential decoding module. The sampling is done in the center of the eye-diagram. In the symbol detection and differential decoding module, we first recover the symbols from the incoming samples, then make the differential decoding of symbols. To recover the symbols, the I and Q components are added to a complex signal. Now, according to the original signal constellation, a decision must be made about which symbol in our complex sample must be mapped. This decision has two parts. First, an amplitude decision is made to determine to which ring the sample belongs.

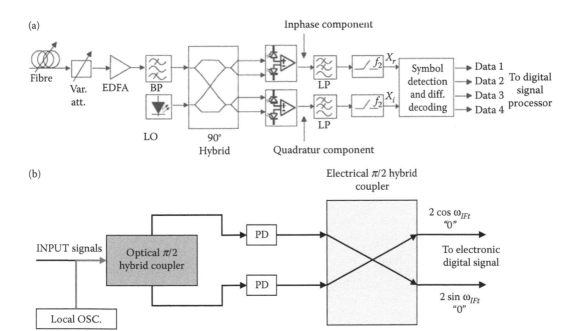

FIGURE 5.48 Coherent receiver for star 16-QAM (a) homodyne and heterodyne I and Q detection model and (b) heterodyne model with optical and electrical pi/2 hybrid couplers-electrical detection of I and Q components.

After this, a phase decision takes place to determine to which of eight possible symbols on a ring our sample belongs.

For amplitude decisions, a known bit sequence for receiver (training sequence) is used and an amplitude threshold a_{th} is defined according to

$$a_{th} = \{ \max 1 \leq k \leq n|s_1(k)| + \min 1 \leq k \leq m|s_2(k)|\}/2 \qquad (5.22)$$

where $s_1(k)$ is the kth complex received samples of symbols on the inner ring and n is the whole number of symbols on the inner ring. $s_2(k)$ is the kth complex received sample of symbols on the outer ring and m is the whole number of symbols on the outer ring. If the amplitude of one sample is larger than the threshold, the symbol is decided to be on the upper ring, otherwise on the inner ring. For the phase decision, the complex plane is divided into eight equi-phase intervals. According to which interval the phase of sample falls in, an index from zero to seven is assigned to this sample. The next steps are differential decoding and mapping. The amplitude differential decoding and phase differential decoding are done separately. From their results, the symbol detection and after it symbol to bit mapping are done in the inverse to that of the encoding in the transmitter.

Alternatively, the detection can be conducted with a heterodyne receiver that use only one single $\pi/2$ hybrid optical coupler, detected by the two photodiodes, and then coupled through a $\pi/2$ electrical hybrid coupler to detect the I and Q components for the phase and amplitude reconstruction of the received signals, as shown in Figure 5.48b. The phase estimation can then be estimated by processing I and Q signals in the electronic domain, as described in the next section.

Because of the two levels of the Star 16-QAM, there must be an amplitude detection sub-system that can be implemented using a single photo-detector, followed by an electronic pre-amp, as shown in Figure 5.49. An electronic processor would be able to determine the position of the received signals on the constellation, and hence, decoding can be implemented without any problem. The transmission performance presented in this article would not be affected. Only the technological implementation would be affected, and hence, the electronic noise or optical noises contribution to the receiver can be considered.

5.4.6.4 Coherent Detection Receiver with Phase Estimation

The method and structure of this receiver is almost the same as for the previous receiver shown in Figure 5.48. The difference here is that a phase estimation is done before the phase decision in the symbol detection and differential decoding module. The dispersion of the single mode optical fiber is purely phase effects and thus causes phase rotation and thus results as phase decision error. The effect of dispersion on star 16-QAM format is shown in Figure 5.50.

The left plot (a) in the figure shows the sampled input signal into the fiber and the plot on the right side shows the sampled output of the fiber with a chromatic dispersion of 300 ps/nm. The

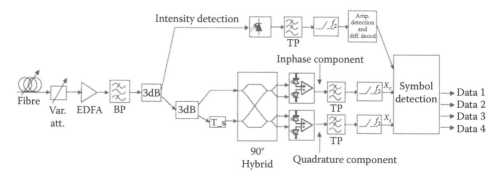

FIGURE 5.49   Direct detection receiver for star 16-QAM.

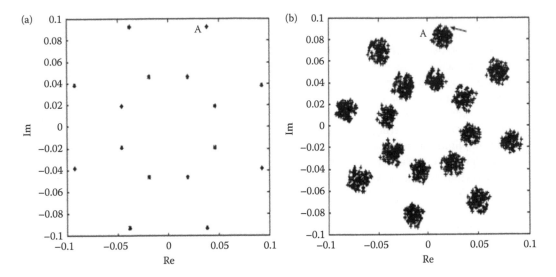

FIGURE 5.50 Signal constellation before (a) and after (b), propagation through the optical fiber link-phase rotation.

fiber is considered as linear in this simulation. A comparison of point A in both figures shows that this point is spread and rotated due to dispersion. The spreading causes phase and amplitude distortion, while rotation causes only phase distortion. To solve the problem of phase rotation, the following phase estimation method is implemented. Generally, this phase estimation method is for phase synchronization between LO with linewidth and signal to replace the optical phase lock loop.

The incoming signal after sampling can be described as

$$c(k) = Ae^{j(\Phi' tot(k) \pm mod(k))} \tag{5.23}$$

where Φ'_{tot} is the phase distortion due to the dispersion and noise and ϕmod is the signal phase that must be recovered. Now ϕ_{tot} must be eliminated from $c(k)$. Φ_{mod} values are $\pi/8$, $3\pi/8$, $5\pi/8$, $\pi/8$, $9\pi/8$, $11\pi/8$, $13\pi/8$, and $15\pi/8$. If this phase values are multiplied by 8 then:

$$c_8(k) = c^8(k) = A^8 e^{j(8' tot(k)+8\_mod(k))} = A^8 e^{j(8' tot(k)+-)} = -A^8 e^{j(8' tot(k))} \tag{5.24}$$

and from this we obtain:

$$\Phi'_{tot}(k) = 1/8 arg(-c_8(k)) \tag{5.25}$$

$\Phi'_{tot}(k)$ is the estimated phase for $\phi'_{tot}(k)$. In the simulations in this work $arg(c_8(k))$ is filtered the filter makes average between 20 neighbor symbols) to avoid the phase jumps from symbol to symbol. The filter order of 20 is not optimized for each CD.

Now the signal phase $\phi_{mod}(k)$ can be estimated as

$$\phi_{mod}(k) = arg(c(k)) - \phi'_{tot}(k) = \Phi_{mod}(k) + \phi'_{tot}(k) - \phi'_{tot}(k). \tag{5.26}$$

After this phase estimation, the signal decision takes place with the same method as for amplitude decision case.

5.4.6.5 Direct Detection Receiver

The block diagram of the direct detection receiver is shown in Figure 5.49. After the optical filter, the signal is split into two branches via a 3 dB coupler. We name these two branches the intensity branch and the phase branch. In the phase branch, the phase differential demodulation is done in the optical domain. The signal and the delayed signal at Ts (symbol duration) are sent into the $(2 \times 4)\,90°$ hybrid and after that, again into balanced detectors. At the output of the balanced detectors, the inphase and quadrature component of the demodulated and differential decoded and received signal can be derived. After electrical filtering, the signal is sampled and then sent into the symbol detection module. In the amplitude branch, the amplitude is determined and differentially decoded. After the photo diode, the signal is low-pass filtered and sampled and then fed into the amplitude detection and differential decoding module, a well-known optical coherent structure can be used to accomplish the amplitude decision and differential decoding. At the end, the in phase and quadrature component and the amplitude branch are send to the symbol detection module for further processes as symbol detection and symbol to bit mapping.

5.4.6.6 Coherent Receiver without Phase Estimation

In this section, the required OSNR at 10^{-4} BER (using Monte Carlo simulation) is determined for the coherent and incoherent direct detection receivers. For each detection method, the optimum RR is obtained to minimize the OSNR at BER $= 10^{-4}$ (the BER is determined via Monte-Carlo simulations). After that, the dispersion tolerance (DT) at 2 dB OSNR penalty at BER $= 10^{-4}$ is determined. OSNR penalty is only in this work 2 dB and can have other values. The OSNR penalty is defined as the OSNR difference in dB between the OSNR of back-to-back case and the OSNR of other CD values. DT is the CD interval that can be achieved with a certain OSNR penalty. In practice, DT describes how much dispersion (residual dispersion) a system can tolerate with a OSNR penalty smaller than 2 dB. The simulations in this work are done via the simulation tool. The simulations are done for the linear channel and for the non-linear channel. The average input power of non-linear fiber in this section is set at 5 dBm.

5.4.6.6.1 Linear Channel

In Figure 5.51b, the optimum RR (RR_{opt}) can be seen for each CD. RR_{opt} is the RR, which minimize the OSNR for the given CD. The optimum RR changes here nearly linear with CD and can be expressed as

$$RR_{opt} = -0.002|CD| + 1.92 \quad for \ 50\,ps/nm \leq |CD| \leq 300\,ps/nm \qquad (5.27)$$

RR_{opt} increases with CD because increasing of CD means increasing of phase rotation due to dispersion. This causes more phase detection errors and thus a higher OSNR requirement. Increasing of RR reduces the phase error probability but increases the amplitude error probability. RR_{opt} is the best tradeoff between phase errors and amplitude errors for each CD.

In the back-to-back case, RR_{opt} is around 1.87. The theoretical value obtained for RR_{opt} is 1.77. The difference occurs because in the introduced coherent receiver, the symbol detection is done first, followed by the differential decoding. In the case of differential decoding before symbol detection, the optimum RR is around 1.77, as expected. To determine the RR_{opt} for each CD, RR is changed for each CD. The RR value yields the smallest OSNR is RR_{opt}. The RR step (determines the RR accuracy) in simulations is 0.05. The characteristic for other CDs is similar to that in Figure 5.51a. To compare the OSNRs, the reference in this work is the OSNR from the back-to-back case. In case of residual dispersion is of interest, the system performance changes with the change of RR. The simulation results of three different RR can be seen in Figure 5.57. As shown in Figure 5.57a, the OSNR performance for the back-to-back case is decreased if (in simulated interval) the RR from 1.87 decreases. A degradation of 6.5 dB is determined if the RR from 1.87 decreases to 1.32. From the other side, the Dispersion Tolerance (DT) at 2 dB OSNR penalty can be increased. The DT for

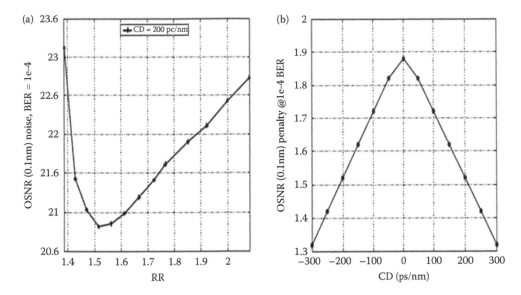

FIGURE 5.51 OSNR vs. RR (left)(a) and optimum RR vs. CD (right)(b) for coherent receiver without phase estimation.

RR = 1.87 is 220 ps/nm and for RR = 1.32 is 460 ps/nm. To understand the reason for this behavior, Figure 5.51a should be considered again. For each CD, if the RR decreases from the RR_{opt} value, the OSNR increases rapidly. That means RR = 1.32 for the back-to-back case has increased the OSNR, but CD = 300 ps/nm has the minimum OSNR for this RR. The result is that the OSNR for the back-to-back case increases and without phase estimation in linear channel, CD = 300 ps/nm decreases. This effect causes a larger DT at a certain OSNR penalty. In practical systems, according to higher requirement for OSNR or DT, the RR can be chosen.

On the left in Figure 5.52a, it can be seen that the required OSNR for |CD| = 300 ps/nm makes a jump compared to other CDs. With |CD| = 350ps/nm, it is not possible to reach a BER of 10^{-4}. The reason is that |CD| = 350 ps/nm is the limit of the system. For this CD, it is not possible to transmit error-free, even without noise due to phase rotations caused by dispersion (phase detection error). The signal constellation of received signal after sampling with CD = 350 ps/nm can be seen in Figure 5.53, for example, some of received symbols of A are over the phase threshold line and they generate the detection errors. A typical eye diagram at the output of the coherent receiver at the limit of the distortion is shown in Figure 5.54. Phase estimation can be implemented in the digital signal processor. It is noted that the signal constellation is rotated uniformly because of the property of the single mode fiber that is purely phase distortion. In the processing of the constellation it is best if the reference frame of the phasor diagram is rotated to align with the constellation, thus, it may simplify the phase estimation process at $\pi/8$ and its multiple values.

5.4.6.6.2 Non-linear Effects

The optimum *RR* in case of a non-linear channel for different CDs is shown in Figure 5.55. As mentioned for the linear channel, here the RR_{opt} also decreases with increase of CD. The difference from Figure 5.51 is that the diagram is not symmetric. The reason for this is the interaction between dispersion and SPM. For a positive CD, the dispersion and SPM have a constructive interaction, which results in a better OSNR performance. For a negative CD, the dispersion and SPM have destructive interaction and it results in a much worse OSNR performance. This effect can be appreciated when contrasting Figure 5.51b with Figure 5.55. The curve slope for the negative CD is higher in a non-linear channel. It means the phase distortion is higher in the non-linear channel

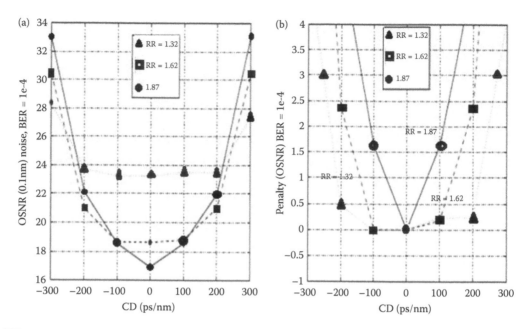

FIGURE 5.52   OSNR vs. CD (a) and OSNR penalty vs. CD (b) for coherent receiver.

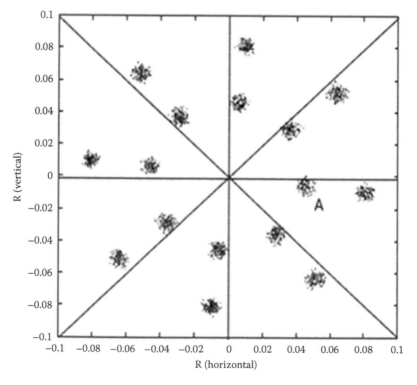

FIGURE 5.53   Received signal constellation with CD = 350 ps/nm without noise for coherent receiver without phase estimation in linear channel.

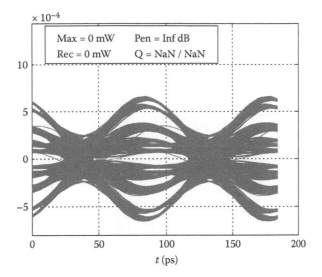

FIGURE 5.54 Typical Eye diagram at the output of the coherent receiver under significant distortion limit of the 2A-8P Start 16-QAM.

(Figure 5.56). For the positive CD, the slope in non-linear channel is lower and it means the phase distortion is less. The simulation results for different *RR* are shown in Figure 5.57. It can be seen again that the required OSNR for the back-to-back case increases with decrease of *RR*. Similarly, with the same reason as for the linear case, the dispersion tolerance at 2 dB OSNR penalty increases as well (Figure 5.58).

5.4.6.7 Remarks

The design of a Start 16-QAM modulation scheme is proposed for coherent detection for ultra-high-speed ultra-high-capacity optical fiber communications schemes. Two amplitude levels and eight phases (2A-8P 16-QAM) are considered to offer simple transmitter and receiver configurations, and, at the same time, the best receiver sensitivity at the receiver. An optical SNR of about 22 dB

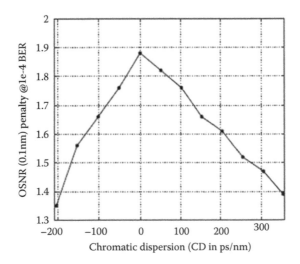

FIGURE 5.55 RR_{opt} vs. CD in non-linear channel for coherent receiver without phase estimation.

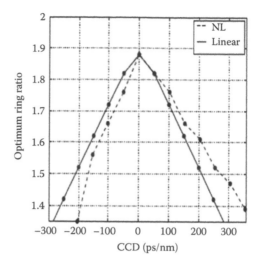

FIGURE 5.56 *RR*$_{\text{opt}}$ comparison in linear and non-linear channel for coherent receiver without phase estimation.

is required for the transmission of Star 16-QAM over an optically amplified transmission dispersion–compensated link. Dispersion tolerance of 300 ps/nm is possible with 1.0 dB penalty of the eye-opening at 40 Gb/s bit rate or 10 Giga-symbols/second with an OSNR of 18 dB. An OSNR could be about 22 dB for 107 Gb/s bit rate and a symbol rate of 26.75 G symbols/second. The transmission link consists of several spans of total 1000 kilometer dispersion compensated optically amplified

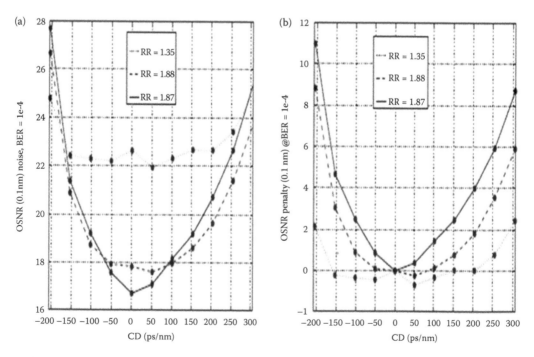

FIGURE 5.57 OSNR vs. CD (a) and OSNR penalty vs. CD (b) for coherent receiver without phase estimation in non-linear channel. Note: No equalization only phase estimation processing of *I* and *Q* receives components in electrical domain.

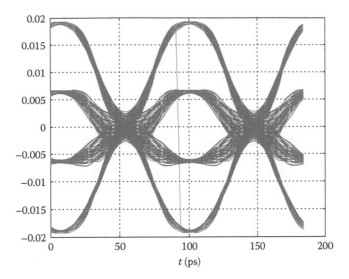

FIGURE 5.58 Eye diagram of the 16QAM detected at the output of a balanced receiver. Bit rate of 40 Gb/s and baud rate of 10 Gb/s.

transmission link. Optical gain of the in-line optical amplifiers is set to compensate for the attenuation of the transmission and compensating fibers with a noise figure (NF) of 3 dB.

The optical transmitter and receivers incorporating commercially available coherent receivers are structured and sufficient for engineering of the optical transmission terminal equipment for the bit rate of 107 Gb/s and a symbol rate of 26.3 GSy/s (Sy=Sympol or Baud=Bd). Furthermore, electronic equalization of the receiver phase shift keying signals can be done using blind equalization that would improve the dispersion tolerance much further. For a symbol rate of 10.7 Gb/s, this dispersion tolerance for 1 dB penalty would reach 300 km of standard SMF. This electronic equalization can be implemented without any difficulty at the 10.7 GSy/s. For 107 Gb/s bit-rate, similar improvement of the dispersion tolerance can be at 26.5 GSy/s, provided that the electronic sampler can offer more than 50 Gig-Samples/s sampling rate.

5.5 OTHER MULTI-LEVEL AND MULTI-SUB-CARRIER MODULATION FORMATS FOR 100 GB/S TRANSMISSION

Numerous technologies have been introduced in recent years to cope with the ever-growing demand for transmission capacity in optical communications. Although the optical single-mode fiber offers enormous bandwidth in the order of magnitude of 10 THz, efficient exploitation of bandwidth started to become an issue a couple of years ago. Moreover, limited speed of electronics and electro-optic devices such as modulators and photo receivers are considered as bottleneck for further increase of data rate based on binary modulation.

For all these reasons, optical modulation formats offering a high ratio of bits per symbol are an essential technology for next generation's high-speed data transmission. This way, data throughput can be increased while required bandwidth in the optical domain as well as for electronics is kept on a lower level.

Based on the demand for transmission technologies offering high ratio of bits per symbol, two promising candidates for achieving a data rate of 100 Gb/s per optical carrier are discussed, namely, optical orthogonal frequency division multiplexing and 16-ary multi-level modulation.

In this section, these are introduced. Performance is analyzed by means of numerical simulation and by experiment.

5.5.1 MULTI-LEVEL MODULATION

Optical modulation formats incorporating four or eight bits per symbol were investigated in numerous contributions in the last couple of years (e.g., DQPSK [13] and 8-DPSK [13]). However, to carry out the step from 10 Gb/s to 40 Gb/s data rate using devices designed for 10 Gsymb/s, a number of 16 bits are required per symbol. The main challenge is to find the optimal combination of ASK and DPSK formats. Several approaches, which are reviewed in References 2 and 5, are possible:

The simplest structure can be an extension of a 30 Gb/s 8-DPSK by an additional phase modulator resulting in 40 Gb/s 16-DPSK. That is, in the complex plane, 16 symbols are placed onto a unit circle as shown in Figure 5.59a. Depending on the current bit at the data input of the additional phase modulator, the 8-DPSK symbol is shifted by $\pi/8$ in case of a "1," while in case of a "0," the incoming phase of the symbol is preserved. Although it seems simple, experimental implementations have shown that the phase stability of the modulators and corresponding demodulators is very critical. Thus, experimental setup must be stabilized. Moreover, 16-DPSK is sub-optimum regarding exploitation of the full area that the complex plane offers. That is the ratio of symbol distance and signal power is low resulting in poor receiver sensitivity.

Similar behavior can be found for the other extreme case of 16-ASK, as shown in Figure 5.59b. Here, the 16 symbols are placed onto the positive real axis. The symbol distance is extremely narrow, thus resulting in poor sensitivity as well.

Much improved performance can be achieved by combining the amplitude and phase shift keying modulation of ASK and DPSK, the M-ary ADPSK as described above. There are a number of combinations of the M-ary ADPSK. One approach is the extension of an 8-DPSK by two rings of ASK levels. Thus, the 16 symbols appear as two rings in the complex plane with eight symbols per ring, as shown in Figure 5.59c. Alternatively, this topology is known as star-16 QAM. A second structure is given by combining the DQPSK with four-level ASK (or, equivalently, M-ary ADPSK) resulting in four rings with four symbols each, as analyzed in Section 4 and shown in Figure 5.59d. Both structures utilize effectively the complex plane. However, both structures require that the sensitivity or the magnitude (diameters) of the rings must be optimized to compromise the sensitivity performance of

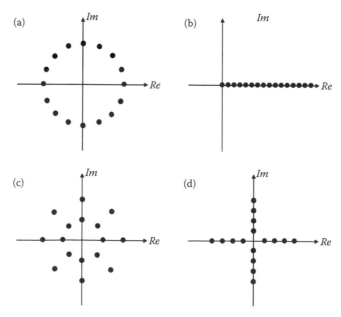

FIGURE 5.59 Constellation of symbols in the complex plane for (a) 16-DPSK; (b) 16-ASK; (c) Star 16-QAM; and (d) 16ADPSK.

the DPSK and the ASK geometrical distribution. The inner ring must be of sufficient size to enable distinction of the different phases of the symbols on this ring. They are limited by the non-linear effects of the transmission fiber and the noises contributed by the receiver and the inline optical amplifiers. Hence, the distance between the constellations are limited by these two limits.

A strategy to mitigate this tradeoff was introduced in [13] by using a special pulse shaping called inverted RZ. For binary ASK in conjunction with inverse RZ, a "0" is encoded as temporary breakdown of the optical power, while for a "1," the optical power remains at high level. Using this pulse shaping, for the M-ary ADPSK format the four levels of the QASK part are transmitted by means of four different values for temporary decay of optical power. The DQPSK part, however, is transmitted by modulating the phase of the signal in the time slot between, that implies that in the transmitter the phase of the signal can be detected while the signal has maximum power.

Measurement results for this modulation format are depicted in Figure 5.60, where the BER is plotted as a function of the OSNR measured within a 0.1 nm optical filter bandwidth. The main outcome is the fact that the DQPSK-part is insignificantly disturbed by an additional QASK part. Moreover, even after transmission over 75 km of standard single-mode fiber, the DPSK part shows a very low penalty. The QASK component inherently shows low performance due to low symbol Euclidean distance. Improvement might be achievable by optimizing the duty cycle of inverse RZ and the ring ratio. A simulated eye diagram of the 16-square QAM is shown with the constellation before and after the optical transmission link of two optically spans with 2 km SSMF mismatch are shown in Figure 5.40.

5.5.2 Optical Orthogonal Frequency Division Multiplexing (oOFDM)

Orthogonal frequency division multiplexing (OFDM) is a transmission technology that is primarily known from wireless communications and wired transmission over copper cables. It is a special case of the widely known frequency division multiplexing (FDM) technique for which digital or analog data is modulated onto a certain number of carriers and transmitted in parallel over the same transmission medium. The main motivation for using FDM is the fact that due to parallel data transmission in frequency domain, each channel occupies only a small frequency band. Signal distortions originating from frequency-selective transmission channels, the fiber chromatic dispersion, can be minimized. The special property of OFDM is characterized by its very high spectral efficiency. While

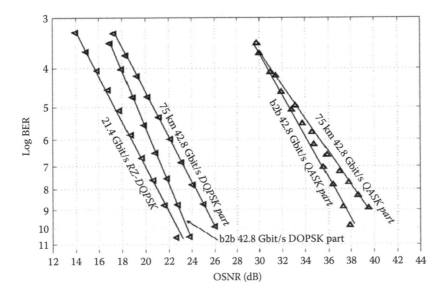

FIGURE 5.60   Measurement BER results for 16-ary Inverse RZ ADPSK modulation format transmission.

for conventional FDM, the spectral efficiency is limited by the selectivity of the bandpass filters required for demodulation, OFDM is designed such that the different carriers are pair wise orthogonal. This way, for the sampling point the inter-carrier interference (ICI) is suppressed although the channels can overlap spectrally.

Orthogonality is achieved by placing the different RF-carriers onto a fixed frequency grid and assuming rectangular pulse shaping. In this special case, the OFDM signal can be described as the output of a discrete inverse Fourier transform with the parallel complex data symbols as input. This property has been one of the main driving aspects for OFDM in the past since modulation and demodulation of a high number of carriers can be realized by simple digital signal processing (DSP) instead of using many local oscillators in transmitter and receiver. Recently, OFDM has become an attractive topic for digital optical communications. It is just another example of the current tendency in optical communications to consider technologies that are originally known from classical digital communications. Using OFDM appears to be very attractive since the low bandwidth occupied by a single OFDM channel increases the robustness toward fiber dispersion drastically allowing the transmission of high data rates of 40 Gb/s and more over hundreds of kilometers without the need for dispersion compensation. In the same way as for modulation formats like DPSK or DQPSK that were introduced in recent years, the OFDM challenge for optical system engineers is to adapt a classical technology to the special properties of the optical channel and the requirements of optical transmitters and receivers.

Thus, two approaches have been recently reported. An intuitive approach introduced by Lorente et al. [6] makes use of the fact that the wavelength-division multiplexing (WDM) technique itself already realizes data transmission over a certain number of different carriers. By means of special pulse shaping and carrier wavelength selection, the orthogonality between the different wavelength channels can be achieved resulting in the so-called orthogonal WDM technique (OWDM). However, this way, the option of simple modulation and demodulation by means of discrete Fourier transforms (DFT) cannot be utilized as this kind of digital signal processing is not available in the optical domain.

As an alternative and popular method, generation of an electrical OFDM signal by means of electrical signal processing followed by modulation onto a single optical carrier. This approach is known as oOFDM. Here, the modulation is a two-step process: First, the electrical OFDM signal already is a broadband bandpass signal, which is then modulated onto the optical carrier. Second, to increase data throughput, oOFDM can be combined with WDM, resulting in multi-Tb/s transmission system as shown in Figure 5.61. Nevertheless, oOFDM itself offers different options for implementation. An important issue is optical demodulation that can be realized by means of direct detection (DD) or coherent detection (CD) using a local oscillator. The DD is preferable because of its simplicity. However, for DD, the optical intensity must be modulated. Because the electrical OFDM signal is quasi-analog with zero mean and high peak-to-average ratio, most the optical power must be wasted for the optical carrier (i.e., an additional DC-value of the complex baseband signal) resulting in low receiver sensitivity. For CD, in addition, the bandwidth efficiency is twice as high as for DD since pure intensity modulation inherently a double-sideband signal is generated. For CD, a complex optical I-Q modulator composed of two real modulators in parallel followed by superposition with $\pi/2$ phase shift allows for transmission of twice as much data within the same bandwidth. For intensity modulation, the bandwidth efficiency may be increased by suppressing one of the redundant sidebands resulting in oOFDM with single-sideband (SSB) transmission. First, the serial data can be converted to parallel streams. These parallel data sequences are then mapped to QAM constellation in the frequency domain, then by IFFT converted it back to the time domain. The time domain signals are in the I- and Q-components which are then fed into I and Q optical modulator. This optical modulation can be DQPSK or any other multi-level modulation sub-system. At the end of the optical fiber transmission, I and Q components are detected by direct detection or coherent detection. For coherent detection, a 2×4 90-degree hybrid coupler is used to mix the polarizations of local oscillator and that of the received signals. The outputs of the couplers are then fed into the balanced optical receivers.

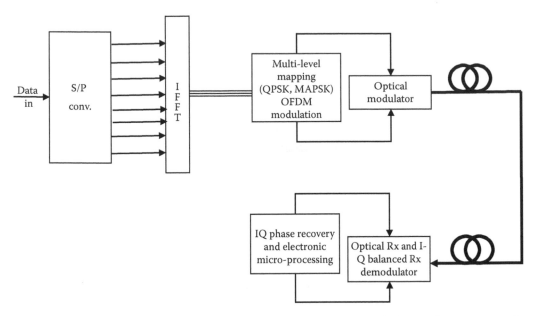

FIGURE 5.61 Schematic diagram of oOFDM transmitter of a long haul optical transmission system using multi-level modulation formats.

The mixing of the local laser source and that of the signals preserves the phase of the signals which are then processed by a high speed electronic processor. For direct detection, the I and Q components are detected differentially, the amplitude and phase detection are then compared and processed similarly as for the coherent case.

To show the robustness of oOFDM towards fiber dispersion and fiber non-linearity, numerical simulations are carried out for a data stream of 42.7 Gb/s data rate. The number of OFDM channels can be varied between $N_{min} = 256$ and $N_{max} = 2048$. A guard interval of 12 ns can be inserted, a strategy belonging inherently to OFDM technology that ensures orthogonality of the different channels in case of a transmission channel with memory. For the optical modulation, intensity modulation using a single Mach–Zehnder modulator in conjunction with SSB filtering and direct detection was implemented. The non-linear optical transmission channel consisted of eight 80 km non-DCF spans of SSMF. As a criterion for performance, required OSNR for a BER of 10^{-3} (Monte Carlo) is measured. Using FEC, after decoding this is transferred into a BER below 10^{-9} depending on the specific code.

Figure 5.62 shows the required OSNR as function of fiber launch power for different values of N. The most important result is that transmission is possible over 640 km over SSMF without any dispersion compensation. It can be explained by the fact that even for the lowest value of $N_{min} = 256$, each sub-channel occupies a bandwidth of approximately 42.7 GHz/256 = 177 MHz resulting in high robustness towards fiber dispersion.

The principal difficulties of oOFDM are that the pure delay due to the variation of the refractive index of the fiber with respect to the optical frequency lead to bunching of the sub-channels, and hence the increase of the optical power, and thus unexpected SPM may occur in a random manner.

5.5.3 100 Gb/s 8-DPSK_2-ASK 16-Star QAM

5.5.3.1 Introduction

A multi-level modulation scheme enables the transmission baud rate to be reduced and the spectral efficiency to be obtained. Another significant advantage of this modulation scheme is to reduce the

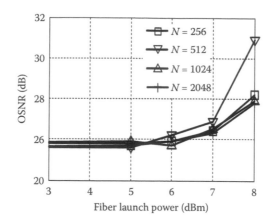

FIGURE 5.62 Simulation result for 42.7 Gb/s oOFDM transmission over 640 km of standard single-mode fiber; OSNR required for BER $= 10^{-3}$ (Monte Carlo simulation) as function of fiber launch power. (From C. Wree, J. Leibrich, W. Rosenkranz, RZ-DQPSK format with high spectral efficiency and high robustness towards fiber non-linearities, *Proc. ECOC 2002*, Copenhagen, Denmark, ECOC 2002 September 2002, paper 9.6.6. extracted with permission.)

requirement of high-speed processing electronics. This is of particular interest for high-speed optical transmission systems.

This section investigates a multi-level modulation scheme that has eight phase and two amplitude levels. This scheme, which is named 8-DPSK_2-ASK, effectively utilizes four bits per one symbol for transmission, in which the first three bits are for coding phase information while the coding of the amplitude levels is implemented with the fourth bit. As a result, the transmission baud rate is equivalently a quarter of the bit rate from bit pattern generator.

This section is organized as follows: Section 5.2 presents detailed description of the optical transmitter for generating 8-DPSK_2-ASK signals. In Section 5.3, the detailed configuration of the receiver is provided. The configuration of the optical transmitter and receiver is referred from those reported by Djordjevic and Vasic [7]. Section 5.4 provides study on dispersion tolerance and transmission performance of the 8-DPSK_2-ASK scheme. Finally, a short summary of the report is provided.

5.5.3.2 Configuration of 8-DPSK_2-ASK Optical Transmitter

There have been several different configurations of an optical transmitter for generating multi-phase/level optical signals with the use of amplitude or phase modulators arranged in either serial or parallel configurations. However, the optical transmitters reported in References 8–10 require a pre-coder with high complexity. On the contrary, the configuration reported by Ivan et al. [9] utilizes the Grey mapping technique to differentially encode the phase information and this significantly reduce the complexity of the optical transmitter. In addition, as elaborated in more detail in Section 5.3, this pre-coding technique enables the detection scheme using the *I-Q* demodulation techniques, as equivalently in coherent transmission systems.

The optical transmitter of the 8-DPSK_2-ASK scheme employs the *I-Q* modulation technique with two Mach–Zehnder intensity modulators (MZIMs) in parallel, and a $\pi/2$ optical phase modulator, as shown in Figure 5.63. At each kth instance, the absolute phase of transmitted lightwaves $\underline{\theta_k}$ is expressed as: $\theta_k = \theta_{k-1} + \Delta\theta_k$ where θ_{k-1} is the phase at $(k-1)$th instance and $\Delta\theta_k$ is the differentially coded phase information. The encoding of this $\Delta\theta_k$ for generating 8-DPSK_2-ASK modulated optical signals (4 bits per one transmitted symbol) follows the well-known Grey mapping rules (Table 5.5). This grey mapping phasor diagram is shown in Figure 5.64. The phasor is normalized with the maximum energy on each branch, that is, $E_1/2$.

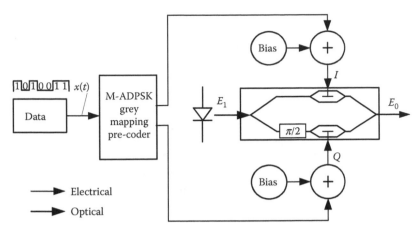

FIGURE 5.63   Optical transmitter configuration of the 8-DPSK_2-ASK modulation scheme.

The amplitude levels are optimized in order that the Euclidean distances d_1, d_2, and d_3 are equal, i.e $d_1 = d_2 = d_3$. After derivation, we obtain: $r_1 = 0.5664$. The I and Q field vectors corresponding to Grey mapping rules from the M-ADPSK pre-coder (see Figure 5.63) are provided in Table 5.6.

The above-described transmitter configuration can be replaced with a dual-drive MZIM. The explanation and derivation for generating 8-DPSK_2-ASK optical signals are also based on the phasor diagram of Figure 5.64. In this case, the output field vector is the summation of two component field vectors, each of which is not only determined by the amplitude, but also by initially biased phases.

5.5.3.3   Configuration of 8-DPSK_2-ASK Detection Scheme

The detection of 8-DPSK_2-ASK optical signals is implemented with the use of two Mach–Zehnder delay interferometric (MZDI) balanced receivers (see Figure 5.65). Several key notes in this detection structure can be stated as (i) The MZDI introduces a delay corresponding to the baud rate. (ii) One arm of MZDI has a $\pi/4$ optical phase shifter while the other arm has an optical phase shift of $-\pi/4$. (iii) The outputs from two balanced receivers are superimposed positively and negatively, which leads to I and Q detected signals respectively. The I and Q detected components are expressed as $I = \mathrm{Re}\left\{E_k E_{k-1}^*\right\}$ and $Q = \mathrm{Im}\left\{E_k E_{k-1}^*\right\}$. (iv) I-Q detected components are demodulated using the

TABLE 5.5
Comparison of DQPSK Transmitter Structures

Parameters for Comparison	Parallel MZIM	Serial MZIM&PM	Single PM	Dual-Drive MZIM
Complexity of circuit design	Complicated in matching of ultra-high frequency electrical paths; high insertion loss Flexible in biasing	Complicated in matching of ultra-high frequency electrical paths; high insertion loss Flexible in biasing	Simple in photonic but complicated in realization of ultra-high frequency signal connections	Simplest but required multi-level voltage switching at symbol rate (microwave speed)
Ability to create MADPSK signal	Not possible. A separate ASK modulator required	Not possible. A separate ASK modulator required	Impossible. A separate ASK modulator required	Dual-drive MZIM acts as ASK and DPSK simultaneously

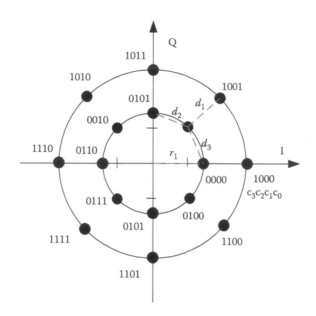

FIGURE 5.64 Grey mapping for optimal 8-DPSK_2-ASK modulation scheme.

popular *I-Q* demodulator in the electrical domain. These detected signals are then sampled and represented as shown in the signal constellation.

5.5.3.4 Transmission Performance of 100 Gb/s 8-DPSK_2-ASK Scheme

Performance characteristics of the 8-DPSK_2-ASK scheme operating at 100 Gb/s bit rate are studied in terms of receiver sensitivity, dispersion tolerance, and the feasibility for long-haul transmission.

TABLE 5.6

I and *Q* field Vectors in 8-DPSK_2-ASK Modulation Scheme Using Two MZIMs in Parallel

Binary Sequence	$(\Delta\theta_k,$ Amplitude)	I_k	Q_k
1000	$(0, 1)$	1	0
1001	$(\pi/4, 1)$	$\sqrt{2}/2$	$\sqrt{2}/2$
1011	$(\pi/2, 1)$	0	1
1010	$(3\pi/4, 1)$	$-\sqrt{2}/2$	$\sqrt{2}/2$
1110	$(\pi, 1)$	-1	0
1111	$(-3\pi/4, 1)$	$-\sqrt{2}/2$	$-\sqrt{2}/2$
1101	$(-\pi/2, 1)$	0	-1
1100	$(-\pi/4, 1)$	$\sqrt{2}/2$	$-\sqrt{2}/2$
0000	$(0, 0.5664)$	$1 * 0.5664$	0
0001	$(\pi/4, 0.5664)$	$\sqrt{2}/2 * 0.5664$	$\sqrt{2}/2 * 0.5664$
0011	$(\pi/2, 0.5664)$	0	$1 * 0.5664$
0010	$(3\pi/4, 0.5664)$	$-\sqrt{2}/2 * 0.5664$	$\sqrt{2}/2 * 0.5664$
0110	$(\pi, 0.5664)$	$-1 * 0.5664$	0
0111	$(-3\pi/4, 0.5664)$	$-\sqrt{2}/2 * 0.5664$	$-\sqrt{2}/2 * 0.5664$
0101	$(-\pi/2, 0.5664)$	0	$-1 * 0.5664$
0100	$(-\pi/4, 0.5664)$	$\sqrt{2}/2 * 0.5664$	$-\sqrt{2}/2 * 0.5664$

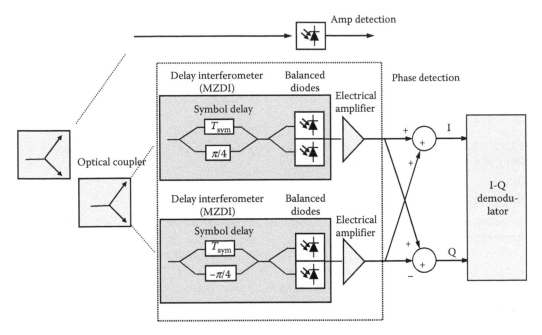

FIGURE 5.65 Detection configuration for the 8-DPSK_2-ASK modulation scheme.

Bit error rates (BER) are the pre-Forward Error Correct (FEC) BERs and the pre-FEC limit is conventionally referenced at 2e-3. In addition, the BERs are evaluated by the Monte Carlo method.

5.5.3.5 Power Spectrum

The power spectrum of 8-DPSK_2-ASK optical signals is shown in Figure 5.66. It can be observed that the main lobe spectral width is about 25 GHz as the symbol baud rate of this modulation scheme is equal to a quarter of the bit rate from the bit patter generator. The harmonics are not highly suppressed thus, requiring bandwidth of the optical filter to be necessarily large in order not to severely distort signals.

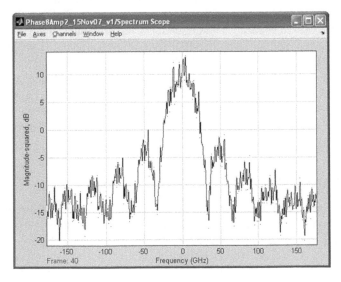

FIGURE 5.66 Power spectrum of 8-DPSK_2-ASK signals.

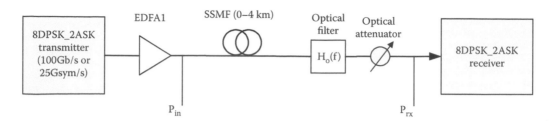

FIGURE 5.67 Set up for the study of receiver sensitivity (back to back) and dispersion tolerance (0–4 km SSMF) for the 8-DPSK_2-ASK modulation scheme.

5.5.3.6 Receiver Sensitivity and Dispersion Tolerance

The receiver sensitivity is studied by connecting the optical transmitter of the 8-DPSK_2-ASK scheme directly to the receiver to make a back-to-back setup (see Figure 5.67). On the other hand, the dispersion tolerance is studied by varying the length of SSMF from 0 km to 5 km ($|D| = 17$ ps/(nm.km)). Received powers are varied by using an optical attenuator. The optical Gaussian filter has BT = 3 (B is approximately 75 GHz). Modeling of receiver noise sources comprises of shot noise, equivalent noise current density of $20\,pA/\sqrt{Hz}$ at the input of the trans-impedance electrical amplifier and dark current of 10nA for each of the two photodiodes in balanced structure. A fifth-order Bessel electrical filter with a bandwidth of BT = 0.8 is used.

The numerical BER curves of the receiver sensitivity for cases of 0–5 kilometer SSMF are shown in Figure 5.68. The receiver sensitivity of the 8-DPSK_2-ASK scheme is approximately −18.5 dBm at BER = 1e-4. The receiver sensitivity at BER = 1e-9 can be obtained by extrapolating the BER curve of the 0 kilometer case. The power penalty versus residual dispersion results are then obtained and plotted in Figure 5.69. It is realized that the 2-dB penalty occurs for the residual dispersion of approximately 60 ps/ nm or equivalently to 3.5 km SSMF.

5.5.3.7 Long-Haul Transmission

The long-haul transmission performance of this modulation format is conducted over 10 optically-amplified and fully compensated spans and each span is composed of 100 km SSMF and 10 km DCF100 (Sumitomo fiber). As a result, the length of the transmission fiber link is 1100 km. This long-haul range is selected to reflect the distance between Melbourne and Sydney. The wavelength of interest is 1550 nm and the dispersion at the end of the transmission link is fully compensation. The simulation set up is shown in Figure 5.69. Additionally, the fiber attenuation because of SSMF and DCF is also fully compensated by using two EDFAs with optical gains as depicted in Figure 5.70. These EDFAs have Noise Figure (NF) set at 5 dB.

Numerical transmission BERs are plotted against received powers in Figure 5.71 and compared to the back-to-back BER curve. It can be observed that the BER curve of 1100 km follows a linear trend and feasibly reaches 1e-9 if extrapolated as shown in Figure 5.71. It should be noted that this transmission performance can be significantly improved with the use of high-performance forward error coding (FEC) scheme.

5.6 REMARKS AND FURTHER DISCUSSIONS

The ever-increasing bandwidth hungry in telecommunication networks based mainly on optical fiber communication technology indicate that low bandwidth efficient modulation formats such as ASK would no longer satisfy the transmission capacity demands, and new advanced optical modulation schemes should replace ASK soon. Advanced optical modulation schemes, especially the multi-level amplitude and phase schemes presented above, are able: (i) to provide long reach, error-free, and high transmission capacity, (ii) to have high bandwidth efficiency (no. bits/Hz parameter), (iii) to push the

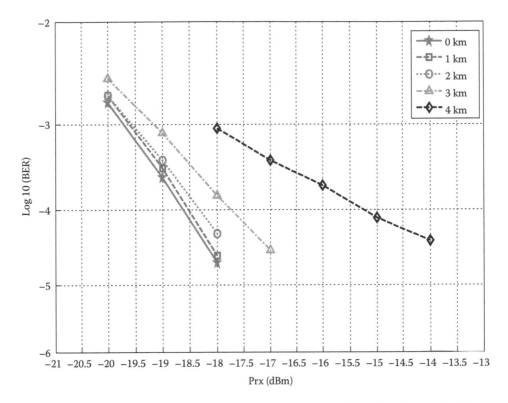

FIGURE 5.68 Plot of BER versus receiver sensitivity (back to back) and dispersion tolerance (0–4 km SSMF) for the 8-DPSK_2-ASK modulation scheme.

bit rate well above that could be offered by electronic technology, for example, 100 Gb/s with the detection at the symbol rate. (iv) to tolerate dispersion and non-linearity, and (v) to maximally utilize the existing optical network infrastructure.

Current developments of photonic technology have enabled the use of differential phase modulation and demodulation in optical domain. Presently, BDPSK and DQPSK formats have received a great attention due to their improvement in the receiver sensitivity as compared with ASK. Furthermore, the RZ and CSRZ formats would assist the combat with non-linear effects. However, as long as transmission capacity and bandwidth efficiency are concerned, MADPSK modulation formats would be a better performance in trading off for its complexity in the receiver structures.

Alternatively, there are other possible multi-level modulation schemes that would offer further improvement of optical transmission performance [11].

5.6.1 Offset MADPSK Modulation

The binary DPSK and quaternary DPSK (DQPSK) is just a special case of a more general class of differential phase modulation formats that maps data bits into phase difference between neighboring symbols. This phase difference $\Delta\phi_i$ can be described as

$$\Delta\phi_i = \theta + \frac{2\pi(i-1)}{M}, \quad i = 1, 2, \ldots M - 1 \tag{5.28}$$

where θ is the initial phase, M is total number of phase levels. This class of formats is called offset DPSK (O-DPSK) and denoted as θ-M-DPSK. Specifically, with $\theta = 0$, $M = 2$ or $M = 4$

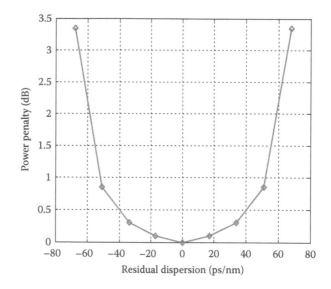

FIGURE 5.69   Power penalty due to residual dispersions for the 8-DPSK_2-ASK modulation scheme.

θ-M-DPSK become 0-2DPSK or 0-4-DPSK ,which are the conventional binary DPSK and DQPSK mentioned above.

ODPSK has been used to transmit over satellite non-linear channel because its phase transition is smooth and can avoid a 180° phase jump. As fiber medium also exhibits non-linearity, this modulation format attracts our attention as a candidate, together with ASK, for creating a new multi-level modulation format, possibly termed as offset MADPSK.

5.6.2   MULTI-LEVEL AMPLITUDE-MINIMUM SHIFT KEYING (MAMSK) MODULATION

Minimum shift keying (MSK) is a form of OQPSK with sinusoidal pulse weighting. In MSK, data bits are first coded into bipolar signals ± 1, which are then separated into $V_I(t)$ and $V_Q(t)$ streams consisting of even and odd bits, respectively. In the next stage, $V_I(t)$ and $V_Q(t)$ are used to modulate a carrier f_c to create a MSK signal $s(t)$ which can be presented as

$$s(t) = V_I(t)\cos\left(\frac{\pi t}{2T}\right)\cos\left(2\pi f_c t\right) + V_Q(t)\cos\left(\frac{\pi t}{2T}\right)\sin\left(2\pi f_c t\right) \qquad (5.29)$$

MSK is a well-known modulation format in wireless communications with efficient spectral characteristics due to high compactness of its main lobe as compared with DPSK, and high suppression of the side lobes. These characteristics indicate that MSK signal is highly dispersion tolerant. A

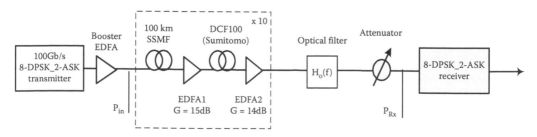

FIGURE 5.70   Transmission set up of 1100 km optically amplified and fully compensated fiber link.

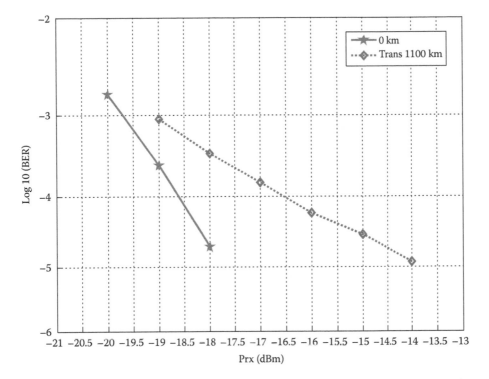

FIGURE 5.71 Plot of BER versus receiver sensitivity for 8DPSK 2ASK modulation format transmission.

combination of MSK and ASK into multi-level amplitude-minimum shift keying (M-AMSK) modulation would even improve the transmission performance without increasing the complexity of the detection scheme.

Multi-level techniques for 100 Gb/s are given, particularly the OFDM with multi-carrier and multi-level amplitude modulation with orthogonality between adjacent channels are proven to be cost effective and appropriate for current electronic technologies. Two interesting approaches to achieve data transmission of 40 Gb/s and beyond (e.g., 100 Gb/s Ethernet) based on low symbol rate are discussed. On one hand, oOFDM can combine a large number of parallel data streams into one broadband data stream with high spectral efficiency. Simulated results are shown for different values of the number of parallel data streams in a non-linear environment. On the other hand, 16-ary modulation formats enable 40 Gb/s transmission with 10 GSym/s (i.e., 100 Gb/s with 25 GSym/s). For a special case, namely inverse RZ 16ADQPSK, measurement results are demonstrated.

5.6.3 STAR QAM COHERENT DETECTION

These last two sections of this chapter describe two optical transmission schemes using coherent and incoherent transmission and detection techniques, the 2A-8P Star QAM and 8DPSK 2ASK 16-Star QAM for 100 Gb/s or 25 G Symbols/sec.

First, the design of a Start 16-QAM modulation scheme is proposed for coherent detection for ultra-high-speed ultra-high-capacity optical fiber communications schemes. Two amplitude levels and eight phases (2A-8P 16_QAM) are considered to offer significant simple transmitter and receiver configurations and at the same time the best receiver sensitivity at the receiver. An optical SNR of about 22 db is required for the transmission of Star 16-QAM over optically amplified transmission dispersion–compensated link. Dispersion tolerance of 300 ps/nm is possible with 1 dB penalty of the eye-opening at 40 Gb/s bit rate or 10 Giga-symbols/s with an OSNR of 18 dB. AN OSNR could be about 22 dB for 107 Gb/s bit rate and a symbol rate of 26.75 G symbols/sec. The transmission link consists of several spans of total 1000 km dispersion compensated optically amplified transmission link. Optical gain of the in-line optical amplifiers are set to compensate for the attenuation of the transmission and compensating fibers with a noise figure of 3 dB. The optical transmitter and receivers incorporating commercially available coherent receiver are structured and sufficient for engineering of the optical transmission terminal equipment for the bit rate of 107 Gb/s and a symbol rate of 26.3 Gbauds/s.

Furthermore, electronic equalization of the receiver phase shift keying signals can be done using blind equalization that would improve the dispersion tolerance much further. For a symbol rate of 10.7 Gb/s, this dispersion tolerance for 1 dB penalty would reach 300 km of standard SMF. This electronic equalization can be implemented without any difficulty at the 10.7 Gbauds/s. For 107 Gb/s bit rate similar improvement of the dispersion tolerance can be at 26.5 Gbauds/s provided that the electronic sampler can offer more than 50 Gig-sampling rate.

Second, the transmitter and receiver configurations for generating 8-DPSK_2-ASK optical signals and direct detection are described as well as the transmission performances. In addition, performance characteristics of this modulation format at 100 Gb/s (equivalently 25 GBd) has also been investigated in terms of the receiver sensitivity, dispersion tolerance the long-haul transmission performance. The simulation results show that 8-DPSK_2-ASK is a promising modulation for very high-speed (100 Gb/s) and long-haul optical communications.

REFERENCES

1. L.N. Binh, Monash Optical Communication System Simulator. Part I: Ultra-long Untra-High Speed Optical Fiber Communication Systems, Manual for Technical Development, Monash University, Melbourne, Australia, 2004.
2. S. Hayase et al., Proposal of 8-state per symbol (binary ASK and QPSK) 30-Gb/ optical modulation demodulation scheme, *Proc. European Conf. on Optical Communication*, paper Th2.6.4, pp. 1008–1009, ECOC 2003, Rimini, Italy, September 2003.
3. K. Sekine et al., 40 Gb/s 16-ary (4 bit/symbol) optical modulation/demodulation scheme, *IEEE Electron. Lett.*, vol. 41, no. 7, March 2005.
4. L.N. Binh, *Advanced Digital Optical Communications*, 2nd ed., Boca Raton, FL., USA: CRC Press, 2015.
5. C. Wree et al., *Coherent Receivers for Phase-Shift Keying Transmission*, Discovery Semiconductors Inc.
6. R. Llorente, J.H. Lee, R. Clavero, M. Ibsen, and J. Martí, Orthogonal wavelength-division-multiplexing technique feasibility evaluation, *J. Lightw. Technol.*, 23, pp. 1145–1151, March 2005.
7. I.B. Djordjevic and B. Vasic, Multilevel coding in M-ary DPSK/differential QAM high-speed optical transmission with direct detection, *IEEE Journal of Lightwave Technology*, vol. 24, pp. 420–428, 2006.
8. M. Serbay, C. Wree, and W. Rosenkranz, Experimental investigation of RZ-8DPSK at 3x 10.7Gb/s, in LEOS'05, 2005.
9. M. Seimetz, M. Noelle, and E. Patzak, Optical systems with high-order DPSK and star QAM modulation based on interferometric direct detection, *IEEE Journal of Lightwave Technology*, vol. 25, p. 1515–1530, 2007.
10. H. Yoon, D. Lee, and N. Park, Performance comparison of optical 8-ary differential phase-shift keying systems with different electrical decision schemes, *Optic Express*, vol. 13, pp. 371–376, 2005.
11. S. Walklin and J. Conradi, Multilevel signaling for increasing the reach of 10Gb/s lightwave systems, *IEEE Journal of Lightwave Technology*, vol. 17, no. 11, pp. 2235–2248, November 1999.
12. C. Wree, J. Leibrich, W. Rosenkranz, RZ-DQPSK format with high spectral efficiency and high robustness towards fiber nonlinearities, *Proc. ECOC 2002*, Copenhagen, Denmark, ECOC 2002 September 2002, paper 9.6.6.
13. L.N. Binh, *Digital Optical Communications*, Boca Raton, FL., USA: CRC Press, 2008.

6 Incoherent and Coherent Optical Receivers

Detection of optical signals can be carried out at the optical receiver by direct conversion of optical signal power to electronic current in the photodiode and then electronic amplification. This chapter gives the fundamental understandings of coherent detection of optical signals that require the mixing of the optical fields of the optical signals with that of the local oscillator, a high-power laser so that its beating product would result in the modulated signals preserving its phase and amplitude characteristics in the electronic domain. Optical pre-amplification in coherent detection can also be integrated at the front end of the optical receiver.

6.1 INTRODUCTION

With the exponential increase in data traffic, especially due to the demand for ultra-broad bandwidth driven by multi-media applications, cost-effective ultra-high-speed optical networks have become highly desired. It is expected that Ethernet technology will not only dominate in access networks, but will also become the key transport technology of next-generation metro/core networks. Currently, 100 Gigabit Ethernet (100 GbE) is considered to be the next logical evolutionary step after 10 Gigabit Ethernet (10 GbE). Based on the anticipated 100 GbE requirements, 100-Gbit/s data rate of serial data transmission per wavelength is required. To achieve this data rate while complying with current system design specifications such as channel spacing, chromatic dispersion, and polarization mode dispersion (PMD) tolerance, coherent optical communication systems with multi-level modulation formats will be desired because they can provide high spectral efficiency, high receiver sensitivity, and potentially high tolerance to fiber dispersion effects [1]. Compared to conventional direct detection in intensity-modulation/direct-detection (IMDD) systems which only detects the intensity of the light of the signal, coherent detection can retrieve the phase information of the light, and therefore, can tremendously improve the receiver sensitivity.

Coherent optical receivers are important components in long-haul optical fiber communication systems and networks to improve the receiver sensitivity and add extra transmission distance. Coherent techniques were considered for optical transmission systems in the 1980s when the extension of repeater distance between spans was pushed to 60 km instead of 40 km for single mode optical fiber at bit rate of 140 Gb/s. However, in the late 1980s, the invention of optical fiber amplifiers overcame this attempt. Recently, interests in coherent optical communications have attracted significant research activities for ultra-bit rate DWDM optical systems and networks. The motivation has been possible because of (i) the fact that uses of optical amplifiers in cascade fiber spans have added significant noises and helped limit the transmission distance, (ii) the advances of digital signal processors whose sampling rate can reach few tens of Giga-samples/second, allowing the processing of beating signals to recover the phase or phase estimation, (iii) the availability of advanced signal processing algorithms such as Viterbi and Turbo algorithms, and (iv) the differential coding and modulation and detection of such signals may not require optical phase lock loop, hence self-coherent and digital signal processing to recover transmitted signals.

A well-known, typical arrangement of an optical receiver is one in which the optical signals are detected by a photodiode (a pin diode or APD or a photon counting device). The electrons generated in the photodetector are then electronically amplified through a front-end electronic amplifier. The electronic signals are decoded for recovery of their original format. However, when the fields of incoming optical signals are mixed with those of a local oscillator whose frequency can be identical or different to that of the carrier, the phase, and frequency property of the resultant signals reflect those of the original signals. Coherent optical communication systems have also been reviving dramatically due to electronic processing and availability of stable narrow linewidth lasers.

This chapter deals with the analysis and design of incoherent and coherent receivers with direct detection and coherent detection, including optional optical phase locking then mixing of optical signals and that of the local oscillator in the optical domain and thence detected by the opto-electronic receivers following this mixing. Thus, the optical mixing and photodetection devices act as the fundamental elements of a coherent optical receiver. Depending on the frequency difference between the lightwave carrier of the optical signals and that of the local oscillator, the coherent detection can be termed as heterodyne or homodyne detection. For heterodyne detection, there is a difference in the frequency and thus the beating signal is fallen in a passband region in the electronic domain. Thus, all the electronic processing at the front end must be in this passband region. In homodyne detection there is no frequency difference and thus the detection is in the base band of the electronic signal. Both cases would require a locking of the local oscillator and carrier of the signals. An optical phase lock loop is thus treated in this chapter.

This chapter is organized as follows: Section 6.2 describes the detection processes in incoherent reception systems and related noise generation processes in such sub-system. Section 6.3 gives an account of the components of coherent receivers, including the outlined principles of optical coherent detection under heterodyne, homodyne, or intradyne techniques. Section 6.4 gives a brief account of self-coherent reception, that is, the sub-system in which the channel signals are split into two paths and the delayed by one bit period and then combined with its original signals and beating with pother in the photodetector. Section 6.5 gives an account on electronic responses and noise processes. Section 6.6 gives an account of digital-signal-processing-based optical receivers for various coherent detection sub-systems. Although these DSP-based detections can also be applied to direction detection, the analysis and processing are similar so this is not dealt with in this book.

6.2 INCOHERENT OPTICAL RECEIVERS

6.2.1 INCOHERENT RECEIVER IN VARIOUS SYSTEMS

In modern optical and photonic communications systems, optical amplifiers can be incorporated in front of a front-end opto-electronic receiver to from an optical amplified pre-amplifier. This type of optical receiver will be discussed in Chapter 10, which deals with optical amplification.

The ultimate goals of the design of optical receiver is to determine the minimum optical energy in term of the number of photons per bit period required at the input of the photodetector so that it would satisfy a certain optical signal-to-noise ratio for an analog optical system or the sensitivity of a digital optical communications satisfying a certain bit-error-rate. In other words, the minimum optical power required at a certain bit rate in a digital communication system so that the decision circuitry can be detected with a specified bit error rate BER, for example, BER $= 10^{-9}$ or 10^{-12}.

Table 6.1 shows the important requirement for optical receivers in various communications systems from a terrestrial to local and wide area networks to undersea submarine systems.

The electronic amplifier is considered a block diagram rather than the detail circuit configuration. However, for completeness, a section is dedicated to the design of the electronic front end amplifier. This is essential that if a generic approach for the design of optical receivers, this is the ultimate objective of this chapter.

6.2.2 INCOHERENT RECEIVER COMPONENTS

The design of an optical receiver depends on the modulation format of the signals and thus transmitted through the transmitter. It is dependent on the modulation in analog or digital, Gaussian or exponential pulse shape, on–off keying or multiple levels, etc.

TABLE 6.1

Optical Receiver Characteristics and Their Relative Importance in Typical Optical-Fiber Communications Systems

	Features	Undersea Optical Communications	Terrestrial Communications and WAN	Point-to-Point (Access Networks)	LAN	Modulation and Reception
High receiver sensitivity	Maximum repeater spacing	Critical	Moderate critical	Moderate	Moderate	Critical
Wide dynamic range	Flexible and convenient systems configurations	Moderate	Moderate critical	Moderate critical	Critical	Critical
Bit rate transparency	Variable bit rate operation	Not required	Not required	Desirable	Desirable	Multiplication factor
Bit rate dependency	Flexible	Accommodated by the use of appropriate line codes and scrambling/descrambling		Desirable	Desirable	Moderate
Fast acquisition time	Short preamble bit sequence	Not required	Not required	Moderate critical	Moderate critical	Critical to get sampling accuracy

Optical signals discussed in this chapter are involved with an intensity modulated and direct detection (IM/DD) ON/OFF keying (OOK) systems. They are thus treated with an assumption that lightfaces arrived at the receiver are assumed to be polarization independent and their polarization plays no part in the degradation of the optical signals, except due to the polarization mode dispersion which would be independently treated in the optical fiber (Chapter 11). Coherent optical systems have been discussed in many other textbooks on optical communications where the polarization and the coherence of the optical source plays a very significant part in the design and implementation of the optical fiber communications systems. These coherent optical communications are of great interest to the research and development community in the early to mid-1980s, but no longer now, due to the invention of the Erbium-doped optical amplifiers (EDFA) in the late 1980s.

Figure 6.1 shows the schematic diagram of a digital optical receiver. An optical receiver front end consists of a photodetector for converting lightwave energy into electronic currents, an electronic pre-amplifier for further amplification of the generated electronic current, usually in voltage form, a main amplifier for further voltage amplification, a clock recovery circuitry for regenerating the timing sequence, and a voltage-level decision circuit for sampling the waveform for the final recovery of the transmitted and received digital sequence. Therefore, the opto-electronic pre-amplifier is followed by a main amplifier with an automatic control to regulate the electronic signal voltage to be filtered and then sampled by a decision circuit with synchronization by a clock recovery circuitry.

Inline fiber optical amplifier can be incorporated in front of the photodetector to form an optical receiver with an optical amplifier front end to improve its receiving sensitivity. This optical amplification at the front end of an optical receiver will be treated in this chapter dealing with optical amplification processes.

The structure of the receiver consists of three parts: the front-end section, the linear channel of the main amplifier and AGC or linear amplification stage, and the data recovery or digital signal processing (DSP) in association with an analog-to-digital converter (ADC) [2] section. These sections are

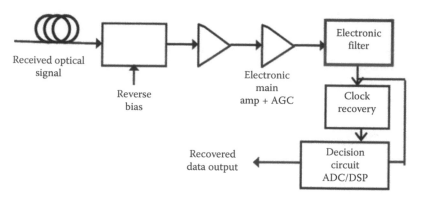

FIGURE 6.1 Schematic diagram of the digital optical receiver IM/DD.

described in order below (Figure 6.2). Currently, the integration of these electronic amplifiers in photonic integrated transceiver circuits to form compact module [3], as shown in Figure 6.3.

6.2.2.1 Photodiodes

A photodetector detects and converts the optical input power into an electric current output. The ideal photodetector would be highly quantum-efficient, and, ideally, it would add no noises to the received signals, respond uniformly to all signals with different wavelengths within the transparent windows of optical fibers around 1300 nm and 1550 nm, and, finally, it would not be saturated and behave linearly as a function of signal amplitude. There are several different types of photodetectors that are commercially available. Of the semiconductor-based types, the photodiode is used almost exclusively for high-speed fiber optic systems because of its compactness and extremely high bandwidth. The two most commonly used types of photodiodes are the PIN and avalanche photodiode (APD). Detailed reviews of these photodiodes have been presented in well-known published literature [4–6]. The fundamental characteristics of these two device types are briefly described for completeness (Figure 6.4).

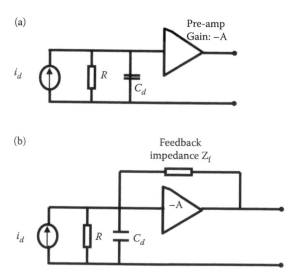

FIGURE 6.2 Schematic diagram of electronic pre-amplifier in an optical receiver (a) high impedance (HI) optical front end receiver and (b) transimpedance (TI) optical front end receiver. The current source represents the electronic current generated in the photodetector. C_d = photodiode capacitance.

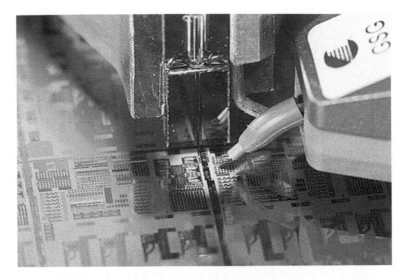

FIGURE 6.3 Integrated electro-photonic circuits as optical transceiver with integrated optic sections and micro-electronic section.

6.2.2.1.1   P-I-N Photodiode

A pin photodtector consists of three regions of semiconductors, the heavily doped p r, the intrinsic, and the n doped sections. It is essential that the *p-n* junction is reversed-biased to cause mobile electrons and holes moving away from the junction, thus increasing the width of the depletion layer and hence a high electric field is produced by the immobile charges at both ends. If lightwaves are evident on the PIN surface and absorbed in the "*high-field depleted*" region, the electron-hole pairs generated will move at saturation-limited velocity. There is no great contribution if photons fall in other regions. Thus, since the width of p region, W_p is inversely proportional to its doping concentration, decreasing the doping level to intrinsic behavior would widen, hence the depleted region is now wider and compatible with the absorption region to produce large photon-generated current. A p^+ region must be added to make good ohmic contact. Wider intrinsic region gives higher quantum efficiency, η but a lower response rate (Figure 6.5).

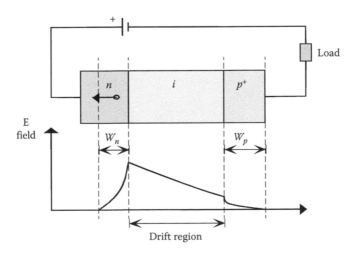

FIGURE 6.4   Schematic diagram of a *p-i-n* photodiode.

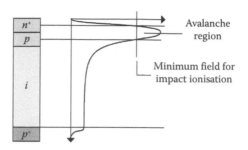

FIGURE 6.5 Schematic diagram of an avalanche photodiode (APD).

6.2.2.1.2 Avalanche Photodiodes (APD)

High field multiplication region is added adjacent to the lightly doped depletion region—intrinsic as in contrast with the PIN structure. Photons absorbed in the intrinsic region produce electron-hole pairs. These pairs drift to the high-field region whereby they gain sufficient energy (E field must be above the threshold for impact ionization to occur) to ionize bound electrons in the valence band upon colliding with them. Thus, the avalanche effect occurs only when the electric field existed in the high-field multiplication region is above the threshold for impact ionization. A guard-ring is usually added to prevent leakage surface current, I_{surf}.

6.2.2.1.3 Quantum Efficiency and Responsivity

Two most important characteristics of a photodetector are its quantum efficiency and its response speed. These parameters depend on the material band gap, the operating wavelength, the doping, and thickness of the p, i, and n regions of the device. The quantum efficiency, η is the number of electron-hole carrier pairs generated per incident photon of energy $h\nu$ and is given by

$$\eta = \frac{\text{number of electron pairs generated}}{\text{number of inciden photons}} = \frac{I_p/q}{P_0/h\upsilon} = \frac{I_p h\upsilon}{P_0 q} \tag{6.1}$$

where I_p is the photocurrent and P_0 is the incident optical power. The performance of a photodiode is often characterized by its responsivity $\Re$ which is related to the quantum efficiency by

$$\Re = \left(\frac{I_p}{P_0}\right)G = \left(\frac{\eta q}{h\upsilon}\right)G = \left(\frac{\eta q \lambda}{hc}\right)G \tag{6.2}$$

where G is the APD *average multiplication factor* and $G = 1$ for non-APD case. The optical power is then given by

$$P_0 = \frac{i_s}{G\Re} \tag{6.3}$$

where it indicates that for the same incident optical power, APD could produce higher photocurrent than PIN. Thus, based on Equations 6.2 and 6.3, we could model PIN and APD.

6.2.2.2 High-Speed Photodetectors

When the speed of the photodetectors is increased, for example, to be used in high-speed systems above 10 Gb/s, pin detectors are normally used and the effective receiving area is smaller in order to reduce the junction capacitance. This decreased area then forces the detector to have lower breakdown voltage.

6.3 INCOHERENT DETECTION AND NOISES

The front-end sub-system is the most important component because noise contributed by this part severely affects the overall performance of the system. The design of an amplifier front-end depends on the sensitivity required and the bandwidth of the system. For example, an optical front-end amplifier for an optical time domain reflectometer (OTDR) would require a very high sensitive front end, while its bandwidth is not critical. On the other hand, a front end optical amplifier for a high bit rate optical communication requires a wide bandwidth amplifier, as well as a reasonably high sensitivity and low noise contribution. As observed in Table 6.1, optical front-end amplifiers and demands on their characteristics with its limit are given.

In general, optical amplified front ends are required for two types of optical systems: the short link in local area or metropolitan area networks and the terrestrial link systems. The former requires high data accumulation rate and the latter requires a fast acquisition rate and reasonable sensitivity.

Thus, electronic pre-amplifiers can take one of two most typical structures: the high impedance amplifier (HIA) and the transimpedance amplifier (TIA) front end. The difference is that in TI configuration there is a shunt feedback from the output to the input. This TIA configuration offers a wide bandwidth and the HIA provides a high sensitivity due its noise, and is almost negligible compared with the TIA model. The TIA and HIA configurations are shown in Figure 6.2.

A photodetector (PD) is represented by a current source and a capacitor in parallel. The noise current (quantum shot noise) is not shown. A diode capacitance due the capacitance dropped across the depletion region of the pn junction is also included. In fact, this diode capacitance C_d is a critical component for the design of the PD.

The HIA front end would have very high impedance Z_i as the input impedance of the electronic amplifier. This high impedance would limit the bandwidth of the overall amplifier. Usually Z_i would be a few hundred of ohms. A TIA front end offers lower receiver sensitivity, but the bandwidth is increased by a factor of A, the amplifier linear gain.

6.3.1 LINEAR CHANNEL

The linear channel in optical receivers consists of a main electronic amplifier and a low-pass filter or an equalizer whose objectives are to reduce the noise or distortion in the channel without incurring further intersymbol interference (ISI). The pre-amplifier and main amplifier have a bandwidth normally wider than that of the system. Thus, in calculating the total noise contribution, the noise spectra density must be integrated over the electronic amplifier bandwidth.

6.3.2 DATA RECOVERY

Once the optical signals are detected and amplified, the data streams are sampled with respect to the time interval recovered by a clock recovery circuit, normally by using a surface acoustic wave filter if the bit rate is high, and then amplified and fed into a triggering circuit. In several high-speed optical receivers, its components are usually integrated in a chip or by using a hybrid circuit if the bit rate is higher than 28 Gbits/s [3].

6.3.3 NOISES IN PHOTODETCETORS

The most important noise source in the phtotodetcetor is the fundamental quantum or shot noise caused by the intrinsic fluctuations in the photo-excitation of electronic carriers. In any interval of T seconds, the probability of exactly N primary electrons are generated is given by the time-varying Poisson distribution

$$P[N, (t_0, t_0 + T)] = \frac{\Lambda^N e^{-\Lambda}}{N!} \tag{6.4}$$

where

$$\Lambda = \int_{t_0}^{t_0+T} \lambda(t)dt \quad \text{where } \lambda(t) = \frac{\eta}{\hbar \nu} p(t) + \lambda_d$$

with λ_d is the number of electrons generated under dark condition, or the dark current electrons of the PD, thus Λ is the average number of primary electrons produced during the interval.

For avalanche photodetector (APD), the internal avalanche multiplication factor is also a random process. The variation of the avalanche gain gives rise to excess noise. The probability of distribution of the avalanche gain depends upon the types of APD. In particular, it is a function of the ratio of hole ionization probability to that of the electron, the factor k. The extra noise is represented by an excess noise factor $F(g)$ given by

$$\langle g^2 \rangle = F(g)G^2 \quad \text{with } G = \langle g \rangle \tag{6.5}$$

The term $F(g)$ is usually determined by experimental works and can be approximated as

$$F(G) \simeq kG + \left(2 - \frac{1}{G}\right)(1 - k) \tag{6.6}$$

and usually $F(G)$ can be estimated by

$$F(G) \simeq G^x \tag{6.7}$$

For Ge APD $x \sim 1$ while for well-designed Si APD $x \sim 0.4$ or 0.5.

6.3.4   RECEIVER NOISES

Before proceeding to the systems calculations to determine the performance of optical receivers, it is necessary that the noises generated in the photodetector and the pre-amplifier front end are considered.

Readers can, in fact, consider investigating the system calculation and returning to the noise calculation if they are taking either the total equivalent noise spectral density at the input of the detector or its noise. This section describes all noise mechanisms related to the photodetection process including the electronic noise associated with the receiver.

Shot noises and thermal noises are the two most significant noises in optical detection systems. Shot noises are generated by either quantum process or electronic biasing. The noises are specified in noise spectral density that is square of noise current per unit frequency (in Hz). Thus, the noise spectral density is to be integrated over the total amplifier bandwidth to obtain the equivalent noise currents.

6.3.4.1   Shot Noises

Electrical shot noises are generated by the random generation of streams of electrons (current). In optical detection, shot noises are generated by (i) biasing currents in electronic devices and (ii) photo currents generated by the photodiode.

For a bias current I, the spectral current noise density S_I given by

$$S_I = \frac{d(i_I^2)}{df} = 2qI \quad \text{in A}^2/\text{Hz}$$ (6.8)

where q is the electronic charge. The current i_I represents the noise current due to the biasing current I.

6.3.4.1.1 Quantum Shot Noise

The average current $\langle i_s^2 \rangle$ generated by the photodetector by an optical signal with an average optical power P_{in} is given by

$$S_Q = \frac{d\langle i_s^2 \rangle}{df} = 2q\langle i_s^2 \rangle \quad \text{in A}^2/\text{Hz}$$ (6.9)

In the case that the APD is used, then the noise spectral density is given by

$$S_Q = \frac{d\langle i_s^2 \rangle}{df} = 2q\langle i_s^2 \rangle\langle G_n^2 \rangle \quad \text{in A}^2/\text{Hz}$$ (6.10)

With $\langle G_n^2 \rangle$ is the average avalanche gain of the detector. It is noted here again that the dark currents generated by the photodetector must be included to the total equivalent noise current at the input after it is evaluated. These currents are generated even in the absence of the optical signal. The dark currents can be eliminated by cooling the photodetector to at least below the temperature of liquid nitrogen (77°K).

6.3.4.1.2 Thermal Noise

At a certain temperature, the conductivity of a conductor varies randomly. This random movement of electrons generates a fluctuating current even in the absence of an applied voltage. The thermal noise of a resistor R is given by

$$S_R = \frac{d(i_R^2)}{df} = \frac{2k_BT}{R} \quad \text{in A}^2/\text{Hz}$$ (6.11)

where k_B is the Boltzmann's constant, T is the absolute temperature (in °K), and R is the resistance in ohms and i_R denotes the noise current due to resistor R.

6.3.4.2 Noise Calculations

In this section, the design of a receiver for use in a digital communication system and the methods for noise calculations are described. Binary and multi-level operations are also given.

A schematic representation of an optical pre-amplifier of the receiver and an electronic analog equalizer for the detection of optical digital modulated signals is shown in Figure 6.6. An equalizer is considered to extend the bandwidth of the receiver to the range of several GHz for ultra-high speed operations of these digital optical receivers. This is similar to earlier design considerations of optical receivers when multi-mode fibers were used in the first generation of optical fiber communications systems [7–9]. The noise sources and small amplification circuit model can be simplified as shown in Figure 6.7, in which all the noise sources of the electronic amplifier and the PD are presented and grouped into total noise current sources at the input and output of the amplifier. These sources and then transferred to a total equivalent current source as seen in the input of the amplifier. Thus, it is very straightforward to find out the signal to noise ration and contribution of noises from the electronic amplification process and the quantum shot noise process in the detection of optical signals.

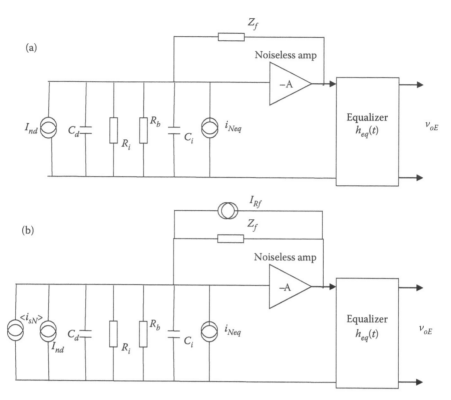

FIGURE 6.6   Optical receiver small-signal model and associate noise sources.

Our goal is to obtain an analytical expression of the noise spectral density equivalent to a source looking into the electronic amplifier including the quantum shot noises of the photodetector. A general method for deriving the equivalent noise current at the input is by representing the electronic device by a Y-equivalent linear network, as shown in Figure 6.7. The two current noise sources $d(i_N^2)$ and $d(i_N''^2)$ are representing the summation of all noise currents at the input and at the output of the Y-network. This can be transformed into a Y-circuit with the noise current at the input as follows:

The output voltages V_0 of Figure 6.7a can be written as

$$V_0 = \frac{i_N'(Y_f - Y_m) + i_N''(Y_i + Y_f)}{Y_f(Y_m + Y_i + Y_o + Y_L) + Y_i(Y_o + Y_L)} \qquad (6.12)$$

and for Figure 6.7b

$$V_0 = \frac{(i_N')_{eq}(Y_f - Y_m)}{Y_f(Y_m + Y_i + Y_o + Y_L) + Y_i(Y_o + Y_L)} \qquad (6.13)$$

thus, comparing these two equations we can deduce the equivalent noise current at the input of the detector is

$$i_{Neq} = i_N' + i_N'' \frac{Y_i + Y_f}{Y_f - Y_m} \qquad (6.14)$$

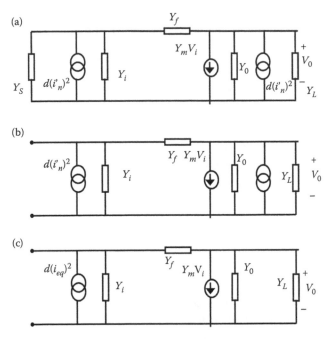

FIGURE 6.7 Small equivalent circuits including noise sources (a) Y-parameter model representing the ideal current model and all current noise sources at the input and output ports, (b) with noise sources at input and output ports, and (c) with a total equivalent noise source at the input.

Then reverting to mean square generators for a noise source we have

$$d(i_{Neq})^2 = d(i'_N)^2 + d(i''_N)^2 \left| \frac{Y_i + Y_f}{Y_f - Y_m} \right|^2 \tag{6.15}$$

It is therefore expected that if the Y-matrix of the front end low-noise amplifier is known, the equivalent noise at the input of the amplifier can be obtained using Equation 6.15.

6.4 INCOHERENT DECTRECTION PERFORMANCE CALCULATIONS FOR BINARY DIGITAL OPTICAL SYSTEMS

As described above, the noises of the optical receivers consist of the thermal noises and quantum shot noises due to the bias currents and the photocurrent generated by the photodetector with and without the optical signals. Thus, this quantum shot noise is strongly signal dependent.

These noises degrade the sensitivity of the receiver, and thus a penalty can be estimated. Another source of interference that would also result in signal penalty is the intersymbol interference (ISI). The goal of this section is to obtain an analytical expression for the receiver sensitivity of the optical receiver in an IM/DD pulse-code-modulated system.

6.4.1 SIGNALS RECEIVED

If the received signal power is

$$p(t) = \sum_j a_j h_p(t - jT_B) \tag{6.16}$$

T_B is the bit period. The average output voltage is thus given by

$$\langle v_o \rangle = \Re \langle G_n \rangle \left[\sum_j a_j \frac{1}{T_B} \int\limits_{-T_b/2}^{T_b/2} h_p(t - jT_B) dt \right] R_I A \qquad (6.17)$$

$h_p(t-jT_B)$ is the impulse response of the system evaluated at each time interval. R_I is the input resistance of the overall amplifier of the system including the front end and linear channel amplifier. It is assumed that the overall amplifier has a flat gain response A over the band width of the system.

$$a_j \Big|_{t=0} = \begin{cases} b_o \approx 0 \\ b_1 \end{cases} \qquad (6.18)$$

with b_o is the energy when a transmitted "0" is received and b_1 is the energy for a transmitted "1" is received. The summation over a number of periods is necessary to take into account the contribution of adjacent optical pulses.

We now must distinguish between two cases when a "0" or a "1" is transmitted and received.

Case (a): OFF or a transmitted "0" is received

Using Equations 6.17 and 6.18, we have

$$\langle v_o \rangle_0 = v_{oo} = \Re \frac{b_o}{T_B} G R_I A \cong 0 \qquad (6.19)$$

with the total equivalent noise voltage at the output, v_{NTo}^2 is

$$v_{NTo}^2 = v_{NA}^2 = i_{Neq}^2 R_I^2 A^2 \qquad (6.20)$$

Case (b): ON transmitted "1" received

In this case, the average signal voltage at the output is received as

$$\langle v_o \rangle_1 = v_{01} = \Re \frac{b_1}{T_b} \langle G_n \rangle R_I A \qquad (6.21)$$

with a total noise equivalent mean voltage at the output is given by

$$v_{NT1}^2 = v_{oSN}^2 + v_{NA}^2 \qquad (6.22)$$

where v_{oSN}^2 is the signal-dependent shot noise. v_{NA}^2 is the amplifier noise at the output and given by

$$v_{NA}^2 = i_{Neq}^2 R_I^2 A^2 B \qquad (6.23)$$

The signal-dependent noise is in fact the quantum shot noise and is given by

$$v_{oSN}^2 = \int\limits_0^B 2q \langle i_s \rangle_1 \langle G_n^2 \rangle R_I^2 A^2 df \qquad (6.24)$$

where B is the 3dB bandwidth of the overall amplifier, $\langle i_s \rangle_1$ is the average photocurrent current received when a "1" was transmitted. This current is estimated as follows:

$$\langle i_s \rangle_1 = \sum_{-\infty}^{\infty} \Re \frac{b_1}{T_B} \int_{-T_b/2}^{T_b/2} h_p(t - jT_B)dt \tag{6.25}$$

or

$$\langle i_s \rangle_1 = \Re \frac{b_1}{T_B} \int_{-\infty}^{\infty} h_p(t)dt \tag{6.26}$$

Using normalization with $\int_{-\infty}^{\aleph} h_p(t)dt = 1$, (6.26) becomes

$$\langle i_s \rangle_1 = \Re \frac{b_1}{T_B} \tag{6.27}$$

6.4.2 PROBABILITY DISTRIBUTION

The optical systems under consideration are typical IM/DD systems in which the optical energy of each pulse period is equivalent to that of at least a few hundred photons. This number is large enough so that a Gaussian distribution of the probability density is warranted.

The probability density function of a "0" transmitted and received by the optical receiver is thus given by and illustrated in Figure 6.8.

$$p[v_o|\text{"0"}] = \frac{1}{\left(2\pi v_{NT0}^2\right)^{1/2}} \exp\left[\frac{-(v_0 - v_{00})^2}{2v_{NT0}^2}\right] \tag{6.28}$$

and similarly for a "1" transmitted

$$p[v_o|\text{"1"}] = \frac{1}{\left(2\pi v_{NT1}\right)^{1/2}} \exp\left[\frac{-(v_0 - v_{01})^2}{2v_{NT1}^2}\right] \tag{6.29}$$

The total probability of error or a bit-error-rate (BER) is defined [10, 11] as

$$BER = p(1)p(0/1) + p(0)p(1/0) \tag{6.30}$$

where $p(1)$ and $p(0)$ are the probabilities of receiving "1" and "0," respectively, and $p(1/0)$ and $p(0/1)$ are the probabilities of deciding "1" when "0" is transmitted and vice versa.

(a)   (b)

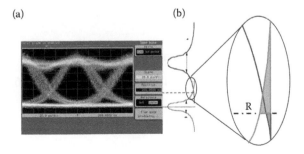

FIGURE 6.8   (a) Time-dependent fluctuating signal in the receiver in response to the random digital bit stream and (b) the probability density distribution as a function of the output voltage level and the decision level for recovery of the transmitted signals.

In a binary bit stream, "1" and "0" are likely to occur equally, that is $p(1) = p(0) = 0.5$, then Equation 6.30 a becomes

$$BER = \frac{1}{2}[p(0/1) + p(1/0)] \tag{6.31}$$

Thus, as equal probability of transmitting "0" and "1" are assumed then for a decision voltage level of d as indicated in Figure 6.5 the total probability of error P_E is the summation of the errors of deciding "0" or "1" by integrating the probability density function over the shaded regions, thus given by

$$BER = P_E = \frac{1}{2}\int_d^\infty p[v_0|\text{"0"}]dv_0 + \frac{1}{2}\int_{-\infty}^d p[v_0|\text{"1"}]dv_0 \tag{6.32}$$

Substituting for the probability distribution using Equation 6.30a and b leading to

$$BER = \frac{1}{2\sqrt{\pi}}\int_{\frac{d-v_{00}}{v_{NT0}}}^\infty e^{-\frac{x^2}{2}}dx + \frac{1}{2\sqrt{\pi}}\int_{\frac{v_{01}-d}{v_{NT1}}}^d e^{-\frac{x^2}{2}}dx \tag{6.33}$$

The functions in Equation 6.33 have the standard form of the complementary error function $Q(\alpha)$ defined as

$$Q(\alpha) = \frac{1}{\sqrt{2\pi}}\int_\alpha^\infty e^{-x^2/2}dx \tag{6.34}$$

then Equation 6.36 becomes

$$BER = \frac{1}{2}\left[Q\left(\frac{d}{v_{NT0}}\right) + Q\left(\frac{v_{01}-d}{v_{NT1}}\right)\right] \tag{6.35}$$

The $Q(\delta)$ function is a standard function and this curve is shown in Figure 6.9. It is noted that for a $BER = 10^{-9}$, the value of δ is about six, which is the normal standard for communications at bit rate of 155 Mbps to 40 Gb/s. However, in laboratory demonstration, it requires a BER of 10^{-12} that is corresponding to a δ of 7.

6.4.3 Minimum Average Optical Received Power

Again, using the condition $p[v_o/\text{"0"}] = p[v_o/\text{"1"}]$ we have from Equation 6.35

$$\frac{d}{v_{NT0}} = \frac{v_{01}-d}{v_{NT1}} \equiv \delta \tag{6.36}$$

Now assuming $v_{00} = 0$, hence

$$BER = Q(\delta) \tag{6.37}$$

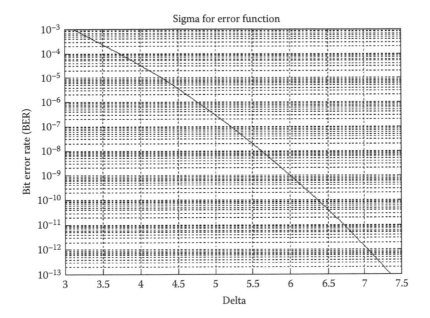

FIGURE 6.9   BER as a function of the δ (delta) parameter. Sometimes δ is assigned as the system Q Magnum function.

Thus, by eliminating the variable d from Equation 6.36 we obtain

$$\delta = \frac{v_{01} - \delta v_{NT0}}{v_{NT1}} \tag{6.38}$$

Or, alternatively,

$$v_{01} = \delta(v_{NT0} + v_{NT1}) \tag{6.39}$$

substituting v_{01}, v_{NT1}, v_{NT0} we have

$$\Re \frac{b_1}{T_b} \langle G_n \rangle R_I A = \partial \left\{ \left[2\Re q \frac{b_1}{T_B} \langle G_n^2 \rangle B R_I^2 A^2 + v_{NA}^2 \right]^{1/2} + v_{NA} \right\} \tag{6.40}$$

However, the amplifier noise voltage v_{NA} can is given by $v_{NA} = i_{Neq} R_I A B^{1/2}$. Thus, by substituting this noise voltage and eliminating $R_I A$ and solving for the energy essential for the "1" transmitted and received at the photodetector, the energy b_1 can be written as

$$b_1 = \frac{q}{\Re} \left[\delta^2 G^x + \frac{2\delta i_{Neq} T_B}{qG} \right] \tag{6.41}$$

where we have used the approximation $\langle G_n \rangle = G$ and $\langle G_n^2 \rangle = \langle G \rangle^{2+x}$ with x is the factor dependent on the ionization ration in an APD. For pin photodetector, $G = 1$.

The optical receiver sensitivity, denoted as RS, can thus be defined as

$$\text{Receiver sensitivity} = RS = 10 \text{Log}_{10} \frac{P_{av}}{P_0} \quad \text{in } dBm \tag{6.42}$$

where $P_{av} = b_1 / T_b$ and $P_0 = 1.0\,\text{mW}$ (the reference power level for evaluating power in dBm). This is the optical receiver sensitivity, which is defined as the minimum optical power required for the receiver to operate reliably with a BER below a specified value.

In Equation 6.45, there are two terms clearly specifying the dependence of the signal dependence δ and the amplifier noise contribution. The term b_1 represents the optical energy of the "1" required for the optical receiver to detect with a certain bit error rate. The term $q/\mathcal{R}$ in the RHS of Equation 6.44 is equivalent to optical power required to generate one electron, or the number of photons required at the optical receiver to generate one electron. Typical measure for the optical receiver is the number of photon required for it to operate with a specified BER. Thus, the term $[\delta^2 G^x]$ indicates the number of photons required to overcome the opening of the eye at a particular BER, and the term $\dfrac{2\delta i_{Neq}T_b}{qG}$ indicates the number of photons required to overcome the noises of the electronic pre-amp at a specific BER. With an avalanche photodetector, the gain of the avalanche gain does also contribute to an increase of the noise level and hence higher power. However, the current avalanche gain would compensate for this.

We note that the analysis here has not taken the effects of pulse shape and the transfer function characteristics of the receiver into account. Thus, in this case an equivalent current noise power as seen from the input port of the electronic pre-amplifier would simplify the analysis of the optical receiver sensitivity or SNR. The effects of the pulse shape and transfer function of the receiver are considered in the next sections.

6.4.3.1 Fundamental Limit: Direct Detection

Referring to Equation 6.44, we can see that in the case that there are no electronic noises then the first term is the number of photons in the quantum limit of direct detection receiver. Thus, we have

$$n_p = [\delta^2 G^x] \tag{6.43}$$

For a PIN detector and a BER of 1e-9, the number of photons of the quantum limit is 36.

EXAMPLE

For an ASK modulated optical signal at a bit rate of 10 Gb/s, estimate the number of photon energy required for the incoming symbol "1" so that a BER of 10^{-9} can be achieved.

6.4.3.2 Equalized Signal Output

Recalling the optical signal $p(t)$ fallen on the input of the PD as

$$p(t) = \sum_{k=-\infty}^{\infty} a_k h_p(t' - kT_B) \tag{6.44}$$

where T_B is signal bit period, and $h_p(t)$ is the input optical shape constrained by

$$\int_{-\infty}^{+\infty} h_p(t)dt = 1 \tag{6.45}$$

It is noted that this function represents the optical power variation as a function of time and thus it is always positive.

Now neglecting the DC component due to dark current, the signal at the output of the equalizer can be written as

$$\langle v_{0E}(t) \rangle = \mathcal{R}Gp(t) * h_{fe}(t) * h_{eq}(t) \tag{6.46}$$

where $h_{fe}(t)$ is the impulse response of the optical receiving sub-system and $h_{eq}(t)$ is the impulse response of the electronic equalizer and * indicates the convolution.

The optical receiver is usually dominated by the pole contributed by the diode capacitance C and associated resistance of the network. It can be shown without difficulty that (refer to Figure 6.6) the frequency domain transfer function is given by

$$H_{fe}(\omega) = -\frac{R_f}{1 + j\omega \frac{R_f C}{A}} \tag{6.47}$$

where A is the open loop voltage gain of the electronic pre-amplifier, R_f is the feedback resistance, and C is the total input equivalent capacitance at the input port of the amplifier. Thus, knowing the input optical shape, the equalizer transfer function for a given pulse shape can be determined by multiplying the transfer function of the receiver and that of the equalizer. It is noted that the 3 dB bandwidth of a feedback receiver is given by $A/R_f C$ which is about A times the bandwidth of receiver without feedback. Thus, the demand on the equalizer is much less for this case.

6.4.3.3 Photodiode Shot Noise

The output noise power due to the random fluctuation of the multiplied Poisson nature of the current $i_s(t)$ produced by the PD can be written [12] as

$$\langle i_{sN}^2 \rangle = \int_{-\infty}^{+\infty} qG^{2+x}\Re \left\{ \sum_{k=-\infty}^{\infty} a_k h_p(t' - kT_B) + \lambda_d \right\} h_I^2(t - t')dt' \tag{6.48}$$

where $h_I(t) = h_{fe}(t) * h_{eq}(t)$ is the overall *current* impulse response of the optical pre-amplifier and the equalizer in cascade λ_d is the electrons/sc. generated by the PD when no lightwave is shined on the PD. It is important to note that the quantum shot noise power depends upon the signal levels $\{a_k\}$ and the time t that is the noise is signal dependent and non-stationary.

Without the loss of generality, we can consider the output quantum shot noise at the instant $t = 0$, then by replacing $t = 0$ and $t' = \tau$, (6.51) becomes

$$\langle i_{sN}^2(0) \rangle = \int_{-\infty}^{+\infty} qG^{2+x}\Re \left\{ \sum_{k=-\infty}^{\infty} a_k h_p(\tau - kT)h_I^2(-\tau) \right\} d\tau + \int_{-\infty}^{+\infty} qG^{2+x}\Re\lambda_d h_I^2(\tau)d\tau \tag{6.49}$$

The first term is contributed by the signal amplitude, the signal-dependent shot noise and the second term indicates the contribution of the dark current of the PD.

The shot noise power due to the signal sequence $\{a_k\}$ at the time $t = 0$ can thus be rewritten as

$$\langle i_{sN}^2(0) \rangle = qG^{2+x}\Re \left\{ \sum_{k=-\infty}^{\infty} a_k (h_p * h_I^2)(-kT_B) \right\} \tag{6.50}$$

$$\langle i_{sN}^2(0) \rangle = qG^{2+x}\Re \left\{ \sum_{k=-\infty}^{\infty} a_k (h_p * h_I^2)(-kT_B) \right\} = qG^{2+x}\Re a_0 g(0) + qG^{2+x}\Re \left\{ \sum_{\substack{k=-\infty \\ k \neq 0}}^{\infty} a_k g(-kT_B) \right\}$$

with $(h_p * h_I^2)(-kT_B) = g(-kT_B)$

$$\tag{6.51}$$

The first term is the quantum shot noise due the pulse under consideration and the second term is due to the preceding and succeeding signal pulses. The effects of the succeeding pulses on the noise at $t = 0$ is likely since there are some delays due to signal travelling through the system. This dependence of noise power on preceding and succeeding pulses is similar to intesymbol interference. However, it is important to note that this effect cannot be eliminated by equalizing the pulse to obtain zero ISI, since $h_0(kT) = 0$ does not imply that $g(kT) = 0$. The exact estimation of the noise power contributed to the considered pulse $\langle i_{SN}^2(0) \rangle$ requires statistical knowledge of the distribution of the pulse sequence. This is commonly unknown and thus some criteria must be used. The worst case that would contribute the largest noise effect is when the sequence $\{a_k\}$ has the adjacent symbols of maximum values b_M. In this case, we have

$$\langle i_{SN}^2(0) \rangle = qG^{2+x}\Re a_0 g(0) + qG^{2+x}\Re \left\{ \sum_{\substack{k=-\infty \\ k \neq 0}}^{\infty} b_M g(-kT_B) \right\} \tag{6.52}$$

The frequency domain of the noise power can be estimated as

$$\left\langle [i_{SN}^F(0)]^2 \right\rangle = qG^{2+x}\Re \left\{ \frac{b_M}{T_B} \sum_{\substack{k=-\infty \\ k \neq 0}}^{\infty} b_M G\!\left(\frac{2\pi N}{T_B} \right) - \frac{b_M - a_0}{2\pi} \int_{-\infty}^{\infty} G(\omega) d\omega \right\} \tag{6.53}$$

where

$$G(\omega) = \Im[g(t)]$$

$$G(\omega) = \frac{1}{2\pi} H_p(\omega)[H_I(\omega) * H_I(\omega)] \tag{6.54}$$

$$\sum_{k=-\infty}^{\infty} g(-kT_B) = \frac{1}{T_B} \sum_{n=-\infty}^{\infty} b_M G\!\left(\frac{2\pi N}{T_B} \right)$$

The worst case total noise power is therefore given by

$$\left\langle [i_{SN}^F(0)]^2 \right\rangle = \langle i_{sd}^2 \rangle + \langle i_{SN}^2(0) \rangle \tag{6.55}$$

With $\langle i_{sd}^2 \rangle$ is shot noise due to dark current of the PD.

6.4.3.4   Total Output Noises and Pulse Shape Parameters
It is advantageous to normalize the input optical pulse and the output equalized pulse as follows:

$$h_p'(t) = T_B h_p(tT_B)$$
$$h_o'(t) = T_B h_o(tT_B) \tag{6.56}$$

Or equivalently in the frequency domain

$$H_p'(f) = H_p\!\left(\frac{2\pi f}{T_B} \right)$$

$$H_o'(f) = \frac{1}{T_B} H_o\!\left(\frac{2\pi f}{T_B} \right) \tag{6.57}$$

These temporal and spectral functions depend only on the shapes of the input and output pulses not on the epoch T. The normalization can be achieved by setting:

$$H'_p(0) = 1 \tag{6.58}$$

While the requirement on $H'_o(f)$ for zero intersymbol interference is

$$\sum_{-\infty}^{\infty} H'_0(f + k) = 1 \quad \text{for } |f| < 0.5 \tag{6.59}$$

Thus, all the output noises can be expressed in term of these normalized functions. We can thus obtain from Equation 6.57:

$$H_I\left(\frac{2\pi f}{T_B}\right) = \frac{T_B}{\Re G} H'(f) \ldots \ldots \text{with} \ldots H'(f) = \frac{H'_o(f)}{H'_p(f)} \tag{6.60}$$

The normalized effects of the pulse shaping are given as

$$I_1 = \int_{-\infty}^{\infty} H'_p[H' * H'(f)] df$$

$$I_2 = \int_{-\infty}^{\infty} |H'(f)|^2 df \tag{6.61}$$

$$I_2 = \int_{-\infty}^{\infty} |H'(f)|^2 f^2 df$$

The term I_1 influences the quantum shot noise of the photodetector from the optical signal received at the input, the term I_2 influences the dark current shot noises and the thermal noises due to the input resistance of the pre-amplifier, and the term I_3 influences the frequency dependent parts of the transfer function of the electronic pre-amp.

6.4.4 MULTI-LEVEL PERFORMANCE

From the probabilistic distribution for multi-level and Gaussian distribution assumption, one can derive the BER for such multi-level direction as

$$BER_Q = \frac{1}{N\log_2(N)} \frac{1}{2} \sum_{i=1}^{N} \left[erfc\left(\frac{|\hat{\mu}_i - I^{th}_{i=low}|}{\hat{\sigma}_i\sqrt{2}}\right) + erfc\left(\frac{|\hat{\mu}_i - I^{th}_{i=high}|}{\hat{\sigma}_i\sqrt{2}}\right) \right] \tag{6.62}$$

where $\hat{\mu}_i$ is the average value of the received symbols that were transmitted as I^{th}_i and $\hat{\sigma}^2_i$ its variance. Both are computed and obtained at the receiver before the hard decision. The quantities $I^{th}_{i=low}$ and $I^{th}_{i=high}$ are the optimum decision threshold levels between I_{i-1} and I_i and between I_{i+1} and I_i, respectively. From this analytical BER_Q obtained using solely the first two statistical moments of all levels $\hat{\mu}_i$ and $\hat{\sigma}_i$, one can compute the equivalent Q-factor to give the Q-factor for PAM 4 signaling.

6.5   RECEIVER FRONT END NOISES

6.5.1   FET FRONT-END OPTICAL RECEIVER

As an example of the estimation of noises at the output of the optical receiver, we illustrate by an example of an optical receiver with a photodetector followed by a FET front end. Referring to Figure 6.6 for a FET front end, we have the following parameters:

$$C = C_a + C_d \tag{6.63}$$

$$R_p = R_b R_i/(R_b + R_i) \tag{6.64}$$

where R_b is the bias resistance, R_i is the input resistance of the amplifier at mid-band, C_d is the diode capacitance under reverse bias, and C_a is the input capacitance of the amplifier.

The spectral density of the thermal noises due to the bias resistance and the feedback resistance R_f are given by

$$S_R = \frac{2k_B T}{R_f}$$
$$S_{R_b} = \frac{2k_B T}{R_b} \tag{6.65}$$

where θ is the temperature and k is the Boltzmann's constant. Then the noise power at the equalizer output is given by

$$\langle i_{RN}^2 \rangle = 2k_B T \left(\frac{1}{R_f} + \frac{1}{R_b}\right) \cdot \int_{-\infty}^{\infty} |H_I(2\pi f)|^2 df = 2k_B T \left(\frac{1}{R_f} + \frac{1}{R_b}\right) \cdot \left(\frac{1}{\Re G}\right)^2 I_2 T_b \tag{6.66}$$

T_b is the bit period and T is the absolute temperature. The noise due to the trans-conductance of the FET is given by

$$\langle i_{g_m N}^2 \rangle = \frac{2k_B T \Gamma P}{g_m} \cdot \frac{1}{2\pi} \int_{-\infty}^{\infty} |H_I(2\pi f)|^2 \left(\frac{1}{R_f^2} + \omega^2 C^2\right) d\omega$$
$$= \frac{2k_B T \Gamma P}{g_m} \cdot \left(\frac{1}{\Re G}\right)^2 \left(\frac{T_b I_2}{R_f^2} + \frac{(2\pi C)^2}{T_b} I_3\right) \tag{6.67}$$

where g_m is the transconductance of the FET, $\Gamma \sim 0.7$ is the factor related materials, and 0.7 for GaAs. P is the imperfection factor of the FET device.

The quantum shot noise of the photodetector is given by

$$\langle [i_{SN}^F(0)]^2 \rangle = qG^{2+x}\Re \left\{ \frac{b_M}{T} \sum_{\substack{k=-\infty \\ k \neq 0}}^{\infty} b_M G\left(\frac{2\pi N}{T}\right) - \frac{b_M - a_0}{2\pi} \int_{-\infty}^{\infty} G(\omega)d\omega \right\}$$
$$= \frac{q}{\Re} G^x \left[a_0 I_1 + b_M \left(\sum\nolimits_1 - I_1\right)\right] \tag{6.68}$$

$$\text{with} \dots \sum\nolimits_1 = \sum_{-\infty}^{\infty} H_p'(k)[H'(k) * H'(k)]$$

6.5.2 BJT Front-End Optical Receiver

Bipolar junction transistor (BJT) can also be employed as the front-end amplification stage of the optical receivers. A typical circuit design of such a receiver is shown in Figure 6.10 in which three BJTs have been used with a direct coupled pair and a peaking stage to extend the bandwidth of the overall gain and then a shunt feedback to form a trans-impedance configuration.

6.5.3 Noise Generators

Electrical shot noises are generated by the random generation of streams of electrons (current). In optical detection, shot noises are generated by (i) biasing currents in electronic devices and (ii) photo currents generated by the photodiode.

A biasing current I generates a spectral current density S_I given by

$$S_I = \frac{d\langle i_I^2 \rangle}{df} = 2qI \quad \text{in A}^2/\text{Hz} \tag{6.69}$$

where q is the electronic charge. The quantum shot noise $\langle i_s^2 \rangle$ generated by the PD by an optical signal with an average optical power P_{in} is given by

$$S_Q = \frac{d\langle i_s^2 \rangle}{df} = 2q\langle i_s^2 \rangle \quad \text{in A}^2/\text{Hz} \tag{6.70}$$

In the case that the APD is used, then the noise spectral density is given by

$$S_Q = \frac{d\langle i_s^2 \rangle}{df} = 2q\langle i_s^2 \rangle \langle G_n^2 \rangle \quad \text{in A}^2/\text{Hz} \tag{6.71}$$

It is noted here again that the dark currents generated by the PD must be included to the total equivalent noise current at the input after it is evaluated. These currents are generated even in the absence of the optical signal. These dark currents can be eliminated by cooling the PD to at least below the temperature of liquid nitrogen (77°K).

At a certain temperature, the conductivity of a conductor varies randomly. The random movement of electrons generates a fluctuating current even in the absence of an applied voltage. The thermal

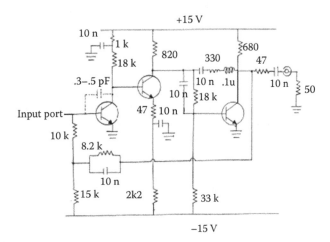

FIGURE 6.10 Design circuit of the electronic pre-amplifier for the balanced optical amplifier. All BJTs are either of type Phillips BFR90A or BFT24.

noise spectral density of a resistor R is given by

$$S_R = \frac{d(i_R^2)}{df} = \frac{4k_BT}{R} \quad \text{in A}^2/Hz \tag{6.72}$$

where k_B is the Boltzmann's constant, T is the absolute temperature (in °K) and R is the resistance in ohms and i_R denotes the noise current due to resistor R.

6.5.4 EQUIVALENT INPUT NOISE CURRENT

Our goal is to obtain an analytical expression of the noise spectral density equivalent to a source looking into the electronic amplifier, including the quantum shot noises of the PD. A general method for deriving the equivalent noise current at the input is by representing the electronic device by a Y-equivalent linear network as shown in Figure 6.3. The two current noise sources $di_n'^2$ and $di_n''^2$ represent the summation of all noise currents at the input and at the output of the Y-network. This can be transformed into a Y-network circuit with the noise current referred to the input as follows:

The output voltages V_0 as referred to Figure 6.7 can be written as

$$V_0 = \frac{i_N'(Y_f - Y_m) + i_N''(Y_i + Y_f)}{Y_f(Y_m + Y_i + Y_o + Y_L) + Y_i(Y_o + Y_L)} \tag{6.73}$$

and then using the equivalent model, we have

$$V_0 = \frac{(i_N')_{eq}(Y_f - Y_m)}{Y_f(Y_m + Y_i + Y_o + Y_L) + Y_i(Y_o + Y_L)} \tag{6.74}$$

thus, comparing these two equations we can deduce the equivalent noise current at the input of the detector is

$$i_{Neq} = i_N' + i_N'' \frac{Y_i + Y_f}{Y_f - Y_m} \tag{6.75}$$

then reverting to mean square generators for a noise source we have

$$d(i_{Neq})^2 = d(i_N')^2 + d(i_N'')^2 \left| \frac{Y_i + Y_f}{Y_f - Y_m} \right|^2 \tag{6.76}$$

It is therefore expected that if the Y-matrix of the front end low-noise amplifier is known, the equivalent noise at the input of the amplifier can be obtained by using Equation 6.76.

For a given source, the input noise current power of a BJT front end can be found by (see also the Appendix of Chapter 7)

$$i_{Neq}^2 = \int_0^B d(i_{Neq})^2 = a + \frac{b}{r_E} + cr_E \tag{6.77}$$

where B is the bandwidth of the electronic pre-amplifier, r_E is the emitter resistance of the front end transistor of the pre-amplifier. a, b, and c are the parameters depending on the circuit elements and amplifier bandwidth. Hence, an optimum value of r_E can be found hence an optimum biasing current is to be set for the collector current of the BJT such that i_{Neq}^2 is at minimum.

$$r_{Eopt} = \sqrt{\frac{b}{c}} \ldots \text{hence} \ldots \rightarrow i_{Neq}^2 \Big|_{r_E = r_{Eopt}} = a + 2\sqrt{bc} \qquad (6.78)$$

If two types of BJT are considered as Phillips BFR90A and BFT24, then a good approximation of the equivalent noise power can be found as

$$a = \frac{8\pi B^3}{3} \left\{ r_B C_s^2 + (C_s + C_f + C_{tE})\tau_T \right\}$$

$$b = \frac{B}{\beta_N} \qquad (6.79)$$

$$c = \frac{4\pi B^3}{3} (C_s + C_f + C_{tE})^2$$

The theoretical estimation of the transistors can be derived from the measured scattering parameters as given by the manufacturer as

BFR90A

$r_{Eopt} = 59\,\Omega$

$I_{Eopt} = 0.44\,\text{mA}$

$$I_{eq}^2 = 7.3 \times 10^{-16}\,\text{A}^2 \longrightarrow \frac{I_{eq}^2}{B} = 4.9 \times 10^{-24}\,\text{A}^2/\text{Hz} \qquad (6.80)$$

$$I_{eq} = 27\,\text{nA} \longrightarrow \frac{I_{eq}}{\sqrt{B}} = 2.21\,\text{pA}/\sqrt{\text{Hz}}$$

BFT24

$r_{Eopt} = 104\,\Omega$

$I_{Eopt} = 0.24\,\text{mA}$

$$I_{eq}^2 = 79.2 \times 10^{-16}\,\text{A}^2 \longrightarrow \frac{I_{eq}^2}{B} = 6.1 \times 10^{-24}\,\text{A}^2/\text{Hz} \qquad (6.81)$$

$$I_{eq} = 30.2\,\text{nA} \longrightarrow \frac{I_{eq}}{\sqrt{B}} = 2.47\,\text{pA}/\sqrt{\text{Hz}}$$

Note that the equivalent noise current depends largely on some not well defined values such as the capacitance, the transit times and the base spreading resistance and the short circuit current gain β_N. The term $I_{eq}/\sqrt{B}$ is usually specified as the noise spectral density equivalent referred to the input of the electronic pre-amplifier.

6.5.5   HEMT Matched Noise Network Pre-Amplifier

A third-order noise matching network (shown in Figure 6.11) has been obtained by Park [13] to tailor for 10 GHz low-noise optical receiver. This network is inserted in our pre-amplifier modeling and the electronic circuit is connected as shown in Figure 6.12.

For high receiving bit rate, we need to design a pre-amplifier with high-frequency response. In electronics, the drain-to-source HEMT parasitic capacitance, C_{ds} is negligible small at high frequency. Furthermore, we could reduce the complexity of solving this circuit by considering the

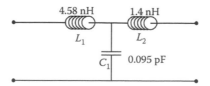

FIGURE 6.11   Third-order noise matching network.

Miller effect capacitance. Thus, we can construct the small signal high-frequency response equivalent circuit for the electronic circuit and the simplified equivalent circuit is shown in Figure 6.12.

The above equivalent circuit Figure 6.13 can be further simplified, as shown in Figure 6.14.

Thus, the required pre-amplifier's transfer function can be obtained in the following:

$$i_3 = \left(\frac{sC_{in}}{s^3 L_2 C_1 C_{in} + sC_1 + sC_{in}} \right) \cdot i_1 \tag{6.82}$$

$$i_1 = \left(\frac{Z_{in}}{Z_{in} + Z_A} \right) \cdot i_s \tag{6.83}$$

By substituting Equations 6.83 into Equation 6.82, we obtain the transfer function of the HEMT matched noise network pre-amplifier,

$$H_{\text{HEMT}} = \frac{V_{out}}{i_s} = -g_m R_{out} \left(\frac{1}{s^3 L_2 C_1 C_{in} + sC_1 + sC_{in}} \right) \cdot \left(\frac{Z_{in}}{Z_{in} + Z_A} \right) \tag{6.84}$$

where

$$R_{out} = \frac{R_d R_l}{R_d + R_l} \tag{6.85}$$

$$Z_{in} = \frac{R_g}{sR_g C_d + 1} \tag{6.86}$$

$$Z_A = sL_1 + \frac{\left(\dfrac{1}{sC_1} \right) \cdot \left(sL_2 + \dfrac{1}{sC_{in}} \right)}{\dfrac{1}{sC_1} + \dfrac{1}{sC_{in}} + sL_2} = \frac{s^4 L_1 L_2 C_1 C_{in} + s^2 L_1 C_1 + s^2 L_1 C_{in} + s^2 L_2 C_{in} + 1}{s^3 L_2 C_1 C_{in} + s(C_1 + C_{in})} \tag{6.87}$$

We can describe the noise theoretical understandings and the equivalent referred to input noise current in the following sections (Figures 6.15 and 6.16).

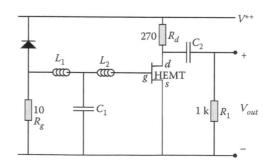

FIGURE 6.12   Electronic circuit of HEMT noise matched network pre-amplifier.

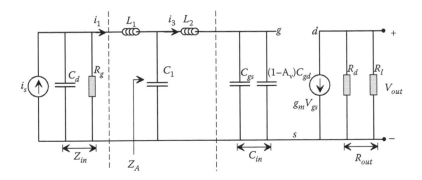

FIGURE 6.13 Equivalent circuit of HEMT noise matched network pre-amplifier.

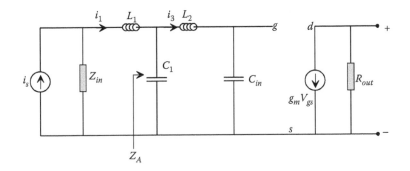

FIGURE 6.14 Simplified equivalent circuit of HEMT noise matched network pre-amplifier.

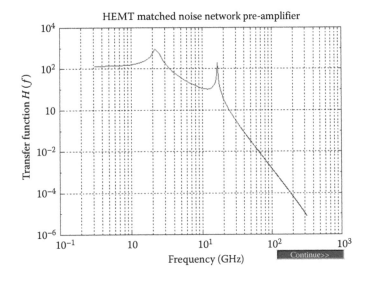

FIGURE 6.15 Frequency response of the HEMT noise matched network pre-amplifier.

The circuit diagram shown in Figure 6.12 can be represented with reflection coefficients given in Figure 6.17. We denote the reflection coefficient of Z_{ph}, Z_s and Z^*_{opt} by Γ_{ph}, Γ_s and Γ^*_{opt}, which are normalized to Z_0 and defined by

$$\Gamma_x = \frac{Z_x - Z_0}{Z_x + Z_0} \tag{6.88}$$

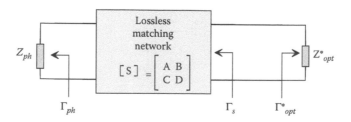

FIGURE 6.16 Lossless matching network with source and load impedances.

where $x = ph$, s, and opt. It is useful to obtain the S-parameters of the third-order noise matching network defined in the matrix form in the following (Equation 6.8).

$$S = \begin{bmatrix} S_{11} & S_{12} \\ S_{21} & S_{22} \end{bmatrix} \qquad (6.89)$$

From Figure 6.17, we find Z_1, which is given by

$$Z_1 = sL_1 + \left(sC_1 + \frac{1}{1 + sL_2}\right)^{-1} = \frac{s^3C_1L_1L_2 + s^2C_1L_1 + s(L_1 + L_2) + 1}{s^2C_1L_2 + sC_1 + 1} \qquad (6.90)$$

Applying the circuit theory of the T-Network and by substituting Equation 6.89, we obtain

$$S_{11} = \frac{Z_1 - 1}{Z_1 + 1} = \frac{s^3C_1L_1L_2 + s^2(C_1L_1 - C_1L_2) + s(L_1 + L_2 - C_1)}{s^3C_1L_1L_2 + s^2(C_1L_1 + C_1L_2) + s(L_1 + L_2 + C_1) + 2} \qquad (6.91)$$

From circuit theory, S_{21} is given by

$$S_{21} = 2\sqrt{\frac{r_1}{r_2}}\left(\frac{V_2}{e_1}\right) \qquad (6.92)$$

From Figure 6.17, we can obtain

$$\frac{V_2}{e_1} = \frac{1}{s^3C_1L_1L_2 + s^2(C_1L_1 + C_1L_2) + s(L_2 + L_1 + C_1) + 2} \qquad (6.93)$$

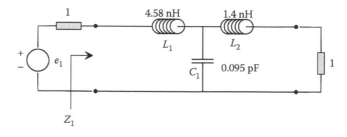

FIGURE 6.17 Third-order noise matching network to find the S-parameters.

Hence, by substituting Equation 6.93 into Equation 6.91, we have

$$S_{21} = \frac{2}{s^3 C_1 L_1 L_2 + s^2(C_1 L_1 + C_1 L_2) + s(L_2 + L_1 + C_1) + 2} \qquad (6.94)$$

Thus, using the symmetrical property, the other two S-parameters can be obtained as follows:

$$S_{22} = S_{11} = \frac{s^3 C_1 L_1 L_2 + s^2(C_1 L_1 - C_1 L_2) + s(L_1 + L_2 - C_1)}{s^3 C_1 L_1 L_2 + s^2(C_1 L_1 + C_1 L_2) + s(L_1 + L_2 + C_1) + 2} \qquad (6.95)$$

$$S_{12} = S_{21} = \frac{2}{s^3 C_1 L_1 L_2 + s^2(C_1 L_1 + C_1 L_2) + s(L_2 + L_1 + C_1) + 2} \qquad (6.96)$$

By using Equation 6.88, we can derive all the required reflection coefficients Table 6.2. First, consider the equivalent photodiode circuit shown in Figure 6.18.

Then, Z_{ph} can be found as

$$Z_{ph} = R_s + \frac{1}{sC_d} \qquad (6.97)$$

From Equation (6.97), we have

$$\Gamma_{ph} = \frac{Z_{ph} - Z_0}{Z_{ph} + Z_0} \qquad (6.98)$$

Then, we substitute Z_{ph} from (6.97):

$$\Gamma_{ph} = \frac{sC_d(R_s - 1) + 1}{sC_d(R_s + 1) + 1} \qquad (6.99)$$

Knowing the S-parameters and Γ_{ph}, we can obtain Γ_s as

$$\Gamma_s = S_{22} + \frac{S_{12}S_{21}\Gamma_{ph}}{1 - \Gamma_{ph}S_{11}} \qquad (6.100)$$

TABLE 6.2
Noise Parameters of the Packaged HEMT with 0.3 μm Gate Length

Frequency	Γ_{opt}		F_{min}	R_n
(GHz)	(Mag)	(Ang)	(dB)	(Ω)
2	0.79	30	0.33	29
4	0.73	59	0.35	21
6	0.68	87	0.44	14.5
8	0.63	119	0.55	9.5
10	0.59	139	0.66	6
12	0.55	164	0.75	4

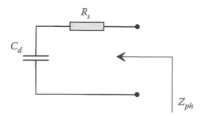

FIGURE 6.18 Equivalent circuit for photodiode.

and

$$\Gamma_{opt} = \Gamma_s = S_{22} + \frac{S_{12}S_{21}\Gamma_{ph}}{1 - \Gamma_{ph}S_{11}} \tag{6.101}$$

The equivalent input noise current density appearing across the photodiode junction capacitance can be expressed in terms of reflection coefficients [13] as

$$i_{eq}^2 = 4kTR_s(\omega C_d)^2 \frac{\left(1 - \left|\Gamma_{opt}^*\right|^2\right)}{G_M\left|1 - \Gamma_s\Gamma_{opt}^*\right|^2} \left\{ F_{min}(1 - |\Gamma_s|^2) + \frac{4R_n}{Z_0}\frac{|\Gamma_s - \Gamma_{opt}|^2}{|1 + \Gamma_{opt}|^2} \right\}\Delta f \tag{6.102}$$

where F_{min} and R_n can be obtained from the given data in Table 4.2 and, G_M is the transducer power gain of the lossless matching network. Our goal is to minimize the equivalent input noise current density. This can be done by minimizing the photodiode junction capacitance C_d, the series resistance R_s of photodiode, the minimum noise figure F_{min} and the noise resistance R_n, and by maximizing the transducer power gain. Among these parameters in Equation 6.102, only the transducer power gain G_M and the output reflection coefficient Γ_s are related to the noise matching network and these may be optimized by design. The transducer power gain can be expressed in terms of reflection coefficients [13] and is defined by

$$G_M = 1 - \left|\frac{(1 + \Gamma_s)(1 - \Gamma_{opt}) - (1 - \Gamma_s)(1 + \Gamma_{opt})}{(1 + \Gamma_s)\left(1 - \Gamma_{opt}^*\right) + (1 - \Gamma_s)\left(1 + \Gamma_{opt}^*\right)}\right|^2 \tag{6.103}$$

The transducer power gain has its maximum value of unity when $\Gamma_s = \Gamma_{opt}$. This condition is equivalent to the output admittance of the noise matching network being matched to the complex conjugate of the optimum source admittance (Figure 6.19).

6.5.6 REMARKS

This section has addressed the considerations of optical receivers in which the electronic currents are generated when the optical energy/power from the modulated optical signals is absorbed by the photosensitive regions directly. The noise generation process is described and related small signal models of electronic pre-amplifier. An equivalent noise current, as seen at the input port of the electronic pre-amplifier, is presented, as well as method to derive it for the front-end amplifier. Both FET and BJT types are given. Furthermore, the effects of the pulse shape on the output signals are also given related to the transfer function of electronic pre-amplifier. A noise matching network at the front end of the optical pre-amplifier would reduce the noise effects at high frequency, and this is crucial in the design of ultra-wide band optical receiver.

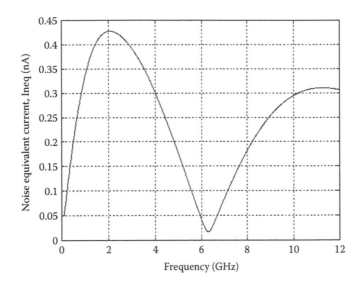

FIGURE 6.19   Equivalent input noise current density of the HEMT noise matched front-end receiver.

These noise models can be integrated into coherent receivers, which will be described in this chapter. For coherent optical receiver, a local oscillator, a powerful laser, mixes the optical signals with its output field and gives a beating signal in the photodetector due to its square law property. The phase and amplitude of the optical signals are preserved in the electronic domain and thus processing in the electronic domain can be performed to extract the digital and analog property of the original signals.

REFERENCES

1. A. H. Gnauck and P. J. Winzer, Optical phase-shift-keyed transmission, *IEEE J. Lightwave Technol.*, 23, pp. 115–130, 2005.
2. Cristian Prodaniuc; Nebojsa Stojanovic, F. Karinou, Z. Qiang, and R. Llorente, Performance comparison between 4D Trellis coded modulation and PAM-4 for low-cost 400 Gbps WDM optical networks, *IEEE Journal of Lightwave Technology*, vol. 34, no. 22, pp. 5308–5316, Nov. 15, 15 2016.
3. L. Zimmermann, Silicon photonics in optical communications, *Semicon Europa* 2013, Munich, 2013.
4. R.C. Alferness, Waveguide electrooptic modulators, *IEEE J. Microwave Theory and Tech.*, vol. MTT-30, pp. 1121–1137, 1982.
5. R.C. Booth, LiNbO-integrated optic devices for coherent optical fiber systems, *Thin Solid Films*, vol. 126, pp. 167–176, 1985.
6. S. Ritchie and A.G. Steventon, The potential of semiconductors for optical-integrated circuits, in *Proc. Conf. Digital Optical Circuit Tech.*, NATO, France, vol. 362, pp. 1111–1120.
7. S. D. Personick, *Bell Syst. Tech. J.*, vol. 52, pp. 843–86, 1973.
8. S. D. Personick, Optical detectors and receivers, *IEEE Journal of Lightwave Technology*, vol. 26, no. 9, pp. 1005–1020.
9. J. L. Hullet and T. V. Muoi, A feedback receive amplifier for optical transmission systems, *IEEE Trans. Com.*, vol. 24, no. 10, pp. 1180–1185, 1976.
10. R. Dogliotti, G. Luvison, and G. Pirani, Error probability in optical fiber transmission systems, *IEEE Trans. Inf. Theory*, vol. IT-25, no. 2, pp. 170–178, 1979.
11. S. D. Personick, *Optical Fiber Transmission Systems*, N.Y.: Plenum Publications, 1980.
12. S. D. Personick., Receiver design for digital fiber optic communications systems, *Bell Syst. Tech. J.*, vol. 52, pp. 843–886, Jul–Aug. 1973.
13. M. S. Park and R. A. Minasian, Ultralow noise 10 Gb/s p-i-n-HEMT optical receiver, *IEEE Photonics Technology Letters*, vol. 5, no. 2, Feb. 1993.

7 Coherent Optical Receiver

7.1 INTRODUCTORY REMARKS

Late in the first decade of the 21st millennium and then again in early 2010, ultra-broadband transmission systems of 100 Gbps employing coherent reception in association with digital signal processing emerged. The digital coherent optical receiver was first invented and patented by Taylor [1]. This has sparked tremendous interest and practical implementation of coherent reception. Coherent reception was initially developed in the 1980s when the single mode optical fiber (SMF) was invented following the multi-mode guiding of lightwaves in circular dielectric waveguide [2,3]. Since the collapse of the ".com" in the early 2000s, the slowdown in optical communications systems has allowed for research and development communities to think of higher speed techniques and PDM-QPSK modulation schemes under DSP-based coherent reception was proven to be the 100 Gbps/channel as the basic rate of the new fiber fixed line communications systems.

The design of an optical receiver depends on the modulation format of the signals and how they are transmitted through the transmitter. The modulation of the optical carrier can be in the form of amplitude, phase, and frequency. Furthermore, the phase shaping also plays a critical role in the detection and the bit error rate of the receiver and thence the transmission systems. In particular, it is dependent on the modulation in analog or digital, Gaussian or exponential pulse shape, on–off keying or multiple levels, etc.

Figure 7.1 shows the schematic diagram of a digital coherent optical receiver which is similar to the direct detection receiver but with an optical mixer at the front end. Figure 7.2 shows the small signal equivalent circuits of such receiver front end. However, the phase of the signals at base or passband of the detected signals in the electrical domain would remain in the generated electronic current and voltages at the output of the electronic pre-amplifier. An optical front end is an optical mixer combining the fields of the optical waves of the local laser and the optical signals so as the envelope of the optical signals can be beating with each other to a product with summation of the frequencies and the difference of the frequencies of the lightwaves. Only the lower frequency term which is fallen within the absorption range of the photodetector is converted into the electronic current preserving the phase and amplitude of the modulated signals.

Thence, an optical receiver front end, very much the same as that of the direct detection, is connected following the optical processing front end consisting of a photodetector for converting lightwave energy into electronic currents, an electronic pre-amplifier for further amplification of the generated electronic current, and followed by an electronic equalizer for bandwidth extension, usually in voltage form, a main amplifier for further voltage amplification, a clock recovery circuitry for regenerating the timing sequence, and a voltage-level decision circuit for sampling the waveform for the final recovery of the transmitted and received digital sequence. Therefore, the opto-electronic preamplifier is followed by a main amplifier with an automatic control to regulate the electronic signal voltage to be filtered and then sampled by a decision circuit with synchronization by a clock recovery circuitry.

The inline fiber optical amplifier can be incorporated in front of the photodetector to form an optical receiver with an optical amplifier front end to improve its receiving sensitivity. This optical amplification at the front end of an optical receiver will be discussed in this chapter dealing with optical amplification processes.

The structure of the receiver consists of four parts: The optical mixing front, the front end section, the linear channel of the main amplifier and AGC (if necessary), and the data recovery section. The optical mixing front end performs the summation of the optical fields of the local oscillator and that of

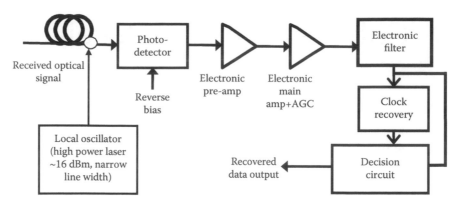

FIGURE 7.1 Schematic diagram of a digital optical coherent receiver with an additional local oscillator mixing with the received optical signals before detected by an optical receiver.

the optical signals. Polarization orientation between these lightwaves is very critical in order to maximizing the beating of the additive field in the photodiode. Depending on the frequency, difference is finite or null between these fields the resulting electronic signals derived from the detector, the electronic signals can be in the base band or the pass band, and the detection technique is termed as heterodyne or homodyne techniques, respectively.

7.2   COHERENT DETECTION

Optical coherent detection can be distinguished by the "demodulation" scheme in communications techniques in association with the following definitions: (i) coherent detection is the mixing between two lightwaves or optical carriers, one is information bearing lightwaves and the other a local oscillator with an average energy much larger than that of the signals; while (ii) demodulation refers to the recovery of baseband signals from the electrical signals.

A typical schematic diagram of a coherent optical communications employing guidedwave medium and components is shown in Figure 7.1 in which a narrow band laser incorporating an optical isolator cascaded with an external modulator is usually the optical transmitter. Information is fed via a microwave power amplifier to an integrated optic modulator, commonly used $LiNbO_3$ or EA types. The coherent detection is principal feature of coherent optical communications which can be further distinguished with heterodyne and homodyne techniques depending whether there is a difference or not between the frequencies of the local oscillator and that of the carrier of the signals. A local oscillator

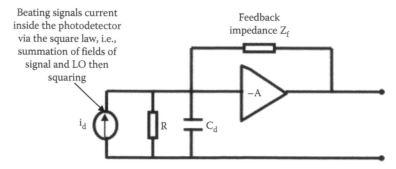

FIGURE 7.2 Schematic diagram of electronic pre-amplifier in an optical receiver of trans-impedance (TIA) electronic amplifier at the front end. The current source represents the electronic current generated in the photodetector due to the beating of the local oscillator and the optical signals. C_d = photodiode capacitance.

is a laser source whose frequency can be tuned and approximately equivalent to a monochromatic source, a polarization controller would also be used to match its polarization with that of the information-bearing carrier. The local oscillator and the transmitted signal are mixed via a polarization maintaining coupler and then detected by coherent optical receiver. Most of the previous coherent detection schemes are implemented in a mixture of photonic domain and electronic/microwave domain.

Coherent optical transmission returns into the focus of research in the mid-2000s. One significant advantage is the preservation of all the information of the optical field during detection, leading to enhanced possibilities for optical multi-level modulation. This section investigates the generation of optical multi-level modulation signals. Several possible structures of optical M-ary-PSK and M-ary-QAM transmitters are shown and theoretically analyzed. Differences in the optical transmitter configuration and the electrical driving lead to different properties of the optical multi-level modulation signals. This is shown by deriving general expressions applicable to every M-ary-PSK and M-ary-QAM modulation format and exemplarily clarified for Square-16-QAM modulation.

Coherent receivers are distinguished between synchronous and asynchronous. Synchronous detection requires an optical phase lock loop (OPLL) which recover the phase and frequency of the received signals to lock the local oscillator (LO) to that of the signal to measure the absolute phase and frequency of the signals relative to that of the LO. Thus, synchronous receivers allow direct mixing of the bandpass signals and the base band; thus, this technique is termed as homodyne reception. For asynchronous receivers, the frequency of the LO is approximately the same as that of the receiving signals and no OPLL is required. In general, the optical signals are first mixed with an intermediate frequency (IF) oscillator which is about two to three times the 3 dB passband. The electronic signals can then be recovered using electrical PLL at a lower carrier frequency in the electrical domain. The mixing of the signals and a local oscillator of an IF frequency is referred as heterodyne detection.

If no LO is used for demodulating of the digital optical signals, then differential or self-homodyne reception may be utilized, or classically termed as autocorrelation reception process or self-heterodyne detection.

Coherent communications have been an important technique in the 1980s and the early 1990s, but then research was interrupted by the advent of optical amplifiers in the late 1990s that offered up to 20 dB gain without difficulty. Nowadays, coherent systems have emerged again because of the availability of digital signal processing and low-priced components, the partly relaxed receiver requirements at high data rates, and several advantages that coherent detection provides. The preservation of the temporal phase of the coherent detection enables new methods for adaptive electronic compensation of chromatic dispersion. When concerning WDM systems, coherent receivers offer tunability and allow channel separation via steep electrical filtering. Furthermore, only the use of coherent detection permits to converge to the ultimate limits of spectral efficiency. To reach higher spectral efficiencies the use of multi-level modulation is required. Concerning this matter, coherent systems are also beneficial because all the information of the optical field is available in the electrical domain. That way complex optical demodulation with interferometric detection—which has to be used in direct detection systems—can be avoided and the complexity is transferred from the optical to the electrical domain. Several different modulation formats based on the modulation of all four quadrature of the optical field were proposed in the early 1990s, and described the possible transmitter and receiver structures and calculated the theoretical BER performance. However, a more detailed and practical investigation of multi-level modulation coherent optical systems for today's networks and data rates is missing so far.

Currently, coherent reception has attracted significant interest for the following reasons: (i) the received signals of the coherent optical receivers are in electrical domain which is proportional to that in the optical domain. This, in contrast to the direct detection receivers, allows exact electrical equalization or exact phase estimation of the optical signals. (ii) Using heterodyne receivers, DWDM channels can be separated in the electrical domain by using electrical filters with sharp roll of the passband to the cut-off band. Presently the availability of ultra-high sampling rate digital signal processors (DSP) allows users to conduct filtering in the DSP in which the filtering can be changed with easy.

However, there are disadvantages that coherent receivers would suffer: (i) coherent receivers are polarization sensitive, and that requires polarization tracking at the front end of the receiver, (ii) homodyne receivers require OPLL and electrical PLL for heterodyne that would need control and feedback circuitry, optical or electrical, which may be complicated, and (iii) for differential detection the compensation may be complicated due to the differentiation receiving nature.

In a later chapter when some advanced modulation formats are presented for optically amplified transmission systems, the use of photonic components are extensively exploited to take advantage of the advanced technology of integrated optics and planar lightwave circuits. Modulation formats of signals depend on whether the amplitude, the phase, or the frequency of the carrier is manipulated as mentioned in Chapter 2. In this chapter, the detection is coherently converted to the intermediate frequency range in the electrical domain and the signal envelop. The down converted carrier signals are detected and then recovered. Both binary level and multi-level modulation schemes employing amplitude, phase, and frequency shift keying modulation are described in this chapter.

Thus, coherent detection can be distinguished by the difference between the central frequency of the optical channel and that of the local oscillator. Three types can be classified (i) heterodyne when the difference is higher than the 3 dB bandwidth of the base band signal; (ii) homodyne when the difference is nil; and (iii) intradyne with the frequency difference fallen within the base band of the signal.

It is noted that in order for maximizing the beating signals at the output of the photodetector that polarizations of the LO and the signals must be aligned. In practice, this can be implemented but best by polarization diversity technique (Figure 7.3).

7.2.1 Optical Heterodyne Detection

The basic configuration of optical heterodyne detection is shown in Figure 7.4. The local oscillator whose frequency can be higher or lower than that of the carrier, is mixed with the information-bearing carrier, thus allowing down or up conversion of the information signals to the intermediate frequency range. The down-converted electrical carrier and signal envelope is received by the photodetector. This combined lightwave is converted by the PD into electronic current signals that are filtered by an electrical bandpass filter (BPF), and then demodulated by a demodulator. A low pass filter is also used to remove higher-order harmonics of the non-linear detection photodetection process, the square-law detection. Under the use of an envelope detector, the process is asynchronous, hence the name term asynchronous detection. If the down-converted carrier is recovered and then mixed

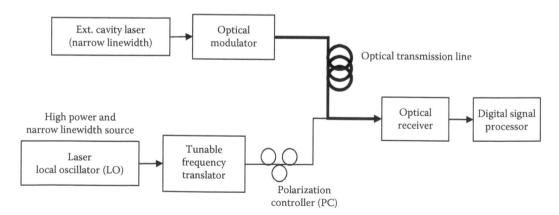

FIGURE 7.3 Typical arrangement of coherent optical communications systems [4,5] LD/LC is a very narrow linewidth laser diode as a local oscillator without any phase locking to the signal optical carrier. PM coupler is polarization maintaining fiber coupled device, *PC* = polarization controller.

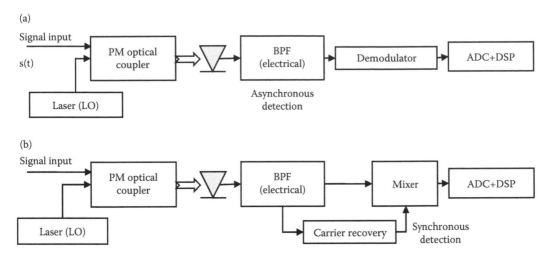

FIGURE 7.4 Schematic diagram of optical heterodyne detection—(a) asynchronous and (b) synchronous, receiver structures. BPF is a bandpass filter. PD = photodiode; ADC = analog to digital converter; DSP = digital signal processor.

with IF signals, this is synchronous detection. It is noted that the detection is conducted at the intermediate frequency range in electrical domain, hence the needs for controlling the stability of the frequency spacing between the signal carrier and the local oscillator. That means the mixing of these carriers would result in an IF carrier in the electrical domain prior to the mixing process or envelope detection to recover the signals.

The coherent detection thus relies on the electric field component of the signal and the local oscillator. The polarization alignment of these fields is critical for the optimum detection. The electric field of the optical signals and the local oscillator can be expressed as

$$E_s(t) = \sqrt{2P_s(t)}\cos\{\omega_s t + \phi_s + \varphi(t)\} \qquad (7.1)$$

$$E_{LO} = \sqrt{2P_L}\cos\{\omega_{LO}t + \phi_{LO}\} \qquad (7.2)$$

where $P_s(t)$ and P_{LO} are the instantaneous signal power and average power of the signals and local oscillator, respectively, $\omega_s(t)$ and ω_{LO} are the signal and local oscillator angular frequencies, ϕ_s and ϕ_{LO} are the phase including any phase noise of the signal and the local oscillator, and $\psi(t)$ is the modulation phase. The modulation can be amplitude with the switching on and off (amplitude shift keying —ASK) of the optical power or phase or frequency with the discrete or continuous variation of the time-dependent phase term. For discrete phase it can be phase shift keying (PSK), differential PSK (DPSK) or differential quadrature PSK—DQPSK, and when the variation of the phase is continuous we have frequency shift keying if the rate of variation is different for the bit "1" and bit "0."

Under an ideal alignment of the two fields, the photodetection current can be expressed by

$$i(t) = \frac{\eta q}{h\upsilon}\left[P_s + P_{LO} + 2\sqrt{P_s P_{LO}}\cos\{(\omega_s - \omega_{LO})t + \phi_s - \phi_{LO} + \varphi(t)\}\right] \qquad (7.3)$$

where the higher frequency term (the sum) is eliminated by the photodetector frequency response, η is the quantum efficiency, q is the electronic charge, h is Plank's constant, and υ is the optical frequency.

Thus, the power of the local oscillator dominates the shot noise process and, at the same time, boosts the signal level, hence enhancing the signal to noise ratio. The oscillating term is the beating between the local oscillator and the signal and the signal proportion with the amplitude is the square root of the product of the power of the local oscillator and the signal.

The electronic signal power S and shot noise N_s can be expressed as

$$S = 2\mathfrak{R}^2 P_s P_{LO}$$
$$N_s = 2q\mathfrak{R}(P_s + P_{LO})B$$
$$\mathfrak{R} = \frac{\eta q}{h\nu} = \text{responsivity}$$

(7.4)

where B is the 3 dB bandwidth of the electronic receiver. Thus, the optical signal-to-noise ratio (OSNR) can be written as

$$OSNR = \frac{2\mathfrak{R}^2 P_s P_{LO}}{2q\mathfrak{R}(P_s + P_{LO})B + N_{eq}}$$

(7.5)

where N_{eq} is the total electronic noise equivalent power at the input to the electronic pre-amplifier of the receiver. From this equation, we can observe that if the power of the local oscillator is significantly increased so that the shot noise dominates over the equivalent noise, and, at the same time, increasing the signal to noise ratio SNR, the sensitivity of the coherent receiver can only be limited by the quantum noise inherent in the photodetection process. Under this quantum limit, the $OSNR_{QL}$ is given by

$$OSNR_{QL} = \frac{\mathfrak{R}P_s}{qB}$$

(7.6)

7.2.2 ASK Coherent System

Under the ASK modulation scheme, the demodulator in Figure 7.4 is an envelope detector (in lieu of the demodulator) followed by a decision circuitry. That is, the eye diagram is obtained and a sampling instant is established with a clock recovery circuit. While the synchronous detection would require a locking between the carrier frequencies be obtained, and then tuned to the local oscillator frequency according to track the frequency component of the signal. The amplitude demodulated envelope can be expressed as

$$r(t) = 2\mathfrak{R}\sqrt{P_s P_{LO}} \cos(\omega_{IF})t + n_x \cos(\omega_{IF})t + n_y \sin(\omega_{IF})t$$
$$\omega_{IF} = \omega_s - \omega_{LO}$$

(7.7)

The intermediate frequency (IF), ω_{IF}, is the difference between those of the LO and the signal carrier, and n_x and n_y are the expected values of the orthogonal noise power components which are random variables.

$$r(t) = \sqrt{[2\mathfrak{R}P_s P_{LO} + n_x]^2 + n_y^2} \cos(\omega_{IF}t + \Phi)t \quad \text{with}\ldots \Phi = \tan^{-1}\frac{n_y}{2\mathfrak{R}P_s P_{LO} + n_x}$$

(7.8)

7.2.3 Envelop Detection

The noise power terms can be assumed to follow a Gaussian probability distribution and are independent with each other with a zero mean and a variance σ, the probability density function (PDF) can thus be given as

$$p(n_x, n_y) = \frac{1}{2\pi\sigma^2} e^{\frac{-(n_x^2 + n_y^2)}{2\sigma^2}}$$

(7.9)

with respect to the phase and amplitude, this equation can be written as [5]

$$p(\rho, \phi) = \frac{\rho}{2\pi\sigma^2} e^{\frac{-(\rho^2 + A^2 - 2A\rho\cos\phi)}{2\sigma^2}} \tag{7.10}$$

where

$$\rho = \sqrt{\left[2\Re\sqrt{P_s(t)P_{LO}} + n_s(t)\right]^2 + n_y^2(t)} \tag{7.11}$$

$$A = 2\Re\sqrt{P_s(t)P_{LO}}$$

The PDF of the amplitude can be obtained by integrating the phase amplitude pdf over the range of 0–2π and given as

$$p(\rho) = \frac{\rho}{\sigma^2} e^{\frac{-(\rho^2 + A^2)}{2\sigma^2}} I_0\left\{\frac{A\rho}{\sigma^2}\right\} \tag{7.12}$$

where I_0 is the modified Bessel's function. If a decision level is set to determine the "1" and "0" level, then the probability of error and the bit error rate (BER) can be obtained if assuming an equal probability of error between the "1s" and "0s" is equal, as

$$BER = \frac{1}{2}P_e^1 + \frac{1}{2}P_e^0 = \frac{1}{2}\left[1 - Q\left(\sqrt{2\delta}, d\right) + e^{-\frac{d^2}{2}}\right] \tag{7.13}$$

where Q is the Magnum function and δ is given by

$$\delta = \frac{A^2}{2\sigma^2} = \frac{2\Re^2 P_s P_{LO}}{2q\Re(P_s + P_{LO})B + i_{N_{eq}}^2} \tag{7.14}$$

when the power of the local oscillator is much larger than that of the signal and the equivalent noise current power, then this SNR becomes

$$\delta = \frac{\Re P_s}{qB} \tag{7.15}$$

The physical representation of the detected current and the noises current due to the quantum shot noise and noise equivalent of the electronic pre-amplification can be seen in Figure 7.5, in which the

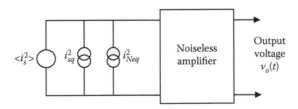

FIGURE 7.5 Equivalent current model at the input of the optical receiver, average signal current and equivalent noise current of the electronic pre-amplifier as seen from its input port.

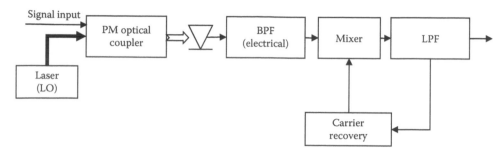

FIGURE 7.6 Schematic diagram of optical heterodyne detection for PSK format.

signal current can be general and derived from the output of the detection scheme, that from a photodetector or a back-to-back pair of photodetectors of a balanced receiver for detecting the phase difference of DPSK, DQPSK, or CPFSK signals and converting to amplitudes.

The BER is optimum when setting its differentiation with respect to the decision level δ, an approximate value of the decision level can be obtained as

$$d_{opt} \cong \sqrt{2 + \frac{\delta}{2}} \Rightarrow BER_{ASK-e} \cong \frac{1}{2} e^{-\frac{\delta}{4}} \tag{7.16}$$

7.2.4 Synchronous Detection

ASK can be detected using synchronous detection*, and the BER is given by

$$BER_{ASK-S} \cong \frac{1}{2} erfc \frac{\sqrt{\delta}}{2} \tag{7.17}$$

7.2.5 PSK Coherent System

Under the phase shift keying modulation format the detection is similar to that of Figure 7.4 for heterodyne detection (see Figure 7.6), but after the BPF, an electrical mixer is used to track the phase of the detected signal. The received signal is given by

$$r(t) = 2\Re\sqrt{P_s P_{LO}} \cos [(\omega_{IF})t + \varphi(t)] + n_x \cos (\omega_{IF})t + n_y \sin (\omega_{IF})t \tag{7.18}$$

The information is contained in the time-dependent phase term $\varphi(t)$.

When the phase and frequency of the voltage control oscillator (VCO) are matched with those of the signal carrier, the received electrical signal can be simplified to

$$r(t) = 2\Re\sqrt{P_s P_{LO}} a_n(t) + n_x$$
$$a_n(t) = \pm 1 \tag{7.19}$$

Under the Gaussian statistical assumption, the probability of the received signal of a "1" is given by

$$p(r) = \frac{1}{\sqrt{2\pi\sigma^2}} e^{-\frac{(r-u)^2}{2\sigma^2}} \tag{7.20}$$

---

* Synchronous detection is implemented by mixing the signals and a strong local oscillator in association with the phase locking of the local oscillator to that of the carrier.

Furthermore, the probability of the "0" and "1" are assumed to be equal. We can obtain the BER as the total probability of the received "1" and "0" as

$$BER_{PSK} = \frac{1}{2} erfc(\delta) \tag{7.21}$$

7.2.6 DIFFERENTIAL DETECTION

As observed in the synchronous or coherent detection, a carrier-recovery circuit is necessary that can be normally implemented by using a an OPLL. This OPLL can complicate the overall receiver structure. To overcome this complexity, it is possible to detect the signal by a self-homodyne process by beating the carrier of one bit period to that of the next consecutive bit, this is called the differential detection. The detection process can be modified, as shown in Figure 7.7, in which the phase of the IF carrier of one bit is compared with that of the next bit and a difference is recovered to represent the bit "1" or "0." This requires a differential coding at the transmitter and an additional phase comparator for the recovery process. In later chapters, on differential PSK the differential decoding is implemented in photonic domain via a photonic phase comparator in form of a MZ delay interferometer (MZDI) with a thermal section for tuning the delay time of the optical delay line. The BER can be expressed as

$$BER_{DPSK-e} \cong \frac{1}{2} e^{-\delta} \tag{7.22}$$

$$r(t) = 2\Re\sqrt{P_s P_{LO}} \cos\left[\pi A_k s(t)\right] \tag{7.23}$$

where s(t) is the modulating waveform and A_k represents the bit "1" or "0." This is equivalent to the baseband signal and the ultimate limit is the BER of the baseband signal.

The noise is dominated by the quantum shot noise of the local oscillator with its square noise current is given by

$$i^2_{N-sh} = 2q\Re(P_s + P_{LO}) \int_0^\infty |H(j\omega)|^2 d\omega \tag{7.24}$$

where $H(j\omega)$ is the transfer function of the receiver system, normally a trans-impedance of the electronic pre-amp and that of a matched filter. As the power of the local oscillator is much larger than the signal and integrating over the dB bandwidth of the transfer function, this current can be approximated by

$$i^2_{N-sh} \simeq 2q\Re P_{LO}B \tag{7.25}$$

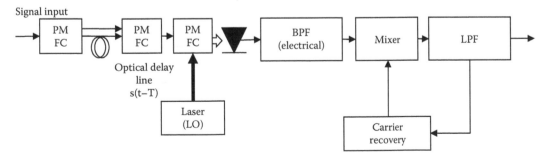

FIGURE 7.7 Schematic diagram of optical heterodyne and differential detection for PSK format.

Hence, the signal-to-noise ratio (power) SNR is given by

$$SNR \equiv \delta \simeq \frac{2\Re P_s}{qB} \tag{7.26}$$

The BER is the same as that of a synchronous detection and is given by

$$BER_{\text{homodyne}} \cong \frac{1}{2} erfc\sqrt{\delta} \tag{7.27}$$

The sensitivity of the homodyne process is at least 3 dB better than that of the heterodyne, and the bandwidth of the detection is half of its counter part due to the double sideband nature of the heterodyne detection.

7.2.7 FSK COHERENT SYSTEM

The nature of FSK is based on the two frequency components that determine the bits "1" and "0." There are a number of formats related to FSK, depending on whether the change of the frequencies representing the bits is continuous or non-continuous, the FSK or CPFSK modulation formats. For non-continuous FSK, the detection is usually performed by a structure of dual frequency discrimination, as shown in Figure 7.8, in which two narrow band filters are used to extract the signals. For CPFSK, both the frequency discriminator and balanced receiver for PSK detection can be used. The frequency discrimination is preferred compared to the balanced receiving structures because it would eliminate the phase contribution by the local oscillator or optical amplifiers which may be used as an optical pre-amp.

When the frequency difference between the "1" and "0" equals to a quarter of the bit rate, the FSK can be termed as the minimum shift keying (MSK) modulation scheme. At this frequency spacing, the phase is continuous between these states.

7.3 OPTICAL HOMODYNE DETECTION

Optical homodyne detection matches the transmitted signal phases to that of the local oscillator phase signal. A schematic representation of the optical receiver is shown in Figure 7.9. The field of the incoming optical signals is mixed with the local oscillator whose frequency and phase are locked with that of the signal carrier waves via a phase lock loop. The resultant electrical signal is then filtered and thence a decision circuitry.

7.3.1 DETECTION AND OPTICAL PLL

Optical homodyne detection requires the phase matching of the frequency of the signal carrier and that of the local oscillator. This type of detection would give a very sensitive, in principle, of

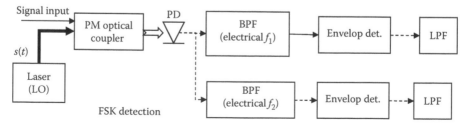

FIGURE 7.8   Schematic diagram of optical homodyne detection of FSK format.

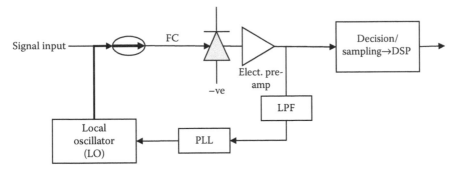

FIGURE 7.9   General structure of an optical homodyne detection system. FC = fiber coupler, LPF = low pass filter, and PLL = phase lock loop.

9 photons/bit. Implementation of such system would normally require an optical PLL whose structure of a recent development [6] is as shown in Figure 7.10. The local oscillator frequency is locked into the carrier frequency of the signals by shifting it to the modulated sideband component via the use of the optical modulator. A single sideband optical modulator is preferred. However, a double sideband may also be used. This modulator is excited by the output signal of a voltage controlled oscillator whose frequency is determined by the voltage level of the output of an electronic bandpass filter conditioned to meet the required voltage level for driving the electrode of the modulator. The frequency of the local oscillator is normally tuned to the region such that the frequency difference, with respect to the signal carrier, is fallen within the passband of the electronic filter. When the frequency difference is zero, there is no voltage level at the output of the filter, and thus, the optical PLL has reached the final stage of locking. The bandwidth of the optical modulator is important because it can extend the locking range between the two optical carriers.

Any frequency offset between the LO and the carrier is detected and noise filtered by the low pass filter (LPF). This voltage level is then fed to a voltage control oscillator to generate a sinusoidal wave that is then used to modulate an intensity modulator modulating the lightwaves of the LO. The output spectrum of the modulator would exhibit two sideband and the LO lightwave. One of these components would then be locked to the carrier. A closed loop would ensure a stable locking. If the intensity modulator is biased at the minimum transmission point and the voltage level at the output of the VCO

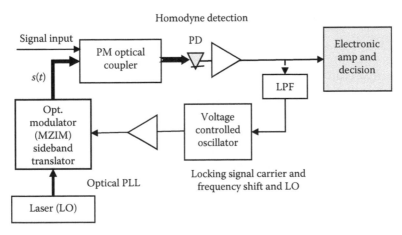

FIGURE 7.10   Schematic diagram of optical homodyne detection—electrical line (dashed) optical line (continuous and solid) using an OPLL.

is adjusted to $2V_\pi$ with driven signals of $\pi/2$ phase shift with each other, then we would have carrier suppression and sideband suppression. This eases the confusion of the closed loop locking.

Under a perfect phase matching, the received signal is given by

$$i_s(t) = 2\Re\sqrt{p_s P_{LO}}\cos\{\pi a_k s(t)\} \tag{7.28}$$

where a_k takes the value ± 1, and $s(t)$ is the modulating waveform. This is a baseband signal, and thus the error rate is the same as that of the baseband system.

The shot noise power induced by the local oscillator and the signal power can be expressed as

$$i_{NS}^2 = 2q\Re(p_s + P_{LO})\int_0^\infty |H(j\omega)|\,d\omega \tag{7.29}$$

where the transfer function is $|H(j\omega)|$ of the receiver whose expression, if under a matched filtering, can be

$$|H(j\omega)|^2 = \left[\frac{\sin(\omega T/2)}{\omega T/2}\right]^2 \tag{7.30}$$

where T is the bit period. Then the noise power becomes:

$$i_{NS}^2 = q\Re(p_s + P_{LO})\frac{1}{T} \simeq \frac{q\Re P_{LO}}{T}$$

$$\text{when}\ldots p_s \ll P_{LO} \tag{7.31}$$

Thus, the signal-to-noise ratio is

$$SNR = \frac{2\Re p_s P_{LO}}{q\Re P_{LO}/T} = \frac{2p_s T}{q} \tag{7.32}$$

thence, the bit error rate is

$$P_E = \frac{1}{2}erfc\left(\sqrt{SNR}\right) \rightarrow BER = erfc\left(\sqrt{SNR}\right) \tag{7.33}$$

7.3.2 Quantum Limit Detection

For homodyne detection, a super quantum limit can be achieved. In this case, the local oscillator is used in a very special way that matches the incoming signal field in polarization, amplitude, and frequency, and is assumed to be phase-locked to the signal. Assuming that the phase signal is perfectly modulated such that it acts in-phase or counter-phase with the local oscillator, the homodyne detection would give a normalized signal current of

$$i_{sC} = \frac{1}{2T}\left[\mp\sqrt{2n_p} + \sqrt{2n_{LO}}\right]^2 \ldots \text{ for}..0 \le t \le T \tag{7.34}$$

Assuming further that $n_p = n_{LO}$ is the number of photon for the LO for generation of detected signals, then the current can be replaced with $4\,n_p$ for the detection of a "1" and nothing for a "0" symbol.

7.3.3   LINEWIDTH INFLUENCES

7.3.3.1   Heterodyne Phase Detection

When the linewidth of the light sources is significant,, the intermediate frequency deviates due to a phase fluctuation and the probability density function is related to the this linewidth conditioned on the deviation $\delta\omega$ of the intermediate frequency. For a signal power of p_s, we have the total probability of error is given as

$$P_E = \int_{-\infty}^{\infty} P_C(p_s, \partial\omega)p_{IF}(\partial\omega)\partial\omega \tag{7.35}$$

The probability density function of the intermediate frequency under a frequency deviation can be written as [7]

$$p_{IF}(\partial\omega) = \frac{1}{\sqrt{\Delta\upsilon BT}} e^{-\frac{\partial\omega^2}{4\pi\Delta\upsilon B}} \tag{7.36}$$

where $\Delta\nu$ is the is full linewidth at FWHM of the power spectral density and T is the bit period.

7.3.3.2   Differential Phase Detection with Local Oscillator

7.3.3.2.1   DPSK Systems

The DPSK detection requires an MZDI and a balanced receiver either in the optical domain or in the electrical domain. If in the electrical domain, then the beating signals in the PD between the incoming signals and the LO would give the beating electronic current, which is then split and one branch is delay by one bit period and then summed up. The heterodyne signal current can be expressed as [8]

$$i_s(t) = 2\Re\sqrt{P_{LO}p_s}\cos(\omega_{IF}t + \phi_s(t)) + n_x(t)\cos\omega_{IF}t - n_y(t)\sin\omega_{IF}t \tag{7.37}$$

The phase $\phi_s(t)$ is expressed by

$$\phi_s(t) = \varphi_s(t) + \{\varphi_N(t) - \varphi_N(t+T)\} - \{\varphi_{pS}(t) - \varphi_{pS}(t+T)\} - \{\varphi_{pL}(t) - \varphi_{pL}(t+T)\} \tag{7.38}$$

The first term is the phase of the data and takes the value 0 or π. The second term represents the phase noise due to shot noise of the generated current, and the third and fourth terms are the quantum shot noise due to the local oscillator and the signals. The probability of error is given by

$$P_E = \int_{-\pi/2}^{\pi/2} \int_{-\infty}^{\infty} p_n(\phi_1 - \phi_2)p_q(\phi_1)\partial\phi_1\partial\phi_1 \tag{7.39}$$

where $p_n(.)$ is the probability density function of the phase noise due to the shot noise, and $p_q(.)$ is for the quantum phase noise generated from the transmitter and the local oscillator [9].

The probability or error can be written as

$$p_N(\phi_1 - \phi_2) = \frac{1}{2\pi} + \frac{\rho e^{-\rho}}{\pi} \sum_{m=1}^{\infty} a_m \cos(m(\phi_1 - \phi_2))$$

$$a_m \sim \left\{ \frac{2^{m-1}\Gamma\left[\frac{m+1}{2}\right]\Gamma\left[\frac{m}{2}+1\right]}{\Gamma[m+1]} \left[I_{m-1/2}\frac{\rho}{2} + I_{(m+1)/2}\frac{\rho}{2}\right] \right\}^2 \tag{7.40}$$

where $\Gamma(.)$ is the gamma function and is the modified Bessel function of the first kind. The PDF of the quantum phase noise can be given as [10]

$$p_q(\phi_1) = \frac{1}{\sqrt{2\pi D\tau}} e^{\frac{\phi_1^2}{2D\tau}} \tag{7.41}$$

where D is the phase diffusion constant, the standard deviation from the central frequency given as

$$\Delta v = \Delta v_R + \Delta v_L = \frac{D}{2\pi} \tag{7.42}$$

is the sum of the linewidth (FWHM) of the transmitter and that of the local oscillator [11]. Substituting Equations 7.41 and 7.40 into Equation 7.39, we obtain

$$P_E = \frac{1}{2} + \frac{\rho e^{-\rho}}{2} \sum_{n=0}^{\infty} \frac{(-1)^n}{2n+1} e^{-(2n+1)^2 \pi \Delta v T} \left\{ I_{n-1/2} \frac{\rho}{2} + I_{(n+1)/2} \frac{\rho}{2} \right\}^2 \tag{7.43}$$

This equation gives the probability of error as a function of the received power. The probability of error is plotted against the receiver sensitivity and the product of the linewidth with the bit rate (or the relative bandwidth of the laser line width and the bit rate) shown in Figure 7.11 for DPSK modulation format at 140 Mb/s bit rate and the variation of the laser linewidth from 0 to 2 MHz.

7.3.3.3 Differential Phase Detection under Self-Coherence

Recently, the laser linewidth requirement for DQPSK modulation and differential detection for DQPSK has also been studied, and no local oscillator being used means self-coherent detection. It has been shown that, for the linewidth of up to 3 MHz, the transmitter laser would not significantly influence the probability of error, as shown in Figure 7.12 [8]. Figure 7.13 shows the maximum linewidth of a laser source in a 10-GSymbols/s system. The loose bound is to neglect linewidth if the impact is to double the BER with the tighter bound being to neglect linewidth if the impact is a 0.1-dB SNR penalty.

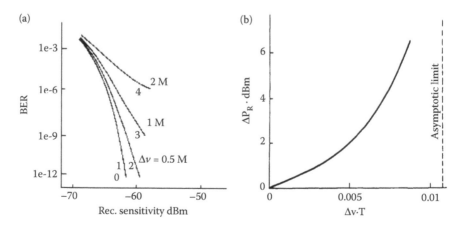

FIGURE 7.11 (a) Probability of error versus receiver sensitivity with linewidth as a parameter in MHz. (b) Degradation of optical receiver sensitivity at BER = 1e-9 for DPSK systems as a function of the linewidth and bit period—bit rate = 140 Mb/s. (Extracted from G. Nchoson, *Elect Lett.*, vol. 20/24, pp. 1005–1007, 1984.)

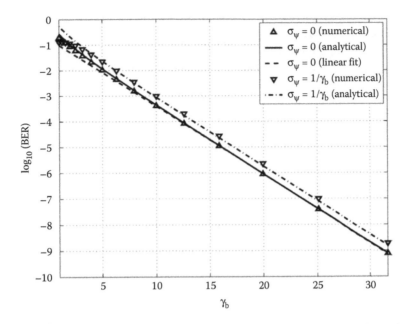

FIGURE 7.12 Analytical approximation (solid line) and numerical evaluation (triangles) of the BER for the cases of zero linewidth and that required to double the BER. The dashed line is the linear fit for zero linewidth. Bit rate 10 Gb/s per channel. (Extracted from S. Savory and T. Hadjifotiou, *IEEE Photonic Tech Lett.*, vol. 16, no. 3, pp. 930–932, 2004.)

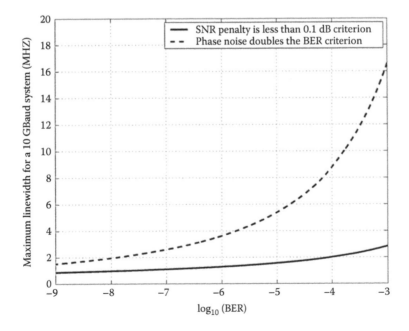

FIGURE 7.13 Criteria for neglecting linewidth in a 10-GSymbols/s system. The loose bound is to neglect linewidth if the impact is to double the BER with the tighter bound being to neglect linewidth if the impact is a 0.1-dB SNR penalty. Bit rate 10 GSy/s. (Extracted from G. Nchoson, *Elect Lett.*, vol. 20/24, pp. 1005–1007, 1984.)

7.3.3.4 Differential Phase Coherent Detection of Continuous Phase FSK Modulation Format

The probability of error of CPFSK can be derived by taking into consideration the delay line of the differential detector, the frequency deviation, and phase noise [12]. Similar to Figure 7.8 [13], the differential detector configuration is shown in Figure 14a, and the conversion of frequency to voltage relationship in Figure 14b. If heterodyne detection is employed, then a bandpass filter is used to bring the signals back to the electrical domain.

The detected signal phase at the shot noise limit at the output of the low-pass filter (LPF) can be expressed as

$$\phi(t) = \omega_c t + a_n \frac{\Delta\omega}{2}\tau + \varphi(t) + \varphi_n(t)$$

$$\text{with}\ldots\ldots \omega_c = 2\pi f_c = (2n+1)\frac{\pi}{2\tau} \tag{7.44}$$

where τ is the differential detection delay time, $\Delta\omega$ is the deviation of the angular frequency of the carrier for the "1" or "0" symbol; $\varphi(t)$ is the phase noise due to the shot noise, $n(t)$ is the phase noise due to the transmitter and the local oscillator quantum shot noise takes the values of ± 1, the binary data symbol.

Thus, by integrating the detected phase from $-(\Delta\omega/2)\tau \rightarrow \pi - (\Delta\omega/2)\tau$, we obtain the probability of error as

$$P_E = \int_{-\frac{\Delta\omega}{2}\tau}^{\pi - \frac{\Delta\omega}{2}\tau/2} \int_{-\infty}^{\infty} p_n(\phi_1 - \phi_2) p_q(\phi_1) \partial\phi_1 \partial\phi_1 \tag{7.45}$$

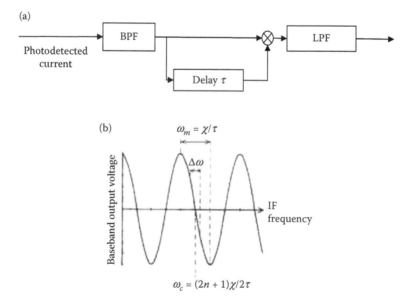

FIGURE 7.14 (a) Configuration of a CPFSK differential detection and (b) frequency to voltage conversion relationship of FSK differential detection. (Extract from K. Iwashita and T. Masumoto, *IEEE J. Lightwave Tech.*, vol. LT-5/4, pp. 452–462, 1987. (Figure 1).

Similar to the case of DPSK system, substituting Equations 7.40 and 7.41 into Equation 7.45, we obtain

$$P_E = \frac{1}{2}\frac{\rho e^{-\rho}}{2}\sum_{n=0}^{\infty}\frac{(-1)^n}{2n=1}e^{-(2n+1)^2\pi\Delta v\tau}\left\{I_{n-1/2}\frac{\rho}{2}+I_{(n+1)/2}\frac{\rho}{2}\right\}^2 e^{-(2n+1)^2\pi\Delta v\tau}\cos\{(2n+1)\alpha\}$$

(7.46)

$$\alpha = \frac{\pi(1-\beta)}{2}.. \text{ and }..\beta = \frac{\Delta\omega}{\omega_m} = 2\pi\tau/T_0$$

where ω_m is the deviation of the angular frequency with m as the modulation index, and T_0 is the pulse period or bit period. The modulation index parameter b is defined as the ration of the actual frequency deviation to the maximum frequency deviation. Figure 7.15 shows the dependence of degradation of the power penalty to achieve the same BER as a function of the linewidth factor $\Delta v\tau$ and the modulation index β.

7.3.3.5 Optical Intra-Dyne Detection

Optical phase diversity receivers combine the advantages of the homodyne with minimum signal processing bandwidth and heterodyne reception with no optical phase locking required. The term diversity is well known in radio transmission links that describes the transmission over more than one path. In optical receivers, the optical paths consist of paths of different polarized waves and those due to the $\pi/2$ orthogonal phases. In intradyne detection, the frequency difference, the intermediate frequency or, the LOFO (local oscillator frequency offset), between the LO and the central carrier is non-zero, and lies within the signal bandwidth of the baseband signal, as illustrated in Figure 7.16 [14]. Naturally, the control and locking of the carrier and the local oscillator can not be exact, sometimes due to jittering of the source, and, most of the time, the laser frequency is locked stably by oscillating the reflection mirror, hence the central frequency is varied by a few hundreds of kilohertz. Thus, intradyne coherent detection is more realistic. Furthermore, the digital signal processor in modern coherent reception systems would be able to extract this difference without much difficulty in the digital domain [15]. Obviously, the heterodyne detection would require large frequency range of operation of electronic devices, while homodyne and intradyne reception require simpler electronics. Differential or non-differential format can be used in DSP-based coherent reception. For differential-based reception, the differential decoding would gain advantage when there are slips in the cycles of bits due to walk-off of the pulse sequences over very long transmission non-compensating fiber line.

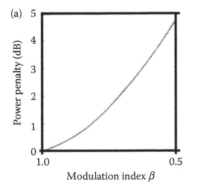
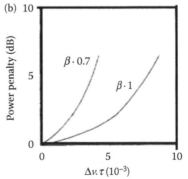

FIGURE 7.15 (a) Dependence of receiver power penalty at BER of 1e-9 on modulation index β (ratio between frequency deviation and maximum frequency spacing between f_1 and f_2) and (b) receiver power penalty at BER 1e-9 as a function of product of beat bandwidth and bit delay time-effects excluding LD phase noise. (Extracted from K. Iwashita and T. Masumoto, *IEEE J. Lightwave Tech.*, vol. LT-5/4, pp. 452–462, 1987, Figures 2 and 3.)

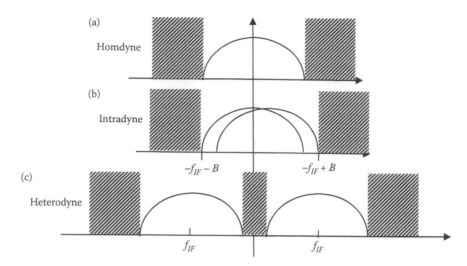

FIGURE 7.16 Spectrum of coherent detection (a) homodyne, (b) intradyne, and (c) heterodyne.

The diversity in phase and polarization can be achieved by using a $\pi/2$ hybrid coupler that splits the polarization of the LO and the received channels and mixing with $\pi/2$ optical phase shift, then the mixed signals are detected by a balanced photodetectors. This diversity detection is described in the next few sections (see also Figure 7.24).

7.4 SELF-COHERENT DETECTION AND ELECTRONIC DSP

The coherent techniques described above would offer significant improvement, but face a setback because of the availability of stable local oscillator and an OPLL for locking the frequency of the local oscillator and that of the signal carrier.

Digital signal processors (DSP) have been widely used in wireless communications and play key roles in the implementation of DSP-based coherent optical communication systems. DSP techniques are applied for coherent optical communication systems to overcome the difficulties of optical phase-locking loop (PLL), and to also improve the performance of the transmission systems in the presence of fiber degrading effects including chromatic dispersion, PMD, and fiber non-linearities.

Coherent optical receivers have the following advantages: (i) the shot-noise limited receiver sensitivity can be achieved with a sufficient local oscillator (LO) power; (ii) closely spaced WDM channels can be separated with electrical filters having sharp roll off characteristics; and (iii) the ability of phase detection can improve the receiver sensitivity compared with the IMDD system [16]. In addition, any kind of multi-level phase modulation formats can be introduced by using the coherent receiver. While the spectral efficiency of binary modulation formats is limited to 1 bit/s/Hz/polarization (which is called the Nyquist limit), multi-level modulation formats with N bits of information per symbol can achieve up to the spectral efficiency of N bit/s/Hz/polarization. Recent research has focused on M-ary phase-shift keying (M-ary PSK) and even quadrature amplitude modulation (QAM) with coherent detection, which can increase the spectral efficiency by a factor of $\log_2 M$ [17–19]. Moreover, for the same bit rate, since the symbol rate is reduced, the system can have higher tolerance to chromatic dispersion and PMD.

However, one of the major challenges in coherent detection is to overcome the carrier phase noise when using a local oscillator (LO) to beat with the received signals to retrieve the modulated phase information. Phase noise can result from lasers, which will cause a power penalty to the receiver sensitivity. A self-coherent multi-symbol detection of optical differential M-ary PSK is introduced to improve the system performance; however, higher analog-to-digital conversion resolution and more digital signal-processing power are required as compared to a digital coherent receiver [20]. Furthermore, differential encoding is also necessary in this scheme. As for the coherent receiver, initially, an optical phase-locked loop (PLL) is an option to track the carrier phase with respect to the LO carrier in homodyne detection. However, an optical PLL operating at optical wavelengths in combination with distributed feedback (DFB) lasers may be quite difficult to be implemented because the product of laser linewidth and loop delay is too large [21]. Another option is to use electrical PLL to track the carrier phase after down converting the optical signal to an intermediate frequency (IF) electrical signal in a heterodyne detection receiver as mentioned above. Compared to heterodyne detection, homodyne detection offers better sensitivity and requires a smaller receiver bandwidth [22]. On the other hand, coherent receivers employing high-speed analog-to-digital converters (ADCs) and high-speed baseband digital signal processing (DSP) units are becoming increasingly attractive, rather than using an optical PLL for demodulation. A conventional block Mth power phase estimation (PE) scheme is proposed in References 16, 21 to raise the received M-ary PSK signals to the Mth power to estimate the phase reference in conjunction with a coherent optical receiver. However, this scheme requires non-linear operations, such as taking the Mth power and the $\tan^{-1}(\cdot)$, and resolving the $\pm 2\pi/M$ phase ambiguity, which incurs a large latency to the system. Such non-linear operations would limit further potential for real-time processing of the scheme. In addition, non-linear phase noises always exist in long-haul systems due to the Gordon-Mollenauer effect [23], which severely affect the performance of a phase-modulated optical system [24]. The results in Reference 25 show that such Mth power PE techniques may not effectively deal with non-linear phase noise.

The maximum likelihood (*ML*) carrier phase estimator derived in Reference 26 can be used to approximate the ideal synchronous coherent detection in optical PSK systems. The *ML* phase estimator requires only linear computations, and thus it is more feasible for online processing for real systems. Intuitively, one can show that the *ML* estimation receiver outperforms the Mth power block phase estimator and conventional differential detection, especially when the non-linear phase noise is dominant, thus significantly improving the receiver sensitivity and tolerance to the non-linear phase noise. The algorithm of *ML* phase estimator is expected to improve the performance of coherent optical communication systems using different M-ary PSK and Quadrature Amplitude Modulation (QAM) formats. The improvement by DSP at the receiver end can be significant for the transmission systems in the presence of fiber degrading effects, including chromatic dispersion, PMD, and non-linearities for single channel and DWDM systems.

7.5 ELECTRONIC AMPLIFIERS: RESPONSES AND NOISES

7.5.1 INTRODUCTORY REMARKS

Electronic amplifier as a pre-amplification stage of an optical receiver plays a major role in the detection of optical signals so that optimum signal-to-noise ratio (*SNR*) and thence the optical *SNR* (*OSNR*) can be derived based on the photodetector responsivity. Under coherent detection, the amplifier noises must be much less than that of the quantum shot noises contributed by the high power level of the *LO*, which is normally about 10 dB above that of the signal average power.

Thus, this section introduces electronic amplifiers for wideband signals applicable to ultra-high speed, high gain and low noise trans-impedance amplifiers (TIA). We concentrate on differential input TIA, but address the detailed design of a single-input, single-output with noise suppression technique in Section 7.7 with design strategy for achieving stability in the feedback amplifier as well as low noise and wide bandwidth. Electronic noise of the pre-amplifier stage is defined as the total equivalent input noise spectral density, that is, all the noise sources (current and voltage sources) of all elements of the amplifier are referred to the input port of the amplifier, and thence an equivalent current source is found, thence the current density is derived. Once this current density is found, the total equivalent at the input can be found when the overall bandwidth of the receiver is determined. When this current is known, and with the average signal power, we can obtain the SNR at the input stage of the optical receiver, thence the OSNR. On the other hand, if the OSNR is required at the receiver is determined for any specific modulation format, then with the assumed optical power of the signal available at the front of the optical receiver and the responsivity of the photodetector we can determine the maximum electronic noise spectral density allowable by the pre-amplification stage, and hence the design of the amplifier electronic circuit.

The principal function of an opto-electronic receiver is to convert the received optical signals into electronic equivalent signals, then amplification and sampling and processed to recover properties of the original shapes and sequence. At first, the optical domain signals must be converted to electronic current in the photodetection device, the photodetector of either *p-i-n* or avalanche photodiode (APD) in which the optical power is absorbed in the active region and both electrons and holes generated are attracted to the positive and negative biased electrodes, respectively. Thus, the generated current is proportional to the power of the optical signals, hence the name "square law" detection. The *p-i-n* detector is structured with a *p+* and *n+*: Doped regions sandwiched the intrinsic layer in which the absorption of optical signal occurs. A high electric field is established in this region by reverse biasing the diode, and thence electrons and holes are attracted to either sides of the diode, thus generation of current. Similarly, an APD works with the reverse biasing level close to the reverse breakdown level of the *pn* junction (no intrinsic layer) so that electronic carriers can be multiplied in the avalanche flow when the optical signals are absorbed.

This photo-generated current is then fed into an electronic amplifier whose transfer impedance must be sufficiently high and low noise so that a sufficient voltage signal can be obtained and then further amplified by a main amplifier, a voltage gain type. For high-speed and wideband signals, trans-impedance amplification type is preferred as they offer wideband, much wider than high impedance type, though the noise level might be higher. With trans-impedance amplifiers (TIA), there are two types: The single input single output port and two differential inputs and single output. The output ports can be differential with a complementary port. The differential input TIA offers much higher trans-impedance gain (Z_T) and wider bandwidth as well. This is contributed to the use of a long-tail pair at the input and hence reasonable high input impedance that would ease the feedback stability [27–29].

In Section 7.2, a case study of coherent optical receiver is described from design to implementation including the feedback control and noise reduction. Although the corner frequency is only a few hundreds of MHz, with limited transition frequency of the transistors, this bandwidth is remarkable. The design is scalable to ultra-wideband reception sub-systems.

7.5.2 WIDEBAND TRANS-IMPEDANCE AMPLIFIERS (TIA)

Two type of trans-impedance amplifiers (TIA) are described. They are distinguished by whether one single input or differential inputs exist. which are dependent on the use of a differential pair or long tail pair or a single transistor stage at the input of the TIA.

7.5.2.1 Single Input Single Output

We prefer to treat this section as a design example and experimental demonstration of a wideband and low noise amplifier in a separate annex, the Annex 2 of Reference 30. In tthe next section, the differential input TIA is treated with large transfer impedance and reasonably low noise.

7.5.2.2 Differential Inputs, Single/Differential Output

An example circuit of the differential input transfer impedance amplifier is shown Figure 7.17, in which a long tail pair or differential pair is employed at the input stage. Two matched transistors are used to ensure the minimum common mode rejection and maximum differential mode operation. This pair has very high input impedance, and thus the feedback from the output stage can be stable. Thus, the feedback resistance can be increased until the limit of the stability locus of the network pole is reached. This offers the high transfer impedance Z_T and wide bandwidth. Typical Z_T of

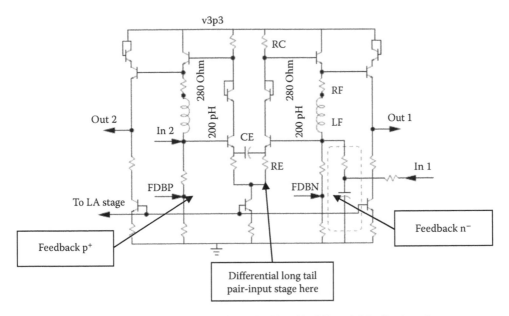

FIGURE 7.17 A typical structure of a differential TIA [29] with differential feedback paths.

3000 to 6000Ω can be achieved with 30 GHz 3 dB bandwidth (see Figure 7.19), which can be obtained as shown in Figures 7.18 and 7.19. The chip image of the TIA can be seen in Figure 7.18a [30–32]. Such TIA can be implemented in either InP or SiGe material. The advantage of SiGe is that the circuit can be integrated with a high-speed Ge-APD detector and ADC and DSP. On the other hand, if implemented in InP, then high-speed *p-i-n* or APD can be integrated and RF interconnected with ADC and DSP. The differential group delay may be serious and must be compensated in the digital processing domain.

7.5.2.3 Amplifier Noise Referred to Input

There are several noise sources in any electronic system that include thermal noises, shot noises, and quantum shot noises, especially in opto-electronic detection. Thermal noises occur because the

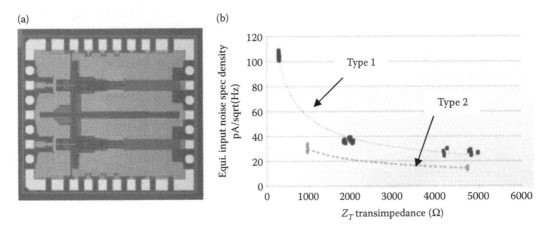

FIGURE 7.18 Differential amplifiers: (a) chip level image and (b) referred input noise equivalent spectral noise density. (Inphi TIA 3205 [type 1] and 2850 [type 2]). (Courtesy from Inphi inc., Technical information on 3205 and 2850 TIA device, 2012.)

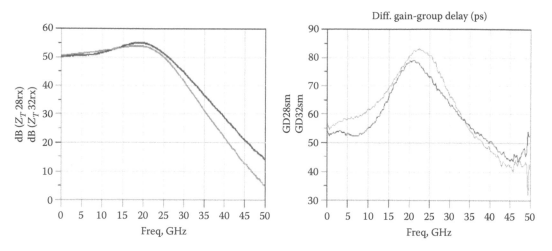

FIGURE 7.19   Differential amplifier: frequency response and differential group delay.

operating temperature is well above the absolute temperature at which no random movement of electrons occur and the resistance of electronic element. This type of noise is dependent on ion temperate. Shot noises are due to the current flowing and random scattering of electrons, thus this type of noise depends on the strength of the flowing currents such as biasing current in electronic devices. Quantum shot noises are generated because of the current emitted from opto-electronic detection processes, which are dependent on the strength of the intensity of the optical signals or sources imposed on the detectors. Thus, this type of noise is dependent on signals. In the case of coherent detection, the mixing of the local oscillator (LO) laser and signals normally occur with the strength of the LO much larger than that of signal average power. Thus, the quantum shot noises are dominated by that from the LO.

In practice, an equivalent electronic noise source is the total noises as referred to the input of electronic amplifiers which can be measured by measuring the total spectral density of the noise distribution over the whole bandwidth of the amplification devices. Thence, the total noise spectral density can be evaluated referred to the input port. For example, if the amplifier is a transimpedance type, then the transfer impedance of the device is measured first then the measure voltage spectral density at the output port can then be referred to the input. In this case, it is the total equivalent noise spectral density. The common term employed and specified for transimpednce amplifiers is the total equivalent spectral noise density over the midband region of the amplifying device. The midband range of any amplifier is defined as the flat gain region from DC to the corner 3 dB point of the frequency response of the electronic device.

Figure 7.20 illustrates the meaning of the total equivalent noise sources as referred to the input port of a two port electronic amplifying device. A noisy amplifier with an input excitation current source, typically a signal current generated from the PD after the optical to electrical conversion, can be represented with a noiseless amplifier and the current source in parallel with a noise sources whose strength is equal to the total equivalent noise current referred to the input. Thus, the total euquivalent current can be found by taking the product between this total equivalent current noise spectral density and the 3 dB bandwidth of the amplifying device. Thus, one can find the signal-to-noise ratio at the iutput of the electronic amplifier given by

$$SNR = \frac{\text{square\_of\_current\_generated}}{\text{suqare\_of\_current\_generated} + \text{total\_equivalent\_noise\_current\_power}} \tag{7.47}$$

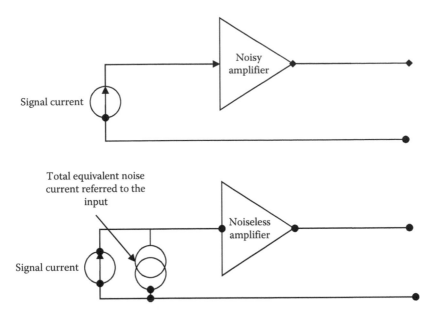

FIGURE 7.20   Equivalent noise spectral density current sources to be represented so that amplifier is noiseless.

From this SNR referred at the input of the electronic front end, one can estimate the eye opening of the voltage signals at the output of the amplifying stage, which is normally required by the ADC to sampling and conversion to digital signals for processing. Thus, one can then estimate the require OSNR at the input of the photodetector and hence the launched power required at the transmitter over several span links with certain attenuation factors.

Detailed analyses of amplifier noises and their equivalent noise sources, referred to input ports, are given in the Annex 2 of Reference 33. It is noted that noises have no flowing direction as they always add and no substraction, thus the noises are measured as noise power and not as a current. Electrical spectrum analyzers are commonly used to measured the total noise spectral density, or the distribution of noise voltages over the spectral range under consideration, which is defined as the noise power spectral density distribution.

7.5.3 INTEGRATED PHOTODETECTOR AND WIDEBAND HIGH TRANS-IMPEDANCE AMPLIFIER

Currently, the integration of electronic in photonic integrated circuit to form electro-photonic integrated circuits (e-PIC) is very critical to create the footprint small enough to size such as CFP4 [34]. Thus, the deposition of Ge on Si on insulator (SOI) on Si substrate is very critical, and hence the coupling of the lightwaves from fibers becomes very attractive, as shown in Figure 7.21. The PD can be formed to create balanced detector pair followed by SiGe wideband amplifier [35], as shown in Figure 7.22. The design of broadband amplifier is based on the differential amplifier stage at the input. This stage is commonly known as the long tail pair [36] with a design strategy that has a high impedance and then followed by a low-impedance input stage with appropriate feedback control by shunt emitter feedback or shunt feedback from collector to base. This differential stage is followed by a linear amplification stage with peaking. The peaking is a resonant circuit providing a complex pole-pair to extend the 3 dB bandwidth by appropriate adjusting of the inductors and capacity of the circuits. One can see the inductors given in Figure 7.22, which commonly exist in the connection wires, for example, the bonding between electronic stages or the parasitic capacitances. The non-peaking and peaking frequency responses of the amplifier whose schematic representation can be seen for extension of passband, as shown in Figure 7.23. The broadband design follows the technique reported by Cherry and Hooper [37].

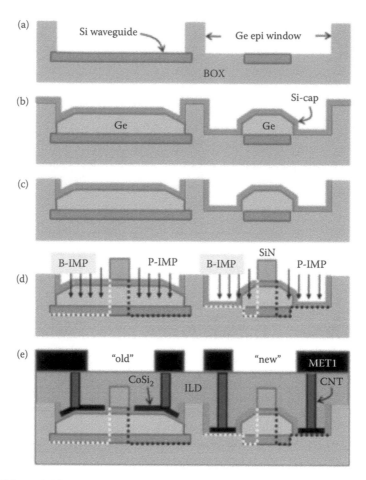

FIGURE 7.21 Schematic illustration by cross sections perpendicular to the direction of light incidence of Ge PD on SOI silicon photonic circuits by (a) fabrication flow of Ge lateral *p-i-n* photodiodes on Si optical waveguide with former layout (left → "old") and new layout (right → "new"), etching a window in an isolator layer stack with a planarized, some 100 nm thick silicon oxide layer; (b) selective Ge epitaxy (about 400 nm) followed by non-selective Si deposition; (c) silicon chemical mechanical polishing (Si-CMP); (d) formation of a 600 nm wide silicon nitride (SiN) pedestal followed by self-aligned implantation steps; and (e) $CoSi_2$ formation, contact (CNT) formation, first metal layer (MET1) deposition and structuring. BOX denotes the 2 μm thick buried SiO_2 layer beneath the 220 nm thick Si waveguide.

To enhance the Rx bandwidth, the TIA can be excited with the measured S-parameters model of the Ge-PD, that is, the current source and noise models given in Section 5.2.2, optimizing for the bandwidth while monitoring the group delay and the gain ripples. The dominant pole at the input transfer impedance (TI) stage is controlled by the photodiode capacitance C_d, the TIA input capacitance and resistance, C_{in} and R_{in}, respectively. To compensate for C_d, a series inductor (L_1) between the photodiode and the TIA can be used to create a pi-matching network to extend the bandwidth. Two factors control the value of L_1, which shows that beyond a certain value of L_1 the group delay variation will be very high and the bandwidth would decrease. The second factor is the practical realization of the inductor, as the self-resonance frequency of the inductor is inversely proportional to its value. For the latter factor, electromagnetic simulation is necessary. These two factors were considered when choosing the inductance value of L_1 to be 450 pH at low frequencies. Using this inductor, the bandwidth can be enhanced from 32 to 53 GHz, as shown in Figure 7.23a where the normalized S_{21} is plotted, comparing the cases with and without bandwidth enhancement. To improve the gain

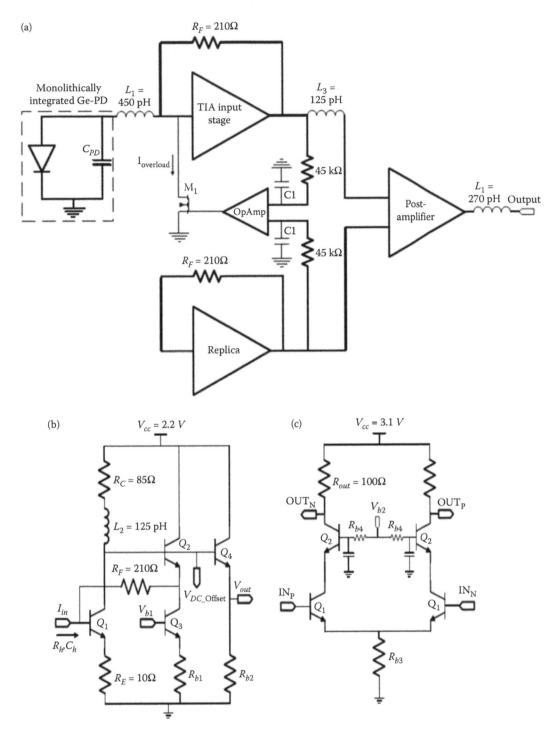

FIGURE 7.22 Block diagram of an integrated monolithic wideband amplifier by SiGe and PD on SOI Si photonic platform.: (a) block diagram, (b) TIA input stage—Cherry—Hooper configuration, and (c) linear amplification poast stage. (Extracted from M. H. Eissa et al., A wideband monolithically integrated photonic receiver in 0.25-μm SiGe:C BiCMOS technology, *Proc. 42nd European Solid-State Circuits Conference, ECOC 2016*, Duesseldorf Germany, 12–15 Sept. 2016)

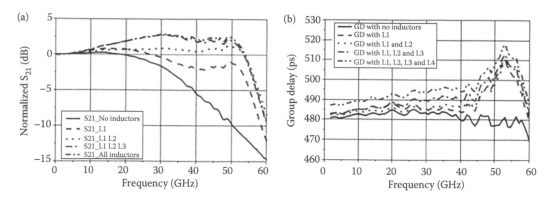

FIGURE 7.23 Simulated 3dB passband of the wideband TIA by monolithic integration of PD and TIA on SOI. (a) Frequency response $S_{21}(f)$ of the amplifier gain and (b) group delay variation, with respect of inductor peaking inductors. (Extracted from H. Tran, F. Pera, D. S. McPherson, D. Viorel, and S. P. Voinigescu, *IEEE Journal of Solid State Circuits*, vol. 39, no 10, pp. 1680–1689, 2004.)

flatness across the band, two additional peaking inductors have been used. The first one (L_2) is a shunt peaking inductor at the load of the TI stage (Rc), which enhances the gain roll off by 2.5 dB, achieving a nearly flat response. The second one is an interstage inductor (L_3) between the TI stage and the post-amplifier that compensates for the input capacitance of the amplifier. L_3 also creates an extra gain peaking of 2 dB account for any extra losses that might degrade the inductors quality factor. Stability checks have been done for the internal nodes, especially for the output of Q4 to ensure that the inductive loading of the emitter-follower is not causing any instabilities. The effects of inductors L_2 and L_3 on the bandwidth extension are shown in Figure 7.23a. In Figure 7.23, the group delay of the whole receiver with and without the bandwidth enhancement inductors is plotted. In the case of no inductors, a group delay variation of ± 2.2 ps is simulated for a 3-dB bandwidth of 34 GHz. It increased to ± 15 ps, after adding the inductors, for 3-dB bandwidth of 53 GHz can be reached. Output matching can be done to ensure the broadband behavior of the amplifier. Although the post-amplifier is resistively loaded, the output matching degrades with frequency due to the effect of the output capacitance of the transistor Q_2 and the pad capacitance. To improve the output matching, a series inductor (L_4) is employed to compensate for this parasitic capacitance. This is important to avoid standing waves and reflections due to long connections and on board transitions.

7.6 DIGITAL SIGNAL PROCESSING SYSTEMS AND COHERENT OPTICAL RECEPTION

7.6.1 DSP-Assisted Coherent Detection

Over the years since the introduction of optical coherent communications in the mid-1980s, the invention of optical amplifiers has left coherent reception behind until recently when long-haul transmission suffered from non-linearity of dispersion compensating fibers (DCF) and SSMF transmission line because of its small effective area. Furthermore, the advancement of DSP in wireless communication has contributed to the application of DSP in modern coherent communication systems. Thus, the name "DSP-assisted coherent detection" that is, when a realtime DSP is incorporated after the opto-electronic conversion of the total field of the local oscillator and that of the signals the analogue received signals are sampled by a high-speed ADC, and then the digitalized signals are processed in a DSP. Currently, realtime DSP processors are intensively researched for practical implementation. The main difference between realtime and off line processing is that the realtime processing algorithm must be effective due to limited time available for processing.

When polarization division multiplexed (PDM) QAM channels are transmitted and received, polarization and phase diversity receivers are employed. The schematic representation of such receiver are shown in Figure 7.24a. Furthermore, the structures of such reception systems incorporating DSP with the diversity hybrid coupler in optical domain are shown in Figure 7.24b–d. The polarization diversity section with the polarization beam splitters at the signal and local oscillator inputs facilitate the demultiplexing of polarized modes in the optical waveguides. The phase diversity using a 90° optical phase shifter here allows the separation of the inphase ($I-$) and quadrature ($Q-$) phase components of QAM channels. Using a 2×2 coupler also enables balanced reception using PDP-connected back-to-back, and hence a 3 dB gain in the sensitivity occurs. Section 2.7 of Chapter 2 describes the modulation scheme QAM using I-Q modulators for single polarization or dual polarization multiplexed channels.

7.6.2 DSP-BASED RECEPTION SYSTEMS

The schematic representation of a synchronous coherent receiver based on digital signal processing is shown in Figure 7.25. Once the polarization and the I- and Q-optical components are separated by the hybrid coupler, the positive and negative parts of the I- and Q-optical components are coupled into balanced opto-electronic receivers, as shown in Figure 7.24b. Two photodiodes (PD) are connected back-to-back so that a push-pull operation can be achieved, hence a 3 dB betterment as compared to a single PD detection. The current generated from the back-to-back connected PDs is fed into a transimpedance amplifier so that a voltage signal can be derived at the output. Furthermore, a voltage-gain amplifier is used to boost these signals to the right level of the analog to digital converter (ADC) so that sampling can be conducted and conversion of the analog signals to digital domain occur. These digitized signals are fetched into digital signal processors (DSP) and processing in the "soft domain" can be conducted. Thus, a number of processing algorithms can be employed in this stage. A number of processing algorithms are employed to compensate for linear and non-linear distortion effects because of optical signal propagation through the optical guided medium to recover the carrier pohase and the clock rate for resampling of the data sequence, etc. Chapter 6 will describe in detail the fundamental aspects of these processing algorithms. Figure 7.25 shows a schematic representation of possible processing phases in the DSP incorporated in a DSP-based coherent receiver. The soft processing of the optical phase locking as described in Chapter 5 is necessary to lock the frequencies of the local oscillator and that of the signal carrier to with certain limit within which the algorithms for clock recovery can function, for example, within ± 2 GHz.

7.6.3 COHERENT RECEPTION ANALYSIS

7.6.3.1 Sensitivity

At an ultra-high bit rate, the laser must be externally modulated so the phase of the lightwave is conserved along the fiber transmission line. The detection can be direct detection, self-coherent, or homodyne and heterodyne. The sensitivity of the coherent receiver is also important for the transmission system, especially the PSK scheme [38] under both homodyne and heterodyne transmission techniques. This section gives the analysis of receiver for synchronous coherent optical fiber transmission systems. Consider that the optical fields of the signals and local oscillator are coupled via a fiber coupler with two output ports 1 and 2. The output fields are then launched into two photodetectors connected back-to-back and then the electronic current is amplified using a TI type and further equalized to extend the bandwidth of the receiver. Our objective is to obtain the receiver penalty and its degradation due to imperfect polarization mixing and unbalancing effects in the balanced receiver. A case study of the design, implementation, and measurements of an optical balanced receiver electronic circuit and noise suppression techniques is given in Section 7.5.

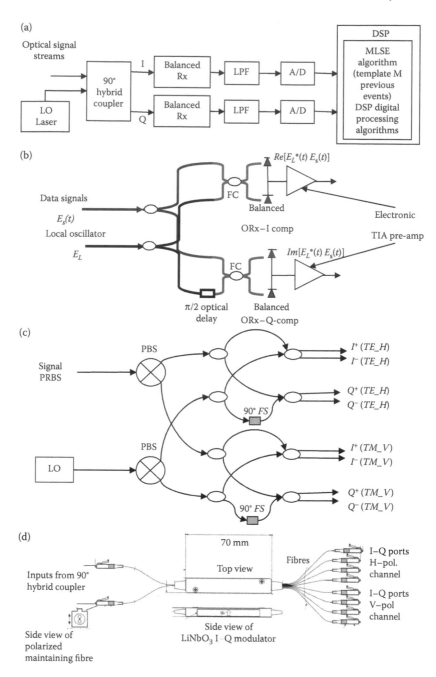

FIGURE 7.24 Scheme of a synchronous coherent receiver using digital signal processing for phase estimation for coherent optical communications. (a) Generic scheme, (b) detailed optical receiver using only one polarization phase diversity coupler, (c) hybrid 90° coupler for polarization and phase diversity, and (d) typical view of a hybrid coupler with two input ports and 8 output ports of structure in (c). TE_V, TE_H = transverse electric mode with vertical (V) or horizontal (H) polarized mode, TM = transverse magnetic mode with polarization orthogonal to that of the TE mode. FS = phase shifter; PBS = polarization beam splitter; MLSE = Maximum Likelihood Phase Estimation. (Adapted from (a) S. Zhang et al., A comparison of phase estimation in coherent optical PSK system, *Photonics Global '08*, Paper C3-4A-03. Singapore, December 2008; (b) S. Zhang et al., Adaptive decision-aided maximum likelihood phase estimation in coherent optical DQPSK system, *OptoElectronics and Communications Conference (OECC) '08*, Paper TuA-4, pp. 1–2, Sydney, Australia, July 2008.)

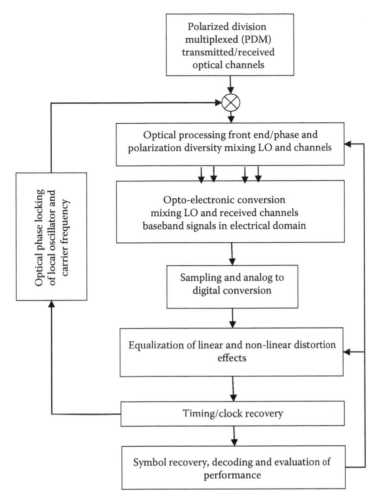

FIGURE 7.25 Flow of functionalities of DSP processing in a QAM coherent optical receiver with possible feedback control.

The following parameters are commonly used in analysis:

E_s	Amplitude of signal optical field at the receiver
E_L	Amplitude of local oscillator optical field
P_s, P_L	Optical power of signal and local oscillator at the input of the photodetector
$s(t)$	The modulated pulse
$\langle i_{NS}^2(t) \rangle$	Mean square noise current (power) produced by the total optical intensity on the photodetector
$\langle i_s^2(t) \rangle$	Mean square current produced by the photodetector by $s(t)$
$S_{NS}(t)$	Shot noise spectral density of $\langle i_s^2(t) \rangle$ and local oscillator power
$i_{Neq}^2(t)$	equivalent noise current of the electronic pre-amplifier at its input
$Z_T(\omega)$	Transfer impedance of the electronic pre-amplifier
$H_E(\omega)$	Voltage transfer characteristic of the electronic equalizer followed the electronic pre-amplifier

The combined field of the signal and local oscillator via a directional coupler can be written with separate polarized field components as

$$E_{sX} = \sqrt{K_{sX}}E_S \cos(\omega_s t - \phi_{m(t)})$$

$$E_{sY} = \sqrt{K_{sY}}E_S \cos(\omega_s t - \phi_{m(t)} + \delta_s)$$

$$E_{LX} = \sqrt{K_{LX}}E_L \cos(\omega_L t) \qquad (7.48)$$

$$E_{LY} = \sqrt{K_{LY}}E_L \cos(\omega_L t + \delta_L)$$

$$\phi_{m(t)} = \frac{\pi}{2}K_m s(t)$$

where $\phi_m(t)$ represents the phase modulation, K_m is the modulation depth, and $K_{sX}\,K_{sY}\,K_{LX}\,K_{LY}$ are the intensity fraction coefficients in the X- and Y-direction of the signal and local oscillator fields, respectively.

Thus, the output fields at ports 1 and 2 of the FC in the X-plane can be obtained using the transfer matrix as

$$\begin{bmatrix} E_{R1X} \\ E_{R2X} \end{bmatrix} = \begin{bmatrix} \sqrt{K_{sX}(1-\alpha)}\cos(\omega_s t - \phi_{m(t)}) & \sqrt{K_{LX}\alpha}\sin(\omega_L t) \\ \sqrt{K_{sX}\alpha}\sin(\omega_s t - \phi_{m(t)}) & \sqrt{K_{LX}(1-\alpha)}\cos(\omega_L t) \end{bmatrix}\begin{bmatrix} E_s \\ E_L \end{bmatrix} \qquad (7.49)$$

$$\begin{bmatrix} E_{R1Y} \\ E_{R2Y} \end{bmatrix} = \begin{bmatrix} \sqrt{K_{sY}(1-\alpha)}\cos(\omega_s t - \phi_{m(t)}) & \sqrt{K_{LY}\alpha}\sin(\omega_L t + \delta_L) \\ \sqrt{K_{sY}\alpha}\sin(\omega_s t - \phi_{m(t)}) & \sqrt{K_{LY}(1-\alpha)}\cos(\omega_L t) + \delta_L \end{bmatrix}\begin{bmatrix} E_s \\ E_L \end{bmatrix} \qquad (7.50)$$

where α defined as the intensity coupling ratio of the coupler. Thus, the field components at ports 1 and 2 can be derived by combining the X and Y components from Equations 7.49 and 7.50, thence the total power at ports 1 and 2 are given as

$$P_{R1} = P_s(1-\alpha) + P_L\alpha + 2\sqrt{P_s P_L \alpha(1-\alpha)K_p}\sin(\omega_{IF}t + \phi_{m(t)} + \phi_p - \phi_c)$$

$$P_{R2} = P_s\alpha + P_L(1-\alpha) + 2\sqrt{P_s P_L \alpha(1-\alpha)K_p}\sin(\omega_{IF}t + \phi_{m(t)} + \phi_p - \phi_e + \pi)$$

with $\quad K_p = K_{sX}K_{LX} + K_{sY}K_{LY} + 2\sqrt{K_{sX}K_{LX}K_{sY}K_{LY}}\cos(\delta_L - \delta_s) \qquad (7.51)$

$$\phi_p = \tan^{-1}\left[\frac{\sqrt{K_{sX}K_{LY}}\sin(\delta_L - \delta_s)}{\sqrt{K_{sX}K_{LX}} + \sqrt{K_{sY}K_{LY}}\cos(\delta_L - \delta_s)}\right]$$

ω_{IF} is the intermediate angular frequency and equals to the difference between the frequencies of the local oscillator and the carrier of the signals. ϕ_e is the phase offset and $\phi_p - \phi_e$ is the demodulation reference phase error.

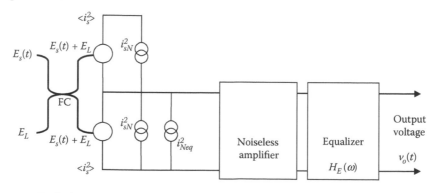

FIGURE 7.26 Equivalent current model at the input of the optical balanced receiver under coherent detection, average signal current and equivalent noise current of the electronic pre-amplifier as seen from its input port and equalizer. FC = fiber coupler.

In Equation 7.51, the total field of the signal and the local oscillator are added and then taking the product of the field vector and its conjugate to obtain the power. Only the term with a frequency fallen within the range of the sensitivity of the photodetector would produce the electronic current. Thus, the term with the sum of the frequency of the wavelength of the signal and local oscillator would not be detected and only the product of the two terms would be detected as given.

Now assuming a Binary *PSK* (*BPSK*) modulation scheme, the pulse has a square shape with amplitude $+1$ or -1, the *PD* is a *p-i-n* type, and the PD bandwidth is wider than the signal 3 dB-bandwidth followed by an equalized electronic pre-amplifier. The signal at the output of the electronic equalizer or the input signal to the decision circuit is given by

$$\hat{v}_D(t) = 2K_H K_p \sqrt{P_s P_L \alpha (1-\alpha) K_p} \int_{-\infty}^{\infty} H_E(f) df \int_{-\infty}^{\infty} (t) dt \sin\left(\frac{\pi}{2} K_m\right) \cos(\phi_p - \phi_e)$$

$$\rightarrow \hat{v}_D(t) = 2K_H K_p \sqrt{P_s P_L \alpha (1-\alpha) K_p} \sin\left(\frac{\pi}{2} K_m\right) \cos(\phi_p - \phi_e)$$

(7.52)

$K_H = 1$ for homodyne; $K_H = 1/\sqrt{2}$ for..heterodyne

For a perfectly balanced receiver $K_B = 2$ and $\alpha = 0.5$, otherwise $K_B = 1$. The integrals of the first line in Equation 7.52 are given by

$$\int_{-\infty}^{\infty} H_E(f) df = \frac{1}{T_B} \quad \because H_E(f) = \sin c(\pi T_B f)$$

$$\int_{-\infty}^{\infty} s(t) dt = 2T_B$$

(7.53)

$V_D(f)$ is the transfer function of the matched filter for equalization and T_B is the bit period. The total noise voltage as a sum of the quantum shot noise generated by the signal and the local oscillator and the total equivalent noise of the electronic pre-amplifier at the input of the pre-amplifier, at the output of the equalizer is given by

$$\langle v_N^2(t) \rangle = \frac{[K_B \alpha S'_{IS} + (2 - K_B) S_{Ix} + S_{IE}] \int_{-\infty}^{\infty} |H_4(f)|^2 df}{K_{IS}^2}$$

(7.54)

or

$$\langle v_N^2(t) \rangle = \frac{[K_B \alpha S'_{IS} + (2 - K_B) S_{Ix} + S_{IE}]}{K_{IS}^2 T_B}$$

For homodyne and heterodyne detection, we have

$$\langle v_N^2(t) \rangle = \Re q \frac{P_L}{\lambda T_B} [K_B \alpha S'_{IS} + (2 - K_B) S_{Ix} + S_{IE}]$$

(7.55)

where the spectral densities S'_{IX}, S'_{IE} are given by

$$S'_{IX} = \frac{S_{IX}}{S'_{IS}}$$

$$S'_{IE} = \frac{S_{IE}}{S'_{IS}}$$

(7.56)

Thus, for the receiver sensitivity for binary PSK and equi-porobable detection and Gaussian density distribution, we have

$$P_e = \frac{1}{2} erfc\left(\frac{\delta}{\sqrt{2}}\right) \tag{7.57}$$

with δ given by

$$P_e = \frac{1}{2} erfc\left(\frac{\delta}{\sqrt{2}}\right) \text{ with } \dots \delta = \frac{\hat{v}_D}{2\sqrt{\langle v_N^2 \rangle}} \tag{7.58}$$

Thus, using Equations 7.55 and 7.58 and 7.52, we have the receiver sensitivity in linear power scale as

$$P_s = \langle P_s(t) \rangle = \frac{\Re q \delta^2}{4\lambda T_B K_H^2} \frac{[K_B \alpha S'_{IS} + (2 - K_B)S_{Ix} + S'_{IE}]}{\eta K_p (1 - \alpha)\alpha K_B^2 \sin^2\left(\frac{\pi}{2} K_m\right)\cos^2(\phi_p - \phi_e)} \tag{7.59}$$

7.6.3.2 Shot Noise-Limited Receiver Sensitivity

In the case that the power of the local oscillator dominates the noise of the electronic pre-amplifier and the equalizer, then the receiver sensitivity (in linear scale) is given as

$$P_s = \langle P_{sL} \rangle = \frac{\Re q \delta^2}{4\lambda T_B K_H^2} \tag{7.60}$$

This shot-noise limited receiver sensitivity can be plotted as shown in Figure 7.27.

7.6.3.3 Receiver Sensitivity under Non-Ideal Conditions

Under non-ideal conditions, the receiver sensitivity departs from the shot noise limited sensitivity and is characterized by the receiver sensitivity penalty PD_T as

$$PD_T = 10Log\frac{\langle P_s \rangle}{\langle P_{sL} \rangle} \quad \text{dB} \tag{7.61}$$

$$PD_T = 10Log_{10}\left[\frac{K_B \alpha S'_{IS} + (2 - K_B)S_{Ix} + S_{IE}}{K_B \alpha}\right]$$
$$- 10Log_{10}[K_B(1 - \alpha)] \tag{7.62}$$
$$- 10Log_{10}\left([\eta][K_p]\sin^2\left(\frac{\pi}{2} K_m\right)\cos^2(\phi_p - \phi_e)\right)$$

where η is the local oscillator excess noise factor.

The receiver sensitivity is plotted against the ratio f_B/λ for the case of homodyne and heterodyne detection, and is shown in Figure 7.27a. The power penalty of the receiver sensitivity against the excess noise factor of the local oscillator is shown in Figure 7.27b. Receiver power penalty can be deduced as a function of the total electronic equivalent noise spectral density, and as a function of the rotation of the polarization of the local oscillator. It can be found in Reference 39. Furthermore, in Reference 40, the receiver power penalty and the normalized heterodyne center frequency can be

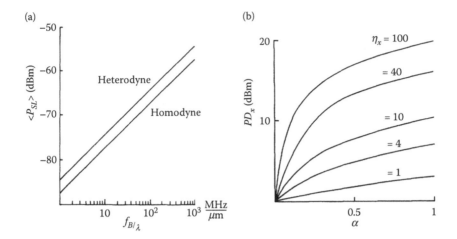

FIGURE 7.27 (a) Receiver sensitivity of coherent homodyne and heterodyne detection, signal power versus bandwidth over the wavelength. (b) Power penalty of the receiver sensitivity from the shot noise limited level as a function of the excess noise of the local oscillator. (Extracted from I. Hodgkinson, *IEEE J. Lightw. Tech.*, vol. 5, no.4, pp. 573–587, 1987, Figures 1 and 2.)

varied as a function of the modulation parameter and as a function of the optical power ratio at the same polarization angle can also be found.

7.6.3.4 Digital Processing Systems

A generic structure of the coherent reception and digital signal processing system is shown in Figure 7.28 in which the digital signal processing system is placed after the sampling and conversion from analogue state to digital form. Obviously, the optical signal fields are beating with the local oscillator laser whose frequency would be approximately identical with the signal channel carrier. The beating occurs in the square law photodetectors, that is, the summation of the two fields are squared and the product term is decomposed into the difference and summation term, thus only the difference term is fallen back into the baseband region and amplified by the electronic pre-amplifier, which is a balanced differential transimpedance type.

If the signals are complex, then there are real and imaginary components which form a pair. The other pair comes from the other polarization mode channel. The digitized signals of both the real and imaginary parts are processed in real time or offline. The processors contain the algorithms to combat

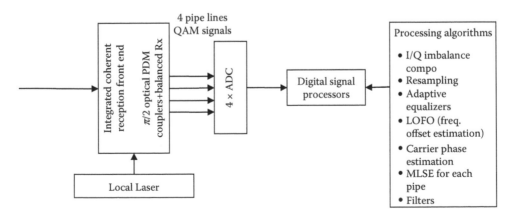

FIGURE 7.28 Coherent reception and digital signal processing system for PDM QAM scheme.

TABLE 7.1

Milestones of Progresses of Linewidth Resolution

10 μm — 1971	800 nm (.80 μm) — 1989- UV	90 nm — 2002: electron	14 nm — approx. 2014: X-ray lithography
3 μm — 1975	lithography	lithography	10 nm — approx. 2016: x-ray litho
1.5 μm — 1982	600 nm (.60 μm) — 1994	65 nm — 2006	7 nm — approx. 2018: x-ray litho
1 μm — 1985	350 nm (.35 μm) — 1995	45 nm — 2008	5 nm — approx. 2020: x-ray lithography
	250 nm (.25 μm) — 1998	32 nm — 2010	
	180 nm (.18 μm) — 1999	22 nm — 2012	
	130 nm (.13 μm) — 2000		

a number of transmission impairments such as the imbalance between the inphase and the quadrature components created at the transmitter, the recovery of the clock rate and timing for resampling, the carrier phase estimation for estimation of the signal phases, adaptive equalization for compensation of propagation dispersion effects using MLSE, etc. These algorithms are built into the hardware processors or memory and loading to processing sub-systems.

The sampling rate must normally be twice that of the signal bandwidth to ensure to satisfy the Nyquist criteria. Although this rate is very high for 25 G–32 GSy/s optical channels, Fujitsu ADC has reached this requirement with sampling rate of 56 G–64 GSa/s, as depicted later near the end of this chapter in Figure 7.40.

The linewidth resolution of the processing for semiconductor device fabrication has been progressed tremendously over the year in an exponential trend, as shown in Table 7.1. This progress could be made because of the successes in the lithographic techniques using optical at short wavelength such as the UV, the electronic optical beam and x-ray lithograpohic with appropriate photoresist such as SU-80 would allow the line resolution to reach 5 nm in 2020. If we plot the trend in a log-linear scale as shown in Figure 7.29, a linear line is obtained meaning that the resolution is reduced exponentially. When the gate width is reduced the electronic speed would increased tremendously, at 5 mm the sped of electronic CMOS device in SiGe would reach several tens of GHz. Regarding the high-speed ADC and DAC, the clock speed is increased by parallel delayed and summation of all the digitized digital line to form a very high-speed operation. For example, for Fujitsu 64 GSa/s DAC or ADC, the applied clock sinusoidal waveform is only 2 GHz. Figure 7.30 shows the progresses in the speed development of Fujitsu ADC and DAC.

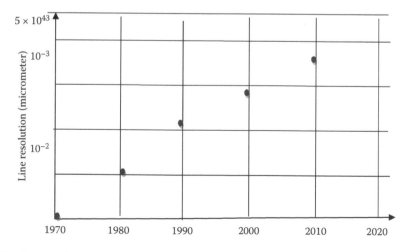

FIGURE 7.29 Semiconductor manufacturing with resolution of line resolutiun.

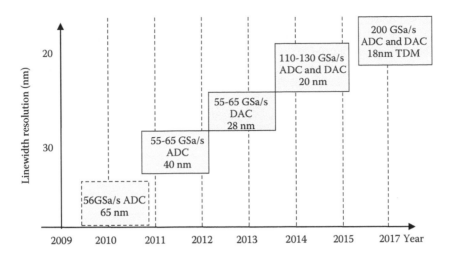

FIGURE 7.30 Evolution of ADC and DAC operating speed with corresponding line width resolution.

7.6.3.4.1 Effective Number of Bits (ENOB)

7.6.3.4.1.1 Definition A effective number of bits (ENOB) is a measure of the quality of a digitised signal. The resolution of a digital-to-analog oranalog-to-digital converter (DAC or ADC) is commonly specified by the number of bits used to represent the analog value, in principle giving 2^N signal levels for an N-bit signal. However, all real signals contain a certain amount of noise. If the converter is able to represent signal levels below the system noise floor, the lower bits of the digitised signal only represent system noise and do not contain useful information. ENOB specifies the number of bits in the digitised signal above the noise floor. Often, ENOB is used as a quality measure also for other blocks like sample-and-hold amplifiers. This way, analog blocks can also be easily included to signal-chain calculations, as the total ENOB of a chain of blocks is usually below the ENOB of the worst block.

Thus, we can represent the ENOB of a digitalized system by writing

$$ENOB = \frac{SINAD - 1.76}{6.02} \tag{7.63}$$

where all values are given in dB, and SINAD is the ratio of the total signal including distortion and noise to the wanted signal; the 6.02 term in the divisor converts decibels (a $\log_{10}$ representation) to bits (a $\log_2$ representation); the 1.76 term comes from quantization error in an ideal ADC [40].

This definition compares the SINAD of an ideal ADC or DAC with a word length of ENOB bits with the SINAD of the ADC or DAC being tested. The signal-to-noise and distortion ratio (SINAD) is a measure of the quality of a signal from a communications device, often defined as

$$SINAD = \frac{P_{sig} + P_{noise} + P_{distorion}}{P_{noise} + P_{distorion}} \tag{7.64}$$

where P is the average power of the signal, noise, and distortion components. SINAD is usually expressed in dB and is quoted alongside the receiver sensitivity to give a quantitative evaluation of the receiver sensitivity. Note that with this definition, unlike SNR, a SINAD reading can never be less than 1 (i.e., it is always positive when quoted in dB).

When calculating the distortion, it is common to exclude the DC components. Due to widespread use, SINAD has collected a few different definitions. SINAD is calculated as one of (i) the ratio of

(a) total received power, that is, the signal to (b) the noise-plus-distortion power. This is modeled by the equation above. (ii) The ratio of (a) the power of original modulating audio signal, that is, from a modulated radio frequency carrier to (b) the residual audio power, that is, noise-plus-distortion powers remaining after the original modulating audio signal is removed. With this definition, it is now possible for SINAD to be less than one. This definition is used when SINAD is used in the calculation of ENOB for an ADC.

Example: Consier the following are measurements of a 3-bit unipolar D/A converter with reference voltage $V_{ref} = 8$ V:

Digital input	000	001	010	011	100	101	110	111
Analog output (V)	−0.01	1.03	2.02	2.96	3.95	5.02	6.00	7.08

The offset error in this case is −0.01 V or −0.01 LSB as 1 V = 1 LSB in this example. The gain error is $(7.08 + 0.01/7) − (7/1) = 0.009$ LSB with LSB stands for the least significant bits. Correcting the offset and gain error, we obtain the following list of measurements: (0, 1.03, 2.00, 2.93, 3.91, 4.96, 5.93, 7) LSB. This allows the INL and DNL to be calculated: INL = (0, 0.03, 0, −0.07, −0.09, −0.04, −0.07, 0) LSB, and DNL = (0.03, −0.03, −0.07, −0.02, 0.05, −0.03, 0.07, 0) LSB.

Differential non-linearity (DNL): For an ideal ADC, the output is divided into $2N$ uniform steps, each with Δ width, as shown in Figure 7.31. Any deviation from the ideal step width is called differential non-linearity (DNL) and is measured in number of counts (LSBs). For an ideal ADC, the DNL is 0LSB. In a practical ADC, DNL error comes from its architecture. For example, in a SAR ADC, DNL error may be caused near the mid range due to mismatching of its DAC.

Integral non-linearity (INL) is a measure of how closely the ADC output matches its ideal response. INL can be defined as the deviation in LSB of the actual transfer function of the ADC from the ideal transfer curve. INL can be estimated using DNL at each step by calculating the cumulative sum of DNL errors up to that point. In reality, INL is measured by plotting the ADC transfer characteristics. INL is popularly measured using either (i) best fit (best straight line) method or (ii) end point method.

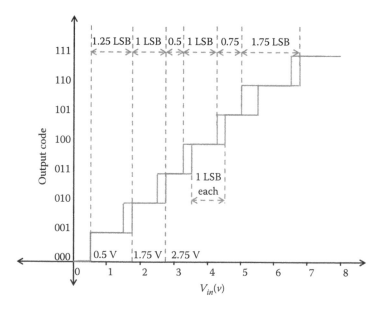

FIGURE 7.31 Representation of DNL in a transfer curve of an ADC.

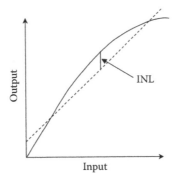

FIGURE 7.32 Best Fit INL.

Best Fit INL: The best fit method of INL measurement considers offset and gain error. One can see in Figure 7.32 that the Ideal transfer curve considered for calculating best-fit INL does not go through the origin. The ideal transfer curve here is drawn such that it depicts the nearest first-order approximation to the actual transfer curve of the ADC.

The intercept and slope of this ideal curve can lend us the values of the offset and gain error of the ADC. Quite intuitively, the best fit method yields better results for INL. For this reason, many times this is the number present on ADC datasheets.

The only real use of the best fit INL number is to predict distortion in time variant signal applications. This number would be equivalent to the maximum deviation for an AC application. However, it is always better to use the distortion numbers than INL numbers. To calculate the error budget, end-point INL numbers provide a better estimation. Also, this is the specification which is generally provided in datasheets. So, one has to use the same instead of end-point INL.

End-point INL: The End-Point method provides the worst case INL. This measurement passes the straight line through the origin and maximum output code of the ADC. (Refer to Figure 7.6). As this method provides the worst case INL, it is more useful to use this number as compared to the one measured using best fit for DC applications. This INL number would be typically useful for error budget calculation. This parameter must be considered for applications involving precision measurements and control Figure 7.33.

The absolute and relative accuracy can now be calculated. In this case, the ENOB absolute accuracy is calculated using the largest absolute deviation D, in this case 0.08 V:

$$D = \frac{V_{ref}}{2^{ENOB}} \rightarrow ENOB = 6.64 \, \text{bits} \tag{7.65}$$

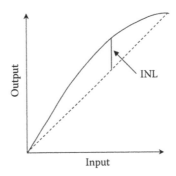

FIGURE 7.33 Endpoint INL.

The ENOB relative accuracy is calculated using the largest relative (INL) deviation d, in this case 0.09 V.

$$d = \frac{V_{ref}}{2^{ENOB}} \rightarrow ENOB_{rel} = 6.47 \, \text{bits} \tag{7.66}$$

For this kind of ENOB calculation, note that the effective number of bits can be larger or smaller than the actual number of bits. When the ENOB is smaller than the ANOB, this means that some of the least significant bits of the result are inaccurate. However, one can also argue that the ENOB can never be larger than the ANOB because you always should add the quantization error of an ideal converter which is ± 0.5 LSB. Different designers may use different definitions of ENOB.

7.6.3.4.1.2 High-Speed ADC and DAC Evaluation Incorporating Statistical Property The effective number of bits of an analogue-to-digital converter (ADC) is considered as the number of bits that an analogue signal can be converted to its digital equivalent by the number of levels represented by the modulo-2 levels which are reduced due to noises contributed by electronic components in such convertor. Thus, only an effective number of equivalent bits can be accounted for. Hence, the term effective number of bits (ENOB) is proposed.

As shown in Figure 7.34b, a real analog to digital (ADC) can be modeled as a cascade of two ideal ADCs and additive noise sources and an automatic gain control (AGC) amplifier [41,42]. The quantized levels are thus equivalent to a specific effective number of bits (ENOB) as far as the ADC is

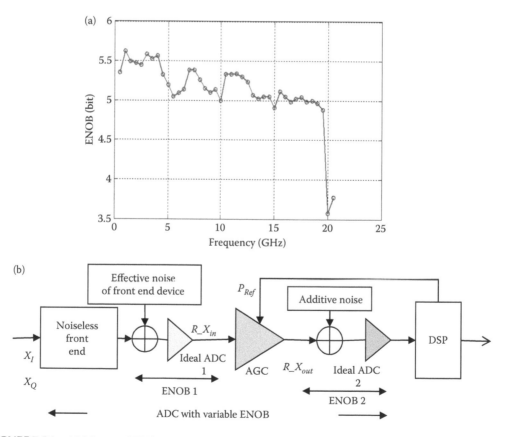

FIGURE 7.34 (a) Measured ENOB frequency response of a commercial real-time DSA of 20 GHz bandwidth and sampling rate of 50 GSa/s and (b) Deduced ADC model of variable ENOB based on experimental frequency response of (a) and spectrum of broadband signals.

operating in the linear non-saturated region. If the normalized signal amplitude/power surpasses unity, the saturated region, then the signals are clipped. The decision level of the quantization in an ADC normally varies following a normalized Gaussian probability density function, thus we can estimate the RMS noise introduced by the ADC as

$$RMS\_noise = \sqrt{\int_{-\frac{LSB}{2}}^{\frac{LSB}{2}} \int_{-\infty}^{\infty} \frac{x^2 \frac{1}{\sqrt{2\pi\sigma^2}}\exp\left(-\frac{(x-y)^2}{2\sigma^2}\right)}{LSB} dxdy} = \sqrt{LSB^2/12 + \sigma^2} \qquad (7.67)$$

where σ is the variance; x is the variable related to the integration of decision voltage; similarly, y for integration inside one LSB. Given the known quantity $LSB^2/12$ by the introduction of the ideal quantization error, σ^2 can be determined via the Gaussian noise distribution. We can thus deduce the ENOB values corresponding to the levels of Gaussian noise as

$$ENOB = N - \log_2\left(\frac{\sqrt{LSB^2/12 + (A\sigma)^2}}{LSB/\sqrt{12}}\right) \qquad (7.68)$$

where A is the RMS amplitude derived from the noise power. According to the ENOB model, the frequency response of ENOB of the DSA is shown in Figure 7.34a with the excitation of the DSA by sinusoidal waves of different frequencies. As observed the ENOB varies with respect to the excitation frequency, in the range from 5 to 5.5. Having known the frequency response of the sampling device, then what is the ENOB of the device when excited with broadband signals? This indicates the different resolution of the ADC of the receiver of the transmission operating under different noisy and dispersive conditions, thus an equivalent model of ENOB for performance evaluation is essential. We note that the amplitudes of the optical fields arrived at the receiver vary depending on the conditions of the optical transmission line. The AGC has a non-linear gain characteristic in which the input sampled signal power level is normalized with respect to the saturated (clipping) level. The gain is significantly high in the linear region and saturated in the high level. The received signal R_X_{in} is scaled with a gain coefficient according to $R\_X_{out} = R\_X_{in}/\sqrt{P_{in\_av}/P_{Ref}}$ where the signal averaged power P_{in_av} is estimated and the gain is scaled relative to the reference power level P_{Ref} of the AGC, then a linear scaling factor is used to obtain the output sampled value R_X_{out}. The gain of the AGC is also adjusted according to the signal energy, via the feedback control path from the DSP (see Figure 7.34b). Thus, new values of ENOB can be evaluated with noise distributed across the frequency spectrum of the signals, by an averaging process. This signal-dependent ENOB is now denoted as $ENOB_S$.

7.6.3.5 Impact of ENOB on Transmission Performance

Shown in Figure 7.35a is the BER variation with respect to the OSNR under back-to-back (B2B) transmission using the simulated samples at the output of the 8-bit ADC with $ENOB_S$ and full ADC resolution as parameters. The difference is due to the noise distribution (Gaussian or uniform). Figure 7.35b depicts the variation of BER versus OSNR with $ENOB_S$ as the variation parameter in case of offline data with ENOB of DSA shown in Figure 7.1a. Several more tests were conducted to ensure the effectiveness of our ENOB model. When the sampled signals presented to the ADC of different amplitudes, controlled and gain non-linearly adjusted by the AGC, different degree of clipping effect would be introduced. Thus, the clipping effect can be examined for the ADC of different quantization levels but with identical $ENOB_S$, as shown in Figure 7.36a for the B2B experiment. Figure 7.36b through e shows, with BER as a parameter, the contour plots of the variation of the

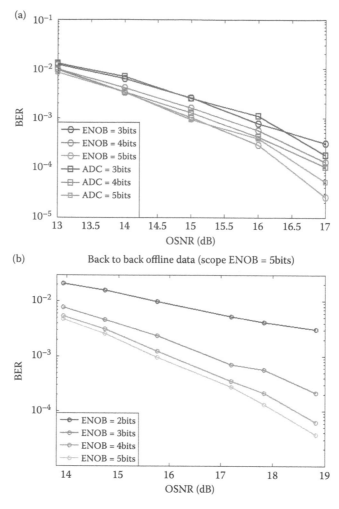

FIGURE 7.35 (a) B2B performance with different $ENOB_S$ values of the ADC model with simulated data (8-bit ADC), (b) OSNR versus BER under different simulated $ENOB_S$ of offline data obtained from experimental digital coherent receiver. (From B. N. Mao et al., Investigation on the ENOB and clipping effect of real ADC and AGC in coherent QAM transmission system, *Proc. ECOC 2011*, Geneva, 2011.)

adjusted reference power level of the AGC and $ENOB_S$ for the cases of 1500 km long haul transmission of full CD compensation and non-CD compensation operating in the linear (0 dBm launch power in both links) and non-linear regimes of the fibers with the launch power of 4 and 5 dBm, respectively. When the link is fully CD compensated, the non-linear effects further contribute to the ineffectiveness of the ADC resolution, and hence the moderate AGC freedom in the performance is achieved. While in the case of non-CD compensation link (Figure 7.36d and e), the dispersive pulse sampled amplitudes are lower with less noise allowing the resolution of the ADC higher via the non-linear gain of the AGC, thus an effective phase estimation and equalization can be achieved. We note that the offline data sets employed prior to the processing using $ENOB_S$ to obtain the contours of Figure 7.36, produce the same BER contours of 2e-3 for all cases. Hence a fair comparison when the $ENOB_S$ model is used. The opening distance of the BER contours indicates the dynamic range of the $ENOB_S$ model, especially the AGC. It is obvious from Figure 7.36a through e that the dynamic range of the model is higher for non-compensating than for full CD compensated transmission and even for the case of B2B. However, for non-linear scenario for both cases, the requirement for $ENOB_S$ is higher for

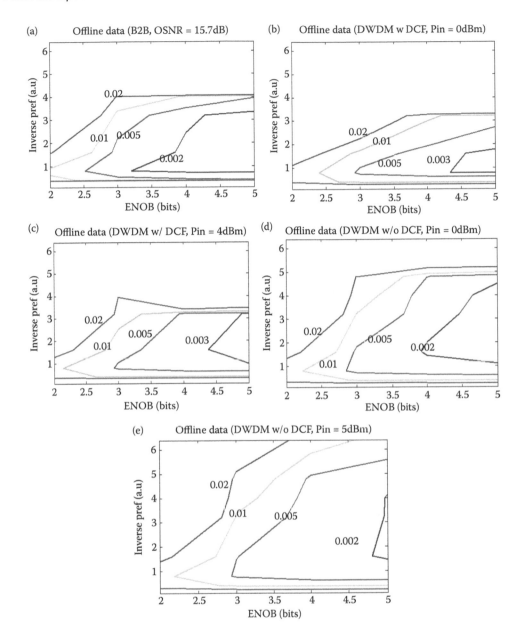

FIGURE 7.36 Comprehensive effects of AGC clipping (inverse of P_{Ref}) and $ENOB_S$ of coherent receiver, experimental transmission under (a) B2B, (b) DWDM linear operation with full chromatic dispersion (CD) compensation and (b) linear, (c) non-linear regions with 4 dBm launch power, (d) non-CD compensation and (d) linear, and (e) non-linear region with launch power of 5 dBm.

the dispersive channel (Figure 7.36c and e). This may be due to the cross phase modulation effects of adjacent channels, hence more noise.

7.6.3.6 Digital Processors

The structures of the DAC and ADC are shown in Figures 7.37 and 7.38, respectively. Normally, there would be four DACs in an IC in which each DAC section is clocked with a clock sequence that is derived from a lower frequency sinusoidal wave injected externally into the DAC. Four units are required for the inphase and quadrature phase components of QAM modulated polarized

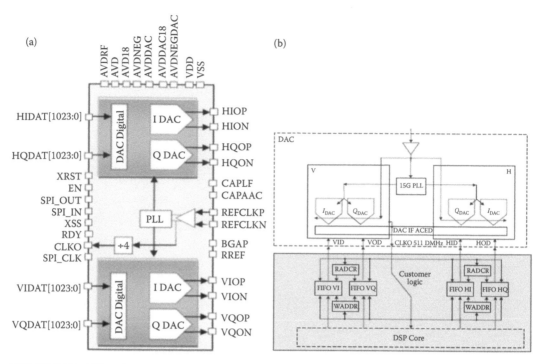

FIGURE 7.37 Fijitsu DAC structures for 4 channel PDM_QPSK signals (a) schematic diagram and (b) processing function.

channels, thus the notations of I_{DAC} and Q_{DAC} shown in the diagram. Similarly, the optical received signals of PDM-QAM (polarization division multiplexed–quadrature amplitude modulation) would be sampled by a four-phase sampler and then converted to digital form into four groups of I and Q lanes for processing in the DSP sub-system. Due to the interleaving of the sampling clock waveform, the digitalized bits appear simultaneously at the end of a clock period that is, sufficient long so that the sampling number is sufficient large to achieve all samples. For example, shown in Figure 7.38, 1024 samples are achieved at a periodicity corresponding to 500 MHz cycle clock for 8- bit ADC. Thus, the

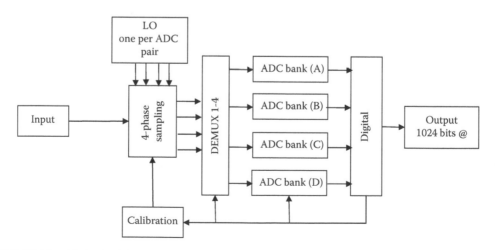

FIGURE 7.38 ADC principles of operations (CHAIS-charge mode interleave sampling).

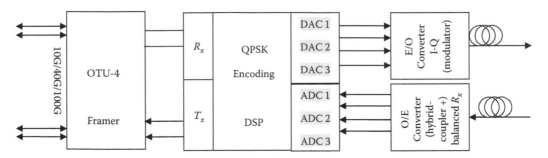

FIGURE 7.39 Schematic representation of typical structure of ADC and ADC transceiver sub-systems for PDM-QPSK modulation channels.

clock has been slowed down by a factor of 128, or alternatively the sampling interval is $1/(500$ MHz.128$) = 1/64$ GHz sec. The sampling is implemented using a CHAIS (CHArged mode Inter-leaved Sampler).

Figure 7.39 shows a generic diagram of an optical DSP based transceiver employing both DAC and ADC under Q_{PSK} modulated or QAM signals. The current maximum sampling rate of 64 GSa/s is available commercially. An IC image of the ADC chip is shown in Figure 7.40.

7.7 CONCLUDING REMARKS

This chapter has described the principles of both incoherent (Chapter 5) and coherent reception and associated techniques with noise considerations and main functions of the DSP. Furthermore, the matching of the local oscillator laser and that of the carrier of the transmitted channel is very important for effective coherent detection, if not degradation of the sensitivity of results. The ITU standard requires that for digital processing based coherent receiver the frequency offset between the LO and the carrier must be within the limit of ± 2.5 GHz. Furthermore, in practice it is expected that in network and system management the tuning of the LO is to be done remotely and automatic locking of the LO with some prior knowledge of the frequency region to set the initial frequency. Thus this action briefly describes the optical phase locking the local oscillator source for intradyne coherent reception sub-system.

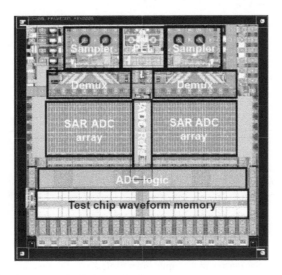

FIGURE 7.40 An ADC sub-systems including a dual convertor structure.

APPENDIX 7A: NOISE EQUATIONS

Referring to the small signal and noise model given in Figures 7A.1 and 7A.2, the noises generated in a transistor can be expressed as

$$di_1^2 = 4k_B T g_B df$$

$$di_2^2 = 2qI_c df \simeq 2k_B T g_m df$$

$$di_3^2 = 2qI_B df \simeq 2k_B T(1 - \alpha_N)g_m df \qquad (7A.1)$$

$$di_4^2 = shot..noise..of..diodes..and..thermal..noise..of..bias.in\ g_B..resistors$$

where g_B is the base conductance, I_C is the collector bias current, and I_B is the base bias current. From nodal analysis of the small signal equivalent circuit given in Figure 7A.2 ,we can obtain the relationship

$$\begin{bmatrix} Y_s + g_B & -g_B & 0 \\ -g_B & g_B + y_1 + y_f & -y_f \\ 0 & g_m - y_f & y_f + y_2 \end{bmatrix} \begin{bmatrix} V_1 \\ V_2 \\ V_3 \end{bmatrix} = \begin{bmatrix} i_{eq} - i_{N1} \\ i_{N1} + i_{N3} \\ i_{N2} \end{bmatrix} \qquad (7A.2)$$

Hence, V_3, V_2, and V_1 can be found by using Euler's rule for the matrix relationship

$$V_3 = \frac{\Delta_{13}}{\Delta}(i_{eq} - i_{N1}) + \frac{\Delta_{23}}{\Delta}(i_{N1} + i_{N3}) + \frac{\Delta_{33}}{\Delta}(i_{N2}) \qquad (7A.3)$$

The noise currents as referred to the input are

$$d(i_1'^2) = \left|\frac{Y_s}{g_B}\right|^2 dI_1^2 = \omega_2 C_s^2 r_B 4kT df \qquad (7A.4)$$

$$d(i_1'^2) = \left|(Y + y_1 + y_f) + \frac{Y_s}{g_B}(y_1 + y_f)\right|^2 \left|\frac{1}{y_f - g_m}\right|^2 = \left|\frac{1}{y_f - g_m}\right|^2 dI_2^2$$

$$= \left(\frac{1}{\beta_N r_E^2} + \omega^2\left(C_0^2 - \frac{2C_B r_B}{\beta_N r_E}\right) + \omega^4 2C_s r_B C_B^2\right) 2kT r_E df \qquad (7A.5)$$

$$d(i_3'^2) = \left|\frac{Y_s}{g_B} + 1\right|^2 dI_3^2 = (\omega C_s^2 r_B^2 + 1)\frac{2kT}{\beta r_E} df \qquad (7A.6)$$

in which we have assumed that $\omega C_f \ll g_m$.

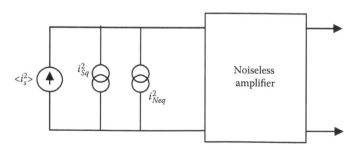

FIGURE 7A.1 Equivalent noise current at the input and noiseless amplifier model, i_{Sq}^2 is the quantum shot noise which is signal dependent, i_{Neq}^2 is the total equivalent noise current referred to the input of the electronic amplifier.

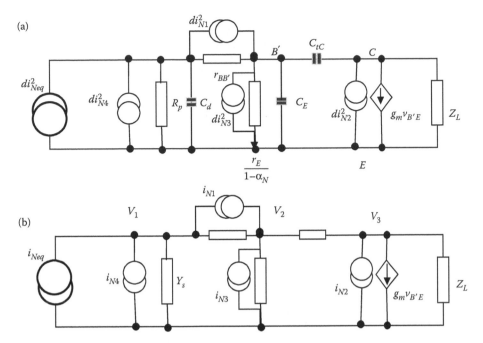

FIGURE 7A.2 (a) Approximated noise equivalent and small signal model of a BJT front end, (b) generalized noise and small signal model circuit. Note that $r_B = r_{sd} + r_{BB}$ and $C_d = C_p + C_i$ with $r_{BB'}$ is the base resistance, r_{sd} is the diode resistance, C_d is the photodiode capacitance and C_i is the input capacitance.

REFERENCES

1. M. Taylor, Coherent optical detection and signal processing method and system, Patent US 20040114939 A1, 17. Juni 2004.
2. T. Okoshi, *Coherent Optical Fiber Communications*, Tokyo: Ohmsha, 1986.
3. F. Doer, Coherent optical QPSK intradyne system: Concept and digital receiver realization, *IEEE J Lightw Tech.*, vol. 10, no. 9, pp. 1290–1296, 1992.
4. R. C. Alferness, Guided wave devices for optical communication, *IEEEJ. Quantum. Elect.*, vol. QE-17, pp. 946–959, 1981.
5. W. A. Stallard, A. R. Beaumont, and R. C. Booth, Integrated optic devices for coherent transmission, *IEEE J Lightwave Tech.*, vol. LT-4, no. 7, pp. 852–857, July 1986.
6. V. Ferrero and S. Camatel, Optical phase locking techniques: An overview and a novel method based on single sideband sub-carrier modulation, *Optics Express*, vol. 16, no. 2, pp. 818–828, 2008.
7. I. Garrett and G. Jacobsen, Theoretical analysis of heterodyne optical receivers for transmission systems using (semiconductor) lasers with non-negligible linewidth, *IEEE J. Lightw Tech.*, vol. LT-3/4, pp. 323–334, 1986.
8. G. Nchoson, Probability of error for optical heterodyne DPSK systems with quantum phase noise, *Elect. Lett.*, vol. 20/24, pp. 1005–1007, 1984.
9. S. Shimada, *Coherent Lightwave Communications Technology*, London: Chapman and Hall, 1995, p. 27.
10. Y. Yamamoto and T. Kimura, Coherent optical fiber transmission system, *IEEE J. Quantum Elect.*, vol. QE-17, pp. 919–934, 1981.
11. S. Savory and T. Hadjifotiou, Laser linewidth requirements for optical DQPSK optical systems, *IEEE Photonic Tech Lett.*, vol. 16, no. 3, pp. 930–932, March 2004.
12. K. Iwashita and T. Masumoto, Modulation and detection characteristics of optical continuous phase FSK transmission system, *IEEE J. Lightwave Tech.*, vol. LT-5/4, pp. 452–462, 1987.
13. K. Iwashita and T. Masumoto, Modulation and detection characteristics of optical continuous phase FSK transmission system, *IEEE J. Lightwave Tech.*, vol. LT-5/4, pp. 452–462, 1987. (Fig.1).
14. F. Derr, Coherent optical QPSK intradyne system: Concept and digital receiver realization, *IEEE J. Lightw.Tech.*, vol. 10, no. 9, pp. 1290–1296, 1992.

15. Gabriella Bosco, Ivan N. Cano, Pierluigi Poggiolini, Liangchuan Li, and Ming Chen, MLSE-based DQPSK transmission in 43Gb/s. DWDM long-haul dispersion managed optical systems, *IEEE J. Lightw. Technology*, vol. 28, no. 10, May 15, 2010.

16. D.-S. Ly-Gagnon, S. Tsukamoto, K. Katoh, and K. Kikuchi, Coherent detection of optical quadrature phase-shift keying signals with carrier phase estimation, *J. Lightwave Technol.*, vol. 24, pp. 12–21, 2006.

17. E. Ip and J. M. Kahn, Feedforward carrier recovery for coherent optical communications, *J. Lightwave Technol.* vol. 25, pp. 2675–2692, 2007.

18. L. N. Binh, Dual-ring 16-star QAM direct and coherent detection in 100 Gb/s optically amplified fiber transmission: Simulation, *Opt. Quantum Electron.*, vol. 40, 707, 2008. Accepted December 2008. Published on line December 2008.

19. L. N. Binh, Generation of multi-level amplitude-differential phase shift keying modulation formats using only one dual-drive Mach–Zehnder interferometric optical modulator, *Opt. Eng.*, vol. 48, no. 4, pp. 045005, 2009.

20. M. Nazarathy, X. Liu, L. Christen, Y. K. Lize, and A. Willner, Self-coherent multisymbol detection of optical differential phase-shift-keying, *J. Lightwave Technol.*, vol. 26, pp. 1921–1934, 2008.

21. R. Noe', PLL-free synchronous QPSK polarization multiplex/diversity receiver concept with digital I&Q baseband processing, *IEEE Photon. Technol. Lett.*, vol. 17, pp. 887–889, 2005.

22. L. G. Kazovsky, G. Kalogerakis, and W.-T. Shaw, Homodyne phase-shift-keying systems: Past challenges and future opportunities, *J. Lightwave. Technol.*, vol. 24, pp. 4876–4884, 2006.

23. J. P. Gordon and L. F. Mollenauer, Phase noise in photonic communications systems using linear amplifiers, *Opt. Lett.*, vol. 15, pp. 1351–1353, 1990.

24. H. Kim and A. H. Gnauck, Experimental investigation of the performance limitation of DPSK systems due to nonlinear phase noise, *IEEE Photon. Technol. Lett.*, vol. 15, pp. 320–322, 2003.

25. S. Zhang, P. Y. Kam, J. Chen, and C. Yu, Receiver sensitivity improvement using decision-aided maximum likelihood phase estimation in coherent optical DQPSK system, *Conference on Lasers and Electro-Optics/Quantum Electronics and Laser Science and Photonic Applications Systems Technologies, Technical Digest (CD)*, Optical Society of America, San Jose, USA, 2008, paper CThJJ2.

26. P. Y. Kam, Maximum-likelihood carrier phase recovery for linear suppressed-carrier digital data modulations, *IEEE Trans. Commun.*, vol. COM-34, pp. 522–527, June 1986.

27. E. M. Cherry and D. A. Hooper, *Amplifying Devices and Amplifiers*, NY: J. Wiley, 1965.

28. E. Cherry and D. Hooper, The design of wide-band transistor feedback amplifiers, *Proc. IEE*, vol. 110, no. 2, pp. 375–389, 1963.

29. N. M. S. Costa and A.V. T. Cartaxo, Optical DQPSK system performance evaluation using equivalent differential phase in presence of receiver imperfections, *IEEE J. Lightwave Tech.*, vol. 28, no. 12, pp. 1735–1744, June 2010.

30. L. N. Binh, *Noises in Optical Communications and Photonic systems*, Boca Raton, FL, USA: CRC Press, 2017.

31. H. Tran, F. Pera, D. S. McPherson, D. Viorel, and S. P. Voinigescu, 6-kΩ, 43-Gb/s differential transimpedance-limiting amplifier with auto-zero feedback and high dynamic range, *IEEE Journal of Solid State Circuits*, vol. 39, no 10, pp. 1680–1689, 2004.

32. Inphi inc., Technical information on 3205 and 2850 TIA device, 2012.

33. L. N. Binh *Noises in Optical Communications and Photonic Systems*, Boca Raton, FL, USA: CRC Press, 2017.

34. CFP MSA CFP4 Hardware Specification, Revision 0.1, 2 March 2014, http://www.cfp-msa.org/Documents/CFP-MSA_CFP4_HW-Spec-rev0.1.pdf access date: 2017-02-19

35. M. H. Eissa, A. Awny, G. Winzer, M. Kroh, S. Lischke, D. Knoll, L. Zimmermann, D. Kissinger, and A. C. Ulusoy, A wideband monolithically integrated photonic receiver in 0.25-μm SiGe:C BiCMOS technology, *Proc. 42nd European Conference on Optical Communications, ECOC 2016*, Duesseldorf, Germany, pp. 487–490, September 12–15, 2016.

36. E. Cherry and D. Hooper, *Amplifying Devices and Low Pass Amplifier Design*, N.Y.: J. Wiley, 1968.

37. E. M. Cherry and D. E. Hooper, The design of wide-band transistor feedback amplifiers, *Proc. IEE*, vol. 110, no. 2, pp. 375–389, 1963.

38. (a) S. Zhang, P. Y. Kam, J. Chen, and C. Yu, A comparison of phase estimation in coherent optical PSK system, *Photonics Global '08, Paper C3-4A-03*. Singapore, December 2008; (b) S. Zhang, P. Y. Kam, J. Chen, and C. Yu, Adaptive decision-aided maximum likelihood phase estimation in coherent optical DQPSK system, *OptoElectronics and Communications Conference (OECC) '08*, Paper TuA-4, pp. 1–2, Sydney, Australia, July 2008.

39. I. Hodgkinson, Receiver analysis for optical fiber communications systems, *IEEE J. Lightw. Tech.*, vol. 5, no. 4, pp. 573–587, 1987.
40. http://en.wikipedia.org/wiki/ENOB—cite_note-3. Accessed on September 2011.
41. N. Stojanovic, An algorithm for AGC optimization in MLSE dispersion compensation optical receivers, *IEEE Trans.Circ.Sys. I*, 55, 2841–2847, 2008.
42. B. Mao, N. Stojanovic, C. Xie, L. Binh, and M. Chen, Investigation on the ENOB and clipping effect of real ADC and AGC in coherent QAM transmission system, *Proc. ECOC 2011*, Geneva, 2011.

8 Multi-Carrier OFDM Optical Modulation

This chapter deals with the optical transmission employing orthogonal frequency division multiplexing (OFDM) techniques in which several sub-carriers are employed to carry partitioned data for transmission over the fiber channel. The channels are frequency multiplexed and orthogonal. The orthogonality comes easily from the generation of the FFT and IFFT of the data sequence in the electronic domain. These sub-carriers in the baseband are then modulating the optical carrier under some modulation formats that are most appropriate for optical fiber channels. The narrow band of the data of sub-channels ensure the low broadening of the pulses and effectively combat the impairments expected from the fiber. This advanced modulation and oOFDM has emerged over the last few years as one of the most promising candidates for long-haul optically amplified communications systems and networks and for low-cost metro and access optical networks. The OFDM modulation technique is applied to generate the discrete multi-tone (DMT) modulation for access network delivery (Chapter 9).

The main reasons are that IFFT and FFT can be implemented in the electronic digital domain at low cost due to availability of ultra-high sampling rate ADC and DSP, hence the frequency shifting of sub-carriers of the OFDM signals.

8.1 INTRODUCTION

Orthogonal frequency division multiplexing (OFDM) is a transmission technology that is primarily known from wireless communications and wired transmission over copper cables. It is a special case of the widely known frequency division multiplexing (FDM) technique for which digital or analog data is modulated onto a certain number of carriers and transmitted in parallel over the same transmission medium. The main motivation for using FDM is the fact that, due to parallel data transmission in frequency domain, each channel occupies only a "small" frequency band. Signal distortions originating from frequency-selective transmission channels, the fiber chromatic dispersion, can be minimized. The special property of OFDM is characterized by its very high spectral efficiency. While for conventional FDM, the spectral efficiency is limited by the selectivity of the bandpass filters required for demodulation, OFDM is designed such that the different carriers are pair wise orthogonal. This way, the sampling point the inter-carrier interference (ICI) is suppressed although the channels can overlap spectrally.

8.1.1 PRINCIPLES OF OOFDM: OFDM AS A MULTI-CARRIER MODULATION FORMAT

8.1.1.1 Spectra

OFDM is a multi-carrier transmission technique and bed on the use of multiple frequencies to simultaneously transmit multiple signals in parallel form. Each sub-channel is assigned with a sub-carrier that is within the range allowable for an optical channel within the DWDM Optical transmission system. Unlike the normal frequency division multiplexing technique in which the spectra of the sub-channels are separated by a guard band such that there is no overlapping between them, in OFDM, the term orthogonality comes from the property that adjacent channels are orthogonal, that is, they are perpendicular or the dot product of the channel is zero as shown in Figure 8.1a and b, respectively. This allows the overlapping of the spectra of adjacent channels without creating any cross talks between them [1]. The frequency domain representation of an OFDM symbol in the

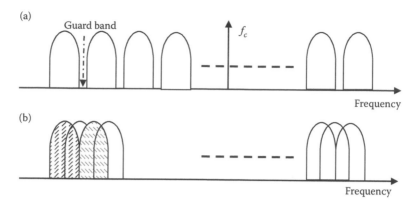

FIGURE 8.1 Multi-carrier modulation technique (a) spectrum of FDM sub-carriers and (b) spectrum of OFDM sub-carriers.

sampled discrete plane is shown in Figure 8.2. Thus, a schematic arrangement of the generation/mapping of OFDM symbols and reception/demapping can be shown in Figure 8.3.

The entire bandwidth allocated for a wavelength channel may be occupied. The data source is distributed over all sub-carrier. Thus, each sub-carrier transports a small amount of the information. By this lowering of the bit rate per sub-carrier channel, the ISI due to the distortion of the channel can be significantly reduced.

8.1.1.2 Orthogonality

The complex baseband OFDM signal $s(t)$ can be written as

$$s(t) = \sum_k \sum_{n=0}^{N-1} a_n(k)g_n(t - kT) \tag{8.1}$$

where k is the time index, N is the total number of sub-carriers, $T = NT_s$ is the stretched OFDM symbol period because of the conversion from serial to parallel, and T_s is the sampling period. $a_n(k)$ is the kth data symbol of the nth sub-carrier and $g_n(t)$ is the baseband data pulse given by

$$\begin{aligned} g_n(t) &= e^{j2\pi f_n}g(t) \\ f_n &= f_c \pm n\Delta f \end{aligned} \tag{8.2}$$

including the sub-carrier component of frequency shifted from the central carrier frequency by n amounts of the frequency spacing between the sub-carriers. $G(t)$ is the pulse shaping function which normally is very close to Gaussian after propagating through the optical MZIM.

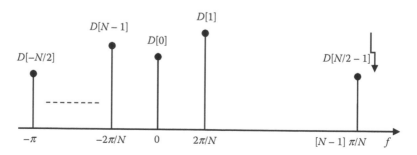

FIGURE 8.2 Frequency domain representation of an OFDM symbol.

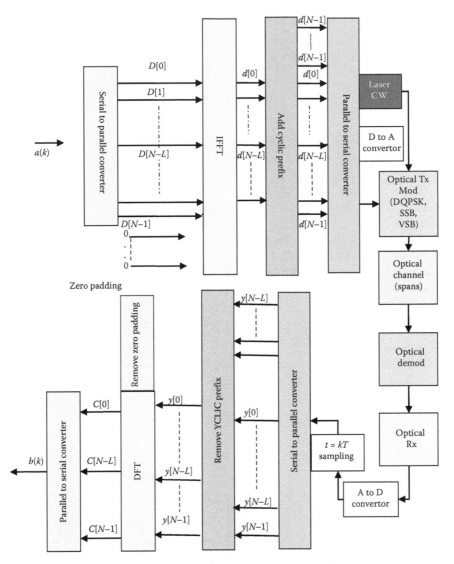

FIGURE 8.3 Block diagram of the sampled discrete time model of OFDM system using N-point FFT and IFFT in flow of sampled signals and incorporating blocks of transmission link consisting of laser, transmitter, and receiver.

The orthogonality of the channels requires:

$$\int_{-\infty}^{\infty} f_n(t)f_m^*(t)dt = \frac{1}{T}\int_{-T/2}^{T/2} f_n(t)f_m^*(t)dt \begin{cases} 1 & n = m \\ 0 & n \neq m \end{cases} = \delta_0[n - m] \qquad (8.3)$$

where the asterisk * indicates the complex conjugate. Thus, it requires that the selection of the carrier frequency and the pulse shaping function in such a way that the orthogonality can be achieved.

8.1.1.3 Sub-Carriers and Pulse Shaping

For orthogonality, each sub-carrier should take an integer number of cycles over a symbol period T, and the number of cycles between adjacent sub-carriers differs by exactly unity. That means that the frequencies of the sub-carriers are multiples of each other.

The pulse shapes of the signals can take the form of rectangular or sin c function and can be written as

$$g(t) = \frac{\frac{4\alpha t}{T}\left\{\cos\left(1+\alpha\right)\frac{\pi t}{T} + \sin\left(1+\alpha\right)\frac{\pi t}{T}\right\}}{\frac{\pi}{T}\left(1 - \left[\frac{4\alpha t}{T}\right]^2\right)} \tag{8.4}$$

where α is the roll off factor, similar to the raise cosine function.

For the rectangular pulse, we have

$$g(t) = \begin{cases} 1 & |t| \leq \dfrac{T}{2} \\ 0 & \text{otherwise} \end{cases} \tag{8.5}$$

However, this "brick wall" pulse shape would be very difficult to be realized in practice, and the raise cosine shape would be preferred.

The orthogonality can be achieved by setting $\Delta f = 1/T$. For simplicity, we can assume that the signal would take the form with the shape function is the same for all sub-carrier channels

$$s(t) = \sum_{n=0}^{N} a_n g_n(t - kT) = \sum_{n=0}^{N} a_n g(t) e^{j2\pi(f_c + n\Delta f)} \tag{8.6}$$

where a_n is the data stream fed into each sub-carrier channel from the output of the serial to parallel converter as shown in Figure 8.4. Taking the Fourier transform of Equation 8.6, we have

$$S(f) = \sum_{n=0}^{N} a_n G(f - (f_c - n\Delta f)) \tag{8.7}$$

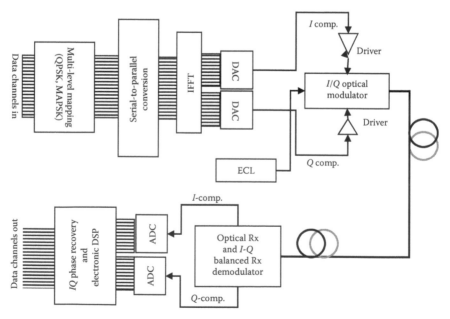

FIGURE 8.4 Schematic diagram of oOFDM transmitter of a long haul optical transmission system using OFDM and IQ multi-level modulation formats. Separate polarized channels PDM are not shown. I = inphase, Q = quadrature phase, ECL = external cavity lasers, DSP = digital signal processing, and IFFT Inverse Fast Fourier Transform.

which is the summation of all sin c functions if the pulse shape is that of rectangular shape. These spectral functions sin c would cross over at the null of each other and their peaks. On the other hand, if the temporal function of the data of each sub-carrier channel, then the frequency spectrum of the channel would be rectangular and thus they can be separated. Thus, we can have

$$g(t) = \frac{1}{T} \sin c\left(\pi \frac{t}{T}\right) = \Delta f \sin c(\pi \Delta f t) \rightarrow G(f) = \text{rect}\left(\frac{f}{\Delta f}\right) \tag{8.8}$$

with $\Delta f = 1/T$.

Therefore, the spectrum of OFDM signal takes the form

$$S(f) = \sum_{n=0}^{N-1} a_n \text{rect}\left(\frac{f - f_c + n\Delta f}{\Delta f}\right) \tag{8.9}$$

which is made up by a number of rectangular spectra separated by Δf.

In practice, the availability of rectangular time pulse shape would not be possible except with certain raise cosine features, while for sin c, shape pulses can be generated without difficulty by using a transmission filter in the electrical domain. The spectrum of OFDM signal is thus the summation of the sin c shape spectra of all the sub-channels. If the number of sub-carrier channel is sufficiently high, then one would obtain a flat rectangular spectrum [2].

The orthogonal sub-channels can be generated using IFFT or FFT normally available from digital signal processors.

8.1.1.4   OFDM Receiver

In a conventional FDM system, the sub-channels are normally separated by filtering using banks of filters and demodulators. For orthogonal signals, the sub-channels can be separated by correlation techniques [3,4].

8.1.2   FFT- and IFFT-Based OFDM Principles

To recover or demodulate the sub-carrier channels by using a bank of filters or local oscillator to tune to separate the sub-channels, it is very costly, when the number of sub-channel is very large. Recently, the speed of digital signal processing has increased significantly. Certainly, the FFT and IFFT are the main features of these processors. The main feature of the FFT and IFFT is that the sampling and their sampled transformed sub-channels are orthogonal. Thus, they can be effectively used ion the transmitter and receiver of OFDM systems. The OFDM discrete model is shown in Figure 8.4. The signal is defined in the frequency domain at the transmitter. It is a sampled digital signal and is defined such that the discrete Fourier spectrum exists only at the discrete frequencies. Each OFDM sub-channel corresponds to one element of this discrete Fourier spectrum. The magnitude and phase of the carrier depends on the data transmitted. The serial-stream input data sequence is converted into parallel that would then be represented as the components of the frequency spectra, $D[n]$ *with equally spaced frequencies* $f_n = 2\pi n/N$ for $n = 0,1 \dots N$. Thus, $D[0]$ represents the DC component of the frequency spectrum of OFDM signals. The frequency representation of the signal are shown in Equation 8.10.

An N-point inverse FFT block (the DSP processor) then generates the time domain components whose prefix. Thus, the signal at this stage is in analog form, as shown in Figure 8.4, padded with zeros for a cyclic discrete Fourier transform. They are then converted from parallel for into serial form and used for modulating the optical modulator. The passband optical signals are then transmitted over the channel.

The definition of the N-point inverse DFT is

$$d[k] = \text{IDFT}\{D[n]\} = \frac{1}{N}\sum_{n=0}^{N-1}D[n]e^{j2\pi k/N} \quad n, k = 0, 1\ldots N-1 \tag{8.10}$$

where N is the symbol length. After the pulse shaping of a rectangular function of

$$\text{rect}\left[\frac{k}{N}\right] = \begin{cases} 1 & \text{for } k = 0, 1\ldots N-1 \\ 0 & \text{otherwise} \end{cases} \tag{8.11}$$

the sequence $d[n]$ in the time domain is given by

$$d[k] = \frac{1}{N}\sum_{n=0}^{N-1}D[n]e^{j2\pi k/N}\text{rect}\left[\frac{k}{N}\right] \quad n, k = 0, 1\ldots N-1$$

$$= \frac{1}{N}\sum_{n=0}^{N-1}D[n]g_n[k] \tag{8.12}$$

$$\text{with} \quad g_n[k] = e^{j2\pi k/N}\text{rect}\left[\frac{k}{N}\right]$$

$g_n[k]$ can be considered as the baseband pulse which is considered to be orthogonal and satisfying the condition.

$$g_n[k] = e^{j2\pi k/N}\text{rect}\left[\frac{k}{N}\right] \tag{8.13}$$

Thus, a rectangular or sin c pulse shaper can be inserted between the serial to parallel and the IFFT blocks and then multiplied by the sub-carrier frequency, then superimposed and multiplied by a factor $1/N$.

At the receiver, after the photodetection and electronic pre-amplification, the noisy time domain OFDM samples are converted to its corresponding frequency domain symbols by an N-point FF. The coefficients of an N-point DFT are given by

$$C[n] = \text{DFT}\{y[k]\} = \sum_{n=0}^{N-1}y[k]e^{j2\pi k/N} \quad n, k = 0, 1\ldots N-1$$

$$= \sum_{-\infty}^{+\infty}y[k]e^{j2\pi k/N}\text{rect}\left[\frac{k}{N}\right] \tag{8.14}$$

An N-point DFT is equivalent to a bank of N-orthogonal "matched filters" matched to the corresponding "baseband pulses" in IDFT, followed by samplers that sample once per N symbols. The impulse of the matched filter can be written as

$$g_n^*[-k] = e^{j2\pi k/N}\text{rect}\left[\frac{-k}{N}\right] \tag{8.15}$$

This follows the logical sequence of

$$C[n] = y[k] * g_n^*[-k] = y[n] * e^{j2\pi k/N} \text{rect}\left[\frac{k}{N}\right] \quad n, k = 0, 1 \ldots N - 1$$

$$= \sum_{-\infty}^{+\infty} y[k] e^{j2\pi k/N} \text{rect}\left[\frac{-(k-u)}{N}\right]$$

(8.16)

After sampling at the instant $k = 0$, the frequency spectral component $C[n]$ becomes

$$C[n] = \sum_{-\infty}^{+\infty} y[u] e^{j2\pi u/N} \text{rect}\left[\frac{u}{N}\right]$$

(8.17)

$$\text{with} \ldots u \to k \to \quad C[n] = \sum_{-\infty}^{+\infty} y[u] e^{j2\pi u/N} \text{rect}\left[\frac{k}{N}\right] = \sum_{k=0}^{N-1} y[k] e^{j2\pi k/N}$$

which is the DFT of $y[k]$.

8.2 oOFDM TRANSMISSION SYSTEMS

Orthogonality is achieved by placing the different RF-carriers onto a fixed frequency grid and assuming rectangular pulse shaping. For OFDM, the signal can be described as the output of a discrete inverse Fourier transform block using the parallel complex data symbols as input. This property has been one of the main driving aspects for OFDM in the past since modulation and demodulation of a high number of carriers can be realized by simple digital signal processing (DSP) instead of using many local oscillators in transmitter and receiver. Recently, OFDM has become attractive for digital optical communications [5,6]. Using OFDM appears to be very attractive because the low bandwidth occupied by a single OFDM channel increases the robustness toward fiber dispersion drastically allowing the transmission of high data rates of 40 Gb/s and more over hundreds of kilometers without the need for dispersion compensation [2]. In the same way as for modulation formats like DPSK or DQPSK that were introduced in recent years, the challenge for OFDM for optical system engineers is to adapt a classical technology to the special properties of the optical channel and the requirements of optical transmitters and receivers.

However, Reference 7 shows that an optical SSB modulation can assist the combat of fiber dispersion. SSB can be achieved by driving the MZIM with two microwave signals $\pi/2$ phase shift with each other or by optical filtering (VSB), as shown in Chapters 2 and 9. However, the truly SSB transmitter using dual drive and $\pi/2$ phase shift is preferred to preserve the energy contained within the bands of the signals. In Optical SSB, the phase information can be preserved after the square-law detection of the photodetector and the CD is limited by reducing the optical spectral bandwidth by a factor of 2 [5].

Two approaches have been reported recently. An intuitive approach introduced by Lorente et al. [6] makes use of the fact that the wavelength-division multiplexing (WDM) technique itself already realizes data transmission over a certain number of different carriers. By means of special pulse shaping and carrier wavelength selection, the orthogonality between the different wavelength channels can be achieved resulting in the so-called orthogonal WDM technique (OWDM). However, in this way the option of simple modulation and demodulation by means of discrete Fourier transforms (DFT) cannot be utilized as this kind of digital signal processing is not available in the optical domain.

The basic OFDM transmitter and receiver configurations are given in Figure 8.3 [8]. Data streams (e.g., 1.0 Gb/s) are mapped into a 2D signal point from a point signal constellation such as QAM. The complex-valued signal points from all sub-channels are considered as the values of the discrete

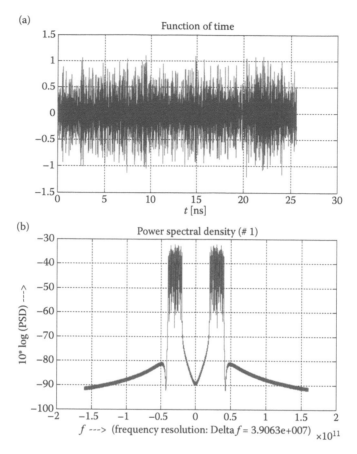

FIGURE 8.5 Signals of an optical FFT/IFFT-based system (a) typical time domain OFDM signals (b) power spectral density of OFDM signal with 512 sub-carriers a shift of 30 GHz for the line rate of 40 Gb/s and QPSK modulation. *Note*: electrical mixers for I and Q components recovery in electrical domain.

Fourier transform (DFT) of a multi-carrier OFDM signal. The serial to parallel converter arrange the sequences into equivalent discrete frequency domain. By selecting the sufficiently large number of sub-channels, the OFDM symbol interval can be made much larger than the dispersed pulse width in a single-carrier system, resulting in an arbitrary small inter-symbol interference (ISI). The OFDM symbol, shown in Figure 8.5, is generated under software processing, as follows: input QAM symbols are zero-padded to obtain input samples for inverse fast Fourier transform (IFFT), the samples are inserted to create the guard band, and the OFDM symbol is multiplied by the window function which can be represented by a raised cosine function. The purpose of cyclic extension is to preserve the orthogonality among sub-carriers even when the neighboring OFDM symbols partially overlap due to dispersion.

The principles of FFT and IFFT for OFDM symbol generations are shown in Figure 8.3. Section 3.1.15 of Chapter 3 has also outlined the principles of PFDM transmission system. A system arrangement of the OFDM for optical transmission in a laboratory demonstration is shown in Figure 8.5. The data sequences are arranged in the sampled domain, to the frequency domain, and then to the time domain representing the OFDM waves which would look like analogue waveforms as shown in Figure 8.5a and b (see also Section 3.1.15 Figure 3.37). Each individual channel at the input would carry the same data rate sequence. These sequences can be generated from an arbitrary waveform generator. The multiplexed channels are the combined and converted to time domain using the IFFT module and then converted to the analog version via the two digital to analog

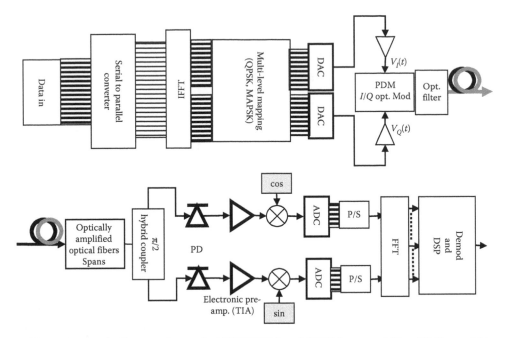

FIGURE 8.6 Schematic diagram of an optical FFT/IFFT-based OFDM system. S/P and P/S = serial to parallel conversion and vice versa. PD = photodetector, TIA = trans-impedane amplifier (see also Figure 3.37 of Chapter 3 for more details). Note electrical demodulation of inphase and quadrature components.

converters. These orthogonal data sequences are then used to modulate I- and Q-optical waveguide sections of the electro-optical modulator to generate the orthogonal channels in the optical domain. Similar decoding of I and Q channels is performed in the electronic domain after the optical transmission and optical-electronic conversion via the photodetector and electronic amplifier. Figure 8.6 illustrate similar transmission systems with the demodulation of the I and Q channels in the electrical domain while these functions are implemented in the hybrid optical coupler in the optical domain in which the $\pi/2$ phase shift for the I and Q can be done by a $\pi/2$ shift in the optical domain by using pyro-optic effects via an electrode heating to change the refractive index of the optical path of either I or Q optical signals. This is commonly done in the PLC (planar lightwave circuit) optical hybrid couple.

In comparison to other optical transmitters, the oOFDM transmitter will require a DSP and DAC for shaping the pulse spectrum and then the IFFT operation to generate the OFDM analog signals to modulate the optical modulator.

The performance of an OFDM coherent and transmission under back-to-back and over 1000 km of SSMF spans without DCM (dispersion compensation modules) as shown in Figure 8.7a and b, respectively, and indicates the resilience of OFDM format to dispersion with an OSNR of 8 dB for a BER of 1e-3 which is the error free rate under FEC integration. Note that polarization multiplexing is employed in the systems with both polarized channels are detected and extracted for BER as a function of the OSNR (measured over 0.1 nm spectral width).

In OFDM, the serial data sequence, with a symbol period of T_s and a symbol rate of $1/T_s$, is split up into N-parallel sub-streams (sub-channels).

As an alternative method [9], generation of an electrical OFDM signal by means of electrical signal processing followed by modulation onto a single optical carrier [10,11]. This approach is known as oOFDM. Here, the modulation is a two-step process: first, the electrical OFDM signal already is a broadband bandpass signal which is then modulated onto the optical carrier. Second, to increase data throughput, oOFDM can be combined with WDM resulting in multi-Tb/s transmission system, as shown in Figure 8.4. Nevertheless, oOFDM itself offers different options for implementation. An

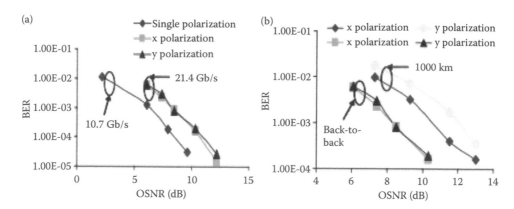

FIGURE 8.7 BER versus OSNR under coherent reception. (Extracted with permission from A. Ali, Orthogonal frequenc division multiplexing in optical transmission systems with high spectral efficiency, Master Thesis Dissertation, Technische Facultaet, University of Kiel, Kiel, Germany, 2007.) (a) back-to-back and (b) 1000 km non-DCM optically amplified spans.

important issue is optical demodulation that can be realized either by means of direct detection (DD) or coherent detection (CD) including a local laser oscillator. DD is preferable because of its simplicity. However, for DD the optical intensity must be modulated. Because the electrical OFDM signal is quasi-analog with zero mean and high peak-to-average ratio, most the optical power should be wasted for the optical carrier. That means there is an additional DC-value of the complex baseband signal, resulting in low receiver sensitivity. For chromatic dispersion (CD), in addition the bandwidth efficiency is twice as high as for DD since a double-sideband signal is generated for pure intensity modulation. For CD, a complex optical I-Q modulator composed of two real modulators in parallel followed by superposition with $\pi/2$ phase shift allows for transmission of twice as much data within the same bandwidth. For intensity modulation, the bandwidth efficiency may be increased by suppressing one of the redundant sidebands resulting in oOFDM with single-sideband (SSB) transmission. First, the serial data in can be converted to parallel streams. These parallel data sequences are then mapped to QAM constellation in the frequency domain, then by IFFT converted it back to the time domain. The time domain signals are in I- (in-phase) and Q-(quadrature) components that are then fed into an I-Q optical modulator. This optical modulation can be DQPSK or any other multi-level modulation sub-system. At the end of the optical fiber transmission, I and Q components are detected by direct detection or coherent detection. For coherent detection, a 2×4 $\pi/2$ hybrid coupler is used to mix the polarized optical fields of the local oscillator and that of the received signals. The outputs of the couplers are then fed into balanced optical receivers. The mixing of the local laser source and that of the signals preserves the phase of the signals which are processed by a high-speed electronic digital processor. For direct detection, I and Q components are detected differentially, the amplitude and phase detection are compared, and are then processed similarly to the coherent case.

To show the robustness of oOFDM toward fiber dispersion and fiber non-linearity, numerical simulations are carried out for a data stream of 42.7 Gb/s data rate. The number of OFDM channels can be varied between $N_{min} = 256$ and $N_{max} = 2048$. A guard interval of 12 ns can be inserted, a strategy belonging inherently to OFDM technology that ensures orthogonality of the different channels in case of a transmission channel with memory. For the optical modulation, intensity modulation using a single Mach–Zehnder modulator in conjunction with SSB filtering and direct detection was implemented. The non-linear optical transmission channel consisted of eight 80 km non-DCF spans of SSMF. MZIM with linearization should be used [12]. As a criterion for performance, required OSNR for a BER of 10^{-3} (Monte Carlo) is measured. Using FEC, after decoding the BER is equivalent an error rate below 10^{-9} depending on the specific code.

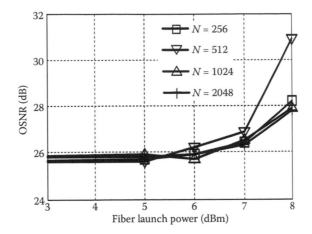

FIGURE 8.8 Simulation result for 42.7 Gb/s oOFDM transmission over 640 km of standard single-mode fiber; OSNR required for BER = 10^{-3} as function of fiber launch power. (Extracted with permission from A. Ali, Investigations of OFDM transmission for direct detection optical systems, Dr. Ing Dissertaion, Albretchs Christian Universitaet zu Kiel, 2012.)

Figure 8.8 shows the required OSNR as function of fiber launch power for different values of N. Transmission is error free (1e-9) over 640 km of SSMF or a dispersion factor of about 1100 ps/nm without dispersion compensation. It can be explained by the fact that even for the lowest value of $N_{min} = 256$, each sub-channel occupies a bandwidth of approximately 42.7 GHz/256 = 177 MHz resulting in high robustness toward fiber dispersion.

The principal difficulties of oOFDM are that the pure delay is due to the variation of the refractive index of the fiber with respect to the optical frequency lead to bunching of the sub-channels and hence the increase of the optical power, thus unexpected SPM may occur in a random manner.

8.2.1 IMPACTS ON NON-LINEAR MODULATION EFFECTS ON oOFDM

The MZIM is biased at quadrature point where the power transfer characteristic is linearizable. By means of low modulation depth, the non-linear distortions due to the MZM can be considered to be arbitrarily small, hence leading to low ratio of useful power to carrier power, and thus low sensitivity. Therefore, the modulation depth is a compromise between these constraints.

Figure 8.9 shows [13,14] results for the BER versus the OSNR under back-to-back transmission with the BER is obtained by Monte-Carlo simulation. The OSNR for BER = 10^{-3} is plotted versus normalized driving voltage, as shown in Figure 8.9. The normalization is performed such that minimum and maximum optical output power can be obtained for instantaneous input voltages of −0.5 and 0.5, respectively. Beyond these values, clipping occurs and intermodulation distortion also exists, due to MZM characteristic. The OFDM signal is analog having nearly Gaussian amplitude distribution [4]. In good approximation, the peak voltage is within an interval from plus to minus the triple of the RMS voltage. This relation is used to create the lower from the upper of the two horizontal axes. The clipping threshold obtained for a peak voltage of ± 0.5 is also given. Finally, the impact of MZM non-linearity, which is within the range of acceptable sensitivity, obviously does not depend significantly on N, and is identified by means of the dashed line.

Figure 8.9 depicts the non-linear resilience of the fiber transmission link including 8 × 80 km SSMF without dispersion compensation. To investigate the impact of linear factor one by one, the MZM is driven in the quasi-linear range with a normalized effective voltage swing of ≈0.15 resulting in a back-to-back OSNR of ≈25.5 dB. This value is significantly above those reported, for example, in Reference 15, which is due to incoherent detection and higher bandwidth. For low values of launch power, Figure 8.9 shows the robustness of oOFDM toward fiber dispersion, as the OSNR penalty achieved with

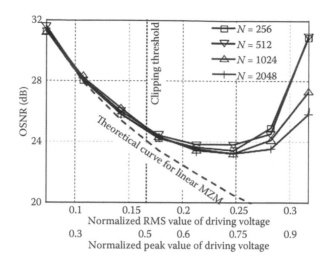

FIGURE 8.9 Required OSNR for BER $= 10^{-3}$ as function of driving voltage swing fed to MZIM electrodes. (Extracted with permission from A. Ali, Investigations of OFDM transmission for direct detection optical systems, Dr. Ing Dissertaion, Albretchs Christian Universitaet zu Kiel, 2012.)

≈ 11000 ps/nm accumulated dispersion is negligible. In the non-linear regime, however, penalty increases rapidly. Beyond the 8 dBm launch power at the fiber input, the BER does not fall below 10^{-3} because of strong signal distortion. The optical power consists of a strong DC component and a weaker AC component carrying the OFDM-signal. Since only the AC component results in signal distortion, the acceptable launch power found in this contribution is higher than for coherent detection.

Figure 8.10 shows the eye-opening penalty of modulation formats NRZ-ASK and OFDM with a guard interval of 25, 50, and 100 ps. It shows clearly the superiority of the OFDM over ASK at 42.7 Gb/s bit rate*.

Except for the value for $N = 512$ at a launched power of 8 dBm, attributed to limited simulation accuracy, the result does not depend on N. Increasing N decreases the separation between the sub-channels, and fiber non-linearity is expected to cause strong XPM and FWM. However, with increasing N the power per sub-channel is decreased. Apparently, these two aspects have been cancelled out each other so that equal performance for all N is achieved.

The impact of non-linear modulator characteristic and Kerr effect shows that for an uncompensated oOFDM 8×80 km fiber link with a varying number N of sub-carriers, both impairments are independent of N. Obviously, oOFDM is quite robust toward the specific non-linear impairments in fiber-optic transmission systems.

8.2.2 DISPERSION TOLERANCE

The variation of the eye-opening penalty as a function of the ratio of the guard interval and the bit period with the fiber SSMF length as a parameter is obtained as shown in Figure 8.11. This indicates the guard interval is very critical for different length of the fibers, and thus the bit rate and the number of sub-carriers.

8.2.3 RESILIENCE TO PMD EFFECTS

Reference 14 reports the first experiment of PMD impact on fiber non-linearity in coherent optical OFDM (CO-OFDM) systems. The optimal Q value at 10.7 Gb/s has been improved by 1 dB

---

* Simulation developed under MOVE_IT (MATLAB) and simulation platform provided by M. Ali of Technische Facultaet, Universit of Kiel.

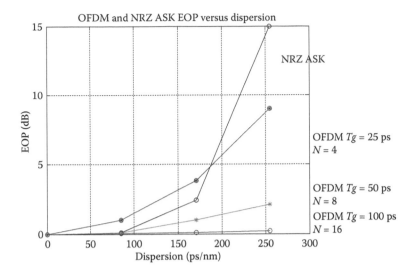

FIGURE 8.10 Eye-opening penalty of OFDM as compared with NRZ-ASK with different guard interval time of 25, 50, and 100 ps for 42.7 Gb/s bit rate.

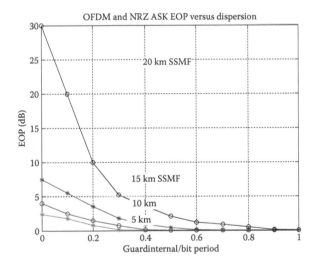

FIGURE 8.11 Eye-opening penalty of OFDM against the ratio of the guard interval and bit period with fiber length as a parameter, $N = 8$.

after introduction of DGD of 900 ps with the variation of the Q-value against the launched power as shown in Figure 8.12. A total of 900 ps is equivalent to 22 symbol period of the baud rate of 25 GB. This is a very substantial resilience rate. Note that with both polarized channels employed as in the PDM systems, the two polarized channels received can be employed in a 2×2 MIMO (multiple input multiple output) processing technique to improve the BER and minimize the required OSNR as shown in Figure 8.7b.

8.3 OFDM AND DQPSK FORMATS FOR 100 GB/S ETHERNET

For 100 Gb/s Ethernet, the decomposition of the bit rate to symbol rates by multi-level modulation formats as described in Chapter 7. The OFDM offers significant lower symbol rate due to the

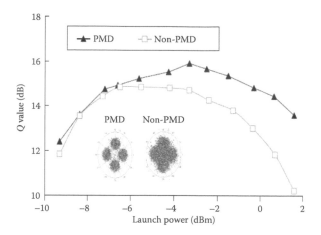

FIGURE 8.12 System Q as a function of launch power for PMD-supported and non-PMD-supported OFDM QPSK transmission systems. (Extracted with permission from Y. Ma et al., Characterization of non-linearity performance for CO-OFDM signals under the influence of PMD, *Electron. Lett.*, September issue, 2007.) Note the rotation of the signal constellation.

decomposition into several sub-channels. Reference 8 reported the transmission performance of RZ-DQPSK formats over 75 km of SSMF as well as RZ-ASK with OSNR of 22 and 36 dB, respectively.

On the other hand, the OFDM can combine a large number of parallel data streams into one broadband data stream with high spectral efficiency and would offer significant reduction of bit rate into symbol rate. This would challenge the use of multi-level modulation formats. However, there are still several issues to be examined by oOFDM such as high peak to average carrier ratio or FWM effects and other non-linear effects.

OFDM schemes offer the following advantages to optical long haul, metro and access networks: Different users can be assigned with different OFDM sub-carriers within one OFDM band of N-sub-carriers as shown in Figure 8.13a. This assignment is very much suitable for the case of massive MIMO antenna in which each antenna element would be assigned with a specific channel. These MIMO antennae would carry wireless OFDM signals and this oOFDM transport technique would be most suitable for the convergence between wireless and optical transport layers. Furthermore, each channel can be assigned different slots for the case of TDM (see Figure 8.13b), as well as different wavelength (see Figure 8.13c)

Regarding speed and distance: 100 Gb/s/λ downstream and same for upstream. 100 km for access, 1000 km for metro and 5–10,000 km for long haul.

On flexibility: OFDM offers adaptive modulation and FEC on sub-carrier basis, dynamic bandwidth allocation in time and frequency, transparency to arbitrary services, and optically transparent ONUs.

On cost efficiency: OFDM offers colorless architecture, stable, accurate DSP-based operation, and non-disruptive to legacy optical data networks (ODN).

On access passive optical networks (PON): Orthogonal Frequency Division Multiplexing (OFDM) PON: very well-suited for future PON systems, transparent to emerging heterogeneous applications, highly flexible, dynamic bandwidth allocation, and non-disruptive to legacy ODN.

Recent demonstrations of ultra-high-speed OFDMA PON by NEC and Huawei OFDM with aggregate speed of 108 Gb/s/ downstream and upstream, feasible on class C+ ODN (30+ dB power budget), and highly dispersion tolerant (60–100 km transmission)

However, OFDM requires advanced DSP which has been under intense research and development. DSP complexity: IFFT/FFT dominates, ~log(N) scaling and optimized algorithms.

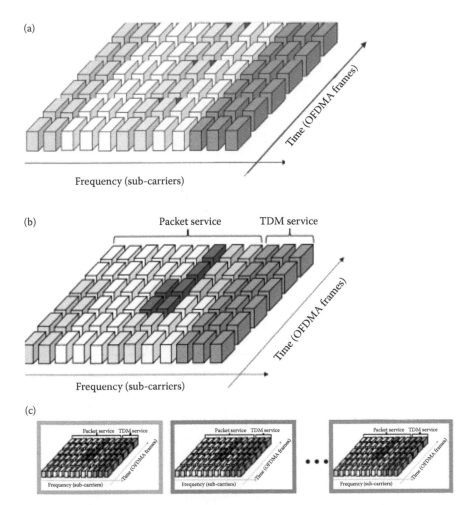

FIGURE 8.13 OFDM in different assignments of channels and sub-carriers and wavelength (a) different sub-carriers to different channels; (b) different slots to TDM channels; and (c) different wavelength for OFDM symbols.

8.4 OFDM SSB IN DATA CENTER NETWORKING

A comb laser can also be integrated with external modulators and modulation formats such as single side band (SSB) ODM, as shown in Figure 8.14. The multi-carrier source injects multi-carrier light-waves into an external dual-drive modulator and the split through different paths MZDI and the transmitting through a length of fiber (see Figure 8.15). At the receiving end, the modulated channels are filter by OBPF (optical band pass filter) and the recovered in the electronic domain. An AWG (arbitrary waveform generator) is used to provide a software development of the modulation scheme OFDM with two outputs of $\pi/4$ phase shift with each other so that SSB OFDM (Figure 8.16) can be generated and the analog output is then used to modulate a dual drive Mach–Zehnder modulator. Electrical amplifiers are added to provide sufficient voltage level for NRZ modulation in optical domain to the electrodes of the modulator.

The devices used in these transmission experiments are single-section Q-Well and Q-Dash-based lasers [4,17,18]. The multi-carrier source was grown by gas source molecular beam epitaxy (GSMBE) on an S-doped (001) InP substrate. The Q-Well device was 1-mm long and consists of a single InGaAsP strained quantum well embedded in 214-nm thick InGaAsP barriers. The active

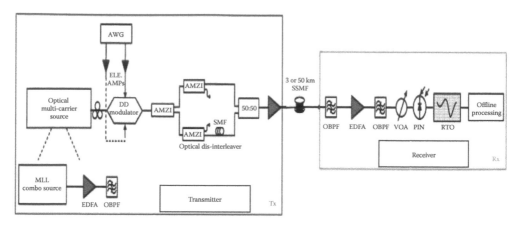

FIGURE 8.14 Schematic representation of the MLL Terabit/s transmitter (Tx) which employs IM- and DD-based receiver (Rx).

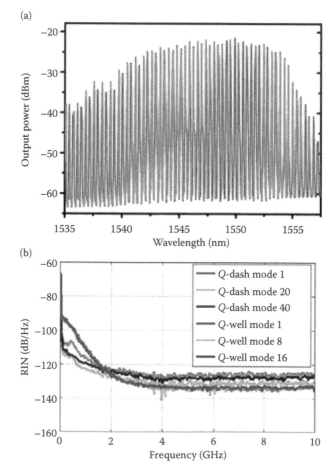

FIGURE 8.15 Typical characteristics of comb source (a) spectral modes of the source and (b) relative intensity noise (RIN) of the multi-carrier comb source.

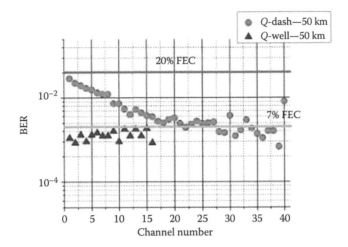

FIGURE 8.16 SSB OFDM performance of quantum well and quantum dash comb sources under modulation SS-OFDM and after 50 km SSMF transmission with the FEC BER as the base line.

region of the Q-Dash device consists of nine layers of InAs Q-Dashes separated by InGaAsP barriers, and Q-Dash as-cleaved laser used here had a total length of 950 Pm, corresponding to a repetition frequency of 44.7 GHz. The lasers were temperature controlled throughout the measurements; the Q-Well at 20°C, while the Q-Dash at 25°C using a TEC module. The Q-Dash laser exhibited the characteristic square-shaped emission spectra [4,7] with a spectral bandwidth more than 1.7 THz. The Q-Well PMLL have narrower spectral bandwidth (~0.5 THz) compared to the Q-Dash laser. However, the Q-Well device has a higher output power (~10 dBm) compared to the Q-Dash PMLL (~6 dBm), which, coupled with narrower spectral bandwidth, results in higher power per mode (and better optical carrier-to-noise ratio (OCNR)). Characterization of the amplitude noise is critically important, as semiconductor mode-locked lasers can exhibit significant mode-partition noise (MPN), which could impair system performance. The RIN of these devices was determined for each individual spectral mode and for the entire emission spectrum.

8.5 CONCLUDING REMARKS

OFDM is strongly suitable as a modulation technique for optical digital communications, especially for lowering the ultra-high speed 100 Gb/s Ethernet to much more acceptable rate per sub-channel. As a multi-carrier channels with natural orthogonality from the FFT and IFFT, it is much more efficient than the use of single carrier as seen in other advanced modulation formats described in most chapters of this book. By closely spacing all of the sub-channels via orthogonality, one can increase the spectral efficiency.

The pulse shaping function can certainly influence the orthogonality of the sub-channels and minimizing the cross talks between channels. Sin c function in the time domain would be the best choice. The guard bands influence the transmission performance of OFDM transmission.

OFDM is shown to be more resilient to non-linearity and PDM, as well as CD. Coherent OFDM offers higher receiver sensitivity or lower eye opening penalty than the direct detection OFDM without much more complexity, as the processing of the received optical sequence can be implemented in the electronic domain.

The trends for oOFDM in optical networking are as follows:

- DSP-based system cost can be significantly and rapidly reduced by component integration and mass production.
- Optimized OFDM algorithms reused from wireless and wireline building blocks.

- Next-generation 100 Gb/s long-haul fiber transmission will be heavily DSP-based 50+ GSa/s, 2 channel ADC chips and 10–30 GS/s, 2 channel DAC chips must be commercially available.
- Intensive on-going effort in parallelized, real-time DSP architectures.
- Aggressive 100 Gb/s DSP development for fiber transmission expected to have favorable effect on ultra-high-speed OFDM-PON for access and metro networks.

The key challenges to be tackled rest in

- DSP 2 channel ADC/DAC, DSP processor key components.
- Favorable technology trends in terms of cost profile.
- 41.25 Gb/s real-time, variable-rate receiver for WDM-OFDM and higher bit rates to 100 Gbps must be developed.
- The main difficulties with OFDM are the sampling rate which leads to heat dissipation and hence not preferred in data centers or long haul transmitter when the baud rate is very high.

REFERENCES

1. R. V. Nee and R. Prasad, *OFDM for Wireless Communications*, Norwood, USA: Artech House Publication, 2009.
2. J. Proakis, *Digit. Commun.*, New York, NY: McGrawHill, 2002.
3. L. Hanzo, S. X. Ng, T. Keller, and W. Webb, *Quadrature Amplitude Modulation: From Basics to Adaptive Trellis Coded, Turbo-Equalized Space-Tme Coded OFDM, CDMA and MC-CDMA Systems*, 2nd ed., Chichester, England, IEEE Press, J. Wiley and Sons, Ltd., 2004.
4. L. Hanzo, M. Münster, B. J. Choi, and T. Keller, *OFDM and MC-CDMA for Broadband Multi-User Communications, WLANs and Broadcasting*, Wiley, 2003.
5. M. Sieben, J. Conradi, and D. E. Dodds, Optical single-sideband transmission at 10 Gb/s using electrical dispersion compensation, *IEEE J. Lightwave Technol.*, vol. 17, no. 10, pp. 2059–2068, 1999.
6. R. Llorente, J. H. Lee, R. Clavero, M. Ibsen, and J. Martí, Orthogonal wavelength-division-multiplexing technique feasibility evaluation, *J. Lightw. Technol.*, vol. 23, pp. 1145–1151, March 2005.
7. A. J. Lowery and J. Armstrong, Otrthogonal—frequency—division multiplexing for dispersion compensation in long haul WDM systems, *Optics Express*, vol. 14, no. 6, March 2006.
8. A. Ali, Investigations of OFDM transmission for direct detection optical systems, Dr. Ing Dissertaion, Albretchs Christian Universitaet zu Kiel, 2012.
9. Leibrich, J., A. Ali, and W. Rosenkranz, Optical OFDM as a promising technique for bandwidth-efficient high-speed data transmission over optical fiber, *Proceedings of the12th International OFDM-Workshop 2007, InOWo 2007*, pp. 1–5, Hamburg, Germany, August 29–30, 2007.
10. A. J. Lowery, L. Du, and J. Armstrong, Orthogonal frequency division multiplexing for adaptive dispersion compensation in long haul WDM systems, *Proc. OFC 2006*, Anaheim, USA, March 2006, paper PDP39.
11. W. Shieh and C. Athaudage, Coherent optical orthogonal frequency division multiplexing, *Elect. Lett.*, vol. 42. pp. 587–589, May 2006.
12. X. J. Mcng, A. Yacoubian, and J. H. Bechtel, Electro-optical pre-distortion technique for linearization of Mach-Zehnder modulators, *Elect. Lett.*, vol. 37, no. 25, pp. 1545–1547, 2001.
13. A. Ali, J. Leibrich, and W. Rosenkranz, Impact of nonlinearities on optical OFDM with direct detection, *Proc. ECOC 2007*, pp. 1–2, Berlin, Germany, 2007.
14. A. Ali, Orthogonal frequenc division multiplexing in optical transmission systems with high spectral efficiency, Master Thesis Dissertation, Technische Facultaet, University of Kiel, Kiel, Germany, 2007.
15. S. L. Jansen et al., *Proc. OFC 2007, paper PDP15*, LA, USA, 2007.
16. Y. Ma, W. Shieh, and X. Yi, Characterization of nonlinearity performance for coherent optical OFDM signals under the influence of PMD, *Electron. Lett.*, September issue, pp. 943–945, 2007.
17. F. Lelarge et al., Recent advances on InAs/InP quantum dash based semiconductor lasers and optical amplifiers operating at 1.55 μm, *IEEE J. Sel.Topics Quant. Electron.*, vol. 13, no. 1, pp. 111–124, 2007.
18. V. Vujicic, C. Calo, R. Watts, F. Lelarge, C. Browning, K. Merghem, A. Martinez, A. Ramdane, and L. P. Barr, Quantum dash mode-locked lasers for data centre applications, *IEEE J. Sel. Top. Quantum Electron.*, vol. 21, no. 6, p. 111058, November/December 2015.

9 Modulation for Short-Reach Access Optical Transmission

9.1 INTRODUCTORY REMARKS

9.1.1 ECONOMICAL ASPECTS IN ACCESS NETWORK DELIVERY

Currently, delivery of massive information capacity to users and communities of users is very important to meet the challenges of 5G technological network engineering. Economical solutions are required, thus the uses of lower bit rates, such as those of 10 GBd to more than 25 GBd and then 56 GBd are considered to be fairly reasonable, leading to 100G, 400G, and even 1000 GBbps. Therefore, the use of advanced modulation formats can help to reduce the number of components required for short reach optical communication modules at 100 Gb/s (4 × 25 GBd) and beyond. The main reason is that for lower-bandwidth photonic and electronic devices, fabrication technology can be used, such as 110 nm resolutions instead of 64 nm resolution fabrication processes. Furthermore, the proliferation of bandwidth-intensive services and cloud computing has driven the speed of data communication links to higher and higher data rates. The 100 Gbps bit rate has been standardized, while discussion has been actively carried out for 400 Gbps and 1.0 Tbps data links. The short reach area includes the last mile delivery to users and intercommunications of data within data centers and interdata or severs centers.

Unlike long haul optical communication systems where externally modulated high-order modulation formats with coherent detection receivers are commonly used, direct modulated lasers with direct detection are likely to be employed for short reach system implementation. The performances of pulse amplitude modulation (PAM), carrier-less amplitude/phase modulation (CAP), and direct detection orthogonal frequency-division multiplexing (OFDM) for short-reach optical communication systems (SROCS), are given and their possible uses for future high-capacity SROCS are explored in this chapter.

9.1.2 WAVELENGTH CHANNELS AND MODULATION TECHNIQUES

Although multiple wavelength channels can be used (e.g., 4 × 25.8 Gbps for 100 Gbps link) for implementing transceiver modules for such communication systems, the demand for system integration to reduce system cost requires the number of wavelengths to be kept to the minimum. One possible approach to realize this is through the use of more advanced modulation formats to increase spectral efficiency by increasing the data rate for a given transmission system bandwidth. High-order modulation formats such as quadrature phase shift keying (QPSK), M-ary quadrature amplitude modulation (M-QAM or QAM-M-ary), and orthogonal frequency-division multiplexing (OFDM) with coherent detection and a digital signaling processing (DSP) algorithm have been studied extensively in recent years for long haul optical communication systems. External modulation using IQ modulators is typically employed at the transmitter, while at the receiver, optical hybrid and local oscillators together with DSP algorithms implemented using application-specific integrated circuits (ASICs) are employed for signal detection.

Dispersion and polarization mode dispersion (PMD), as well as other linear impairments can be compensated via DSP algorithms; spectral efficiency as high as $10\,b/s/Hz$ can be achieved, and coherent transceivers supporting 100 Gb/s (with 28 or 56 GBd) per channel using polarization-multiplexed QPSK are widely available for commercial deployment. However, for the short to medium term, coherent detection technique is unlikely to be used for short reach systems. The use of IQ modulator and optical hybrid will not only increase system cost, but also increase transceiver footprint and, as a result, increase the difficulty of system integration. The complicated coherent detection algorithm will also significantly increase the power consumption of the transceiver modules, which can be an important concern in SROCS. For short reach communication systems, low-cost light sources such as vertical cavity surface emitting lasers (VCSELs), directly modulated lasers (DMLs), and Integrated externally modulated lasers (IEMLs) are likely to be used, and signal will be directly detected at the optical receiver. As a result, no optical phase information will be available. Because of these issues, we need to study solutions that can increase system spectral efficiency while using DML, EML, or VCSEL lasers and direct detection. A number of schemes have been investigated. These include pulse amplitude modulation (PAM) [1,2], carrier-less amplitude/phase modulation (CAP) [3–5], direct detected orthogonal frequency-division multiplexing (OFDM), and discrete multi-tone (DMT) modulation [6,7]. Performance as well as power consumption have been studied and compared. In this chapter, we review the characteristics of these modulation schemes, especially for SROCS and highlight some significant applications in network deployment.

This chapter is organized as follows: The next section gives a brief system schematic description of the SRCOS and some typical performance of the system. Then the most up-dated PAM-4 employed for SROCS is described in detail, including concepts, transmission, and technology and measurement techniques using most modern pieces of equipment. The bit rate can reach 112 Gbps with 56 GBd employing direct modulation VCSEL.

9.2   HIGH-ORDER PAM TRANSMISSION SYSTEMS

Figure 9.1 shows schematic representation of a direct modulation transmission systems employing DSP in the transmitter and receiver. As with any common optical transmission system, the transmitter consists of direct modulation of the laser or by an external modulator integrated with the laser and pre-emphasis can be implemented via the DSP and DAC. While at the receiver, the photodetector (PD) can be a P-I-N type or APD (avalanche PD), and then a TIA and Linear amplifier in association with ADC and DSP.

Theoretical and typical experimental performance of the PAM-m ($m = 2, 4, 8$ or 16) are shown in Figure 9.2a and b. We note that the required increase in optical power would be 6 dB whenever a modulo-2 order of modulation is used. This will put a demand on the output optical power of the laser at the transmitter.

If WDM transmission is required that is very common with 4-lane at 25 Gb/s each lane to offer 100G module system, then WDM multiplexer (MUX) and demultiplexer (Demux) must be used at the transmitter and receiver, respectively. These are not shown in Figure 9.1, but readers are expected to understand. For the spectral regions for access systems, both the O-band and C-band are useful. For the O-band, the loss is higher but dispersion broadening is much less and hence less demanding on DSP pressing time. However, lower power would be at the receiver and thus the available OSNR may not be sufficient if the total equivalent noise at the input of the receiver is too high.

9.3   BEYOND 50G VCSEL MODULATED VCSEL SHORT REACH

Short range optical links are important for interconnecting servers or clusters in data centers, for example, access links between cloud centers to user node points. Economic factors are important in these scenarios and direct modulated vertical cavity surface emitting lasers (VCSELs) operating in the first fiber spectral windows 850 and 1300 nm (O-band) are commonly employed together with multi-mode or single mode fibers (or few mode fibers at these wavelength regions). Typical transmission medium can be multi-mode fiber (MMF). While it has a limited bandwidth distance product due to modal and chromatic dispersion (CD), the large core diameter (LCD) results in a

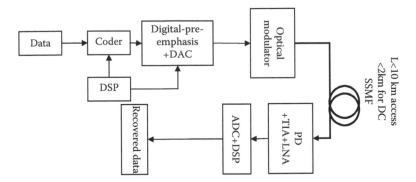

FIGURE 9.1 Typical digital optical communication systems for short haul using DSP and optical modulation plus opto-electronic reception and DSP. Legends: DC = data center, SSMF = standard single-mode-fiber; DSP = digital signal processor; ADC = analog to digital convertor, DAC, digital to analog converter; TIA = trans-impedance amplifier; LNA = linear amplifier.

large alignment tolerance and consequently helps keep the transceiver cost low. Furthermore, VCSELs offer several unique advantages that made them popular in data communications (Datacom) applications. They are energy efficient, as seen by the energy dissipation of 73 fJ/bit at 40 Gbps and 95 fJ/bit at 50Gbps reported in Reference 8. VCSELs can also be directly modulated at very high data rates, demonstrated at as high as 57Gbps using on-off keying (OOK) without equalization [9] and 71Gbps using OOK with transmitter and DSP receiver equalization [10]. Higher bit rate can be achieved by employing higher order modulation such as pulse amplitude modulation of order n (PAM-n). The next level is PAM-4 in order to double the effective bit rate of a transmission system.

9.3.1 VCSEL Structure and Gain

Consider the typical VCSEL structure in Figure 9.1. In this example, the gain region consists of three quantum wells placed in a one-wavelength-thick spacer region. On either side of the spacer region, there are extremely high reflectivity mirrors made from semiconductors that have total reflectivity greater than 99.5%. The mirrors are fabricated from altering material layers that are $\lambda/4$ thick to form a Distributed Bragg Reflector (DBR). Superimposed on the structure of a schematic is the electric field distributed across the vertical direction of the VCSEL (Figure 9.3)

The total cavity length of the VCSEL, L_{eff} is expressed as the sum of the active region thickness plus the so-called penetration depth of the electric field into the DBR structure. Finally, the longitudinal mode spacing of VCSEL can be expressed as

$$\Delta\lambda = \frac{\lambda^2}{2n_{eff}L_{eff}} \tag{9.1}$$

where n_{eff} is the average index of refraction of the mirrors. Note that the exact penetration depth, which may be different for the n and p mirrors can be found from the mirror properties. Typical longitudinal mode spacing of a VCSEL is 30 nm, while the longitudinal mode spacing of a typical cleaved cavity edge-emitting laser (EEL) may be only 0.3 nm as given in Chapter 2. Now, when the gain profile of the active region is taken into account, and the stop band of the DBR is also considered, it is easy to see why a VCSEL will only lase in a single longitudinal mode. This is shown in the schematically in Figure 9.2. One can see from this picture that the longitudinal mode spacing of the VCSEL is approximately the same width as the stop band of the DBR, and are significantly removed from the center of the cavity. The dip in the DBR spectrum is the Fabry-Perot resonance of the entire VCSEL structure and is measured by white light reflectance

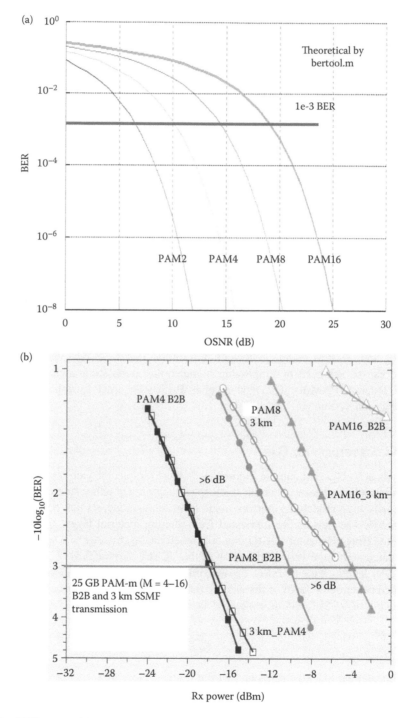

FIGURE 9.2 BER versus OSNR (a) and receiver power (dBm) (b) for PAM2, 4, 8 and 16 under back-to-back and 3km transmission. B2B = back to back, SSMF = standard single mode fiber.

of the entire structure. In addition, there is negative gain for the other cavity modes. In contrast, the longitudinal mode spacing of an EEL can be very small, and there is essentially very little difference in cavity reflectance and gain among the several center modes. A type of EEL can be made to emit in a single wavelength by utilizing distributed feedback (DFB) to the laser cavity. This is

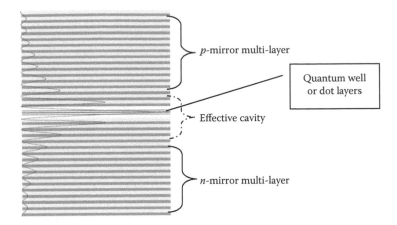

FIGURE 9.3   Schematic view of VCSEL and Electric Field (light grey-oscillating.)

typically done by using lithography to define a grating structure in the laser to provide wavelength selective feedback (Figure 9.4).

9.3.2   TRANSVERSE LASER MODES

Most communications-grade FPs and DFBs are made to be single-transverse mode by defining a ridge waveguide structure to limit the lateral extent of the electric field. Sometimes it is advantageous to allow more than one transverse laser mode to exist in an EEL, especially when high power operation is desired. The vertical extent of the field is generally defined by a refractive index guiding structure in the epitaxial growth. One such structure is the commonly used graded index separate confinement hetero-structure (GRINSCH) laser. In a VCSEL, the transverse laser modes are controlled by several parameters, including the lateral oxidation or proton implantation used to steer current into the active region, the self-heating of the active region, and particular design of the oxidation boundary.

9.3.3   SINGLE MODE VCSEL

The easiest VCSEL to conceptualize is a single transverse mode (STM) VCSEL. An STM VCSEL can be fabricated by reducing the lateral extent of the fundamental optical mode such that it is the only

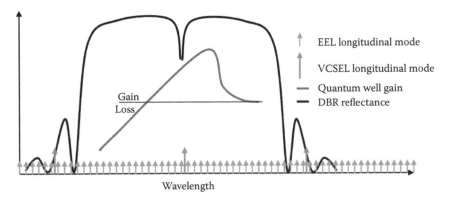

FIGURE 9.4   Schematic representation of DBR reflectance (observed from the top of the VCSEL), mode spacing and quantum well gain curve. The mode spacing of the EEL is not to scale, and is much finer than depicted here.

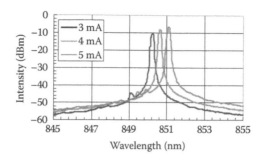

FIGURE 9.5 Typical optical spectrum of a single mode VCSEL as a function of driving current.

allowable mode of the cavity. Typically, this is done using a lateral oxidation layer to form an aperture less than 5 μm in extent. This is shown schematically in Figure 9.5. Most VCSELs do not have a method to control the polarization of the optical mode, which may allow for coexistence of two or more polarization modes. The polarizations are slightly different in wavelength, typically less than 0.1 Angstroms or less. A typical optical spectrum of a single mode VCSEL is shown in Figure 9.5. Please note that the optical linewidth of a single mode laser cannot be measured on a typical optical spectrum analyzer. Accurate measurements of the optical linewidth must be done with a Fabry–Perot interferometer or other means. The typical optical linewidth of AOC single mode lasers is less than 100 MHz.

9.3.4 VCSEL Equivalent Circuit Model

An equivalent small-signal circuit model of the VCSEL can be represented, as shown in Figure 9.6a, in which C_p is the parasitic capacitance between the ground and the driving electrode. C_m is the laser diode junction capacitance and R_i is the intrinsic junction resistance due to accumulation of the carriers, R_{pDBR} and R_{nDBR} are the resistances of the grating p-region and n-regions, respectively. The intrinsic frequency response of a VCSEL can be represented by [11]

$$H_{inf}(f) = \frac{p(f)}{i_a} = C\frac{f^2}{f_r^2 - f^2 + j\gamma\frac{f}{2\pi}} \tag{9.2}$$

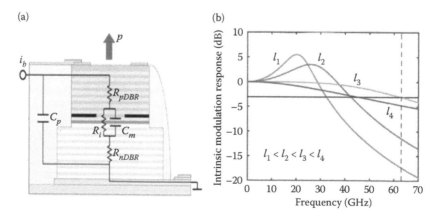

FIGURE 9.6 (a) Equivalent small-signal circuit model of VCSEL and (b) small-signal frequency response.

with

$f_r = resonance.frequency$; $\gamma = damping\_const$
$C = arbitrary\_cons\tan t$.

the resonance frequency and the damping factor are given by

$$f_r = D.\sqrt{I - I_{th}}; \quad D = \frac{1}{2\pi}\sqrt{\frac{\eta_i\gamma v_g}{qV_a}\frac{\delta v_g/\delta\Delta n}{\chi}} \tag{9.3}$$

$$\gamma = Kf_r^2 + \gamma_0; \quad K = 4\pi^2\left(\tau_p + \frac{\dfrac{\varepsilon\chi}{\delta v_g/\delta\Delta n}}{\chi}\right) \tag{9.4}$$

and the damping limited band width can be derived as

$$f_{3dB} = \frac{2\pi\sqrt{2}}{K} \tag{9.5}$$

where f is the frequency variable, f_r is the resonance frequency of the VCSEL, γ is the damping rate the inductor to make a resonant combined circuit and is not shown but understood to be the inductance of the gold bonding wire. v_g denotes the group velocity, τ_p is the photon lifetime, and ϵ is the permittivity of the active area. The frequency response shown in Figure 9.6b indicates the 3 dB passband of the VCSEL depends on the driving current. Thus, by increasing current would lead to increasing temperature (self-heating) thence the saturation of photon density and resonance frequency, therefore the thermally limited bandwidth can be observed. Controlling the biasing and driving current I3 to the VCSEL may increase 3 dB bandwidth as shown in Figure 9.6b to about 60 GHz. One other way to increase the effective modulation bandwidth is by optical injection to VCSEL by an external laser acting as the master laser to add a peaking resonance to the combined resonant cavity as described in Chapter 2.

As we can see in Equation 9.2, the damping coefficient is a function of the photon lifetime in the lasing layer (Quantum well QW or quantum dot QD), the carrier group velocity and the rate of the change with respect to the electron concentration (See also Figure 9.7). The 3dM bandwidth of the VCSEL by this damping resonance is inversely proportional with respect to the magnitude of damping coefficient and thus one can control the damping effect of the VCSEL to extend the passband to maximum value, which can reach about 25 GHz as reported so far [11].

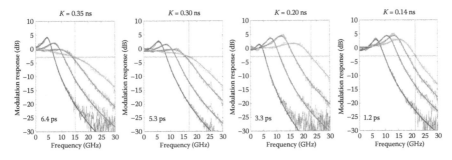

FIGURE 9.7 Bandwidth variation (enhancement) as a function of damping coefficient, K, under different driving current such that $I_1 < I_2 < I_3 < I_4$ (different grey colors as indicated in Figure 9.6).

9.3.5 EFFECTS OF OPTICAL MODE ON DATA TRANSMISSION

The change of wavelength and mode hopping in optical data links is an important factor in single-mode fiber data links because of chromatic dispersion. In multi-mode fiber data links, there is no appreciable chromatic dispersion, and the effect that must be considered is the change in optical bandwidth of the fiber. Other effects like mode partition noise must also be considered. These are taken together as the "K" factor in common link budget rise-time budget analysis techniques such as those employed in Ethernet analysis. Typical values used can very conservatively be taken as $K = 0.7$, for the link analysis.

9.3.6 MODAL EFFECTS ON VCSEL EYE DIAGRAM QUALITY

The optical energy in a VCSEL can move from mode to mode, and switch polarization during an optical pulse, or even randomly in time and temperature. Significant mode competition effects can also be observed in VCSELs as the device is coming to equilibrium. A model was previously developed that demonstrated excellent agreement between theoretical calculations and measured results [12]. The most important conclusion reached in that modeling exercise was that there can be significant pulse shape distortion caused by modal selective coupling in a VCSEL. However, if all of the VCSEL modes are equally coupled, then there is no degradation in the optical performance of the VCSEL due to mode competition. Figure 9.10 contains three pictures which are repeated below for clarity. In oxide VCSELs, the effect of mode selective coupling is generally seen as overshoot in the optical signal. This can be readily observed by decoupling a TOSA (Transmitter Optical Sub-Assembly) in the direction of the fiber axis.

9.3.7 MULTI-LEVEL DIRECT MODULATION FORMAT

Multi-level PAM was used also to improve the transmission reach in the MMF. PAM-4 transmission over up to 200 m of OM4+ MMF at 50 Gbps data rate was demonstrated in Reference 13, 104 Gbps transmission over 50 m was demonstrated using two wavelengths, 850 and 880 nm, carrying a 52 Gbps 4-PAM signal each. Multi-level PAM-m presents a set of unique challenges in link, VCSEL and driver design [14]. This section we present the PAM-4 especially for high speed transmission link for short reach applications.

The electronic circuit driver can be seen in Figure 9.8b in which the base material is InP-DHBT (high-electron mobility bipolar transistor) of 250 nm gate length and a cut-off transition frequency of

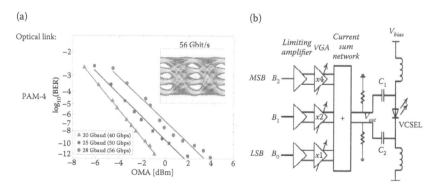

FIGURE 9.8 Performance of PAM-4 up to 56 Gbps (28 GBd) of VCSEL under MMF transmission link (a) GER versus sensitivity (OMA) and (b) electronic circuit of driver. (Extract from presentation of A. Larsson et al. VCSEL design and integration for high-capacity optical interconnects, *Proc. SPIE OPTO Photonic West 2017*, San Francisco, 2017.)

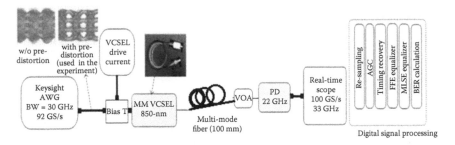

FIGURE 9.9   112 Gb/s PAM-4 over MMF (OM4) experimental set up platform.

350 GHz. The bandwidth of the VCSEL is 24 and 22 GHz for the photo-receiver. The PAM-4 eye diagram is inserted in Figure 9.8a.

Similarly, Karinou et al. [15,16] have demonstrate the transmission of VCSEL modulated at 56 GBd over MMF with DSP-based Optical receiver as shown in Figure 9.9. The transmission performance is shown in Figure 9.10.

VCSEL is a transverse multi-mode device operating at the 850-nm wavelength. The threshold current is 1.0 mA. The optimum bias operating current and peak-to-peak drive voltage are optimized for the back-to-back (B2B) case and the 100-m OM4 transmission, respectively. For B2B case the optimum conditions are found to be $I_{drive} = 8$ mA and $V_{pp} = 500$ mV, whether for the transmission case over 100 m OM4 MMF (multi-mode fiber) $I_{drive} = 6$ mA and $V_{pp} = 600$ mV. The 112 Gb/s optical PAM-4 signal is then launched over 100-m OM4 fiber and is received by a 22-GHz 850-nm photo-detector (PD). The optical power is varied in front of the PD in order to perform BER measurements for the B2B case and after 100-m transmission. The output photo-current is captured on a real-time scope with channel bandwidth of 33 GHz bandwidth and a sampling rate of 100 GS/s. The data are stored and processed offline using digital signal processing (DSP) algorithms implemented at the receiver.

On the other hand, EML (external modulator laser) module was used to modulate and transmission over 20 Km of single mode fiber (SSMF) in association with DSP-based optical receiver at a baudrate of 28 GBd and PAM-4 modulation format. The EML is a laser incorporating an

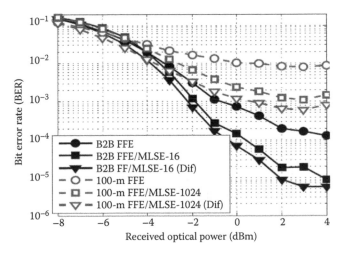

FIGURE 9.10   BER vs. received optical power for 112 Gb/s PAM-4 back-to-back and transmission over 100-m MMF under different post-equalization schemes.

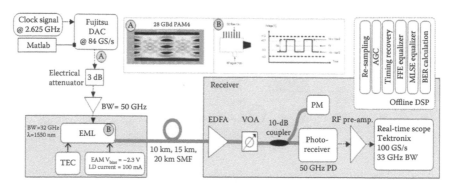

FIGURE 9.11 Experimental platform for 20 km standard single mode fiber transmission by 56 Gbps PAM-4.

external in-line modulator which normally is an electro-absorption (EA) modulator type. The received power is plotted versus the BER of such 20 km SSMF transmission is shown in Figures 9.11 and 9.12.

The PAM-4 modulation format and DSP post-processing indicate that the module for data center interconnect are possible to handle the high capacity operation of data center in 5G generation networking. Direct modulation is preferred as compared to coherent transmission because of its low cost and simplicity on integration of different sub-modules such as the micro-controller unit, the photonic integrated circuits and the high speed micro-electronic drivers and trans-impedance amplifiers and linear amplifiers.

EE note also that the electronic—optical delay time is dependent on the amplitude level during the VCSEL as observable in the eye diagram inserted in Figure 9.8 that would create error in the recovery of the data channels due to misalignment of the sampling instant. Thus, a correct circuit should be added or DSP processing must take into account of this delay amplitude-dependence when VCSEL is employed. Such delay difference has not been observed in EML modulation.

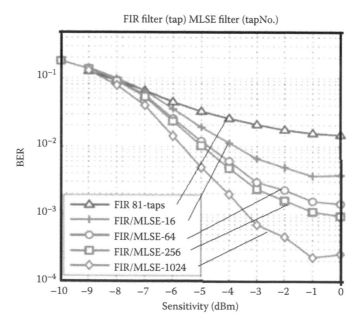

FIGURE 9.12 20 km SSMF transmission of 56G PAM-4 signals: BER versus received optical power.

9.3.8 PAM-4 Measurement Challenges

PAM-4 (Pulse Amplitude Modulation, 4-level) from a time domain perspective has four digital amplitude levels (−3, −1, 1, and 3), as shown in Figure 9.13a.

PAM-4 has an advantage over non-return-to-zero (NRZ) in that for each level ("symbol") there are two bits of information providing twice as much throughput for the same Baud rate (28 GBaud PAM-4 = 56 Gb/s). From a frequency domain perspective, PAM-4 requires half the bandwidth of that of NRZ. In the PAM-4 eye view (Figure 9.2), one can see three vertical eyes created by the four levels. Unlike NRZ, where the decision level is fixed to 0 V for a differential signal, the three slicer levels used by a PAM-4 receiver can be adaptive or time varying.

The use of multi-level signaling for PAM-4 has entirely changed what has been expected in Ethernet test in the past. Newly developed technology is required to accomplish implementation of PAM-4 components and serial links with changes to system test as more complex transmit and receive circuit designs are required to address PAM-4 challenges. PAM-4 technical challenges include a shift from saturating output stages to linear IO behavior in order to achieve multiple-levels. New chip designs face the challenge of managing the size of the integrated circuits (ICs) supporting PAM-4 which have increased nearly 30%. The larger IC (integrated circuit) size is due to additional linear drivers and detectors and has resulted in up to 35% increase in power requirements as well. For Ethernet chips with large IO counts, the problems are even greater.

Data transmission advancement using PAM-4 includes many new design and test challenges. Research will continue to determine how to address many PAM-4 challenges such as

- *Clock Recovery*: Finite rise time acting on different transition amplitudes creates inherent inter symbol interference (ISI) and makes clock recovery much more difficult. Transition time of the PAM-4 data signal can create significant horizontal eye closure due to switching jitter, which is dependent on the rise and fall time of the signal. Transition qualified phase detectors are needed to look at analog levels for clock recovery. Whether direct detection (comparators), which require a lot of power, or digitizing ADCs, which are expensive, are used is still to be determined.
- *Decision Feedback Equalization (DFE)*: DFE is used to calculate a correction value that is added to the logical decision threshold and results in the threshold shifting up or down so new logical decisions can be made on the waveform based upon the new equalized threshold level. The technology required to manage DFE and the possible addition or use of other equalizers for PAM-4 multi-levels is still to be determined.
- *Reduction of Signal to Noise Ratio (SNR)*: The PAM-4 signal has 1/3 the amplitude of that of a similar NRZ signal (SNR loss of ~9.5 dB) due to level spacing and is more susceptible to noise. However, it is possible that the lower PAM-4 insertion loss compensates for the 9.5 dB loss in SNR due to reduced signal amplitude in PAM-4 signaling.

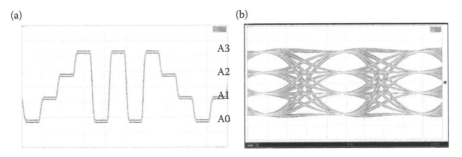

FIGURE 9.13 Time domain and eye-diagram of PAM-4. (a) Time-domain signals of 4 levels (A0-A3) of PAM-4. (b) Eye diagram of PAM-4 at 28GBd.

9.3.9   PAM-4 Theoretical Simulation/Characteristics

Typical PAM-4 communications system configurations for transmitter (Tx), channel, and receiver (Rx) can be chip-to-chip, chip-to-module (electrical or optical), or electrical backplane. Development of these high-performance components, modules, and networks requires sophisticated test and measurement techniques for interface interoperability and test validation. For NRZ 400G design, there are decades of experience to base new development on. Unlike NRZ, PAM-4 is not commonly simulated. This makes augmenting PAM-4 system simulation very valuable.

New considerations for tests and end-to-end link simulation will be essential to verify PAM-4 compliance and ensure interoperability. During simulation and test at the higher data rates it is also important to keep in mind that test instruments must be more capable than what they are intended to measure. For example, 25 Gb/s is 40ps and increasing speed to 56 Gb/s relates to 18ps, so the higher link jitter is at or beyond the capability of some instruments.

The challenge of device design validation and test increase as the data rates increase not only because of noise susceptibility, but also because of test probe or fixture interference. As cables and test fixtures are inserted between the device under test and test equipment, one may not expect to create a notable influence on the signal. However, at higher data rate transmissions, noise and interference become significant factors. Device test interference can occur from simple insertion of measurement instruments. For example, random noise can be induced from the device transmitter and then it root-sum-square adds to the intrinsic random noise from a measuring oscilloscope. In this case, it is important to know what the oscilloscope noise is for the bandwidth and gain being measured so it can be removed or de-embedded from the signal. Signal loss even in test cables at higher speeds is significant and may require de-embedding.

Typically, engineers design test fixtures with a controlled impedance environment so the fixture effects can be minimized, but it is impossible to completely eliminate skin effect series trace loss, dielectric shunt loss and inductive or capacitive impedance discontinuities of the fixture channel. Test fixture deficiencies cause signal loss and reflection with increased effects at higher frequencies. A few different approaches used to remove the test fixture effects from the measurement include direct measurement (pre-measurement process) and de-embedding (post-measurement processing). Direct measurement uses specialized calibration standards that are inserted into the test fixture and measured. De-embedding uses models of the test fixture [17] and mathematically removes the fixture characteristics from the overall measurement. De-embedding methods require that an accurate characterization model of the fixture be obtained first. Examples of de-embedding methods include "simulation-based" and "calibration-based." The simulation-based method simulates the waveform before the fixture is fabricated. The calibration-based method is used to eliminate fixture effect after the fixture is available [18] for use during the measurement process. Simulation modeling is enabled for PAM-4 designs in the arbitrary waveform generators (AWG). Simulations help to provide evaluation and performance prediction of a specific portion of a link or a complete end-to-end link design. Complex behaviors of NRZ transmitter and receiver signals can be modeled using the algorithmic modeling interface (AMI) standard. AMI is a behavioral model first defined in the input/output (IO) buffer information (IBIS) 5.0 specification, and provides both passive channel characteristics and SerDes functionalities [17]. The current AMI standard supports NRZ signaling, assuming the Tx signal has two levels and the receiver slicer reference is at 0V. The simulation flow for PAM-4 link can be seen in Figure 9.14 with a four-level input waveform and output waveform with clock data recovery (CDR) times and eye diagram with delay shift dependent on the modulated amplitude in Figure 9.15.

Measurement challenges for PAM-4 are similar and the key challenges facing engineers today are clock recovery (CR), eye skew, and noise. Transmitter (Tx) or output testing verifies that factors such as eye parameters and jitter measurements meet or exceed standard requirements and most often uses an oscilloscope. Receiver (Rx) or input testing verifies that the correct bits are detected with

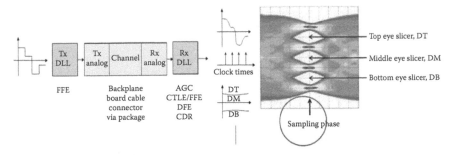

FIGURE 9.14 Block simulation of PAM-4 transmission link (a) schematic representation and (b) exemplar of PAM-4 eye and sampling. Legend: FFE = feed-forward equalization, DFE = decision feedback equalization, AGC = automatic gain control, MT = decision top, DM = decision middle, DB = decision bottom, Tx = Optical transmitter, Rx = optical receiver.

impairments from a worst-case channel traditionally using a bit error ratio tester (BERT) as the principal measurement tool. System testing verifies the Tx/Rx device works in a system under all conditions and can use many different measurement tools such as oscilloscopes, BERTs, protocol analyzers, etc. With PAM-4, there are new complexities that increase measurement challenges. Test and measurement in these early stages of PAM-4 links is helping to build an understanding of the causes and mechanisms of artifacts that impair link error performance. Impaired links can be related to implementation of clock recovery, and closed eye challenges such as skew, compression and non-linearity. As the PAM-4 technology continues to advance and knowledge of PAM-4 challenges and solutions grows, new measurements will emerge to characterize transmitter outputs and new stress impairments will be developed to test and characterize receiver inputs.

For the higher 56 Gb/s data rate, equalization is mandatory. Due to the increased signal rate, the channel the signal travels through distorts the signal at the receiver. The result can be seen using an oscilloscope and show a partially or completely closed eye diagram (Figure 9.6) that will prevent the receiver's ability to extract the clock and/or data. Equalization needs to be applied to reopen the eye diagram correcting for the inter-symbol interference (ISI) and recover the clock or data. Equalization methods used include feed forward equalization (FFE) for the transmitter and continuous-time-linear equalizer (CTLE), decision feedback equalizer (DFE) and CDR for the receiver. Clock recovery tracks low-frequency jitter and is utilized by either a real-time oscilloscope or a sampling oscilloscope when characterizing a transmitter. Current commercial oscilloscopes include software that can be

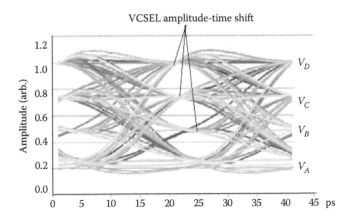

FIGURE 9.15 Simulated PAM-4 eye diagram with time delay shift dependent on amplitude of modulation (VCSEL).

used to model equalization such as CTLE, DFE, and FFE. An example is the *Keysight N5461A* Infiniium *Serial Data Equalization Software*, which allows users to choose which equalizer, or combination, to implement.

PAM-4 receivers can be impaired by time skew from eye to eye resulting from the optical output of a vertical cavity surface emitting laser (VCSEL) in a fiber optic transceiver driven by a PAM-4 signal (see Figure 9.15). VCSELs turn on faster when driven with a higher voltage, causing the lower transition eye to lag the upper two eyes. The alignment of the upper, middle and lower eyes is defined in PAM-4 standards. At high baud rates, the skew can be great enough to require different detection sampling times for each transition eye. More complex PAM-4 receiver designs will be needed to address the sampling delay per eye. Higher data rate transmissions cause a change in signal characteristics and additional error correction techniques are needed to maintain an open eye. Non-linearity or amplitude compression can alter the eye height of different transition eyes causing a linearity error due to a lower signal to noise ratio of the lower transitions.

Transmitter testing verifies that the output parameters meet or exceed the standard requirements including eye parameters and jitter measurements. The types of measurements required to characterize the transmitter output will continue to grow and improve as PAM-4 development progresses. According to Clause 94 in IEEE 802.3-2014, PAM-4 transmitters must be capable of generating simple test patterns such as JP03A and JP03B. The pattern defined as JP03B is used to measure random jitter (RJ), uncorrelated deterministic jitter or periodic jitter (PJ), and even-odd (F/2) jitter.

- JP03A Test Pattern (repeating {0,3} sequence)
- JP03B Test Pattern ({0,3} is repeated 15 times then, {3,0} is repeated 16 times) Also defined as a 62-bit clock pattern with phase reversal.

PAM-4 linearity is a very important parameter that can be tested with the standard Transmitter Linearity Test Pattern (Figure 9.13a). Each of the four PAM-4 levels can be measured for mean voltage, power if optical, or noise. The pattern is also used to analyze the transmitter linearity by measuring minimum signal level (S_{min}), effective symbol levels (ES_1 and ES_2), and level separation mismatch ratio (RLM)

- Transmitter Linearity Test Pattern (repeating 160 symbol patterns with a sequence of 10 symbol values, each 16 UI in duration)
- 10 consecutive symbols mitigates the impact of ISI on "Level" measurements (V_A, V_B, V_C, V_D) (See Figure 9.15)

Level, the mean "thickness," and skew are also measured (Figure 9.16). There are two different types of PAM-4 receivers being considered for development; a slicer level-based receiver where eye height/width are important, or an ADC (analog-to-digital-converter)-based receiver where level variation in constellation form is most important. The quaternary test pattern can be used to verify transmitter outputs for either receiver design.

Keysight or Tektronix can provide instrument solutions for the test and measurement of transmitter signals. Transmitter (Tx) characterization can be performed by real-time or equivalent-time (sampling) oscilloscopes. PAM-4 signals have a lower SNR than NRZ, so added noise as a result of the oscilloscope becomes an important consideration. Because of their hardware architecture, sampling oscilloscopes have an inherent noise floor that is very low compared to real-time oscilloscopes having comparable bandwidth, and as a result, will often yield the most accurate characterization of a PAM-4 signal. The Keysight 86100D with 86108B (Figure 9.10) is a sampling oscilloscope that provides high signal fidelity because of its wide bandwidth (>50 GHz), ultra-low time-base jitter (RJ<45 fs RMS), low noise floor, (<800 μV) together with an integrated clock recovery that operates on NRZ and PAM-4 signals to 32GBdd. With Keysight 86100D-9FP PAM-N Analysis software, the 86100D DCA-X provides PAM-4 measurements such as level mean/thickness/skew, eye mean/height/width/skew, and linearity. For optical PAM-4 systems, the Keysight 86100D DCA-X with 86105D-281 provides up to two optical channels per module. The DCA can also be configured to

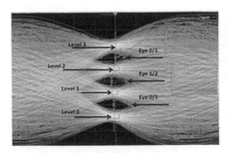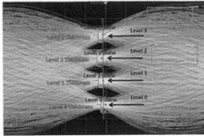

FIGURE 9.16   Eye centric measurements (left) and level centric measurements (right).

analyze up to 16 electrical PAM-4 signals simultaneously using the N1045A 60 GHz Electrical or N1055A 50 GHz TDR/TDT modules.

Transmitter characterization can also be measured using a real-time oscilloscope such as the Keysight DSO/DSAZ634A (Figure 9.17), which is best for troubleshooting and single shot events. The real-time oscilloscopes provide the fastest sample rate (160 GSa/s), large record length (deep single-shot memory), and do not require repetitive signals to generate pattern waveforms. New PAM-4 algorithms are required for real-time oscilloscopes that use software for clock recovery. Software for the real-time scopes can capture PAM-4 measurements such as eye width/height, BER, and equalization (FFE/CTLE). N8827A/B PAM-4 analysis software for Infiniium [17] real-time oscilloscopes accurately and quickly characterizes PAM-4 electrical signals described in IEEE 802.3bj Clause 94 [19]. The PAM-4 analysis software also addresses future measurement needs as outlined in developing standards such as OIF-CEI-56G and IEEE 400G. Data pattern generation can be achieved by using a second DUT, BERT, or arbitrary waveform generator.

9.3.10   PAM-4 PATTERN GENERATION METHOD

There are two approaches for generating PAM-4 patterns for receiver testing. The most common is based on traditional BERT NRZ pattern generators (Figure 9.18). Two channels of pattern aligned data streams, representing the most and least significant bits, are combined together to create the PAM-4 signal (Figure 9.18c). The output representing the least significant bit (LSB) is attenuated 6 dB relative to the other output. A delay equal to the attenuation path is added in the output representing the most significant bit (MSB), and the two signals are summed together using an RF power divider. In practice, two attenuators are often used, a 10 dB and a 3 dB. The attenuation in both paths

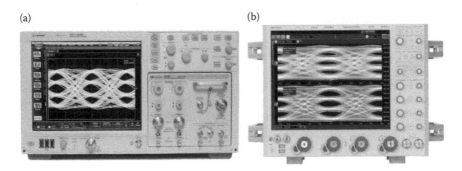

(a)   (b)

FIGURE 9.17   (a) Sampling oscilloscope (Keysight 86100D Infiniium DCA-X Wide-Bandwidth Oscilloscope.) and (b) Real-time sampling oscilloscope for measurement of PAM-4 signals (DSAZ634A Infiniium 63 GHz). Equipment type: Keysight Inc.

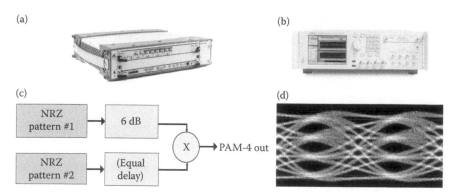

FIGURE 9.18 Commercial equipment for generation of multi-level signals. (a) Keysight AWG system. (b) Tektronix AWG. (c) Generating PAM-4 by passive coupling and splitting. (d) Output of SHF 6-bit DAC generated PAM-4 eye diagram as monitored by a wideband sampling oscilloscope.

reduces the effect of reflections from mismatch in the transmission lines, which cause problems in PAM-4 systems. 10 dB is used in the LSB output, as 9 dB attenuators are not commonly available. The 1 dB error is corrected with the amplitude controls in the pattern generator outputs. As with most choices, there are tradeoffs between the two patterns generation methods. The combining outputs of NRZ BERTs described above allows long patterns, limited only by the length of user memory in the BERT. It also has potentially faster rise and fall times, assuming proper selection of the attenuators and RF power divider. Faster rise times are important for using the generator as a golden transmitter emulator, but not for receiver input testing, as channels are generally used to slow the edges down even further. This approach can be expensive, as two high-performance pattern generator outputs are required to generate a single channel of PAM-4 pattern. It is also noted that the skew of different paths of PAM-4 pattern must be tuned to ensure the synchronization of the two NRZ sequence.

9.3.11 REMARKS

As Ethernet technology progresses from 100G to 400G, new test technology, measurement capabilities, and ability to accurately simulate PAM-4 designs for Tx and Rx hardware are mandatory. The change from NRZ to PAM-4 multi-level signaling presents many new design and measurement challenges. This section addresses the techniques for generation and measurements of PAM-4 signals, in particular the direct modulation of VCSEL and features of its generated and detected eye diagram, the delay amplitude dependence leading to correction to sampling time skew to compensate for this potential problem.

9.3.12 DMT MODULATION FORMATS

Direct detected OFDM (DD-OFDM), also known as DMT, is another attractive scheme for low-cost and short-reach communication [20]. The output after inverse fast Fourier transform (IFFT) in a DMT scheme is real-valued, which makes the IQ modulation onto an RF or optical carrier unnecessary, so it reduces the system cost. As a kind of multi-carrier modulation technique, DMT shows high spectrum efficiency, flexible multi-level coding, and obvious tolerance to ISI.

Figure 9.19 shows the schematic diagram of the DMT system. The original bit sequence is first fed to an encoder, which maps blocks of bits into complex symbols. A high-order coding scheme can be used here. Then the serial complex data is divided into parallel streams, and each stream is

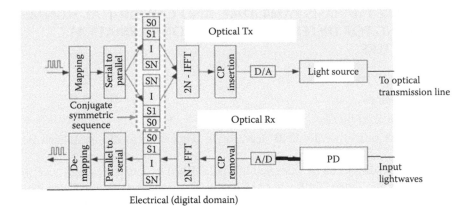

FIGURE 9.19   Schematic representation of DMT modulation transmission systems.

modulated onto N sub-carriers. To get a real-valued time-domain output, a new 2N points sequence is constructed, and the second N points is the conjugate symmetric sequence of the first half. After 2N point IFFT, the output signal is of real value. The cyclic prefix is then inserted for each DMT symbol. The generated DMT signal is passed through a D/A converter (hence looking like random analog signals) and subsequently used for optical modulation. At the receiver, direct detection can be employed. After A/D conversion, the cyclic prefix is removed and followed by 2N points FFT. Only the first half of the 2N points output after FFT is necessary for signal decoding. There are several issues in DMT modulation systems. The most important one is the high peak-to-average power ratio (PAPR) for multi-carrier modulation. This shortcoming would bring serious non-linear distortion in the electrical and optical domains, especially at the driving power amplifier and in the fiber link. It would also reduce the dynamic range of a DMT system. Figure 9.20 shows the PAPR by the parameter CCDF (complementary cumulative distribution function) against PAPR (PAPR at the origin)

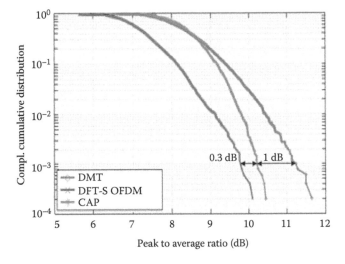

FIGURE 9.20   PAPR performance comparison for different schemes. The number of sub-carriers for the OFDM scheme is 256, and there are four modulation levels.

9.4 DIGITAL PRE-EMPHASIS PAM4, DMT, AND CAP: DIGITAL SIGNAL PROCESSING FOR DIFFERENT MODULATION FORMATS AT TRANSMITTERS

9.4.1 Principles

9.4.1.1 PAM Schemes

9.4.1.1.1 Unipolar and Bipolar PAM

- Unipolar PAM suffers ~3 dB SNR degradation comparing to the un-utilized polar PAM (Figure 9.21)
- The ~3 dB degradation can be recovered in a polar PAM system if a balance receiver is used. This scheme is also called DPSK since MZ modulator provides the "digital" phase modulation if it is biased at null instead of quadrature. No coherent detection is needed if data is differentially pre-coded.

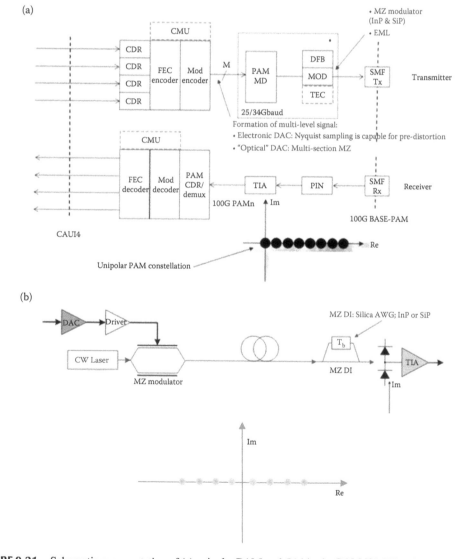

FIGURE 9.21 Schematic representation of (a) unipolar PAM and (b) bipolar PAM [21,22] and physical devices.

PAM can be implemented by simple Tx (single MZ modulator, EML, or DML). The required SNR increases as multiplicity increases; to keep the low-noise implementation scaled up to high multiplicity is also a challenge.

For low-cost, practical implementation, an electric data at symbol rate $\leq$25.78 GBaud with multiplicity m $\leq$4, PAM-4 is desirable.

9.4.1.1.2 What Other Modulation Formats?

Alternatively, there are possible three modulation schemes considered to be attracting R&D engineers to pursue including (i) QAM-16; (ii) CAP-16; and (iii) PolMux-QAM-16. What are the required detection schemes associated with the modulation formats? They can be non-coherent or direct detection and coherent receptions and transmission. The effective symbol rates can be tabulated as shown in Table 9.1. This implies that the technologies to be employed in the formation of the 100G modules are very important.

Many systems have been proposed to maximize the system performance for arbitrary channels. The performance is measured by the baud rate and the bit-error rate (BER) at the receiver end. Discrete multi-tone (DMT) [5,6] and carrier-less amplitude/phase modulation (CAP) [7,8] are two viable techniques for high-speed digital transmission over metal links, as well as in the optical access networks.

9.4.1.2 CAP Modulation

Originally, integrated optical communications networks could carry many kinds of symbol rates on the same fiber to access services. Utilization of the potentially under loaded optical networks is a good starting point for flexibility in wired-access internet at moderate speed in the 1990s, using twisted pair telephone loops to transmit high-speed data is a cost-driven choice. This led to the introduction of the digital optical access for transmission of digital information over the access optical loops and data center centric networking.

Carrier-less amplitude/phase modulation (CAP) is a multi-level and multi-dimensional modulation technique proposed by Bell Labs [23]. This is briefly given in the next sub-section. In contrast to quadrature amplitude modulation (QAM), CAP does not use a sinusoidal carrier to generate two orthogonal components. CAP uses two orthogonal signature waveforms to modulate the data in two dimensions. At the receiver, two filters are used to reconstruct the signal from each component. CAP is especially attractive for access networks and short-range links as it employs low complexity electronics, and allows for narrow channel spacing due to the high spectral efficiency, hence energy efficient. As an example, we have used four-level encoding for each dimension generating the so-called CAP-16. Previous publication demonstrated eight levels encoding CAP-64 [24], three orthogonal components 3D-CAP [25], and working with bit-rate up to 40 Gbps [26].

9.4.1.3 DMT Modulation

The DMT line code is the discrete-time version of the known multi-tone modulation (DMT). Although multi-tone modulation is very expensive in terms of complexity, DMT is a practical implementation. On the other hand, the CAP line code deals with the channel as one tone and leaves the

---

TABLE 9.1

Modulation Format, Symbol Rate, and Number of Lanes for 100 Gbps Module

Format	PAM-8	PAM-16	QAM-16	CAP-16	PM_QAM-16
Symbol rate (GBd)	34.38	25.78	25.78	25.78	12.89
Number of data lanes	1	1	2	2	4

problem of retrieving the original data dependent on the receiver processing ability. Different adaptive equalization algorithms can be deployed to achieve this processing power. Although implementation of adaptive equalizers is rather expensive, the overall CAP complexity is considerably less than that of the DMT.

The understanding of the digital communication environment as a multi-rate trans-multiplexer (MRTM) offers many new frontiers to have been explored. The notation of multiple-access topologies, such as time-division multiple access (TDMA) and frequency-division multiple access (FDMA), can be expressed directly as special cases of the general MRTM. It has been shown that the traditional CAP system can be expressed as a 2D MRTM system.

9.4.1.4   CAP Transceiver

The original CAP is a bandwidth-efficient 2D passband transmission scheme. The basic idea of the CAP system is to use two signals as signature waveforms to modulate two data streams. Bandwidth efficiency is achieved in two steps. The first step involves multi-level encoding of the data stream.

Using four-level encoding for each dimension can generate the so-called CAP-16. Bandwidth efficiency can also be achieved by using efficient signature waveforms. The theoretical limit of that parameter is achieved when using Nyquist signaling.

The baseband Nyquist signal has a bandwidth where is the symbol rate, while for a passband signal. Efficient shaping necessitates using signature signals that occupy more than one symbol period in time. Extending the signature waveform can compress the frequency-domain characteristics of the signal and lead to overlapping signatures of successive symbols. The design of the signature waveforms should ensure no inter-symbol interference (ISI) between consecutive symbols, as well as no crosstalk between symbols in each dimension. Another factor considered in designing the signature waveforms is the excess bandwidth defined as

$$\alpha = \frac{W - 1/T}{1/T} \tag{9.6}$$

where W is the symbol rate and T is the actual bandwidth used for the baseband signal. In practice, different signals can be used to meet those criteria. Examples for these are the raised-cosine signal and the square root raised-cosine signals.

Two-dimensional signaling has the advantage that it can retain the same bandwidth efficiency for a passband signal. The two orthogonal signals used as signature waveforms are modulated versions of a baseband signal, and are given by

$$f_1(t) = g(t) \cos\left(2\pi f_c(t)\right)$$
$$f_2(t) = g(t) \sin\left(2\pi f_c(t)\right) \tag{9.7}$$

where $g(t)$ is the baseband signal and f_c is a frequency that is larger than the largest frequency of the baseband signal. The pair is called a Hilbert pair. Figure 9.1a of Reference 27 shows the time domain modulated raised-cosine signature waveforms and the normalized frequency characteristics for a symbol rate of 25 MHz with excess bandwidth of 20%. It shows the corresponding receiver filters for PR of the transmitted sequence. The structure of the CAP transceiver is shown in Figure 9.22a. The data stream is scrambled into two symbol streams, and each is modulated with the corresponding signature waveform.

The receiver is implemented in an adaptive fashion to invert the channel and the signature filters and retrieve the original sequence of symbols. Many topologies can be used to implement the receiver, such as the linear equalizer and the decision-feedback equalizer. The major challenge in designing the receiver is to guarantee PR of the original sequences. We use a fractionally spaced linear equalizer (FSLE) [28] and a training sequence to update the filter taps adaptively via the least mean square (LMS) algorithm. If the QAM line code is used instead of CAP, a cross-coupling would

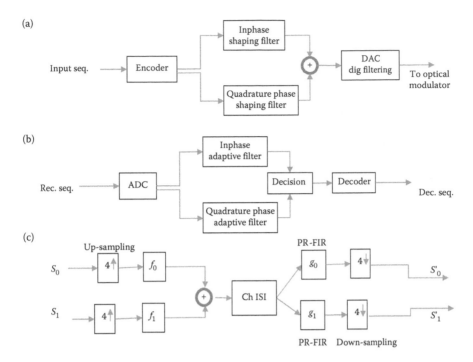

FIGURE 9.22 CAP System transmitter and receiver, trans-multiplexer schematic representation of (a) transmitter; (b) receiver; and (c) multi-rate trans-multiplexer (Transmux).

be needed to perform the complex filtering, but for the CAP system only a simple filtering operation is needed for each dimension.

The transmitter, as well as the receiver, are implemented in a digital domain. The transmitter signature filters are implemented as fixed finite-impulse response (FIR) filters. To implement the system with that topology, the sampling rate of the implementation must be high enough to prevent aliasing effects. For that implementation, the input symbol sequences are up-sampled to match the implementation sampling rate, and in turn, the receiver output is down-sampled to the original symbol rate. The performance of the CAP-16 system is found to be acceptable in terms of system throughput and receiver BER for the transceiver.

Arbitrary waveform generator (AWG) and analogue to digital convertor (ADC) are now commonly employed in the system, and digital signal processing can be applied at both transmitter and receiver sides. In this section, we describe the digital signal processing (DSP) required for different modulation formats at both ends of the transmission system.

9.4.2 DSP for PAM-4 Signal

Figure 9.1 shows the DSP flow chart for PAM-4 signal at the transmitter and the receiver sides. A 2^{16} de Bruijn bit sequence was used for bit to symbol mapping and the generation of PAM-4 signal. To achieve a bit rate of 112 Gbps (100 Gbps excluding Ethernet and FEC overhead), a baud rate of 56 Gbps is required for PAM-4. In the experimental setup [29], the sampling rate of AWG is set to be 63 GSa/s. Therefore, the symbol stream was up-sampled by a factor of 1.125 (63/56) samples/bit. To avoid aliasing, a raised cosine (RC) pulse-shaping filter with a roll-off factor of 0.12 was used for pulse shaping. Because the bandwidth of both the transmitter and the receiver is much smaller than signal bandwidth, the up-sampled signal was pre-emphasized by using an inverted linear filter (Figure 9.23).

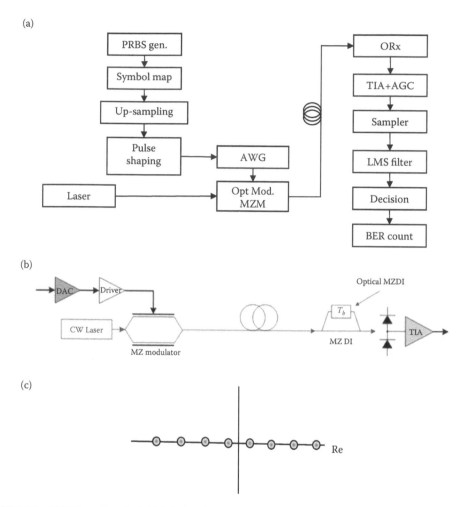

FIGURE 9.23 DSP flow chart for PAM-4 signal at transmitter and receiver sides. At receiver side, the sampled signal is normalized via AGC and resampled to two samples/symbol. A training symbol aided least mean square (LMS) algorithm was first used to initialize the equalizer taps. After the taps are fully converged, the equalizer will switch to a decision directed mode. After symbol decision, the bit error rate was calculated by error counting. (a) Schematic representation of DSP PAM-4 signaling at Tx, Rx and transmission and (b) details of optical transmission structure (c) constellation. Legend: AGC = automatic gain control; BER =bit-error-rate; MZ DI: passive optics; Specs can be relaxed for single-ended Rx.

MZDI is a passive integrated optic device with some thermal tuning and can be implemented in SiP [30], and the PIN cost is relatively low comparing to Tx, this can be fabricated by InP or Ge-SiP deposited on top of Si waveguide. The most challenging task is to create the I-Q modulator in integrated Si photonic platform. However, the I-Q modulator is also known as the single side band (SSB) modulator, which can be generated by using Hilbert transformed electrical data sequence or in generation in the optical domain by an optical $\pi/2$ phase shift via a dual-drive single MZ modulator.

For a single MZ with dual drives, each arm generates phase modulations:

$$E_{out} = \frac{E_{in}}{2}\left[e^{jV_1} + e^{jV_2}\right] \tag{9.8}$$

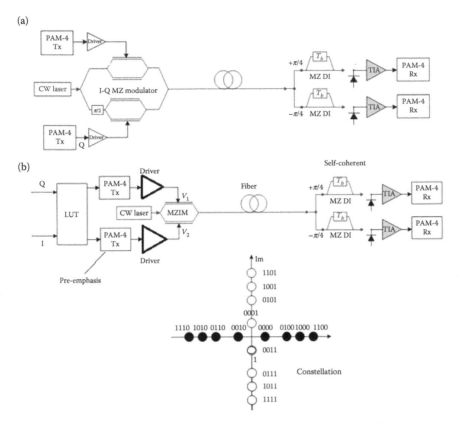

FIGURE 9.24 Differential QAM-16 optical signaling schematic representation (a) without pre-emphasis and (b) with pre-emphasis.

An example given in Figure 9.24 shows how to generate the sub-set of a constellation along the real and imaginary axes with a circle $r = 0.4E_i$, and thus an M-ary quadrature amplitude modulation can be realized by the different dual-drive voltages V_1 and V_2 with a single MZ modulator. The combination of MSBs and LSBs can be used to determine amplitude and phase-domain, respectively.

9.4.3 POL-MUX-N-ARY-QAM

Polarization multiplexed system (e.g., QPSK) is the baseline for 100G long-haul transmission. However, the polarization mixture at the Rx side is problematic for direct detection without the use of expansive polarization rotator, thence coherent detection is required to find the inverse Jone's Matrix PM-QAM-n and PM-PAM-n are NOT cost effective for client-side optics (Figures 9.25 through 9.27).

9.4.4 DSP FOR CAP-16 SIGNAL

Figure 9.2 shows the DSP flow chart for CAP-16 signal. At transmitter side, a 2^{16} de Bruijn bit sequence is used for bit-to-symbol mapping and the generation of inphase and quadrature signals. After the IQ separation, the two signals continue to two shape filters. The impulse responses of two filters in orthogonality or Hilbert transform pair can be expressed as

$$h_I(t) = g(t)\cos(2\pi f_C(t))$$
$$h_Q(t) = g(t)\sin(2\pi f_C(t))$$

(9.9)

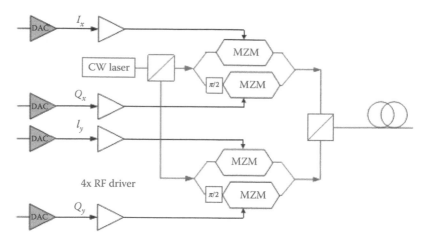

FIGURE 9.25 Pol-Mux QAM transmitter.

A square-root-raised-cosine shaping filter with a roll-off coefficient of $\alpha = 0.06$ is used as the baseband impulse response. The center frequency fc is given by

$$fc = \frac{Bd}{2}(1 + \alpha) + \Delta f \tag{9.10}$$

here, Bd is the baud-rate and is 28 GBd/s to achieve a bit rate of 112 Gb/s (100 Gbit/s excluding Ethernet and FEC overhead) with CAP-16 format. Δf is the frequency offset and is set to be 0.1 GHz. After the shaping filter, the signal are resampled and pre-emphasized by an inverted linear filter.

We note that at 100 Gbps composing of four wavelength lanes, the FIR filter implementation is more difficult. Furthermore, the system performance and relation to Tx/Rx parameters require further investigation. At receiver side, the signal samples are first sent into two matched filters. The two filters are the time-reversed version of the shaping filters at the transmitter in order to separate the in-phase and quadrature components. Then the signal is re-sampled at two samples per symbol. Here, a hybrid equalizer can be used for channel equalization. A modified cascaded multi-modulus algorithm

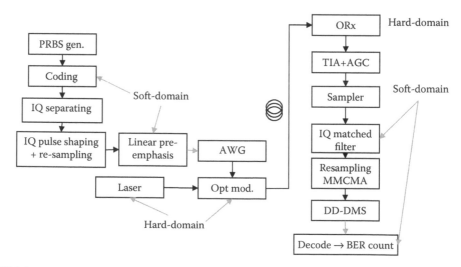

FIGURE 9.26 DSP flow for CAP-16 signaling.

(MCMMA) can be used for pre-convergence [31]. Decision directed least-mean square (DD-LMS) can also be used as the second stage for fine equalization. After the equalization, the signal is decoded and the bit error rate (BER) is calculated by an error counting.

9.4.5 DSP FOR DMT SIGNAL

Figure 9.3 shows the DSP flow chart for DMT system. At transmitter side, the DMT signals are encoded using a 2^{16} de Bruijn bit sequence with the following parameters: FFT size is 512 with Hermitian symmetry, cyclic prefix (CP) is eight, and the number of data sub-carriers is 217.

For every data frame, 21 training symbols are inserted at the beginning of each data frame, which consists of one symbol for symbol frame synchronization and 20 symbols for channel estimation. The aggregate data rate is 111.8 Gbps, which results in a useful bit rate of 100 Gbps excluding Ethernet and FEC overheads. The generated DMT signal is pre-emphasized by using an inverted linear filter and loaded into an AWG operated at 63 GSa/s with eight bits resolution.

At receiver side, after normalization and re-sampling, symbol synchronization is realized by training symbols. Thus, CPs are removed and the data symbols on each sub-carrier are obtained by FFT operation. A one-tap equalizer is implemented to compensate for channel distortions. Finally, the recovered symbols are decided and decoded to obtain the bit sequence. The bit error rate (BER) can be calculated by error counting.

9.4.6 REMARKS

Possible modulation formats and the associated detection scheme under single optical carrier under high-level modulation formats, and their potential application for the future 100G are briefly given, including transmitters of DSP-PAM4, DSP-based CAP, and DSP- PMD (polarization multiplexing division) are investigated as a high possibility for Ethernet in metro or access networking:

1. Extension of unipolar PAM to polar PAM can be realized by detection by a balanced Rx, which could result in ~3 dB SNR improvement (high cost).

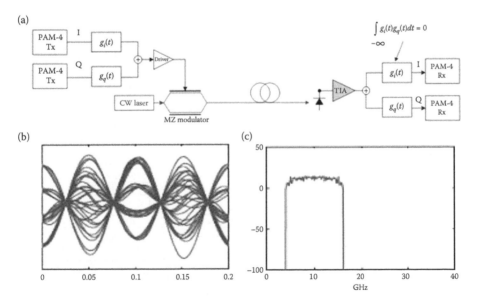

FIGURE 9.27 Signaling modulation scheme of CAP-16. PAM-4 orthogonal signals and single drive of a dual-drive MZIM (a) Schematic representation of system transmission with signal flow for CAP-16; (b) eye diagram, and (c) spectrum. Note: Quadrature and Inphase signaling in electrical domain.

2. QAM can be implemented by an I-Q modulator or a dual-drive single MZ modulator (a) dual-drive MZ modulator has been demonstrated at symbol rate = 25 GBd with both integrated SiP and InP, which is capable of QAM-16 implementation; (b) QAM-16 with single-ended Rx has a SNR performance better than PAM-8; and (c) MZ delay interferometer (MZ DI) specs for single-ended Rx can be relaxed. Integrated Si Photonics-based PIN Rx and passive MZ DI have been demonstrated that can provide further cost reduction.

3. PM-PAM or PM-QAM are not suitable because of polarization mixing at Rx, which requires coherent Rx to remove the polarization related degradation.

4. Detailed study for CAP-16 at both system level and component level are needed. Further Link budget simulation and experiment to compare QAM-16 vs. CAP-16 is a good starting point to differentiate the cost and feasibility between the two. The flow of the DSP processing is shown in Figure 9.28.

9.4.7 NON-DSP HARDWARE PRE-EMPHASIS FOR EQUALIZATION

9.4.7.1 Electronic Feed-Forward Equalizer

This section summarizes the report on the first experimental demonstration of 100 Gbit/s 3-level duo-binary (DuoB) optical transmission enabled by inhouse newly developed SiGe BiCMOS transmitter/receiver ICs. Operated in real-time, we demonstrate a 100 Gbit/s data-rate over 2 km SSMF without DSP. This pre-emphasis electronics are critical for current access networks, especially in data center interconnects and access fiber loops. Similarly, we report on the pre-equalizer given by ADI.

Currently, the evolution from 100 Gbit/s Ethernet to the rate of 400 Gbit/s is under discussion within the IEEE P802.3bs 400 Gigabit Ethernet Task Force. Among different approaches, the four-lane 100 Gbit/s scheme is particularly attractive for 500 m and 2 km SMF (single mode fiber) applications, as it allows lower lane counts and offers higher spatial efficiency.

In research, several experiments of NRZ or PAM-4 transmissions at 100–112 Gbit/s were reported. However, most of them required heavy offline digital signal processing (DSP) to compensate the limited bandwidth of the O/E components. In reference, a real-time 112 Gb/s PAM-4 optical link over 2 km standard SMF (SSMF) can be demonstrated, with a BiCMOS transceiver (including CDR) power consumption of ~8.6 W

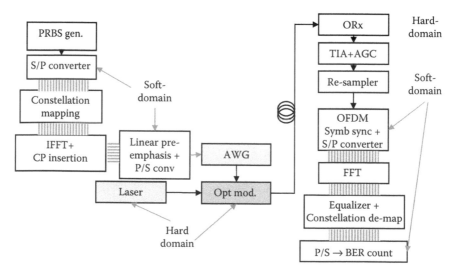

FIGURE 9.28 DSP flow chart for DMT at T_X and Rx and the optical hardware transmission system. AWG = arbitrary waveform generator; P/S = parallel to serial convertor; S/P = serial to parallel convertor; BER = bit error rate counter.

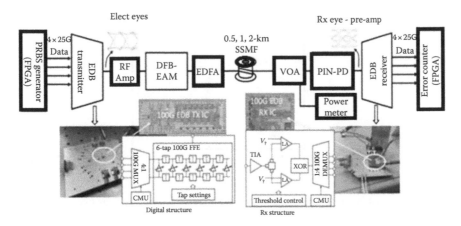

FIGURE 9.29 Duo-binary 211 Gbps and non-DSP hardware equalizer—setup.

The experimental setup is illustrated in Figure 9.29. A Xilinx FPGA board is used and adopted to generate four electrical 25 Gbit/s NRZ decorrelated 2^7-1 PRBS signal streams. An EDB (electrical Duo-Binary) Tx (Transmitter) IC (integrated circuits) multiplexes $4 \times 25\,G$ streams into a serial 100 Gbit/s, and pre-equalizes the serialized NRZ signal. The pre-emphasized signal was amplified by a 50 GHz RF amplifier and then used to drive a C-band 100 GHz electro-absorption modulated laser, (EML). In the experiment, the integrated distributed feedback laser (DFB) in the EML emitted at 1548.7 nm and the output optical power of the EML was around 0 dBm. After transmission over an SSMF, the received optical signal was detected by a PIN photodiode (PD) and a custom EDB Rx (Receiver)) IC. An Erbium-doped fiber amplifier (EDFA), a variable optical attenuator (VOA) and a power meter were used before the PIN-PD to adjust/record the received optical power for measurement purpose. The EDFA can be removed from the setup when a sufficient higher-power EML is available. The PIN-PD is a high-speed InP-based O/E (optical to electrical) converter which can be packaged prototype with a responsivity of 0.5 A/W. The sub-sequent DuoB Rx IC demodulates the three-level eyes with two separate threshold levels. The demodulated signal is then deserialized on-chip into 4×25 GBd NRZ outputs for real-time error detection, which is implemented in the same FPGA board as for PRBS generation. Both EDB Tx and Rx ICs are fabricated in a 0.13 um SiGe BiC-MOS technology. The Tx IC consists of a 4-to-1 MUX and a six-tap feed forward equalizer (FFE) that can compensate the frequency roll-off of the fiber limited bandwidth and the receiver responses.

The Tx IC occupies an area of 1555 µm × 4567 µm including IOs (input/output) and dissipates about 1 W. The measured frequency response of the E/O/E (electrical/Optical/Electrical) components excluding the EDB ICs is shown in Figure 9.30. The frequency responses of various fibers

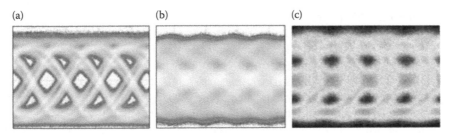

FIGURE 9.30 Measured 100-Gbit/s (a) eye-diagram at the EDB Tx IC output without pre-emphasis FFE, (b) eye-diagram at the PIN-PD output with pre-emphasis at Tx by a real-time six-tap FFE, and (c) received eye diagram.

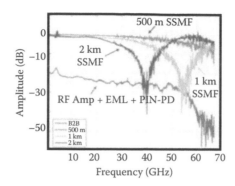

FIGURE 9.31 Measured frequency responses of the B2B optical link and various fibers.

of 500 m, 1 and 2 km were measured and plotted in the same figure. As can be seen in Figure 9.31, in the optical back-to-back (B2B) case the bandwidth is mainly limited by the electrical amplifier and interfaces.

For fiber operating in C-band, the response severely degenerates the flatness of the frequency response, especially at 2 km. Together with the bandwidth-limited components and the dispersive fiber, the real-time FFE in the TX is employed to create an equivalent channel response that transforms the NRZ from the parallel to serial convert to serial sequence output into a three-level DuoB at the Rx input.

The EDB RX IC consists of a trans-impedance (TIA) input stage, two parallel level-shifting limiting amplifiers, an XOR and a 1-to-4 DEMUX. The EDB RX can be reconfigured into a NRZ Rx with 4 demuxed outputs, which is also used in the experiments for NRZ signaling comparison. The FFE is preferably done in TXs to avoid noise emphasis. Without the need of complex DSP, the Rx IC occupies 1926 um × 2585 um including IOs and consumes below 1.2W.

As shown in the inserts of Figure 9.29, both EDB Tx and Rx ICs are flip-chipped bonding on test PCBs. Before performing the optical transmission experiment, the EDB TX/Rx ICs have been first verified B2B pure electrically. The electrical B2B measurement was performed continuously over 1 h, and measured bit-error rate (BER) was less than 10^{-12} for 100 Gbps DuoB signal, which revealed a very stable performance of the EDB Tx/Rx ICs. Optical performance of the transmission link can be evaluated using the experimental setup shown in Figure 9.29. To compare the performance, we first evaluated both NRZ and the proposed three-level DuoB at various rates in optical B2B configuration. Figure 9.30a shows the electrical 100 Gbps NRZ eye-diagram after the Tx IC without FFE, which can then be open. However, the received optical eye at the PIN-PD (Figure 9.30b) was completely closed due to the component bandwidth limitation, as indicated in Figure 9.31. As shown in Figure 9.3c, by enabling the real-time six-tap FFE in the EDB Tx IC, the received optical eye has been shaped into three-level DuoB and two separate eyes were clearly formed after the PIN-PD (p-type intrinsic-n-type photodiode). For fair comparison, the Tx FFE is always enabled in the experiment and separately optimized for each rate. The measured BER curves for 50 Gbit/s and 70 Gbit/s NRZ signals are shown in Figure 9.32 [32]. The optical power shown here is measured at the input of the PIN-PD (after the VOA). The NRZ signaling works relatively well in B2B up to 70 Gbit/s. However, after 2km of SSMF, we noticed a decreased performance at 70 Gbit/s (4.4 dB penalty compared to the B2B). As clearly indicated by the received eye-diagram, the CD introduces significant inter-symbol interference (ISI) and closes the NRZ eye. We can utilize this ISI effect and create three-level DuoB signal by optimizing the Tx FFE. The optimized three-level DuoB eye in Figure 9.32a shows two widely open eyes. The measured BER for 70 Gbit/s EDB link over 2 km can reach a BER of ~1e-10, which is well below the hard-decision forward error correction (HD-FEC) with 7% overhead (BER = $3.8.10^{-3}$) or G-FEC

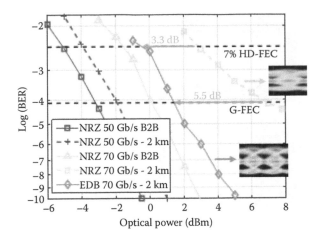

FIGURE 9.32 Measured NRZ BER curve up to 70 Gbit/s and 2 km SSMF, compared to 70 Gbit/s EDB modulation.

threshold (BER $= 8.10^{-5}$). In addition, using EDB modulation, we obtained a sensitivity improvement of 3.3 dB at 7% HD-FEC limit (or 5.5 dB at G-FEC limit) with respect to the 70 Gbit/s NRZ transmission over 2 km distance.

Thence, the next feature illustrates the performance of 100 Gbit/s EDB measurements under different fiber lengths [33]. The resulting eye-diagrams for B2B, 500 m, 1 km, and 2 km SSMFs are depicted in Figure 9.33a. The realtime BER measurements are shown in Figure 9.5b. For 500 m applications, no noticeable penalty was observed. Longer than 2 km SSMF, 100 Gbit/s EDB signal can still be received with a BER less than 7% HD-FEC (hard decision forward error correction) threshold.

9.4.7.2 Analog Device (ADI) Equalizers

A similar ADI has implemented a 32 Gbps, Dual Channel, Advanced Linear Equalizer (ALE) as shown in Figure 9.34.

The HMC6545 supports data rates from DC up to 32 Gbps Protocol and data rate agnostic low latency ($<$170 ps); integrated AGC with differential sensitivity of $<$50 mV; Up to 20 dB programmable multiple unit interval input; equalization extended chromatic and polarization mode dispersion tolerance; programmable differential output amplitude control of up to 600 mV; single 3.3 V supply eliminating external regulators; wide temperature range from $-40°$C to $+95°$C; 5 mm $\times$ 5 mm, 32-lead LFCSP package; applications to 40 Gbps or 100 Gbps DQPSK direct detection receivers; short and long reach CFP2 and QSFP+ modules; CEI-28G MR and CEI-25G LR 100 GE line cards; 16 Gbps and 32 Gbps Fibre Channel; Infiniband 14 Gbps FDR and 28 Gbps EDR rates; signal conditioning for backplane and line cards; and Broadband test and measurement equipment.

9.5 MODULE

9.5.1 28 GBd Module

To reduce the cost compact modules are required. Current modules are shown in Figure 9.35. Transceiver can be shown in Figure 9.36 with a block diagram of a four lanes 100G transmitter and four lanes receiver, a microcontroller unit that controls the biasing of the modulators and provide voltage supplies to the electronics of the receiver. The block diagram of a transceiver of 100 Gbps with four lanes is shown in Figure 9.36. There are clearly three main sections: (a) transmitter; (b) receivers; and (c) micro-control units. The transmitter consists of 4 × 2 and can be included in association with electronic derivers 8G TOSAs in which an external modulation laser (EML). A pre-emphasis electrical

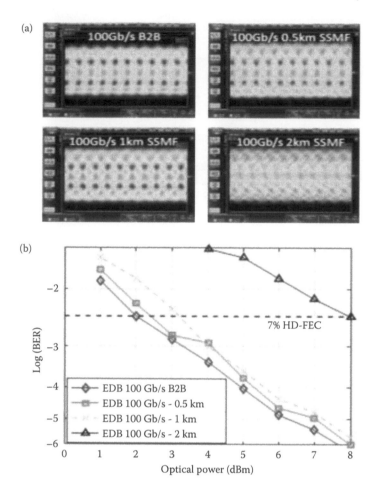

FIGURE 9.33 Measured (a) eye-diagrams of 100 Gbit/s EDB transmission at different fiber lengths and (b) BER curves of 100 Gbit/s EDB transmission at different fiber lengths.

equalizer or filter can be employed to pre-distort the signals to compensate for fiber chromatic distortion. The four optical channels are then multiplexed into one line to launch into the transmission fiber link (Figure 9.37).

Similar in a reverse manner, the inline incoming multiplexed channels are demultiplexed into four parallel optical lines via ROSAs and the detected via PDs and then amplified via TIAS and Post-amplification (PA) and then possibly equalized by a feed forward equalizer.

The MCU is used to control the biasing of the modulators and then a separate DC converter to supply power to the power rail of the transmitters and receivers.

Electrical and electronic circuit design, including module layout, can follow the diagram shown in Figure 9.38, particularly the 3D modeling to obtain the S-parameters of the electrical circuits and then transfer the design to the electro-photonic circuits and opto-electronics.

9.5.2   112Gbps PAM-4 Time Domain Multiplexing Transceiver Chipset in 0.18 μm SiGe Bi-CMOS

9.5.2.1   Transmitter Structure Driver Plus FFE Integrated Driver

The simplified block diagram of the 112 Gbps 4-PAM transmitter is shown in Figure 9.39. The data interface lanes accept 28 Gbps single-ended NRZ data streams which are converted to differential

Functional block diagram

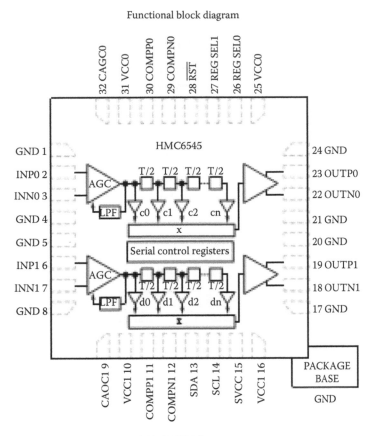

FIGURE 9.34 Functional block diagram of ADI HTMC.

signals internally with skew adjustment, before retiming. A synchronous 28 GHz clock signal is also used for all internal multiplexers and retimers. Digitally programmable phase shifters ensure proper clock and data alignments for all retimers and multiplexers. Phase shifters comprise multi-stage vector summation of quadrature clock signals. Furthermore, the 56 GHz internal clock signal is generated by doubling the 28 GHz clock. The two multiplexed 56 Gbps NRZ data streams (labeled A and B on Figure 9.1) are replicated internally (C and D) before full-rate retiming. The replica paths are processed through an extra latching stage to provide a half-UI data delay. The main and delayed signals are then processed through a bank of variable gain amplifiers whose gain can be adjusted from -2 to $+2$V/V. An active linear summation block combines the weighted main and delayed 56 Gbps data

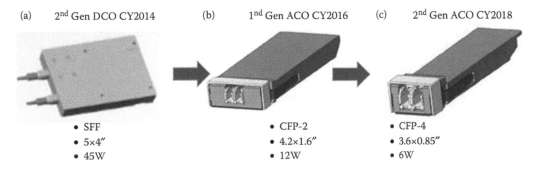

FIGURE 9.35 Typical modules (a) SFF, (b) CFP-2, and (c) CFP-4.

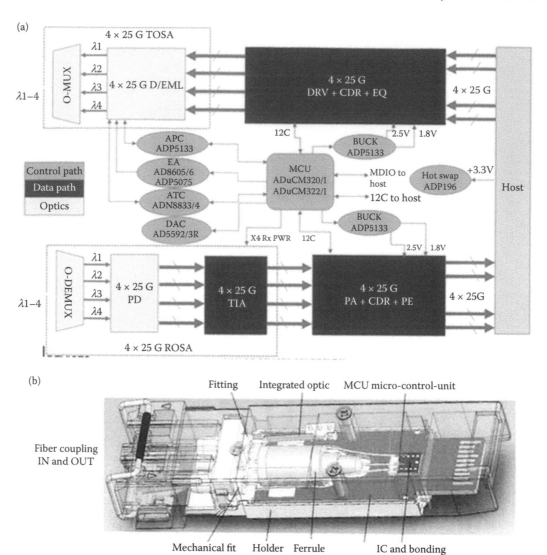

FIGURE 9.36 100G CFP-4/QSFP28 LR4 Functional Block Diagram (in courtesy of Analog Devices Inc.) (a) block diagram of module; and (b) picture of module. TOSA = transmitter optical sub-assembly; ROSA = receiver optical sub-assembly; TIA = trans-impedance amplifier; PA = post-amplification; CDR = chromatic dispersion; EML = external modulator laser; PD = photodetector; MCU = micro-controller unit; DAC = digital to analog converter.

streams. This results in a 4-PAM output signal with programmable levels and feed-forward equalization (FFE). FFE coefficients can be adjusted by changing the gain of the VGAs in the main path. The PAM-4 output signal from the DAC is amplified up to 3.5Vp-p differential by using a distributed amplifier (DA). Each segment of the four-stage DA comprises emitter follower differential cascode stages with programmable offset cancellation.

9.5.2.2 Receiver Structure

The simplified block diagram of the 112 Gbps 4-PAM receiver is shown in Figure 9.40. An input VGA provides a constant input amplitude to the signal distribution network. A bank of four slicers with adjustable threshold levels followed by a chain of limiting amplifiers are used to detect the

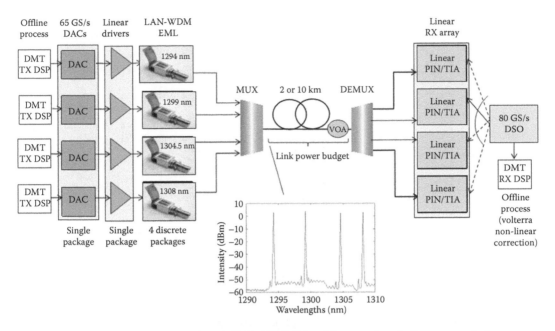

FIGURE 9.37   Schematic representation of 65 GBd NRZ O-band four wavelength 10 Km transmission systems for 250 Gbps access networks using DMT modulation format.

4-PAM signal levels. The resulting thermometer-coded 56 Gbps NRZ data streams (MSB, LSB+, and LSB− in Figure 9.38) are retimed at full rate before thermometer to binary conversion where the MSB signal selects the appropriate LSB+ or LSB− signals through a high-speed selector.

The 56 Gbps MSB and LSB signals are converted to four 28 Gbps data lanes through a pair of demultiplexers. A set of selectors are used to choose between the data output or quarter-rate recovered clock. All output signals are retimed at 28 GHz before arriving at the output drivers. The clock and data recovery block (CDR) uses a replica of the MSB signal and a 56 Gbps phase detector for clock

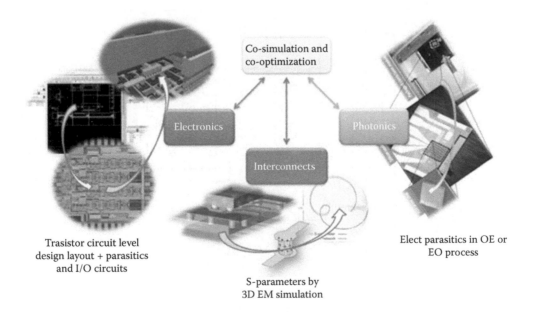

FIGURE 9.38   Design cycle for electrical, electronic circuits, and module lay out including photonic sections.

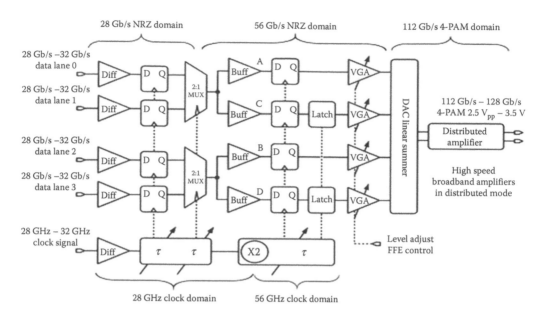

FIGURE 9.39 Simplified block diagram of the 112 Gb/s large-swing 4-PAM transmitter with FFE section, TDM of 4x 28 GB, thus 28 G, 56, then 112 G structure [34].

recovery. A dual VCO architecture ensures a wide locking range and immunity against temperature and process variations. A built-in second order PLL with a 1/16 ratio is also implemented to bring the VCO frequency within the locking rage of the CDR. If available, the receiver can also use a coherent external 28 GHz clock signal. All clock paths are equipped with phase shifter to align the 56 and 28 GHz clock signal to data transitions. Phase shifter setting, CDR loop bandwidth, gain control and threshold levels are all set digitally through a serial digital interface.

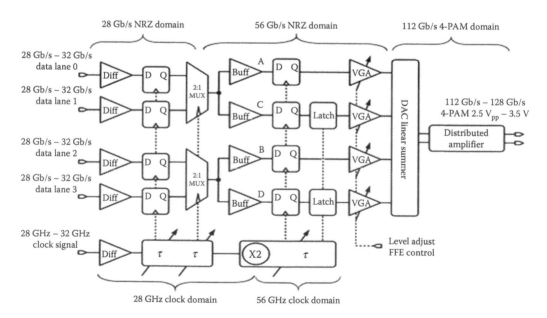

FIGURE 9.40 Simplified block diagram of the 112 Gbps PAM-4 receiver with clock and data recovery. VGA = voltage gain automatic; LA = linear amplifier; SEL = selective switch; D, Q = input and output of high speed flip-flop.

9.5.2.3   Fabrication and Packaging

The transceiver chipset is implemented using the 180 nm SiGe BiCMOS process with its transition frequency and maximum frequency of f_T/f_{Max} of 240/270 GHz, respectively. The employed fab process offers 180 nm CMOS transistors and six layers of Al metallization. The transmitter and receiver ICs occupy an area of 2.6×2 mm² and 3.5×2.8 mm², respectively (Figure 9.41). Both ICs operate from a single 3.3 V power supply and consume 6W (receiver) and 2W (transmitter). Figure 9.41 also shows a packaged dual-transmitter IC, as well as a single packaged receiver IC. Both ceramic packages use leads for 28 Gbps IOs, as well as power supply and clock signals. All 112 Gbps signals are handled through GPPO interfaces at the edge of the package (Figures 9.42 through 9.45).

9.5.3   PAM-4 DOUBLE AND SINGLE-SIDEBAND DuoB MODULATION FORMATS FOR UP TO 80 KM SSMF TRANSMISSION

Short-reach and access networks really require transmission links whose transmission distance can be flexible so as to provide flexibility in information delivery to clients. This section summarizes our recent progress on high baud-rate duobinary-PAM-4 system. Dual-drive MZM is used as a complex signal modulator with relative low insertion loss and high bandwidth to overcome the influence of dispersion and generation of single sideband signal. Furthermore duobinary (DB) and pre-emphasis at the transmitter side to extend the transmission reach are demonstrated to rove the effectiveness of the modulation scheme under direct reception.

9.5.3.1   Introductory Remarks

With the quick increasing capacity demands of data center and wireless communication technology, 100 Gbit/s and beyond systems with advanced modulation schemes have drawn lots of attentions. Four-level pulse amplitude modulation (PAM-4) has been selected as the standard modulation scheme of 400GE because of its moderate bandwidth requirement and power consumption. Lots

FIGURE 9.41   Images of the (a) transmitter die, (b) receiver die, (c) dual packaged transmitter, and (d) packaged receiver.

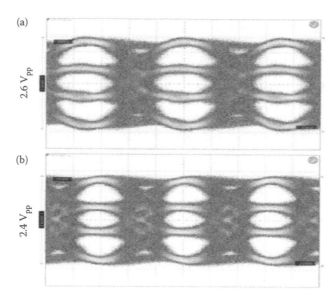

FIGURE 9.42 Measured packaged electrical 4-PAM output eye diagrams at (a) 56 GBaud/s and (b) 64 GBaud/s.

of investigations have been carried out; however, most of them are in O-band. PAM-4 can be a general modulation scheme if it can also support transmission in C-band. The most challenging issue for direct detection (DD) system with double sideband (DSB) configuration in C-band is chromatic dispersion (CD) that can induce serious power fading problem and degrade the system performance greatly. That is why dispersion compensation fibers are needed when only electro-absorption laser (EML) is used at transmitter side. With the progress of integrated photonics technology, we can choose some complex signal modulator with some extra cost, such as dual-drive MZM (DDMZM),

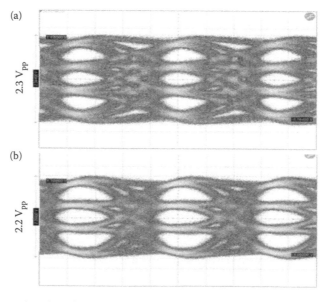

FIGURE 9.43 Measured packaged electrical 4-PAM output eye diagram through a 28 GHz 4th order LPF (a) before and (b) after equalization.

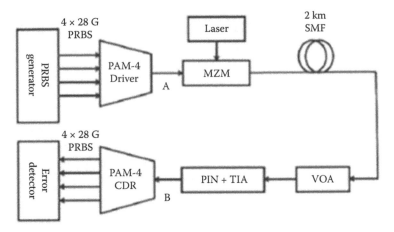

FIGURE 9.44 Schematic representation of optical transmission system set-up. Note on electrical mux and demux to recover 28 GB channels.

to reduce the influence of CD. Meanwhile, PAM-4 signal can work with single sideband (SSB) condition because its Hermitian property or $\pi/2$ phase shift of one branch with respect to each other, which can double the spectral efficiency (SE) of PAM-4 system.

In this section, the transmission distance and bit rate of up to 180 Gbit/s (90 GBaud) PAM-4 signals transmission are described. A performance comparison between conventional PAM-4 and duobinary (DB)-PAM-4 systems is carried out and the results show the advantage of DB-PAM-4. Then afterwards, the SSB-DB-PAM-4 and dispersion pre-compensated DSB-DB-PAM-4 are demonstrated.

9.5.3.2 Complex Signal Generation Using Dual Drive MZ Modulator (DDMZM)

The DDMZM consists of two independent phase modulator and the output signal can be written as [35]:

$$E_{out} = e^{j\omega t}\left(e^{j\frac{V_1}{V_\pi}\pi} + e^{j\frac{V_2}{V_\pi}\pi}\right) = 2e^{j\omega t}\cos\left(\frac{V_1 - V_2}{2V_\pi}\pi\right)e^{j\frac{V_1+V_2}{2V_\pi}\pi} \qquad (9.11)$$

where V_1 and V_2 are the electrical signals added to the two branches and V_π is the voltage needed to induce π phase shift in each branch. The driving voltage signals (V_1 and V_2) include two parts: direct

(a)   (b)   (c)

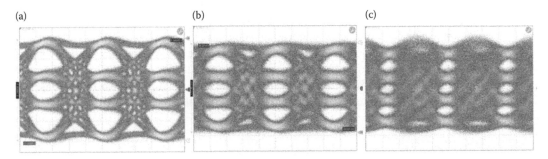

FIGURE 9.45 Measured 56 GBaud, 4-PAM eye diagrams: (a) Electrical 2.2V_{pp} differential signal at point A depicted in Figure 9.44; (b) electrical signal at point B in Figure 9.44 with 0-km of SSMF; and (c) electrical signal after 2-km of SSMF at point B in Figure 9.44. Note: no amplitude dependence due to the use of MZM instead of VCSEL.

current (DC) part and radio frequency (RF) part. The DC voltages will induce a constant phase shift between upper and lower branches to adjust the bias point. The RF signals carry the information to be sent. Then we can write the drive signals as

$$V_1 = V_{b1} + s_1(t)$$
$$V_2 = V_{b2} + s_2(t)$$

$$(9.12)$$

where V_{b1} and V_{b2} are the DC bias voltages for the upper and lower branch electrodes, respectively, and $s_1(t)$ and $s_2(t)$ are the RF signals. When the upper and lower branches are biased with $V_{b1}=V_\pi/4$ and $V_{b1} = -V_\pi/4$, the DDMZM works at quadrature point and the phase shift value between the two branches is $\pi/2$ to satisfy the Hilbert transform condition for sideband suppression. In this case, the output signal can be written as

$$E_{out} = e^{j\omega t} e^{j\pi/4} \left(e^{j\frac{s_1}{V_\pi}\pi} - e^{j\frac{s_2}{V_\pi}\pi} \right)$$

$$(9.13)$$

If the RF signal is small, the modulator can linearly convert the signal from electrical to optical domain that is defined by the following expression [32]:

$$E_{out} = e^{j\omega t} e^{j\pi/4} \left(1 + j\frac{s_1}{V_\pi}\pi - j + \frac{s_2}{V_\pi}\pi + o_2 \right)$$

$$(9.14)$$

For PAM-4 signal, at back-to-back (B2B) condition, the drive signals are added with push-pull method ($s_1 = -s_2$). This stands for that the signal we transmitted is $j(s_1/V_\pi)\pi - (s_1/V_\pi)\pi$, which will align the signal added and the optical carrier and get the maximum beating signal between the optical carrier the modulated optical signal.

9.5.3.3 DSB-DB-PAM-4 System

9.5.3.3.1 Experimental Platform

Figure 9.46 shows the setup we used for the DSB-DB-PAM4 measurement. A digital to signal converters (DAC) with 92 GSa/s and a 30 GHz bandwidth was used, and the outputs were amplified to drive the two branches of the 30 GHz DDMZM with quadrature bias points. After transmission through the SSMF link, the optical signal was converted to an electrical signal by a 50 GHz optical receiver with a conversion gain of 150 V/A. Finally, the signal was captured by a 160 GSa/s digital sampling oscilloscope with a 63 GHz cut-off frequency. A pseudo-random bit sequence with a length of $2^{15}-1$ bits was used for the bit-to-PAM-4 symbol mapping. The generated PAM-4 symbols were pre-coded and up-sampled. Bandwidth and dispersion pre-compensation was then carried out. The resulting signal was resampled to 92 GSa/s, and its real and imaginary parts were loaded into the DAC. At the receiver side, the captured signal was first re-sampled to 180 GSa/s. Timing recovery was used to remove the timing offset and jitter. A Volterra filter was then used to remove most of the linear and non-linear distortions, followed by MLSE to remove the residual distortion. Finally, the bit error rate (BER) is determined, to evaluate system performance.

9.5.3.3.2 Experimental Results and Discussion

To compare the system performance of PAM-4 and DB-PAM-4, we modify the end to end transfer function to be flat (PAM-4) and cosine profile (DB-PAM-4) in frequency by bandwidth pre-compensation. The recovered eye-diagrams and performance are shows in Figures 9.47 and 9.48, respectively. One can see that the DB-PAM-4 signal shows much clearer eye-diagram and also better

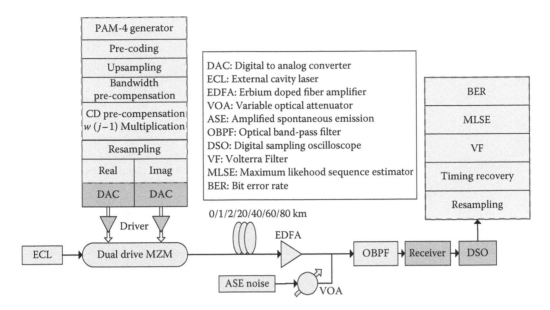

FIGURE 9.46 Experimental setup of DSB-DB-PAM-4 system.

performance, which can support 180 Gbit/s DB-PAM-4 signal transmission over 2 km fiber without dispersion pre-compensation. This shows that DB-PAM-4 is a better choice based on our setup to reach 180 Gbit/s signal generation and transmission.

If dispersion pre-compensation is implemented then the reachable distance can be extended to 80 km, as shown in Figure 9.49. One can note that there is an extra penalty after the signal transmitted through the fiber even though the dispersion pre-compensation is assisting. This is mainly caused by the fact that the DDMZM is a polar modulator and there are some residual distortions. The penalty increases with the fiber length.

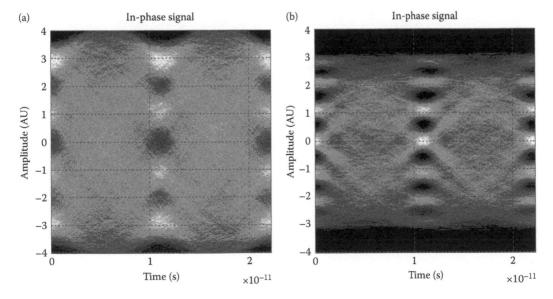

FIGURE 9.47 Recovered eye-diagrams of PAM-4 (a) and DB-PAM-4 (b) signals.

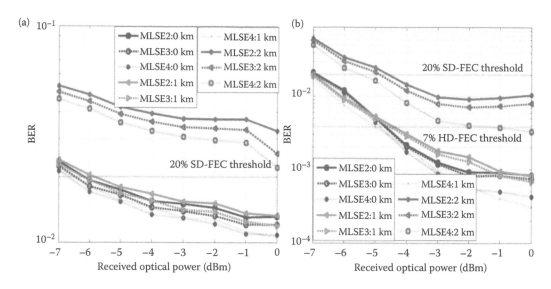

FIGURE 9.48   Performance of PAM-4 (a) and DB-PAM-4 (b) systems.

9.5.3.4   SSB-DB-PAM-4 System

9.5.3.4.1   Experimental Setup

To increase the SE (spectral efficiency) of the system, we can remove one sideband of the DB-PAM-4 signal, which, in theory, has the same SE with that of 16QAM. The setup of SSB-DB-PAM-4 system is shown in Figure 9.50. The hardware is the same as DSB-DB-PAM-4 system. The main difference is that Hilbert transform is carried out at transmitter side to generate the SSB-DB-PAM-4 signal in DSP. Figure 9.51 shows the optical spectrum of the signals with 100 and 50 GHz optical demux. With 50 GHz demux, we can achieve SE up to 3.8 bit/s/Hz, which is the highest reported SE in PAM-4 systems. Figures 9.52 and 9.53 show the system performance with FFE at receiver side and with FFE and MLSE, respectively. The insets are the recovered eye-diagrams. One can see that with the increase

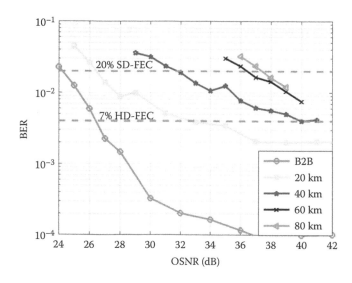

FIGURE 9.49   Performance of DB-PAM-4 with dispersion pre-compensation over different length of fibers [36].

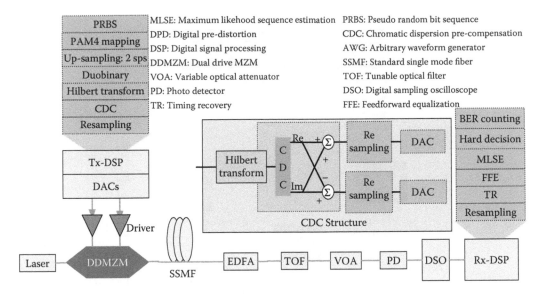

FIGURE 9.50 Experimental setup and offline DSP of SSB-DB-PAM-4 system.

of fiber length, the eye-diagrams become noisier because of the extra distortions. The system can rarely reach below the limit of SD-FEC. With the help of MLSE, the performance can be improved and the error floor can be reduced by one order. The reachable distance can reach up to 13 km. Comparing with DSB-DB-PAM-4 signal, the SSB-DB-PAM-4 signal has a slightly worse performance.

9.5.3.5 Remarks

This section describes progresses on high baud-rate DB-PAM-4 system. As high as 90 GBaud DB-PAM-4 signal can be generated by a 92 GSa/s DAC. DB-PAM-4 system shows better performance than conventional PAM-4 system. If dispersion pre-compensation is assisting with DDMZM, the

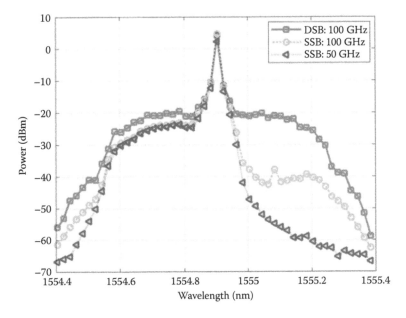

FIGURE 9.51 Optical spectrum of 90 GBaud DB-PAM-4 signals.

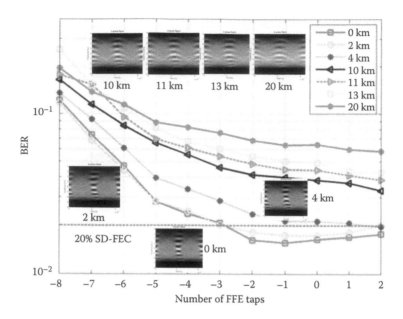

FIGURE 9.52   System performance with just FFE at Rx [37].

reachable distance is extendable to 80 km to cover the metro point to point scenario. SSB-DB-PAM-4 is also feasible to greatly increase the SE and capacity of the system.

9.6   MULTI-TERA-BITS/S OPTICAL ACCESS TRANSPORT TECHNOLOGY

Tremendous efforts have been developed for multi-Tbps over ultra-long distance and metro and access optical networks. With the exponential increase demand on data transmission, storage and

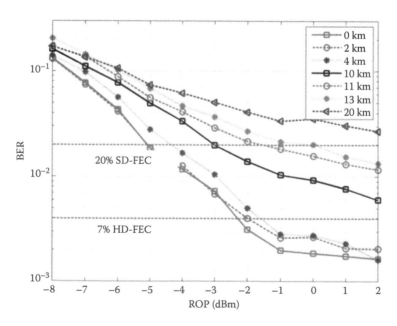

FIGURE 9.53   System performance with FFE and MLSE.

serving, especially the 5G wireless access scenarios, the optical Internet networking has evolved to data-center-based optical networks pressuring on novel and economical access transmission systems. This section describes the following: (1) experimental platforms and transmission techniques employing band-limited optical components operating at 10G for 100G based at 28G baud. Advanced modulation formats such as PAM4, DMT, duo-binary, etc. are reported and their advantages and disadvantages are analyzed so as to achieve multi-Tbps optical transmission systems for access inter- and intradata-centered-based networks; (2) integrated multi-Tbps combining comb laser sources and micro-ring modulators meeting the required performances for access systems are reported. Ten-sub-carrier quantum dot com lasers are employed in association with wideband optical intensity modulators to demonstrate the feasibility of such sources and integrated micro-ring modulators (see Chapter 3) acting as a combined function of demultiplexing/multiplexing and modulation, hence compactness and economy scale. Under the use of multi-level modulation and direct detection at 56 GBd an aggregate of higher than 2 Tbps and even 3 Tbps can be achieved by interleaved two comb lasers of 16 sub-carrier lines; and, finally (3) the fundamental designs of ultra-compact flexible filters and switching integrated components based on Si photonics for multi-Tera-bps active interconnection are presented. Experimental results on multi-channels transmissions and performances of optical switching matrices and effects on that of data channels are proposed.

9.6.1 INTRODUCTORY REMARKS

The emergence of data center (DC) centric networks have exerted tremendous pressure on traditional optical networks to be flattened and simplified, as well as increasing the network capacity and the contents to be delivered to end users. The capacity of the access networks would reach several Tera-b/s when 4K video for high definition and ultra-fast (faster than fast) delivery over fixed lines or wireless mobile 5G networks expected to be coming toward the end of this decade. The evolutionary network is schematically shown in Figure 9.54 in which several Peta-b/s capacities would be transported in the long-haul core and metropolitan core networks via several clouds at the edge, as

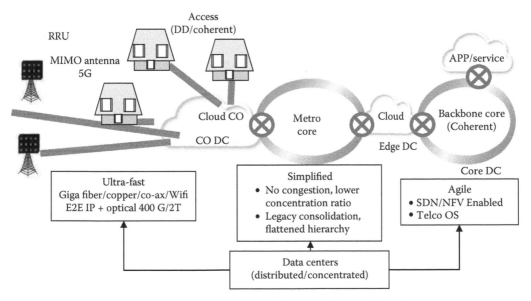

FIGURE 9.54 Evolutionary network modernization strategy. CO-Cloud = central office cloud, DC = data center. 100 Gbps access transmission systems.

well as at local distributed nodes. The exchanges of traditional telecom networks are expected to locate Access clouds in their transformed flatten networks.

It is thus expected that multi-Tbps would be required in the newly evolutionary networks to be deployed in the near future. The economy scale, as well as the flexibility of such access transport networks or systems, is a very important factor so that bandwidth on demand can be offered to end users. This paper thus addresses the following: First, the transmission and direct detection of four optical lanes of different and packed wavelength channels each carrying 25–28 Gbps depending whether forward error coding (FEC) is required. Furthermore, several of such 100G modules would result into accumulated capacity reaching Tbps (Tera-bps). Second, the 28 GBd can be employed to produce 56 Gbps via the modulation format PAM-4 produce 56 Gbps per channel are reported. LThe employment of comb sources and special modulators such as micro-ring types, to achieve multi-Tbps is given in both coherent and non-coherent reception. Lastly, we propose some integrated optical switching systems to simplify the routing of ultra-fast optical channels to the end user locations.

9.6.2 100G Access Systems and Networks

A typical access transmission system is shown above in Figure 9.55 [38], in which a low cost VCSEL with limited bandwidth of 18GHz is used under a baud (Bd) rate of 56 GBd data sequence is fed to directly modulate the VCSEL. Shown also in this configuration is the T-bias and the PAM4 data sequence obtained SHF-10001 attenuator 6 dB and a high speed combiner. The PAM4 eye diagram in the electrical domain is shown as the far left oscilloscope image in Figure 9.55, followed by the optical output eye in back-to-back (B2B), transmitted through a length of standard single mode fiber (SSMF), before feeding into the processor so as to evaluate the bit-error-rate (BER) after passing through the algorithm MLSE (minimum least square estimation). The performance of such system is shown in Figure 9.56. Under FEC of 20% and 7%, we can expect that such 56 Gbps PAM4 can be transmitted over few to some tens of kilometers of SSMF.

Alternatively, another low-cost solution for access networks can be achieved by using components developed for 10G and operating at 28G with DSP processing and pre-emphasis at the transmitter so as to produce systems of 100 Gbps via 4-optical wavelength transmitters employing EML (external modulation laser) in which four lasers are integrated with electro-absorption modulators. The transmission performance for such optical transmission is shown in Figure 9.56b and c. It is believed that the footprint of such transponders employing 56 Gbps (PAM4) or 4 × 28 Gbps (NRZ) are the least

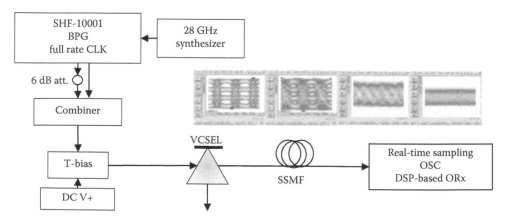

FIGURE 9.55 Typical access transmission system employing PAM4 direct detection and associate digital signal processing. Legend: SSMF = standard single mode fiber; VCSEL = vertical cavity stimulated emission laser; DSP = digital signal processor; att. = attenuator; CLK = clock.

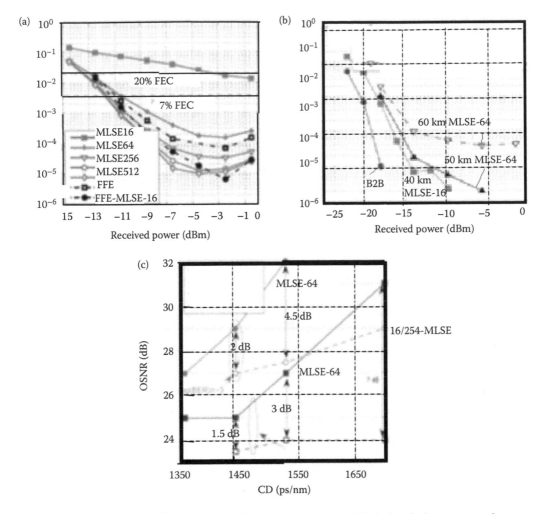

FIGURE 9.56 Performance of low-cost transmission system using band-limited optical components for metro-access (a) BER versus received optical power for PAM-4 28 GBd; (b) BER versus received optical power for NRZ 28G using 10G components; and (c) OSNR versus residual chromatic dispersion in SSMF for NRZ 28 Gbps. Legend: MLSE-16 = MLSE equalization with 16 states, similar for 64, 256, or 512 states.

costly to deploy into the access networks for DC centric networks or flattened traditional telecom networks.

9.6.3 Tbps Access Transmission and Optical Routing Technology

It is well known that the limitation of the electronic components is around 56 Gbps with bandwidth to around 35 GHz, especially when digital signals are generated such as in high speed DAC and ADC. Noise contribution in these signal generators must also be considered, as they grow as a cubic function of the frequency. These noises can then reduce the allowable optical signal-to-noise ratio (OSNR), and thus the amplitude levels when multi-level modulation format is employed. Multi-level level modulation is required to increase the bit rate or the number of bits per symbol so that 100G or 400G can be obtained. In our case, we report the 112 Gbps via 56 GBd PAM4 modulation format signals. Furthermore, in order to achieve Tbps, super-channels are required that mean that parallel channels are simultaneously generated and transmitting. We propose here the use of comb laser sources and equivalent modulators and multiplexers via the device known as the micro ring modulators (MRM).

9.6.3.1 Direct Detection

The MRM can act as an optical filter, as well as a modulation device. In comb generation, one can generate multiple sub-carriers from one original source in the frequency domain or in the time domain a very short pulse repetitive sequence followed by a periodic Fabry–Perot filter, as its free spectral range meets the required frequency spacing in a DWDM grid spectrum. In the case of flex grid networks, the periodically spaced grid can be occupied differently in multiple equally spaced wavelength grids. Figure 9.58 lays out the paths to generation of comb lines in these two domains. Furthermore, there are a number of techniques for generating comb lines in the frequency domain via different optical processing structures, including (i) frequency shifting recirculating optical loop with single sideband (SSB) shift of the guided lightwaves in the circulating cavity, (ii) phase modulation and carrier suppression return zero for shifting and then generating, (iii) modulating integrated in a laser cavity, and (iv) short pulse generation and then supercontinuum via crystal waveguide then filtering of determined frequency spacing by a Fabry–Perot filter and then an optical amplification to the desired level of the comb lines. Figure 9.59 shows the spectra of comb sources generated by frequency shift recirculating loop and combined phase and amplitude modulation in the non-linear region. Furthermore, the modulated spectrum of the comb lines of Figure 9.59a is given in (c). The five sub-carrier comb source if modulated to produce 400 Gbps per line then the aggregate bit rate would be 2.0 Tbps. We have demonstrated that by shifting left and shifting right the comb lines can cover the whole C-band with frequency shifting of 50 GHz which is one frequency shift applied by the frequency synthesizer to the optical modulator incorporated inside the recirculating loop as shown in Figure 9.57.

In access networks, the 1300 nm can be employed to avoid the dispersion effects of the SSMF. One such comb laser in the 1310 nm is shown in Figure 9.60a. The cob laser is fabricated using quantum dot structures in InGaAs with flat broadband mirror at the two end of the laser cavity [39]. A comb line can be filtered as shown in Figure 9.60b. Such a comb laser in the 550 nm region can be generated by an embedded modulator in a quantum dot laser cavity as reported in References 40 and 41. These comb lasers would offer the generation of modulated channels reaching a total capacity of 9.4 Tbps over the entire C-band with 100 Gbps per channel and 94 sub-carriers of 50 GHz spacing. This band has now been extended to 94 wavelength carriers as optical amplifiers have been available and optimized over the C-band.

Employing the 1310 nm comb laser produced by Innolume GmbH and modulating using an external modulator whose 3 dB bandwidth of 35 GHz with the modulation format PAM4, we obtain 112 Gbps per carrier, so the total aggregate capacity is 13×112 Gb/s or 1.45 Tbps can be produced by this comb laser. The 1310 nm ensure that the transmission is attenuation limited and not dispersion limited. Figure 9.61 shows the spectra of the modulated comb line under matching of the filter center

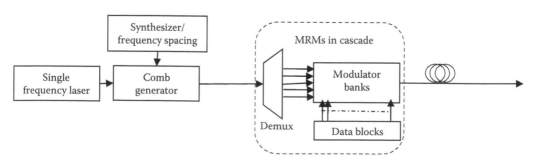

FIGURE 9.57 Schematic representation of multi-Tbps optical transmitter.

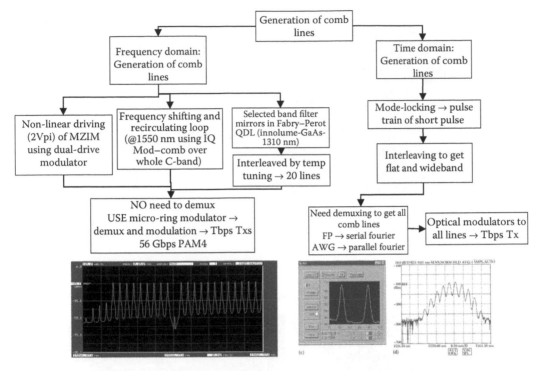

FIGURE 9.58   Methods of generation of comb sources in the frequency and time domain.

wavelength with that of the carrier comb line. The PAM4 eye diagram under B2B are shown in Figure 9.62. The BER is estimate at 1e-5. After 2 km transmission this is about 1e-3 under equalization using MLSE. The BER is similar for all 13 comb lines indicating that the chromatic dispersion is not critical for this Tbps transmission systems. Thus, they are most suitable, with the employment of FEC, for low cost Tbps access or data center environments.

9.6.3.2   Tbps Coherent Reception Systems

The coherent transmission technology using polarization multiplexing (PDM) and M-ary quadrature amplitude modulation (M-QAM) has been deployed throughout the long haul and metropolitan core networks. The flattening of the telecom networks under multi-Tbps have put tremendous pressure on network carriers to transform several fiber rings topologies into mesh topology in the metro-core

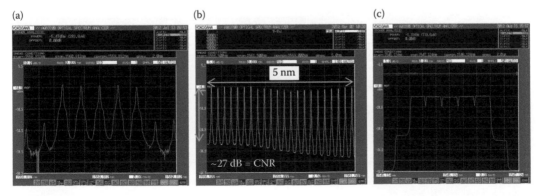

FIGURE 9.59   Comb spectrum generated by (a) frequency shift recirculating loop; (b) phase and amplitude modulation in non-linear overdrive region; and (c) modulated spectrum of comb laser of (a).

(a)   (b)

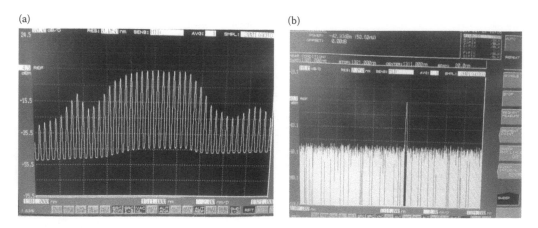

FIGURE 9.60 Spectrum of comb laser at 1310 nm region (a) all comb lines and (b) a filtered comb line.

(a)   (b)

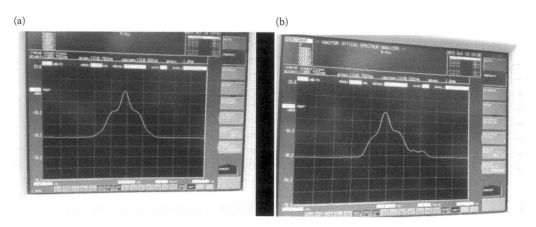

FIGURE 9.61 Modulated spectrum of PAM4 56 GBd (a) filtered at center of spectrum and (b) misaligned filtered modulated spectrum at PAM6 56 GBd.

(a)   (b)

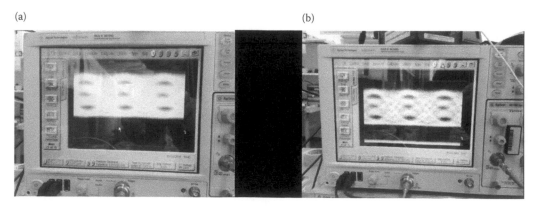

FIGURE 9.62 Eye diagram at output of transmitter PAM4 56 GBd of a filtered channel of the comb modulated channels (a) and (b).

networks and close to the access users as much as possible. Although it is more complicated in terms of additional optical hybrid coupler and local oscillators in the coherent reception sub-systems, it offers much better operating margins and higher capacity of at least doubling by the PDM. It is also more expensive, but this is expected to be lowered with the integrated Si photonic technology under current intense development and research [42–45]. A number of significant integrated photonics based on Si CMOS technology are now commercially available.

This section gives an overview of our demonstration of multi-Tbps transmission system based on coherent transmission and reception techniques with application to access or metro or even long haul networks A general multi-Tbps optical experimental platform is shown in Figure 9.63. The comb laser source is generated by using the frequency-shift circulating technique described above. The comb source is demultiplexed into individual comb lines that are fed into a pulse shape Nyquist transmitter through which the Nyquist sampling rate and roll off is enforced on the optical modulator so that the modulated spectrum can reach close to the sharp roll off. In this way, the optical channels can be packed close together to minimize the crosstalk, as well as efficient in the spectrum, hence higher capacity for all channels transmitted over the C-band. An arrangement of the PDM QAM transmitter is shown in Figure 9.64. The spectrum of super-channel 2 Tbps is shown in Figure 9.65. Alternatively, the micro-ring modulator in Si integrated photonic platform, can be used as the Tbps transmitter as shown in Figure 9.66. The transmission performance of such channel over 3500 km is shown in Figure 9.65 and its transmission performance over 1600 km of optically amplified SSMF spans in Figure 9.67c. Typical spectrum and eye diagram of the Nyquist channel are shown in Figure 9.67a and b. This length of transmission the Tbps over this length allow the shorter transmission with a number of reconfigured add/drop mux (ROADM) expected to be in metro network, especially add/drop into access networks. It is noted that the coherent reception sub-system consists of an optical hybrid coupler through which the mixing of a comb laser acting as the multiple line local oscillator beating

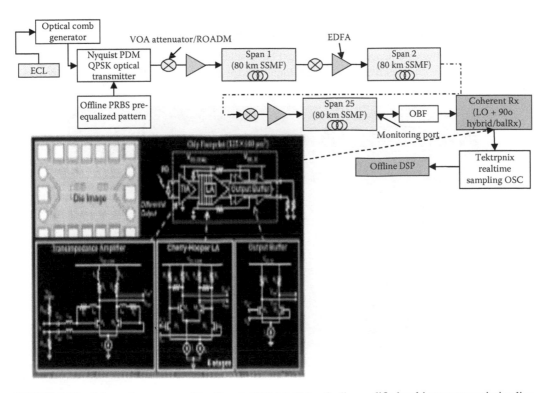

FIGURE 9.63 Schematic structure of the Nyquist PDM-QAM optically amplified multi-span transmission line. Insert is the advanced differential trans-impedance (TIA) optoelectronic receiver by CMOS 90 nm technology.

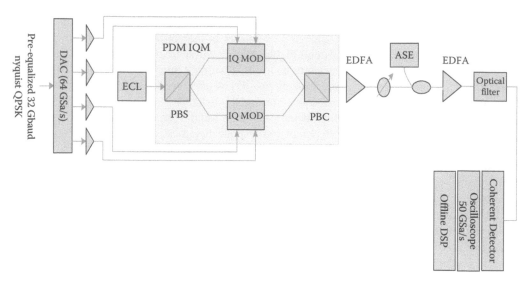

FIGURE 9.64 Experimental setup of 128 Gbps Nyquist PDM-QPSK transmitter and B2B performance evaluation.

with all received channels in the photodetector after the optical demux. The balanced photodetectors are connected to the differential TIA so as to obtain the highest output voltage level signals which are compatible for feeding into the ADC and then to the sampling oscilloscope and digitally processed. The BER is sufficient for further FEC to achieve error free. Such Tbps superchannels can be integrated to transmit over multi-core fibers to give a total aggregate capacity reaching a few Peta-bps.

9.6.4 Optical Interconnect for Multiple Tbps Access Networks

Optical add-drop mux for access networks must be available from cloud to end users in the multi-Gbps per user or even several Tbps to MIMO antenna for the future 5G wireless access networks.

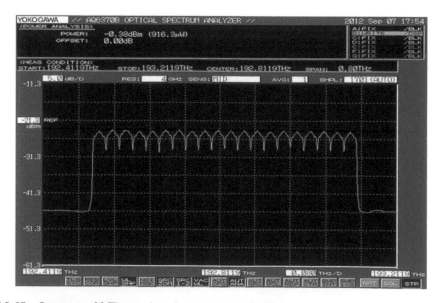

FIGURE 9.65 Spectrum of 2 Tbps optical channels using non-linear comb generator with two probe channels and 18 dummy channels.

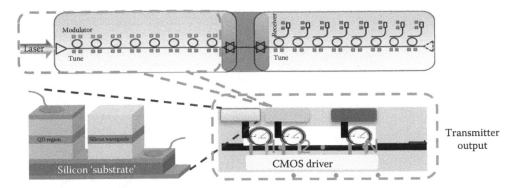

FIGURE 9.66 Tbps transmitter employing micro-ring modulator as simultaneous mux/demux and modulation device.

We assume here that the multi-Tbps access networks accept channels of at least 100 Gbps, to 400 Gbps and the super-channel 1 Tera, 2 T to 10 Tbps and 100 Tbps (see Chapter 2 on 112 GBd by optical injection locking direct modulation). Thus, the bandwidth of channels can be variable, hence the flexi grid WDM. This type of networks would be installed in the next coming modernization of optical networking. These networks are under deployment by carriers such as Telefonica, Deutsch Telecom, etc. We propose an all-optical interconnection for multi-Tbps capacity network between metro-core and metro access environment, as shown in Figure 9.68a. It consists of an optical kernel which composes of a matrix of cross waveguides and micro-ring modulators/switches which can not only demultiplexing or multiplexing and routing as well. Thus the lightwave paths can be routed and wave length channels can be selected. Such wavelength switching and routing in three directions is shown in Figure 9.68b. This structured can be implemented in Si integrated photonic circuits with a density switching matrix of order 100×100 without much problems on optical waveguide paths but the electrical connections and heating tuning section.

9.6.4.1 Remarks

This section has addressed the issues of access technologies for network capacity reaching several Tbps. Such networks will come in the year 2020 when 5G mobile network and associate clouds,

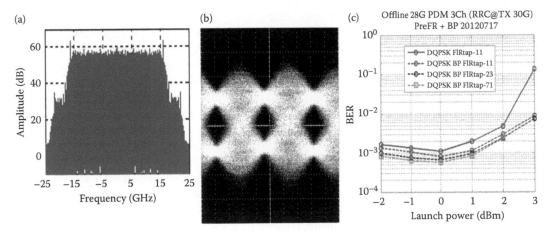

FIGURE 9.67 (a) Typical received spectrum; (b) Nyquist eye diagram; and (c) BER versus launched power (dBm) of 20 channels with central channel as probe channel over transmission distance of 1600 km non-DCF, 20 optically amplified 80 km SSMF spans with VOA inserted in front of EDFA. 28 GBaud PDM-Nyquist QPSK (a) single channel and (b) three channels.

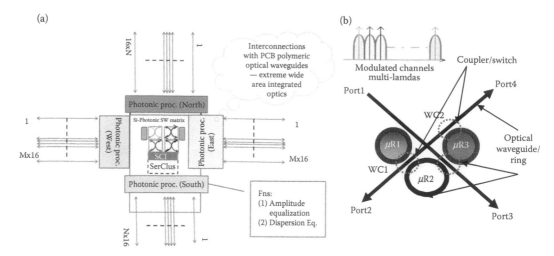

FIGURE 9.68 An optical interconnect consisting of an optical kernel and optical processor in north-south-east west with fibers carrying 25 T to 100 Tbps all in the C-band with 92 wavelength channels or super-channels occupying multiple number of 50G grid spacing.

for example, CRAN (Cloud Radio Access Networks) and 4K video are delivered throughout global Internet. The transmission technologies at 100G per channel and Tbps have been presented and proven that they are appropriate for such multi-Tbps transport networks. Furthermore, we have proposed an optical interconnect structure for add/drop from the metro core to access networks. This optical interconnect can accept and operate with channels employing coherent or non-coherent reception. Tbps transmission systems operating under coherent or direct detection are described using comb generators and multiple wavelength mux/demux and modulation. There is no doubt that such multi-Tbps access networks will be extensively deployed soon. DMT. CAP and multi-level modulation schemes can offer much advantages in the multi-Tbps access technology.

9.6.5 SUPER-CHANNEL COHERENT TECHNOLOGIES TO IMPROVE NETWORK EFFICIENCY

New high-speed digital signal processors (DSPs) and their integration with coherent optical receivers can significantly enhance optical performance while offering improved flexibility and programmability. These photonic engines power the latest generation of high-speed optical transponders, achieving record capacity, spectral efficiency, and optical reach. With these programmable, multi-modulation transponders, carriers can adjust the transponder capacity to the physical characteristics of an optical path. This enables them to maximize network capacity and efficiency on each route.

Previous generations of coherent transponders typically relied on a single line rate (100G) and a single modulation type (polarization-division multiplexed quadrature phase-shift keying, or PDM-QPSK) for all applications. The 100G coherent transponders represented significant technical advancements in high-speed optics and yielded large capacity improvements over 10G wavelengths.

However, these single-modulation transponders were a one-size-fits-all solution. While the 100G transponders are a good, general-purpose fit for most applications, they demanded compromises on shorter optical paths and very long routes. On shorter optical paths, for example, in metro or regional applications, each optical channel is capable of transporting 200–250G of capacity. Therefore, the use of 100G PDM-QPSK transponders on metro/regional routes underutilizes the true capabilities of the optical network. On ultra-long haul routes, 100G PDM-QPSK modulated wavelengths may not be the best fit because they may require costly regeneration nodes.

New generations of DSPs and transponders eliminates the need for such compromises by supporting programmable, multiple modulation modes. These advanced features enable carriers to optimize capacity, optical reach, and spectral efficiency for each optical route. Such capabilities, collectively known as Superchannel Coherent Technologies, are now ready for deployment in the field.

Superchannel Coherent Technologies offer high potentialities to vastly improve the flexibility and performance of optical transponders. As understanding given in Section 6, Super Coherent Technologies encompass four areas: multiple modulation formats, multiple baud rates, flexible spectrum, and advanced coding.

Multi-modulation balances capacity and reach. In optical networking, modulation is used to convert digital ones and zeros into symbols in the optical domain that are transmitted over fiber span or optically amplified span transmission line. In most cases, the modulation formats increase overall capacity by supporting the encoding and transmission of multiple bits per symbol.

Modern coherent optical networks rely on phase modulation, primarily PDM-QPSK, to encode and transmit 100G coherent wavelengths. QPSK modulation encodes two bits per symbol, combined with two different polarization modes (i.e., PDM), resulting in a total of four bits transmitted. QPSK modulation produces the familiar constellation diagram, as shown in Chapter 3 and in Sections 3.5 of this chapter.

Higher-complexity modulations such as 8-QAM and 16-QAM increase overall capacity by encoding even more bits per symbol provided optical power, hence OSNR (optical signal to noise ratio) higher to satisfying constellation point distance, for example, 4 dB for PAM4 in contrast to NRZ. With PDM-16QAM modulation, for example, it is possible to send 200G of traffic over the same optical channel used to transport 100G PDM-QPSK modulated signals. The higher capacity that comes with these more advanced modulation types also results in shorter optical reach due to the closer spacing of the constellation points. For many metro or regional applications, the optical reach is sufficient and the higher modulation formats provide double the capacity per wavelength.

The tradeoff between capacity and optical reach for different optical modulation types. Simpler modulation methods, such as QPSK, are well suited for long-haul and ultra-long-haul routes. Higher-complexity modulations are ideal for high-capacity routes at metro and access reach distances. The new super coherent DSPs support programmable, multiple modulation modes. The multiple modes enable carriers to optimize both the capacity and optical reach of the transponder for each optical path and application.

9.6.5.1 Baud Rate Flexibility

The capacity of an optical channel is based on its modulation type and baud rate. The modulation type defines how many bits are encoded per symbol, and the baud rate measures how fast the symbols are transmitted. Carriers can increase channel capacity by using more complex modulation formats, which encode more bits per symbol, by using faster baud rates or by combining both techniques. One limitation of more complex modulations is shorter optical reach, as described previously. An increased baud rate combined with the same modulation type enables longer reach and higher capacity. Baud rate flexibility is a powerful Super-channel Coherent Technology on next-generation DSPs. The maximum baud rate is limited by the speed of the electronics within the coherent DSP. For current-generation DSPs and 100G coherent transponders, the industry uses DSP baud rates of approximately 32 Gbaud. Total channel capacity is simply the baud rate (32 Gbaud) multiplied by the number of bits per symbol (2 bits/symbol) multiplied by the two polarization modes for PDM-QPSK modulation, resulting in a line rate of approximately 128 Gbps. The 128-Gbps rate is the actual line rate of a nominal 100G signal, including soft-decision forward error correction (SD-FEC) bits and Optical Transport Network (OTN) framing.

Newer superchannel coherent DSPs support higher baud rates, as well as the ability to program both the baud rate and modulation type to optimize the capacity and optical reach of each optical path. One key application higher baud rates enable is transport of 200G using 8QAM modulation. Carriers have traditionally used PDM-16QAM modulation to transport 200G, which limits its use to shorter distances. Higher baud rates enable 200G to be transported using PDM-8QAM modulation, so carriers get the benefit of both higher capacity and longer optical reach.

A network study by Nokia compared the percentage of optical routes suitable for 200G 16QAM and 200G 8QAM wavelengths, based on a European network model and a North American network model, as shown in Figure 9.69. The study found that 16QAM modulation posed significant limitations, especially on North American routes where only 40% of the routes could be supported. It also found that 200G 8QAM was suitable on 100% of the European optical routes and more than 92% of all North American optical routes. Higher baud rates and flexibility to program baud rates are key Super Coherent Technologies enabling more efficient, cost-effective optical networks.

Flexible spectrum groups wavelengths closer together as described in Section 9.6. Traditional WDM systems support approximately 90 channels using fixed 50-GHz spaced channels, following the ITU-defined WDM grid pattern. However, grouping these wavelengths slightly closer together can provide up to 120 usable channels which offer a 30% capacity enhancement. Carriers require a few new techniques to group these wavelengths closer together and maintain the same optical performance.

Newer-generation DSPs support a feature called Tx pulse shaping, as described in Section 9.4, sometimes referred to as Nyquist filtering. This feature shapes and compresses each wavelength into a slightly smaller bandwidth. With Tx pulse shaping, carriers can transport 100G coherent signals over a 37.5-GHz channel, as opposed to the 50-GHz channel on traditional WDM networks.

Carriers can combine flexible grid channel spacing and Tx pulse shaping techniques to create multi-carrier superchannels. A superchannel is a grouping of two or more carrier wavelengths.

A superchannel enables the inner wavelengths to be grouped closer together instead of transporting each wavelength in individual 50-GHz channels. The inner sub-carrier channels are less exposed to optical network transmission impairments, such as passband narrowing. The combination of Tx pulse shaping and grouping superchannel sub-carriers closer together provides improved spectral efficiency, in other words, more bits transmitted for every Hertz of spectrum. Carriers can use the spectrum savings to carry additional channels and capacity.

Advanced coding is the secret sauce. Super Coherent Technologies include advanced coding methods and techniques, many of which are considered highly proprietary by vendors. However, some of the common techniques being developed in the industry include stronger SD-FEC (soft decision forward error correction), coded modulation, and symbol encoding/decoding flexibility.

SD-FEC is a powerful error detection and correction algorithm implemented on current-generation DSPs. Each vendor has its own proprietary SD-FEC algorithm. In general, these algorithms provide about 11dB net coding gain. Some vendors are implementing stronger second-generation SD-FEC

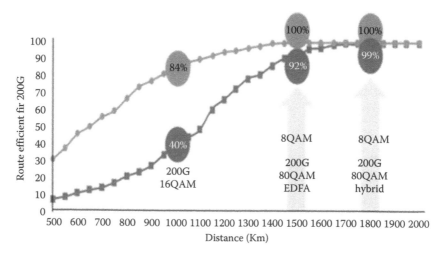

FIGURE 9.69   Applicability of 200G 8QAM on WDM networks. (After Nokia.)

algorithms, sometimes referred to as ultra-SD-FEC. These new algorithms provide higher net coding gains, resulting in longer optical reach compared to existing SD-FEC implementations.

SD-FEC is applied and processed at the frame layer, within the electrical domain inside coherent DSPs. It is also possible to apply additional error correction information to the optical layer, by way of the modulated symbols themselves. This technique is referred to as coded modulation. The use of coded modulation formats along with the newer ultra-SD-FEC (SD = software-decision FEC = forward error correction) algorithms offers substantial performance improvements. Finally, there are different techniques for encoding and decoding the modulated symbols. These techniques offer different performance levels based on each method. As described previously, coherent optical networks use modulation to encode digital ones and zeros into symbols transmitted on the fiber. The result is a phase modulated constellation pattern, like the sample 100G QPSK modulation shown in Section 9.6.

The encoding and decoding of these symbols occur in the transmitter, coherent receiver, and DSP, based on either differential encoding or absolute encoding of the phase. The two encoding methods determine how the coherent receiver tracks and decodes the modulated constellation phases and converts the symbols back into ones and zeros. Each encoding method, differential or absolute, offers advantages depending on the application.

Superchannel coherent DSPs implement robust symbol decoding including advanced polarization tracking, phase tracking, and cycle-slip mitigation algorithms to prevent fiber impairments and network disruptions from causing burst errors in the transmission. Naturally occurring phenomena, such the lightning strike-induced state-of-polarization changes described in aerial fibers in access networks, the lightning and wind generated cable sagging stress influence on coherent optical transmission in aerial fiber, can be largely resolved by advanced polarization and phase-tracking DSP algorithms. Ideally, super coherent DSPs will offer differential and absolute encoding options, along with a broad range of multi-modulation and baud rate options.

The latest generation of DSPs includes several new and powerful enhancements, collectively referred to as Superchannel Coherent Technologies. This collection of technologies includes multiple modulation modes, multi-baud rate, Tx pulse shaping, superchannels, and advanced coding techniques. When combined, Super Coherent Technologies offer a vast improvement in capacity and optical reach compared to current-generation transponders. They also offer the flexibility to program and tune optical transponder capacity and optical reach to each specific optical path. These new super coherent DSPs, and their corresponding optical transponders, improve optical performance, increase network capacity, enhance spectral efficiency, and lower the overall transport cost per bit for optical networks.

REFERENCES

1. J. L. Wei et al., Performance and power dissipation comparisons between 28 Gb/s NRZ, PAM, CAP and optical OFDM systems for data communication applications, *J. Lightwave Tech.*, vol. 30, no. 20, pp. 3273–3280, Oct. 2012.
2. A. Ghiasi and B Welch, IEEE 802.3bm Fiber Optic Task Force, Sept. 2012.
3. J. D. Ingham et al., 40 Gb/s carrierless amplitude and phase modulation for low-cost optical data communication *links*, *Proc. OFC/NFOEC*, Los Angeles, USA, paper OThZ3, Mar. 2012.
4. L. Tao et al., Experimental demonstration of 10 Gb/s multi-level carrierless amplitude and phase modulation for short range optical communication systems, *Optics Express*, vol. 21, no. 5, pp. 6459–6465, Mar. 2013.
5. R. Rodes et al., Carrierless amplitude phase modulation of VCSEL with 4 bit/s/Hz spectral efficiency for use in WDM-PON, *Optics Express*, vol. 19, no. 27, pp. 26551–26556, Dec. 2012.
6. T. Gui et al., Auto bias control technique for optical OFDM transmitter with bias dithering, *Optics Express*, vol. 21, no. 5, pp. 5833–5841, 2013.
7. Y. Bao et al., Nonlinearity mitigation for high-speed optical OFDM transmitters using digital pre-distortion, *Optics Express*, vol. 21, no. 6, pp. 7354–7361, 2013.

8. Haglund, E., P. Westbergh, J. S. Gustavsson, E. P. Haglund, A. Larsson, M. Geen, and A. Joel, 30 GHz bandwidth 850 nm VCSEL with sub-100 fJ/bit energy dissipation at 25–50 Gbit/s, *Electron. Lett.* vol. 51, no. 14, pp. 1096–1098, 2015.

9. Westbergh, P. E. P. Haglund, E. Haglund, R. Safaisini, J. S. Gustavsson, and A. Larsson, High-speed 850 nm VCSELs operating error free up to 57 Gbit/s, *Electron. Lett.* vol. 49, no. 16, pp. 1021–1023, 2013.

10. Kuchta, D. M., Rylyakov, A. V., Doany, F. E., Schow, C. L., Proesel, J. E., Baks, C. W., Westbergh, P., Gustavsson, J. S., and Larsson, A., A 71 Gb/s NRZ modulated 850 nm VCSEL-based optical link, *IEEE Photon. Technol. Lett.* vol. 27, no. 6, pp. 577–580, 2015.

11. A. Larsson, J.S. Gustavsson, P. Westbergh, E. Haglund, E.P. Haglund, E. Simpanen, T. Lengyel, K. Szczerba, and M. Karlsson, VCSEL design and integration for high-capacity optical interconnects, *Proc. SPIE OPTO Photonic West 2017*, vol. 10109, San Francisco, 2017. doi: 10.1117/12.2249319.

12. J. Tatum et al., High speed characteristics of VCSELs, *Fabrication, Testing, and Reliability of Semiconductor Lasers II, Proceedings of the SPIE*, vol. 3004, pp. 151–159, 1997.

13. Motaghian, R. and C. Kocot, 104 Gbps PAM-4 Transmission over OM3 and OM4 Fibers using 850 and 880 nm VCSELs, in [Proc. CLEO], paper SW4F.8, 2016.

14. K. Sczerba, T. Lengyel, Z. He, J. Chen, P. A. Andrekson, M. Karlsson, H. Zirath, and A. Larson, High speed optical interconnects with 850 nm VCSELs and advanced modulation formats, *Proc. SPIE*, vol. 10122, pp. 10–220G1, 2017.

15. F. Karinou, N. Stojanovic, and C. Prodaniuc, 56 Gb/s 20-km transmission of PAM-4 signal employing an EML in C-band without in-line Chromatic Dispersion Compensation, *Proc. ECOC2016, 42nd European Conference and Exhibition on Optical Communications*, September 18–22, pp. 860–862, 2016, Dusseldorf, Germany.

16. F. Karinou, N. Stojanovic, C. Prodaniuc, Z. Qiang, and T. Dippon, 112 Gb/s PAM-4 optical signal transmission over 100-m OM4 multimode fiber for high-capacity data-center interconnects, *Proc. ECOC2016, 42nd European Conference and Exhibition on Optical Communications*, September 18–22, pp. 860–862, 2016, Dusseldorf, Germany.

17. Keysight Technologies, PAM-4 Design Challenges and the Implications on Test: Application note, 2014.

18. Keysight Technologies, Flex DCA software package, 2014.

19. IEEE P802.3bj 100 Gb/s Backplane and Copper Cable Task Force, http://www.ieee802.org/3/bj/.

20. L. Tao, J. Yu, and J. Zhang, Reduction of intercarrier interference based on window shaping in OFDM RoF systems, *IEEE Photon. Tech. Lett.*, vol. 25, no. 9, pp. 851–54, May 2013.

21. H. Schmeckebier, G. Fiol, C. Meuer, D. Arsenijević, and D. Bimberg, Complete pulse characterization of quantum-dot mode-locked lasers suitable for optical communication up to 160 Gbit/s, *Optics Express*, vol. 18, no.4, pp. 3415–3425, 2010.

22. Yole Inc, *A view on the Silicon Photonics market*, report 2014.

23. C. Doerr et al., Monolithic Rx including MZ DI and PIN: Signal analysis, OFC post-deadline paper, PDP23, 2008.

24. A. Gnauck and P. Winzer, Optical phase-shift-keyed transmission, *J. Lightw. Technol.*, vol. 23, no. 1, pp. 115–130, Jan. 2005.

25. W. Y. Chen, G. H. Im, and J. J. Werner, Design of digital carrierless AM/PM transceivers, AT&T and Bellcore contribution to ANSI T1E1.4/92-149, 1992.

26. M. Wieckowski, J. B. Jensen, I. Tafur Monroy, J. Siuzdak, and J. P. Turkiewicz, 300 Mbps transmission with 4.6 bit/s/Hz spectral efficiency over 50 m PMMA POF link using RC-LED and multi-level carrierless amplitude phase modulation, *Proc. Optical Fiber Conference and National Fiber OPtic Engineer Conference (OFC/NFOE'2011)*, OSA Technical Digest (CD) (Optical Society of America 2011) paper NTuB8.

27. A. Shalash and K. K. Parhi, Multidimensional carrierless AM/PM systems for digital subscriber loops, *IEEE Trans. Com.* vol. 47, no. 11, pp. 1655–1667, 1999

28. J. D. Ingham, R. Penty, I. White, and D. Cunningham, 40 Gb/s carrierless amplitude and phase modulation for low-cost optical datacommunication links, *Optical Fiber Communication Conference and Exposition (OFC/NFOEC)*, Los Angeles, CA, USA, paper OThZ3, 2011.

29. A. F. Shalash and K. K. Parhi, Multidimensional carrierless AM/PM systems for digital subscriber loops, *IEEE Transactions on Communications*, vol. 47, no. 11, p. 1655, November 1999.

30. J. Proaki and M. Salehi, *Digital Communications*, 5th edition (Irwin Electronics & Computer Engineering) 5th Edition, NY: McGrawHill, 2016.

31. K. Zhong, X. Zhou, T. Gui, L. Tao, Y. Gao, W. Chen, J. Man, L. Zeng, A. Pak T. Lau, and C. Lu, Experimental study of PAM-4, CAP-16, and DMT for 100 Gb/s short reach optical transmission systems, *Optics Express* vol. 23, no. 2, p. 1178, 26 Jan 2015.

32. X. Yin et al., First demonstration of real-time 100 Gbit/s 3-level DuoB transmission for optical interconnects, *Proc. Optical Fiber Conference OFC'2017*, PDP, Anaheim, 2017.

33. K. Voigt, , L. Zimmermann, G. Winzer, and C. Schubert, Performance of 40-Gb/s DPSK demodulator in SOI-technology, *IEEE PTL* vol. 20, no. (8), 614–616, 2008.

34. B.T. Smith, D. Feng, H. Lei, D. Zheng, J. Fong, and M. Asghari, Fundamentals of Silicon Photonic Devices, http://www.mellanox.com/related-docs/whitepapers/KOTURA_Fundamentals_of_Silicon_Photonic_Devices.pdf.

35. L. Tao, Y. Wang, Y. Gao, A. P. T. Lau, N. Chi, and C. Lu, Experimental demonstration of 10 Gb/s multi-level carrier-less amplitude and phase modulation for short range optical communication systems, *Optics Express* vol. 21, no. 5, 6459–6465, 2013.

36. F. Y. Gardes, D.J. Thomson, N.G. Emerson, and G.T. Reed, 40 Gb/s silicon photonics modulator for TE and TM polarizations, *Optics Express*, vol. 19, no. 12, pp. 11804–11814, June 2011.

37. Y.A. Vlasov, Silicon Integrated Nanophotonics: Road from Scientific Explorations to Practical Applications, CLEO'2012, Plenary paper, 2012.

38. S. Shahramian, J. Lee, J. Weiner, R. Aroca, Y. Baeyens, N. Kaneda, and Y-K. Chen, A 112Gb/s 4-PAM Transceiver Chipset in 0.18 μm SiGe BiCMOS Technology for Optical Communication Systems, in *IEEE Compound Semiconductor Integrated Circuit Symposium (CSICS)*, IEEE, New Orleans, LA, USA, October. 2015. http://ieeexplore.ieee.org/lpdocs/epic03/wrapper.htm?arnumber=7314464.

39. Q. Zhang, N. Stojanovic, C. Xie, C. Prodaniuc, and P. Laskowski, Transmission of single lane 128 Gbit/s PAM-4 signals over an 80 km SSMF link, enabled by DDMZM aided dispersion pre-compensation, *Optics Express*, vol. 24, no. 21, p. 24580, 2016.

40. Q. Zhang, N. Stojanovic, C. Prodaniuc, C. Xie, M. Koenigsmann, and P. Laskowski, Single-lane 180 Gbit/s PAM-4 signal transmission over 2 km SSMF for short-reach applications, *Opt.Lett.* vol. 41, no. 19, pp. 4449–4452, 2016.

41. Q. Zhang, N. Stojanovic, J. Wei, and C. Xie, Single-lane 180 Gbit/s DB-PAM-4-signal transmission over an 80 km DCF-free SSMF link, *Opt. Lett.* vol. 42, no. 4, pp. 883–886, 2017.

42. Q. Zhang, N. Stojanovic, T. Zuo, L. Zhang, C. Prodaniuc, F. Karinou, C. Xie, and E. Zhou, Single-lane 180 Gb/s SSB-duobinary-PAM-4 signal transmission over 13 km SSMF, *OFC* 2017, Tu2D.2.

43. F. Karinou et al., Directly PAM4 modulated 1530 nm VCSEL enabling 56 Gb/s data center interconnects, *IEEE Photonic Tech Lett.*, vol. 27, pp. 1872–1874, 2014.

44. Innolume GmbH, Dortmund Germany @ http://innolume.de/_pdfs/Comb/LD-1310-COMB-8.pdf.

45. D. Huhse, M. Schell, D. Bimberg, and I. S. Tarasov, Generation of electrically wavelength tunable (Delta lambda = 40 nm) singlemode laser pulses from a 1.3 μm Fabry–Perot laser by self-seeding in a fibre-optic configuration, *Electronics Letters*, vol. 30, no. 2, 2014.

10 Pulse Shaping Modulation and Superchannels

This chapter presents the analytic technique and an experimental platform by pulse shaping modulation in order to conduct Tera-bps optical transmission whereby the generation of multi-carrier lightwaves and modulation techniques incorporating pulse shaping of data sequences. The optical channels are polarized division multiplexed and follow the complex signal formats QPSK (PDM-QPSK). The pulse shaping leads to a closely spaced channel spectrum so as to achieve the most effective spectral density. Total bit rates of 1.12 and 2.24 Tbps are demonstrated and transmission of superchannels over 3500 km of optically amplified SSMF multi-spans with BER of 2×10^{-3} (hard FEC qualified) for 28 GBd Nyquist QPSK and 2×10^{-2} (soft FEC qualified) at 32 GBd. Further 20% FEC can be employed to achieve an extension of transmission distance of up to 2500 km.

The pulse shaping modulation Tbps superchanel platform consists of the following principal subsystems: (i) a digital signal processing (DSP)-based digital-to-analog converter-polarization division multiplexing (DAC-PDM) Nyquist pulse shaped modulation QPSK optical transmitter incorporating a comb-generator as the multi-carrier light source; (ii) comb-generators using recirculating frequency shifting techniques and non-linear driving of optical modulator to achieve N × 5 sub-carriers and thence by modulation achieving the superchannel with a total capacity of 1 Tbps or 2 Tbps; and (iii) DSP algorithms are developed for DAC generation of Nyquist optical channel as well as or off line processing and forward error correction (FEC) coding and decoding to further improve the sensitivity of the DSP-based optical coherent receiver. A comparison of Tbps transmission system employing two different techniques of comb-generation, thence Nyquist sub-channels, is also given.

10.1 INTRODUCTION

10.1.1 Overview

PDM-QPSK has been exploited as the 100 Gbps long-haul transmission commercial systems, the optimum technologies for 400 GE or Tera-E* transmission for next-generation optical networking have now attracted significant interest for deploying ultra-high-capacity information over the global internet backbone networks. In this chapter, the development of the hardware platform of 1 Tbps transmission systems is critical for proving the design concept and conducting the field trial on telecom carriers' client networks, such as those of Deutsch Telecom, Vodafone or Telefonica. In Chapter 9, we have presented high-level design and a number of generic options for delivery of Tbps over optically amplified multi-span link. Nyquist QPSK has been elected as the most effective format for delivery of high spectral efficiency and is effective in transmission and equalization at the transmitter and receiver.

High speed and sampling rate DAC are available and enable several digital processing techniques at the transmitter to deal with impairments created by the transmission media, such as the digital pre-equalization or pre-emphasis to pre-distort the transmitter pulse sequence. Furthermore, the pulse shaping, such as Nyquist or raise-cosine shaping functions, can be used to compact the pulse spectrum to improve the eye opening and packing adjacent channels.

---

* GE = Giga-bit Ethernet; Tera-E = Tera-bit Ethernet.

Thus, in this chapter, we describe in detail the design and an experimental platform for delivery of Tbps transmission system employing Nyquist-QPSK at symbol rate of 28–32 GSa/s and 10 sub-carriers. The generation of sub-carriers has been demonstrated using either recirculating frequency shifting (RFS) or non-linear driving of an IQ modulator to create five sub-carriers per main carrier, thus two main carriers are required. Techniques for evaluation of the performance of the Tbps transmission system are described in detail.

Nyquist pulse shaping modulation is used to effectively pack the spectral distribution of the multiplexed channels. A DAC with sampling rate varied from 56 GS/s to 64 GS/s is used for generating Nyquist pulse shape. Equalization of the transfer functions of the DAC and optical modulators is also coded.

10.1.2 CHAPTER ORGANIZATION

This chapter is organized as follows: Section 10.2 gives an overview of the system requirements, the techniques for pulse shaping and modulation, and thence, a generic architecture of the Tbps transmission systems is given. Section 10.3 gives further details of the multi-channel arrangement so that a Tbps superchannel can be achieved, with special focus given to the arrangement of comb-generation of sub-carriers and methods for modulation of these carriers and pulse shaping so that spectral efficiency can be achieved. Section 10.4, then, gives details of key hardware sub-systems of the Tbps transmission systems, including the three principal blocks, the comb-generators and DAC-based pulse shaping modulation sub-system, as well as the coherent receiving and extraction of the probe channels for evaluation of the transmission performance. Technical details of the pattern generation from DAC are also given. Section 10.5 summarizes the transmission performance of the Tbps transmission systems, especially the 1 and 2 Tbps using RCFS and non-linear generation techniques. Finally, Section 10.6 gives some conclusions. The fiber technical parameters of G.652 are given in Chapter 14 or Appendix 10A.1 (considerable as a detailed appendix or annex). Further technical details on optical modulators are given in Chapter 12. Appendix 10B.1 also gives more details of ADC.

10.2 SYSTEM REQUIREMENTS, SPECIFICATIONS, AND PULSE SHAPING MODULATION

10.2.1 REQUIREMENTS AND SPECIFICATIONS

Transmission distance: As the next generation of backbone transport, the transmission distance should be comparable to the previous generation, namely the 100 Gbps transmission system. As the most important requirement, we require that the 1 Tbps transmission for long haul should be 1500–2000 km, for metro application ~300 km.

CD (chromatic dispersion) tolerance: As standard single mode fiber (SSMF), fiber CD factor/coefficient 16.8 ps/nm is the largest among the current deployed fibers, CD tolerance should be up to 30,000 ps/nm at the central channel whose wavelength is approximated at 1550 nm. At the edge of C-band, this factor is expected to increase by about 0.092 ps/(nm^2 km) or about 32,760 ps/nm at 1560 nm and 26,400 ps/nm at 1530 nm.*

Polarization mode dispersion (PMD) tolerance: The worst case of deployed fiber with 2000 km would have a differential group delay (DGD) of 75 ps. The PMD (meaning all-order DGD) tolerance is 25 ps. State of polarization (SOP) rotation speed: According to the 100 Gbps experiments, SOP rotation can be up to 10 KHz; we take the same spec as 100G system.

Modulation format: PDM-QPSK for long-haul transmission; PDM-16QAM for metro application.

Spectral efficiency: Compared to 100G system with an increasing factor of two. Nyquist-WDM and CO-OFDM can fulfill this. However, it depends on technological and economical

___
* See Appendix for technical specification of Corning fiber G.652 SSMF.

requirements that would determine the suitability of the technology for optical network deployment.

10.2.2 Comparison of Nyquist_N_Tbps Transmission System

Table 10.1 gives a comparison of two different Tera-bps superchannels under different comb-generation techniques incorporating pulse shaping with variable roll-off coefficients and requirements on the digital processing systems as well as possible applications. Various parameters of the transmission platform are also specified.

10.2.3 Superchannels by Pulse Shaping Modulation for Tera-bps Transmission

Single channel 1 Tbps transmission is impossible to realize using only one carrier or sub-carrier and modulation. If we keep the non-FEC baudrate close to 25 GBaud, we need a PDM-4096QAM modulation scheme to achieve 1 Tbps per single lightwave carrier. This is impossible under the constraint of maximum laser amplitude power available to date. Furthermore, as the OSNR requirement increases exponentially to the modulation format level, it is impossible to reach the transmission distance with limited laser power and the threshold of the non-linear level of the transmission fiber. Even PDM-16QAM needs, theoretically, 7 dB more OSNR than PDM-QPSK to reach the same distance. To solve this problem, one can either increase the baudrate or employ more sub-carriers, that is, a superchannel with multiple sub-carriers. Increasing the baudrate leads to higher demand on the bandwidth for O/E components, and is thus challenging to technology advances and cost. Employing a high spectral efficient superchannel with sub-channel bandwidth of 50 GHz or below seems to be the most favorable choice.

Considering overhead of OTN framer and FEC of either 7% or 20% overhead, the total bit-rate can thus be 28 G × 4 × 10 to obtain 1.12 Tbps or 32 G × 4 × 10 for 1.28 Tbps (Figure 10.1).

For a multiple sub-carrier transmission, we would need multi-carrier modulation techniques and corresponding multi-carrier demodulation and coherent detection. Currently, there are a few methods and apparati to generate multi-carriers. The most commonly used is using a single ECL laser source and a recirculating shift loop and an MZM modulated by a RF signal. The frequency of the RF signal is the spacing of the sub-channels. Tx-DSP and DAC are needed for shaping the signal pulse, for example, Nyquist pulse shaping, and pulse shaping for compensating non-ideal O/E components transfer function and CD pre-compensation, etc.

An interleaver can be used to separate the sub-carriers for individual sub-channel modulation. At the output, all the signals of sub-channels are multiplexed together to form a superchannel 1.12 Tbps optical signal (Figure 10.2)

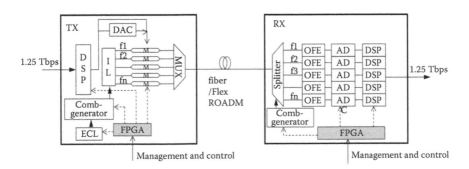

FIGURE 10.1 Architecture of a superchannel 1 Tbps transmission platform using DSP-based coherent reception systems.

TABLE 10.1

1 Tbps Off Line System Specification

Parameter Technique	Channel RCFS Comb-Gen	Channel Non-linear Comb-Gen	Some Specs	Remarks
Bit rate	1,2...N Tbps (whole C-band)	1 Tbps, 2, ... N Tbps	~1.28 Tbps @28–32 GB	20% OH for OTN, FEC
Number of ECLs	1	$N \times 2$		
Nyquist roll-off α	0.1 or less	0.1 or less		DAC pre-equalization required
Baud rate (GBauds)	28–32	28–32	28, 30 or 31.5 GBaud	Pending on FEC coding allowance
Transmission distance	2500	2500	1200(16 span)~2000 km (25 spans) 2500 km (30 spans)	20%FEC req. for Long haul application
			500 km	Metro application
Modulation format	QPSK/16QAM	QPSK/16QAM	Multi-carrier Nyquist WDM PDM-DQPSK/QAM	For long haul
			Multi-carrier Nyquist WDM PDM-16QAM	For metro
Channel spacing			4×50 GHz	For long haul
			2×50 GHz	For long haul
Launch power	≪0 dBm if 20 Tbps used		~ −3 to 1 dBm lower if $N > 2$	Depending on QPSK/16QAM and Long haul /metro can be different
B2B ROSNR @ 2e − 2 (BOL) (dB)	14.5	14.5	15 dB for DQPSK 22 dB for 16QAM	1 dB hardware penalty 1 dB narrow filtering penalty
Fiber type	SSMF G.652 (or G.655)	SSMF G.652 (or 655)	G.652 SSMF	
Span loss	22	22	22 dB (80 km)	
Amplifier	EDFA (G > 22 dB); NF < 5 dB		EDFA (OAU or OBU)	
BER	2e − 3	2e − 3	Pre-FEC 2e − 2 (20%) or 1e − 3 classic FEC (7%)	
CD penalty (dB)			0 dB @ +/−3000 ps/nm <0.3 dB @ +/−30,000 ps/nm;	16.8 ps/nm/km and 0.092 ps/(nm^2.km)
PMD penalty (DGD)			0.5 dB @75 ps, 2.5 symbol periods	
SOP rotation speed	10 kHz	10 kHz	10 kHz	
Filters cascaded penalty			<1 dB @12 pcs WSS	
Driver linearity	Required	Required	THD <3%	16QAM even more strict

The principal question is how can we create superchannels under laboratory conditions to achieve the proof of concept of the Tbps transmission? That is, how do we generate superchannels, transmission and detection/demodulation at the receiving end? Thus, one of the main principles of this chapter is to describe techniques employed to generate superchannels and probe channel to evaluate the performance of the transmission systems for Tbps superchannel.

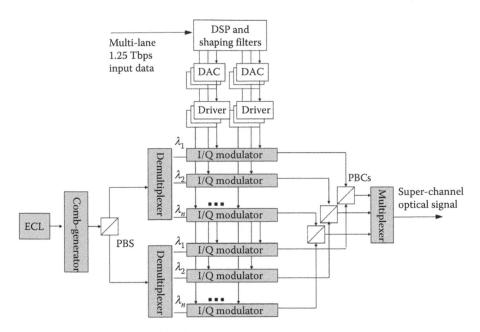

FIGURE 10.2   General architecture of a superchannel transmitter for coherent reception systems.

10.3   MULTI-CARRIER NYQUIST PULSE SHAPING MODULATION UNDER TRANSMISSION SYSTEM

In Nyquist WDM, the goal is to place different optical sub-channels as close as possible to their baud-rate, the Nyquist rate, by minimizing channel crosstalk through the formation of approximately close to rectangular spectrum shape of each channel by either optical filtering or electrical filtering and equalization.

10.3.1   NYQUIST SIGNAL GENERATION USING DAC BY EQUALIZATION IN FREQUENCY DOMAIN

The rectangular spectrum has a $\sin c$, i.e., $\sin x/x$ time domain impulse response. At the sampling instants $t = kT$ ($k = (1,2 \ldots N)$ is a non-zero integer), its amplitudes reach zero. That implies that at the ideal sampling instants, the ISI from neighbouring symbols is negligible, or free of intersymbol interference (ISI). Figure 10.3 depicts such a Nyquist pulse and its spectrum for a single channel or multiple channels. Note that the maximum of the next pulse raise is the minimum of the previous impulse of the consecutive Nyquist channel.

Considering one sub-channel carrier 25 GBaud PDM-DQPSK signal, then the resulting capacity is 100 Gbps for a sub-channel, hence the ability to reach 1 Tbps 10 sub-channels would be required. To increase the spectral efficiency, the bandwidth of these 10 sub-channels must be packed densely together. The most likely technique for packing the channel as close as possible in the frequency with minimum ISI is the Nyquist pulse shaping, which is described later in this section. Thus, the name Nyquist-WDM system is coined. However, in practice, a "brick wall" like spectrum shown in Figure 10.3, is impossible to obtain and is not an ideal solution because non-ISI pulse shape should be found, and the raise cosine pulse with some roll off property condition though to be met.

The raised-cosine filter is an implementation of a low-pass Nyquist filter, i.e., one that has the property of vestigial symmetry. This means that its spectrum exhibits odd symmetry about $1/2T_s$,

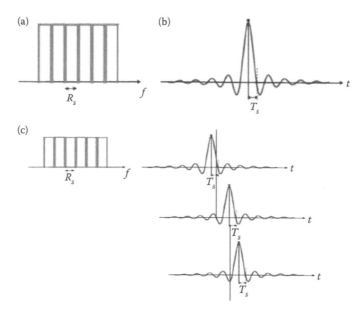

FIGURE 10.3 A superchannel Nyquist spectrum and its corresponding "impulse" response (a) spectrum, (b) impulse response in time domain of a single channel, and (c) sequence of pulse to obtain consecutive rectangular spectra. A superposition of these pulse sequences would form a rectangular "brick wall like" spectrum.

where T_s is the symbol-period. Its frequency-domain representation is 'brick-wall-like' function, given by

$$H(f) = \begin{cases} T_s & |f| \leq \dfrac{1-\beta}{2T_s} \\ \dfrac{T_s}{2}\left[1 + \cos\left(\dfrac{\pi T_s}{\beta}\left\{|f| - \dfrac{1-\beta}{2T_s}\right\}\right)\right] & \dfrac{1-\beta}{2T_s} < |f| \leq \dfrac{1+\beta}{2T_s} \\ 0 & \text{otherwise} \end{cases} \tag{10.1}$$

with $0 \leq \beta \leq 1$

This frequency response is characterized by two values: β, the roll-off factor and T_s, the reciprocal of the symbol-rate in Sym/s, that is $1/2T_s$ is the half bandwidth of the filter. The impulse response of such a filter can be obtained by analytically taking the inverse Fourier transformation of Equation 10.1, in terms of the normalized sin c function, as

$$h(t) = \sin c\left(\frac{t}{T_s}\right)\frac{\cos\left(\pi\beta t/T_s\right)}{1 - (2(\pi\beta t/T_s))^2} \tag{10.2}$$

where the roll-off factor, β, is a measure of the excess bandwidth of the filter, i.e., the bandwidth occupied beyond the Nyquist bandwidth as from the amplitude at $1/2T$. Figure 10.4 depicts the frequency spectra of raise cosine pulse with various roll-off factors. Their corresponding time domain pulse shapes are given in Figure 10.4b.

When used to filter a symbol stream, a Nyquist filter has the property of eliminating ISI, as its impulse response is zero at all nT (where n is an integer), except when $n = 0$. Therefore, if the transmitted waveform is correctly sampled at the receiver, the original symbol values can be recovered completely. However, in many practical communications systems, a matched filter is used at the

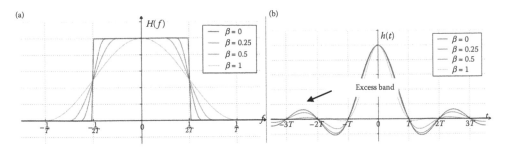

FIGURE 10.4 (a) Frequency response of raised-cosine filter with various values of the roll-off factor β. (b) Impulse response of raised-cosine filter with the roll-off factor β as a parameter.

receiver to minimize the effects of noises. For zero ISI, the net response of the product of the transmitting and receiving filters must equate to $H(f)$, thus we can write:

$$H_R(f)H_T(f) = H(f) \tag{10.3}$$

Or, alternatively, we can rewrite that

$$|H_R(f)| = |H_T(f)| = \sqrt{|H(f)|} \tag{10.4}$$

The filters that can satisfy the conditions of (10.4) are the root-raised-cosine filters. The main problem with root-raised-cosine filters is that they occupy a larger frequency band than that of the Nyquist sin c pulse sequence. Thus, for the transmission system, we can split the overall raise cosine filter with root raise cosine filter at the transmitting and receiving ends, provided the system is linear. This linearity is to be specified accordingly. An optical fiber transmission system can be considered to be linear if the total power of all channels is under the non-linear SPM threshold limit. When it is over, this threshold, a weakly linear approximation, can be used.

The design of a Nyquist filter influences the performance of the overall transmission system. Oversampling factor, selection of roll-off factor for different modulation formats, and FIR Nyquist filter design are key parameters to be determined. If taking into account the transfer functions of the overall transmission channel including fiber, WSS and the cascade of the transfer functions of all O/E components, the total channel transfer function is more Gaussian-like. To compensate this effect in the Tx-DSP, one would need a special Nyquist filter to achieve the overall frequency response equivalent to that of the rectangular or raise cosine with roll-off factor shown in Figure 10.5 (Figure 10.6).

10.3.2 FUNCTIONAL MODULES OF NYQUIST PULSE SHAPER

A generic schematic representation of the functional modules of Nyquist-WDM system is described in Figure 10.2. Two special features should be taken into account. At the Tx side, pulse shaping is so defined that the overall transfer function from driver to the ADC on the Rx side should ideally be rectangular or NRZ-form. The implementation can be either using look-up-table or Nyquist low-pass filter in frequency domain or time domain. Alternatively, the optical spectrum of the modulated lightwaves at the output of the optical modulator can be obtained via the port of the optical spectrum analyser (OSA). This is then equalized to achieve "brick wall" like spectrum. Thus, the required spectrum on the DAC to achieve this equalization is known and then used to modify the DAC driving output voltage levels at (H_I^+, H_I^-) and (H_Q^+, H_Q^-) and (V_I^+, V_I^-) and (V_Q^+, V_Q^-) where the equalized optical spectra can be obtained for the two polarized modes of the linearly polarized mode to be launched into the SSMF optically amplified transmission lines.

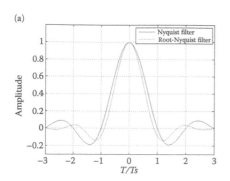 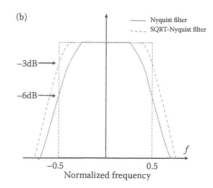

FIGURE 10.5 (a) Impulse and (b) corresponding frequency response of sin c Nyquist pulse shape or root raise cosine (RRC) Nyquist filters.

For each sub-channel, 4 × DACs are needed to convert the discrete digital signal to analog signal (X_I, X_Q, Y_I, Y_Q). To match the voltage amplitude (power) requirement of the I-Q modulator, 4 × RF broadband drivers are required to provide appropriate RF signal amplitude for driving the MZM modulator so as to obtain the phase difference between the in-phase and quadrature phase constellation points. At the output of the transmitter, all sub-channels are multiplexed together to form a superchannel (Figure 10.7).

There is another way to generate the Nyquist signal, which is suitable when Tx-DSP is not available. The Nyquist spectrum is generated by an optical rectangular (or raise-cosine roll-off) filter as shown in Figure 10.8. However, in practice, it is very difficult to achieve a sharp filter roll off without much phase fluctuation at the edges of the filter.

At the receiver, multi-carrier parallel demodulation is used. At first, the superchannel will be separated into individual sub-channels. Each sub-channel will be demodulated like a single channel reception: OFE down-converts the optical spectrum to the baseband (Homodyne coherent demodulation), the four-lane analog signals are converted into digital signals domain using four ADCs. The digitized signals are processed by the ASIC containing the software processing modules for CD compensation, for timing recovery, for MIMO FIR filtering, for carrier recovery, etc. (Figure 10.9).

10.3.3 DSP ARCHITECTURE

Single sub-channel DSP can be either using the traditional FDEQ+TDEQ architecture or pure FDEQ architecture (Figure 10.10).

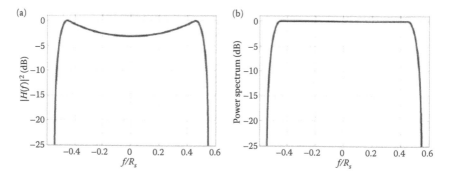

FIGURE 10.6 (a) Desired Nyquist filter for spectral equalization. (b) Output spectrum of the Nyquist filtered QPSK signal.

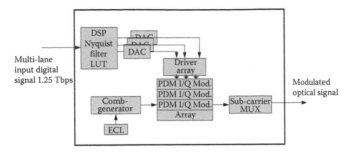

FIGURE 10.7 Transmitter architecture of a Nyquist-WDM system with electrical Nyquist filter.

10.4 KEY HARDWARE SUB-SYSTEMS OF THE TBPS SUPERCHANNEL TRANSMISSION PLATFORM

This section briefly outlines the key hardware of the sub-systems of the transmission platform for Tbps superchannels, consisting of the generation of multi-sub-carriers, the comb-generator; the pulse shaper, the digital to analog convertor, the optical modulators, and the optical transmission line.

10.4.1 COMB-GENERATION TECHNIQUES

10.4.1.1 Recirculating Frequency Shifting

The main modules of a multi-carrier generator (comb-generator) are illustrated as follows (Figure 10.11):

1. A continuous-wave (CW), External Cavity Laser (ECL), or multi-laser bank of ECLs whose line width is sufficiently narrow, possibly in order of less than 100 KHz), can be employed as the master lightwave carrier/carriers.
2. An MZM for pulse shaping (CSRZ), which generates channels whose spectral output is flat.
3. An MZM used as phase modulator, the phase control signal coming from a RF signal generator.
4. RF generator whose frequency of the generated sinusoidal wave determines the frequency spacing between the carriers of the sub-channels.
5. Sinusoidal signal generator.
6. $\pi/2$ phase shifter performs the Hilbert transform to generate single-side band (SSB) shifted sub-carrier.

10.4.1.2 Non-Linear Excitation Comb-Generation and Multiplexed Laser Sources

Under the condition that the modulator, an IQ modulator, can be driven such that the amplitude swings to $2V_\pi$ the generation of first- and second-order frequency shifting components can be formed

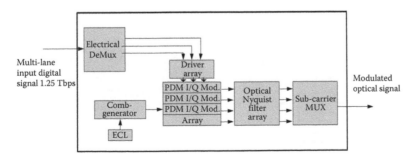

FIGURE 10.8 Schematic representation of the Nyquist optical transmitter architecture with optical Nyquist filter.

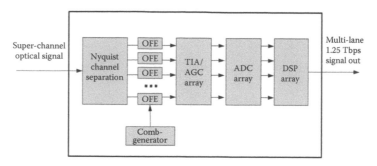

FIGURE 10.9   Receiver architecture of Nyquist-WDM system.

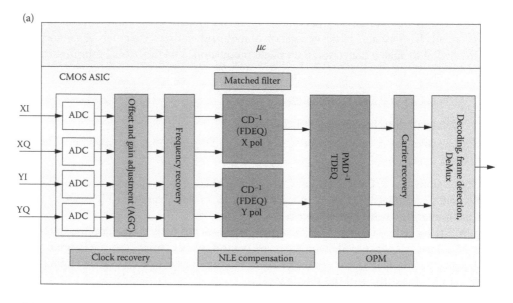

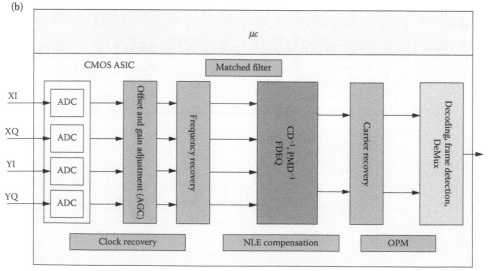

FIGURE 10.10   (a) DSP architecture, FDEQ+TDEQ and (b) pure FDEQ.

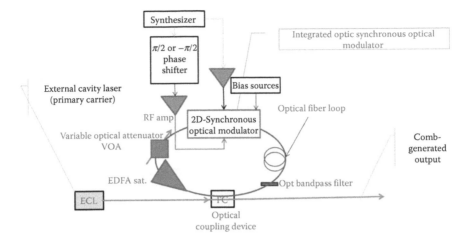

FIGURE 10.11   Block diagram of a recirculating frequency shifting comb-generator.

(Figure 10.12). We operate the modulator in the region such that no suppression of the primary carrier can be achieved, and thus, five sub-carriers can be generated from one main carrier. Using two main carriers, we can generate 10 sub-carriers and hence the modulation of these sub-carriers can create 10×100 Gb/s or 1.0 Tbps superchannels.

10.4.2   COMB-GENERATORS

10.4.2.1   RCFS Comb-Generator

The general path flow of the experiment platform is shown in Figure 10.13. There are two separate paths that need to be noted (in order of importance). The main structure is the optical fiber or integrated optic loop in which besides the optical path interconnecting all optical components which shift the frequency of the lightwaves and tapping part of its energy out to the output port. The optical coupler is used to inject the lightwaves from a sources operating at CW mode, and tap the frequency shifted lightwaves out to the output port.

An optical modulator incorporated in the loop performs the frequency shifting. Single side band operation is used by RF phase delay by pi/2 with respect to each other, hence the Hibert

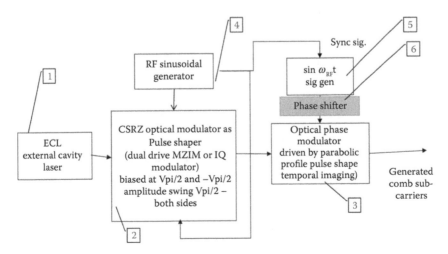

FIGURE 10.12   Comb-generation using non-linear driving condition on MZIM I/Q modulator.

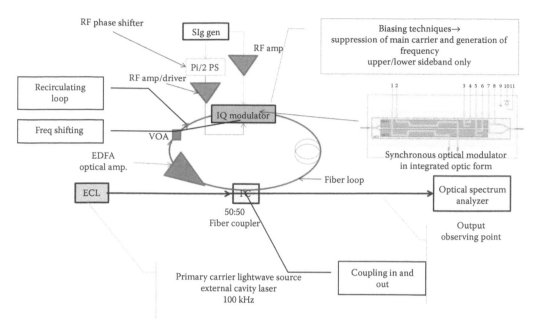

FIGURE 10.13 Principle of the experimental setup of the comb-generator using recirculating frequency shifting.

transformation and the suppression of the right or left side band accordingly, depending on the relative phase shift between the in-phase and quadrature phase electrodes of the IQ modulator. The insert of Figure 10.13 is the integrated optic plan view of the IQ-modulator waveguide and electrodes. There are two child interferometers with overlaid electrodes that perform the suppression of the carrier or sideband by biasing and driving condition by RF waves (see Figure 10.13). Note that electrical phase shifters are inserted in both arms of the RF paths before applying to the electrodes to create pi/2 phase shift, thence suppression of one sideband. See Figure 10.14b for the comb-generator already packaged.

In addition, there is a parent interferometric optic structure in which both child interferometers are positioned. A pair of electrodes is also employed on one branch of this interferometer so that phase shifting between the emerging lightwaves from the child interferometers can be altered with respect to each other. A variable optical attenuator is also used to ensure that the total loop gain is less than unity so that lasing effects would not occur. This also used to adjust the gain equalization due to different gain factor.

A sinusoidal RF wave generator is also required. However, in practice, such an oscillator is simple provided that the frequency is known. It is noted that the output optical port of the comb-generator must be connected with an angled PM fiber connector so that no feedback of optical comb waves back to the recirculating loop. If not, back scattering would then create interference effects and noise disturbance on modulated signals would be observed, hence much higher BER.

Referring to Figures 10.16 and 10.17, we can observe that: (i) The spectra of the generated sub-carriers are equalized over a wide region of 5 nm, carrier noise ratio (CNR) reaches 27–32 dB and (ii) The spacing between two adjacent sub-carriers can be varied by tuning the excitation RF frequency, shown is 28 GHz spacing.

10.4.2.2 Comb-Generation by Non-Linear Driving of Optical Modulators

The schematic structure of the non-linear comb-generator is shown in Figure 10.14, and its packaged version is shown in Figure 10.15. As can be observed, the pincipal component of this

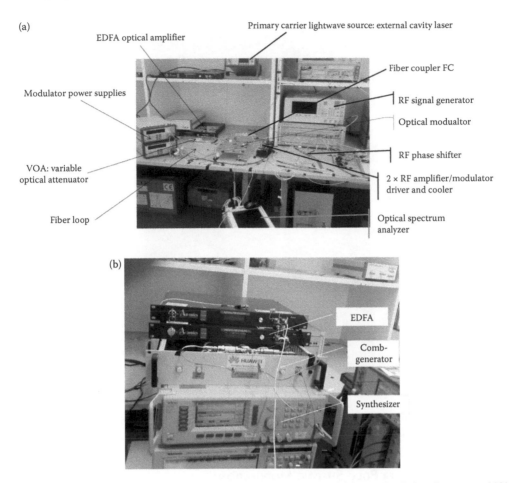

FIGURE 10.14 Laboratory setup of a comb-generator prototype employing recirculating frequency shifting technique (a) comb-generator under development and (b) packaged comb-generator with external EDFA and signal synthesizer.

comb-generator is the IQ modulator driven into non-linear region, mainly over doubling the voltage level such that it covers twice the range of the voltage required for shifting the phase of pi. The packaged comb-generator is shown with two RF phase shifter so that the harmonic level can be adjusted to obtain equalized amplitudes with respect to the main carrier. WSS cn be used to split selected channels and equaliztion of the channel power level. In practice, it is required to have all channels of the same power level lauched into the first spans and then all cascaded optically amplified fiber spans (Figures 10.16 and 10.17). A typical spectrum of a 1×5 sub-carrier comb-generator is shown in Figure 10.18 in which the fundamenatl/primary carrier is located at the middle of the spectrum and the first and second harmonics on the two side of the bands can also be identified without much difficulty. The modulation of such five sub-channels can be seen in Figure 10.18. However, if one channel can carry only Nyquist QPSK of 112 Gb/s then there must be another primary carrier position at a distance from the first primary carry so that 10 sub-carriers can be created, and thence, 1 Tbps channel can be generated via the modulation of the sub-carriers. The LiNbO$_3$ modulator is transparent at these spectral regions and the electro-optic coefficients are quite constant as well, so the modulation of these sub-carriers would result into more or less the same phase modulation. The spectrum of such 10 channels created from 10 sub-carriers is depicted in Figure 10.19.

(a)

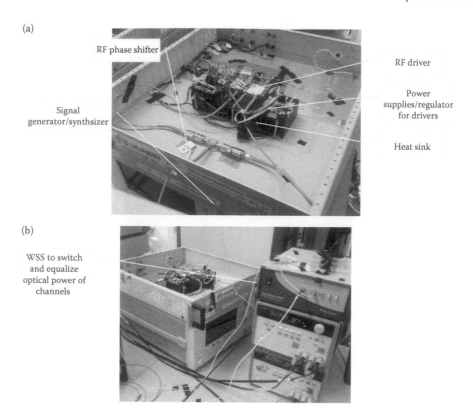

FIGURE 10.15 Non-linear comb-generator (a) structure including optical modulator and RF phase shifters, two branches to RF port inputs and (b) overall packaged comb-generator with power supply and signal generator.

10.4.3 MULTI-CARRIER PULSE SHAPING MODULATION

This section describes the generation and modulation of superchannels by (i) generation of multi-subcarriers from primary carriers by either non-linear driving of optical modulators or (ii) recirculating frequency shifting techniques. The formation of Tbps superchannels can be $10 \times 100G$ (1 Tbps) or $20 \times 100G$ (2 Tbps). It is noted that $5 \times 200G$ can also be used and the 200G can be generated by using the high level modulation format, for example, 16QAM instead of 4-QAM such as QPSK.

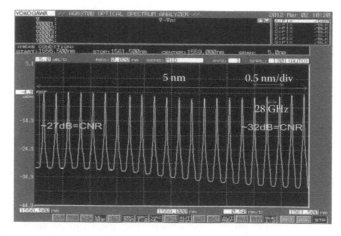

FIGURE 10.16 Spectrum of generated multi-carriers.

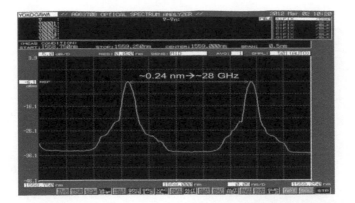

FIGURE 10.17 Measured sub-channel spacing between two adjacent channels.

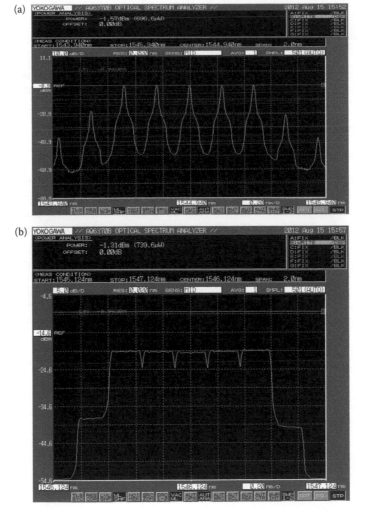

FIGURE 10.18 Spectrum of non-linear comb-generator 1×5 sub-carriers (a) sub-carriers at 30 GHz spacing and (b) modulated sub-channels.

(a)

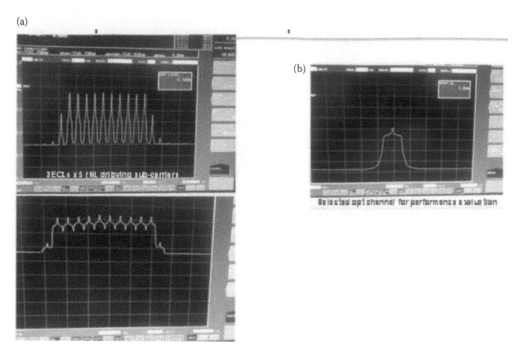

FIGURE 10.19 Spectrum of 10 sub-carriers produced by non-linear comb-generation from two primary carriers and modulated channel (a) group of 10 sub-carrier under modulation and (b) single channel under modulation.

10.4.3.1 Generation of Multi-Sub-Carriers for Tbps Superchannels

10.4.3.1.1 Recirculating Frequency Shifting Technique

Multi-carrier modulation using a comb-generator is constructed and tested in the laboratory for advanced optical communication systems of Huawei ERC in Munich, Germany. The schematic representation of this comb-generator is shown in Figure 10.20, in which a recirculating loop is the basic configuration incorporating an optical modulator, an optical amplifier to compensate the insertion

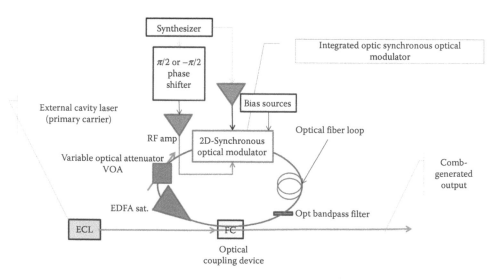

FIGURE 10.20 Schematic structure of the RCFS comb-generator.

loss of various devices and an optical attenuator whose coefficient can be adjusted so that the overall loop gain is slightly less than unity. This loop gain less than unity is required to ensure no lasing would occur. An optical coupler with a coupling ratio of 50:50 or 3 dB split is employed to couple the lightwaves and combed sub-carrier lines into and out of the loop, respectively. A synthesizer, which is a synthsizer/signal generator, is used to generate a sinusoidal wave in RF domain whose frequency determines the frequency shifting of the optical lines of the sub-carriers of the comb-generator.

The optical modulator incorporating in the loop is a synchronous modulator that can be an IQ modulator biased such that synchronization of the interferometers in the two optical paths of the parent MZIM can be maximized. Two arms of the optical modulator are driven with a pi/2 phase shift with respect to each other so that a single side band (SSB) shifting of the lightwave can be achieved at the output of the modulator with the loop open. The spectrum of the SSB lightwave at the output of the modulator is shown in Figure 10.21a. Note that this frequency shifting can be either shifting left or right depending on the relative phase shift between the two arms of the synchronous modulator as shown in Figure 10.21b.

FIGURE 10.21 Image of the RCFS comb-generator including the optical loops and optical amplifiers connected externally, the RF synthesizer and RF ports to be launched into the synchronous modulator. The top optical amplifier (EDFA) is used for amplifying the comb lines for data modulation and thence transmission. (a) front panel and (b) back panel with DC power supply line inputs.

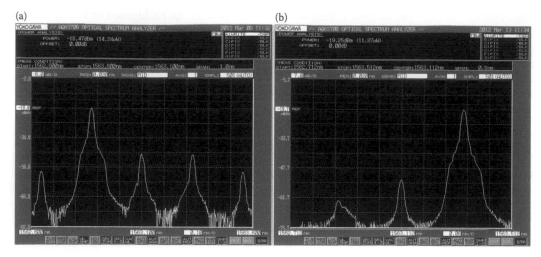

FIGURE 10.22 Spectrum of the SSB lightwave at output of the synchronous optical modulator (a) shifting right in frequency scale and (b) shifting right in frequency.

A synthesizer is required to launch the RF waves to the two arms with RF amplifiers to boost the electrical signals to appropriate level so that maximum modulation depth can be achieved (Figure 10.22). We note that although the modulator 3 dB bandwidth is specified at 23 GHz, it is still possible to operate the modulator at 33 GHz although some RF signal loss would be expected. This is because the driving signal is purely sinusoidal and very narrow band unlike data signals employed to drive the data modulator. We have successfully generated comb lines from 28 GHz to 33 GHz for channel spacing 28, 30, and 33 GHz.

The ECL is the primary lightwave that is used to generate comb sub-channels. This laser output power is set at about 13 dBm, which is the highest power level. A higher power level can be used if EDFA booster amplifier is used. Thus, we note the power level of the SBS line in Figure 10.21 at −10 and −4 dBm for the comb lines in Figure 10.23, which has also been passing through an EDFA at saturation power of 16 dBm that is distributed to all sub-carrier lines. The images of the front panel and back panel of the constructed RCFS comb-generator are shown in Figure 10.21a and b, respectively.

The optical amplifier incorporated in the recirculating ring is operating in the saturation mode only if when the power of the lines is sufficient to boost it into the saturated region otherwise it would operate in the linear region. For this reason, we observe that the frequency shifting lines would not be in the saturated level for the first few sub-carrier lines as seen in Figure 10.23b, the far most right side of the spectrum. The tuning of the DC bias supply voltage to the electrodes is quite sensitive and would be in the 10's of mV range.

One very important point we must make is that the output fiber port must be angled connector type so that no reflection of sub-carrier combed lines can be fed back into the recirculating to avoid the noise oscillation of the generated comb lines. Note that the suppression of the other harmonic lines is more than 25 dB.

The superchannel spectra shown in Figure 10.24a and b display those of the main carrier under modulation using 25 GSym/s QPSK and 28 GSym/s Nyquist QPSK with spacing 50 and 30 GHz, respectively. The roll-off factor for Nyquist signals is 0.1 in optical domain after equalization embedded in electronic domain DAC signals.

A typical spectrum of the generated multi-sub-carriers by the RCFS is shown in Figure 10.23a and b for all sub-carriers and window of selected 20 sub-carriers with amplitude equalization, respectively, while Figure 10.24a and b show the spectra of modulated lightwaves as observed by an OSA under the modulation formats QPSK and QPSK Nyquist at 28 GB/s. Furthermore

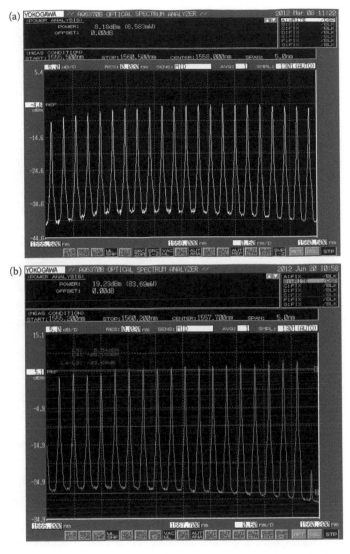

FIGURE 10.23 Spectrum of multi-sub-carriers generated by the RCFS loop. (a) Far left part of the spectrum and (b) near central section of the spectrum.

shown in Figure 10.24c is also the spectrum of the modulated channels of the RCFS combed sub-carriers.

A further note on the RCFS comb-generator is that the RF drivers to both ports of the optical modulator should be a linear driver type so that no harmonic distortion would be generated and interfering with the harmonics of the comb lines. It is noted that the magnitude of the third harmonic is suppressed to more than −30 dB with respect to that of the fundamental, the first and second harmonics.

10.4.3.1.2 Operating the RCFS Comb-Generator

The operation of the RCFS comb-generator should follow these procedures: (i) disconnect the fiber recirculating loop if it is already connected; (ii) connect the output optical port of the synchronous modulator and monitor it via an optical spectrum analyser (OSA); (iii) tune the DC power supplies to the two slave MZIMs of the synchronous modulator such that the suppression of the primary carrier

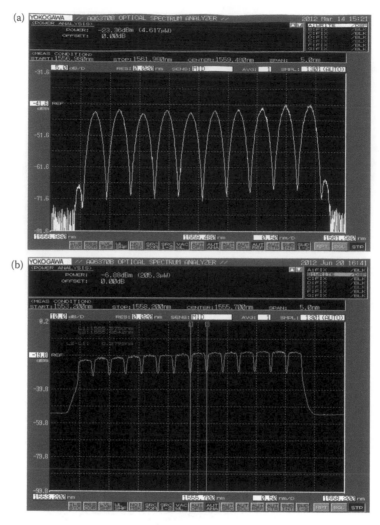

FIGURE 10.24 Spectrum of 10×25 GBaud QPSK superchannel modulation (a) under QPSK with 50 GHz spacing, (b) 14 channels under Nyquist QPSK after transmission over 2000 km optically amplified SSMF spans. (*Continued*)

can be observed and to the minimum value. Once the minimum level is achieved by one electrode pair on a MZIM then change to the other slave MZIM and repeat above so that minimum primary carrier is suppressed; (iv) apply the RF signals into the input of the RF 3 dB splitter and monitor the optical spectrum. Adjust the supplies to the electrodes of MZIM slaves and the electrode applied to the master MZIM so that suppression of one sideband can be observed. Note that the RF level can be tuned so that maximum generation of the sideband line can be achieved; and (v) tune the RF phase shifter to obtain the maximum level of the SSB line and maximum suppression of the other SSB lines. Note that adjustment of the DC supplies to the slave MZIM may be required to suppress the primary carrier.

10.4.3.1.3 Non-linear Modulation for Multi-Sub-carrier Generation

Non-linear driving over $2V_\pi$: Alternating to the superchannel generation using recirculating frequency shifting techniques described above, we have also conducted the generation of higher-order sub-carriers by driving the IQ modulator over its non-linear regions, that is, the amplitude of the RF

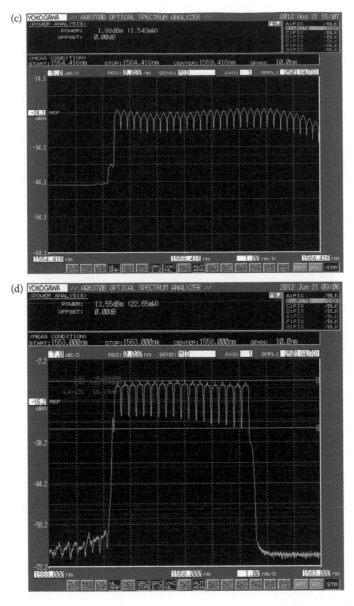

FIGURE 10.24 (Continued) (c) higher than 20 sub-carriers under Nyquist QPSK modulation, and (d) 20 modulated lightwave superchannels after filtering.

waves are covering over at least two $2V_\pi$ of the child MZM sections of the IQ optical modulator. The generation of multi-sub-carriers can be implemented using frequency double and triple by swinging the modulator with an amplitude of V_π and biased at the minimum transmission point, hence frequency doubling and suppression of carrier. This scheme is shown in Figure 10.25a and b.

A group of 10 or 20 sub-carriers can be generated by using two or four primary carriers tuned to appropriate spacing to accommodate the upper and lower higher order sub-carriers. For example, if 30 GHz spacing between sub-carriers, then 150 GHz spacing between two main primary carriers to accommodate for two lower sideband sub-carriers of the higher frequency source and two other upper sideband sub-carriers of the lower frequency light source. Figure 10.26a and b shows the spectra of 1×5 and 4×5 sub-carriers from one and two primary carrier sources respectively. Note that

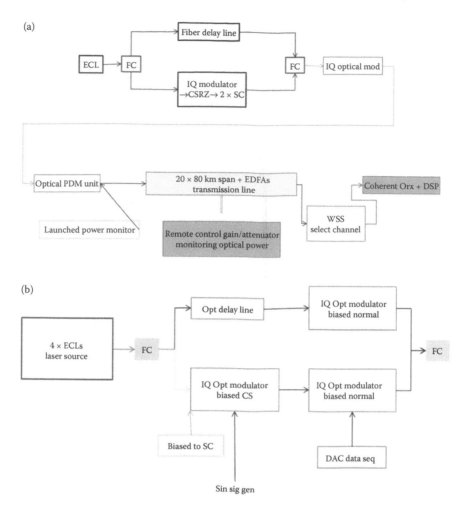

FIGURE 10.25 Schematic representation of optical modulation to generate 4×3 sub-channels to obtain 1 Tbps by CSRZ frequency doubling by one path generating $\times 2$ sub-carriers by CSRZ driving the modulator and then combined with non-frequency shifting optical path (a) 1×3 generation and (b) 4×3 for 1 Tbps channel formation.

for the spectra of 20 sub-carriers, one sub-carrier is band-stopped by passing through an optical notch (or band-stopped) filter whose centered frequency is set such that it falls exactly at the frequency of the sub-carrier. By doing this, one can extract sub-carriers arbitrary and injected carriers from different sources so that independent channels can be generated to ensure the decorrelation of adjacent channels and thence the measured of channels can reflect the true performance of independent channels. It is also noted that the third-order sub-carrier was suppressed to more than 30 dB in case the seven sub-carriers to be generated the RF driving signal power level can be used; however, we doubt that the modulator would stand the high voltage level developed across the electrical-optical transfer characteristics of the IQ modulator.

CSRZ and combined paths: Another possibility to generate 1 Tbps the combined four lasers are split into two optical paths. One path is fed into a CSRZ modulator to generate 2×4 sub-carriers with frequency doubling or +1 first and minus first-order. These eight sub-carriers are then combined with non-frequency shifted primary carriers, thus the combined optical path would produce a set of 10 sub-carriers which are then modulated by an IQ modulator driven via the DAC output ports to obtain Nyquist QPSK channels. The combined optical path can also be fed into a WSS (wavelength selective

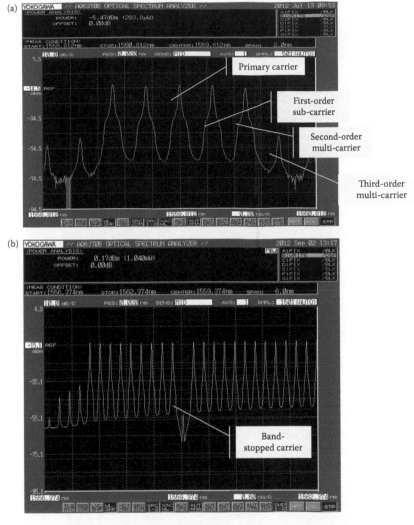

FIGURE 10.26 Spectrum of sub-carriers generated from a primary laser source by non-linear driving of IQ modulator (a) 1 × 5 and (b) 4 × 5 sub-carriers with one band-stopped.

switch) to separate two independent streams of carriers that are then fed into two IQ Modulators driven by independent PRBS from the DAC output ports to produce decorrelated modulated optical channels.

Remarks: Although both schemes described have been tested and implemented, the scheme using 4 × 5 to obtain 20 sub-carriers is finally selected to feed into optical modulators to obtain 10 or 20 Tbps for the demonstration platform to field trials with Deutsche Telecom and Vodafone, respectively.

10.4.3.2 Supercomb-Generation Modulated Channels as Dummy

Some sub-carriers of the super comb lines generated by the RCFS comb-generators can be used as dummy channels. The generated comb lines can be filtered by band-stopping of the wavelength selective switch device, which is shown in Figure 10.27. One sub-carrier is filtered out and the rest of the super comb lines are then fed to a PDM-IQ-modulator so that PDM QPSK channels can be generated as shown in Figure 10.28.

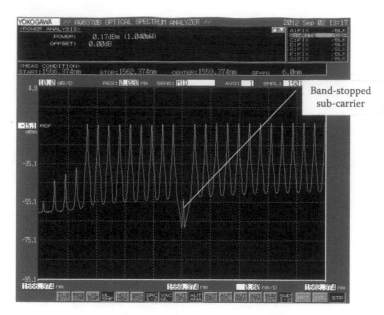

FIGURE 10.27 Band-stopped sub-carrier of a multi-sub-carrier lines used for modulation and generation of dummy channels.

10.4.3.3 2 and 1 Tbps Optical Transmitter at Different Symbol Rates

The 1 and 2 Tbps superchannels can be generated using either RCFS comb-generator or the non-linear comb-generator describe in previous sections (Figure 10.29). The difference between these two schemes are that for the RCFS comb-generated channels the phase and frequencies of all sub-channels are locked to the reference primary carrier. Thus, with the number of dummy channels employed for transmission the synchronization of these channels would lead to high non-linear SPM effects (Figures 10.30 and 10.31).

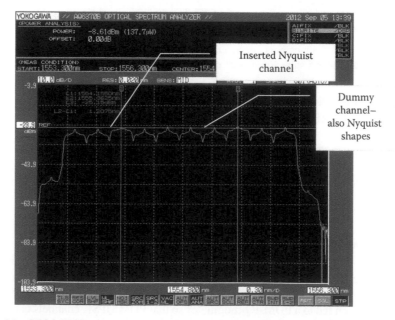

FIGURE 10.28 PDM QPSK channels by modulation for use as dummy channels of 1 Tbps superchannel.

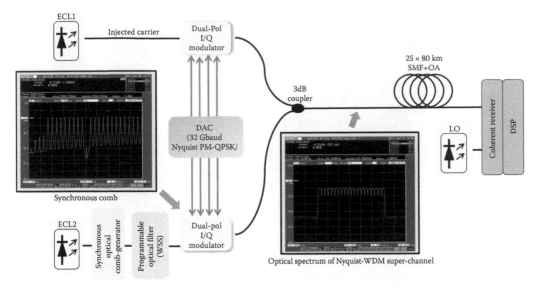

FIGURE 10.29 Arrangement of ECL1 and ECL2 for non-linear comb-generation of 10 sub-carriers in association with Nyquist pulse shaping and modulation for 25 spans of optically amplifier fiber (non-DCF) transmission. The spacing between the lasers allows allocation of equally spaced of 28–32 GHz—totally 4 sub-channels by non-linear comb-generation.

10.4.4 DIGITAL TO ANALOG CONVERTER

10.4.4.1 Structure

The principal sub-system of the superchannel transmission system is the digital signal processing based optical transmitter in which the digital to analog converter (DAC) play the central role in pulse shaping, equalization and pattern generation. An external sinusoidal signal is required to be fed into

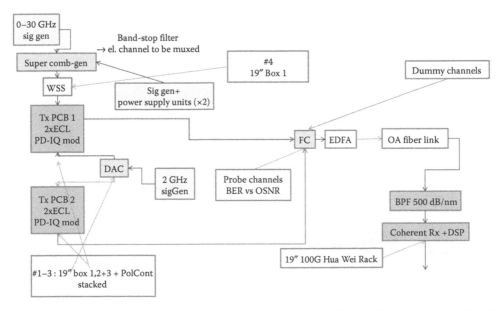

FIGURE 10.30 Schematic representation of structure of Tbps optical transmission systems using 4×5 sub-carrier generation techniques.

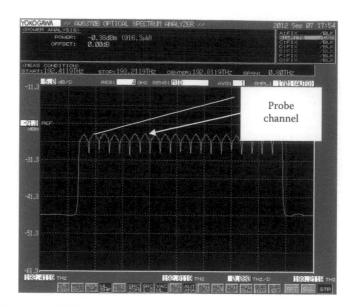

FIGURE 10.31 Spectrum of 2 Tbps optical channels using non-linear comb-generator with two probe channels and 18 dummy channels.

the DAC so that multiple clock source can be generated for sampling at 56 to 64 GSa/s. Thus, the noises and clock accuracy depends on the stability and noise of this synthesizer. An Agilent signal generator model N9310A is employed in our platform. However, a Rohde and Schwartz signal generator with low phase noise is preferred. Four DACs sub-modules are integrated in one IC with four pairs of eight outputs of $(V_I^+, V_Q^+)(H_I^+, H_Q^+)$ and $(V_I^-, V_Q^-)(H_I^-, H_Q^-)$.

10.4.4.2 Generation of I and Q Components

Electrical outputs from the quad DACs are in pair of positive and negative and complementary with each other. Thus, we would be able to form two sets of four output ports from the DAC development board. Each output can be independently generated with offline uploading of pattern scripts. The arrangement of the DAC and PDM-IQ optical modulator is depicted in Figure 10.32. Note that we require two PDM IQ modulators for generation of odd and even optical channels.

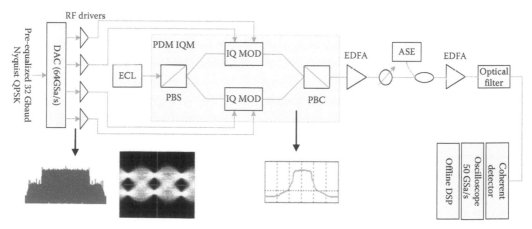

FIGURE 10.32 Experimental setup of 128 Gb/s Nyquist PDM-QPSK transmitter and B2B performance evaluation.

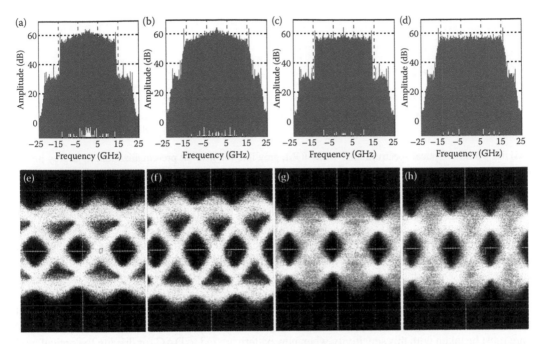

FIGURE 10.33 Spectrum (a) and eye diagram (e) of 28 Gbuad RF signals after DAC without pre-equalization, (b) and (f) for 32 Gbaud, Spectrum (c) and eye diagram (g) of 28 Gbuad RF signals after DAC with pre-equalization, (d) and (h) for 32 Gbaud.

As Nyquist pulse shaped sequences are required, a number of pressing steps are conducted (i) characterization of the DAC transfer functions and (ii) Pre-equalization in the RF domain to achieve equalized spectrum in the optical domain, that is, at the output of the PDM IQ modulator.

The characterization of the DAC is conducted by launching to the DAC sinusoidal wave at different frequencies and measured the waveforms at all eight output ports. As observable in the inserts of Figure 10.32, the electrical spectrum of the DAC is quite flat, provided that pre-equalization is done in the digital domain launching to the DAC. The spectrum of the DAC output without equalization is shown in Figure 10.33a and b. The amplitude spectrum is not flat because of the transfer function of the DAC as given in Figure 10.34, which is obtained by driving the DAC with sinusoidal waves of different frequencies. This shows that the DAC acts as a low pass filter with the amplitude of its passsband gradually decrease when the frequency is increased. This effect can come from the number of samples is reduced when the frequency is increased as the sampling rate can only be set in the range

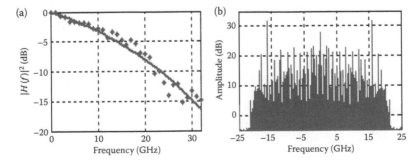

FIGURE 10.34 (a) Frequency transfer characteristics of the DAC. Note the near linear variation of the magnitude as a function of the frequency and (b) noise spectral characteristics of the DAC.

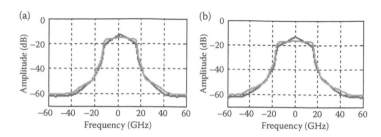

FIGURE 10.35 Optical spectrum after PDM IQM, grey line for without pre-equalization, light grey line for with pre-equalization (a) 28 GSym/s and (b) 32 GSym/s.

of 56–64 GSa/s. The equalized RF spectra are depicted in Figure 10.33c and d. The time domain waveforms corresponding to the RF spectra are shown in Figure 10.33e and f, and thence 10.33g and h for the coherent detection after the conversion back to electrical domain from the optical modulator via the real time sampling oscilloscope Tektronix DPO 73304A or DSA 720004B. Furthermore, the noise distribution of the DAC shown in Figure 10.34b indicates that the sideband spectra of Figure 10.33 come from these noise sources.

It is noted that the driving conditions for the DAC are very sensitive to the supply current and voltage levels, which are given in Appendix 10B.1 with resolution of down to $1e - 3$ V (Figure 10.35). Care must be taken with this sensitivity when new pattern are fed to DAC for driving the optical modulator. Optimal procedures must be conducted with the evaluation of the constellation diagram and BER derived from such constellation. However, we believe that the new version of the DAC supplied from Fujitsu Semiconductor Pty Ltd of England in Europe have somehow overcome these sensitive problems. We still recommend that care should be taken and inspection of the constellation after the coherent receiver must be done to ensure that error free in the B2B connection. Various time-domain signal patterns obtained in the electrical time domain generated by DAC at the output ports are illustrated in Figure 10.36. Obviously, the variations of the in-phase and quadrature signals give raise to the noise, hence blurry constellations.

10.4.4.3 Optical Modulation

Modulation of the PDM IQ modulator can be implemented using RF signals of the output pairs of the DAC. Two sets of RF output ports are employed to provide the inphase and quadrature signals to the modulators traveling wave electrode pair. DC voltage power supplies are also applied to DC biasing

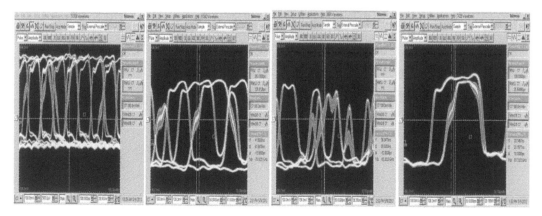

FIGURE 10.36 Various DAC time domain waveforms at output ports of DAC where non-optimal driving conditions are applied.

Detailed arrangement of comb-gen using NL driving of IQ modulator
for 2ECL × 5sc→ 10 × 28G × 2 × 2 → 1.12 Tbps

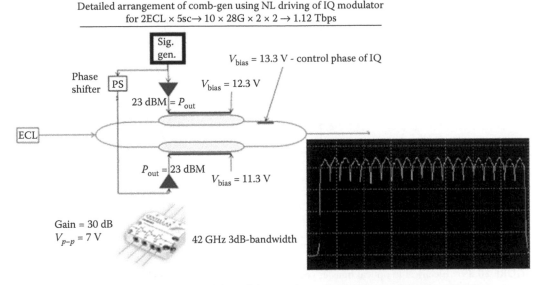

FIGURE 10.37 Plane view of the PDM IQ modulator and associated bias and RF driving circuitry.

connections. The integrated structure of the PDM IQ modulator is shown in Figure 10.37. There are a number of DC electrodes for biasing the IQ modulator, as shown in Figure 10.38. This set is duplicated for polarized modulators integrated on the same $LiNbO_3$ chip. The S21 parameter frequency response shows that the 3 dB optical bandwidth of the modulator is 23 GHz. This is sufficiently wide for 28 and 32 GB digital sequence. The electrical to optical transfer characteristics is shown in Figure 10.39. We note that as the DC bias electrodes are separate from the RF travelling wave

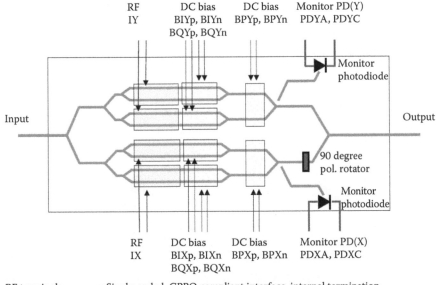

RF terminals	: Single-ended, GPPO compliant interface, internal termination.
DC bias terminals	: separated from RF section, push-pull configuration
photodiodes	: Located at each QPSK modulator sections, the phase of the photocurrent inverted to the optical output from QPSK modulator.

FIGURE 10.38 Schematic representation of optical waveguides and electrodes of the PDM IQ modulator. Two polarization dependent IQ modulators in parallel.

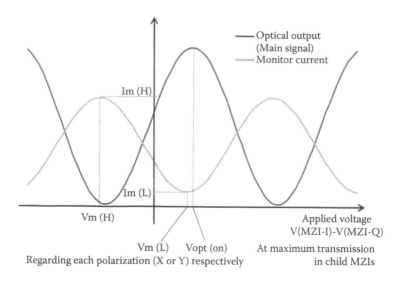

FIGURE 10.39   Electrical to optical transfer characteristics of the PDM IQ modulator*.

electrode, thus the voltage level is quite high due to the short length of the DC electrode pair. The electrode V_π may reach around 15–20 V when using these electrodes, so care should be taken to avoid the damage of these electrodes, that is, the air break down electric field of about 50 kV/m across the electrode spacing of only about 5 μm.

10.4.4.4   Synchronization

Care must be taken to ensure that synchronization of the pulse sequences at the output of the DAC ports occur, otherwise the constellation of the QPSK structure would be noisy and create a worse BER. This can be done by programmable delay time in the data sequence to be loaded into the DAC.

10.4.4.5   Modular Hardware Platform

The overall packed comb-generator using non-linear driving of the modulator is shown in Figure 10.40 with the comb packaged box on top and signal generator plus power supplies placed in the bottom and staked separately. While the comb-generator using RCFS structure is shown in Figure 10.41 in which two EDFA are placed on the top, one for optical compensation inside the recirculating loop

FIGURE 10.40   Non-linear comb-generator—view of with and without power supply connection.

---

* Fujitsu optical components: http://www.fujitsu.com/jp/group/foc/en/products/optical-devices/100gln/ access date: September 2017.

FIGURE 10.41 RCFS optical comb-generator.

and the other for amplifying the signals power to be coupled to the WSS or the IQ modulator (Figure 10.42). As noted in the above section, the total optical power must be distributed to all comb lines thus typical power level for each comb lines of the RCFS comb is about −5 to −7 dBm depending on the number of comb lines generated from the recirculating loop. Note that EDFAs are of PM type, including input and output fiber patch cords.

10.5 NON-DCF OPTICALLY AMPLIFIED MULTI-SPAN TRANSMISSION LINE

10.5.1 TRANSMISSION PLATFORM

The schematic structure of the transmission system is shown in Figure 10.43 consisting of an optically amplified multi-SSMF span fiber transmission incorporating variable optical attenuators, and optical

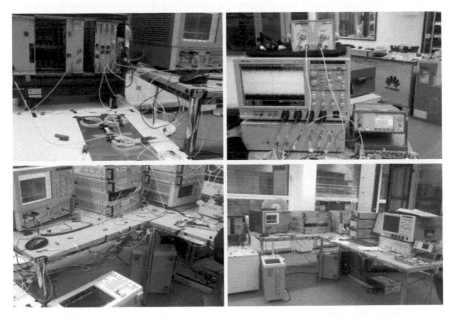

FIGURE 10.42 Overview of the Nyquist WDM QPSK transmission platform.

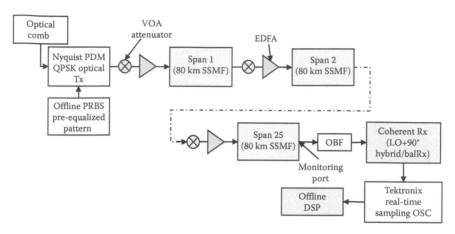

FIGURE 10.43 Schematic structure of the Nyquist QPSK PDM optically amplified multi-span transmission line.

transmitters and coherent optical receiver with offline DSP. Optical bandpass filter (OBPF) is employed using either the D-40 demux or Yenista sharp roll off (500 dB/nm) to extract the sub-channel of the superchannel whose performance is to be measured, normally in term of BER versus OSNR.

The multi-span optically amplified fiber transmission line consists of two main parts: The fiber spools and banks of optical amplifiers and VOA, as shown in Figure 10.44a and b, respectively. The launched power to the first span is to be adjusted and thence at the VOA and amplifiers into each span as the saturation level of EDFA varying with respect to the total launched power. Thus, to achieve the same launched power into each span, the remote control is employed to adjust these input launched power span by span, especially when the launched power is varied to obtain different OSNR. Back-to-back (B2B) is measured by inserting noises at different level to the signal power monitoring port. A D-40 demultiplexer is also available and incorporated in the 19″ rack so that a sub-channel of the superchannel can be extracted for measuring the transmission or B2B performance (Figure 10.45).

The transmitter employs Nyquist QPSK, as described in a previous section (see Section 10.4.4), with the channel spacing and baud rate can being varied and optimization of the DAC. The optical receiver is a coherent type with an external local oscillator whose wavelength tuned to the right wavelength location of the sub-channel whose performance is to be measured. In practice, this LO must be remotely/automatically locked to the wavelength of the sub-channel.*

10.5.2 PERFORMANCE

10.5.2.1 Tbps Initial Transmission Using Three Sub-Channel Transmission Test

During the initial phase of the development of the Tbps transmission system, we considered a test of transmission of three channels only for transmission over 2000 km optically amplified multi-span, as described in the previous section, to ensure that the signal quality can be achieved using Nyquist QPSK with interference because of linear cross talk due to overlapping of sub-channels as they are packed so close to each other using Nyquist criteria. We set up the generation of three sub-channels by using an ECL source split 50:50 into two branches, one fed through a CSRZ optical modulator driven with a sinusoidal signal source of frequency equal to the sub-channel spacing, and the

---

* LN Binh et alia, "Optical PLL super combed carrier cloning: circuit and methodology for locking superchannel coherent receivers," patent under preparation and approval.

(a)

(b)

FIGURE 10.44 Fiber spools (a) to be connected to optical amplifiers and variable optical attenuator and (b) arranged in bank in 19″ rack.

other branch would then be modulated with Nyquist QPSK format combined with the two CSRZ carriers that are also modulated by Nyquist QPSK but with different random pattern. Thus, we do have three sub-channels that are decorrelated and the BER versus OSNR can be obtained to ensure that the effects of overlapping can be justified. The schematic representation of this arrangement is shown in Figure 10.46. The transmission performance measured with BER against launched power and

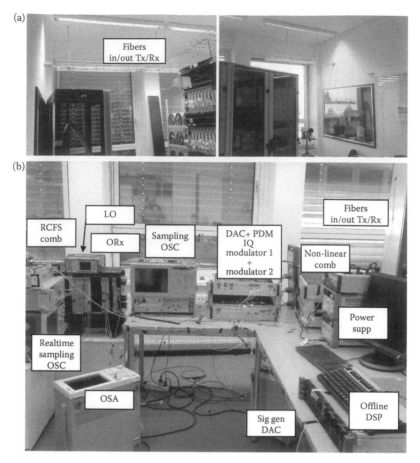

FIGURE 10.45 Fiber lines (a) running from optically amplified multi-fiber span transmission line to transmitter and receiver platform (b).

sub-channel spacing of 28 and 30 GHz with 28 GBauds Nyquist QPSK transmitted over 1600 km line is shown in Figure 10.47.

The BER is optimum at launched power of −1 to 0 dBm for single channel and probe channel which is the central of the three channels. This proves that the Nyquist pulse shaping can offer the same performance close to that of a single channel transmission over 1600 km optically amplified multi-span and non-dispersion compensating fiber line (Figure 10.48). These performance results gave much confidence in the pulse shaping and for proceeding to superchannel of 1 and 2 Tbps over a single dual polarized mono-mode fiber.

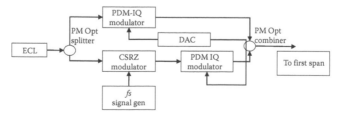

FIGURE 10.46 Schematic representation of generation of three sub-carrier channel transmitter for test transmission in the initial phase for Tbps transmission system. PM = polarization maintaining.

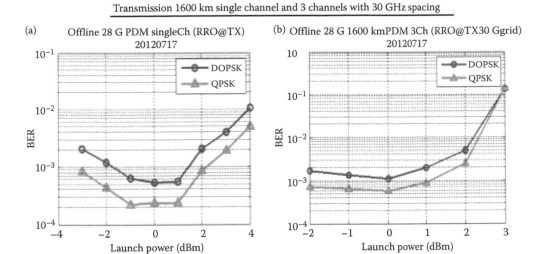

FIGURE 10.47 BER versus launched power (dBm) using three channels with central channel as probe channel over transmission distance of 1600 km non-DCF, 20 optically amplified 80 km SSMF spans with VOA inserted in front of EDFA. 28 GBaud PDM-Nyquist QPSK (a) single channel and (b) three channels.

Furthermore, we use optical filter at the receiver that is before feeding into the hybrid coupler the optical filter (the DeMux D-40 or a sharp roll off Yenistar filter with 500 dB/nm) is employed to extract the probe channel (Figure 10.49). The transmission performance for a five-channel superchannel is shown in Figure 10.50. In general, the sharp roll-off filter would offer 1 dB improvement over the DeMux D-40 that commonly has a parabolic-like roll-off filtering characteristic.

10.5.2.2 N-Tbps (N = 1,2 ... N) Transmission

We can now extend the transmission of Nyquist pulse shaped sub-channels with the total capacity of all channels of 1.12 Tbps and thence 2.24 Tbps. For 1.12 Tbps per second, we use nine sub-channels plus one probe channel, all of which are modulated and pulse shaped satisfying Nyquist criteria with

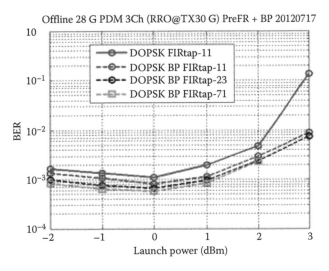

FIGURE 10.48 1600 km transmission Nyquist QPSK processed with FIR 11 to 23 taps: BER versus OSNR. Number of sub-channels is three generated and modulated as shown in Figure 10.47.

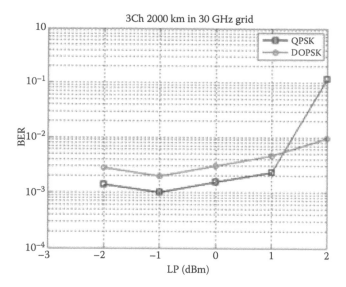

FIGURE 10.49 Three-channel test with central channel as probe over 2000 km transmission by simulation 31.8 GHz channel spacing 28 GBaud QPSK with processing QPSK and DPQSK (to optimize for cycle slippage).

raise cosine filter and pre-equalization. Thus, we have 10 channels generated using RCFS or non-linear comb-generation. One channel is suppressed by using band stop filter of a WSS and then combined with another Nyquist shaped QPSK channel independently driven by a decorrelated sequence generated from DAC ports. This inserted channel acts as the probe channel, hence, the one select to measure the performance of transmission for any sub-channel of the superchannel as required instead of all independently driven sub-channels because of the limitation of independent sequence from one DAC sub-system. Figure 10.51 shows the generic scheme for generating such probe channel within the superchannel. We can select whichever channel spectral location to insert the probe channel by tuning the source wavelength and the WSS depending on whether the RCFS or

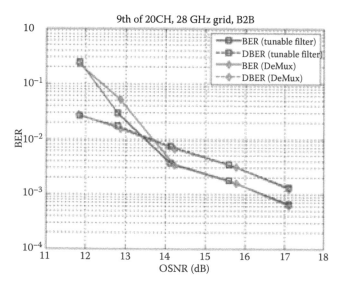

FIGURE 10.50 Performance of channel nine of the superchannel transmission over 2000 km with D-40 demux or Yenistar filter.

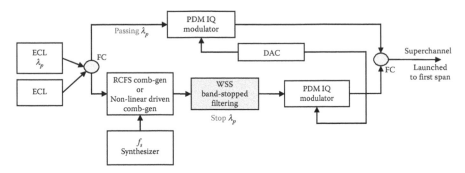

FIGURE 10.51 Generation of $(N-1)$ sub-carrier dummy channels and plus-one probe channel for Tbps superchannel transmission systems.

non-linear conb-generator is employed. Figure 10.52 shows the B2B performance of 10 sub-channels (2×5 by non-linear driving modulator described above) under Nyquist pulse shaping and PDM QPSK modulation format. Individual inphase and quadrature phase components are also processed to see the effects of one component on other channels. Thus, total BER of QPSK is the total sum of all these individual components. DQPSK processing offer 1 dB better in OSNR for the same BER and BER of 1e − 3 achieved for OSNR of 15.2 and 16 dB, respectively, for QPSK and DQPSK, respectively. The main reasons for evaluating all individual inphase and quadrature received signals are to ensure that the DAC generated signals at the ports $(V_I^+, V_Q^+)(H_I^+, H_Q^+)$ and $(V_I^-, V_Q^-)(H_I^-, H_Q^-)$ enforcing no penalties on the coherent detected signals. With this scheme for generation of 1 and 2 Tbps, it is not difficult to extend to N_Tpbs provided that the RCFS comb-generator is used or one has to employ $N/5$ ECLs if non-linear comb-generation technique is resorted. The principal problems that we would face in this case would be the non-linear distortion effects that will be described in the coming section.

Using non-linear comb-generator, 10 sub-channels are generated, with one channel eliminated by passing through the WSS set by band-stopped filtering. These sub-carriers are then modulated and combined with the probe channel then amplified launched in to the first span. The B2B performance

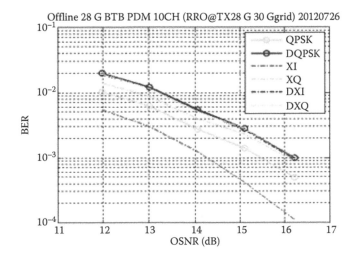

FIGURE 10.52 10 sub-channels of 112 Gb/s 28 GBauds Nyquist QPSK superchannel generation and transmission using non-linear driven comb-generator. 28 GBauds, 30 G grid sub-channel spacing, 10 sub-channels with one probe channel.

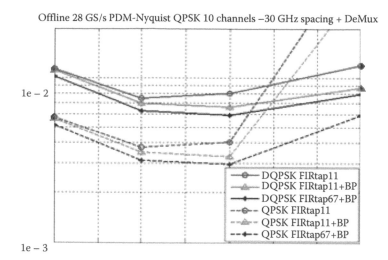

Offline 28 GS/s PDM-Nyquist QPSK 10 channels −30 GHz spacing + DeMux

FIGURE 10.53 BER vs. launched power (−2 dBm to 1.5 dB with 0.5 dBm per division in horizontal scale) 28 GBauds Nyquist QPSK with 30 GHz spacing with and without back propagation additional processing and 11 tap FIR and 2000 km transmission Non-DCF fiber multi-span line probe channel selected using D40 demux.

is shown in Figure 10.52. The channel spacing is 30 GHz with 28 Gbauds as the symbol rate, the BER obtained is 1e − 3 at 14.5 dB. The transmission performance of the 10 sub-channel superchannel over 200 km of optically amplified non-DCF SSMF multi-span transmission line is shown in Figures 10.52 and 10.53 in which B2B with four in-phase and quadrature components of the PDM-28G Nyquist QPSK are analysed with their BER plotted against OSNR at the initial phase prior to transmission and then over the complete link. The attenuation pers span is 22 dB and the EDFA stages are optimized so that the launched power into each span can be kept the same at each span. The optimum launched power is −1 to 0 dBm with a BER of 2e − 3 with 11 taps Fir filter and back propagation to moderately compensate for fiber non-linearity.

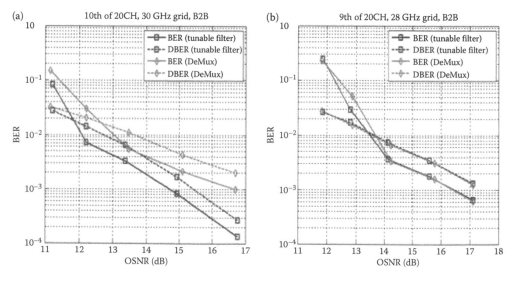

FIGURE 10.54 Transmission performance of (a) Channe 19 and (b) channel 20 of the >2 Tbps superchannel with channel spacing of 30 GHz and 28 Gbauds Nyquist QPSK roll off factor 0.1 using D-40 demux and Yenistar sharp roll off filter (500 dB/nm).

The back propagation is conducted by propagating the received sampled signals at the receiver, then converted into the optical domain level and the propagation though span by span with the non-linear coefficient equal and in opposite sign with those of the SSMF. The back propagation distance per span is about 22 km as the effective length of the fiber under non-linear SPM effects (Figures 10.54 through 10.56).

10.5.2.3 Tbps Transmission Incorporating FEC at Coherent DSP Receiver

10.5.2.3.1 Coding Gain of FEC and Transmission Simulation

Simulations are conducted to estimate the gain and BER against OSNR with error coding and non-coding to assisting with the experimental performance (Figures 10.57 through 10.61). The simulation proves the coding gain for Nyquist QPSK of 0.1 roll off factor, as shown in Figure 10.62. Under coding, we could see that the OSNR requires for $1e-3$ may be reduced down to 11 dB. Thus, an extra margin of 4 dB can be gained to allow the extension of the transmission reach of Nyquist QPSK, an extra 500 km. Figure 10.62a and b shows the effects of 30 and 28.5 GHz sub-channel spacing on the BER of the Nyquist QPSK with LDPC and significant improvement of FEC on their performance. Similarly, for Figure 10.62c and d relates to 33 and 32.5 sub-channel spacing with 32 GBauds. Figure 10.62e and f then proves the improvement when BCJR additional coding is superimpose to obtain further coding gain. Figure 10.62g displays all gain curves into one graph and Figure 10.62h and i shows the spectra of sub-channels, the odd channels only.

10.5.2.4 MIMO Filtering Process to Extend Transmission Reach

The filter structure incorporated in the DSP processing is shown in Figure 10.63. Due to whitening noise effects in the optical amplification stages in the link, the root-raised cosine filter is used at the receiver for match filtering. That is the complete filtering process in the link form a complete Nyquist filter and satisfying the Nyquist shaping criteria. Following the Rx imperfections compensation and CD compensation, a complex butterfly FIR structure is applied to compensate for PMD. Each of four complex FIR filters is realized by a butterfly structure of corresponding real FIR filters. The recursive constant modulus algorithm (CMA) LMS algorithm continuously updates the filter taps, which

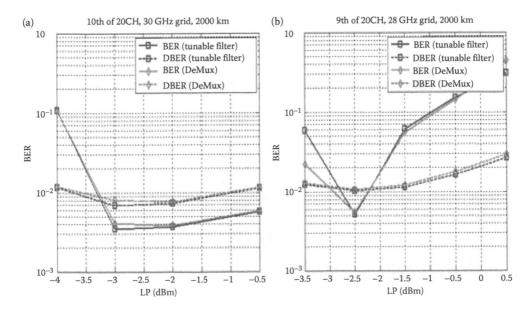

FIGURE 10.55 BER versus launched power of superchannel and probe channels are channel No. 9 (a) and No. 20 (b), transmission distance 2000 km multi-span optically amplified non-DCF line.

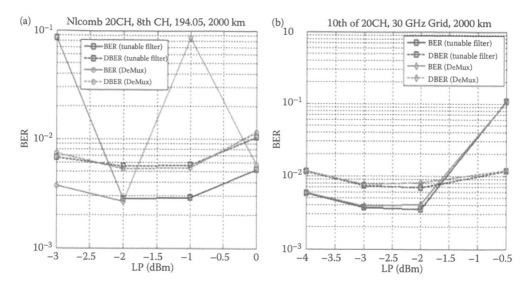

FIGURE 10.56 Five-channel non-linear comb-gen sub-carriers transmission over 2000 km: BER versus OSNR using 40 channel demultiplexer or sharp roll off optical filter (a) test channel is channel No. 8 and (b) Channel No. 10.

guarantee the initial convergence and tracking of time-variant channel distortions. In the steady state, the complex butterfly structure is a digital, real representation of the inverse impulse response determined by the tap coefficients.

The output signals from the FIR equalization stage (x' and y') at time k, are related to the input signal vectors (x and y) containing samples $k - L + 1$ to k by

$$x'(k) = h_{xx} \cdot \mathbf{x}(k) + h_{yx} \cdot \mathbf{y}(k)$$
$$y'(k) = h_{yy} \cdot \mathbf{y}(k) + h_{xy} \cdot \mathbf{x}(k) \tag{10.5}$$

where $h_{xx}, h_{xy}, h_{yy}, h_{yx} >$ are the $T/2$-spaced tap vectors (T is symbol period), for the FIR filter, and the dot "$\cdot$" denoting the vector dot product. The length L of the tap vector is equal to the

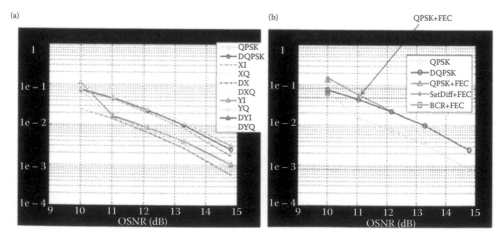

FIGURE 10.57 Soft differential FEC processing for Nyquist QPSK, BER versus OSNR. 28 Gbauds with individual components of QPSK, i.e., real and imaginary parts (inphase and quadrature phase components), B2B set up (a) without FEC and (b) with FEC.

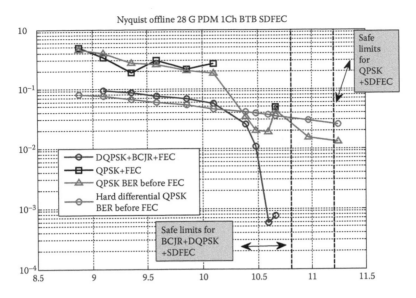

FIGURE 10.58 FEC improvement BER versus OPSNR for Nyquist QPSK with BCJR and without FEC.

impulse response of the distorted transmission medium to be compensated. Initial equalizer acquisition is performed on the first several thousand symbols depending on the "learning/training" process. These symbols are subsequently discarded in the error counting. The equalizer tap vectors are then updated continuously throughout the processing of the data set in order to track channel changes.

The MIMO filter length in the commercial 100G receivers without tailoring for Nyquist transmission is usually set between seven and 11 as a tradeoff between complexity and requirements since the measured mean DGD in real long-haul optical links is around 25 ps. Further increasing the FIR complexity would enhance the gain in case of a longer pulse response to be employed due to the limited transfer function of the transmission system. This is verified in our experimental

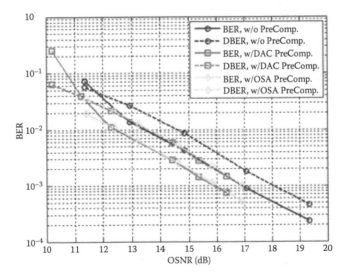

FIGURE 10.59 32 GBauds PDM Nyquist QPSK transmission, BER versus OSNR with and without electronic compensation in DAC, back-to-back scenario.

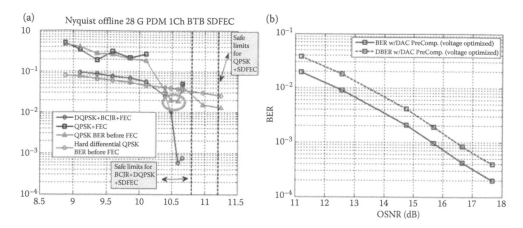

FIGURE 10.60 Performance of transmission systems under (a) optimization and (b) non-optimized DAC generation of random sequence at 28 G and 32 GBauds, back-to-back scenario, modulation format Nyquist QPSK.

platform with the BER against OSNR, the launched power and number of taps as depicted in Figure 10.64a–d. It is noted that the tap number higher than nine does not offer any gain improvement in performance as depicted in Figure 10.64. The MIMO filter tap length extending effects in commercial WDM optical transmission experiments (a) BTB performance, L stands for the filter tap length, (b) 1500 km transmission line, (c) 2000 km transmission line, and (d) MIMO filter with tap length of 23 convergence results. h_{xx} for the in-phase real part of h_{xx}, h_{xx} for the quadrature imaginary part of h_{xx}.

In our Nyquist experimental platform, the pulse sequence shaped by a Nyquist square root filter (RRC) of a roll-off factor of 0.1 generating by a time-domain FIR filter with 65 taps. Simulation results do not show any penalty caused by our FIR tap settings. However, hardware imperfections likely require filter pulse response of more FIR taps to achieve the best performance. Furthermore, the convergence of the CMA algorithm would fail for FIR with longer tap length, as shown in

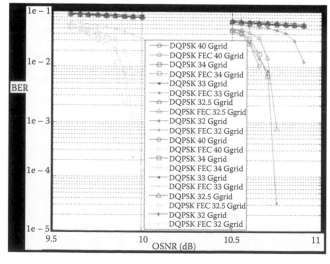

FIGURE 10.61 32 GB Nyquist QPSK roll-off = 0.1 with sub-channel spacing as parameter and transmission over 2000 km non-DCF optically amplified spans under scenarios of with and without FEC 20%, far left and far right of the graph, respectively.

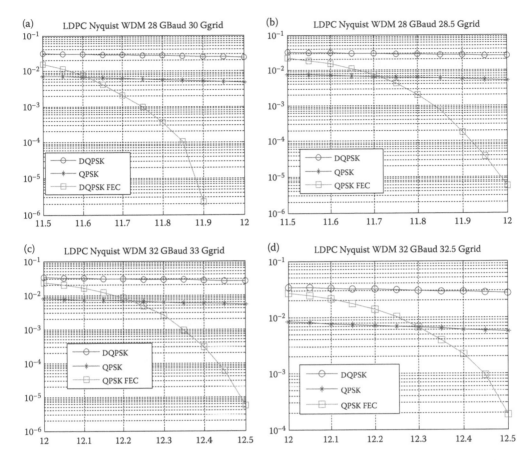

FIGURE 10.62 Simulation of Nyquist QPSK with and without error coding gain (a and b) LDPC coding and gain at 28 GHz grid. *(Continued)*

Figure 10.65. With seve taps, we are able to acquire the channel while the FIR filter and no convergence with 23 and 43. Therefore, the conventional CMA method is used and the performance is shown in Figure 10.65a. The coarse step applies fast learning via larger values of a weighting factor μ_1 and a forgetting factor α_1. The final step uses smaller values of these two parameters and improves the final performance.

To solve the convergence problem, we perform the acquisition procedure in two steps, with smaller and larger FIR filter lengths (Figure 10.66). First, a shorter FIR filter with a smaller number of taps L_1, for example, less than nine, and larger μ_1 and α_1 can be used in the initial step to find the main values of the starting taps. After the pre-convergence phase, the filter is extended with the reiterative tap values obtained in the first procedure, while the extended taps are set to null in association with smaller values of μ_2 and α_2. This method assures the filter convergence.

Using this new algorithm, the B2B performance can be improved by up to 0.7 dB at BER of 10^{-3} with a tap extension from 7 to 23, as shown in Figure 10.65a. The performance of new algorithm is further verified over the transmission in 1500 and 2000 km links. The tap extension from seven to 23 enables a Q gain of almost 0.5 dB at two optimum launch power of -1 and 0 dBm. To test the MIMO filter performance improvement by extending the filter length, we checked the filter of length 83. The improvement could only be about 0.1 dB indicating that further increasing the complexity brings only a negligible gain. Thus, a filter length of 23 is the FIR filter limit.

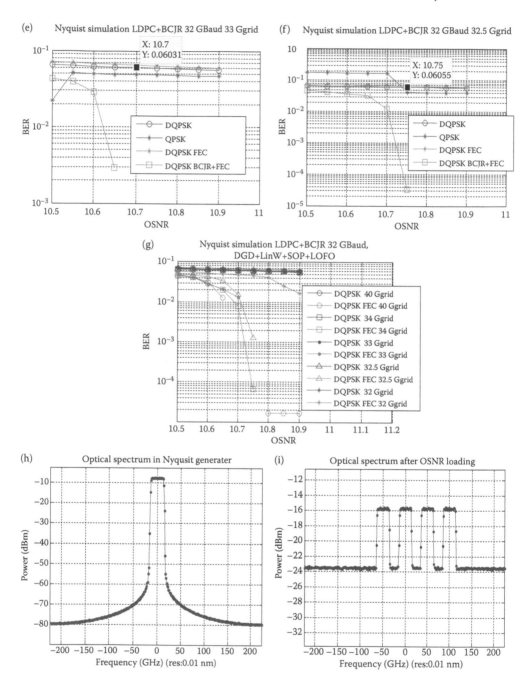

FIGURE 10.62 (Continued) (c and d) LDPC Nyquist QPSK with 33 GHz grid; (e and f) BCJR+LDPC and 3 GHz grid; (g) Nyquist QPSK with and w/o coding gain with grid –frequency spacing as parameters (32–40 GHz) as summary of (a–f); and (h and i) Spectra of channels before and after ASE noise loading.

10.6 MULTI-CARRIER SCHEME COMPARISON

A preliminary comparison and analysis of Nyquist QPSK transmission scheme for superchannel Tbps transmission over long-haul and expected metro networks is shown in Table 10.2.

Hardware complexity: The transmitter side of all schemes requires a comb-generator which can be using one ECL, then employing RCFS technique to generate N sub-carriers of locked phase and

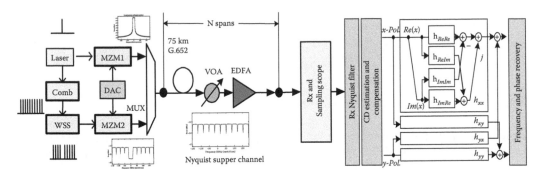

FIGURE 10.63 Nyquist superchannel experimental setup and receiver structure; each of four complex MIMO filters consists of four real FIR filters.

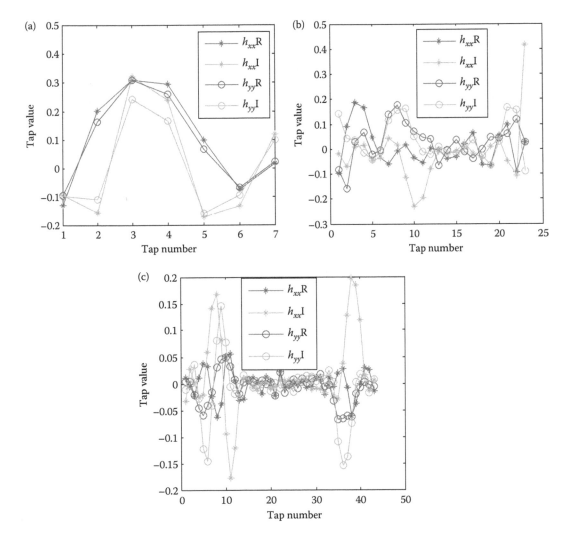

FIGURE 10.64 MIMO filter tap length extending effects in Nyquist WDM optical transmission experimental results. (a) tap length 7, (b) tap length 23, and (c) tap length 43, $h_{xx}R$ for the real part of h_{xx}, and $h_{xx}I$ for the imaginary part of h_{xx}.

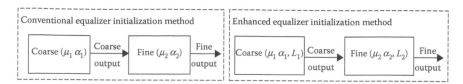

FIGURE 10.65 Conventional and enhanced MIMO filter tap convergence algorithm.

frequency, or a multipled factor of five sub-carriers per ECL using non-linear driving method. Furthermore, a set of parallel PDM I/Q modulators and DAC with eight ports of positive and complementary signals, an optical sub-channel DeMux and a Mux would be necessary for generating independent sets of inphase and quadrature signals for modulating the PDM-IQ modulators. At the receiver side, a standard coherent receiving system can be used with insertion of sharp optical filters at the front end of the optical hybrid coupler. This is required to separate the sub-channels, and may require a set of optical filters, one for each sub-channel. Thus, a technical solution should be provided to avoid this complex and expensive solution. This is the set back of the Nyquist WDM technique.

DSP complexity: For Nyquist-WDM, because the Nyquist filtering results in ISI, at the receiver side one needs sequence estimation algorithms such as MAP and MLSE. For Nyquist QPSK and different comb-generated sub-carriers and sub-channels, FIR with lower number of taps would be suitable, so that the complexity can be acceptable.

ECLs: For RCFS comb-generation, there needs only one ECL source but additional modulator and optic components are required while for non-linear driving comb-gen the number of ECLs would be increased accordingly. To cover the whole C-band with superchannels of 30 GHz spacing, there would be around 12 ECLs. This number may be high, especially when they are to be packaged in the same line card.

Spectral efficiency: All Nyquist pulse shaping schemes can enhance the SE by a factor of about two, the overlapping between sub-channels would create minimum cross talk if the roll-off factor less than 0.1 and the third-order harmonics of the sub-channels and comb-generated sub-carriers are more than 30 dB below the primary carrier.

Tx-DSP and DAC: Ideally, CO-OFDM may not need DSP, but for compensation of components and transmission impairments, it is preferred to use Tx-DSP. The other two schemes must use Tx-DSP and DAC to shape the pulse or generate the designed signal.

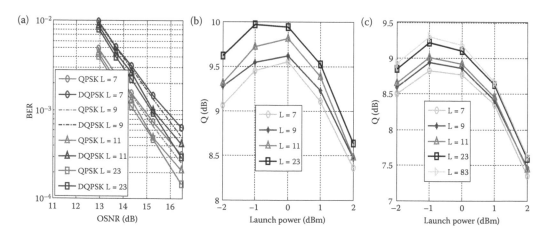

FIGURE 10.66 Enhanced MIMO filter tap length extending performance (BER versus OSNR) in Nyquist WDM optical transmission experiments. (a) BTB, (b) 1500 km transmission results, and (c) 2000 km transmission.

TABLE 10.2

Comparisons of Tbps Transmission Schemes

	Nyquist-WDM	CO-OFDM	eOFDM
Hardware complexity	Tx: similar Rx: multi-OFE	Tx: similar Rx: need OFFT	Tx: similar Rx: multi-OFE
DSP complexity	More complex, possibly because of compensation—but no higher degree of complexity, possibly more processing time required for CPU	Normal	Normal
SE (theoretically)	Similar	Similar	Similar
Tx-DSP + DAC	Essential	Possibly not	Essential
ADC sampling rate	$1.2 \times$ Baud	$2 \times$ Baud	$2 \times$ Bandwidth
Bandwidth requirement on O/E components	Depending on sub-channel spacing	Depending on sub-channel spacing	Depending on sub-channel spacing
Special requirements	DSP for sequence estimation (MAP, MLSE)	Orthogonal channel separation	Cyclic prefix and guard band for OFDM symbols
Flexibility	Medium	Medium	High

ADC sampling rate: This depends on channel spacing, ideally by a factor of $2\times$ the single-sided signal bandwidth. As Nyquist-WDM narrow filters the two-sided signal bandwidth to only 1.3 of Baud rate; for the same baud rate it requires the lowest sampling rate of ADC. 56 to 64 GS/s ADC and DAC are available from Fujitsu which allow the generation of random sequence for 28 GB/s to 32 GB/s.

Bandwidth requirements on O/E (O/E = optical/Electrical) components: Similar to the sampling rate, Nyquist-WDM needs the lowest bandwidth assuming the same baudrate. The other two schemes are similar.

Flexibility: All three schemes have flexibility in number of sub-channels, modulation formats and bandwidth of sub-channels, whilst e-OFDM (e-OFDM = electrical OFDM) can adjust some parameters in electrical domain by means of Tx-DSP, it has more flexibility than the other two schemes.

10.7 CONCLUDING REMARKS

N-Tbps optical transmission are experimentally demonstrated to cover all the possible schemes for the emerging technology. We have conducted and covered all possibilities of pulse shaping and spacing between sub-channels with sharp roll off, as well as the extending the spectral region over the whole C-band. Nyquist pulse shaping modulation is described and applied to generating superchannel for spectral compactness hence minimizing interchannel cross talk and increasing effective occupied spectrum. Thence, ultra-dense WDM (UDWDM) channels are created by modulating sub-carriers obtained from the comb-generations by RCFS or non-linear driving techniques. The target is to have a best-in-class 1 Tbps Nyquist-WDM demonstration system that has been developed and its performance is proven under field demonstrations over the Deutsche Telecom and Vodafone installed transmission systems in Europe. We have completed the development of the solution of 10×28 GB/s (or 30 or 32 GB/s) PDM Nyquist QPSK with 28 or 30, or 32 GHz spacing between adjacent sub-channels. The transmission performance achieved so far have reached 2000 km with BER of $1e-3$ with an OSNR of 14.7 dB. This performance allows to confidently conduct field trial, with FEC coding to obtain further coding gain to combat additional ASE noises from an additional 500 km with 6 spans of 80 km length and six additional optical amplifiers to extend the transmission line to 2500 km reach.

In summary, this chapter has described the Tbps transmission systems that function well in the laboratory environment, as well as in installed fiber multi-span lines. Field trials are proven of the systems employing pulse shaping modulation techniques. The transmission performance from the Tbps transmission systems in laboratory environment and field trials is available and confirmation/comparison of the technical feasibility of Tbps optical coherent communication systems, and more importantly the decision to provide the production of such Tbps super-channel transmission systems. These are for optical core or long-haul networks. Chapter 9 has also given a short-haul multi-Tbps system for optical access networks.

APPENDIX 10A: TECHNICAL DATA OF STANDARD SINGLE MODE OPTICAL FIBER

10A.1   CORNING® SMF-28™ OPTICAL FIBER

10A.1.1   Product Information
PI1036
   Issued: April 2001
   Supercedes: March 2001
   ISO 9001 Registered

10A.2   CORNING® SINGLE-MODE OPTICAL FIBER

10A.2.1   The Standard for Performance
Corning® SMF-28™ single-mode optical fiber has set the standard for value and performance for telephony, cable television, submarine, and utility network applications. Widely used in the transmission of voice, data, and/or video services, SMF-28 fiber is manufactured to the most demanding specifications in the industry. SMF-28 fiber meets or exceeds ITU-T Recommendation G.652, TIA/EIA-492CAAA, IEC Publication 60793-2 and GR-20-CORE requirements.

Taking advantage of today's high-capacity, low-cost transmission components developed for the 1310 nm window, SMF-28 fiber features low dispersion and is optimized for use in the 1310 nm wavelength region. SMF-28 fiber also can be used effectively with TDM and WDM systems operating in the 1550 nm wavelength region.

10A.2.2   Features ad Benefits
- Versatility in 1310 and 1550 nm applications.
- Outstanding geometrical properties for low splice loss and high splice yields.
- OVD manufacturing reliability and product consistency.
- Optimized for use in loose tube, ribbon, and other common cable designs.

10A.2.3   The Sales Leader
Coming SMF-28 fiber is the world's best selling fiber. In 2000, SMF-28 fiber was deployed in over 45 countries around the world. All types of network providers count on this fiber to support network expansion into the twenty-first century.

10A.2.4   Protection and Versatility
SMF-28 fiber is protected for long-term performance and reliability by the CPC™ coating system. Corning's enhanced, dual acrylate CPC coatings provide excellent fiber protection and are easy to work with CPC coatings are designed to be mechanically stripped and have an outside diameter of 245 μm. They are optimized for use in many single- and multi-fiber cable designs including loose tube, ribbon, slotted core, and tight buffer cables.

10A.2.5   Patented Quality Process

SMF-28 fiber is manufactured using the Outside Vapor Deposition (OVD) process, which produces a totally synthetic ultra-pure fiber. As a result, Corning SMF-28 fiber has consistent geometric properties, high strength, and low attenuation. Corning SMF-28 fiber can be counted on to deliver excellent performance and high reliability, reel after reel. Measurement methods comply with ITU recommendations G.650, IEC 60793-1, and Bellcore GR-20-CORE.

10A.2.6   Optical Specifications

10A.2.6.1   Attenuation

Standard Attenuation Cells		
Wavelength	Attenuation Cells (dB/km)	
(nm)	Premium[a]	Standard
1310	≤ 0.35	≤ 0.40
1550	≤ 0.25	≤ 0.30

[a] Lower attenuation available in limited quantities.

10A.2.6.2   Point Discontinuity

No point discontinuity greater than 0.10 dB at either 1310 nm or 1550 nm.

10A.2.6.3   Attenuation at the Water Peak

The attenuation at 1383 ± 3 nm shall not exceed 2.1 dB/km.

Attenuation vs. Wavelength		
Range (nm)	Ref. λ (nm)	Max. α Difference (dB/km)
1285–1330	1310	0.05
1525–1575	1550	0.05

The attenuation in a given wavelength range does not exceed the attenuation of the reference wavelength (λ) by more than the value α.

Attenuation with Bending			
Mandrel Diameter (mm)	Number of Turns	Wavelength (nm)	Induced Attenuation[a] (dB)
32	1	1550	≤ 0.50
50	100	1310	≤ 0.05
50	100	1550	≤ 0.10

[a] The induced attenuation due to fiber wrapped around a mandrel of a specified diameter.

10A.2.6.4   Cable Cutoff Wavelength (λ_{ccf})

$$\lambda_{ccf} \le 1260\,\text{nm}$$

10A.2.6.5   Mode-Field Diameter
  $9.2 \pm 0.4\,\mu\text{m}$ at $1310\,\text{nm}$
  $10.4 \pm 0.8\,\mu\text{m}$ at $1550\,\text{nm}$

10A.2.6.6   Dispersion
Zero dispersion wavelength (λ_0):

$$1302\,\text{nm} \le \lambda_0 \le 1322\,\text{nm}$$

Zero dispersion slope (S_0):

$$\le 0.092\ \text{ps/(nm}^2 \cdot \text{km)}$$

$$\text{Dispersion} = \text{D}(\lambda) :\approx \frac{S_0}{4}\left[\lambda - \frac{\lambda_0^4}{\lambda^3}\right]\text{ps/(nm} \cdot \text{km)}$$

$$\text{for } 1200\,\text{nm} \le \lambda \le 1600\,\text{nm}$$
$$\lambda = \text{operating wavelength}$$

10A.2.6.7   Polarization Mode Dispersion

Fiber Polarization Mode Dispersion (PMD)	Value (ps/$\sqrt{\text{km}}$)
PMD link value	≤ 0.1[a]
Maximum individual fiber	≤ 0.2

[a] Complies with IEC SC 86A/WG1, Method 1, September 1997.

The PMD link value is a term used to describe the PMD of concatenated lengths of fiber (also known as the link quadrature average). This value is used to determine a statistical upper limit for system PMD performance.

Individual PMD values may change when cabled. Corning's fiber specification supports network design requirements for a 0.5 ps/$\sqrt{\text{km}}$ maximum PMD.

APPENDIX 10B: DAC OPERATING CONDITIONS

10B.1   POWER CABLE SETUP INSTRUCTION SHEET

1. The first task is to unplug the PSU board from the DK board.
2. Some DK boards have a soldered connection across all three pins of link LK4 (on lower side of PCB). This should be removed if present.
3. Others have a three-pole jumper plug on LK4. This should be removed.
4. Place a standard two pole jumper across pins two and three of LK4 if one is not already present.
5. Next, connect the cables supplied to the board, and set the power supplies for the following voltages.

Rail	Voltage (V)	Current Limiter Setting (mA)
AVDDACNSY	1.248	3000
1V8A	1.95	2000
1V8B	1.82	2000
0V9A	0.998	3000
0V9B	1.015	2500
AVDNEGDAC	−0.936	800
AVDNEGNSY	−0.973	800
AVDEDAC	2.04	3000
5V_SYN	5.0	500

In the wiring harness supplied, red is +ve, black is common ground, blue is −ve wrt ground. All connectors are labelled with their rail names.

Note that the DAC RF supply has been taken from the 0V9B rail, so that is why it is higher than 0.9 V. All other 0.9 V loads have been placed on 0V9A via links on the DK board.

The following power-up sequence is suggested;

1. output_high(5V SYN);   //DK board fan and 5 V rail to clock source module
    delay_ms(20);
2. output_high(1V8A);   //Feeds 1V8A rail, supplying VDDE rail to DAC via LK2
    delay_ms(20);
3. output_high(1V8B);   //Feeds 1V8B rail, supplying AVDEDACNSY rail to DAC via LK1. Also AVDESFIS51 rail to DAC via LK3.

    delay_ms(20);
4. output_high(0V9A);   //Feeds 0V9A rail, which supplies AVDDAC/AVDDAC_B rail to DAC via LK5. Also VDDI rail to DAC via LK4. Also AVDISFIS51

Section III

11 Single Mode Planar Optical Waveguide

11.1 INTRODUCTION

The term "integrated optics" was first coined by Miller in 1969 [1], as an analogy of lightwave to the electronic integrated circuits. The present term used in the industry is planar lightwave circuits (PLC). Since that proposed term, there has been tremendous progress. Various integrated optical devices have been researched, developed, and deployed in practical optical transmission systems and networks. Extensive surveys have been given over the years [2,3] and even defined and described on the Internet [4,5]. Both linear and non-linear integrated optics [6] have been exploited.

Recent experimental demonstrations have pushed the information transmission bit rate per channel to 100 Gb/s [7–9] with the multiplexing of several wavelength channels reaching to tens of Tera-bits/sec [10–12]. At such speed, the needs of optical modulation, switching, optical pre-processing using non-linearity of integrated optical devices, and the compensation of dispersion of single mode optical fibers using fiber or integrated optic components and equalization of the losses of fibers by optical amplification. Prior to the availability of Erbium-doped fiber amplifiers (EDFA), attempts to increase the span distance between repeaters of amplitude modulated single fiber transmission systems by employing coherent techniques [13,14] in which a narrow linewidth and high power is used to mix with the received signals to improve the receiver sensitivity. This linewidth requirement on the local oscillator laser limits the deployment of such coherent transmission in real practice. Recently, coherent optical communications have attracted much interest as the possible technique to further increase the transmission spans [15,16]. Naturally, the modulation formats such as amplitude shift keying, phase shift keying, and frequency shift keying are employed to reduce the effective signal band width to minimize the dispersion effects in single mode optical fibers. The detection of ultra-high-speed optical signals can be in the direct or coherent detection. This attracts the employment of integrated optical devices in the optical transmitters, receivers, and online components.

The basic component of these integrated photonic devices is the optical waveguide formed by a thin film or diffused waveguiding layer structure on some substrate. From the mode propagation of point of view, design optimization requires accurate estimation of the propagation constant, thence dispersion characteristics, mode size, and group velocities, depending on the types of applications. These requirements led us to develop simple, accurate, and efficient methods of analysis or single mode waveguides. This is the motivation of the first chapter of this book.

Most waveguides found in integrated optics, especially lithium niobate types, are fabricated by a diffusion process that is commonly formed by diffusion of impurities into various substrate ferroelectric materials such as lithium niobate and lithium tantalate. Complementary error function is usually used to represent the distribution of the impurity from surface of the substrate into the depth of the substrate [17] and the time of diffusion of the metallic impurities. A Gaussian profile is expected when the diffusion time is sufficiently long [18], which is used to fit the experimental values of Se into CdS crystal. In addition, various profiles can be used to form optical waveguides using molecular beam epitaxy (MBE), metallic organic chemical vapor deposition (MOCVD) techniques as in the waveguiding structures for laser diodes of separate confinement of the heterojunction [19].

Exact analytical solutions are available for the step [20], exponential [21], hyberbolic secant [22], clad linear [23], clad-parabolic [24], and Fermi [25] profiles. In general, approximate analytical or exact numerical methods are required to analyze general classes of profiles. For some practical profiles, universal charts describing mechanism of waveguiding have already been presented by several authors. These have been obtained by variational analysis [26], WKB [27], and multi-layer staircase [28]. However, these curves are only accurate for multi-mode waveguides, except for the last two references. Single mode planar optical waveguides are very important for integrated optic circuits for applications in advanced ultra-high-speed optical communications, see, for example,e the pioneering woks by Karotsky [29]. However, methods of analysis are confined predominantly to those originally used in multi-mode waveguides. In the single-mode regime, the variational and WKB methods are expected to perform poorly. In the former, the solutions are strongly affected by the choice of trial fields. In the latter, more accurate prediction of the phase changes at the turning point are required. In the Runge–Kutta outward integration methods, instability is caused by the solution and error increasing at large "x" [30]. This problem is resolved by approximating the fields at sufficient depth in the waveguide by an evanescent field [31]. However, this requires the knowledge of the location of solution matching.

Thus, this chapter recognizes that any general technique of analysis must be numerical in nature due to the more stringent accuracy requirements for single mode waveguides. In the case that the single mode waveguide is used as a non-linear interaction medium, the phase matching is very important and an accurate estimation of the dispersion curves plays a very important part in the conversion efficiency [32,33].

This chapter is organized as follows: Section 11.2 gives the formation of the problems for solving single mode optical waveguides in which all parameters are expressed in terms of normalized quantities. A novel relationship between a newly defined mode spot size and the normalized waveguide parameter is described. This is very close to the definition of the mode spot size as defined for single mode optical fibers that is commonly specified. Section 11.3 describes the simplest method for calculating the propagation constant and the model field. The stepping function recurrence method of integration originally developed by Killingbeck [34,35] is extensively used. The method is applied for asymmetrical optical waveguide structures.

For diffused waveguides or graded index profile distribution of the refractive index from the surface of the waveguide to the deepest position, two widely used methods for waveguide modal analysis are given. The variational method with a simple Hermite-Gaussian field was first introduced by Korotsky [29] to calculate the mode spot size in a diffused channel waveguide. In Section 11.4, we show that the method to estimate the modal characteristics of all diffused waveguide profiles and shows that is inaccurate and computationally intensive for the calculation of the dispersion characteristics, but very close estimation of the mode spot size.

11.2   FORMATION OF PLANAR SINGLE MODE WAVEGUIDE PROBLEMS

A planar dielectric waveguide with the geometry shown in Figure 11.1 can support modes with two polarizations. These are the transverse electric (TE) and transverse magnetic (TM) field guided modes. In practice, polarized modes can be excited. Provided certain boundary conditions are met, these modes are bounded and propagating along the z-axis, each with a unique effective phase velocity. If we allow for the uniformity of the refractive index and geometrical dimension in the propagation direction, the phase velocity is only a function of the transverse index profile. In this section, we consider the simplest configuration with index variation only in the −x direction. Note that the notation of the coordinate system follows the right hand rule (RHS).

11.2.1   TE/TM Wave Equation

The wave equation for the guided modes stems from the Maxwell equations given in the appendix together with the associate constituent relations. We consider the steady state solutions in the dielectric medium free from any sources and losses. By omitting the common factor $e^{j(\omega t - \beta z)}$ from the equations, we can write, for a medium characterized with the refractive index $n(x)$, the

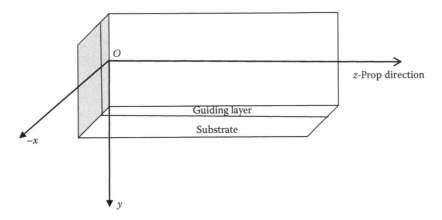

FIGURE 11.1   Schematic structure of a planar dielectric waveguide.

well-known wave equation:

$$\frac{d^2 E_y}{dx^2} + \left[k_0^2 n^2(x) - \beta^2\right]E_y = 0$$

and   (11.1)

$$\frac{d^2 H_y}{dx^2} + \left[k_0^2 n^2(x) - \beta^2\right]H_y = \frac{1}{n^2(x)}\frac{d^2 n^2(x)}{dx^2}\frac{dH_y}{dx}$$

where k_0 = wave number in free space; β = propagation constant of the wave along the z-axis; $E_y(x)$ = transverse electric field; and H_y = transverse magnetic field.

11.2.2   CONTINUITY REQUIREMENTS AND BOUNDARY CONDITIONS

In practical waveguides, it is common to expect that at least one region of dielectric continuity is encountered by the optical fields. At the location of the discontinuity, the wave equations are not valid. However, the identity of the modes is preserved by matching the fields and their derivatives on either side of the dielectric discontinuity. For the TE modes, these boundary conditions impose the continuity E_y and (dE_y/dy) across the interface. For the TM modes, we require the continuity H_y and $1/(n^2(x))(dH_y/dy)$. In addition, the bound modes satisfy the conditions that E_y, H_y vanishes at $x = \infty$. Together, they give rise to the eigenvalue equation from which the propagation constant can be calculated.

Note that the principal object of the eigenvalue equation is to estimate the maximum value of the propagation constant so that the dependence of this propagation parameter on the optical frequency/wavelength is minimum so that there is minimum dispersion of the waves at different wavelength that is normally found for the other spectral components of a modulated lightwave channel in optical communication system. A maximum value of the propagation constant along the z-direction means that the direction of the wave vector is close to the propagation axis.

11.2.3   INDEX PROFILE CONSTRUCTION

For the purpose of computation and analysis, it is customary to write the refractive index profile in the general form as

$$n^2(x) = \begin{cases} n_s^2\left[1 + 2\Delta S\!\left(\dfrac{x}{d}\right)\right] & x \geq 0 \\ n_c^2 & x < 0 \end{cases}$$

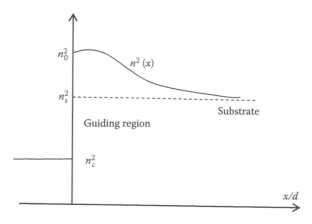

FIGURE 11.2 Square of the refractive index distribution profile for an asymmetrical waveguide.

where

$$\Delta = \frac{n_0^2 - n_s^2}{2n_s^2} \tag{11.2}$$

Δ is the profile height and n_c, n_0, and n_s are the refractive indices of the cover, the guiding layer and the substrate, respectively. $S(x)$ is the profile shape function and d is the diffusion depth of a graded index distribution. Figure 11.2 shows a typical representation of the graded index profile. It turns out that further normalization of the shape profile can be represented as

$$S\left(X = \frac{x}{d}\right) = \frac{n^2(x) - n_s^2}{n_0^2 - n_s^2} \tag{11.3}$$

This definition of the profile shape is unaffected by the symmetry of waveguide. In symmetric waveguide structure with an axis of symmetry at $x = 0$, these equations are equally valid.

11.2.4 NORMALIZATION AND SIMPLIFICATION

The presence of the non-zero term in the RHS of Equation 11.1 complicates the analysis. It is identically zero for a step index profile. The exact solutions of this equation are available for the exponential, hyperbolic secant, and an inverted-x profile [36]. However, for smooth profiles normally encountered in practice, several authors [37] found by perturbation analysis that the RHS of the equation can be neglected. The fundamental mode of an infinite parabolic profile face a 44% error in the group velocity of the profile shape guided mode.

Following Kolgenik and Ramaswamy [38], we can introduce the normalized parameter for the waveguide as follows

$$V = \frac{dk_0}{n_0^2 - n_s^2}$$

$$A = \frac{n_s^2 - n_c^2}{n_0^2 - n_s^2}; \quad \text{and} \tag{11.4}$$

$$b = \frac{n_e^2 - n_s^2}{n_0^2 - n_s^2}$$

where A is defined as the asymmetry factor, $b =$ normalized propagation constant $n_e = \beta/k_0$ is the effective refractive index of the guided mode along the propagation axis; and V is the normalized frequency.

For a guided mode, it requires that

$$n_s < n_e < n_0 \tag{11.5}$$

Thus, the normalized propagation constant must satisfy $B < 1$. The real advantages of the normalization comes in the analysis and design optimization point of view. Substituting the normalized parameters into the wave Equation 11.1, we obtain:

$$\frac{d^2\varphi}{dx^2} + V^2[S(X) - b]\varphi = 0 \tag{11.6}$$

where $\varphi(X) \equiv E_y$; H_y for the TE and TM modes, respectively. The propagation for the TM modes are accurately represented by those of the TE modes, except at cut-off or for waveguide at large symmetry [39]. However, if extreme accuracy or mode splitting is required, then Equation 11.6 can be modified to the changes involving only a slight modification of the boundary condition.

11.2.5 MODAL PARAMETERS OF PLANAR OPTICAL WAVEGUIDES

Solution of the wave Equation 11.6 together with the boundary conditions enables the determination of various optical parameters. The following are the most commonly used for the design of single optical guidedwave devices:

11.2.5.1 Mode Size

The mode size Γ_a for an asymmetrical field is defined as the full-width half maximum (FWHM) power intensity. For a full description of the field, the peak position of the intensity I_p and the field asymmetrical factor Γ_1/Γ_2, defined with respect to I_p are required as shown in Figure 11.3. The knowledge of the mode size is critical to match with that of the single mode optical fiber for inline integration with fiber transmission systems. These parameters are defined in term of the optical power of the field due to practical reasons because the mode size is normally monitored using a CCD camera through which the intensity of the mode field is converted into the charge current and displayed or digitized for data processing.

11.2.5.2 Propagation Constant and Effective Refractive Index

The variation of the normalized propagation constant b as a function of the normalized frequency parameter V is normally required for the design and characterization of optical waveguide. For

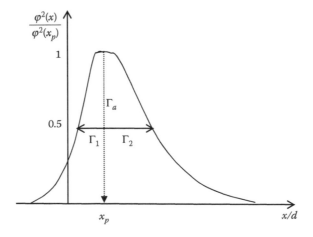

FIGURE 11.3 The minimum set of parameters required for characterization of fundamental mode field in an asymmetrical planar waveguide.

example, this relationship specifies the diffusion depth required once the mode index at a specific operating wavelength is given. The purpose of this section is to present the relation of the modal field to the propagation constant.

If we integrate Equation 11.6 with respect to X, then we have

$$\int_{-\infty}^{+\infty} \left\{ \frac{d^2\varphi}{dx^2} + V^2[S(X) - b]\varphi \right\} dX = 0$$

$$\int_{-\infty}^{+\infty} \varphi'' dX + V^2 \int_{-\infty}^{+\infty} \left\{ [S(X) - b]\varphi(X) \right\} dX \tag{11.7}$$

where the dashes denote the derivative with respect to $X>$ if we further impose the condition that $\varphi(-\infty) = \varphi(+\infty) = 0$, the integral in the left becomes zero and we obtain

$$b = \frac{\int_{-\infty}^{+\infty} S(X)\varphi(X)dX}{\int_{-\infty}^{+\infty} \varphi(X)dX} \tag{11.8}$$

This equation indicates that the dispersion characteristics for an arbitrary index profile must be a smooth curve. The rule of refractive index used by Tien [40] to explain a host of new wave phenomena in integrated optical waveguides that light tends to propagate in the region where the refractive index is largest. In the context of planar optical waveguides with arbitrary index profile, this rule suggests that for a given profile at a given frequency, the mode field adjusts itself so that maximum value of b is achieved. This corresponds to the minimum phase velocity allowed for the mode. This rule is a direct statement of Fermat's law in ray optics and is a special case of a generalized rule in quantum mechanics formulated in the form of well-known Feymann's path integral of which Maxwell's equations are also satisfied [41].

11.2.5.3 Waveguide Dispersion and Spot Size

A second and potentially useful relation between b and the modal field can be established from the stationary expression for b. This can be obtained from Equation 11.6 after multiplication by φ and taking the integration with respect to X from negative to positive infinitive. After integrating the results by parts and imposing the boundary conditions $\varphi(-\infty)\varphi'(-\infty) = \varphi(+\infty)\varphi'(\infty) = 0$, we obtain the well-known stationary relation

$$bV^2 = \frac{V^2 \int_{-\infty}^{+\infty} S(X)\varphi^2(X)dX - \int_{-\infty}^{+\infty} \varphi^2(X)dX}{\int_{-\infty}^{+\infty} \varphi^2(X)dX} \tag{11.9}$$

This is the basic equation for the variational analysis. It possesses the unique property that for any trial fields which satisfy the boundary conditions above, the quotient remains stationary provided the mismatch between the trial and actual fields is small [42].

Thus, we can write

$$\frac{d(bV^2)}{dV^2} = \frac{\int_{-\infty}^{+\infty} S(X)\varphi^2(X)dX}{\int_{-\infty}^{+\infty} \varphi^2(X)dX} \tag{11.10}$$

that is,

$$\frac{1}{2}\left(bV + V\frac{d(bV)}{dV} \right) = V\frac{\int_{-\infty}^{+\infty} S(X)\varphi^2(X)dX}{\int_{-\infty}^{+\infty} \varphi^2(X)dX} \tag{11.11}$$

Taking the derivative second time and using Equations 11.9 and 11.11, we obtain

$$\frac{1}{2}\left(V\frac{d^2(bV)}{dV^2} + 2\frac{d(bV)}{dV}\right) = V\frac{d}{dV}\left(b + \frac{2}{V^2 W_m}\right) + \frac{1}{2}\left(b + \frac{d(bV)}{dV}\right) \quad (11.12)$$

where we have defined a new spot size parameter as

$$W_m = \frac{2\int_{-\infty}^{+\infty}\varphi^2(X)dX}{\int_{-\infty}^{+\infty}[\varphi'(X)]^2 dX} \quad (11.13)$$

Furthermore, algebraic manipulation leads to the simple relationship

$$V\frac{d^2(bV)}{dV^2} = 4\frac{d}{dV}\left(\frac{2}{VW_m^2}\right) \quad (11.14)$$

This relation is analogous to the relation between Petermann's spot size and the waveguide dispersion in single mode optical fibers [43]. The preceding analysis was first performed by Sansonetti [44] who inspired Petermann to define a new spot size in the characterization of single mode optical fibers from spot size measurement.

The relation by Equation 11.14 can be found for optical waveguide with a profile following Hermit Gaussian variational field of

$$\varphi(X) = \begin{cases} A_0\alpha_0^{1/2}e^{-\alpha_0 X^2/2}; & X \geq 0 \\ 0; & X < 0 \end{cases} \quad (11.15)$$

$$W_m^2 = \frac{4}{\alpha_0^2} = \Gamma_a^2 \quad (11.16)$$

where α_0 is the variational spot size parameter and A_0 is a constant. The RHS of Equation 11.16 corresponds to Γ_a as defined by Korotsky [45] which has been defined above. Although α_0 is an approximate mode spot size, several experiments [45] show excellent agreement between the theoretical and experimental values W_m and Γ_a.

11.3 APPROXIMATE ANALYTICAL METHODS OF SOLUTION

Despite the availability of direct numerical integration methods for the analysis of optical waveguides, approximate analytical solutions are still being used, improved, and sought after. We have to strike the balance between accuracy and simplification. Three well-known methods of analysis are described in this section. They are valid for single mode planar optical waveguides.

The variational method [45] is applicable only to the fundamental mode of the asymmetrical waveguide due to the form of the trial field. The equivalent profile method is valid only for symmetrical waveguides because it requires the field to be monotonously decreasing. The WKB method can be used in both cases (see Appendix). We group the methodological approaches into symmetry and asymmetry. In Section 3.1, the analytical formulae for a number of widely used profiles are obtained. We explore the improvements to the WKB method and limitations. In Section 3.2, we compare the accuracy of the equivalent profile moment methods using a step and a cosh reference profile [46,47]. The WKB method may not work at all for the analysis of the single mode optical waveguides.

11.3.1   Asymmetrical Waveguides

11.3.1.1   Variational Techniques

The variational method is based on the substitution of a TE_0 mode look-alike trial field into the stationary expression of the normalized propagation constant b given in Equation 11.8. The shape of the field is then adjusted to maximize b for all values of V (see Equation 11.9). The mathematical procedure is given in Reference 48.

Following Korotky et al. [45] and Riviere et al. [49], a trial solution can be proposed that closely fits the form of the TE_0 field

$$\varphi(X) = \begin{cases} \sqrt{\alpha_0}e^{-\frac{\alpha_0 X^2}{2}} & X \geq 0 \\ 0 \ldots X < 0 \end{cases} \tag{11.17}$$

Note that one drawback of the form of this field is that it vanishes at $X = 0$. For single-mode optical waveguides, this condition is only for guides with very large asymmetry. However, only a single parameter needs to be optimized, thus the optimization scheme is simple. In the following, it is shown that there exist closed form formulae for several profiles.

11.3.1.1.1   Eigen Value Equation

If we substitute Equation 11.17 and the derivative of this trial field into Equation 11.9, a simpler expression is obtained, after some tedious algebra, as

$$b = I_1 - \frac{3\alpha_0}{2V^2} \tag{11.18}$$

where

$$I_1 = 4\alpha_0 \left(\frac{\alpha_0}{\pi}\right)^{1/2} \int_0^{\infty} S(X)X^2 e^{-\alpha_0 X^2} dX$$

is the only profile dependent expression. The correct value of α_0 is obtained by noting that b must be stationary with respect to α_0, thus

$$\frac{db}{d\alpha_0} = 0 = \frac{dI_1}{d\alpha_0} - \frac{3}{2V^2} \tag{11.19}$$

Substituting this α_0 into Equation 11.18, we obtain the eigenvalues given in Table 11.1.

11.3.1.1.2   Fundamental Mode Cut-Off Frequency

The lowest-order mode in an asymmetric optical waveguide has a non-zero cut-off frequency for $A \neq 0$. Thus, we can set $b = 0$ and $V = V_c$ in Equations 11.18 and 11.19 to obtain the desired cut-off frequencies. Since V_c appears in both equations, one can solve simultaneously for α_0 initially

---

TABLE 11.1

Optimum Value of α_0 for Selected Asymmetrical Clad-Diffused Waveguide

Profile	$S(X)$	α_0
Gaussian	e^{-X^2}	$(\alpha_0 + 1)\left[\dfrac{(\alpha_0 + 1)}{\alpha_0}\right]^{1/2} - V = 0$
Exponential	e^{-X}	$4V^2\sqrt{\alpha_0} - 3\pi(\alpha_0 + 1)^2 = 0$
Complementary error function erfc	$Erfc(X)$	$2(4\alpha_0 + 1)\sqrt{\dfrac{\alpha_0}{\pi}} - (1 + 6\alpha_0)\left[1 - erfc\dfrac{1}{2}\sqrt{\alpha_0}\right]e^{\frac{1}{4\alpha_0}} - \dfrac{12\alpha_0^3}{V^2}$

TABLE 11.2

Optimum Value of α_0 and Fundamental Cut-Off Normalized Frequency Parameters of exp and erfc Profiles

Profile	Parameter	Equation
Exponential	α_0	$\dfrac{(2\alpha_0^2 + 5\alpha_0 - 2)}{16\alpha_0(2\alpha_0 + 1)}\left[1 - erfc\dfrac{1}{2\sqrt{\alpha_0}}\right] + e^{\frac{1}{4\alpha_0}} + 1 = 0$
	V_c	$\left(\dfrac{3\alpha_0}{2} + \dfrac{1}{\sqrt{\alpha_0}}\right)\left[\left(1 + \dfrac{1}{2\alpha_0}\right)e^{\frac{1}{4\alpha_0}}\left(1 - erfc\dfrac{1}{2\sqrt{\alpha_0}}\right)\right] = V_c^2$
Complementary error function erfc	α_0	$\dfrac{(\alpha_0 + 1)}{\sqrt{\alpha_0}}\tan^{-1}\sqrt{\alpha_0} - \dfrac{\alpha_0}{(\alpha_0 + 1)} - 1 = 0$
	V_c	$V_c^2 = \dfrac{3\pi(\alpha_0 + 1)^2}{4\sqrt{\alpha_0}}$

before substituting back to obtain the cut-off value for the V-parameter, V_c. It happens that the cut-off V_c for a Gaussian profile is given as $V_c = 1.9741$. Table 11.2 tabulates the analytical expressions for the cut-off frequency of the exponential and complementary error function profiles.

The computation of the propagation constant and the mode cut-off frequency for TE_0 mode that requires numerical integration would be tedious. Fortunately, the commonly encountered graded profile waveguide shown in Table 11.1 only involves root-search for the estimation of α_0. There is no existing method to estimate the probable range of α_0 as a function of V. Thus, the consuming process in the computation of is the correct estimation of the interval for root search algorithm. Nevertheless, one would be interested in the instigation of the accuracy of the results over a selected range of V. The estimated values of the propagation constant b are calculated and tabulated in for a number of profile distribution with a set of specific parameters $n_c = 1.0$, $n_s = 2.177$, and $\Delta = 0.043$. This is a typical profile structure for air cover diffused waveguide profile in $LiNbO_3$ or $LiTaO_3$ substrate. The trial field distribution is a Hermite-Gaussian. The corresponding cut-off frequencies at the cut-off limit are given for TE_0 in Table 11.3.

The values of the propagation constants are as expected. The accuracy of the variational field fits to the actual field improves with increasing frequency. At large V, the field in the cover decreases rapidly. Similarly, the evanescent field in the substrate follows similar trend. Thus, the mode field is confined within the guiding region and its shape is accurately modeled by a Hermite Gaussian function. This behavior was first observed experimentally by Kiel and Auracher [50]. This observation leads to the motivation for the pioneering work of Korotky et al. [45] in the use of this simple trial field. An earlier method by Taylor requires of up to 21 terms in the variational field involving parabolic cylinder functions [51]. A simple relationship between α_0 and the mode spot size Γ_a can be

TABLE 11.3

Cut-Off Frequencies of the TE_0 Mode

Profile	Exponential		Gaussian		Complementary Error Function	
	V_c (var.)	V_c (exact)	V_c (var.)	V_c (exact)	V_c (var.)	V_c (exact)
	1.563	1.087	1.974	1.433	2.839	2.085
α_0 at V_c		0.143		0.500		0.697

TABLE 11.4

b–V **Data for Selected Profiles Calculated with a Hermite–Gaussian Trial Field (Variational Method; Exact = Analytical Expression)**

	Exponential		Gaussian		Complementary Error Function	
V	b (var.)	b (exact)	b (var.)	b (exact)	b (var.)	b (exact)
2	0.066	0.105	–	–	–	–
3	0.193	0.299	0.216	0.275	0.015	0.068
4	0.289	0.321	0.370	0.413	0.121	0.169
5	0.362	0.390	0.476	0.510	0.213	0.255
10	0.560	0.578	0.719	0.732	0.477	0.497
100	–	0.897	0.970	0.971	0.883	0.885

obtained as [49]:

$$\Gamma_a = \frac{1.555}{\sqrt{\alpha_0}} \tag{11.20}$$

The mode size Γ_a can be obtained directly without using numerical computing. However, Korotky et al. found good agreement between experimental and theoretical results [52], and our analytical results given here in Table 11.4 show that there are substantial discrepancies in the propagation constant for single mode optical waveguides. He means that the mode spot size Γ_a may not be so dependent on the frequencies. Experimentally speaking, the variation of the wavelength of the guiding waveguide and detection is due to the sensitivity of the spot size image monitoring device. This must be taken into account for the measurements of the full-width half mark of the image. To investigate this possibility, Γ_a is plotted versus the normalized frequency V parameter shown in Figure 11.4. This step is also taken to examine the behavior of α_0 near the cut-off frequency of the

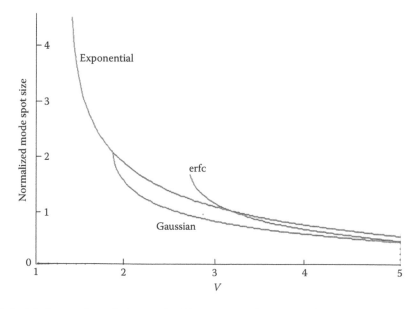

FIGURE 11.4 Mode spot size calculated using Hermite–Gaussian trial field for different profiles.

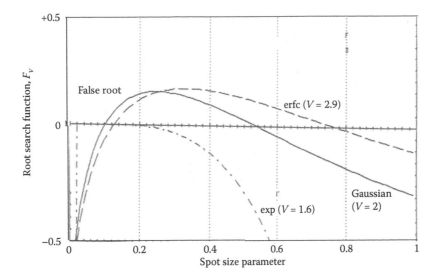

FIGURE 11.5 Multiple roots of the root search function of the variational method.

fundamental mode. The difficulty in the calculation of the cut-off frequencies given in Table 11.3 can be observed. This is due to the volatility of the confinement of the mode near cut-off. This is a well-known phenomenon in optical fibers [53]. Figure 11.4 shows the variation of Γ_a with respect to V over a range of frequencies including at $V = V_c$ for the profiles of exponential, Gaussian and complementary error shapes. Two sets of data and curves are given to notice the method of using the root-search algorithm to compute the optimum spot size parameters α_0. More than one root can exist in the search interval. The smooth set of curves given in Figure 11.3 is obtained by choosing only the negative going cross-over of the curves given in Figure 11.4. The kinks observed in Figures 11.5 and 11.6 are obtained when choosing the smaller and incorrect root. The propagation constant computed from this false zero is much smaller and can be negative. Thus, it is preferred to operate the

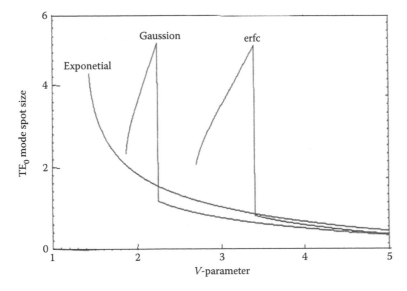

FIGURE 11.6 Mode spot size of TE_0 mode as a function of V for profiles of exponential, Gaussian and complementary error function estimated using variational method.

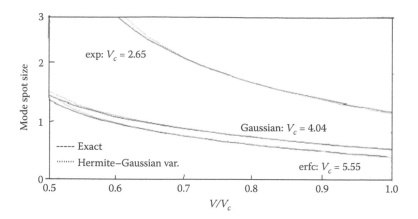

FIGURE 11.7 Spectral variation of mode spot size: accuracy of Hermite–Gaussian trial field fitting for single mode diffused clad profiles. Single mode diffused-clad profiles $n_c = 1.0$; $n_s = 2.177$; $\Delta = 0.043$ ($A = 20$).

waveguide far from the cut-off region so that the mode spot size is not strongly dependent to the V-parameter. This scenario is important for the case when planar optical waveguides are used as an optical amplifier [52], for example, Erbium-doped LiNbO3 waveguide, the wavelength of the pump beam is far from the operating wavelength region and may be close to the cut-off. This must be taken into account. If not, then the fluctuation of the mode spot size would alter the amplification gain of the amplifier. In practice, the refractive index profile could never be modeled by any form of analytical function and an equivalent profile may be used.

The correct spot size behavior computed numerically using the numerical algorithm given here follows similar trend, as shown in Figures 11.7 and 11.8. The variational spot size is superimposed on these curves for comparison, the agreement is remarkable. The wavy curves in Figure 11.8 are caused by numerical noises. The tolerance on each plot is 1%>. Such accuracy is achieved because of the definition described here for the spot size that does not take into account of the tails of the field. This is where serious agreement between the exact and the Hermite Gaussian occurs. This explains the discrepancies in the normalized propagation constant b as estimated by this method.

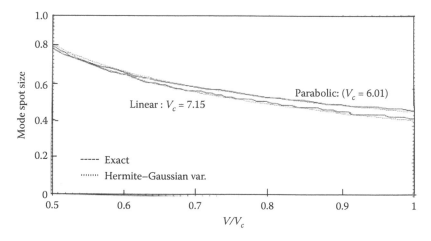

FIGURE 11.8 Accuracy of Hermite Gaussian trial field fitting for single-mode clad power law profiles. Single mode clad-power-law profiles with $n_c = 1.0$; $n_s = 2.177$; $\Delta = 0.043$ ($A = 20$).

11.3.1.2 WKB Methods

The WKB method was first developed by Jeffery [54] and applied to the calculation of energy eigenvalues in quantum mechanics by Wentzel [55], Kramers [56], and Brilluoin [57]. Because of the similarities between problems involving the quantum mechanical potential well and the refractive index profiles of optical waveguides [40], the method can be easily adapted for used in guided wave optics. Marcuse first used the method for proving the eigenvalue equation of asymmetrical graded index optical waveguides [58]. A simplified derivation by Hocker and Burns [59] based on ray optics confirmed Marcuse's results. This is due to the equivalence of the WKB and ray optics formalism [60].

The central problem involved the techniques that lie with the connection of oscillatory and evanescent fields at the turning point where the original WKB solutions are singular. Langer solved the problem by approximating the actual fields there by Airy functions [61]. This is equivalent to replacing the actual profile locally by a linear segment. Its slope and position are implicitly related to the propagation constant in the eigenvalue equation. We have examined the turning point phenomena in detail (see Appendix 11C.1). We thus can state that the WKB method is not limited by the inaccurate phase prediction at the turning point. A more serious limitation is caused by the neglect of the cladding. Coupling effects between the turning point and cladding have been studied in details by Arnold [62]. He found that the cladding effects can be isolated and built into the eigenvalue equation. However the corrections involved a complicated nest of Airy functions and the analytic simplicity of the method is lost.

We take a simpler and more practical approach to account for cladding effects and study the behavior of the WKB errors and found that it is minute for asymmetrical waveguides provided a simple correction is added. More discussions on the improvement of the method presented will be given in appropriate sections.

11.3.1.2.1 Derivation of WKB Eigenvalue Equation

In an asymmetrical of the WKB, the eigenvalue equation is obtained by matching the field and its derivative at the dielectric interfaces. For the WKB method, this is complicated by the fact that the solutions must be matched correctly at the turning point. The Jeffery's solution can be referred as the zeroth-order WKB method and Langer's method with turning point correction as the first-order WKB method (see Figure 11.9). Following Gordon's [63] and Marcuse's [64], one can write for a graded index asymmetrical waveguide as

$$\varphi(X) = \begin{cases} a_0 e^{V\sqrt{A+bX}}; & X \le 0 \\ p^{-1/2}(X)\cos[\phi(X) - \pi/4]; & 0 \le X < X_t^- \\ \left(\dfrac{2\pi\phi}{3p}\right)^{-1/2}\left[J_{1/3}(\phi) + J_{-1/3}(\phi)\right]; & X = X_t^- \\ \left(\dfrac{2\pi\phi}{3p}\right)^{+1/2}\left[I_{1/3}(\phi) + I_{-1/3}(\phi)\right]; & X = X_t^+ \\ \left(\dfrac{p(X)}{4}\right)^{-1/2} e^{-\phi(X)}; & X_t^+ \le X < \infty \end{cases} \qquad (11.21)$$

where $\phi(X) = V\int_X^{X_t}\left|\sqrt{S(X) - b}\right|dX$ and $p(X) = [S(X) - b/(1 - b)]^{1/2}$; a_0 is a constant; I and J are the Bessel's functions representation of the Airy solutions at the turning point, X_t.

The turning point is defined such that

$$S(X_t) = b \qquad (11.22)$$

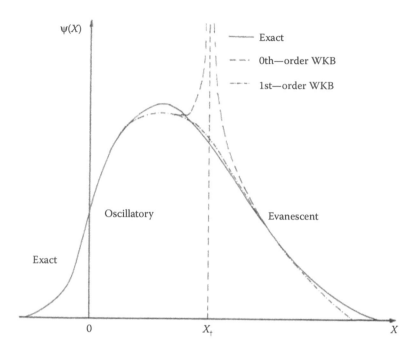

FIGURE 11.9 An illustration of regions of validity of the WKB solutions. The turning point is given by $S(X_t) = B$. The WKB eigenvalue equation is obtained by ensuring that the WKB solutions in the guide are matched to the exact field in the cover (superstrate).

A general proof is given in Appendix 11C.1, which shows that at a turning point, the approximation of the exact field by Airy function is extremely good if

$$S'(X_t) \approx 0 \quad \text{and} \quad S''(X_t) \approx 0 \tag{11.23}$$

Furthermore, the oscillatory solution for $0 \leq X < X_t^-$ and the evanescent field for $X > X_t^+$ are just asymptotic expansions of the Bessel's solutions for $\phi(X) \gg 1$; that is, $V \gg 1 >$ these are just the zeroth-order WKB solutions (see Appendix 11C.1). They must be used in these forms with the correct phase arguments to ensure uniformity of the WKB solutions in the guide and the substrate are already correctly matched.

The eigenvalue equation follows by ensuring the smooth matching of the WKB solution and the exact evanescent field at $X = 0$. The continuity of $\varphi(0)$ and $\varphi'(0)$ gives, after some algebraic manipulations,

$$V \int_0^{X_t} \left| \sqrt{S(X) - b} \right| dX = \left(m + \frac{1}{4} \right) \pi + \tan^{-1} \left(\frac{\sqrt{A + b}}{1 - b} + \delta \right) \tag{11.24}$$

where $\delta = S'(0)/(4V\sqrt{1 - b})$

If setting $d = 0$, then the WKB eigenvalue equation becomes

$$2V \int_0^{X_t} \left| \sqrt{S(X) - b} \right| dX - \frac{\pi}{2} - 2\tan^{-1} \left(\frac{\sqrt{A + b}}{1 - b} + \delta \right) = 2m\pi \tag{11.25}$$

as obtained by Marcuse [65]. For practical multi-mode waveguides, Hocker and Burn [66] claimed that $\tan^{-1}\left(\sqrt{A+b}/(1-b)\right) \approx \pi/2$ since $A \gg 1$, thus leading to the following relationship

$$V \int_0^{X_t} \left|\sqrt{S(X) - b}\right| dX = \left(m + \frac{3}{4}\right)\pi \qquad (11.26)$$

The third term of the RHS of Equation 11.25 exists due to the phase shift under gone by the modal field by the discontinuity at $X = 0$. Equation 11.25 is just the mathematical statement of the similar phase resonance condition of ray optics [67], which states that the phase accumulated along the ray path over one period including reflection as it traverses the guide from $X = 0$ to $X = X_t$ must be a multiple of 2π, if constructive interference is to occur. Constructive interference is essential for maintaining a stable modal pattern. In fact, the phase changes at turning point and the dielectric discontinuity are just $-\pi/2$ and $-2\tan^{-1}\left(\sqrt{A+b}/(1-b)\right)$.

Figure 11.10 illustrates the ray path of the process. To solve b for a given V, the turning point is tuned until the phase resonance condition is met. These results have been derived without the use of WKB formalism by Hocker and Burns [68].

11.3.1.2.2 Limitation of the WKB Method

Three sources of errors inherent in the eigenvalue equation Equation 11.25. It can be easily shown that Equation 11.24 can reduce to the eigenvalue equation for the TE modes of a step index profile

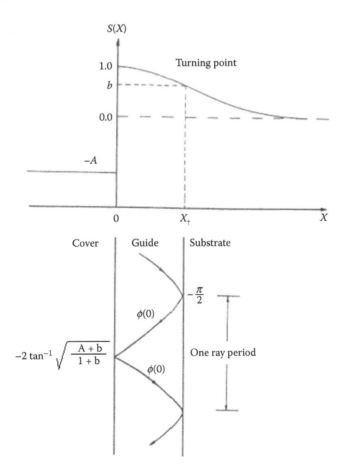

FIGURE 11.10 Ray optic derivation of the WKB eigenvalue equation.

waveguide provide the phase changes can be obtained correctly. For graded index profiles, this may not be the case as described in the next section. Thus, the phase accumulated in the guided as predicted by the WKB is only an approximation. This is due to the representation of the field by an equivalent cosine-like field.

Is the change at a dielectric interface dependent on the slope of the refractive index profile in the second medium at the interface? It is believed that in general $\delta \neq 0$, judging from the exact analysis of the linear-clad profile. This factor was omitted from the results of Marcuse [69]. Furthermore, the phase change at the turning point is estimated from Equation 11.23 without resorting to the correctness of the evanescent field representation of the actual field beyond $X = X_t$. The question is whether one can lump together all sources of errors in the phase into a single error parameter. Thus it is possible to propose, in general

$$V \int_0^{X_t} \sqrt{S(X) - b} . dX = (m + \gamma)\pi \tag{11.27}$$

where γ is the total accumulated phase change at the turning point, γ is normally equal to 3/4.

11.3.1.2.3   Profiles with Analytical WKB Solutions

The WKB integral given in Equation 11.26 is integrable for the step, clad-linear, clad-parabolic, exponential, and Cosh graded-index profiles. With the help of the integration formulae given in Reference 70, the results for the normalized propagation constant b and mode cut-off frequencies are presented in Table 11.5. For other profiles, the integral has to be integrated numerically. In a modern computing facility with ultra-high-speed processors, analytical solutions would give us some insight understanding of the behavior of the wave solution. If numerical integration is conducted for each trial value of the normalized propagation constant b, the turning point change, especially when V becomes very small, b approaches zero and the turning point value becomes very large. Thus, the WKB were not popular before but not now with modern and ultra-high-speed computing systems.

11.3.1.2.4   Ordinary WKB Results

We are faced with two forms of the WKB eigenvalue equation in Equations 11.24 and 11.27. This section studies the performance of this equation over a wide variety of profiles for representative

TABLE 11.5
Equation for Calculating b and V_c via the WKB Method

Profile	b	V_c
Clad-linear	$1 - \dfrac{A_0}{V^{2/3}}$ $A_0 = \left[\dfrac{3\pi}{2}(m+\gamma)\right]^{2/3}$	$\dfrac{3\pi}{2}(m+\gamma)$
Clad-parabolic	$1 - \dfrac{4(m+\gamma)}{V}$	$4(m+\gamma)$
Exponential	$b: \sqrt{1-b} - \sqrt{b}\tan^{-1}\left(\sqrt{\dfrac{1-b}{b}}\right) = \dfrac{\pi(m+\gamma)}{2V}$	$\dfrac{\pi}{2}(m+\gamma)$
Cosh-2	$\left[1 - \dfrac{2}{V}(m+\gamma)\right]^2$	$2(m+\gamma)$
Step	$1 - \left[\dfrac{2(m+\gamma)}{V}\right]^2$	$(m+\gamma)\pi$
Gaussian	—	$(m+\gamma)\pi$

TABLE 11.6

b−V for Clad-Linear Profile and Clad-Parabolic Profile and Exponential Profile

| | Clad-Linear Profile | | | Clad-Parabolic Profile | | | Exponential Profile | | |
| | b (WKB) | | | b (WKB) | | | b (WKB) | | |
V	Enhanced	Ordinary	b (exact)	Enhanced	Ordinary	b (exact)	Enhanced	Ordinary	b (exact)
2	–	–	–	–	–	–	0.1086	0.0831	0.1050
3	–	–	0.0335	0.981	0.000	0.1577	0.23331	0.2054	0.2292
4	0.1333	0.0792	0.1479	0.3081	0.2500	0.3262	0.3249	0.2992	0.3212
5	0.2498	0.2045	0.2500	0.4417	0.4000	0.4475	0.3939	0.3705	0.3903
10	0.5218	0.5001	0.5182	0.7149	0.7000	0.7153	0.5809	0.5658	0.5781
100	0.8945	0.8923	0.8939	0.9705	0.9700	–	0.8974	0.8954	0.8968

values of V. The effect of asymmetry factor on the dispersion was identified by Ramaswamy and Lagu [71] in which they found that for $A > 10$, the error is negligible. However, their conclusion is only valid for multi-mode waveguides. In modern optical waveguides for advanced optical communications systems, single mode optical waveguides are mainly the guidedwave media for applications. The results obtained for single mode optical waveguides prove otherwise when $A = 20$ is employed with the waveguide parameters $n_s = 2.177$ (Lithium niobate as substrate) cover layer $n_c = 1.0$ (air) and $\Delta = 0.043$. Equation 11.26 reaches its asymptotic value when $A \to \infty >$ the exact numerical results have been obtained with an integration step of 0.01. The profile truncation point is set at $X = 10$ for diffused profiles and for $X = 1$ for the clad- power law profiles. The values of b and V for different profiles are calculated and tabulated in Tables 11.6 and 11.7. The improvement of the values of the normalized propagation constant and the V-parameter can be observed and self-explanatory. The value of V is set at region closed to that of the cut-off of the guided mode is TE$_0$

11.3.1.2.5 Enhanced WKB

There are serious drawbacks of both the variational and WKB methods in the computing of the dispersion characteristics of the diffused clad single mode planar waveguides. The cosine-exponential trial field can substantially improve the accuracy of the variational analysis [72]. However, it requires optimization.

TABLE 11.7

b−V for Gaussian Profile and erfc Profile and Cosh^{-2} Profile

| | Gaussian Profile | | | Erfc Profile | | | Cosh-2 Profile | | |
| | b (WKB) | | | b (WKB) | | | b (WKB) | | |
V	Enhanced	Ordinary	b (exact)	Enhanced	Ordinary	b (exact)	Enhanced	Ordinary	b(exact)
2	0.0452	0.0104	0.817	–	–	–	0.1001	0.0625	0.1231
3	0.2538	0.2071	0.2750	0.0575	0.0281	0.0677	0.2908	0.2500	0.3074
4	0.4008	0.3630	0.4133	0.1651	0.1293	0.1695	0.4244	0.3906	0.4357
5	0.3939	0.3705	0.3903	0.5013	0.4712	0.5095	0.2539	0.2198	0.2552
10	0.5809	0.5658	0.5781	0.7301	0.7173	0.7323	0.4991	0.4776	0.4971
100	0.9706	0.9702	0.9707	0.8959	0.8835	0.8852	0.9707	0.9702	0.9708

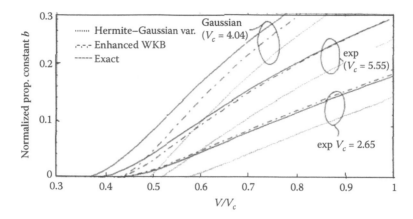

FIGURE 11.11 Dispersion characteristics: comparison of methods of analysis of single mode clad-power-law of Gaussian, exponential distributed profiles with $n_s = 2.17$, $\Delta = 0.043$.

For good field distribution and immunity to the bending of the waveguide, it is anticipated that the waveguide is operating in the region close to the cut-off of the TE_1 mode. Figures 11.11 and 11.12 show the range of applicability of each method for diffused clad as well as clad power-law profiles. Except for the Gaussian profile, all other profiles show that the enhanced WKB method is sufficiently accurate if the operating point lies in the range $0.75 < V/V_c < 1.0$. The discrepancies in the Gaussian profile are caused by the steepness of the profile. Errors caused by the presence of the uniform substrate index should taper off in the stated range of validity. The variational method with a Hermite–Gaussian field cannot be used for the calculation of b for any of these profiles due to the poor overlapping between the trial field and the actual field. The waveguide designs should use an appropriate method for a particular application.

11.3.2 SYMMETRICAL WAVEGUIDES

In this section, we deal mainly with symmetrical optical waveguides and dwell mainly on the equivalent moment methods described above. The WKB method is also discussed briefly. It involves only a slight modification of the previous equations and the entries given in Table 11.8. As presented, the

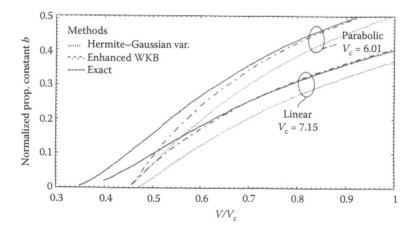

FIGURE 11.12 Dispersion characteristics: comparison of methods of analysis of single mode clad-power-law of parabolic and linear distributed profile with $n_s = 2.17$, $\Delta = 0.043$.

TABLE 11.8

WKB Calculated Cut-Off Frequencies Two Lowest Order Modes TE$_0$ and TE$_1$

Profile	V_{c1}, V_{c2} (WKB)		V_{c1}, V_{c2} (exact)
	Enhanced	Ordinary	
Clad-linear	3.2	3.53	2.46
	7.92	8.24	7.15
Clad-parabolic	2.72	3.0	1.96
	6.72	7.0	6.01
Exponential	1.07	1.18	1.09
	2.64	2.75	2.65
Gaussian	2.14	2.56	1.43
	5.28	5.96	4.04
Erfc	2.32	2.56	2.09
	5.72	5.96	5.55
Cosh-2	1.36	1.50	1.24
	3.36	3.50	3.32

WKB performs poorly in this kind of waveguides. On the other hand, the moment method (based on cosh-2 profile) is accurate in the range of frequencies for single mode operation.

11.3.2.1 WKB Eigenvalue Equation

For a symmetrical optical waveguide, (11.27) becomes

$$2V \int_0^{X_t} \sqrt{S(X) - b}.dX = (m + \gamma_s); \; \gamma_s = \frac{1}{2} \text{ diffused waveguides} \tag{11.28}$$

If the turning point coincides with a dielectric discontinuity, the correct phase shift formula is to be used. For buried modes the complicated expression is given in the Appendix 11B.1. This section is limited to the profiles following diffused or clad-power shape. The factor of two in Equation 11.28 is accounted for the WKB-defined effective guide width now extended from $-X_t$ to X_t, the turning points on both sides of the guiding region. Thus, the formulae in Table 11.5 for the estimation of b and V can be translated to symmetrical optical waveguide by transforming $V \to 2V$ and $\gamma \to \gamma_s$

11.3.2.2 Two-Parameter Profile Moment Method

The profile moment method is related to the variational formalism of optical waveguide problems [73]. The trial field is derived from that of a reference profile where an exact analytical expression is available. It is known that the field distribution of the fundamental mode follows a bell-shape like trend. Thus, by adjusting the V-parameter, a close match to the modal field can be obtained. This condition can be satisfied by monotonously varying the variational parameters.

11.3.2.2.1 Theoretical Basis

The starting point is that for two symmetrical waveguides having the same substrate index, normalized mode propagation constants are related by [74]

$$\beta^2 - \beta_r^2 = \frac{k_0^2 \int_{-\infty}^{\infty} \left[n^2(x) - n_r^2(x) \right] \varphi(x) \varphi_r(x) dx}{\int_{-\infty}^{\infty} \varphi(x) \varphi_r(x) dx} \tag{11.29}$$

where the subscript r indicates the quantities belong to the reference waveguide. The condition for the two waveguides to be equivalent is $\beta = \beta_r$, thus we have

$$\int_{-\infty}^{\infty} [n^2(x) - n_r^2(x)]\varphi(x)\varphi_r(x)dx = 0 \tag{11.30}$$

One can express the product of the field of this equation as a series as

$$\varphi(x)\varphi_r(x) = \sum_{l=0}^{\infty} c_l(k_0)x^{2l} \tag{11.31}$$

where c_l are the frequency-dependent coefficients of the series. Thus (11.30) becomes

$$\sum_{l=0}^{\infty} c_l(k_0) \int_0^{\infty} [n^2(x) - n_r^2(x)]x^{2l}dx = 0 \tag{11.32}$$

Since we impose the condition that these waveguides have the same substrate index, we can write (11.32) in term of the profile shape function $S(X)$ leading to

$$\sum_{l=0}^{\infty} c_l(k_0)[N_{2l} - N_{2lr}] = 0 \tag{11.33}$$

in which N_{2l} can be identified by

$$N_{2l} = 2[n_0^2 - n_s^2]d^{2(l+1)}\Omega_{2l} = 0 \tag{11.34}$$

where Ω_{2l} is defined as

$$\Omega_{2l} = \int_0^{\infty} S(X)X^{2l}dX = 0 \tag{11.35}$$

For profiles that are nearly identical, one can assume that for $\beta = \beta_r$ over the range of k_0 for which the fields are slowly varying, it is sufficient to retain only two terms in the series. Thus, we have

$$\begin{aligned} N_0 &= N_{0r} \\ N_2 &= N_{2r} \end{aligned} \tag{11.36}$$

Expanding these two terms, we have

$$\begin{aligned}{} [n_0^2 - n_s^2]d\Omega_0 &= [n_{0r}^2 - n_s^2]d_r\Omega_{0r} \\ [n_0^2 - n_s^2]d^3\Omega_2 &= [n_{0r}^2 - n_s^2]d_r\Omega_{2r} \end{aligned} \tag{11.37}$$

which can also expressed in term of the normalized frequency V-parameter as

$$\frac{V}{V_r} = \left\{\frac{\Omega_0\Omega_2}{\Omega_{0r}\Omega_{2r}}\right\}^{1/4} \tag{11.38}$$

11.3.2.2.2   Estimation of Normalized Propagation Constant
The normalized propagation constant b can be expressed in term of V and β as

$$b(V) = \left(\frac{d}{V}\right)^2 \frac{n_e^2 - n_s^2}{k_0^2} \tag{11.39}$$

where $n_e = \beta/k_0$ is the effective refractive index of the guided mode or the refractive index of the guided medium as seen by the mode along the z direction. Since $n_e = n_{er}$, thence

$$\frac{b(V)}{b(V_r)} = \frac{\Omega_0}{\Omega_{0r}} \left\{\frac{\Omega_0\Omega_2}{\Omega_{0r}\Omega_{2r}}\right\}^{1/2} \tag{11.40}$$

This equation states that the propagation $b(V)$ of an arbitrary waveguide can be derived from that of a reference waveguide provided that the profile moments and the dispersion relation for $b_r(V_r)$ are known. Table 11.9 lists the three lowest moments of profiles having analytical forms of their shape functions. The profiles listed in this table having step, clad-linear, exponential, and cosh have exact analytical solutions for their propagation constant. Thus, any of these profiles can be employed as a reference profile.

We select two profiles, the step and cosh profiles, for two case studies as follows:

Step Reference Profile [75]

$$V_r\sqrt{1-b_r} = m\frac{\pi}{2} - \tan^{-1}\left(\sqrt{\frac{b_r}{1-b_r}}\right) \quad m = 0, 1, 2\ldots \tag{11.41}$$

where

$$V_r = [3\Omega_0\Omega_2]^{1/4}V; \ b = \left[\frac{\Omega_0^3}{3\Omega_2}\right]^{1/2} b_r \tag{11.42}$$

Cosh Reference Profile [75]

$$b_r = \left\{\left(1 + \frac{1}{4V_r^2}\right)^{1/2} - \frac{1}{V_r}\left(m + \frac{1}{2}\right)\right\}^2 \quad m = 0, 1, 2\ldots \tag{11.43}$$

TABLE 11.9
Profile Moments of Selected Profile Shape

Profile	Ω_0	Ω_2	Ω_4	Ω_4/Ω_2	SDF [see Equation (11.51)]
Step	1	0.333	0.2	0.6	1.0
Clad-linear	0.5	0.0833	0.0033	0.40	0.67
Clad-parabolic	0.667	0.133	0.571	0.43	0.72
Exponential	1	2	24	12	10.08
Gaussian	0.866	0.443	2.659	6	5.04
Erfc(x)	0.564	0.188	0.226	1.2	1.01
Cosh-2(x)	1	0.693	0.823	1.19	1.00

where

$$V_r = \left[\frac{12\Omega_0\Omega_2}{\pi^2}\right]^{1/4} V; \; b = \frac{\pi}{2}\left[\frac{\Omega_0^3}{3\Omega_2}\right]^{1/2} b_r \tag{11.44}$$

These equations are required for calculation of the dispersion relation characteristics.

11.3.2.2.3 Estimation of TE1 Mode Cut-Off

The next higher-order mode is TE_1. The cut-off frequency of this mode is the upper limit of the single mode operation. We can write the product of the guided waves of the reference waveguide and the one to be analyzed as

$$\varphi(x)\varphi_r(x) = \sum_{l=0}^{\infty} c_l(k_0)x^{2l+2} \tag{11.45}$$

and similarly

$$N_2 = N_{2r}; \; N_4 = N_{4r} \tag{11.46}$$

Thus, the relationship between the profile moments and the refractive index can be obtained as

$$\left[n_0^2 - n_s^2\right]d^3\Omega_2 = \left[n_{0r}^2 - n_s^2\right]d_r^3\Omega_2; \; \left[n_0^2 - n_s^2\right]d^5\Omega_4 = \left[n_{0r}^2 - n_s^2\right]d_r^5\Omega_4 \tag{11.47}$$

The estimation of non-profile moment terms using the definition of V for each guide gives the desired mode-cut-off relation after setting $b = 0$.

Step Reference Profile [76]

$$V_c = \frac{\pi}{2}\left(\frac{5\Omega_4}{27\Omega_2}\right) \tag{11.48}$$

Cosh Reference Profile

$$V_c = \sqrt{2}\left(\frac{5\pi^2\Omega_4}{252\Omega_2^3}\right)^{1/4} \tag{11.49}$$

where the cut-offs for the reference profiles have been derived from Equations 11.41 and 11.43.

The propagation constant and the and the cut-off frequency V_c are tabulated in Tables 11.10 and 11.11 for two typical profiles, the step and clad-linear types, and Tables 11.12 through 11.15 are for Clad-parabolic, exponential, Gaussian, and erfc(x) profiles, respectively. Note that the moments of the complementary error function are not listed in Table 11.9, as there are no close forms solutions.

11.3.2.2.4 Choice of methods

There are some interesting insights as observed from the tables:

In the clad-power law profiles, the moment-ESI method consistently gives better results for the propagation constant and the cut-off frequencies of TE_0 mode.

On the other hand, diffused waveguides characterized by non-decreasing higher-order moments of Table 11.9 are more accurately modeled by the cosh profile.

TABLE 11.10
$b-V$ Data for Step Profile

| V | b (moment) | | b (WKB) | b (exact) |
	Step Reference	Cosh Reference		
0.5	$b = b$ (exact)	0.192	<0	0.189
1.0	For all V	0.481	0.383	0.454
1.5		0.697	0.726	0.628
2.0		0.848	0.846	0.725
3.0		>1	0.931	0.849
4.0		>1	0.961	0.902

TABLE 11.11
$b-V$ Data for Clad Linear Profile

| V | b (moment) | | b (WKB) | b (exact) |
	Step Reference	Cosh Reference		
0.5	0.0560	0.0563	<0	0.0561
1.0	0.173	0.177	<0	0.174
1.5	0.286	0.300	0.149	0.290
2.0	0.375	0.404	0.297	0.384
3.0	0.491	0.558	0.464	0.515
4.0	0.558	0.660	0.557	0.579

TABLE 11.12
$b-V$ Data for Clad-Parabolic Profile

| V | b (moment) | | b (WKB) | b (exact) |
	Step Reference	Cosh Reference		
0.5	0.0951	0.0959	<0	0.0952
1.0	0.270	0.280	<0	0.272
1.5	0.419	0.448	0.333	0.423
2.0	0.525	0.580	0.500	0.535
3.0	0.653	0.762	0.667	0.673
4.0	0.721	0.877	0.750	0.751

At low frequencies, both approaches are asymptotically exact.

The eigenvalue Equation 11.41 can be reduced to

$$b(V \to 0) = \left[\left(1 + \frac{1}{4V^2} \right)^{1/2} - \frac{1}{2V} \right]^2 \tag{11.50}$$

TABLE 11.13
$b-V$ Data for Exponential Profile

V	Step Reference	Cosh Reference	b (WKB)	b (exact)
	b (moment)			
0.5	0.142	0.148	0.0205	0.152
1.0	0.263	0.294	0.205	0.317
1.5	0.320	0.387	0.337	0.424
2.0	0.350	0.431	0.426	0.498
3.0	0.377	0.491	0.539	0.593
4.0	0.389	0.525	0.609	0.653

TABLE 11.14
$b-V$ Data for Gaussian Profile

V	Step Reference	Cosh Reference	b (WKB)	b (exact)
	b (moment)			
0.5	0.146	0.148	<0	0.147
1.0	0.342	0.363	0.207	0.354
1.5	0.466	0.520	0.421	0.498
2.0	0.541	0.628	0.549	0.594
3.0	0.620	0.763	0.688	0.709
4.0	0.657	0.842	0.762	0.774

TABLE 11.15
$b-V$ Data for Erfc(x) Profile

V	Step Reference	Cosh Reference	b (WKB)	b (exact)
	b (moment)			
0.5	0.0672	0.0679	<0	0.0678
1.0	0.187	0.194	0.0281	0.193
1.5	0.286	0.306	0.177	0.304
2.0	0.355	0.394	0.293	0.391
3.0	0.436	0.512	0.443	0.509
4.0	0.479	0.586	0.534	0.584

which is just the eigenvalue equation for cosh profile. This comparison is valid only when the profiles have equal volume Ω_0.

For large value V, their dispersion curve split and higher-order moment scaling factors in Equations 11.41 and 11.43 must be used. However, due to different properties of the higher-order moments of the step and cos h profiles, neither can be used to predict each other dispersion characteristics accurately.

The WKB method gives consistently better results at large frequencies. To give an idea of asymptotic range of applicability of the WKB and moment methods, the dispersion curves for the diffused as well clap-power law profile are plotted in Figures 11.13 through 11.16.

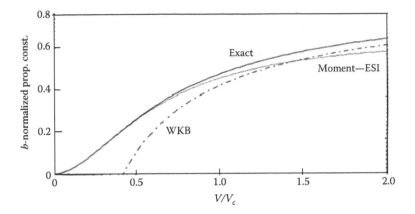

FIGURE 11.13 (Figure 5.6a) Dispersion characteristics: range of applicability of WKB method and moment method. Symmetric clad-linear profile with $A = 20$; $Vc(TE_1$ mode$) = 2.7995$.

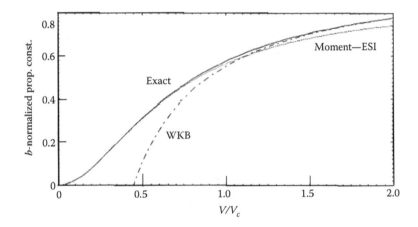

FIGURE 11.14 (Figure 5.6a) Dispersion characteristics: range of applicability of WKB method and moment method. Symmetric clad-parabolic profile with $A = 20$; $Vc(TE_1$ mode$) = 2.330$, $n_c = 2.177$, $\Delta = 0.043$.

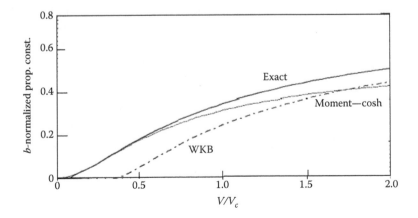

FIGURE 11.15 (Figure 5.6a) Dispersion characteristics: range of applicability of WKB method and moment – cosh method. Symmetric exponential profile with $A = 20$; $Vc(TE_1$ mode$) = 1.2024$, $n_c = 2.177$, $\Delta = 0.043$.

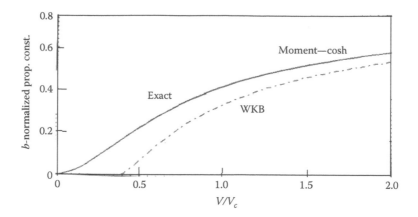

FIGURE 11.16 (Figure 5.6a) Dispersion characteristics: range of applicability of WKB method and moment – cosh method. Symmetric Erfc(x) profile with $A = 20$; $V_c(TE_1$ mode$) = 2.3187$, $n_c = 2.177$, $\Delta = 0.043$.

11.3.2.2.5   A New Method for Profile Classification

Table 11.15 and Figure 11.16 indicate that the dispersion characteristic of the complementary error function profile is well above the expected accuracy of the moment method as calculated from a cosh reference profile. Even at $V = 4.0$, a near perfect agreement is obtained. To account for such observations, a shape derivation factor (SDF) can be proposed as

$$SDF = \frac{\Omega_4/\Omega_2}{\Omega_{4r}/\Omega_{2r}} \tag{11.51}$$

where the subscript r is referred to the reference profile. For clad-profiles, one can chose the step profile as a reference, whereas for diffused waveguides the cosh profile offer much better fit. This SDF parameter is thus entered in Table 11.10. We can see the benefit of this factor for erfc(x) profile, which has a SDF factor of 1.01 as compared to 10 for exponential profile. Thus, the former method offers better accuracy.

11.3.2.3   A New Equivalence Relation for Planar Optical Waveguides

By sketching the spatial distribution of the modal field of the TE_1 mode in a symmetrical waveguide, and the TE_0 field in an asymmetrical waveguide, we can find out why the profile moment method would not work in both cases. However, there is some surprise when $X \geq 0$; the distribution is similar if $A \rightarrow \infty$ as shown in Figure 11.17. For the same profile, if the field distributions are identical then there exists a relation between the dispersion curves of both structures. Figure 11.18 illustrates the correspondence between the modes of both waveguide structures. One can postulate that

$$V_{ca1} = V_{cs2}; \ V_{ca2} = V_{cs4} \tag{11.52}$$

where the left-hand side denotes the cut-offs of the TE_0 and TE_1 modes in the asymmetrical waveguide. $V_{cs2}; V_{cs4}$ are the cut-offs of the TE_1 and TE_3 modes of the corresponding symmetrical waveguide.

Furthermore, it is noted that the separation of the b-V characteristic curves at cut-off of the symmetrical waveguide is nearly uniform. For the step profile, this separation equals $\pi/2$, whereas for graded profiles, this is only approximately true. Thus, we can write

$$V_{cs4} \simeq 3V_{cs2} \tag{11.53}$$

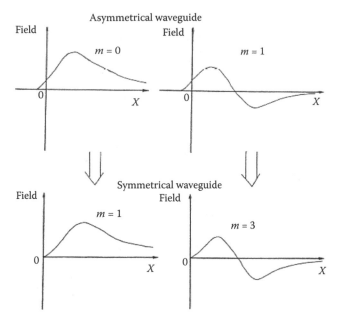

FIGURE 11.17 Correspondence between the modal fields of asymmetrical and symmetrical waveguides with the field distribution of odd modes.

Combining Equations 11.52 and 11.53 leads to

$$\left. \frac{V_{ca2}}{V_{ca1}} \right|_{A \to \infty} \simeq 3 \tag{11.54}$$

This equation allows us to conduct preliminary tests on the postulation above. Tables 11.16 and 11.17 tabulates the ration of the V-parameters at cut-off of the modes TE_0 and TE_1 for $A \sim 20$ and $A \to \infty$ for the profiles listed. It shows that Equation 11.54 does not satisfy for step profile. This may be contributed by the error in the assumptions about the separation of the b-V curve s at cut-off in the Equation 11.53. To see if Equation 11.52 can be satisfied, one can compare the TE_0 mode cut-offs of the symmetrical waveguide as $A \to \infty$. The cut-offs are so close in the last two columns of the table allowing to be considered exactly equal. Thus, this is the new corresponding relationship between the m-modes of an asymmetrical waveguide and the odd $(2m + 1)$th modes of the corresponding symmetrical waveguide.

11.3.3 CONCLUDING REMARKS

The variational method incorporating a Hermite-Gaussian trial field is inaccurate for calculating the $b-V$ dispersion characteristics of single mode optical waveguides. However, it is an accurate and convenient tool for mode spot size calculations provided that the mode is well-confined in the single-spectral range.

The WKB approach is numerically straight forward without any divergence. The enhanced WKB formulae with the correct phase connection at the dielectric interfaces yield very accurate results even for single mode optical waveguides. This is in contrast with a number of published works.

The profile moment method is applicable only to symmetrical profiles. A shape derivation factor (SDF) is defined to allow decision on the choice of reference profile to obtain optimum performance of this method. Foe diffused profiles; the hyperbolic cosh reference profile is required to yield accurate results. For clad profiles, the step reference profile offers accurate results and significantly better when SDF $\to$ 1.

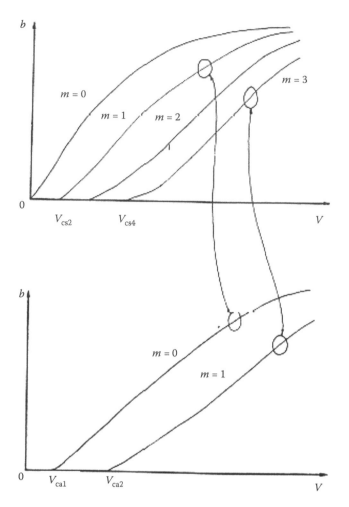

FIGURE 11.18   The m-mode dispersion curves of asymmetrical and symmetrical waveguides from the $(2m + 1)$th mode dispersion curve of the corresponding symmetrical waveguide.

TABLE 11.16

Ratio of TE_0 and TE_1 Mode Cut-Off V-Parameter in Asymmetrical Waveguides (Obtained for Profiles without Analytical Solutions by Forward Recurrence Algorithm with 0.01 Step Size)

	V_{ca2}/V_{ca1}	
Profile	**$A \sim 20$**	**$A \to 20$**
Step	3.33	3.0
Clad-linear	2.91	2.67
Clad-parabolic	3.07	2.78
Exponential	2.43	2.30
Gaussian	2.83	2.57
Erfc(x)	2.65	2.47
Cosh^{-2}	2.68	2.45

TABLE 11.17
Prediction of Odd Mode Cut-Offs in Symmetrical Waveguides from the Mode Cut-Offs of the Corresponding Waveguides

Profile	V_{ca1} (Asymmetrical Waveguide TE$_0$)		V_{ca4} (Symmetrical TE$_1$)
	A ~ 20	**A → 20**	
Step	1.35	1.57	1.57
Clad-linear	2.46	2.80	2.80
Clad-parabolic	1.96	2.26	2.25
Exponential	1.09	1.20	1.20
Gaussian	1.43	1.64	1.64
Erfc(x)	2.09	2.35	2.37
Cosh^{-2}	1.24	1.41	1.41

However, the profile-moment method does not cover the entire single mode region for all profiles. At larger V, its accuracy deteriorates significantly. Neither does the WKB formulae cover the single mode region, a hybrid profile moment-WKB method should be considered. A simple profile characterization factor is presented to assess the applicability of the profile moment method.

From the analysis of the mode field distribution and numerical computation, the m-modes of an asymmetrical waveguide and the $(2m + 1)$ odd modes of its corresponding asymmetrical waveguide are directly related. Computations of their mode cutoffs allow us to establish the exact correspondence for an asymmetrical waveguide with infinite asymmetrical factor.

APPENDIX 11A: MAXWELL EQUATIONS IN DIELECTRIC MEDIA

11A.1   THE MAXWELL EQUATIONS

The general Maxwell equations can be written as

$$\nabla \times E = -j\omega\mu H \tag{11A.1}$$

$$\nabla \times H = J + j\omega\varepsilon E \tag{11A.2}$$

$$\nabla \cdot D = \sigma/\varepsilon_0 \tag{11A.3}$$

$$\nabla \cdot B = 0 \tag{11A.4}$$

Note that in the above equations:

- $j = \sqrt{-1}$.
- $\mu \cong \mu_0$ is the magnetic permeability for non-magnetic materials that normally constitute an optical waveguide.
- $\varepsilon = \varepsilon_0 n^2$ is the dielectric constant of the material where ε_0 is the dielectric constant of free space and n is the refractive index of the materials.
- J is the current density and σ is the surface charge density, which are possible sources.
- The displacement vector D is related to the electric filed via $D = \varepsilon E$ and the magnetic-filed induction B is related to the magnetic field via $B = \mu H$.

In practice, problems of optical waveguide and couplers are often analyzed in the regions that are free of the above sources, that is, $J = 0$, and $\sigma = 0$. In these cases, we have

$$\nabla \times E = -j\omega\mu H \tag{11A.5}$$

$$\nabla \times H = j\omega\varepsilon E \tag{11A.6}$$

$$\nabla \cdot (\varepsilon E) = 0 \tag{11A.7}$$

$$\nabla \cdot H = 0 \tag{11A.8}$$

11A.2 THE WAVE EQUATION

With the usual expression of the time dependent lightwave carrier modulated signals $e^{j\omega t}$, the wave equation can thus be obtained as

$$\nabla x \nabla x \vec{E} - \varepsilon(\omega)\frac{\omega^2}{c^2}\vec{E} = 0 \tag{11A.9}$$

The refractive index of the medium can be related to the permittivity including the non-linear third order effects. Using the relation:

$$\nabla x \nabla x \vec{E} = \nabla(\nabla.\vec{E}) - \nabla^2\vec{E} = -\nabla^2\vec{E}$$

$$\because \nabla.\vec{D} = 0$$

$$\text{then\_the\_wave\_equation} \tag{11A.10}$$

$$\nabla^2\vec{E} - n^2(\omega)\frac{\omega^2}{c^2}\vec{E} = 0$$

11A.3 BOUNDARY CONDITIONS

In the regions free of the sources, we have the following boundary conditions:

- Continuity of the magnetic field and the component of the electric field tangential to the interface, that is,

$$H^{(1)} = H^{(2)}, E^{(1)}_{//} = E^2_{//}; \tag{11A.11}$$

- Continuity of the normal component of the displacement vector, that is,

$$D^{(1)}_{\perp} = D^{(2)}_{\perp} \quad or \quad n_1^2 E^{(1)}_{\perp} = n_2^2 E^{(2)}_{\perp} \tag{11A.12}$$

11A.3.1 Reciprocity Theorems

11A.3.1.1 General Reciprocity Theorem

From the above source-free Maxwell equations, we have for two optical media of dielectric constants ε_1 and ε_2:

$$\nabla \times (\nabla \times E) = \omega^2\varepsilon\mu E \tag{11A.13}$$

$$\nabla \times (\nabla \times H) = \omega^2\varepsilon\mu H \tag{11A.14}$$

Using the above equations and the identity $\nabla \cdot (A \times B) = B \cdot (\nabla \times A) - A \cdot (\nabla \times B)$, we have

$$\nabla \cdot (E_1 \times H_2 - E_2 \times H_1) = j\omega(\varepsilon_2 - \varepsilon_1)E_1 \cdot E_2 \tag{11A.15}$$

and its integral equivalence

$$\frac{\partial}{\partial z}\iint_{A_\infty}(E_1 \times H_2 - E_2 \times H_1) \cdot \hat{z}dA = j\omega \iint_{A_\infty}[\varepsilon_2(x, y) - \varepsilon_1(x, y)]E_1 \cdot E_2 dA \tag{11A.16}$$

11A.3.1.2 Conjugate Reciprocity Theorem

Conjugate reciprocity theorem can be obtained in a similar way as above, except for using the conjugate form of filed expressions. This is particularly convenient in constructing the formulation for the lossless waveguides or couplers, in particular, the expression of the power conservation. Following some algebra, we have

$$\nabla \cdot (E_\mu \times H_\nu^* + E_\nu^* \times H_\mu) = -j\omega(\varepsilon_\mu - \varepsilon_\nu)E_\mu \cdot E_\nu* \tag{11A.17}$$

$$\frac{\partial}{\partial z}\iint_{A_\infty}(E_\mu \times H_\nu + E_\nu^* \times H_\mu) \cdot \hat{z}dA = -j\omega \iint_{A_\infty}(\varepsilon_\mu - \varepsilon_\nu)E_\mu \cdot E_\nu^* dA \tag{11A.18}$$

APPENDIX 11B: EXACT ANALYSIS OF CLAD-LINEAR OPTICAL WAVEGUIDES

The exact analysis of TE modes is guided in an optical waveguide whose refractive index profile follows a clad-linear shape. The profile was first analyzed exactly by Marcuse [77], then treated in full by Adams [78] and applied to the study of low-threshold current laser diode by Chinn [79]. The results presented here are different from these published formulae, as they are expressed in terms of Bessel functions of real positive order. Starting with the eigenvalue equation, we derive the propagation constant and the cut-offs of the waveguide. The treatments of symmetrical and asymmetrical profiles are given separately.

11B.1 Asymmetrical Clad-Linear Profile

11B.1.1 Eigenvalue Equation

The eigenvalue equation is given by [80]

$$\frac{Ai'(\alpha_0) - V^{1/3}\sqrt{A+b}Ai'(-\alpha_0)}{Bi'(\alpha_0) - V^{1/3}\sqrt{A+b}Bi'(-\alpha_0)} = \frac{Ai'(\alpha_1) - V^{1/3}\sqrt{b}Ai(\alpha_1)}{Bi'(\alpha_1) - V^{1/3}\sqrt{A+b}Bi(-\alpha_1)} \tag{11B.1}$$

where $\alpha_0 = (1-b)V^{2/3}$; $\alpha_1 = bV^{2/3}$; AI and Bi are Airy functions and the dash denotes the derivatives with respect to the argument.

Using the relations between the Airy and Bessel functions of Reference 81, we can convert (11B.1) into an immediate form

$$\frac{J_{-2/3}(\gamma_0) - J_{2/3}(\gamma_0) + Q[J_{-1/3}(\gamma_0) + J_{1/3}(\gamma_0)]}{J_{-2/3}(\gamma_0) - J_{2/3}(\gamma_0) + Q[J_{-1/3}(\gamma_0) - J_{1/3}(\gamma_0)]}$$

$$= \frac{-I_{-2/3}(\gamma_1) + I_{2/3}(\gamma_1) + I_{-1/3}(\gamma_1) - I_{1/3}(\gamma_1)}{I_{-2/3}(\gamma_1) + I_{2/3}(\gamma_1) + I_{-1/3}(\gamma_1) + I_{1/3}(\gamma_1)} \tag{11B.2}$$

where the arguments $\gamma_0; \gamma_1$ of the Bessel functions and Q are given as

$$\gamma_0 = \frac{2}{3} V(1-b)^{3/2}; \quad \gamma_1 = \frac{2}{3} V(b)^{3/2}; \quad Q = \left[\frac{A+b}{1-b}\right]^{3/2} \tag{11B.3}$$

In arriving at the Equation 11B.2, we have also used*

$$Ai'(\alpha_1) = -\frac{\alpha_1}{3}(I_{-2/3} - I_{2/3}) \tag{11B.4}$$

Finally, using the recurrence relations for the J and I functions results in the required version of the eigenvalue equation

$$\frac{(2 + 3\gamma_0 Q)J_{1/3}(\gamma_0) - 3\gamma_0 J_{4/3}(\gamma_0)}{(4Q + 3\gamma_0)J_{2/3}(\gamma_0) - 3\gamma_0 J_{5/3}(\gamma_0)} = \frac{(2 + 3\gamma_1 Q)I_{1/3}(\gamma_0) - 3\gamma_1 I_{4/3}(\gamma_1)}{(4 + 3\gamma_1)I_{2/3}(\gamma_1) - 3\gamma_1 I_{5/3}(\gamma_1)} \tag{11B.5}$$

11B.2   Mode Cut-Off

At cut-off, we have $b = 0$ and $Q = [A]^{1/2}\gamma_0 = (2/3)V; \gamma_1 = 0$. The RHS of Equation 11B.5 $\to \infty$ and a substitution of the asymptotic formula for the Bessel functions gives the mode-cut-off for modes with large V_c

$$\frac{2}{3}V_C \sim \left(m + \frac{1}{12}\right)\pi - \tan^{-1}\left(\sqrt{A}\right) \quad m = 0, 1, 2 \ldots \tag{11B.6}$$

11B.3   Symmetrical Waveguuide

11B.3.1   Eigenvalue Equation

The eigenvalue equations for the odd and even modes given in Reference 82 and can be derived from Equation 11B.1. Only the LHS is affected and the eigenvalue equations for these modes are given as

$$\frac{Ai'(-\alpha_0)}{Bi'(-\alpha_0)} = \frac{Ai'(\alpha_1) - V^{1/3}\sqrt{b}Ai(\alpha_1)}{Bi'(\alpha_1) - V^{1/3}\sqrt{A+b}Bi(-\alpha_1)} \quad \text{even\_mode}$$

$$\frac{Ai(-\alpha_0)}{Bi(-\alpha_0)} = \frac{Ai'(\alpha_1) - V^{1/3}\sqrt{b}Ai(\alpha_1)}{Bi'(\alpha_1) - V^{1/3}\sqrt{A+b}Bi(-\alpha_1)} \quad \text{odd\_mode} \tag{11B.7}$$

Similarly, using the relations between the Airy and Bessel functions of Reference 83, we can convert (11B.7) into an immediate form

$$\frac{-2J_{-2/3}(\gamma_0) + 3\gamma_0 J_{4/3}(\gamma_0)}{3\gamma_0 J_{2/3}(\gamma_0)} = \frac{-I_{-2/3}(\gamma_1) + I_{2/3}(\gamma_1) + I_{-1/3}(\gamma_1) - I_{1/3}(\gamma_1)}{I_{-2/3}(\gamma_1) + I_{2/3}(\gamma_1) + I_{-1/3}(\gamma_1) + I_{1/3}(\gamma_1)}$$

$$\frac{-3\gamma_0 J_{1/3}(\gamma_0)}{3\gamma_0 J_{5/3}(\gamma_0) - 4J_{2/3}(\gamma_0)} = \frac{-I_{-2/3}(\gamma_1) + I_{2/3}(\gamma_1) + I_{-1/3}(\gamma_1) - I_{1/3}(\gamma_1)}{I_{-2/3}(\gamma_1) + I_{2/3}(\gamma_1) + I_{-1/3}(\gamma_1) + I_{1/3}(\gamma_1)} \tag{11B.8}$$

---

* Note: the Equation (11B.4) given in Ref. by Abramiwitz and Stegun does not have the negative sign—a vital error.

11B.3.2   Mode Cut-Off

The right-hand side of Equation 11B.5 goes to infinite at cut-off, so we have

$$J_{2/3}\left(\frac{2V_C}{3}\right) = 0 \quad m = 0, 2 \ldots \text{even\_modes}$$

$$J_{2/3}\left(\frac{2V_C}{3}\right) - V_C J_{5/3}\left(\frac{2V_C}{3}\right) = 0 \quad m = 1, 3 \ldots \text{odd\_modes}$$

(11B.9)

Alternatively, these equations can be obtained from Equation 11B.6.

APPENDIX 11C: WKB METHOD, TURNING POINTS AND CONNECTION FORMULAE

11C.1   INTRODUCTION

Consider the scalar wave equation

$$\frac{d^2\varphi(x)}{dx^2} + K^2\varphi(x) = 0 \tag{11C.1}$$

where $K^2 = k_0^2 n^2(x) - \beta^2$ is the transverse propagation constant and $\varphi(x)$ is the modal field in the transverse plane. The characteristic mode factor $e^{j(\omega t - \beta z)}$ is omitted. This version of the wave equation is selected to present the turning points in the subsequent analysis. The turning point is defined by $x_t \rightarrow K(x_t){=}0$.

When the refractive index function $n(x)$ has a certain simple form (11C.1) can be solved explicitly for $\varphi(x)$, the good behavior of this function restricting the axial propagation constant β to discrete values. These are the characteristics of the bound modes. However, in most practical optical waveguides, explicit solutions of the fields are not available and approximation methods of solutions must be developed.

The WKB method is based on an asymptotic expansion in k_0^{-1}, the first term of which leads to geometrical optic results, or the zeroth-order WKB solutions, and higher order terms lead to exact modal solutions. The principal concern of this method lies in the transitional region which connects the oscillatory fields and its evanescent neighbors. These are the turning points of the problem where the semi-classical approximation breaks down. The way in which the WKB solutions valid on either side of the turning point connects, remains as the central problem of the method.

Before proceeding to derive the WKB solutions we assign the following symbols: $\phi(x)$; $\varphi(x)$; $\Phi(x)$ = exact modal field, WKB solution and approximate modal field valid at the turning point.

11C.2   DERIVATION OF THE WKB APPROXIMATE SOLUTIONS

Following established tradition, we postulate a solution of Equation 11.82 in the form of

$$\phi(x) = Ae^{jk_0 S(x)}; \; j = \sqrt{-1}; \; A = cons\tan t \tag{11C.2}$$

Thus, Equation 11C.1 is transformed into the Riccati equation

$$j\frac{1}{k_0}\frac{d^2 S(x)}{dx^2} - \left(\frac{dS(x)}{dx}\right)^2 + \left(n^2(x) - n_e^2\right) = 0 \tag{11C.3}$$

$$n_e \equiv \text{effective\_index\_of\_waveguide\_mode}$$

Now, let $y = S'$ and assume that y admits of a formal series expansion of the form:

$$y = \sum_{n=0}^{\infty} k_0^{-n} y_n \tag{11C.4}$$

Thence

$$y' = S'' == \sum_{n=0}^{\infty} k_0^{-n} y'_n \tag{11C.5}$$

and

$$y^2 = S'^2 = \left\{ \sum_{n=0}^{\infty} k_0^{-n} y'_n \right\}^2 = y_0^2 \left\{ 1 + k_0^{-1} \frac{2y_1}{y_0} + k_0^{-2} \left(\frac{2y_1}{y_0} + \frac{y_1^2}{y_0^2} \right) + \cdots \right\} \tag{11C.6}$$

Therefore, substituting Equations 11C.5 and 11C.6 into Equation 11C.3 and equating the coefficients of the like-power of $k0$ leads to the following recurrence relations:

$$y_0^2 = n^2 - n_e^2 = \frac{K^2}{k_0^2}; \quad jy_0^1 = 2y_1 y_0; \quad jy_0' = 2y_2 y_0 + y_1^2 \tag{11C.7}$$

Thus, we can obtain

$$y_0' = \pm \frac{K}{k_0}; \quad \pm j \frac{K'}{k_0} = \pm \frac{2K}{k_0} y_1 \tag{11C.8}$$

which can be integrated into the form

$$\frac{j}{2k_0} \ln K + C = \frac{1}{k_0} \int_{x_0}^{x} y_1 dx \tag{11C.9}$$

where C is the integration constant.
Hence

$$\phi(x) = A e^{jk_0 S(x)} = A e^{jk_0 \int_{x_0}^{x} \sum_{n=0}^{\infty} k_0^{-n} y_n dx} = A e^{jk_0 \left\{ \int_{x_0}^{x} y_0 dx + k_0^{-1} \int_{x_0}^{x} y_1 dx + k_0^{-2} \int_{x_0}^{x} y_2 dx + \cdots \right\}} \tag{11C.10}$$

$$= A_\pm e^{\pm j \int_{x_0}^{x} K dx + jk_0 C} \quad \rightarrow \quad \varphi(x) = A_\pm K^{-1/2} e^{\pm j \int_{x_0}^{x} K dx}$$

where $A_\pm$ denotes the constants corresponding to the $\pm$ solutions, respectively.
For $K^2 > 0$, we have

$$\varphi_1(x) = DK^{-1/2} \cos \left(\pm j \int_{x_0}^{x} K dx + \delta \right) \tag{11C.11}$$

$$D, \delta \equiv arb.\_constants$$

For $K^2 < 0$, we have

$$\varphi_2(x) = B_{\pm} |K|^{-1/2} e^{\left(\pm \int\limits_{x_0}^{x} |K| dx \right)}$$

Clearly, at the turning point defined by $K^2(x) = 0$, the oscillatory and evanescent fields diverge. Hence, neither form can be retained during the transition from one interval to the other which K^2 changes sign.

Furthermore, a back substitution of the WKB solutions, for example, φ into the original Equation 11C.1 produces an inhomogeneous equation

$$\frac{d^2 \varphi(x)}{dx^2} + K^2 \varphi(x) = W(x)$$

$$W(x) = \frac{3}{4} \left(\frac{K'}{K} \right)^2 - \frac{1}{2} \frac{K''}{K}$$

(11C.12)

For this equation, any point where $K^2(x)$ vanishes is a singular point. However, far from the singularity, higher-order WKB solution can be obtained with Equation 11C.12 as the starting point, and incorporating $XW(x)$ into $K^2 >$ this method contrasts with the straight forward idea using high order recurrence relation to obtain higher-order terms [84].

11C.3   TURNING POINT CORRECTIONS

11C.3.1   Langer's Approximate Solution Valid at Turning Point

For a refractive index profile that is unbounded, we can write, for an nth order zero at $x_0 = 0$ as

$$K^2(x) = (x - x_n)^n f(x)$$

(11C.13)

where $f(x) = \sum_{i=0}^{\infty} C_i (x - x_i)^i$ is a non-vanishing polynomial of x at $x_0 = 0$. For simplicity and without loss of generality, let $x_0 = 0$. Thus, for values of x close to zero we can write

$$K^2(x) = C_0(x)^n$$

(11C.14)

Therefore, in the vicinity of the turning point, we can represent the wave equation by an approximate differential equation

$$\frac{d^2 \phi(x)}{dx^2} + C_0 x^n \phi(x) = 0; \text{ where } \phi \equiv \varphi \text{ at } x = 0$$

(11C.15)

The solutions to Equation 11C.15 are Bessel functions. Thus, it would be necessary to transform the wave equation before setting the condition (11C.14). We can now introduce the Louiville transform

$$\xi = \int\limits_0^x K(x) dx \quad \text{and} \quad \phi = K^{1/2}(x) v$$

(11C.16)

After some algebra, the equation becomes

$$\frac{d^2 v}{d\xi^2} + \left\{ \frac{3}{4} K^{-4} \left(\frac{d^2 K}{dx^2} \right)^2 - \frac{1}{2} K^{-3} \frac{d^2 K}{dx^2} + 1 \right\} v = 0$$

(11C.17)

Now, using the form of the approximate transverse propagation constant in Equation 11C.14 and evaluating the terms in the bracket reduces Equation 11C.17 to the required intermediate form

$$\frac{d^2v}{d\xi^2} + \left\{ 1 + C_0^{-1} x^{-(n+2)} \left(\frac{n^2 + 4n}{16} \right) \right\} v = 0 \qquad (11C.18)$$

Thence, using Equation 11C.16 we obtain

$$\frac{1}{\xi^2} = C_0^{-1} x^{-(n+2)} \left(\frac{n^2 + 2}{2} \right)^2 \qquad (11C.19)$$

Then substituting into Equation 11C.18, we arrive at

$$\frac{d^2v}{d\xi^2} + \left\{ 1 + \frac{1}{\xi^2} \cdot \left(\frac{n(n+4)}{4(n+4)^2} \right) \right\} v = 0 \qquad (11C.20)$$

Now, changing the variable $v \to \xi^{1/2} W$ leads to

$$\frac{d^2v}{d\xi^2} = \xi^{1/2} \frac{d^2W}{d\xi^2} + \xi^{-1/2} \frac{dW}{d\xi} - \frac{1}{4} \xi^{-3/2} W \qquad (11C.21)$$

Finally, substituting Equation 11C.21 back into Equation 11C.20 and multiplying by $\xi^{3/2}$ throughout, the resultant equation is transformed to the desired canonical form:

$$\frac{d^2W}{d\xi^2} + \xi \frac{dW}{d\xi} + \left\{ \xi^2 - \frac{1}{(n+2)^2} \right\} W = 0 \qquad (11C.22)$$

This is the Bessel equation of order $1/(n+2)$ with independent solutions denoted by plus and minus signs

$$W = \left\{ \frac{\pi^2}{4} A_\pm \right\} J_{\pm m}(\xi); \quad \text{with} \quad m = \frac{1}{n+2} \qquad (11C.23)$$

Thus, from $v \to \xi^{1/2} W$ and (11C.16) we can recover the solutions of the original form (11C.15) at the turning point as

$$\phi = K^{-1/2} \xi^{1/2} W(\xi) = \left\{ \frac{\pi^{1/2}}{2^{1/2}} A_\pm \right\} \left(\frac{\xi}{K} \right)^{1/2} J_{\pm m}(\xi) \qquad (11C.24)$$

where $A_\pm$ is an arbitrary constant and ξ is related to $K(x)$ via the Liouville transformation. The relationship between Equations 11C.22 and 11C.15 can be found in standard textbook on Bessel's function [85]. We include it here for interested readers:

To study the behavior of the approximate solutions at the turning point, we obtain the DE satisfied by ϕ. To do this, we represent (11C.24) into the form

$$\phi = G(x) \xi^m J_{\pm m}(\xi); \quad \text{with} \quad G(x) = \left\{ \frac{\pi^{1/2}}{2^{1/2}} A_\pm \right\} K^{-1/2}(x) \xi^{1/2-m} \qquad (11C.25)$$

Differentiating (11C.25) with respect to x, we obtain:

$$\frac{d\phi(x)}{dx} = \frac{\left\{\frac{\pi^{1/2}}{2^{1/2}}A_{\pm}\right\}\xi^{1/2-m}}{G(x)}\left\{\xi^m J'_{\pm m} + m\xi^m J'_{\pm m}\right\} + G'(x)\xi^m J_{\pm m} \tag{11C.26}$$

and $$\frac{d^2\phi(x)}{dx^2} = \xi^m J_{\pm m}G''(x) - \frac{\pi^2}{4}A_{\pm}^4 G^{-3}(x)\xi^{2-3m}J_{\pm m}$$

Using Equation 11C.25, we can write in a more compact form

$$\frac{d^2\phi(x)}{dx^2} = \frac{G''(x)}{G(x)}\phi(x) - K^2\phi(x) \tag{11C.27}$$

Thus, we have the relation

$$K(x) = \frac{d\xi(x)}{dx} = \frac{\pi}{2}A_{\pm}^2\xi^{1-2m}G^{-2}(x) \tag{11C.28}$$

and the Bessel identity

$$\xi^2 J'' + \xi J' = (m^2 - \xi^2)J(\xi) \tag{11C.29}$$

11C.3.2 Behavior of Turning Point

Equation 11C.27 can be recast into

$$\frac{d^2\phi(x)}{dx^2} + \left\{K^2(x) - \theta(x)\right\}\phi(x) = 0; \tag{11C.30}$$

where $\theta(x) = (G''(x)/G(x))$

This is the differential equation that satisfies the approximate Langer's solution ϕ at the turning point. To prove the validity of ϕ, we only must show that $G(x)$ is bounded at $x = 0$, so that the coefficient of ϕ does not possess singularity. To do this, we write

$$\frac{G(x)}{\left\{\frac{\pi^{1/2}}{2^{1/2}}A_{\pm}\right\}} = K^{-1/2}(x)\xi^{1/2-m} = x^{-n/4}f^{-1/2}(x)[I_1(x)]^{1/2-m} \tag{11C.31}$$

Using the expression for K in Equation 11C.14, then

$$I_1(x) = \int_0^x x^{n/2}f^{1/2}(x)dx \tag{11C.32}$$

Thence, integration by parts gives the expanded form

$$I_1(x) = \frac{x^{n/2+1}f^{1/2}}{n/2+1}\left\{1 - \frac{\int_0^x x^{n/2+1}f'(x)f^{-1/2}(x)dx}{2x^{n/2+1}f^{1/2}(x)}\right\} \tag{11C.33}$$

Substituting Equation 11C.33 into Equation 11C.31, we obtain

$$\frac{G(x)}{\left\{\dfrac{\pi^{1/2}}{2^{1/2}}A_\pm\right\}} = \frac{f^{-m/2}}{(n/2+1)^{1/2-m}}[1 - I_2(x)]^{1/2-m} \tag{11C.34}$$

Where the intermediate integral is given by

$$I_2(x) = \frac{\int_0^x x^{n/2+1}f'(x)f^{-1/2}(x)dx}{x^{n/2+1}f^{1/2}(x)} \tag{11C.35}$$

Now, applying l'Hospital rule shows that

$$\lim_{x\to 0} I_2(x) = 0 \tag{11C.36}$$

while we observe that $f(0) \neq 0$ by definition. Hence, $G(x) \neq 0$ at the turning point, the proposition has been proved.

11C.3.3 Error Bound for ϕ Turning Point

To investigate the accuracy with which the DE for Error bound for ϕ represents the exact wave equation at the turning point, one only must compute the value of $\theta(x)$ at $x = 0$. Thus, we find

$$\theta(x) = \frac{G''(x)}{G(x)} = \frac{3}{4}\left(\frac{K'}{K}\right)^2 - \frac{1}{2}\left(\frac{K''}{K}\right) + (m - 1/4)\frac{K^2}{\xi^{1/2}} \tag{11C.37}$$

Note that the first two terms of this equation is the function $W(x)$ defined earlier. Thus, for $m = 1/2$ (i.e., $n = 0$) we have the turning point of order zero and $\phi \to \varphi$. Therefore, it can be concluded that in a region removed from any turning point the WKB solutions are of the same form as Φ and vice versa. This observation is substantially proven by investigations of the asymptotic behavior of ϕ away from the turning point. The proof of this result paves the way for the important connection formulae which is the eventual purpose of this exercise.

Now, using the form of $K^2(x)$ given in Equation 11C.13, we can put $\theta(x)$ in term of $f(x)$ as

$$\frac{3}{4}\left(\frac{K'}{K}\right)^2 = \frac{3}{16}\left\{\left(\frac{f'}{f}\right)^2 + \frac{2n}{x}\left(\frac{f'}{f}\right) + \frac{n^2}{x^2}\right\} - \frac{1}{2}\left(\frac{K''}{K}\right)$$

$$= \frac{1}{8}\left(\frac{f'}{f}\right)^2 - \frac{n}{4}\left(\frac{n}{2} - 1\right)x^{-2} - \frac{n}{4}\left(\frac{f'}{f}\right)x^{-1} - \frac{1}{4}\left(\frac{f''}{f}\right) \tag{11C.38}$$

and

$$\left(m^2 - \frac{1}{4}\right)K^2\xi^{-2} = \left(m^2 - \frac{1}{4}\right)x^n f(x)\frac{1}{I_1^2(x)} \tag{11C.39}$$

Thus, we can rewrite (11C.33) in more suitable form

$$I_1(x) = \left\{x^{\pi/2+1}f^{1/2} - \frac{1}{2}\int_0^x x^{n/2+1}f'(x)f^{-1/2}(x)dx\right\}(n/2+1)^{-1} \tag{11C.40}$$

Since the integral $\to 0$ in the limit of $x \to 0$ we can expand I_1^{-2} in a binomial series as

$$I_1^{-2} = \left(\frac{n}{2}+1\right)^2 \left(x^{\frac{n}{2}+1}f^{1/2}\right)^{-2} \left\{ 1 + \frac{I_3}{x^{\frac{n}{2}+1}f^{-2}} \right\} + \frac{3}{4} \left\{ \frac{I_3^2}{\left(x^{\frac{n}{2}+1}f^{1/2}\right)^2} \right\} + \quad (11C.41)$$

with $\quad I_3 = \int_0^x f^{1/2}f'x^{\frac{n}{2}+1}\,dx$

Integrating by parts again, we can expand I_3 into the form

$$I_3 = \frac{1}{\left(\frac{n}{2}+2\right)} \left\{ x^{\frac{n}{2}+2}f^{-1/2}f' - \int_0^x x^{\frac{n}{2}+2}f^{-1}\left(f^{1/2}f'' - \frac{1}{2}(f')^2 f^{-1/2}\right)dx \right\} \quad (11C.42)$$

$$\therefore I_3^2 = \left(\frac{n}{2}+2\right)^{-2}\left\{ x^{\frac{n}{2}+4}f^{-1}(f')^2 - 2x^{\frac{n}{2}+2}f^{-1/2}f'I_4 + (I_4)^2 + \cdots \right\}$$

where I_4 is the integral expression given in Equation 11C.41. Finally, substituting Equations 11C.42 and 11C.41 into Equation 11C.38, we obtain

$$\left(m^2 - \frac{1}{4}\right)K^2\xi^{-2} = -\frac{n}{16x^2}\left(\frac{n+4}{2}\right) - \frac{2}{8}\frac{1}{x}\left(\frac{f}{f'}\right) - \frac{3}{16}\frac{n}{n+4}\left(\frac{f'}{f}\right)^2$$

$$+ \left[\frac{n}{8}\frac{1}{x^2}\left(\frac{f}{f'}\right) + \frac{3}{8}\frac{n}{n+4}\frac{1}{x}\left(\frac{f'}{f}\right)^2\right]\frac{I_4(x)}{x^{n/2+1}f^{1/2}(x)} + \cdots \quad (11C.43)$$

Therefore, substituting Equations 11C.43 and 11C.38 into Equation 11C.37, we arrive at

$$\theta(x) = \frac{3}{2}\frac{1}{n+6}\left(\frac{n+5}{n+4}\right)\left(\frac{f'}{f}\right)^2 - \left(\frac{f''}{f}\right) + \text{terms\_in\_}x\text{\_and\_higher} \quad (11C.44)$$

Equation 11C.44 shows that if $f(x)$ is a relatively slowly varying function at $x = 0$, then $\theta(0) \to 0$ and Langer's approximate solution Φ is a good approximation to the actual mode field at the turning point. This is a general result of which the expression obtained by Marcuse [86] is a special case.

11C.4 CORRECTION FORMULAE

It is shown above that the DE satisfied by Langer's approximate solution Φ is non-singular. Therefore, unlike the WKB solution Φ is single valued. It is not restricted to yield representations of the solution of the wave Equation 11C.1 in the intervals on one to the other side of the turning point. Therefore, in problems involving a single tuning point, no question of connection formulae arises in association with it. However, in actual waveguides, the simplest scalar wave equation contains two turning points at the minimum. Thus, a single function such as Langer's approximate solution, Φ, valid at a turning point cannot possibly describe the modal fields since essentially there are two regions of evanescent field behavior connected by a region where the field is oscillatory.

Thus, as a first step to obtain a connection formulae, we introduce a new variable

$$t = -\int_0^x |K|(x)dx = j\xi$$

$$j = \sqrt{-1}$$

(11C.45)

so we can express Φ as ϕ_1 - or - ϕ_2 to denote the solutions in regions in which $K^2 > 0$_and_$K^2 < 0$, respectively. We now have

$$\phi_1 = \left\{ \left(\frac{\pi}{2}\right)^{1/2} A_\pm [\xi/K]^{1/2} \right\} J_{\pm m}(\xi); \quad x \geq 0$$

$$\phi_2 = \left\{ \left(\frac{\pi}{2}\right)^{1/2} B_\pm \right\} [t/K]^{1/2} I_{\pm m}(\xi); \quad x < 0$$

(11C.46)

The asymptotic solution of the Bessel functions of large argument are well known and we can write [87]

$$\phi_1^{as} = \phi_1|_{\xi \to \infty} = A_\pm K^{-1/2} \cos\left(\xi \mp \frac{m\pi}{2} - \frac{\pi}{4}\right)$$

$$\phi_2^{as} = \phi_2|_{t \to \infty} = B_\pm |K|^{-1/2} e^{-t+j2\pi\left(\mp m - \frac{1}{4}\right)}$$

(11C.47)

These relations confirm our earlier hypothesis that ϕ_1^{as}; ϕ_2^{as} are just linear combinations of the WKB solutions $\varphi_1 = A_\pm K^{-1/2} e^{\pm j\xi}$_and_$\varphi_2 = B_\pm |K|^{-1/2} e^{-t}$.

We consider the case in which the two turning points are sufficiently far apart so that we can use a turning point of order $n = 1$. Now that the solutions for the linear turning point are just the Airy functions or Bessel functions of order $1/3$, these are

$$\Phi_1 = \left(\frac{\pi}{2}\right)^{1/2} A_\pm \left(\frac{\xi}{K}\right)^{1/2} J_{\pm 1/3}(\xi); \quad x \geq 0$$

$$\Phi_2 = \left(\frac{\pi}{2}\right)^{1/2} B_\pm \left(\frac{t}{|K|}\right)^{1/2} I_{\pm 1/3}(t); \quad x < 0$$

(11C.48)

These expressions are just alternative ways of writing the same solutions. They must be identical. Continuity requirements at $x = 0$ gives

$$B_+ = -A_-$$

$$B_- = A_-$$

(11C.49)

Thus, the asymptotic forms can be written down as

$$\phi_1^{as} = A_\pm K^{-1/2} \cos\left(\xi \mp \frac{\pi}{6} - \frac{\pi}{4}\right)$$

$$\phi_2^{as} = \mp \frac{A_\pm |K|^{-1/2}}{2} e^{-t-j\pi\left(\frac{1}{2} \mp \frac{2}{3}\right)}$$

(11C.50)

To derive the first connection formula, we follow the procedure established by Langer [88] and write

$$\phi_{2+}^{as} + \phi_{2-}^{as} + \phi_{1+}^{as} + \phi_{1-}^{as}$$
$$A_+ = A_-$$

$$\text{thus} \rightarrow |K|^{-1/2}e^{-t} \rightarrow 2|K|^{-1/2}\cos\left(\xi - \frac{\pi}{4}\right)$$

(11C.51)

The arrow indicates that the asymptotic solution on the left goes into the expression on the right as one crosses the turning point. These arrows are irreversible or a small error in the phase of the cosine would be magnified by the positive exponential on the left hand side of Equation 11C.51.

Similarly, we can write down another set of connection formula as

$$|K|^{-1/2}e^{-t} \leftarrow 2K^{1/2}\cos\left(\xi + \frac{\pi}{4}\right)$$

(11C.52)

These formulae suffice for applications involving two turning points.

11C.5   APPLICATION OF CORRECTION FORMULAE

There are three distinct categories of problems encountered in wave propagation in slab dielectric waveguides where the connection formulae obtained above can be applied to yield eigenvalue equation for bound modes. In the WKB context, a bound mode corresponds to a solution which is oscillatory between two turning points beyond which the solution is evanescent. These are treated separately.

11C.5.1   Ordinary Turning Point Problem

The refractive index profile is illustrated in Figure C.1a, where the turning points are at x and x_2. Thus for $x_1 < x \leq x_2$, $K^2 > 0$ and the field is oscillatory. Elsewhere, the field is evanescent. The connection formulae connect solutions on both sides of a turning point and these must be applied at both x_1 and x_2. Correct phase matching of the oscillatory fields yields the WKB eigenvalue equation. This equation contains the implicit prescription of the required propagation constant.

Thus, in region #2a and #2b, we have, respectively,

$$\varphi_{2a} = A_a|K|^{-1/2}e^{-t_1}; \quad \varphi_{2b} = A_b|K|^{-1/2}e^{-t_2}$$

$$\text{with} \quad t_1 = \int_{x_1}^{x}|K|dx; \quad t_2 = \int_{x_2}^{x}|K|dx$$

(11C.53)

By virtue of the first connection formulae Equation 11C.51, the solution in the region #1 which connects with region #2a is

$$\varphi_{1a} = 2A_aK^{-1/2}\cos\left(\xi_1 - \frac{\pi}{4}\right); \quad \text{with} \quad \xi_1 = \int_{x_1}^{x}Kdx$$

(11C.54)

And that which connects the solution in region #2b is

$$\varphi_{1b} = 2A_bK^{-1/2}\cos\left(\xi_2 - \frac{\pi}{4}\right); \quad \text{with}\ldots\xi_2 = \int_{x}^{x_2}Kdx$$

(11C.55)

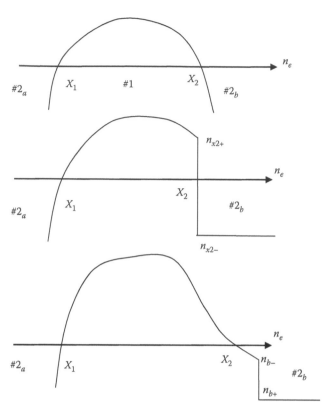

FIGURE 11C.1 (a) Ordinary turning point with caustics at $x = x_1$ and x_2 where $n_2(x) = ne_2$, (b) step discontinuity at $x = x_2$, and (c) buried modes with mode index ne close to step discontinuity at $x = b$.

Now with φ_{1a} and φ_{1b} represent one and only one solution in #1. Thus, this consistency condition implies that

$$A_a = A_b; \quad \xi_1 - \frac{\pi}{4} = \pm\left(\xi_2 - \frac{\pi}{4}\right) \tag{11C.56}$$

The RHS of Equation 11C.56 can be written as, taking the minus sign

$$-\left(\xi_2 - \frac{\pi}{4}\right) = \xi_1 - \frac{\pi}{4} - \eta; \quad \eta = \int_{x_1}^{x_2} K dx - \frac{\pi}{2} \tag{11C.57}$$

which must be a multiple of π to ensure correct phase matching of the WKB oscillatory solution. Equation 11C.57 is expanded to give the familiar form of the WKB eigenvalue equation

$$\int_{x_1}^{x_2} \left[k_0^2 n^2(x) - \beta^2\right]^{1/2} = \left(m + \frac{1}{2}\right)\pi; \quad m = 0, 1, 2 \dots \tag{11C.58}$$

This result fails when $x_1 = x_2$ or when there is an index discontinuity in the vicinity of the turning point.

11C.5.2   Effect of an Index Discontinuity at a Turning Point

Figure 11C.1b illustrates a typical example of a thin film waveguide deposited on a substrate. The turning point at $x = x_2$ coincides with the film/air interface. At $x = x_2$, it is erroneous to apply the connection formulae, since they are derived on the assumption of a linear variation of $K^2(x_2)$. We resort, instead, to the boundary conditions imposed on the fields at $x = x_2$.

If we denote $n(x_{2-})$ and $n(x_{2+})$ as the values of the refractive index just before and after the step, then the standard phase shift at the discontinuity [89] is proportional to δ_{x_2} given by

$$\delta_{x_2} = \tan^{-1}\left(\frac{\beta^2 - k_0^2 n^2(x_{2+})}{-\beta^2 + k_0^2 n^2(x_{2-})}\right)^{1/2} \tag{11C.59}$$

and the WKB eigenvalue equation transforms to

$$\int_{x_1}^{x_{2-}} [k_0^2 n^2(x) - \beta^2]^{1/2} = \left(m + \frac{1}{4}\right)\pi + \delta_{x_2}; \quad m = 0, 1, 2\ldots \tag{11C.60}$$

Note that in the large step $\delta_{x_2} \to \pi/2$ as in the case of a strongly asymmetrical waveguide.

11C.5.3   Buried Modes Near an Index Discontinuity at a Turning Point

The analysis in the previous sub-section treats only the turning point at $x = x_2$ fails directly on top of the index discontinuity. Thus, strictly speaking Equation 11C.60 is only accurate for certain order modes (value of m) which satisfy this condition.

We denote the refractive index just before and after the step at $x = b$ by $n(b^-)$ and $n(b^+)$, respectively. Then in the region $x_2 < x < b$ the step at $x = b$ causes significant reflection of energy. Thus, it is no longer accurate to represent the fields in the region by a single decay exponential. Therefore, we include decaying and growing exponentials and write

$$\varphi_1 = AK^{-1/2}\cos\left(\xi_1 - \frac{\pi}{4}\right); \quad x_1 < x < x_2$$

$$\varphi_{2a} = A|K|^{-1/2}\cos\left(\int_{x_1}^{x_2} K dx\right)e^{t_2} + \frac{1}{2}c\sin\left(\int_{x_1}^{x_2} K dx\right)e^{-t_2}; \quad x_2 < x < b \tag{11C.61}$$

$$\varphi_{2b} = A|K|^{-1/2}e^{\left(\int_{xb}^{x}|K|dx\right)}; \quad x < b$$

where A is a constant. The coefficients of the growing and decaying fields in Equation 11C.61 have been chosen to satisfy the connection formulae at $x = x_2$. Thus, the equation of φ_1 in Equation 11C.61 is redundant as far as the eigenvalue equation is concerned. Continuity requirements on φ_{2a} and φ_{2b} at $x = b$ gives the required eigenvalue equation

$$\int_{x_1}^{x_{2-}} [k_0^2 n^2(x) - \beta^2]^{1/2} = \left(m + \frac{1}{4}\right)\pi + \delta_b; \quad m = 0, 1, 2\ldots \tag{11C.62}$$

where

$$\delta_b = \tan^{-1}\left\{\frac{|K(b^-)| + |K(b^+)|}{|K(b^-)| - |K(b^+)|}\right\}e^{2\int_{x_2}^{b^-}|K|dx}$$

is the corrected overall phase shift of the buried modes [90].

APPENDIX 11D: DESIGN AND SIMULATION OF PLANAR OPTICAL WAVEGUIDES

11D.1 INTRODUCTION

We have outlined in the series of lectures on optical waveguides in which we treat a slab optical wave-guide as the extension of an optical fiber where the structure is restricted to one-dimension and the other dimension of its cross section has been extended to infinity. The optical waveguides considered in this experiment are not a step-index slab type as considered in the theoretical section above but rather they are graded-index, that is, the refractive index is gradually decreased from the core to the cladding region.

This introductory experiment in optical waveguides aims to familiarize potential optical communications engineers with the structures and behavior of the optical field distribution in a number of guided optical waves structures such as straight, bend, and Y-junction, etc. The computer experiment is written in such a way that you can read and perform the preliminary work and experiment in stages.

The objectives of this section are

- To design parameters of slab optical waveguides so that they can support a certain number of guided modes in the single or multi-mode regions.
- To use the fundamental mode of the optical waveguide for observation and measurement of optical fields of several waveguide composite structures.
- To propagate the fundamental optical field through a number of optical waveguide structures such as straight, bend, and Y-junction optical guided waves structures.

11D.2 THEORETICAL BACKGROUND

11D.2.1 Structures and Index Profiles

Optical waveguides are the fundamental element in modern optical communications and photonic signal processing systems. Optical fibers are the guiding medium for optical signal transmission and are formed by a circular core inside a circular cladding region. The mathematics required to represent the electric and magnetic fields components of the guided waves in optical fibers would involve Bessels' functions. These waveguides would be treated in detail in the fourth year course of optical communications engineering. A simplified version of optical fibers is the slab optical waveguide whose structure is shown in Figure 11.1. The cladding regions are the superstrate and substrate would have identical constant refractive index. The guiding region is a slab or thin film layer sandwiched between the cladding regions. The refractive index of the slab region must be higher than that of the cladding and its thickness must be sufficiently thick to support confined (bound) optical guided modes.

The refractive index profile of the step-slab region can be uniform, that is, constant throughout, *or graded where n(x) decreases gradually* from the center of the slab to the cladding. For the sake of simplicity, to obtain an analytical solution of the wave equation representing the guided field, the index profile of our slab structures would have a "$cosh^{-2}$" distribution (graded index profile) given by

$$n^2(x) = n_s^2 + \frac{2n_s \Delta n}{\cosh^2 \dfrac{2x}{h}}$$ (11D.1)

or approximately

$$n(x) = n_s + \frac{\Delta n}{\cosh^2 \dfrac{2x}{h}}$$ (11D.2)

where n_s = cladding refractive index (for the superstrate and substrate), Δn = refractive index difference between the cladding and slab regions and h is the total thickness of the guiding layer (slab thickness). It is convenient to define a normalized parameter V as

$$V = kh(2\,ns\,\Delta n)^{1/2} \tag{11D.3}$$

where $k = 2\pi/\lambda$ and λ is operating wavelength of the optical waves in vacuum.

Note: The expression of the parameter V is identical with that of a circular optical fiber. However, in this experiment, we are dealing with "planar" optical waveguide structures.

11D.2.2   Optical Fields of the Guided TE Modes

Normally, the optical fields in a slab waveguide would consist of two quasi-polarizations TE (transverse electric) and TM (transverse magnetic) where the non-zero electromagnetic field components are (E_y, H_x, H_z) and (E_x, E_z, H_y) for TE and TM modes, respectively. In this experiment, we consider only the behavior of TE modes in slab optical waveguides.

The wave equation for TE modes can be derived from Maxwell's equations; in the case when the refractive index difference is small, the wave equation can be approximated to have a scalar form as

$$\frac{d^2 E_y}{dx^2} - (\beta^2 - n^2 k^2)E_y = 0 \tag{11D.4}$$

When the refractive index distribution of the waveguide structure $n(x)$ has a $\cosh^{-2}$ profile in the slab region and constant in the cladding regions, the field solution of Equation 11D.4 would have an analytical form of

$$Ey(x) = \frac{u_v(2x/h)}{\cosh^2 (2x/h)} \tag{11D.5}$$

where $v = 0,1,2,3\ldots$ subject to the boundary conditions that the field must vanish at a distant very far from the slab-cladding interface. The function $u_v(2x/h)$ would take the following forms:

For even TE modes, $v = 0,2,4,\ldots$

$$
\begin{aligned}
u_v(2x/h) = 1 &- \frac{1}{2}v(2s - v)\frac{\sinh^2(2x/h)}{1.1!} \\
&+ \frac{1}{4}v(v - 2)(2s - v)(2s - v - 2)\frac{\sinh^4(2x/h)}{(1.3.2!)} + \cdots
\end{aligned} \tag{11D.6}
$$

with $\underline{s} = 0.5\,\{(1 + V^2)^{1/2} - 1\}$ the *total number of guided even modes* that this kind of optical waveguide can support.

For odd TE modes, $v = 1,3,5\ldots$

$$u_v(2x/h) = \sinh (2x/h)$$

$$
\left[
\begin{aligned}
1 &- \frac{1}{2}(v - 1)(2s - v - 1)\frac{\sinh^2(2x/h)}{3.1!} \\
&+ \frac{1}{4}(v - 1)(v - 3)(2s - v - 3)\frac{\sinh^4(2x/h)}{(3.5..2!)} + \cdots
\end{aligned}
\right] \tag{11D.7}
$$

with $s = 0.5[(1 + V^2)^{1/2} - 1]$ is the maximum number of guided odd modes.

For lower order modes, we have

$$u_0 = 1 \tag{11D.8}$$

$$u_1 = \sinh \frac{2x}{h} \tag{11D.9}$$

$$u_2 = 1 - 2(s-1)\sinh^2 \frac{2x}{h} \tag{11D.10}$$

$$u_3 = \sinh \frac{2x}{h} \left[1 - \frac{2}{3}(s-2)\sinh^2 \frac{2x}{h} \right] \tag{11D.11}$$

The propagation constant β_υ and the effective indices $(n_{eff} = \beta_\upsilon / k)$ of the υth order modes are given by

$$\beta_\upsilon^2 = n_s^2 k^2 + 4(s-\upsilon)^2/h^2 \tag{11D.12}$$

$$n_{eff}^2 = n_s^2 + (s-\upsilon)^2 (\lambda/\beta_\upsilon)^2 \tag{11D.13}$$

and the normalized propagation constant b is defined as

$$b = \frac{n_{eff}^2 - n_s^2}{n^2 - n_s^2} \tag{11D.14}$$

with n is the refractive index at the center of the guide or approximately

$$b = \frac{n_{eff}^2 - n_s^2}{2n_s \Delta n} \tag{11D.15}$$

Thus, the optical field of a slab optical waveguide can be found if we can specify the following parameters: the slab thickness h; the cladding refractive index; the refractive index difference Δn, and the operating wavelength

11D.2.3   Design of Optical Waveguide Parameters

Choose the parameters n_s, Δn, and h of your waveguide. Some typical refractive indices of certain transparent materials for superstrate and substrate of optical waveguides are

- $n_s = 1.447$ for silica glass at the operating wavelength of 1300 nm. This also is the base material for modern telecommunication optical fibers.
- $n_s = 3.6$ for GaAs semiconductor waveguide at 1300 nm. This is the base material for optical waveguides formed in the resonant cavity and waveguide of semiconductor lasers for optical fiber communications.
- $n_s = 2.2$ for lithium niobate crystal at certain crystal axis as seen by the TE waves. This material is also the base material for optical modulators for optical communications.

Make sure that the chosen parameters would form an optical waveguide that would support no more than four guided modes at the operating wavelength of 1300 nm. Can you design an optical waveguide such that it supports only one guided mode which is the only fundamental mode of the optical waveguide?

In fact, we can plot the $b-V$ curve for $\upsilon = 0, 1, 2,\ldots$ and from this diagram we can design a mono-mode, two-mode... optical waveguides. Notice that only TE modes are considered here.

11D.3 A Simulation of Optical Fields and Propagation in Slab Optical Waveguide Structures

A computer simulation program can be used [90] to study the evolution of optical fields in slab optical waveguides, which form the basic component for several optical wave guiding devices.

To numerically study the behavior of optical waves, particularly the fundamental mode field, in these structures, the whole waveguide region including the slab and cladding regions, that is, W and L, are sliced into several intervals along the propagation direction z, as well as in the vertical direction for numerical calculations.

The field in the first plane, that is, at $z = 0$, can be found by Equation 11D.7. This field would then be propagated to the next plane through a discredited equation by applying the finite difference method to the wave equation with the z-dependence in the Maxwell's equation. This para-axial wave propagation equation is given by

$$2jkn_s \frac{\partial E_y}{\partial z} = \frac{\partial^2 E_y}{dx^2} + k^2 \big[n^2(x, y) - n_s^2 \big] E_y \tag{11D.16}$$

which can be written using the center-finite-difference technique as

$$jkn_s \frac{E_{i,k+1} - E_{i,j}}{\Delta z} = \frac{E_{i-1} - 2E_i + E_{i+1}}{\Delta x^2} + k^2 \big[n_i(x, y) - n_s^2 \big] E_i \tag{11D.17}$$

where $j = (-1)^{1/2}$, the subscripts i and j denote the variation of E with respect to the x and z directions, respectively. That is, the cross section plane is partitioned into several layers with order i and the propagation steps along the z-direction are assigned with order j. The obtained results of the field at location j would then be used as the field initial distribution for propagating through the structure to obtain the optical field of the next plane $j + 1$ and so on. Thus, we can employ an analytical method to obtain the field solution for the optical filed in the transverse plane. A numerical method (the finite difference) is used to study the evolution of the optical field propagating along the optical waveguide structure.

In this section, we are not going to study the finite difference method but the evolution of the optical field in optical waveguides. In the following parts, we will examine the optical field behavior in the structures illustrated in Figure 11D.1.

An additional dimension, the z axis, is now added to the optical waveguide devices. These structures are shown in Figures 11D.2 through 11D.5.

11D.3.1 Lightwaves Propagation in Guided Straight Structures

Please do not hesitate to seek assistance from the demonstrators for procedures in running the simulation package. Typical steps for simulation are: run MATLAB; go to directory STRAIGHT; run FD1, the program for the beam propagation method; choose parameters as prompted by the program

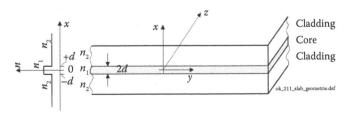

FIGURE 11D.1 Schematic structure of the slab optical waveguide: The guiding layer is sandwiched between the superstrate (upper cladding region) and substrate (lower cladding region) of identical refractive indices.

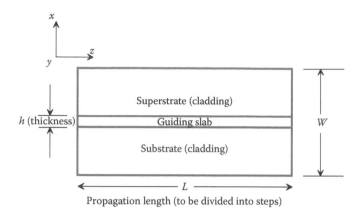

FIGURE 11D.2 (a) Side view of a slab optical waveguide in a straight optical device structure. h is the waveguide thickness, W is the total width of the structure in the transverse plane to be specified for numerical simulation, L is the total length of the device. W is to be divided into several equi-spacing layers for numerical simulation. The length L along the propagation direction is also spitted into several steps for propagation from one plane to the other and so on.

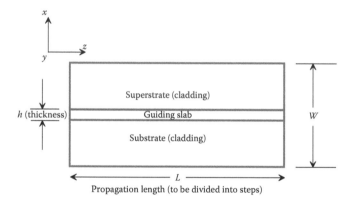

FIGURE 11D.3 Side view of the optical waveguide device using a slab waveguide in bend structure. Notice the bending section. θ is the bend angle, l is the length of the straight section.

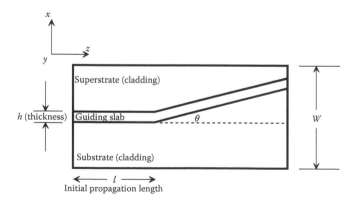

FIGURE 11D.4 Side view of the Y-structure using slab optical waveguide. θ is the half Y-junction angle and l is the straight initial section before splitting.

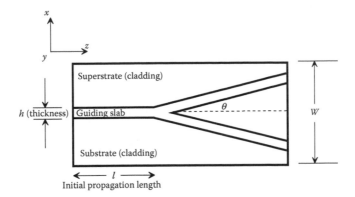

FIGURE 11D.5 Side view of the interferometric structure using slab optical waveguide. θ is the half Y-splitting angle and l is the straight initial section before splitting.

such as (i) waveguide region to be analyzed; (ii) operating wavelength; (iii) slab thickness; (iv) the number of x intervals for optical field and number of propagation steps in the z-direction; (v) the refractive indices of slab and cladding, that is, cladding index and the index difference; and (vi) propagation distance.

a. After successfully obtaining the guiding of optical waves, keep one or two parameter constant (such as the waveguide thickness), vary stepwise other parameters, for example, index difference, wavelength, etc. Observe the evolution of the optical field and plot the 3D guided wave field profile and the field contour. Note: when specifying the number of plane to be plotted by MATLAB the product of the number of intervals in x and z directions must not exceed the MATLAB limit which is 8188 depending on the available computer memory and MATLAB version.

b. Observe the field evolution with respect to the change in refractive index difference, waveguide thickness etc.

An example of the wave guided and propagation in a straight waveguide is shown in Figure 11D.6.

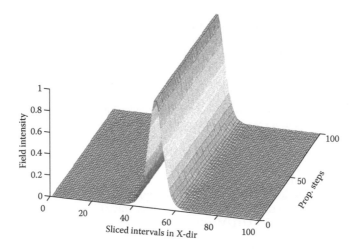

FIGURE 11D.6 Guided mode propagation in a straight slab optical wave guide. Waveguide parameters: cladding refractive index = 2.2, refractive index change 0.02, propagation length 100 micron, step size of 0.5 micron. Step size in x-direction 0.02 micron.

11D.3.2 Lightwaves Propagation in Guided Bent Structures

Similar to the steps as above: (i) choose the most suitable optical waveguide structures of the STRAIGHT structure to enter into the bend structure parameters; (ii) the BEND directory is to be evoked; (iii) additional parameters for this structure are the bend angle θ and the length l of the straight section. Start with a bend angle of about 0.5 degree of arc; (iv) now vary the bend angle in step of 0.5 or 1 degree of arc to about 10 degree of arc, (v) observe the radiation of the guided field at the bend section and observe the guided and radiated optical fields; and (vi) vary the refractive index difference and run the program for the bend angle of about 2–4 degrees and observe the confinement and radiation of the optical field at straight and bend sections.

11D.3.3 Lightwaves Propagation in Y-Junction (Splitter) and Interferometric Structures

As illustrated in Figure 11D.5, the Y-junction or optical splitter is considered as a combination of two identical bend sections. Note that the section right at the Y-junction has a width that is wider than that of the straight section. Thus, the number of guided modes would be higher for this very short section: (i) choose the half angle θ of the Y-junction from 0.5 to 5 degrees of arc, *then* observe the field evolution and measure the field strength, that is, "intensity" or optical power, distribution across the whole device; (ii) one can vary the refractive index difference for a small half angle at the Y-junction to observe the splitting effect. The field behavior at these Y-junctions can be seen in Figure 11D.6, and (iii) then simulation of the propagation of lightwaves through an interferometric optical waveguide structure the Mach–Zhender, in which the input lightguide is spitted into two paths and the combined into one output port.

REFERENCES

1. S. E. Miller, Integrated optics: An introduction, *Bell Sys. Tech. J.*, vol. 48, pp. 2059–2069, 1969.
2. C. R. Doerr and K. Okamoto, Advances in silica planar lightwave circuits, *IEEE J. Lightwave Tech.*, vol. 24, no. 12, pp. 4763–4770, Dec. 2006.
3. P.S. Chung, Waveguide modes, coupling techniques, fabrication and losses in optical integrated optics, *J. Elkect. Electron. Australia*, vol. 5, pp. 201–214, 1985.
4. http://en.wikipedia.org/wiki/Photonic_integrated_circuit. Access date: Dec. 20009; http://electron9.phys.utk.edu/optics421/modules/m10/integrated_optics.htm.
5. R. G. Hunsperger, *Integrated Optics Theory and Technology*, originally published in the series: Advanced Texts in Physics, 6th ed., XXVIII, Berlin: Springer, 2009.
6. L.N. Binh and S.V. Chung, Nonlinear interactions in thin film structures, in *Proceedings of 8th Australian Workshop on Optical Communications*, Adelaide, Session VII, 1983.
7. K. Uchiyama and T. Morioka, All-optical time-division demultiplexing experiment with simultaneous output of all constituent channels from 100Gbit/s OTDM signal, *Elect. Lett.*, vol. 37 no. 10, 642–643, 10th May 2001.
8. M. Nakazawa, T. Yamamoto, and K.R. Tamura, 1.28Tbit/s–70km OTDM transmission using third- and fourth-order simultaneous dispersion compensation with a phase modulator, *Elect. Lett.* vol. 36 no. 24, pp. 2027–2029, 23rd November 2000.
9. C. Schubert, R. H. Derksen, M. Möller, R. Ludwig, C.-J. Weiske, J. Lutz, S. Ferber, A. Kirstädter, G. Lehmann, and C. Schmidt-Langhorst, Integrated 100-Gb/s ETDM receiver, *IEEE J. Lightw. Tech*, vol. 25, no. 1, pp.122–129, Jan. 2007.
10. H. Suzuki, M. Fujiwara, and K. Iwatsuki, Member, application of super-DWDM technologies to terrestrial terabit transmission systems, *IEEE J. Lightw. Tech*, vol. 24, no. 5, pp. 1998–2005, May 2006.
11. J. P. Turkiewicz et al., 160 Gb/s OTDM networking using deployed fiber, *IEEE J. Lightw. Tech*, vol. 23, no. 1, pp. 225–234, Jan. 2005.
12. R. Nagarajan et al., 400 Gbit/s (10 channel x40 Gbit/s) DWDM photonic integrated circuits, *Electronics Letters*, vol. 41 no. 6, 2005.
13. T. Okoshi, Recent advances in coherent optical fiber communications, *IEEE J Lightw. Tech.*, vol. LT-5, pp. 44–52, 1987.

14. LG. Karzovky and O. K. Tonguz, ASK and FSK coherent lightwave systems: A simplified approximate analysis, *IEEE J. Lightw. Tech.*, vol. 8, no. 3, pp.338–351, Mar. 1990.

15. S. Tsukamoto, D.-S. Ly-Gagnon, K. Katoh, and K. Kikuchi, Coherent demodulation of 40-Gbit/s polarization-multiplexed QPSK signals with 16-GHz spacing after 200-km transmission, paper: PDP29, in *Proc. CLEOS 2005*, San Jose, 2005.

16. D.-S. Ly-Gagnon, S. Tsukamoto, K. Katoh, and K, Kikuchi, Coherent detection of optical quadrature phase-shift keying signals with carrier phase estimation, *IEEE. J. Lightw. Tech.*, vol. 24, no. 1, pp. 12–21, Jan. 2006.

17. V. Ramaswamy and R.K. Lagu, Numerical field solution for an arbitrary asymmetrical gradient index planar waveguide, *IEEE J. lightw. Tech.*, vol. LT-1, pp. 408–417, 1983.

18. R. V. Schmidt and IP. Kaminow, Metal-diffused optical waveguides in LiNbO$_3$, *Appl. Physics Letters.*, vol. 25, pp. 458–460, 1974.

19. Streifer W, R. D. Burnham, and D. R. Scifres, Modal analysis of separate-confinement heterojunction lasers with inhomogeneous cladding layers, *Opt Lett.*, vol. 8, pp. 283–285, 1981.

20. D. Marcuse, *Light Transmission Optics*, New York: Van Nostrand, 1972. Chapter 8.

21. E.M. Conwell, Modes in optical waveguides formed by diffusion, *Appl. Phys. Lett.*, vol. 26, pp. 328–329, 1973.

22. H. Kirchoff, The solution of Maxwell equations for inhomogeneous dielectric slabs, *A.E.U.*, vol. 26, pp. 537–541, 1972.

23. D. Marcuse, TE modes of graded index—slab waveguides, *IEEE J. Quant. Elect.*, vol. QE-9, pp. 1000–1006, 1973.

24. A.W. Snyder and J.D. Love, *Optical Waveguide Theory*, London: Chapman and Hall, 1983, Chapter 12.

25. T.R. Chen and Z.L. Yang, Modes of a planar waveguide with Fermi index profile, *Appl. Optics*, vol. 24, pp. 2809–2812, 1985.

26. H.F. Taylor, Dispersion characteristics of diffused channel waveguides, *IEEE. J. Quant. Electronics*, vol. QE-12, pp. 748–752, 1976.

27. G.B. Hocker and W.K. Burns, Mode dispersion in diffused channel waveguides by the effective index method, *Appl. Optics*, vol. 16, pp.113–118, 1975.

28. L. Riviere, A. Yi-Yan, and H. Carru, Properties of single mode planar with Gaussian index profile, *IEEE J. Lightw. Tech.*, vol. LT-3, pp. 368–377, 1986.

29. S.K. Korotky et al., Mode size and method for estimating the propagation constant of single mode Ti: LiNbÔ strip waveguides, *IEEE Trans. Microwave Th. and Tech.*, vol. MTT-30, pp. 1784–1789, 1982.

30. V. Ramaswamy and R.K. Lagu, ibid., 1983.

31. A.N. Kaul, S.L. Hossain, and K. Thyagarajan, A simple numerical method for studying the propagation characteristics of single-mode graded-index planar optical waveguides, *IEEE. Trans. Microw. Th. And Tech.*, vol. MTT_34, pp. 288–292, 1986.

32. L.N. Binh and S.V. Chung, Nonlinear interactions in thin film structures, in *Proc. 8th Aust. Workshop on Opt. Comm., Session VII*, Adelaide, 1983.

33. Binh et al., Design considerations for second harmonic generation in thin film grown by ion beam assisted deposition method, in *Proc. 3rd Laser Conf.*, Melbourne, Session 10A, 1983.

34. J. Killingbeck, A pocket calculator determination of energy eigenvalues, *J. Phys A.*, vol. 10, pp. L09–L103, 1977.

35. R. A. Sammut and C. Pask, Simplified numerical analysis of optical fibers and planar waveguides, *Elect. Lett.*, vol.17, pp. 105–106, 1981.

36. J. D. Love and A. K. Ghatak, Exact solutions for TM modes in graded index slab waveguides, *IEEE J. Quant Elect.*, vol. QE 15, pp. 14–16, 1979.

37. Marcuse D., The effects of the $\nabla 2n$ term on the modes of an optical square-law medium, *IEEE J Quant Elect*, vol. QE-9, pp. 958–960, 1973.

38. H. Kolgenik and V. Ramaswamy, Scaling rule for thin film optical waveguides, *Appl. Opt.*, vol. 13, pp. 1857–1862, 1974.

39. L. N. Binh et al., Design considerations for second harmonic generation in thin film grown by ion-beam assisted deposition method, *3rd Int. Laser Conf.*, Melbourne Australia, Session 10A, 1983.

40. P.K. Tien, Integrated optics and new wave phenomena, *Rev. Mod. Physics*, vol. 49, pp. 361–420, 1977.

41. A. Watson, Physics–where the action is, *New Scientist*, vol. 109, no. 1493, pp.42–44, 30th Jan. 1986.

42. L. Mammel and L.G. Cohen, Numerical prediction of fiber transmission characteristics from arbitrary refractive index profiles, *Appl. Opt.*, vol. 21, pp. 699–703, 1982.

43. Petermann K., Constraints for fundaental mode-size for broadband dispersion-compensated single mode fibers, *Elect. Lett.*, vol. 19, pp. 712–714, 1983.

44. Sansonetti P., Moddal dispersion in single mode-fibers: Simple approximation issued from mode spot size spectral behavior, *Elect. Lett.*, vol. 18, pp. 647–648, 1982.

45. Korotsky et al., Mode size and method for estimating the propagation constant of single mode Ti: LINbO3 strip waveguides, *IEEE Trans. Microw. Th. Techniques*, vol. MTT-30, pp. 1784–1789, 1982.

46. Black R.J. and C. Pask, Slab waveguides characteristics by moments of refractive index profiles, *IEEE J. Quant. Elect.*, vol. QE-20, pp. 996–999, 1984.

47. Ruschin, S., Approximate formula for the propagation constant of the basic mode in slab waveguides pof arbitrary index profiles, *Appl. Opt.*, vol. 24, pp. 4189–4191, 1985.

48. A. W. Snyder and J. Love, *Optical Waveguide Theory*, London, Chapman and Hall, 1983, Chapter 12.

49. L. Riviere, A. Yi-Yan, and H. Carru, Properties of single optical planar waveguides with Gaussian index profiles, *IEEE J. Lightw. Tech.*, vol. 3, pp. 368–377, 1985.

50. R. Kiel and F. Auracher, Coupling of single-mode Ti:diffused LiNbO3 waveguides to single mode fibers, *Opt. Comm*, vol. 30, pp. 23–28, 1979.

51. H.F. Taylor, ibid., 1976.

52. J. D. B. Bradley, M. Costa e Silva, M. Gay, L. Bramerie, A. Driessen, K. Wörhoff, J.-C. Simon, and M. Pollnau, 170 Gbit/s transmission in an erbium-doped waveguide amplifier on silicon, *Optics Express*, vol. LT-17, no. 24, pp. 22201–22208.

53. W.A. Gambling and H. Matsumara, Propagation in radially inhomogeneous single-mode fiber, *Opt Quantum Electronics*, vol. 108, pp. 31–34, 1970.

54. J. Jeffery, On certain approximate solutions of linear differential equations of second order, *Proc. London Math. Soc.*, vol. 23, pp. 428–436, 1923.

55. G. Wentzel, Eine Verrallgemeirneung der QUanttenbedingungen fur die Zwecke der Wellemechanik, *Z. Phys.*, vol. 38, pp. 518–529, 1926.

56. H. A. Kramers, Wellenmechanik und halbzahlig Quantisierung, *Z. Phys.*, vol. 39, pp. 828–840, 1926.

57. L. Brillouin, Remarque's sur la mechanique ondulatoire, *J. de Phys.*, vol. 7, pp. 353–368, 1926.

58. D. Marcuse, ibid., 1973.

59. G. B. Hocker and W. K. Burns, Mode dispersion in diffused channel waveguides of arbitrary index profiles, *IEEE J. Quant Elect.*, vol. QE-11, pp. 270–276, 1975.

60. A. Ankiewicz, Comparison of wave and ray techniques for solution of graded index optical waveguide problems, *Optica Acta*, vol. 25, pp. 361–373, 1978.

61. R. E. Langer, On the connection formulae and the solutions of the wave equation, *Phys Rev.*, vol. 51, pp. 669–676, 1937.

62. J. M. Arnold, Asymptotic analysis of planar and cylindrical inhomogeneous waveguides, *Radio Sciences*, vol. 16, pp. 511–518, 1981.

63. J. P. Gordon, Optics of general guiding media, *Bell Syst. Tech. J.*, vol. 45, pp. 321–332, 1966.

64. D. Marcuse, ibid., 1973.

65. D. Marcuse, ibid., 1973.

66. Hocker and Burns, ibid, 1975

67. P. K. Tien and R. Ulrich, Theory of prism-coupler and thin film lightguides, *J. Opt Soc Am.*, vol. 60, pp. 1325–1337, 1966.

68. Hocker and W.K. Burns, ibid., 1975.

69. Marcuse, ibid., 1973.

70. I. S. Gradshteyn, I. M. Ryzhik, Yu. V. Geronimus, M. Yu. Tseytlin, and A. Jeffrey, 2015 [October 2014]. D. Zwillinger, Moll, Victor Hugo, eds. *Table of Integrals, Series, and Products*, Translated by Scripta Technica, Inc. (8 ed.). Academic Press, Inc. ISBN 978-0-12-384933-5.

71. Ramaswamy and Lagu, ibid., 1983.

72. P. K. Mitra and A. Sharma, Analysis of single mode inhomogeneous planar waveguides, *IEEE J. Lightwave Tech.*, vol. LT-4, pp. 204–212, 1986.

73. Pask and Black 1984.

74. Snyder and Sammut 1979.

75. D. Marcuse, ibid., 1972, Chapter 8.

76. D. Marcuse, ibid., 1972, Chapter 8.

77. D. Marcuse, ibid., 1973.

78. M. J. Adams, *An Introduction to Optical Waveguides (Series: Technology & Engineering)*, New York: John Wiley & Sons, 1981.

79. G. P. Agrawal, *Fiber-Optic Communication Systems 4E W/CD (Wiley Series in Microwave and Optical Engineering)*, New York: John Wiley, 2010.

80. Adams, 1981.

81. M. Abramowitz and I. A. Stegun, *Handbook of Mathematical Functions*, New York: Dover Pub, 1972.
82. Adams, 1981.
83. M. Abramowitz and I. A. Stegun, ibid., 1972.
84. J. L. Dunham, The Wentzel-Brillouin-Kramers method of solving the wave equation, *Phys. Rev.*, vol. 41, p. 713, 1932.
85. Gradshteyn and Ryzhik, 1965.
86. Marcuse ibid., 1973.
87. Abramowitz and Stegun, 1972.
88. R. E. Langer, On the connection formulas and the solutions of the wave equation, *Phys. Rev.* vol. 51, p. 669, 1937.
89. Adams, 1981.
90. L. N. Binh, *Guided Wave Photonics: Fundamentals and Applications with MATLAB® (Optics and Photonics)*, 1st ed, Boca Raton, FL: CRC Press, 2014, See a MATLAB program for simulation.

12 Graded-Index Integrated Modulators

To achieve efficient design of high-speed modulators and switches, especially micro ring resonators, the fabrication of rib-waveguides and diffused waveguide with suitable mode size is essential to minimize total insertion loss (for ring resonator) and to maximize the overlap integral between the guided optical field and the applied modulating field.

In this chapter, FDM is described for analyzing the quasi-*TE* and quasi-*TM* polarized waveguide modes because of its simplicity and plausible accuracy. We employ the semi-vectorial (SV) analysis to automatically take into full account of the discontinuities in the normal electric field components across any arbitrary distribution of internal dielectric interfaces. The eigen-modes of the Helmholz equation can be solved by applying the shifted inverse power iteration method. This method warrants the mode size and its relevant propagation constant. These parameters are important to the design of optical waveguide. The grid size is non-uniform to maximize the accuracy of the optical guided modes and their effective refractive indices. Diffused waveguides and rib waveguides are designed with different parameters to demonstrate the effectiveness of the method and leading to an optimum design of waveguides of optical modulation and micro-ring resonators.

12.1 INTRODUCTION

To achieve efficient design of high-speed modulators and switches, especially micro ring resonators, the fabrication of rib-waveguides and Ti: LiNbO$_3$ waveguide with suitable mode size is essential to minimize waveguide insertion loss and to maximize the overlap integral between the guided optical field and the applied modulating field. Furthermore, the bending or radius of curvature is so important for ring resonator to keep the ring size as small as possible. Extensive studies have been devoted in recent decades on fabricating Ti-diffused LiNbO$_3$ waveguides which couple efficiently to single-mode fibers [1–5]. A major milestone was achieved when a total fiber-waveguide-fiber insertion loss of 1 dB was achieved for z-cut LiNbO$_3$ at 1.3 μm [4,6]. Such low loss was achieved by choosing fabrication parameters to yield a relatively deep, clean diffusion, which simultaneously minimized the fiber waveguide mode mismatch loss and the propagation loss. Suchoski and Ramaswamy [7] have reported on the optimization of fabrication parameters to obtain Ti-LiNbO$_3$ single mode waveguides which exhibit minimum mode size and low propagation loss at 1.3 μm. All these design requirements have led to the significance of the analysis of polarized modes in channel waveguides.

In general, the optical mode of the waveguide is acquired by solving the Helmholtz equation. However, only a few simple waveguide structures can be solved analytically. Therefore, extensive attempts have been made to obtain numerical solutions for a 2D cross section of optical waveguides [6–18]. One method is the approximate modeling of 2D slab waveguide solution successively in both directions, following the method of Marcatilli [8] or the effective-index method (EIM) [9]. However, these methods are not applicable to arbitrarily shaped optical waveguides, nor do they handle waveguide mode near the cut-off region efficiently. A significant number of numerical methods have been proposed to obtain rigorous solutions to the wave equation with pertinent boundary conditions. The most popular techniques are the finite difference method (FDM) [10], finite element method (FEM) [11], and beam propagating method (BPM) [12]. The application of different techniques based on the above methods such as SV E-field FDM [13], SV H-field FDM [14], and Rayleigh quotient solution [15] have been studied and reported. These methods are applicable to arbitrarily shaped optical

515

waveguides. In FEM and FDM, partial differential equations are discretized and then transformed to matrix equations. The calculations of mode indices and optical field distributions are then equivalent to obtaining eigenvalues and eigen-fuctions of the coefficient matrices.

In this paper, we chose FDM to study the quasi-TE and quasi-TM polarized waveguide modes because of its simplicity and plausible accuracy. We have employed the SV analysis [13,14,16], which automatically takes full account of the discontinuities in the normal electric field components across any arbitrary distribution of internal dielectric interfaces. The SV FDM, despite its simplicity and being free from troublesome spurious solutions, has two major disadvantages of being computational intensive and requiring large amount of memory. Hence, it is necessary to introduce the discretization scheme on the non-uniform mesh, in which mesh intervals can be changed arbitrarily depending on waveguide structures. For this reason, we have modeled the waveguide mode with FDM which employs a non-uniform (NU) discretization scheme [16,17]. Such a discretization scheme enables us to increase the size of the problem space so that the field component at the boundary can be assumed to have vanished. The grid spacing increases monotonically with increasing distance from the guiding region. The grid lines can also be aligned with the boundaries of the step index changes in conventional structures such as rib, ridge, and strip-loaded waveguides, as well as quantum well structures. Furthermore, by judiciously placing the grid lines and corresponding cell structure efficiently, we can reduce the required matrix size and hence redundant computer calculations, while preserving the accuracy of the calculations. The non-uniform discretization scheme also enables us to handle waveguide mode near the cut-off region with a relativly simple boundary condition. The eigen-modes of the Helmholz equation is solved by the application of shifted inverse power iteration method. This method warrants both the mode size and its relevant propagation constant, which are both important parameters to the design of optical waveguide.

In Section 12.2, we outline the numerical formulation of the non-uniform finite difference scheme as described above. Both quasi-TE and quasi-TM polarized modes are addressed. We also assess the accuracy of the numerical result of this scheme by computing the effective refractive index of a few rib and slab dielectric waveguides and compare the results with published results. The effect of grid spacing is also investigated. We will also present the effectiveness of the variable grid spacing in dealing with waveguide mode near the cut-off region.

Section 12.3 describes the modeling of Ti-LiNbO$_3$ channel waveguide. The effects of various waveguide fabrication parameters such as the diffusion time, diffusion temperature, thickness, and width of the titanium strips are studied. The accuracy of the numerical model is assessed by comparing our simulations with experimental and simulation results that are reported in several literatures.

Apart from being able to access the accuracy of the final product of our work, which is the SVMM (SV Mode Modeling) computer program, we will also present an overview of its application in modeling Ti-LiNbO$_3$ channel waveguide for optical devices such as modulators and switches.

This part of this work will present an understanding of the numerical formulation involved in the modeling of optical modes in channel waveguides. The robustness of the numerical formulation will enable us to model the optical mode of a channel waveguide with an arbitrary index profile easily and accurately. With recent advancements in computer technology, our work has much to offer in the analysis and design of optical channel waveguides.

12.2   NON-UNIFORM GRID SV POLARIZED FDM

12.2.1   THE PROPAGATION EQUATION

For harmonic wave propagation in the z-direction along a rib or channel waveguide, we consider the fields

$$\mathrm{E}(x, y, z) = (E_x, E_y, E_z) \exp j(\omega t - \beta z) \qquad (12.1)$$

$$H(x, y, z) = (H_x, H_y, H_z) \exp j(\omega t - \beta z) \tag{12.2}$$

$$D = \varepsilon(x, y)E, \ B = \mu H \tag{12.3}$$

where the dielectric constant $\varepsilon(x, y)$ is piecewise constant and the magnetic permeability μ is completely constant throughout the solution domain. The components of the electric and magnetic fields in Equation 12.1 are functions of x and y only. Then, applying the Maxwell equations in the magnetic and charge-free media and take appropriate algebra, we obtain the wave equation:

$$\nabla \times (\nabla \times E) = \nabla(\nabla \cdot E) - \nabla^2 E = \omega^2 \varepsilon \mu E = k^2 n^2 E \tag{12.4}$$

in which $k = \omega(\varepsilon_0 \mu_0)^{1/2} = 2\pi/\lambda$ and $\varepsilon = \varepsilon_0 n^2(x, y)$ with λ being the free space wavelength. With the divergence of $\nabla \cdot D = 0$ and $\nabla \log_e \varepsilon = \nabla \varepsilon / \varepsilon$, we get

$$\nabla \cdot E = -E \cdot \nabla \log_e \varepsilon = -E \cdot \nabla n^2 / n \tag{12.5}$$

This may be substituted into Equation 12.4 to yield the wave equation

$$\nabla^2 E + k^2 E + \nabla(E \cdot \nabla n^2 / n) = 0 \tag{12.6}$$

As $n(x, y)$ is piecewise constant, $\nabla n^2 / n = 0$ and it should be noted that $\nabla n^2 / n$ is undefined at internal dielectric interfaces where $n(x, y)$ is discontinuous. With the assumption that the fields are polarized either perpendicular (quasi-TM) to or parallel (quasi-TE) to the crystal surface and that the major field components of the modes are perpendicular to the direction of the propagation, Equation 12.7 can be reduced to

$$(\nabla_t^2 + k^2 n^2)E = \beta^2 E \tag{12.7}$$

in which $\nabla_T^2 = (\partial^2 / \partial x^2) + (\partial^2 / \partial y^2)$, the transverse Laplacian and β is the propagation constant. This is essentially the Helmholz wave equation.

12.2.2 Formulation of NU Grid Difference Equation

Figure 12.1a shows the grid lines used in the FDM formulation. The grid lines are chosen in such a way that denser grids are allocated around the guiding region, while coarser grids are assigned to regions further away from the waveguide. Boundaries of abrupt index changes are straddled by

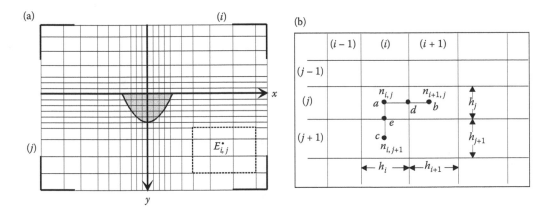

FIGURE 12.1 NU discretized grid for FDM scheme (a) and (b) a magnified portion of grid lattice and cell structure of point i, j.

the grid lines wherever necessary. Figure 12.1b shows the magnified view of a portion of the grid for a more detailed illustration. Each cell point is in the center of each rectangular cell. h_i, and h_j are the horizontal and vertical grid sizes. The refractive index within each cell is assumed to be uniform. $n_{i,j}$ and $n_{i+1,j}$ represent the values of the refractive index of each small cell as an approximation, which are taken from the continuous refractive index profile $n(x, y)$. NU spacing of the grid lines provides some flexibility in setting up the NU grid FDM. The NU discretization with increasing spacing away from the guiding region permits sufficient extension of the boundary. This enables us to assume a Dirichlet boundary condition (metal box) where all fields have vanished.

12.2.2.1 Quasi-TE Mode

For quasi-TE polarized mode, E_y is assumed to be zero. E_x is continuous across the horizontal interfaces, but discontinuous across vertical interfaces. Therefore, the quasi-TE modes are the eigensolutions of the equation

$$\nabla_t^2 E_x + k^2 n^2 E_x = \beta^2 E_x \qquad (12.8)$$

The discontinuity across the vertical interface will need to be considered when formulating the difference equation.

Figure 12.2 illustrates the quasi-TE field discontinuity at the boundary between cells (i, j) and $(i + 1, j)$. Consider the points a, d and b, with d being at the boundary of the dielectric interface. The horizontal axis is the x-axis, while the vertical axis is the electric field amplitude of the respective position of the cell. Assume that the x-axis is pointing towards the east. So, E_E and E_W are the field amplitudes just to the east and the west of the boundary between the cells (i, j) and cell $(i + 1, j)$. $E_{i,j}^v$ is the virtual field in cell (i, j) which is the extension of the actual field $E_{i+1, j}$. In other words, $E_{i,j}^v$ is the field seen by the cell $(i + 1, j)$. Similarly, $E_{i+1,j}^v$ is the extension of $E_{i,j}$. n_E and n_W are the refractive indices just to the east and the west of the boundary. Since we consider a slowly varying index distribution, we assume that n_E and n_W are approximately equal to $n_{i,j}$ and $n_{i+1,j}$, respectively. The boundary conditions between the cells (i, j) and $(i + 1, j)$ are given as follows:

$$n_E^2 E_E = n_W^2 E_W, \Rightarrow n_{i,j}^2 E_E = n_{i+1,j}^2 E_W \qquad (12.9)$$

$$\frac{\partial}{\partial x} E_E = \frac{\partial}{\partial x} E_W = p^+ \qquad (12.10)$$

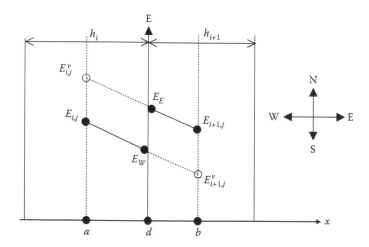

FIGURE 12.2   Quasi-TE electric field discontinuity at the boundary between cells (i, j) and cell $(i + 1, j)$. Solid lines are the actual field profiles along x axis while $E_{i,j}^v$ and $E_{i+1,j}^v$ are virtual fields.

where p^+ represents the field gradient at the boundary between the cells. We can then use the approximate relationship between $E_{i,j}$, $E_{i+1,j}$, $E_{i,j}^v$, $E_{i+1,j}^v$, and obtain the following equations for E_E and E_W:

$$E_{i+1,j} \approx E_E + (h_{i+1}/2) \cdot p^+; \; E_{i,j}^v \approx E_E - (h_i/2) \cdot p^+; \; E_{i+1,j}^v \approx E_W + (h_{i+1}/2) \cdot p^+$$

$$\text{and } E_{i,j}^v \approx E_W - (h_i/2) \cdot p^+ \tag{12.11}$$

where h_i and h_{i+1} are the horizontal lengths of cell (i,j) and $(i+1,j)$. The four equations above are in fact redundant. Therefore, we need only to consider either $E_{i,j}^v$ or $E_{i+1,j}^v$, which we choose $E_{i+1,j}^v$ in our case. The following shows the algebraic manipulation Equation 12.11:

$$p^+ = 2(E_{i+1,j}^v - E_{i,j})/(h_i + h_{i+1}); \; E_{i+1,j}^v = E_{i+1,j} + (E_W - E_E); \; h_{i+1}(E_W - E_{i,j})$$

$$= h_i(E_{i+1,j} - E_E) \tag{12.12}$$

and then

$$E_{i+1,j}^v = \frac{n_{i+1,j}^2(h_i + h_{i+1})E_{i+1} + h_{i+1}\left(n_{i+1,j}^2 - n_{i,j}^2\right)E_{i,j}}{\left(n_{i,j}^2 h_i + n_{i+1,j}^2 h_{i+1}\right)} \tag{12.13}$$

with similar procedure between cell (i,j) and cell $(i+1,j)$, we can obtain

$$p^- = 2\left(E_{i,j} - E_{i-1,j}^v\right)/(h_i + h_{i-1}) \quad \text{and}$$

$$E_{i-1,j}^v = \frac{n_{i-1,j}^2(h_i + h_{i-1})E_{i-1,j} + h_{i-1}\left(n_{i-1,j}^2 - n_{i,j}^2\right)E_{i,j}}{\left(n_{i,j}^2 h_i + n_{i-1,j}^2 h_{i-1}\right)} \tag{12.14}$$

where p^- is now the field gradient at the boundary between the cells $(i-1,j)$ and (i,j). Note that the quasi-TE electric field is continuous in terms of y-direction even if there are discontinuities in refractive index. Therefore $E_{i,j+1}^v = E_{i,j+1}$, $E_{i,j-1}^v = E_{i,j-1}$.

The second derivative can be derived as

$$\frac{\partial^2}{\partial x^2} E_{i,j} = \frac{1}{h_i}[p^+ - p^-] = \frac{1}{h_i}\left[\frac{2\left(E_{i+1,j}^v - E_{i,j}\right)}{h_{i+1} + h_i} - \frac{2\left(E_{i,j} - E_{i-1,j}^v\right)}{h_i + h_{i-1}}\right] \tag{12.15}$$

$$\frac{\partial^2}{\partial y^2} E_{i,j} = \frac{1}{h_j}[p^+ - p^-] = \frac{1}{h_j}\left[\frac{2(E_{i,j+1} - E_{i,j})}{h_{j+1} + h_j} - \frac{2(E_{i,j} - E_{i,j-1})}{h_j + h_{j-1}}\right] \tag{12.16}$$

hence, we get the discrete wave equation as

$$\frac{\partial^2}{\partial x^2} E_{i,j} = \frac{2n_{i-1,j}^2}{h_i \left(n_{i,j}^2 h_i + n_{i-1,j}^2 h_{i-1} \right)} E_{i-1,j} + \frac{2n_{i+1,j}^2}{h_i \left(n_{i,j}^2 h_i + n_{i+1,j}^2 h_{i+1,j} \right)} E_{i+1,j}$$
$$- \left[\frac{2n_{i,j}^2}{h_i \left(n_{i,j}^2 h_i + n_{i-1,j}^2 h_{i-1} \right)} + \frac{2n_{i,j}^2}{h_i \left(n_{i,j}^2 h_i + n_{i+1,j}^2 h_{i+1} \right)} \right] E_{i,j} \tag{12.17}$$

$$\frac{\partial^2}{\partial y^2} E_{i,j} = \frac{1}{h_j} \left[\frac{2(E_{i,j+1} - E_{i,j})}{h_{j+1} + h_j} - \frac{2(E_{i,j} - E_{i,j-1})}{h_j + h_{j-1}} \right] \tag{12.18}$$

Substituting these into the Helmholz equation,

$$C_{i-1,j} E_{i-1,j} + C_{i+1,j} E_{i+1,j} - C_{i,j} E_{i,j} + C_{i,j-1} E_{i,j-1} + C_{i,j+1} E_{i,j+1} = \beta^2 E_{i,j} \tag{12.19}$$

where

$$C_{i-1,j} = \frac{2n_{i-1}^2}{h_i \left(n_{i,j}^2 h_i + n_{i-1,j}^2 h_{i-1} \right)}; \quad C_{i+1,j} = \frac{2n_{i+1}^2}{h_i \left(n_{i,j}^2 h_i + n_{i+1,j}^2 h_{i+1} \right)}; \quad C_{i,j-1} = \frac{2}{h_j(h_j + h_{j-1})};$$
$$C_{i,j+1} = \frac{2}{h_j(h_j + h_{j+1})} \text{ and } C_{i,j} = C_{i-1,j} + C_{i+1,j} + C_{i,j-1} + C_{i,j+1} - k^2 n_{i,j}^2 \tag{12.20}$$

The above equations is essentially an eigenvalue equation of

$$C_{TE} E_{TE} = \beta_{TE}^2 E_{TE} \tag{12.21}$$

in which C_{TE} is a non-symmetric band matrix that contains the coefficient of the above equations, β_{TE}^2 is the TE propagation eigenvalue, and E_{TE} is the corresponding normalised eigenvector representing the field profile $E_x(x, y)$.

12.2.2.2 Quasi-TM Mode

The quasi-TM mode can be formulated in similar fashion. The only difference is that for quasi-TM polarized mode, E_x is assumed to be zero and E_y is continuous across the vertical interfaces but discontinuous across horizontal interfaces. Essentially, the quasi-TM modes are the eigen-solutions of the equation

$$\nabla_t^2 E_y + k^2 n^2 E_y = \beta^2 E_y \tag{12.22}$$

The detailed derivation of the equation can be found in the literature [16]. The following are the derivatives and its relevant difference equations:

$$\frac{\partial^2}{\partial y^2}E_{i,j} = \frac{2n_{i,j-1}^2}{h_i\left(n_{i,j}^2 h_i + n_{i,j-1}^2 h_{j-1}\right)}E_{i,j-1} + \frac{2n_{i,j+1}^2}{h_i\left(n_{i,j}^2 h_i + n_{i,j+1}^2 h_{j+1}\right)}E_{i,j+1}$$
$$- \left[\frac{2n_{i,j}^2}{h_i\left(n_{i,j}^2 h_i + n_{i,j-1}^2 h_{j-1}\right)} + \frac{2n_{i,j}^2}{h_i\left(n_{i,j}^2 h_i + n_{i,j+1}^2 h_{j+1}\right)}\right]E_{i,j} \qquad (12.23)$$

$$\frac{\partial^2}{\partial x^2}E_{i,j} = \frac{1}{h_i}\left[\frac{2(E_{i+1,j} - E_{i,j})}{h_{i+1} + h_i} - \frac{2(E_{i,j} - E_{i-1,j})}{h_i + h_{i-1}}\right] \qquad (12.24)$$

Substituting these into the Helmholtz equation, we get

$$C_{i-1,j}E_{i-1,j} + C_{i+1,j}E_{i+1,j} - C_{i,j}E_{i,j} + C_{i,j-1}E_{i,j-1} + C_{i,j+1}E_{i,j+1} = \beta^2 E_{i,j} \qquad (12.25)$$

where

$$C_{i-1,j} = \frac{2}{h_i(h_i + h_{i-1})} \quad \text{and} \quad C_{i+1,j} = \frac{2}{h_i(h_i + h_{i+1})} \quad C_{i,j-1} = \frac{2n_{j-1}^2}{h_j\left(n_{i,j}^2 h_j + n_{i,j-1}^2 h_{j-1}\right)};$$

$$C_{i,j+1} = \frac{2n_{j+1}^2}{h_j\left(n_{i,j}^2 h_j + n_{i,j+1}^2 h_{j+1}\right)} \quad \text{and} \quad C_{i,j} = C_{i-1,j} + C_{i+1,j} + C_{i,j-1} + C_{i,j+1} - k^2 n_{i,j}^2 \qquad (12.26)$$

12.2.2.2.1 Eigenvalue Matrix

To solve the difference equation, we need first to discretized the problem space. We assume that the space is sliced into *NX* pieces along the *x*-direction and *NY* pieces along the *y*-direction. This will give us a total of $N (=NX \times NY)$ grid points. The refractive index of each cell is then allocated according to the relevant index distribution.

When the Finite Difference Wave Equation is evaluated at a grid point, say $E_{i,j}$, it will yield a five-point linear equation in terms of the *E* field of the immediate neighbors, namely $E_{i-1,j}$, $E_{i+1,j}$, $E_{i,j-1}$, $E_{i,j+1}$, each with its relevant coefficient as shown in Equations 12.32 and 12.37. For a cross-sectional area of a waveguide with *N* such grid points, we would end up with *N* linearly dependent algebraic equations.

We will now scan through the grid points row after row, at the same time relabelling the subscripts of *E* from 1 to *N*. Consider the original grid point (i, j). If the new sequence number is *k*, then Equation 12.36 can be rewritten as

$$p_k E_k + l_k E_{k-1} + r_k E_{k+1} + t_k E_{k-Nx} + b_k E_{k+Nx} = \beta^2 E_k \qquad (12.27)$$

where p_k, l_k, r_k, t_k, b_k are the coefficients $C_{i,j}$, $C_{i-1,j}$, $C_{i+1,j}$, $C_{i,j-1}$, $C_{i,j+1}$, respectively. We can then collect terms and write the equations in a matrix form.

For a 3 × 3 grid of the refractive index profile, we can write the matrix equations as an the eigenvalue equation of the form $[C].[E] = \beta^2[E]$ in which $[C]$ is a non-symmetric band matrix that contains the coefficient of the above equations, β^2 is the propagation eigenvalue, and $[E]$ is the corresponding normalized eigenvector representing the field profile $E(i, j)$. In the next section, we will discuss the approach that we adopt in solving the eigenvalue problem given as

$$
\begin{bmatrix}
p_1 & r_1 & 0 & b_1 & 0 & 0 & 0 & 0 & 0 \\
l_2 & p_2 & r_2 & 0 & b_2 & 0 & 0 & 0 & 0 \\
0 & l_3 & p_3 & r_3 & 0 & b_3 & 0 & 0 & 0 \\
t_4 & 0 & l_4 & p_4 & r_4 & 0 & b_4 & 0 & 0 \\
0 & t_5 & 0 & l_5 & p_5 & r_5 & 0 & b_5 & 0 \\
0 & 0 & t_6 & 0 & l_6 & p_6 & r_6 & 0 & b_6 \\
0 & 0 & 0 & t_7 & 0 & l_7 & p_7 & r_7 & 0 \\
0 & 0 & 0 & 0 & t_8 & 0 & l_8 & p_8 & r_8 \\
0 & 0 & 0 & 0 & 0 & t_9 & 0 & l_9 & p_9
\end{bmatrix}
\begin{bmatrix}
E_1 \\ E_2 \\ E_3 \\ E_4 \\ E_5 \\ E_6 \\ E_7 \\ E_8 \\ E_9
\end{bmatrix}
= \beta^2
\begin{bmatrix}
E_1 \\ E_2 \\ E_3 \\ E_4 \\ E_5 \\ E_6 \\ E_7 \\ E_8 \\ E_9
\end{bmatrix}
\tag{12.28}
$$

There are a few major features of the matrix equation above: (i) this type of matrix is often referred to as tridiagonal matrix with fringes. The order of the matrix is $N \times N$, the square of the total number of grid points. Most of terms in the matrix are zeros; (ii) the matrix is non-symmetrical relative to the diagonal term; (iii) the central three diagonal terms always exist and are always non-zero; (iv) the coefficients $p, l, r, t,$ and b make up the five bands of the matrix, with p being the main diagonal, l and r being the sub-diagonal while t and b the super-diagonal; (v) the sub-diagonal diagonal terms are just one term away from the main diagonal while the upper-diagonal terms are NX terms away from the main diagonal. The distance between main diagonal and the last non-zero super-diagonal band is commonly referred to as the half bandwidth of a band matrix; and (vi) terms such as $l_1, r_N, t_1 - t_{NX}, b_{N-Nx} - b_N$ are missing. This is so since the evaluations of these terms require the E values outside the boundary area, and these values have been assumed zero. Therefore, they need not be represented.

12.2.3 THE INVERSE POWER METHOD

The properties and characteristics of eigenvalue problems are well known and have been addressed rather extensively in many textbooks [18,19]. This section will only provide a brief overview to highlight the more specific points related to our particular approach.

An $N \times N$ matrix A is said to have an *Eigenvector* x and a corresponding *eigenvalue* λ if the following condition is satisfied:

$$
A.x = \lambda x \tag{12.29}
$$

There can be more than one distinct eigenvalue and eigenvector corresponding to a given matrix. The zero vector is not considered to be an Eigenvector at all. The above equation holds only if

$$
\det|A - \lambda I| = 0 \tag{12.30}
$$

which is known as the characteristic equation of the matrix. If this is expanded, it becomes an Nth degree polynomial in λ whose roots are the eigenvalues. This is an indication that there are always N, though not necessarily distinct, eigenvalues. Equal eigenvalues coming from multiple roots are called degenerate. Root-searching in the characteristic equation however, is usually a very poor computational method for finding eigenvalues. There are many more efficient algorithms available in locating the eigenvalues and their corresponding vectors.

Unfortunately, there is no universal method for solving all matrix types. For certain problems, either the eigenvalues or eigenvectors are needed, while others require both. Furthermore, some problems may only need a small number of solutions out of the total N solutions available, while others need all. To complicate the matter even further, the eigen-solutions could be complex, and some matrices can be so ill-behaved that round off errors in computing can lead to a non-convergence of the solution. Therefore, it is of vital importance to be able to choose the right approach in solving an eigen-problem.

Choosing an algorithm often involves the classification of matrix into types like symmetry, non-symmetry, tridiagonal, banded, positive definite, definite Hessenberg, sparse, random, etc. The matrix in our problem is a non-symmetric banded matrix with bandwidth equal to twice the number of columns in the grid profile. It has great sparsity for most of the elements are zeros. Also, we need only a few eigenvalues that correspond to the guided modes of the waveguide. In other words, there are only a limited number of guided modes, hence the number of eigenvalue λ. The number of eigen-solutions required is small compared with the size of the matrix (often in the order of tens of thousands). All these different factors have led to the choice of the approach called the Inverse Iteration Method [18,19].

The basic idea behind the inverse iteration method is quite simple. Let y be the solution of the linear system

$$(A - \tau I) \cdot y = b \tag{12.31}$$

where b is a random vector and τ is close to some eigenvalue λ of A. Then the solution y will be close to the eigenvector corresponding to λ. The procedure can be iterated: replace b by y and solve for a new y, which will be even closer to the true eigenvector. We can see why this works by expanding both y and b as linear combinations of the eigenvectors x_j of A:

$$y = \sum_j \alpha_j x_j \quad \text{and} \quad b = \sum_j \beta_j x_j \tag{12.32}$$

Then we have

$$\sum_j \alpha_j (\lambda_j - \tau) x_j = \sum_j \beta_j x_j \tag{12.33}$$

so that

$$\alpha_j = \frac{\beta_j}{\lambda_j - \tau} \quad \text{and} \quad y = \sum_j \frac{\beta_j x_j}{\lambda_j - \tau} \tag{12.34}$$

If τ is close to λ_n, say, then provided β_n is not accidentally too small, y will be approximately x_n, up to a normalization. Moreover, each iteration of this procedure gives another power of $\lambda_j - \tau$ in the denominator of (16.31). Thus, the convergence is rapid for well-separated eigenvalues.

Suppose at the ith stage of iteration we are solving the equation

$$(A - \lambda_i I) \cdot y = x_i \tag{12.35}$$

where x_i and λ_i are our current guesses for some eigenvector and eigenvalue of interest (we shall see below how to update λ_i). The exact eigenvector and eigenvalue satisfy

$$A \cdot x = \lambda x \rightarrow (A - \lambda_i I) \cdot x = (\lambda - \lambda_i) x \tag{12.36}$$

Since y of Equation 12.35 is an improved approximation to x, we normalize it and set

$$x_{i+1} = \frac{y}{|y|} \tag{12.37}$$

We get an improved estimate of the eigenvalue by substituting our improved guess y in Equation 12.37. By Equation 12.37, the left-hand side is x_i, so calling λ our new value $\lambda_i + 1$, we find $\lambda_{i+1} = \lambda_i + (|x|^2 / |x_i \cdot y|)$.

Although the formulas of Inverse Iteration Method seems to be rather straight forward, the actual implementation can be quite tricky. Most of the computational load occurs in solving the linear system of equations. It would be advantageous if we can solve (12.35) quickly. It is to be reminded that the size of the matrix in our case is dependent upon the total grid size of the problem space. For a typical grid size of 100 by 100, for example, the coefficient matrix would be of size 10,000 by 10,000. The core memory required in a digital computer to store the entire matrix would be phenomenal. A linear system solver such as a routine that is available in LINPACK employs a common LU factorization (Gaussian elimination) plus backward substitution combination algorithm, much like the manual way of solving linear equations. There is an extensive coverage on this topic in Reference 19. We will therefore not discuss it further, except to mention that the LU factorization needs only to be done before the first iteration. When the iteration starts, we already have the steps involved in elimination stored away in an array and only backward substitution is necessary. This approach, even with a storage-optimized mode in the LINPACK routine still has a storage requirement of about 3 × (bandwidth of matrix × matrix size). Even though this would mean a considerable reduction in memory storage, it still amounts to a rather substantial memory size.

Also, the preconditioner that employs the Incomplete Cholesky Conjugate Gradient Method [20] and the Orthominc [20] accelerator have been found to be most stable and converge most quickly for our matrix. On average, the combination of the preconditioner and accelerator enable us to complete a simulation of a typical waveguide in 3–5 min on a PC Pentium 4. The same simulation that incorporates the LINPACK LU decomposition routine would take 25 min on the same computer with a substantially greater amount of memory. Since the zero elements are no longer involved in the calculations, it is understandable that the NSPCG iterative method will perform more efficiently.

By incorporating the NSPCG numerical solver and the Inverse Iterative Method, we have successfully implemented a Mode Modeling Program, SVMM (SV Mode Modeling) capable of modeling channel waveguide of an arbitrary index profile. The Inverse Iterative Method also enables us to model the higher order modes that are supported by the waveguide structure.

12.2.4 Graded-Index Channel Waveguide

The modeling of Ti-LiNbO$_3$ channel waveguide, a graded index waveguide, plays an significant role in the design of the optical modulators and switches. Efficient design of such optical devices requires good knowledge of the modal characteristics of the relevant channel waveguide. Reference 21 outlines the general overview of the waveguide fabrication process. In this section, we will attempt to employ our SVMM program to simulate the waveguide mode of the Ti-LiNbO$_3$ waveguide and compare the results with that of published experimental results. Our objectives, apart from assessing the usefulness of SVMM, is also to understand the key features in the fabrication of Ti-LiNbO$_3$ waveguide for Mach–Zehnder optical modulator (see Figures 12.3 through 12.10).

To achieve our purpose, a good knowledge of the refractive index profile of the diffused waveguide is required. Over the past decades, much work has been done in fabricating low-loss, minimum-mode-size Ti-diffused channel waveguide [1,7,22–25]. From these references, we can gather our knowledge of the diffusion process involved in the fabrication of LiNbO$_3$ waveguide and its relevant diffusion profile. Based on this knowledge, we can then profess to model the modal characteristics of the waveguide by SVMM. The following section shows how SVMM can be used for the design of practical waveguides (see Figures 12.10 through 12.12).

12.2.4.1 Refractive Index Profile of Ti-LiNbO$_3$ Waveguide

When Ti metal is diffused, the Ti-ion distribution spreads more widely than the initial strip width. The profiles can be described by the sum of an error function, while the Ti-ion distributions perpendicular to the substrate surface can be approximated by a Gaussian function [1,7,26]. This of course is true only if the diffusion time is long enough to diffuse all the Ti metal into the substrate. We

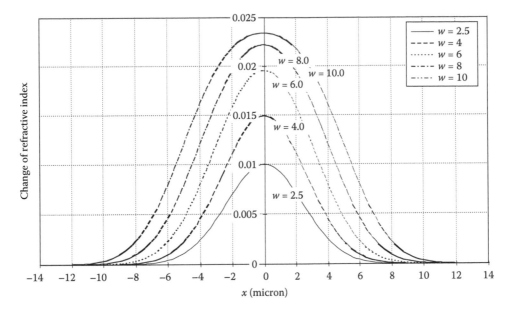

FIGURE 12.3 Lateral diffusion variation with increasing Ti strip width.

consider this case as having the finite diffusant source. However, if the total diffusion time is shorter than needed to exhaust the Ti source, the lateral diffusion profile would take up the sum of the complementary error function, while the depth index profile is given by the complementary function [1,22]. This case is considered to have had a infinite diffusant source [16]. In our study, we would assume that there is sufficient time for the source to be fully diffused because in most practical waveguide, it is undesirable to have Ti residue deposited on the surface of the waveguide because this will increase the propagation loss [26]. This increase in propagation loss is a result of stronger

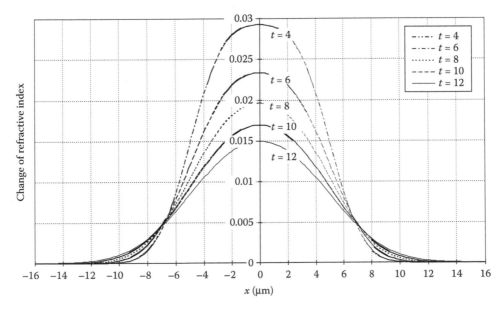

FIGURE 12.4 Lateral diffusion variation with increasing diffusion time.

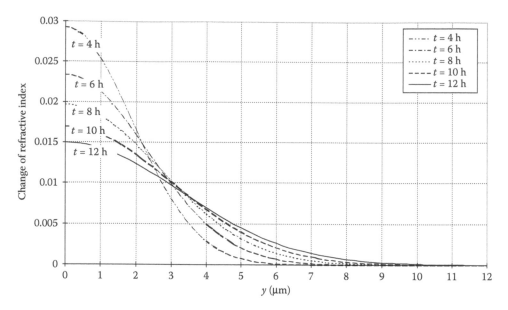

FIGURE 12.5   Depth index variation with increasing diffusion time.

interaction with the LiNbO$_3$ surface (and thus an increased scattering loss) as the modes become weakly guided.

In general, the refractive index distribution of a weakly guiding channel waveguide

$$n(x, y) = n_b + \Delta n(x, y) = n_b + \Delta n_0 \cdot f(x) \cdot g(y) \tag{12.38}$$

where n_b is the refractive index of the bulk (substrate) and $\Delta n(x, y)$ is the variation of the refractive index in the guiding region. $\Delta n(x, y)$ in our diffusion model is essentially a separable function where $f(x)$ and $g(y)$ are the function that describes the lateral and perpendicular diffusion profile,

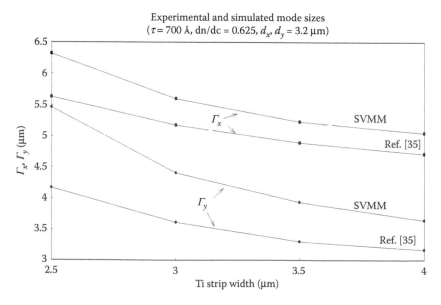

FIGURE 12.6   Experimental and simulated mode spot size as a function of the width of the Ti strip width prior to diffusion.

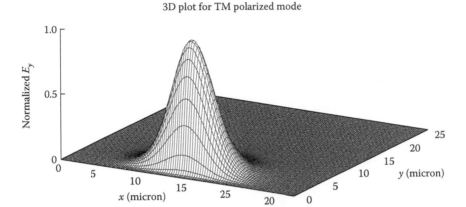

FIGURE 12.7   Typical modal field of a diffused channel waveguide.

while Δn_0 is known as the surface index change after diffusion. The surface index change is defined as the change of refractive index on the substrate just below the centre of the Ti strip. In other words, it is the refractive index when both $f(x)$ and $g(y)$ assume the value of unity.

The variation of the refractive index can be modeled as below [1]:

$$\Delta n(x, y) = \frac{dn}{dc}\tau \int_{-w/2}^{w/2} \frac{2}{d_y\sqrt{\pi}}\exp\left[-\left(\frac{y}{d_y}\right)^2\right]\frac{1}{d_x\sqrt{\pi}}\cdot\exp\left[-\left(\frac{x-u}{d_x}\right)^2\right]du \qquad (12.39)$$

$$= \Delta n_0 \cdot f(x) \cdot g(y)$$

where

$$f(x) = \frac{\dfrac{1}{2}\left[\operatorname{erf}\left(\dfrac{x+\dfrac{w}{2}}{d_x}\right) - \operatorname{erf}\left(\dfrac{x-\dfrac{w}{2}}{d_x}\right)\right]}{\operatorname{erf}\left(\dfrac{w}{2d_x}\right)} \qquad (12.40)$$

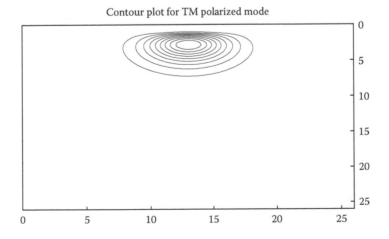

FIGURE 12.8   Contour plot of the modal field of a diffused channel waveguide.

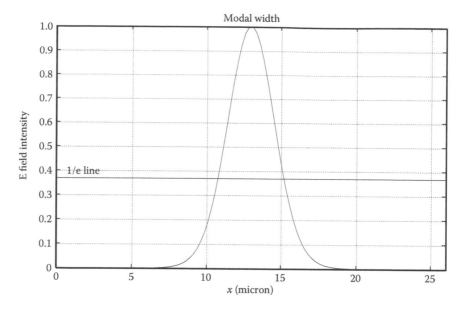

FIGURE 12.9   Horizontal mode profile of a diffused channel waveguide.

$$g(x) = \exp\left[-\left(\frac{y}{d_y}\right)^2\right] \tag{12.41}$$

and

$$\Delta n_0 = \frac{dn}{dc}\frac{2}{\sqrt{\pi}}\frac{\tau}{d_y}\operatorname{erf}\left(\frac{w}{2d_x}\right) \tag{12.42}$$

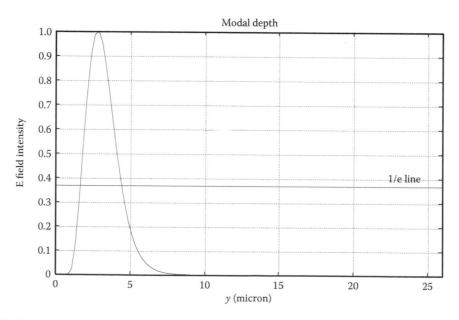

FIGURE 12.10   Vertical mode profile of a diffused channel waveguide.

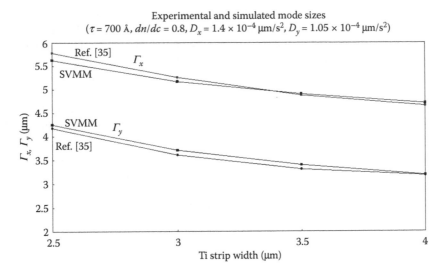

FIGURE 12.11 Comparison of simulated and experimental mode sizes for $\tau = 700$ Å.

with

$$d_x = 2\sqrt{D_x t}, \; d_y = 2\sqrt{D_y t} \tag{12.43}$$

In the above expressions, t is the total diffusion time, c is the Ti concentration, d_x and d_y are the diffusion lengths, and D_x and D_y are the diffusion constants in each direction. τ and w are the initial Ti strip thickness. dn/dc is the change of index per unit change in Ti metal concentration. The change of surface index would approach the value where $\Delta n_0 = (dn/dc)(2/\sqrt{\pi})(\tau/d_y)$. Any increase in the surface index will have to come from a thicker Ti strip, or a decrease in diffusion depth, d_y, which involve an increase or decrease in diffusion temperature. According to the work of Fukuma et al. [26], the diffusion length is very close to one another in both lateral and depth direction (isotropic diffusion) at 1025°C for z-cut crystal. An increase in temperature greater than that would result in a higher diffusion constant in the depth direction and lower value for lateral diffusion and vice versa for diffusion temperature lower than 1025°C. The diffusion length can also be change by monitoring the diffusion

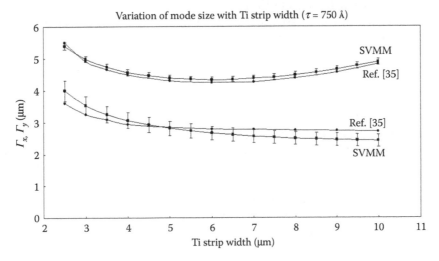

FIGURE 12.12 Comparison of simulated and experimental mode sizes for $\tau = 700$ Å.

time. Essentially, longer diffusion time would mean a lower surface index change as most of the Ti source would be diffused deeper into the substrate. Again, the second model depicts a higher change of surface index since not all the Ti metal is exhausted. The following graphs shows us the variation of the diffusion profile as we vary the initial titanium width and the diffusion time. The fabrication condition and parameters are assumed to have $T = 1025°C$, $\tau = 1100$ Å, $dn/dc = 0.625$, dx and dy are both 2 µm.

We can see in the above graphs that by controlling the width of the initial Ti strip width, we can vary the change of refractive index and the relative size of the channel waveguide, thus enabling us to control the number of mode that can be supported by the waveguide.

In general, a narrow initial Ti width would give a near cut-off mode for the refractive index change would be too small. The optical mode would be weakly confined, thus giving a larger mode size. As we increase the Ti width, the refractive index change would be higher and the waveguide mode would be better confined and has a smaller mode size. However, the mode size would increase with further increase in Ti width due to a larger physical size of the waveguide. The change of surface index can also be controlled by varying the thickness of the Ti strip. As Equation 12.53 implied, the surface index change is proportional to the strip thickness, τ, under diffusion at 1000–1050°C for 6 h would be around 500–800 Å [7]. If the Ti strip is too thin, the refractive index change approaches cut-off conditions. All these characteristics would be illustrated by the next section when we model the waveguide mode with SVMM.

12.2.4.2 Numerical Simulation and Discussion

With the above knowledge of the diffusion profile, we are now in a good position to feed these models into the SVMM program to investigate the modal characteristics of Ti-LiNbO$_3$ waveguide. In this section, we attempt to simulate the experimental work done by P. Suchoski and R.V. Ramaswamy [7] in fabricating minimum mode size low loss Ti-LiNbO$_3$ channel waveguide. We will restrict our analysis to the z-cut y propagating material since this would be the substrate cut for the optical modulator. For this substrate cut, the relevant optical field would be TM polarized, which correspond to the polarization along the extraordinary index axis of the crystal. Hence, the change of refractive index concern would be the extraordinary index, n_e.

In Suchoski's work, the TM polarized mode width and depth, which is defined as $1/e$ intensity full width and full depth, are measured for Ti-LiNbO$_3$ waveguides fabricated under the condition where $T = 1025°C$ for 6 hr. The sample waveguides have Ti thickness ranging from 500 to 1100 Å, and Ti strip widths ranging from 2.5 to 10 µm.

The laser source wavelength is assumed to be at 1.3 µm. The following graphs are extracted from their work. In view of these experimental results, we can see that the mode sizes increase as the Ti strip width is decreased from 4 to 2.5 µm. This increase is more pronounced especially with the thinner Ti films, because the waveguides become closer to cut-off, as thinner Ti film resulted in a lower value of Δn. The TM mode depth and width decrease as the Ti thickness is increased from 500 to 800 Å. However, for 4 µm strip widths, the mode size does not decrease further for Ti films thicker than 800 Å. This is an indication that it is not possible to diffuse any more Ti into the substrate for Ti thickness of more than 800 Å for 6-hour diffusion time. We now proceed to simulate the above experiment with our program. We will focus on Ti thickness that ranges between 700 and 800 Å because it is the thickness that gives minimum mode sizes, which is ideal for the design of optical modulator for maximizing the overlap integral between guided optical mode and applied modulating field. To achieve that, we must first work out the suitable diffusion parameter to be used in our program. Various values of dn/dc has been reported [26]. Measurements reported by Minakata et al. [15] shows the change of extraordinary index n_e per Ti concentration as

$$\frac{dn_e}{dc} = 0.625$$

The nominal values for diffusion constant, D_x and D_y from the work of Fukuma and Noda [26] which were both measured to be $1.2 \times 10^{-4}\,\mu m^2/s$ at the nominated temperature which is 1025°C. This makes both diffusion length of d_x and d_y the value of 2 µm.

With these nominal parameters, we simulate the waveguide with a Ti thickness, τ of 700 Å. The following figures are the result of our simulation compare to the experimental one and some illustrations of the TM mode profile.

As it turns out, the simulated results appear to have overestimated both Γ_x and Γ_y. Such discrepancy is anticipated fabrication of diffused waveguide is subjected to many changes. Various reports [1,7,16,22] have shown that even though the nominal diffusion condition can be very much the same, the measured diffusion parameter can still differ greatly from one another due to possible differences in stoichiometry between different crystals and measurements techniques. Therefore, there would certainly be some uncertainties that lies in fabrication parameters and the application of the refractive index model described in Equation 12.49. Such uncertainty can be compensated by adjusting the value of dn/dc and D_x and D_y. We find that by adjusting the following diffusion parameters where

$$dn/dc = 0.8; \quad D_x = 1.4' \times 10^{-4}\,mm/s^2; \quad and \quad D_y = 1.1' \times 10^{-4}\,mm/s^2, \tag{12.44}$$

our simulation results correspond well within design limit with the experimental work done by Suchoski for the case where the waveguide is well guided. The result is shown in the following figure.

The plots with circular graph markers are extracted from Reference 7 while the one with square graph markers are simulated result. The simulated and measured result matches to within 5%.

Having found the suitable diffusion parameter, we proceed to simulate another experimental result from Reference 26 for $\tau = 750$ Å and Ti strip width, w ranges from 2.5 to 10 µm. The following figure shows the comparison of both simulated and experiment result.

The results show that the mode width, Γ_x correspond well to the experimental result with differences of less than 3%. The mode depth, Γ_y, however matches only to within 8%. Despite the slight discrepancy, the SVMM's result still shows the qualitative characteristic of the diffused waveguide. In the above graph, we can see that the modal width, Γ_x started at a large value and then decreases with wider Ti strip-width before climbing up again to a higher value.

Effectively, the larger initial mode size is due to the lower refractive index change resulted from a much narrower Ti width, thus causing the optical mode to be less confined. As the Ti strip become wider, it gives a higher change of refractive index, hence a better confined optical mode. The mode width however, would increase further as we increase the Ti width simply because of the increase in the physical width of the waveguide. At the same time, the larger physical width would enable the waveguide to support higher order mode.

Figure 12.13 shows a plot of the normalized mode index b which is defined as [27] $b = \left(n_{eff}^2 - n_s^2\right)/(2\Delta n.n_s)$. The waveguide become more strongly guided as we increase the Ti width. At the same time, higher order mode begins to settle in as the strip widths get significantly larger than 6 µm. The following figures shows the higher order modes for the waveguide diffused from a 10 µm Ti strip.

The modal depth, however, continue to decrease as with wider Ti strip width because any wider Ti strip width does not affect the diffusion depth. It only increases the surface index, thus giving a smaller modal depth. The surface index however, will only reach a maximum value as we increase w. Therefore, as we increase the Ti strip width to a certain point, the modal depth cease to decrease further at any apparent rate, as has been shown by both the experimental and simulated result. At this point, lateral diffusion would dominate. It is worth to be reminded that the limiting case of increasing width in Ti width is a planar waveguide.

We continue to simulate with the same diffusion parameter, the experiment where Ti thickness is increased to 800 Å. The following graphs shows the result of the simulation.

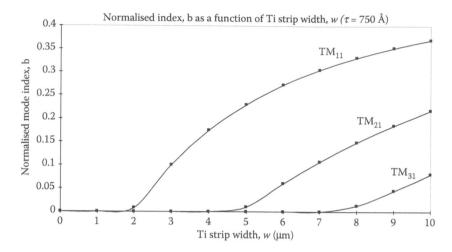

FIGURE 12.13 Normalized mode index, b as a function of Ti strip width, w.

Figure 12.14 shows the mode field distribution in the transverse plane of the guided mode TM_{21} and TM_{31} of the diffused waveguide with the thickness of Ti film of 700 Å and a width of 10 μm.

Figure 12.15 above shows that SVMM overestimates the mode size of the diffused waveguide. As the matter of fact, the thickness of 800 Å, as mentioned before, corresponds to the case where the Ti thickness has just depleted. In other words, the change of surface index is at its highest point for that diffusion time. The mode size would therefore appear to be much smaller compare to those waveguides of which the Ti has been diffused sufficiently longer than the time needed to just deplete all the Ti. This explain why the simulated modal width and depth being larger than the practical one. The following figures summarize the simulated results for waveguide for a range of Ti thickness:

Figure 12.16 shows the variation of the simulated modal width as a function of a range of Ti thickness. Furthermore Figure 12.17 shows the variation of simulated modal width with respect to a range of Ti film thickness prior to the diffusion.

From the graphs, we can see that the modal width and depth increase monotonically with the Ti thickness. It is not difficult to see that from the experimental results that the diffusion model and diffusion parameters is no longer valid when the Ti film exceeds the thickness which is fully diffused, which in this case is around 800 Å. For any thickness beyond that, we will need to resort to another diffusion model. In our case, this isn't necessary because having Ti film thicker than the diffusible amount would means result in scattering loss, thus the insertion loss.

In this section thus far, we have demonstrated how SVMM can be used apart from simulating rib waveguide, to simulate diffused channel waveguide. As a matter of fact, the program can be used to

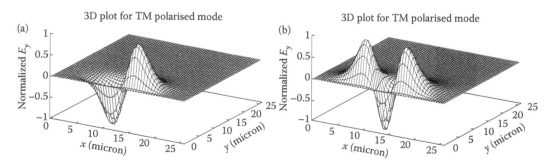

FIGURE 12.14 (a) TM_{21} mode: $\tau = 750$ Å, $w = 10$ μm and (b) TM_{31} mode: $\tau = 750$ Å, $w = 10$ μm.

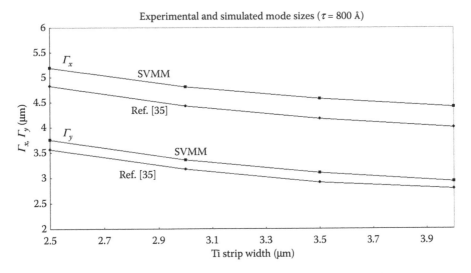

FIGURE 12.15 Comparison of simulated and experimental mode sizes for $\tau = 800$ Å.

simulate waveguide of an arbitrary index profile. Simulation of Ti-LiNbO$_3$ waveguide, however, is not a straightforward matter because fabrication of such a waveguide is subjected to many changes, such as differences in crystal quality, diffusion process, density variations of the deposited titanium films, and differences in measurement techniques. Because of that, we can see many inconsistency in various literature. Fouchet et al. [27] show the relation between refractive index change $\Delta n_{e,o}(Z)$ and Ti concentration $C(Z)$ in the mathematical form of

$$\Delta n_{e,o}(Z) = A_{e,o}(C_o, \lambda) \cdot (C(Z))^{\alpha_{e,o}} \tag{12.45}$$

The expression shows that the proportionality coefficient $A_{e,o}$ depends not only on the wavelength λ, but also on the diffusion parameters, which is characterized by C_o, the Ti surface concentration.

In other words, the diffusion model that we used in our simulation is only a crude representation of the diffused waveguide. To enhance the accuracy of the simulation, we will need to provide a more

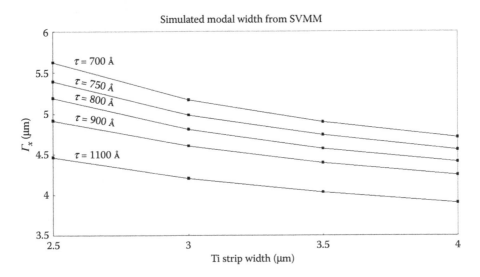

FIGURE 12.16 Simulated modal width Γ_x for a range of Ti thickness, τ.

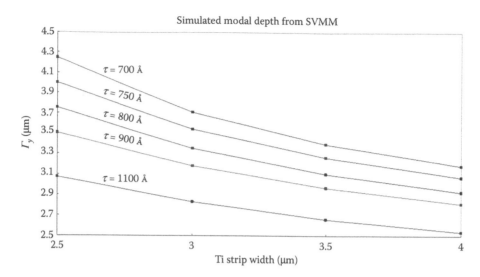

Simulated modal depth from SVMM

Γ_y (µm)

$\tau = 700$ Å

$\tau = 750$ Å

$\tau = 800$ Å

$\tau = 900$ Å

$\tau = 1100$ Å

Ti strip width (µm)

FIGURE 12.17   Simulated modal width Γ_x for a range of Ti thickness, τ.

accurate diffusion model which considers the dispersion relationship of the change in refractive index profile in Ti-LiNbO$_3$.

12.3   REMARKS

In this section, we have described the development of a numerical computing process that is based on a SV Finite Difference analysis to solve the Helmholz equation. The numerical model that we have formulated can model the guided modes in optical waveguides of any arbitrary index profile. A non-uniform mesh allocation scheme is employed in the formulation of the difference equations to free more computer memory for the computation of waveguide regions that bear greater signifi-cance. The accuracy of the simulation package presented here in this chapter, SVMM is assessed by computing the propagation constants and the effective indices of several rib waveguides that have been known to be excellent benchmark waveguide structures. The results of our computation have compared favorably with other published results [7,14,15,26,27]. We then continue to simulate the optical guided modes of diffused optical waveguides in LiNbO$_3$. The computed mode sizes are con-sistent with referenced experimental results. The simulation techniques described in this chapter, however, have shown the inadequacies of the adopted diffusion model for its inability to model the diffused waveguide in a more robust sense. It is suggested that further research to be conducted for a more refine and robust representation of the refractive index profile of Ti-LiNbO$_3$ diffused waveguide. Despite the shortcoming of the diffusion model, this chapter demonstrates the potential of SVMM to be used as an analytical and design tool for graded index integrated optical waveguide.

12.4   TRAVELING WAVE ELECTRODES FOR ULTRA-BROADBAND ELECTRO-OPTIC MODULATION

An integrated electro-optic modulator composes of optical waveguides and electrodes. The integrated optical waveguides would follow structure of 3D waveguides as described in Chapter 4. The elec-trodes must be in planar structure of either coplanar (CPS) or asymmetric coplanar structure (ACPS). This section gives the detailed modeling and design of such electrodes as the modulating electrodes for ultra-high speed broadband optical modulators. The modulation efficiency is also examined with the numerical calculation of the overlap integral between the optical waveguide and the fields generated by the electrodes.

12.4.1 INTRODUCTION

This section gives a detailed formulation and modeling of the traveling wave electrodes for integration into an optical waveguide of an integrated optical modulator whereby the phase modulation is achieved.

Optical transmission at ultra-high bit rate up to 40 Gb/s and multiplexing of several optical channels are becoming the standard deployment of dense division wavelength multiplexing (DWDM) optical networks in certain information transport routes. Various amplitude and phase shift keying modulation formats have been investigated over the last decade to increase the transmission capacity and mitigate the linear dispersion and non-linear induced effects. In ultra-high-speed transmission systems, external modulators play a vital role. Of interest are the modulators in lithium niobate (LiNbO$_3$) and other compound semiconductor-integrated optical devices. The phase modulation can be achieved with single or asymmetric or dual (balanced or symmetric) electrode structures that require precision in the design and fabrication.

Several published works [28,29] have outlined the important steps and considerations involved to achieve efficient design of traveling wave electrodes [30,31]. The principal part of the design calculations were based upon adopted empirical model that was derived from the combination of the quasi-TEM analysis and the finite element method (FEM). The empirical model, despite being impressive in its ability to facilitate the design calculation, does not provide the modulating electric field that is mandatory in calculating the electro-optical overlap integral. More importantly, it does not consider the subtler structural factors, such as the wall angle of the gold-plated electrode. Most analyses, such as the conformal mapping (CM) technique, the FEM [4], or the method of images (MoI) [32], assume an infinitely thin electrode structure and perfect vertical side wall. However, when thick electrodes, typically in the range of 10–20 µm, are fabricated to achieve broadband operation of a 3 dB bandwidth of more than 40 GHz, the assumption of perfectly vertical wall angle is no longer valid. In practical devices, after the gold electroplating stage, the electrode would assume a trapezoidal shape as shown in Figure 12.18.

The wall angle of the electrode has a rather significant influence on the value of the effective microwave index n_m and the electrode characteristic impedance Z. It is therefore important to take into consideration such structural effect of the electrodes in the design of broadband traveling wave electrodes. Chung et al. [4] have modeled the electric field of the electrodes using the FEM but their analysis did not consider the effect of the wall angle. This has led to the limitation of the reported empirical model. Electrode structural factors such as this can only be modeled by a more robust

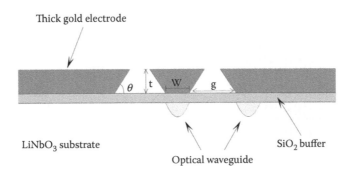

FIGURE 12.18 Cross section structure of an interferometric electro-optic modulator having tilted wall angle of thick electrodes and a dielectric buffer layer; LiNbO$_3$ substrate and air superstrate.

numerical formulation such as the finite element method [2,33] or the FD method [3]. The FD method is selected to solve the anisotropic Laplace equation. Finite differencing is very computationally effective because of its relatively straightforward analysis. In the PC computing environment that has virtual memory support, this scheme can warrant sufficient accuracy in a relatively short time without tedious mathematical manipulation and programming involved as contrast to the FEM. Non-uniform grid allocation can also be used to economize computer storage capacity.

The next section outlines the numerical formulation of the non-uniform mesh FD scheme to solve the anisotropic Laplace equation under the quasi-TEM assumption of the microwave mode. Based on this numerical model, an application program named FD Traveling Wave Electrodes Analysis (FDTWEA) has been developed to compute Z and n_m, the modulating field, E_x and E_y, and the traveling wave and optical wave overlap integral, Γ. Our simulated results will be compared with published results to verify the validity of our FDTWEA simulator. The simulator is also employed to study the effect of the wall tilted angle on the design of electrode structures.

12.4.2 ELECTRIC FIELD FORMULATION

12.4.2.1 Discrete Fields and Potentials

The traveling wave electrodes are miniature transmission where a quasi-TEM wave transmission can be assumed, hence, the electrical fields can be reduced to a 2D electrostatic field distribution, which can be solved by applying the Laplace equation. Since the LiNbO$_3$ substrate and the SiO$_2$ dielectric are involved, we have an anisotropic case. Assuming a Z-cut orientation of the LiNbO$_3$ crystal, then the permittivity tensor is diagonal. Thence the electrostatic potential V is the solution of the anisotropic Laplace equation written as

$$\frac{\partial}{\partial x}\left(\varepsilon_x \frac{\partial V}{\partial x}\right) + \frac{\partial}{\partial y}\left(\varepsilon_y \frac{\partial V}{\partial y}\right) = 0 \tag{12.46}$$

This equation can be numerically solved by a FD method using a non-uniform mesh allocation scheme. Denser mesh is allocated at the edges of the electrode and the buffer layer area so that the effect of the edge field can be modeled more accurately. Points that are further away from the electrode can be modeled with a coarser mesh. Difference equations can be then formulated.

Consider a general electrode structure, as shown in Figure 12.19a. The grid points are placed along dielectric boundary. Since the electrode structures for our consideration involve only three dielectric media, namely, the air, the SiO$_2$ thin-coated layer, and the LiNbO$_3$ substrate. With LiNbO$_3$ being the anisotropic medium, the transition of different dielectric medium occurs only in the y-direction. Therefore, it suffices to analyze the two layers of dielectric medium as shown in Figure 12.19b to formulate the difference equations.

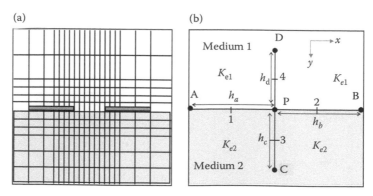

FIGURE 12.19 (a) A simple illustration of the non-uniform grid allocation scheme. (b) The grid points for finite differencing.

Considering a mesh point P along the boundary and four other points surrounding it, namely, A, B, C, and D, the point located between AP, BP, CP, and DP can be labeled 1, 2, 3, and 4 with their respective grid sizes of h_a, h_b, h_c, and h_d. The electrical potentials at each point are V_P, V_A, V_B, V_C, and V_D, respectively. K_{e1} and K_{e2} are the dielectric constants of the media.

$$\text{At 1: } \frac{\partial V}{\partial x} \approx \frac{V_P - V_A}{h_a}; \quad \text{at 2 } \frac{\partial V}{\partial x} \approx \frac{V_B - V_P}{h_b}; \quad \text{at 3 } \frac{\partial V}{\partial y} \approx \frac{V_C - V_P}{h_c}; \quad \text{and at 4 } \frac{\partial V}{\partial y}$$

$$\approx \frac{V_P - V_D}{h_d} \tag{12.47}$$

With the assumption that half of the flux flowing in medium 1 and the other half in medium 2, we have

$$\left(\varepsilon_x = \frac{\partial V}{\partial x}\right)_{PB} = \frac{\frac{1}{2}(Ke_{1x} + Ke_{2x}) \cdot (V_B - V_P)}{h_b} \quad \text{along segment } PB \tag{12.48}$$

$$\left(\varepsilon_x = \frac{\partial V}{\partial x}\right)_{AP} = \frac{\frac{1}{2}(Ke_{1x} + Ke_{2x}) \cdot (V_P - V_A)}{h_a} \quad \text{along segment } AP \tag{12.49}$$

$$\left(\varepsilon_y = \frac{\partial V}{\partial y}\right)_{PC} = \frac{Ke_{2y} \cdot (V_C - V_P)}{h_c} \quad \text{along segment } PC \tag{12.50}$$

$$\left(\varepsilon_y = \frac{\partial V}{\partial y}\right)_{DP} = \frac{Ke_{1y} \cdot (V_P - V_D)}{h_d} \quad \text{along segment } DP \tag{12.51}$$

Using Equation 12.30, the five-point FD relationship between the fields at the grid points can be expressed as

$$K_A V_A + K_B V_B + K_C V_C + K_D V_D - K_P V_P = 0$$

$$\text{with } K_A = \frac{Ke_{1x} + Ke_{2x}}{hb \cdot (ha + hb)}; \quad K_B = \frac{Ke_{1x} + Ke_{2x}}{ha \cdot (ha + hb)}; \quad K_C = \frac{Ke_{2y}}{\frac{1}{2}hc \cdot (hc + hd)}; \tag{12.52}$$

$$K_D = \frac{Ke_{1y}}{\frac{1}{2}hd \cdot (hc + hd)} \quad \text{and } K_P = K_A + K_B + K_C + K_D$$

Similarly, corresponding coefficients and FD relationships can be obtained for the air and the SiO_2 buffer layer and the substrate $LiNbO_3$ regions. Using a sufficiently large problem space, approximately 400×400 µm, we can assume that the electric field along the boundary lines to be infinitesimal. For the potential, a Neumann boundary condition [4,5] can be used to obtain the spatial relationships between the fields at the grid points. The remaining points of the windows are just the permutation of the boundary conditions outlined above. Incorporating the boundary conditions into the FD equations, a set of difference equations can be derived from the mesh grid points in form of

$$[A][u] = [b] \tag{12.53}$$

where [A] is the coefficient matrix, [u] is the vector that contains the potential, V of each grid point, while b is the vector that assumes the right hand side of the Laplace equation which is mostly zero except for the grid points on the electrodes which take up the value of the potential on the relevant electrodes. A typical matrix equation for a problem space can then be generated whose coefficient matrix is a tri-diagonal matrix with fringes. The matrix elements p, a, b, c and d for each point correspond to the coefficient K_p, K_A, K_B, K_c, and K_d, respectively. Take note that for points seven, eight, and nine, which fall on the electrode, $K_p = 1$ while K_A, K_B, K_c, and K_d are all zero. The difference equation can be solved by means of the more conventional Successive Over-Relaxation Method (SOR) [34,35]. This method, however, requires a good initial guess and a good estimate of the relaxation factor to achieve a reasonable rate of convergence.

12.4.2.2 Capacitance, Impedance, and Microwave Index

The above numerical formulation can then be applied to the computation of the operational parameters of the traveling wave electrodes such as the characteristic impedance Z and the microwave effective index n_m which are given as

$$Z = \frac{1}{c\sqrt{CC_0}} \quad \text{and} \quad n_m = \sqrt{\frac{C_o}{C}} \tag{12.54}$$

where C is the capacitance per unit length of the electrode transmission line with the dielectric medium, C_0 is the capacitance per unit length for the air filled medium, with c being the speed of light in vacuum. The capacitances C and C_0 can be determined by calculating the charges on the conductors. The Gauss theorem can be applied to determine the charges and hence it requires the integration of the normal component of the electric flux over a surface enclosing the 'hot' electrode. Forming this surface by lines joining the nodal points parallel to the coordinate directions, at any point P on this surface, we have

$$D_n = \varepsilon E_n = -\varepsilon \frac{\partial V}{\partial n} \tag{12.55}$$

where D_n is the normal component of the electric flux, E_n is the normal component of electric intensity, and the subscript n indicates the orthogonal coordinate. The potential at P may be expressed numerically in terms of the known potentials V_A and V_b on each side of it. For irregular mesh as shown in Figure 12.19, we have,

$$\frac{\partial V}{\partial x} = \frac{V_B - V_A}{h_b + h_a} \quad \text{and} \quad \frac{\partial V}{\partial y} = \frac{V_C - V_D}{h_c + h_d} \tag{12.56}$$

Thus, if using enclosing surface of a square box surround the hot conductor of s straight line segments each containing r nodes, the charge per unit length normal to the cross section is then be given by

$$Q = \varepsilon_r \varepsilon_o l \sum_s \sum_{P=1}^{4'} \left(\frac{\partial V}{\partial n}\right)_P \tag{12.57}$$

The apostrophe indicates the first and last terms in the summation are halved, which is seen to be equivalent to integration by the trapezoidal rule, l is the length of the segment of the integration path. For uniform discretization, l is essentially the grid size h. For our non-uniform scheme, l is assumed to be either $(h_a + h_b)/2$ or $(h_c + h_d)/2$ depending on either a horizontal or vertical line segment over which the summation is taken. The value of the relative permittivity, ε_r depends on which dielectric medium point P falls onto. For instance, when we sum the derivative along the first horizontal line segment ($s = 1$), the segment falls completely in the air, so $\varepsilon_r = 1.0$. If the summation is done on the line segment number three, which is in the LiNbO$_3$ crystal, $\varepsilon_r = \varepsilon_z = 43$. The vertical summation ($s = 2$, 4) however need to be dealt with care because it involves dielectric medium transition. For points located entirely in the air, SiO$_2$ and LiNbO$_3$, the relative permittivities are $\epsilon_a = 1.0$, $\epsilon_b = 3.9$ and $\epsilon_x = 28$, respectively. However, for points fallen on the buffer air interface, we assume half of the flux passing through each medium. For the air-SiO$_2$ interface, $\epsilon_r = (1 + \epsilon_b)/2$, whereas for the SiO$_2$-LiNbO$_3$ interface, $\epsilon_r = (\epsilon_b + \epsilon_x)/2$. The charge capacity is defined as

$$C = \frac{Q}{V_t} \tag{12.58}$$

where V_t is the potential difference between the electrode arms. Similarly, for C_o, the Laplace equation can be solved for the transmission line in the air filled medium without the dielectric medium. The charge capacity and hence the capacitance can then be deduced.

12.4.2.3 Electric Fields E_x and E_y and the Overlap Integral

The electric field generated by the traveling wave electrode can be derived as

$$E_x = \frac{\partial V}{\partial x} = \frac{V_B - V_A}{(h_a + h_b)} \quad \text{and} \quad E_y = \frac{\partial V}{\partial y} = \frac{V_C - V_D}{(h_c + h_d)} \tag{12.59}$$

The overlap integral due to the electro-optic (EO) effects, Γ is defined as [36]:

$$\Gamma = -\frac{\pi n_e^3 r_{33} L}{\lambda} \iint |E_o(x, y)|^2 E_m(x, y)\, dx\, dy \tag{12.60}$$

where $|E_o(x,y)|^2$ and E_m are the square of the normalized total electric field of the guided lightwaves and the electric field generated the electric traveling wave respectively. The choice of E_x or E_y depends on the crystal orientation and the polarization of the optical field to maximize the EO interaction. The normalized optical field intensity profile assumes a Hermitian–Gaussian profile given as

$$|E_o(x, y)|^2 = \frac{4y^2}{w_x w_y^3 \pi} \exp\left[-\left(\frac{x - p}{w_x}\right)^2\right] \cdot \exp\left[-\left(\frac{y}{w_y}\right)^2\right] \tag{12.61}$$

where the $1/e$ intensity width and depth are $2w_x$ and $1.376w_y$, respectively; and p is the peak position of the optical field in the lateral direction. w_x and w_y are strongly dependent on waveguide fabrication parameters. They can be determined either experimentally or by a numerical modeling. For modeling purposes, we use w_x and w_y of 2.5 μm and 2.2 μm, respectively [36]. From the calculated Γ one could determine the voltage V_p defined as the applied voltage to the electrode for a pphase shift change on the lightwave carrier passing through an arm of the MZIM.

12.4.3 ELECTRODE SIMULATION AND DISCUSSIONS

12.4.3.1 Grid Allocation and Modeling Performance

Figure 12.20 shows contour plots of the potentials for CPW, ACPS, and CPS electrode structures obtained from the solution of the Laplace Equations 12.1 through 12.5 in which $w = 10$ μm, $g = 15$ μm, $t = 3$ μm, and $t_b = 1.2$ μm are considered. The electric fields generated by these electrodes are obtained by solving (12.6) and are illustrated in Figure 12.21.

The horizontal field, E_x is strongest in between the gaps, while the vertical field, E_y is strongest along the hot electrodes in all structures. From these plots, we can see that the push pull operation can be achieved most efficiently by the horizontal field, E_x of a CPW structure for the X-cut Y-propagating device. Another configuration the CPS electrode structure would allow us to place the waveguides directly underneath for the vertical field, E_y to exert a maximum EO effects. This, of course, would correspond to a Z-cut device that in turn may suffer frequency chirping on the light wave carrier. Furthermore, the CPS structure, however, exhibits a high propagation loss and is

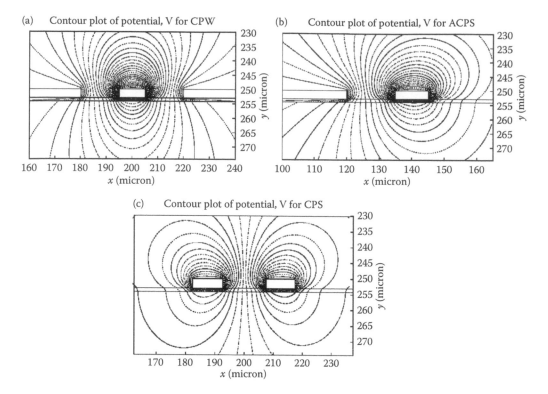

FIGURE 12.20 Contour plot of the Laplace equation solution for electrode structures (a) CPW, (b) ACPS, and (c) CPS.

therefore seldom employed. A common configuration for Z-cut device is to place one waveguide under the hot electrodes and the other at the edge of the ground plane in either the coplanar waveguide (CPW) or ACPS (asymmetric coplanar structure). The vertical field E_y is employed in such configuration. This cannot be considered as a full push-pull operation because the waveguide underneath the ground plane sees only one-half of the field seen by the waveguide under the hot electrode. We can observe that from the contour plots and the field plots, the strongest field exists around the edges of the electrodes, within the SiO_2 buffer layer underneath the hot electrodes and in the gap between the electrodes. Therefore, the denser grids around these areas should be allocated. From our experience, a grid size as small as 0.05 to 0.2 μm at around the electrode edges, the gap and buffer region, up to larger grid size of 8–10 μm for points where the electric field has decayed substantially warrants an accurate calculation result. If the file is not specified, FDTWEA generates one automatically using its default grid allocation scheme suitable for the relevant electrode structure. A problem space of around 400 by 400 μm is sufficient to assume a metallic box boundary condition.

12.4.3.2 Model Accuracy

12.4.3.2.1 General Comparison

To assess the validity of our modeling results, the values of Z and n_m are compared with published experimental results. The comparisons are tabulated in Table 12.1. The simulated results of FDTWEA for the CPW and ACPS electrodes corroborate well with published results. In this model, we determine the Z and n_m values for the buffer layer varying between 0.6 and 0.85 μm that are determined from the fabrication tolerances of several fabricated devices [37]. There seems to be a discrepancy in our calculated Z value of the ACPS structure compared to the one reported by [38] with Z of 35 Ω. We have thus assumed that the calculations in Reference 39 have not considered the thickness

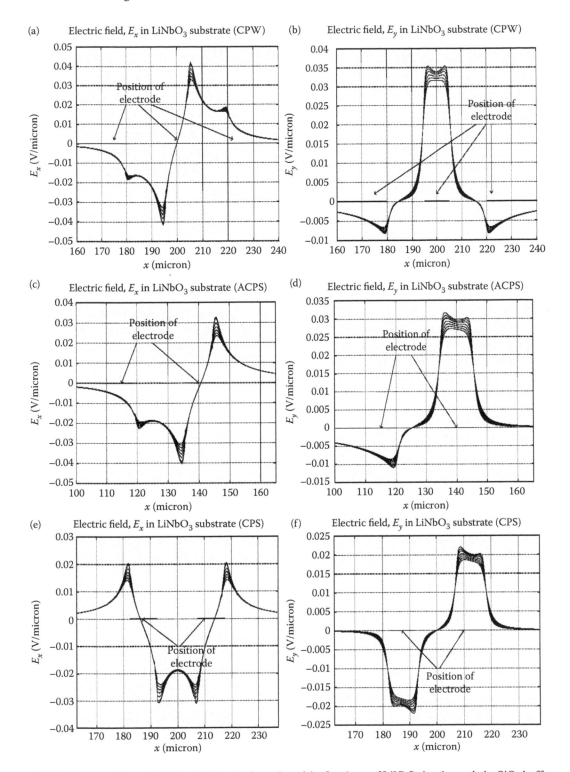

FIGURE 12.21 Electric field distributions in the region of the first 1 μm of LiNbO₃ just beneath the SiO₂ buffer layer [(a) E_x and (b) E_y] of the CPW electrode; [(c) E_x and (d) E_y] of ACPS electrode; and [(e) E_x and (f) E_y] of CPS electrode.

TABLE 12.1

FDTWEA Calculated Values as Compared with Published Results. *W*, *G*, *T*, and *t* Units Are in μm

References	Structure	W	G	T	t_b	Published		FDTWEA	
						n_m	$Z(\Omega)$	n_m	$Z(\Omega)$
[4]	CPW	20	5	3	0.6–0.85	2.7	27	2.867–2.703	27–27.3
—	CPW	48	10	3	0.6–0.85	3.3	24.5	3.358–3.224	24.13–21.4
[38]	ACPS	15	5	4	0	–	~35	3.661	29.83
		15	5	0	0	–	–	4.226	33.15
[4]	ACPS	10	10	1.5	0.1	–	45(cal) 47.1(TDR) 49.8(NA)	3.781	42.7

of the electrodes. The recalculated impedance, with the assumption of an infinitely thin electrode, is about 33.15 Ω, which agrees well with their reported value. The third modeling is performed based on experimental results reported by Reference 32. The simulated results agree closely with their theoretical calculations which are based on the MoI. However, the measured results in Reference 32 differ slightly from that of the theoretical predictions. The characteristic impedance measured by time domain reflectometry (TDR) and scattering parameter network analysis (NA) were 49.8 and 47.1 Ω, respectively. The small discrepancy is possibly due to a slightly thicker SiO₂ buffer layer coated in the experiment.

12.4.3.2.2 FDTWEA versus Spectral Domain Analysis (SDA)

Our simulations are then compared further by emulating the calculations by the SDA as reported in Reference 40. Figure 12.22a and b shows the comparison of the two simulations. Initially, we calculated the parameter based on the electrode parameters of $t = 4$, $t_b = 0$–1.5, $w = 8$, and $g = 30$ μm. Our results, however, have seemed to underestimate both the values of Z and n_m. It was later realized that the SDA may not have taken the thickness of the electrode into account. We recalculate Z and n_m with the assumption of a zero thickness electrodes. This has clearly implied that the SDA does not consider the effect of the electrode thickness. Such limitation does not exist in our present FD analysis.

12.4.3.2.3 FDTWEA versus Finite Element Method

We further verify our calculation program by comparing the value of n_m with the simulation result of very thick electrodes of the FEM. The comparison is shown in Figure 12.23. The calculations are intended to show how the thickness of the electrodes can significantly improve the electrical and optical velocity matching. Thick electrodes that are in the range of 10–20 μm are employed. From Figure 12.22, the plot of the microwave refractive indices versus the thickness of the electrode, we observe that our calculations agree to within 0.5% with its FEM counterpart. In our calculations, a grid size of 0.125 μm is used for points of meshes along the wall of the thick electrodes. Each calculation takes approximately 3–4 min to complete on a standard PC Windows 2000 operating system. Our FD scheme has achieved similar level of accuracy achieved by FEM in a relatively straightforward numerical performance.

So far, it has shown that the FD scheme offers potential and reasonably accurate modeling of the traveling wave electrode parameters, Z and n_m. They are consistent with the simulations and measured data reported in previous works. It has simply verified the validity of our numerical model.

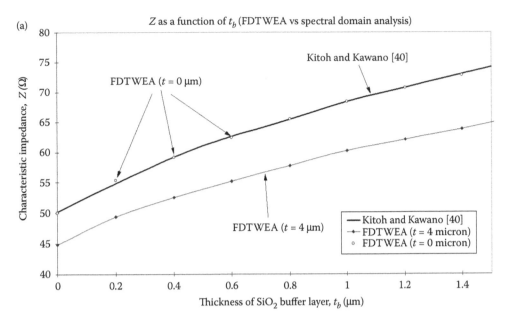

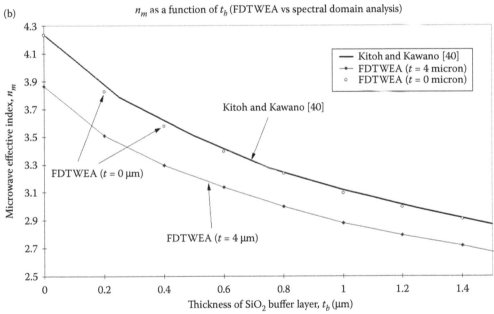

FIGURE 12.22 (a) FDTWEA calculated impedance compared to that of the SDA and (b) calculation of FDTWEA nm compared to that of SDA.

FDTWEA, with its verified precision would thus be a useful tool to provide a good quantitative measure in the design and analysis of traveling wave electrodes.

12.4.4 EO OVERLAP INTEGRAL, Γ

In this section, we show some of the calculation result and discuss a few issues on the interaction between the electrical traveling wave and the guide optical waves, the electrical-optical overlap integral. To compute the overlap integral, we need the normalized optical intensity field profile, which is a

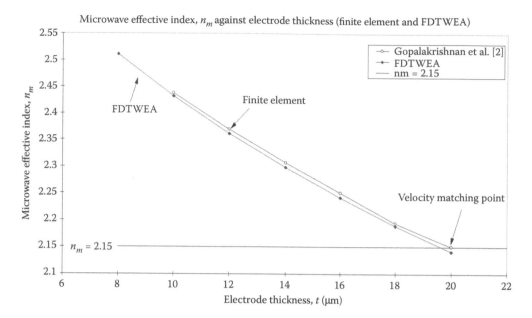

FIGURE 12.23 FDTWEA's calculation of n_m compared with FEM.

Hermitian–Gaussian profile as

$$|E_o(x, y)|^2 = \frac{4y^2}{w_x w_y^3 \pi} \exp\left[-\left(\frac{x - p}{w_x}\right)^2\right] \cdot \exp\left[-\left(\frac{y}{w_y}\right)^2\right] \tag{12.62}$$

The optical mode sizes, w_x and w_y, play very important roles in maximizing the overlap integral Γ. The mode size is defined as twice the $1/e$ modal width and 1.376 times the $1/e$ modal depth. The parameters w_x and w_y can be modeled using the effective index method or the finite difference or element numerical technique and not given here or by even by experimental measurement of the mode size as described in Chapter 3. Furthermore, Γ is also influenced by the relative position of the optical mode with respect to the hot electrodes. All attributes that affect the value of Γ can be easily modeled by FDTWEA. For illustration purposes, we will calculate Γ of the ACPS electrodes and study the effects of various factors that can influence its value.

Figure 12.24 shows the variations of the overlap integral, Γ as the peak position of the optical mode, p shift from one end of the electrodes to the other end. As expected, a tighter confined mode would give a higher value of Γ.

It is thus very critical to fabricate a Ti-LiNbO$_3$ that guide lightwaves with optimum mode spot sizes to maximize the fiber-diffused waveguide coupling and the EO overlapping. Apart from having a tightly confined optical mode, the relative position of the optical waveguide with respect to the electrodes is also critical to provide an optimum EO effect. When narrow electrodes whose width is comparable to the size of the waveguide are designed, it is preferable to place the waveguide directly underneath the electrode to utilize the strong edge field from both side of the electrode. This has been shown in Figure 12.24, in which the maximum Γ is obtained when the position of the optical mode is centered underneath the electrode. However, for much wider electrode structures, the preferred position would be just inside the end of the electrode closer to the gap. When a 30 μm wide electrode is employed, a maximum overlap integral can be achieved by exploiting the higher edge field by placing the waveguide at the center, that is $x \sim 135$ μm. Positioning the optical waveguide right at the center of the electrode would give inefficient EO interaction.

Figure 12.25 shows how the thicker buffer layer can impede the electro-optic effect. Although a thicker buffer layer has the advantage of a lower conductor loss and significantly improves the

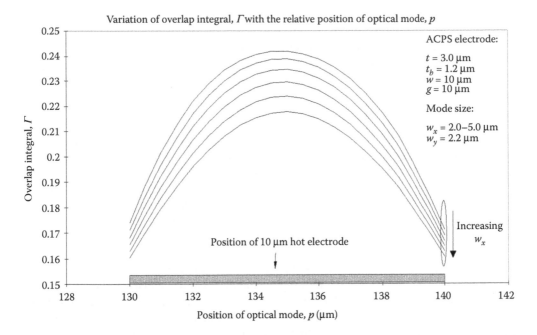

FIGURE 12.24 Variation of Γ as the peak position of the optical mode shift from one end of the hot electrode to another for increasingly wider optical mode.

velocity match, there is a tradeoff involved because the overlap integral will be lower, which can therefore lead to a higher V_π. We can see that without the buffer layer, Γ assumes a much higher value. Unfortunately, having no buffer layer would imply a much higher optical-electrical velocity mismatch and hence reducing the effective operational bandwidth. Essentially, the position of the optical waveguide with respect to the electrode should be considered before fabricating the device. It is also

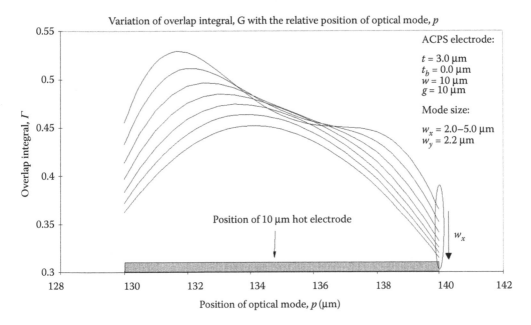

FIGURE 12.25 Variation of Γ as the peak position of the optical mode shift from one end of the hot electrode to another for increasing wider optical mode for electrode with no buffer layer, $t_b = 0$.

important to have a quantitative measure of the effect SiO_2 thickness on the overall performance of the device.

Figure 12.26 confirms that the FDTWEA can provide a measure that would greatly facilitate the design of traveling wave electrodes.

12.4.5 TILTED-WALL ELECTRODE

In previous sections, we have used FDTWEA to analyze various rudimentary design parameters of traveling wave electrodes. In this section, we demonstrate show how it can handle the problem of greater complexity of practically fabricated traveling electrode.

The fabrication of traveling wave electrodes is not straight forward. One way of extending the bandwidth of the device is to employ very thick electrodes that are in the range of 10–20 µm. Such thick electrodes, however, do not always assume a rectangular shape. They are more likely to take up a trapezoidal shape and the contour of the distributed electric fields, as shown in Figure 12.27.

Such geometrical factor cannot be ignored because they certainly have a subtle effect on both the values of Z and n_m. We can see that the wall angle certainly has a substantial effect on the values of Z and n_m as illustrated in Figures 12.28 and 12.29. For the trapezoidal shape electrodes both Z and n_m are reduced. The effect of the wall angle is less severe for thinner electrodes as we can see from the graphs that the difference of the plots for different θ converges as the electrodes become thinner. For thick electrodes, especially those that are greater than 10 µm, the wall angle effect should not be ignored. For example, if we base our design on the assumption of a $\theta = 90°$ rectangular electrode, then we would use a 20 µm thick electrode for the best velocity match. However, if the fabricated electrodes assume a trapezoidal shape with $\theta = 80°$, then we would have overestimated the value of n_m for best velocity matching by about 0.2. This corresponds to almost a 20% loss of the bandwidth

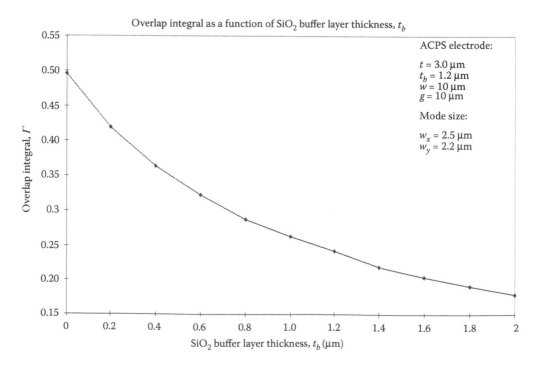

FIGURE 12.26 Γ as a function of the thickness of SiO_2 buffer, t_b.

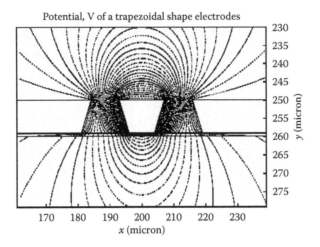

FIGURE 12.27 Potential field distribution of the trapezoidal shape electrode with the contour shown in curves wrapping around the electrodes.

that could have been achieved by a mismatching of 0.01 between the microwave and optical indices. The plots shown in Figure 12.29 suggest that we need only an electrode thickness of around 14 μm to achieve maximum bandwidth, based on the assumption of an 82° electrode wall angle and an electrode thickness of 15 μm. Again, this demonstrates another potential application of the FDTWEA numerical modeling scheme. It thus offers much improved analytical capability beyond most analytical techniques. Undoubtedly, an analytical capability would certainly imply a better design in traveling wave electrodes for ultra-wideband optical modulators, especially for the employment of advanced modulation formats in long haul optically amplified fiber communications systems.

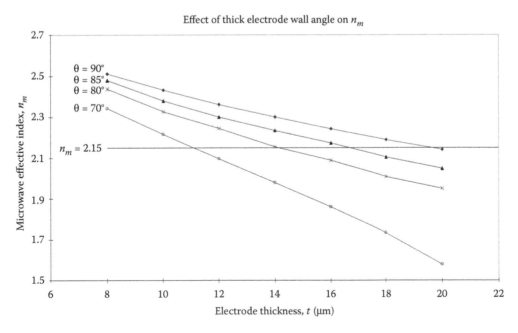

FIGURE 12.28 Electrode microwave impedance n_m as a function of the electrode thickness with the wall angle as a parameter.

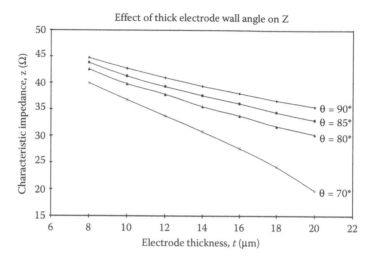

FIGURE 12.29   Characteristic impedance Z (ohms) as a function of electrode thickness with the wall angle as a parameter.

12.4.6   Frequency Responses of Phase Modulation by Single Electrode

In practice, the design and estimation of the frequency response of the microwave electrode and its optical modulation response is very important. Even more important is the measurement technique to determine whether the mismatching of the microwave signals and the lightwave, especially when the optical waves are modulated under amplitude, discrete phase or continuous phase or frequency. Advanced modulation formats have recently emerged as the most effective method in ultra-high-speed optical communications to combat the residual dispersion, non-linear effects [41–44]. The role of the phase modulation of the carrier is critical especially when continuous phase modulation is employed [45]. Thus, the frequency response of phase modulators is briefly investigated both in theory and experiment in this section.

The velocity mismatch between the microwave signals and the optical waves comes from the difference in the effective refractive indices of the optical guided mode and that of the traveling microwave wave along the electrode length. This mismatching reduces the optical bandwidth of the phase modulator. The essential characteristics of the phase modulator are its frequency response which is the phase variation of the optical carrier under the modulation of the travelling wave. Over the years, the measurement of the frequency response of an optical phase modulator has been done in the electrical domain [46], and, until recently, has been a novel method using a Sagnac interferometer to determine the frequency response of phase modulator [47]. This allows us to adapt this measurement technique to characterize the phase responses of a number of optical phase modulator with the set up shown in Figure 12.30 in which the device under test is a phase modulator inserted in a Sagnac interferometer. The length of the ring determines the null of the frequency response. The phase modulation is then converted to amplitude response is due to the interference or phase comparison of the optical path in clockwise and counterclockwise. Thus, any modulation index variation in the modulator would be easily measured. A microwave network analyzer is used to scan the exciting sinusoidal RF wave into the electrode of the optical phase modulator. The optical signals are then detected by the photodiode and a wideband microwave amplifier then fed into port 2 of the network analyzer. The scattering parameter S_{21} is then obtained.

The detected current output of the photodetector is given by

$$i_d = t_f \Re K \sin \left[\sin c(2\pi f_{RF}\tau_{PM})v_{RF}(t+\tau) - v_{RF}(t) \right] \tag{12.63}$$

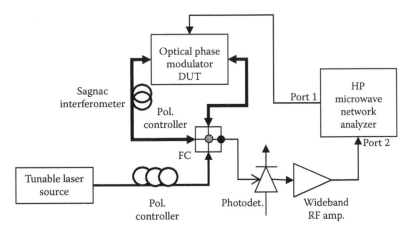

FIGURE 12.30 Schematic representation of a Sagnac interferometer for the measurement of the frequency response of optical phase modulators. Optical fiber line __(bold) dashed; electrical line continuous. DUT = device under test. FC = fiber coupler – 4 ports.

with

$$\sin c(2\pi f_{RF}\tau_{PM}) = \frac{\sin (2\pi f_{RF}\tau_{PM})}{2\pi f_{RF}\tau_{PM}}$$

with f_{RF} is the frequency of the *RF* electrical signal launched into the phase modulator. τ_{PM} is the delay time difference between the travelling times of the lightwaves in the two waveguide sections before entering to the phase modulator. τ is the travelling time of the travelling wave through the electrode. The polarization is assumed to be 1 when the polarization controller aligns the polarized mode launched into the lithium niobate optical waveguide.

Equation 12.61 is obtained by getting the optical field components of the clockwise and counterclockwise with a delay difference depending on the length difference of the fibers. These fields are then modulated with the phase shift via the optical phase modulator. These fields are then entering the fiber coupler. The output field is the interfered field at the output and coupled to the photodetector. This field is then followed a square law and then a current is obtained. Thus, the responsivity $\mathfrak{R}$ of the detector is included in Equation 12.61.

Furthermore, the amplitude conversion from the phase modulation of the two paths can be considered as similar to the case of asymmetric Mach–Zehnder interferometer and the optical amplitude of the interfered optical field does also follow [48]:

$$E_o(t) = \frac{E_{in}}{2}\left[J_o\left(\frac{\pi v_{RF}}{V_{\pi,CW}}\right) - J_o\left(\frac{\pi v_{RF}}{V_{\pi,CCW}}\right)\right] \cos (2\pi f_{RF}t) \qquad (12.64)$$

And the modulation half wave voltage V_π for the *CW* and *CCW* direction is given by

$$V_{\pi,CW} = V_{\pi,CCW}\frac{2\pi f_{RF}\tau}{\sin (2\pi f_{RF}\tau)} \qquad (12.65)$$

A number of optical phase modulators are tested in which the buffer exists or not. We observe that the null and maxima of the frequency response is periodic and due to the time difference of the optical waves before entering the two inputs of the optical phase modulators. No observation of the frequency responses is given in Equation 12.44, but are seen in Equation 12.61. Figure 12.31 shows the frequency response of a non-buffered layer-traveling wave electrode. That means that the

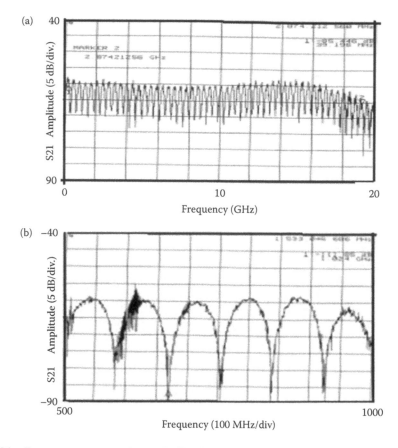

FIGURE 12.31 Frequency response of a non-buffered layer-traveling wave electrode (a) general response and (b) detailed side lobes in amplitude (log scale) versus frequency.

mismatch of the velocities of the travelling wave and optical wave is quite high. While in Figure 12.32 when a SiO_2 buffer layer is used, then the travelling wave can offer a wide band operation and the frequency response is a combined response of Equations 12.61 and 12.44.

12.4.7 REMARKS

In this section, we have formulated a non-uniform FD model that offers an accurate analytical and modeling tool for estimation of the ultra-broadband properties of traveling wave electrodes. Aniso-tropic Laplace equation solved using the FDTWEA gives accurate microwave properties of symmetric and asymmetric electrode structures. A non-uniform grid allocation scheme is implemented for maximizing the memory computing usage without jeopardizing the model accuracy. It is demonstrated that the calculated characteristic impedance Z and microwave effective index n_m are consistent and improvement in accuracy as compared to published results. The FDTWEA is also employed to evaluate the electro-optic overlap integral. Principal features are obtained for optimization of the electro-optic interaction overlap integral. The optimization of Γ assures a low half wave voltage V_π, hence lower driving microwave power. Our calculations have demonstrated consistency with their simulated and measured data. The FDTWEA has also been applied to study the impacts of the trapezoidal shape tilted-wall thick electrode ($>10\,\mu m$) on the electrode characteristic impedance and effective index. The effects of the wall angle are not negligible and should not be ignored when thick electroplated electrode is required to achieve ultra-broadband operation. Quantitatively and qualitatively,

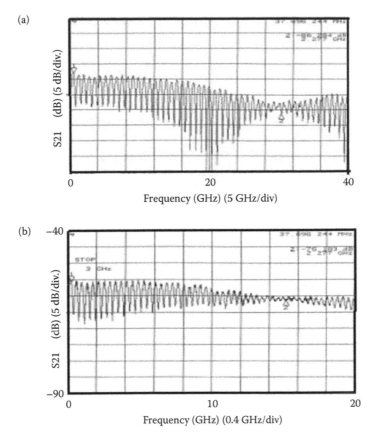

FIGURE 12.32 Frequency response of a SiO_2 buffered layer-traveling wave electrode (a) 40 GHz bandwidth and (b) 10 GHz bandwidth.

our presented model has been shown to be consistent and offers more computationally efficient solutions as compared with other numerical models.

12.5 LITHIUM NIOBATE OPTICAL MODULATORS: DEVICES AND APPLICATIONS

Optical modulators, using acousto-optic, magneto-optic, or electro-optic effects as the principal components for external modulation of lightwaves, have presently played the important role in modern long-haul ultra-high speed optical communications and photonic signal processing systems. $LiNbO_3$ is an ideal host material for such modulators. In this paper we present: (i) a brief overview the electro-optic interaction in $LiNbO_3$ optical waveguides and the travelling electric waves for modulation; (ii) fabrication techniques of optical waveguides and electrodes for the excitation of microwave and millimeter waves, including the prospects of the uses of ultra-thick brick-wall-like type by synchrotron deep x-ray LIGA process; (iii) a comprehensive modeling of the traveling wave electrodes for the design of interferometric optical modulators, including different electrode structures. Tilted and thick and brick-wall-like electroplated electrodes are modeled and confirmed with implemented modulators operating up to 26 GHz in diffused $LiNbO_3$ optical interferometric waveguide structures; and (iv) the design and implementation of $LiNbO_3$ single or dual-drive as modulators for modulating lightwaves in modern transmission systems as ultra-high-speed pulse carvers, generators of modulation formats (NRZ, RZ, CS-RZ, DPSK, DQPSK, etc.) as outlined in Section 12.2, and photonic signal processors.

12.5.1 MZIM and Ultra-High Speed Advanced Modulation Formats

Since the invention of EDFA (erbium-doped fiber amplifiers) and the moderate insertion loss of light-wave coupling, propagation loss in diffused waveguides are no longer the limiting factor for using LiNbO$_3$ MZIM for system applications, especially for ultra-long-haul and ultra-high-speed (40 Gb/s and higher) optical fiber communications. Recently, several advanced modulation formats have been implemented and demonstrated employing single drive, dual-drive, and pairs of MZIMs. There are four typical types of modulation techniques: digital amplitude modulation as commonly known as ASK with NRZ or RZ pulse formats [14], digital differential phase modulation or DPSK or multi-level M-ary PSK, frequency modulation FSK or continuous phase FSM or MSK. The operational principles of these modulators have been described in Section 12.3. This section briefly outlines the demands of these transmitters on the integrated modulators and practical fabrication of such modulators.

12.5.1.1 Amplitude Modulation

Amplitude modulation of the lightwave carrier can be easily implemented by applying an RF signals into the travelling wave electrodes. The bias voltage is critical whether it is position at the phase quadrature, minimum or maximum transmission depending on specific applications. If analog modulation is required such as in the field of microwave photonics and photonic signal processing the MZIM is normally biased at the phase quadrature point. While if a suppression of the carrier is required, then there is a different phase shift of pi between the two arms. For single drive MZIM, then, bias at V_π while for dual drive cases, one could bias both sides with $\pi/2$ and $-\pi/2$ (or alternatively phase quadrature at $\pi/2$ and $3\pi/2$). The biasing at these two operating points can also be used to produce negative gain coefficients in photonic signal processors. For RZ format, a pulse carver is required to generate RZ pulse trains and then followed by a MZIM data modulator. CS-RZ can thence be generated if the pulse carver is biased at the minimum transmission point.

12.5.1.2 Phase Modulation

Phase modulation can offer better performance than amplitude modulation, provided that the detection of the phase of the carrier under the modulation format can be recovered with error free. Previously, coherent technique was used for these types of phase modulation using LiNbO$_3$ MZIM, but it suffers significantly as this technique requires the phase locking of the carriers and hence the line width limitation of the source would deteriorate the signals. However, this can be overcome by the use of different phase shift keying (DPSK) techniques. DPSK can be implemented using parallel MZIMs and M-ary DPSK can be implemented using only one dual drive LiNbO$_3$ MZIM.

12.5.1.3 Frequency Modulation

Optical frequency shift keying can be generated using a pair of parallel LiNbO$_3$ MZM as shown in Figure 12.38. The RF frequencies can be switched in the time domain according to the coding of the data pulses. Again NRZ or RZ formats can be added if a LiNbO$_3$ MZIM pulse carver is added in tandem to the FSK modulators. Novel continuous phase FSM and minimum shift keying (MSK) as an extension of FSK can be developed using MZIMs.

12.5.2 LiNbO$_3$ Modulator Fabrication

Most steps of the fabrication of MZIM can be implemented using standard optical lithography. Optical waveguides are fabricated by using Ti in-diffusion of about 50 nm thick at 1050°C for seven hours. The SiO$_2$ layer is then RF sputtered with thickness variable from 0.6 to 2.0 μm. Electrodes are formed by two main stages. A 20 nm chrome layer is first deposited and the electrode pattern is then created using photographic techniques. A 100 nm thick Au layer is then deposited on top of the *Cr* temporary electrode pattern. The gold layer is then etched to the pattern of the electrode structures. Electroplating of thick Au pattern is then conducted. The thickness of electroplated Au can reach 7–10 μm with a wall angle of about 12–20 degrees without shorting the electrodes. Note that prior to the fabrication of the electrode, optical channel waveguides are tested using the straight optical channels pattern next to the Mach–Zehnder structures to determine the important parameters of the electrodes as designed and simulated. Microwave coupling, fiber pigtails, and mounting are then completed.

The insertion loss S_{21} of our typical fabricated electrodes are simulated and measured as shown in Figure 12.36, which shows the measured RF insertions loss with the upper curve showing the calibrated S_{21} of the microwave set up and the lower curve showing the results for the case when the RF electrodes are included in the measured system. Figure 12.37 shows the scattering parameter S_{21} under optical measurement in which the output optical waves are detected by a wide band optical receiver and its variation with respect to the frequency of the RF waves launched into the modulation electrode. Several modulators have been fabricated. The obtained results indicate that the electrical and optical bandwidths reach a consistent range of about 26–30 GHz. Optical measurement of the packaged modulators has also been obtained and the optical transfer bandwidths are compatible with those obtained in the electrical domain. This concludes that the optical loss is minimum and a good velocity matching between the optical waves and the travelling microwaves. A packaged 26 GHz 3 dB bandwidth MZIM is shown in Figure 12.33.

12.5.3 EFFECTS OF ANGLED-WALL STRUCTURE ON RF ELECTRODES

Since the proposed dielectric optical waveguides in 1966, and thence integrated optics in 1969, the field of integrated photonics has progressed tremendously in parallel with global optical fiber-optic communication networks. Optical communication systems have reached an increasing fast pace in speed and capacity, reaching several Tb/s with 40G OC-192 to 80G OC-768 multi-wavelength transport over ultra-long transmission haul. External modulation is the most important technique for modulating lightwaves at these ultra-high-speed modulators as external modulation of lightwaves [6]. Electro-optic (EO) modulators based on LiNbO$_3$ waveguiding structures have been the most popular and feasible devices in the technological evolution of optical fiber

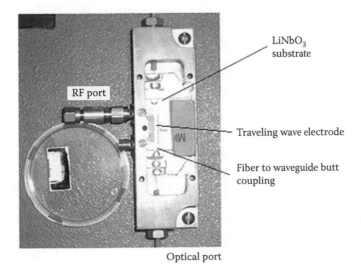

FIGURE 12.33   Fabricated and packaged optical modulator. The substrate and fabricated electrode shown on the left side of the photograph. The package 26 GHz modulator with fiber pigtail shown on the right side.

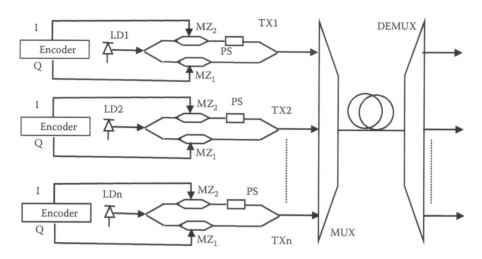

FIGURE 12.34 A complex modulation format transmitter using MZIMs for multi-carrier multiplexed optical fiber transmission system. PS = phase shifter, LD = laser diode – different wavelength, TX = transmitter, Mux = Wavelength multiplexer, DeMUX = wavelength demultiplexer.

communications after the invention of the optical amplification in the early 1990s due to its intrinsic high-speed property, its large EO coefficient, its transparency over the 1550 nm communication spectral window, zero or adjustable frequency chirp, efficient butt-coupling, and, most importantly, relatively simple fabrication technique. Recently, several advanced modulation formats such as non-return-to-zero (NRZ), amplitude shift keying (RZ-ASK), carrier-suppressed RZ (CS-RZ), ASK, and differential phase shift keying (DPSK) can be generated by employing two single-drive interfrometric modulators in tandem or a dual electrode drive device. Figure 12.34 shows a typical optical transmission system employing parallel $LiNbO_3$ MZ intensity modulator ($LiNbO_3$ MZIM) for generation of differential quadrature PSK signal. Thus, the modulators can be used as a pulse carver to generate periodic pulse shape trains, data modulation, and generation and modulation of different phase states of the lightwave carrier. Therefore, it demands the modulators having properties of low drive voltage, low insertion loss, and wideband. It also requires efficient and accurate modeling of the traveling wave electrodes, as well as a method to fabricate thick and vertical brick-wall-like electroplated electrodes, the optical channel waveguides and the interaction of the launched travelling waves in such combined EO modulation system is developed.

The modeling of ultra-broadband $LiNbO_3$ MZIM incorporating travelling wave electrodes has been briefly described in Section 12.6 as they play a crucial role in the optical bandwidth of the devices. The modulating electric field plays critical roles in the EO overlap integral. More importantly, the subtler structural factors such as the wall angle of the gold-plated electrode can be tailored. Most analyses, for example, the conformal mapping technique, the Green function method, or the method of images, either assumes an infinitely thin electrode or assumes the wall angle of the electrode to be 90°. However, when very thick electrodes, typically in the range of 10–20 μm [11], are employed to achieve ultra-broadband property (40 GHz and higher), the 90° wall angle assumption is no longer valid. After the electro-plating process, the trapezoidal shape of the electrode is illustrated in Figure 12.35b. Figure 12.35a shows the schematic plane view of the electro-optic interferometric optical modulators. The wall angle of the electrode alters significantly the effective microwave index n_m and the characteristic impedance Z. It is thus critical to take into considerations such structural defects in the design of travelling wave electrodes. The comparison of the bandwidth of the tilted electrode to that of the brick-wall-like structures is important for increasing the operational bandwidth.

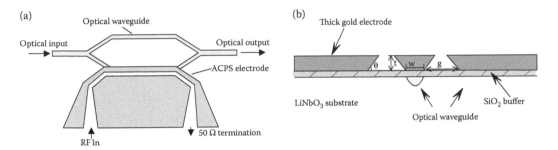

FIGURE 12.35 (a) Schematic diagram of the electro-optic interferometric modulator and (b) the cross-sectional view of wall-angle tilted thick electrodes on optical waveguides.

Our FDTWEA is used to study the effect of the electrode wall angle that is the most practical effects due to fabrication processes. In extending the bandwidth of the device, very thick electrodes of thickness ranges from 10 to 20 μm is used. If conventional Ti-Au electroplating technique is used, such thick electrodes, however, do not assume a perfect brick wall shape, but a trapezoidal shape, as shown in Figure 12.35b. Such a geometrical factor exerts a subtle effect on the values of Z and n_m. The wall-angle of tilted electrodes significantly affects the effective values of Z and n_m. The effect of the wall angle is illustrated in Figure 12.35. The bandwidth difference of the plots for different θ converges as the electrodes become thinner. For thick electrodes, especially when greater than 10 μm, the wall angle effect should not be ignored. For example, if we base our design on the assumption of a $\theta = 90°$ rectangular electrode, then we would use a 20 μm-thick electrode for the best velocity match. However, if the fabricated electrodes actually assume a trapezoidal shape with $\theta = 80°$, we would have overestimated the value of n_m for best velocity matching by about 0.2. This corresponds to almost a 20% reduction of the bandwidth that could have been achieved by a mismatch of 0.01 between the microwave index and the optical propagation velocity. Modeled results indicate that an electrode thickness of around 14 μm can be used to achieve maximum bandwidth. If the deep x-ray lithographic process is employed with a thick polymer layer (e.g., SU-8), a brick wall like polymeric slot structure can be formed. A thick gold layer can then be electroplated and an ideal electrode structure can be fabricated. If such a 20 mμ thick gold electrode can be fabricated this way to give an n_m of 2.15 that matches the optical index and hence a near infinitely wide-bandwidth modulator can be achieved.

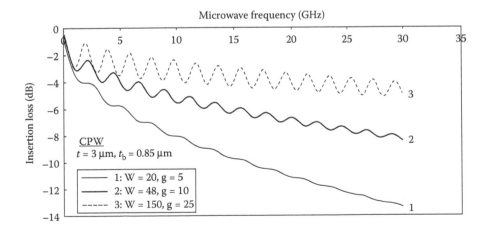

FIGURE 12.36 Insertion loss S_{21} of the RF electrode as a function of the microwave frequency.

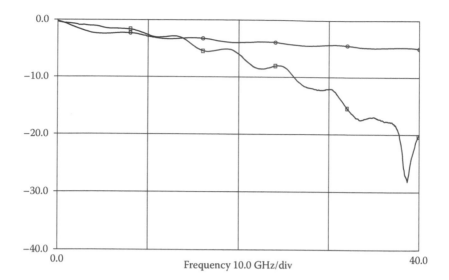

FIGURE 12.37   Electrical (S_{21})/optical response of the implemented interferometric modulator.

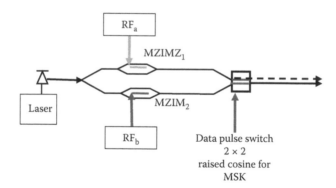

FIGURE 12.38   Schematic representation of a LiNbO$_3$ MZIM for FSK modulation.

12.5.4   REMARKS

An effective numerical model has been briefly presented for modeling of LiNbO$_3$ MZIM, especially the effects of the properties of the travelling electrodes on the EO modulation. The diffused optical waveguides can be designed with ease using the non-uniform FD (finite difference) approach. The characteristic impedance Z and the microwave effective index n_m have been developed and implemented in a 26 GHz bandwidth LiNbO$_3$ MZIM with single drive. The effects of the trapezoidal shape electrode with the wall angle θ, as assumed by most thick electrode ($>10\,\mu$m) fabricated by present electroplating technique. Fabrication technique for ultra-wideband MZIM has been described. A near 30GHz 3 dB BW has been described. Furthermore, we describe how a thick electrode can be fabricated if the deep x-ray synchrotron lithography is used so that an extremely wide-band MZIM can be produced. We have shown that the potentials of LiNbO$_3$ MZIMs in the generation of different digital modulation formats (see also Chapter 2) via the modulating of amplitude, phase, and frequency (by linear phase), as well as in negative gain coefficient photonic processors. Other potential and novel modulation formats such as the continuous phase formats MSK, CPFSK, etc. can be generated using MZIMs and are described in Section 12.3.

REFERENCES

1. M. Minakata, S. Shaito, and M. Shibata, Two dimensional distribution of refractive index changes in Ti diffused LiNbO$_3$ waveguides, *J. Appl. Phys.*, vol. 50, no. 5, pp. 3063–3067, May 1979.
2. G.K. Gopalakrishnan, Burns, W.K., and Bulmer, C.H., Electrical loss mechanisms in traveling wave LiNbO$_3$ optical modulators, *Elect. Lett.*, vol. 28, no. 2, pp. 207–208, 1992; G.K. Gopalakrishnan, and Burns W. K., Performance and modeling of broadband LiNbO$_3$ traveling wave optical intensity modulators. *J. Lightw. Tech.*, vol. 12, no. 10, pp. 1807–1818, 1994.
3. H.E. Green, The numerical solution of some important transmission-line problems, *IEEE Trans. Microw. Th. Tech.*, vol. 13, no. 5, pp. 676–692, 1965.
4. H. Chung, W. S. C. Chang, and E. L. Adler, Modeling and optimization of travelling-wave LiNbO$_3$ interferometric modulators, *IEEE J. Quant. Elect.*, vol. 27, no. 3, pp. 608–617, 1991.
5. S. Ramo and W. Whinery, *Fields and Waves in Communication Electronics*, 2nd ed, Hoboken: J. Wiley & Sons, Inc, 1973.
6. R.C. Alferness et al., Efficient single mode fiber to titanium diffused lithium niobate waveguide coupling for l = 1.32 μm, *IEEE J. Quant. Electron.*, vol. QE-18, pp. 1807–1181, 1982.
7. P.G. Suchoski and R.V Ramaswamy, Minimum mode size low loss Ti:LiNbO$_3$ channel waveguides for efficient modulator operation at 1.3 μm, *IEEE J. Quant. Electron.*, vol. QE-23, no. 10, pp. 1673–1679, Oct. 1987.
8. E. A. J. Marcatilli, Dielectric rectangular waveguide and directional couplers for integrated optics, *Bell Syst. Tech. J.*, vol. 48, pp. 2071–2102, 1969.
9. G. B. Hocker and W. K. Burns, Mode dispersion in diffused channel waveguides by the effective index method, *Appl. Optics*, vol. 16, no. 1, pp. 113–118.
10. M. J. Robertson et al., Semiconductor waveguides: analysis of optical propagation in single rib structures and directional couplers, *IEE Proceedings*, vol. 132, Pt. J, no. 6, pp. 336–342, Dec. 1985.
11. B. M. A. Rahman and J. B. Davies, Vectorial-H finite element solution of GaAs/GaAlAs rib waveguides, *IEE Proceedings-J*, vol. 132, pp. 349–335, 1985.
12. Pao-Lo Liu and Bing-Jin Li, SV beam propagation method for analysing polarised modes of rib waveguides, *IEEE J. Quant. Electron.*, vol. 28, no. 4, pp. 778–782, Apr. 1992.
13. M.S Stern, SV polarised H field solutions for dielectric waveguides with arbitrary index profiles, *IEE Proceedings*, vol. 135, Pt. J, no. 5, pp. 333–338, Oct. 1988.
14. M.S Stern, SV polarised finite difference method for optical waveguides with arbitrary index profiles, *IEE Proceedings*, vol. 135, Pt. J, no. 1, pp. 56–62, Feb. 1988.
15. M. S. Stern, Rayleigh quotient solution of SV field problems for otical waveguides with arbitrary index profiles, *IEE Proc., Optoelectron.* vol. 138, Pt J, pp. 185–190, 1990.
16. C. M. Kim and R.V Ramaswamy, Modeling of graded index channel waveguides using non-uniform finite difference method, *J. Lightwave Technol.*, vol. 7, no. 10, pp. 581–1589, Oct. 1989.
17. S. Seki et al., Two dimensional analysis of optical waveguides with a nonuniform FDM, *IEE Proc. J.*, vol. 138, no. 2, pp. 123–127, Apr. 1991.
18. R. L. Burden and J. D. Faires, *Numerical Analysis*, 4th ed., Boston: PWS-Kent Publishing Company, pp. 492–505, 1989.
19. W. H. Press et al., *Numerical Recipes-the Art of Scientific Computing*, Cambridge University Press, pp. 377–379.
20. T. C. Opp et al., *NSPCG User's Guide Version 1.0—A Package for Solving Large Sparse Linear Systems by Various Iterative Methods*, Austin: Center for Numerical Analysis, The University of Texas.
21. K. S. Chiang, Review of numerical and approximate methods for the modal analysis of general optical dielectric waveguides, *Opt. Quantum Electron.* vol. 26, pp. S113–S134.
22. S. Fouchet et al., Wavelength dispersion of Ti induced refractive index change in LiNbO$_3$ as a function of diffusion parameters, *J. Lightwave Technol.*, vol. Lt-5, no. 5, pp. 700–708, May. 1987.
23. M. Minikata et al., Precise determination of refractive index changes in Ti-diffused LiNbO$_3$ optical waveguides, *J. Appl. Phys.*, vol. 49, no. 9, pp. 4677–4682, Sep. 1978.
24. S.K. Korotky et al., Mode size and method for estimating the propagation constant of single mode Ti: LiNbO$_3$ strip waveguides, *IEEE J Quant. Electron*, vol. QE-18, no. 10, pp. 1796–1801, Oct. 1982.
25. M. R. Spiegel, *Mathematical Handbook of Formulas and Tables, Shaum's Outline Series*, New York: McGraw Hill, 1998.
26. M. Fukuma and J. Noda, Optical properties of titanium-diffused LiNbO$_3$ strip waveguides and their coupling to a fiber characteristics, *Appl. Optics*, vol. 20, no. 4, 15, pp. 591–597, Feb. 1980.

27. A. Sharma and P. Bindal, Analysis of diffused planar and channel waveguides, *IEEE J. Quant. Elect.*, vol. 29, no. 1, pp. 150–156, Jan. 1993.

28. O. G. Ramer, Integrated optic electro-optic modulator electrode analysis, *IEEE J. Quant. Elect.*, vol. QE-18, no. 3, pp. 386–392, 1982.

29. H. J. M. Belanger and Z. Jakubezyk, General analysis of electrodes in integrated optics electrooptic Devices, *IEEE J. Quant. Elect.*, vol. 27, no. 2, pp. 243–251, 1991.

30. D. Marcuse, Optimal electrode design for integrated optics modulators, *IEEE J. Quant. Elect.*, vol. QE-18, no. 3, pp. 393–398, Mar. 1982.

31. C. Sabatier and E. Caquot, Influence of a dielectric buffer layer on the field distribution in an electro-optic guided wave device, *IEEE J. Quant. Elect.*, vol. QE-22, no. 1, pp. 32–37, 1986.

32. W.-C. Chuang, W. Y. Le, J. H. Lieu, and W.-S. Wang, A comparison of the performance of $LiNbO_3$ travelling wave phase modulators with various dielectric buffer layers, *J. Opt. Commun.*, vol. 14, no. 4, pp. 142–148, 1993.; Mares, P. J. and S. L. Chuang, Modeling of self-electro-optic-effect devices, *J. Appl. Phys.*, vol. 74, no. 2, pp. 1388–1389, 15 Jul. 1993.

33. Z. Pantic and R. Mittra, Quasi-TEM analysis of microwave transmission lines by finite element method, *IEEE Trans. Microw. Th. Tech.*, vol. MTT-34, no. 11, pp. 1096–1103, 1986.

34. T. C. Oppe, *NSPCG User's Guide Version 1.0—A Package for Solving Large Sparse Linear Systems by Various Iterative Methods, Center for Numerical Analysis*, Austin, USA: The University of Texas, May 1988.

35. C. M. Kim and R. Ramaswamy, Overlap integral factors in integrated optic modulators and switches, *J. Lightw. Tech.*, vol. 7, no. 7, pp. 1063–1070, 1989.

36. E. L. Wooten and W. S. C. Chiang, Test structures for characterization of electro-optic waveguide modulators in lithium niobate, *IEEE J. Quant. Elect.*, vol. 29, no. 1, pp. 161–170, 1993.

37. L. N. Binh, $LiNbO_3$ optical modulators: Devices and applications, *J. Crystal Growth*, vol. 288, no. 1, pp. 180–187, 2006.

38. S. K. Korotky, G. Eisenstein, R. S. Tucker, J. J. Veselka, and G. Raybon, Optical intensity modulation to 40GHz using a waveguide electro-optic switch, *Appl. Phys. Lett.*, vol. 50, no. 23, pp. 1631–1633, 1987.

39. C. M. Gee, G. D. Thurmond, and H. W. Yen, 17 GHz bandwidth electro-optic modulator, *Appl. Phys. Lett.*, vol. 43, no. 11, pp. 998, 1 Dec. 1983.

40. T. Kitoh and K. Kawano, Modeling and design of $Ti:LiNbO_3$ optical modulator electrodes with a buffer layer, *Elect. Comm. in Japan, Part 2*, vol. 76, no. 1, pp. 25–34, 1993.

41. P. J. Winzer and R-J. Essiambre, Advanced modulation formats for high-capacity optical transport networks, *J. Lightwave Technol.*, vol. 24, no. 12, pp. 4711–4728.

42. I. Morita, Advanced Modulation Format for 100-Gbit/s Transmission, Paper $TuE_{3.3}$, pp. 252–253, OFC, 2008.

43. K. Petermann, Modulation formats for optical fiber transmission, *Book Chapter in Optical Communication Theory and Techniques*, Enrico Forestieri (ed.), Berlin: Springer, 2006.

44. L. N. Binh, *Digital Optical Communications*, Boca Raton: CRC Press, 2008.

45. L. N. Binh and T. L. Huynh, and K. K. Pang, Direct detection frequency discrimination optical receiver for minimum shift keying format transmission, *IEEE J. Lightwave Technol.*, vol. 26, no. 18, pp. 3234–3247, 15 Sept, 2008.

46. W. R. Leeb, A. L. Schotz, and E. Bonek, Measurement of velocity mismatch in travelling wave electro-optic modulators, *IEEE J. Quant. Elect.*, vol. QE-18, no. 1, pp. 14–16, 1982.

47. E. H. W. Chan, and R. A. Minassian, A New Optical phase modulator dynamic response measurement technique, *IEEE J. Lightwave Technol.*, vol. 26, no. 16, pp. 2882–2888, 2008.

48. M. Y. Frankel, and R. D. Esman Optical single sideband Suppression carrier for wideband signal processing, *IEEE J. Lightwave Technol.*, vol. 16, no. 5, pp. 859–863, May 1989.

13 3D Integrated Optical Waveguides for Modulators

Silicon on insulator (SOI) has now been realized as one of the most important candidates for integrated optical systems and its integration with electronic high-speed circuit. The electro-photonic integrated circuits (e-PIC) are important for high-speed transmission systems in which the bandwidth of all sub-systems, such as optical modulators and electronic driving circuits, high-speed photodetectors in association with wideband transimpedance amplifiers (TIA), and linear amplification (LA) stages require the minimization of stray capacitances and inductances due to long cable interconnection. Si-PIC offers the possibility of electronic complementary metal oxide semiconductor (CMOS) circuit integration. SiGe materials can now be grown on Si, and GE material can be employed to establish the photo-detection structure that can be directly coupled on a Si photonic waveguide. Thus, it is only possible when such e-PIC is commonly employed in optical transmissions and networks. In these SOI ePICs, 3D waveguides are the most fundamental guiding structures for guiding and directing lightwaves to and from different regions of e-PIC. Such 3D semiconductor waveguides coupled from an optical fiber can be illustrated are shown in Figure 13.1.

Following the fundamentals of planar optical waveguide described in Chapter 14, this chapter describes the 3D optical waveguides in which the waveguide is restricted in both transverse directions. A simplified analysis of these waveguides, the effective index method and numerical techniques and the finite difference method (FDM), are described and examples are given. In this chapter, we analyze the modes that are guided by 3D waveguides with rectangular geometries using mainly the Marcatilli's method, and the effective index method as analytical techniques.

On the numerical method, we select the simple finite difference method (FDM) as the principal techniques because they are simple and give accurate results for optical waveguides operating in the linear region. We chose the FDM to study the quasi-TE and quasi-TM polarized waveguide modes because of its simplicity and plausible accuracy. We have employed the semi-vectorial analysis which automatically takes full account of the discontinuities in the normal electric field components across any arbitrary distribution of internal dielectric interfaces. The eigenmode of the Helmholz equation is solved by the application of the shifted inverse power iteration method. This method warrants the mode size and its relevant propagation constant, both of which important parameters to the design of optical waveguide. The grid size is non-uniform to maximize the accuracy of the optical guided modes and their propagation constants. Diffused waveguides and rib waveguides are designed with different parameters to demonstrate the effectiveness of the method and leading to an optimum design of waveguides of optical modulation and micro-ring resonators.

13.1 INTRODUCTION

To achieve efficient design of high speed modulators and switches, especially micro ring resonators, the fabrication of rib-waveguides and Ti-LiNbO$_3$ waveguide with suitable mode size is essential to minimize waveguide insertion loss and to maximize the overlap integral between the guided optical field and the applied modulating field. Furthermore, the bending or radius of curvature is important for the ring resonator to keep the ring size as small as possible. Extensive studies have been devoted in recent decades to fabricating Ti-diffused LiNbO$_3$ waveguides which couple efficiently to single-mode fibers [1–5]. A major milestone was achieved when a total fiber-waveguide insertion loss of 1 dB was achieved for z-cut LiNbO$_3$ at 1.3 μm [4]. Such low loss

(a)   (b)   (c)

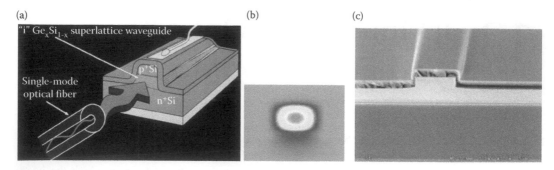

FIGURE 13.1 (a) A 3D optical waveguide coupling lightwaves from an optical fiber; (b) optical field distribution in the transverse plane of a 3D waveguide; and (c) cross-sectional view of a rib 3D waveguide, typically high contrast refractive index difference between guided region and cladding.

was achieved by choosing fabrication parameters to yield a relatively deep, clean diffusion, which simultaneously minimized the fiber waveguide mode mismatch loss and the propagation loss. Suchoski and Ramaswamy [6] have reported on the optimization of fabrication parameters to obtain Ti-LiNbO$_3$ single mode waveguides which exhibit minimum mode size and low propagation loss at 1.3 μm. All these design requirements have led to the significance of the analysis of polarized modes in channel waveguides.

In general, the optical mode of the waveguide is acquired by solving the Helmholtz equation. However, only a few simple waveguide structures can be solved analytically. Therefore, extensive attempts have been made to obtain numerical solutions for a 2D cross section of optical waveguides [7–23]. One method is the approximate modeling of 2D slab waveguide solutions successively in both directions, following the method of Marcatilli [5] or the effective-index method (EIM) [24]. However, these methods are not applicable to arbitrarily shaped optical waveguides, nor do they efficiently handle waveguide mode near the cut-off region. A significant number of numerical methods have been proposed to obtain rigorous solutions to the wave equation with pertinent boundary conditions. The popular techniques are the finite difference method (FDM) [10], finite element method (FEM) [22], and the beam propagating method (BPM) [14]. The application of different techniques based on the above methods, including semi-vectorial E-field FDM [12], semi-vectorial H-field FDM [25], and Rayleigh quotient solution [26], have been studied and reported on. These methods are applicable to arbitrarily shaped optical waveguides. In FEM and FDM, partial differential equations are discretized and then transformed into matrix equations. The calculations of mode indices and optical field distributions are then equivalent to obtaining eigenvalues and eigenfuctions of the coefficient matrices.

In this chapter, we first treat the 3D optical waveguide from an analytical point of view with the representation of two distributions of the refractive index profile by two effective planar profiles. The propagation and mode guiding conditions obtained for these planar waveguides are thus combined for the 3D waveguide.

Sections 13.2 and 13.3 describe the analytical estimation of guided mode using the Marcatilli's method and the effective index method. Section 13.4 outlines the numerical formulation of the non-uniform finite difference scheme. Quasi-TE and quasi-TM polarized modes are addressed. We also assess the accuracy of the numerical result of this scheme by computing the effective refractive index of rib and slab dielectric waveguides. The effect of grid spacing is also investigated. The effectiveness of the variable grid spacing in dealing with waveguide mode near the cut off region is also given. Sections 13.4 and 13.5 give the treatment of the 3D optical waveguides by the finite difference method (FDM) for uniform index regions, the rib waveguide and diffused index profiles, and the diffused optical channel waveguides. Section 13.4 describes the modeling of the 3D optical waveguide with graded index profile, such as the Ti-LiNbO$_3$ channel waveguides. The effects of various waveguide fabrication parameters such as the diffusion time, diffusion temperature, thickness, and width of the titanium strips are studied. The accuracy of the numerical model is assessed by comparing our simulations with experimental and simulation results that are reported in several literatures. Section 13.5 describes the modeling of rib optical waveguides using the same finite difference method.

13.2 MARCATILLI'S METHOD

The cross section of typical 3D waveguides is shown in Figure 13.2, including raised strip or channel waveguide, strip loaded, rib or ridge, and embedded structures with a substrate and an overlay region.

Usually, the raised channel waveguide is formed by depositing a thin film layer, for example, by the molecular chemical vapor deposition (MOCVD) or sputtering; then, if we remove the film material in the outer regions by some means such as dry reactive etching while keeping the film layer in the central portion intact, we have the raised stripe or channel waveguides. The ridge or rib waveguides are similar to the raised strip waveguides except that the film layer on the two sides is partially removed, as shown in Figure 13.2b. If we place a dielectric strip on the top of the film layer, as shown schematically in Figure 13.2c, we have the strip-loaded waveguides. By embedding a high-index bar in the substrate region, we have the buried or embedded strip waveguides (Figure 13.2d). Channel, ridge, strip-loaded, and buried strip waveguides are 3D waveguides with rectangular boundaries. Circular and elliptical fibers, discussed in Chapters 4 and 5, are 3D waveguides with curved boundaries. The refractive index of the 3D waveguide can vary with respect to the distance of depth. In this case, we have graded index channel waveguides such as diffused channel optical waveguides formed by diffusion of impurity into LiNbO$_3$ substrate at temperature around 1000°C. Figure 13.3a shows the plane view of the diffused waveguide by diffusing thin titanium film into LiNbO$_3$ at about 1050°C for 7 h. Figure 13.3b and c displays the field distribution in the transverse plane of the waveguide and the intensity distribution in the horizontal and vertical directions at the maximum intensity points.

In this section, we analyze the modes guided by 3D waveguides with rectangular geometries using mainly the Marcatilli's method, and the effective index method. This chapter consists of five sections.

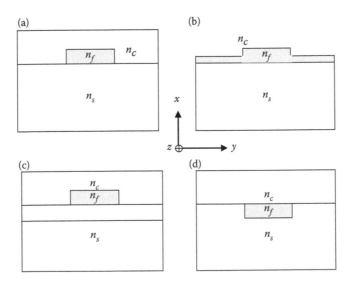

FIGURE 13.2 Rib Channel 3D optical waveguides (a) raised strip or channel, (b) ridge or rib, (c) strip-loaded, and (d) embedded channel.

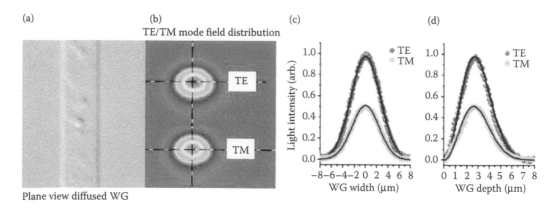

FIGURE 13.3 Diffused waveguide in LiNbO$_3$ (a) plane view of the diffused channel waveguide; (b) optical field distribution of TE and TM modes; (c) intensity distribution of TE/TM modes along the width (horizontal direction); and (d) along depth/vertical direction.

Since fields of 3D waveguides are complicated and difficult to analyze, we begin with a qualitative description.

13.2.1 FIELD AND MODES GUIDED IN RECTANGULAR OPTICAL WAVEGUIDES

13.2.1.1 Mode Fields of H_x Modes

In 2D waveguides, one of the dimensions transverse to the direction of propagation is very large in comparison to the operating wavelength. This is the y-direction in Figure 13.2. The waveguide width in this direction is treated as infinitely large. As a result, fields guided by 2D dielectric waveguides can be classified as transverse electric (TE) or transverse magnetic (TM) modes. For TE modes, the longitudinal electric field component, E_z, is zero, and all other field components can be expressed in terms of H_z. For TM modes, H_z vanishes and all other field components can be expressed in terms of E_z. In 3D optical waveguides, the waveguide width and height are comparable to the operating wavelength. Neither the width nor height can be treated as infinitely large. Thus, neither E_z nor H_z vanishes, except for in some special cases. As a result, modes guided by 3D optical waveguides are neither TE nor TM modes, except for the special cases. In general, they are hybrid modes. A complicated scheme is needed to designate the hybrid modes. Because all field components are present, the analysis for hybrid modes is very complicated. Intensive numerical computations are often required [27].

In many dielectric waveguide structures, the index difference is small. As a result, one of the transverse electric field components is much stronger than the other transverse electric field component. Goell has suggested a physically intuitive scheme to describe hybrid modes [28]. In Goell's scheme, a hybrid mode is labeled by the direction and distribution of the strong transverse electric field component. If the dominant electric field component is in the x (or y) direction, and if the electric field distribution has $p-1$ nulls in the x-direction and $q-1$ nulls in the y-direction, then the hybrid mode is identified as $E_{x,pq}$ (or $E_{y,pq}$) mode. The superscript denotes the direction of the dominant transverse electric mode.

Consider a weakly guiding rectangular optical waveguide with a core of index n1 and surrounded with lower indices n_j with $j = 2, 3, 4,$ and 5. The waveguide cross-section is shown in Figure 13.4.

The rectangular waveguide can be considered to be equivalent to two slab waveguides, one extended in the x-direction and one in the y-direction. That means that the field is confined as a

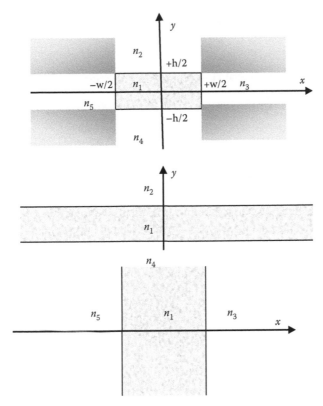

FIGURE 13.4 Model used to analyze E_y modes of a (a) rectangular waveguide, (b) waveguide H, and (c) waveguide W.

mode in the y-direction and the other in the x-direction. This is normally called the hybrid mode. Thus, we can write the field component H_x in the five regions as portioned in Figure 13.4 as follows:

$$H_{x1} = C_1 \cos(\kappa_{x1}x + \phi_{x1}) \cos(\kappa_{y1}y + \phi_{y1})e^{-j\beta z}; \ region 1$$

$$H_{x2} = C_2 \cos(\kappa_{x2}x + \phi_{x2})e^{-j\kappa_{y2}y}e^{-j\beta z}; \ region 2$$

$$H_{x3} = C_3 e^{-j\kappa_{x3}x} \cos(\kappa_{y3}y + \phi_{y3})e^{-j\beta z}; \ region 3 \qquad (13.1)$$

$$H_{x4} = C_4 e^{-j\kappa_{y4}y} \cos(\kappa_{x4}x + \phi_{x4})e^{-j\beta z}; \ region 4$$

$$H_{x5} = C_5 e^{-j\kappa_{x5}x} \cos(\kappa_{y5}y + \phi_{y5})e^{-j\beta z}; \ region 5$$

where C_j, ϕ_{xj}, ϕ_{yj} are the constants to be determined using the boundary conditions, κ_{xj}, κ_{yj} are the propagation constants effective in the x and y transverse directions, respectively. For each region, the propagation constants in the x-, y-, and z-directions. κ_{xj}, κ_{yj}, β must satisfy

$$\kappa_{xj}^2 + \kappa_{yj}^2 + \beta^2 = k^2 n_j^2; \quad j = 1, 2, 3, 4, 5 \qquad (13.2)$$

There are no additional constraints on the transverse propagation constant. In fact, the transverse propagation constant in the regions 2–5 are imaginary, that is, the fields must decay to zero in these regions except for in the rectangular section.

When expressed in terms of κ_{xj}, κ_{yj}, ϕ_x, ϕ_y, Equation 13.1 can be simplified to

$$H_{x1} = C_1 \cos(\kappa_{x1}x + \phi_x)\cos(\kappa_{y1}y + \phi_y)e^{-j\beta z}; \quad region\_1$$

$$H_{x2} = C_2 \cos(\kappa_{x2}x + \phi_x)e^{-j\kappa_{y2}y}e^{-j\beta z}; \quad region\_2$$

$$H_{x3} = C_3 e^{-j\kappa_x x}\cos(\kappa_y y + \phi_y)e^{-j\beta z}; \quad region\_3 \tag{13.3}$$

$$H_{x4} = C_4 e^{-j\kappa_y y}\cos(\kappa_x x + \phi_x)e^{-j\beta z}; \quad region\_4$$

$$H_{x5} = C_5 e^{-j\kappa_x x}\cos(\kappa_y y + \phi_y)e^{-j\beta z}; \quad region\_5$$

Boundary Conditions at the Interfaces

Horizontal Boundary $y = \pm h/2; |x| < w/2$

Along the horizontal boundaries, the tangential components are E_x, E_z, H_x, H_z, and the x-components are ignored, as their amplitudes are extremely small compared to other components. Under the operating condition of the Maxwell equation, we can observe that

- E_z is continuous at the boundary and tangential, implying that $(1/n_j^2)(\partial H_x/\partial y)$.
- Being tangential to the horizontal lines, H_x must be continuous everywhere along the horizontal lines. Thence, the tangential derivative $\partial H_x/\partial x$ and, therefore, H_z, must also be continuous on the horizontal lines. In other words, if H_x is continuous at the horizontal lines, so is the same for H_z.

Thus, all the boundary conditions are met if we have the continuity of the term $(1/n_j^2)(\partial H_x/\partial y)$.

Vertical Boundary $x = \pm w/2; |y| < h/2$

Along this boundary, the tangential components are in the y- and z-direction and the normal direction is x. Only the components E_y and H_x are significant, and E_y is continuous if H_x is continuous. Applying these conditions for the field components at $x = \pm w/2$, we obtain:

$$E_{z1} - E_{z3} = \frac{j\eta_0}{k}\left(\frac{1}{n_1^2}\frac{\partial H_{x1}}{\partial y} - \frac{1}{n_3^2}\frac{\partial H_{x3}}{\partial y}\right) + O(\partial^2)$$

$$= \frac{j\eta_0}{k}\frac{1}{n_1^2}\left(\frac{\partial(H_{x1} - H_{x3})}{\partial y}\right) - \frac{j\eta_0}{n_3}\frac{n_1^2 - n_3^2}{n_1^2}\frac{1}{kn_3}\frac{\partial(H_{x3})}{\partial y} \tag{13.4}$$

The second term of Equation 13.4 can be ignored because of the very small difference in the refractive index terms. Thus, it can be written as

$$E_{z1} - E_{z3} = \frac{j\eta_0}{k}\frac{1}{n_1^2}\left(\frac{\partial(H_{x1} - H_{x3})}{\partial y}\right) + O(\partial^2) \tag{13.5}$$

In other words, the components E_z is continuous if H_x is continuous there.

Transverse Vectors κ_x, κ_y

The transverse momentum vector κ_x can now be determined from the boundary conditions discussed above. One would seek an oscillating behavior of the waves in the waveguide region and exponentially decay to zero in the cladding regions. At $y = \pm h/2$, the continuity of H_x and $(1/n_j^2)(\partial H_x/\partial y)$ leads to

$$C_1 \cos\left(\frac{1}{2}\kappa_y h + \phi_y\right) = C_2 e^{-j\kappa_{y2}h/2}$$

$$-\frac{\kappa_y}{n_1^2}C_1 \sin\left(\frac{1}{2}\kappa_y h + \phi_y\right) = -\frac{j\kappa_{y2}}{n_2^2}C_2 e^{-j\kappa_{y2}h/2} \tag{13.6}$$

Combining these equations we obtain the relation

$$\tan\left(\frac{1}{2}\kappa_y h + \phi_y\right) = -\frac{j\kappa_{y2}n_1^2}{\kappa_y n_2^2} \tag{13.7}$$

From Equation 13.2 we can deduce that

$$j\kappa_{y2} = \sqrt{k^2(n_1^2 - n_2^2) - \kappa_y^2} \tag{13.8}$$

Thus, Equation 13.7 becomes

$$\tan\left(\frac{1}{2}\kappa_y h + \phi_y\right) = \frac{\sqrt{k(n_1^2 - n_2^2) - \kappa_y^2}}{\kappa_y n_2^2} \tag{13.9}$$

Or, alternatively, we have

$$\frac{1}{2}\kappa_y h + \phi_y = q'\pi + \tan^{-1}\left(\frac{\sqrt{k^2(n_1^2 - n_2^2) - \kappa_y^2}}{\kappa_y n_2^2}\right)$$

$$\frac{1}{2}\kappa_y h + \phi_y = q''\pi + \tan^{-1}\left(\frac{\sqrt{k^2(n_1^2 - n_4^2) - \kappa_y^2}}{\kappa_y n_4^2}\right); \quad at \ldots y = -h/2 \tag{13.10}$$

Where q' and q'' and q are integers. Then, eliminating ϕ_y, we can rewrite as

$$\kappa_y h_y = q\pi + \tan^{-1}\left(\frac{\sqrt{k^2(n_1^2 - n_2^2) - \kappa_y^2}}{\kappa_y n_2^2}\right) + \tan^{-1}\left(\frac{\sqrt{k^2(n_1^2 - n_4^2) - \kappa_y^2}}{\kappa_y n_4^2}\right) \tag{13.11}$$

This is the dispersion relation for the TM modes guided in the channel waveguide and also similar to that for a planar waveguide. The two terms on the RHS of Equation 13.11 represent the phase shift, normally called the Goos-Hanchen shift for the "rays" penetrating into the cladding of the guided fields. Thus, similar to this boundary condition and the dispersion relationship, the dispersion characteristics for the transverse vector κ_y can be written as

$$\kappa_x w = p\pi + \tan^{-1}\left(\frac{\sqrt{k^2(n_1^2 - n_3^2) - \kappa_x^2}}{\kappa_x}\right) + \tan^{-1}\left(\frac{\sqrt{k^2(n_1^2 - n_5^2) - \kappa_x^2}}{\kappa_x}\right) \tag{13.12}$$

with p is an integer.

13.2.1.2 Mode Fields of E_y Modes

Like the analysis given for the H_x modes, the E_x modes can be found with the dispersion relation by using the continuity properties of the field components $H_y;(\partial H_y/\partial y)$. We then obtain

$$\kappa_y h = q\pi + \tan^{-1}\left(\frac{\sqrt{k^2\left(n_1^2 - n_2^2\right) - \kappa_y^2}}{\kappa_y}\right) + \tan^{-1}\left(\frac{\sqrt{k^2\left(n_1^2 - n_4^2\right) - \kappa_y^2}}{\kappa_y}\right) \qquad (13.13)$$

$$\kappa_x w = p\pi + \tan^{-1}\left(\frac{n_1^2\sqrt{k^2\left(n_1^2 - n_3^2\right) - \kappa_x^2}}{\kappa_x n_3^2}\right) + \tan^{-1}\left(\frac{n_1^2\sqrt{k^2\left(n_1^2 - n_5^2\right) - \kappa_x^2}}{\kappa_x n_5^2}\right) \qquad (13.14)$$

Equations 13.13 and 13.14 specify the dispersion relationship for the TM modes with a planar waveguide thickness of W.

Thus, the Marcatilli's method is modeled for two equivalent planar waveguides in the horizontal and vertical directions. It corresponds to the dispersion relation (13.11) and (13.12) for TM modes guided by planar waveguide of thickness W. The dominant electric field of E_x modes is in parallel with the horizontal boundaries. Thus, we use the dispersion equation of TE modes guided by waveguide H to determine κ_y. The dominant electric field component of E_x modes is perpendicular to the vertical boundaries of waveguide W. Therefore, we use the dispersion for TM modes guided by the 2D waveguide to evaluate κ_x. With κ_x, κ_y known, the propagation constant can be determined from Equation 13.2.

13.2.1.3 Dispersion Characteristics

As an example, we consider a dielectric bar of index n_1 immersed in a medium with index n_2 as shown in Figure 13.5 with uniform refractive indices in the regions surrounding the channel waveguiding region. To facilitate comparison, we define the normalized frequency parameter V and the normalized guide propagation constant index b, or normalized propagation constant in terms of n_1, n_2, h.

$$V = kh\sqrt{n_1^2 - n_2^2} \simeq \frac{2\pi}{\lambda}hn\sqrt{2\Delta}$$

$$b = \frac{\beta^2 - k^2 n_2^2}{k^2\left(n_1^2 - n_2^2\right)} \qquad (13.15)$$

Thus, the normalized effective refractive index can be evaluated as function of the normalized frequency parameter V to give the dispersion curves as shown in Figure 13.6 in which the curves obtained from finite element method and the Marcatilli methods are also contrasted with agreement.

A numerical evaluation for silica doped with a GeO_2 waveguide and cladding region is pure silica. The relative refractive index of the core and the pure silica cladding is 0.3% or 0.5%. Using the single mode operation given in Figure 13.6, we can select $V = 1$ and using (13.15), then the cross section of the rectangular waveguide is $3 \times 3\ \mu m^2$ for 0.5% relative refractive index and for 0.3% the dimension is 6×6 micron^2, and the refractive index of pure silica is 1.448 for operating wavelength of 1550 nm.

In summary, the procedures for the design of a waveguide, planar or channel structure, for supporting single or multi-modes of the E or H fields are

- Based on the dispersion curves of the E- and H-field modes, that is, the curves representing the V-parameter and the normalized propagation index, determine the desired number of modes in either polarization, then go to the curve getting the corresponding normalized

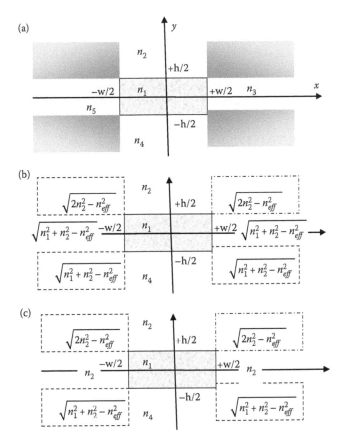

FIGURE 13.5 An embedded channel optical waveguide (a) waveguide structure; and its representation using (b) effective index method and (c) model of the waveguide using Marcatilli method.

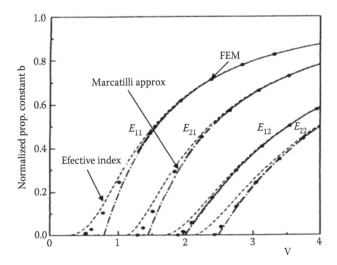

FIGURE 13.6 Dispersion characteristics, dependence of the normalized propagation constant of the guided modes as a function of the parameter V, the normalized frequency: comparison of three numerical, analytical methods for rectangular optical waveguides consisting of uniform core and cladding.

propagation constant index. Thence, the effective refractive index of the guided modes can be determined from this index and hence the expected propagation time of the guided mode over the length of the waveguide. This is important for the design of the multi-mode interference waveguide.

- Continue for other polarization direction, and then combine the two guided solutions to obtain the approximated analytical values for the combined mode.

13.3 EFFECTIVE INDEX METHOD

13.3.1 GENERAL CONSIDERATIONS

Like the Marcatili method discussed in the last section, the effective index method is also an approximate method for analyzing rectangular waveguides. In the Marcatili method, a 3D waveguide (see Figure 13.2) is replaced by two 2D waveguides: waveguides H and W depicted in Figure 13.5. The two 2D waveguides are mutually independent in that the waveguide parameters of the two 2D waveguides come directly from the original 3D waveguide.

To provide a theoretical basis for the effective index method, in lieu of the original 3D waveguide, we consider a pseudowaveguide that can be resolved into waveguides I and II or I' and II'. The pseudowaveguide is chosen such that waveguides I and II, or I' and II', can be easily identified and analyzed. The dispersion of waveguide II or II' is used as an approximation for β of the original 3D waveguide. The structures of these waveguides are shown in Figure 13.7.

Consider the $\underline{E}_y$ modes guided by a 3D waveguide shown in Figure 13.7a. As discussed above, all field components of $\underline{E}_y$ modes can be expressed in terms of H_x, which can be written as $h_x(x, y)e^{-j\beta z}$, where $h_x(x,y)$ is the field distribution in the transverse plane and $n(x,y) = n_j; j = 1-5$. The wave equation in the transverse plane can thus be obtained as

$$\left[\frac{\partial^2}{\partial x^2} + \frac{\partial^2}{\partial y^2} + k^2 n^2(x, y) - \beta^2\right] h_x(x, y) = 0 \tag{13.16}$$

Instead of considering the 3D waveguide problem, we can modify the refractive index distribution so that a pseudowaveguide structure can be obtained as shown in Figure 13.7. The refractive index of the pseudowaveguide can be written as

$$n_{ps}^2 = n_x^2(x) + n_y^2(y) \tag{13.17}$$

Then one can determine $n(x, y)$ by the common method of separation of variables, that is, the distribution $h_x(x, y)$ can be represented as the product of two distribution functions $X(x)$ and $Y(y)$, and the wave equation in the transverse plane can thus be written as

$$\frac{1}{X(x)}\frac{\partial^2 X(x)}{\partial x^2} + \frac{1}{Y(y)}\frac{\partial^2 Y(y)}{\partial y^2} + \left[k^2 n_x^2(x) + k^2 n_y^2(y) - \beta^2\right] h_x(x, y) = 0$$

or $\tag{13.18}$

$$\frac{1}{Y(y)}\frac{\partial^2 Y(y)}{\partial y^2} + k^2 n_y^2(y) = -\frac{1}{X(x)}\frac{\partial^2 X(x)}{\partial x^2} - k^2 n_x^2(x) - \beta^2$$

This equation is physically possible when the two sides equates to nil. So we have

$$\frac{1}{Y(y)}\frac{\partial^2 Y(y)}{\partial y^2} + k^2\left[n_y^2(y) - n_{eff}^2\right] = 0$$

$$-\frac{1}{X(x)}\frac{\partial^2 X(x)}{\partial x^2} - k^2\left[n_x^2(x) + n_{eff}^2\right] - \beta^2 = 0 \tag{13.19}$$

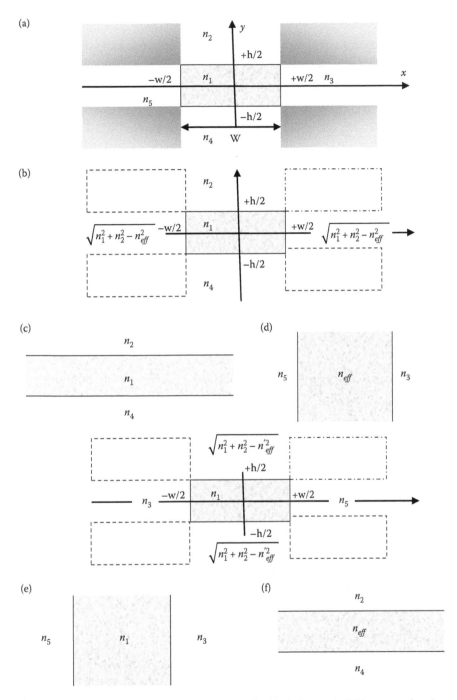

FIGURE 13.7 Model of pseudowaveguides used in the effective index method (a) rectangular channel waveguide, (b) pseudowaveguide, (c) waveguide I, (d) waveguide II, (e) alternate pseudowaveguide, (f) waveguide I', and (g) waveguide II'.

These two equations can be solved subject to the boundary conditions to derive the propagation constant along the z-direction of the waveguide. Thus, the complete solution is the product of the two functions $X(x)$ and $Y(y)$ and the phase term representing the propagation of the field along the z-direction.

13.3.2 A Pseudowaveguide

Consider the waveguide structure shown in Figure 13.7. The refractive index distribution of the channel waveguide core and cladding is shown and given as

$$n_{ps}^2(x,y)=\begin{cases} n_1^2; \quad \ldots region \ldots 1 \\ n_2^2; \quad \ldots region \ldots 2 \\ n_1^2+n_3^2-n_{eff}^2; \quad \ldots region \ldots 3 \\ n_4^2; \quad \ldots region..4 \\ n_1^2+n_5^2-n_{eff}^2; \quad \ldots region \ldots 5 \end{cases} \tag{13.20}$$

This distribution can be considered as the superposition of two distributed functions:

$$\begin{aligned} n_y^2(y) &= \begin{cases} n_2^2; \quad \ldots region \ldots y > h/2 \\ n_1^2; \quad \ldots region \ldots -h/2 \le y \le h/2 \\ n_4^2; \quad \ldots region \ldots y < -h/2 \end{cases} \\ n_x^2(x) &= \begin{cases} n_3^2 - n_{eff}^2; \quad \ldots region \ldots x > w/2 \\ 0; \quad \ldots region \ldots -w/2 \le x \le w/2 \\ n_5^2 - n_{eff}^2; \quad \ldots region..x < -w/2 \end{cases} \end{aligned} \tag{13.21}$$

Similar to the method obtained in the Marcatilli's method, the dispersion relation can be obtained as

$$kh\sqrt{n_1^2-n_{eff}^2}=q\pi+\tan^{-1}\left(\frac{n_1^2}{n_2^2}\frac{\sqrt{n_{eff}^2-n_2^2}}{\sqrt{-n_{eff}^2+n_1^2}}\right)+\tan^{-1}\left(\frac{n_1^2}{n_4^2}\frac{\sqrt{n_{eff}^2-n_4^2}}{\sqrt{-n_{eff}^2+n_1^2}}\right) \tag{13.22}$$

$$kw\sqrt{n_{eff}^2-N}=p\pi+\tan^{-1}\left(\frac{\sqrt{N^2-n_3^2}}{\sqrt{-n_{eff}^2+N^2}}\right)+\tan^{-1}\left(\frac{\sqrt{N^2-n_5^2}}{\sqrt{n_{eff}^2-N^2}}\right) \tag{13.23}$$

Using these dispersion relations, the dispersion characteristics of an embedded channel waveguide with cladding, as shown in Figure 13.5, can be obtained very close to that given in Figure 13.6 as the dashed curves.

13.4 MODE MODELING OF RIB WAVEGUIDES

In every finite difference approach, a few approximations are made and will therefore introduce some error into the final result. The following are a few approximations that are likely to introduce some error in our calculation: (i) the approximation of the full vectorial wave equation by the semi-vectorial one; (ii) the replacement of the differential equation with the difference equations; (iii) discretisation error; (iv) round off error; and (v) the error that are introduced by the NSPCG numerical solver itself.

To assess the accuracy, capability, and limitation of our program, we have calculated fundamental mode indices of three well-known rib waveguide that are often used as a waveguide modeling benchmark. Results of polarized modes have been published [10–22]. The geometry of the rib waveguide is shown in Figure 13.8. Parameters include width of the rib w, height of the rib h, thickness of the guiding layer underneath the rib d, index of the substrate n_s, and index of

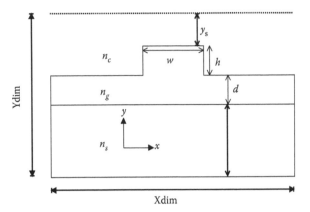

FIGURE 13.8   Typical structure of rib waveguide.

the guiding layer n_g, which are listed in Table 13.1. The refractive index of the air cladding region, n_c is unity.

The three waveguides each have a different characteristic. The first structure has relatively large vertical refractive index steps ($\Delta n = 2.44$ and 0.1) which could, for example, correspond to a GaAs guiding layer bound by air and a $Ga_{0.75}Al_{0.25}$ as a confining layer. In the lateral direction, the rib height is large and the width narrow. This structure, with strong light confinement in lateral and vertical directions, is useful for curved guides, as radiation loss is minimized. This structure does not allow the application of Effective Index Method because the slab outside the rib is cut off.

The second structure shows a weakly guiding feature. In this case, the rib height is much less, allowing the mode to extend laterally. This is particularly useful for directional coupler structures, as strong coupling between adjacent guides will result in short coupling lengths. The guiding layer thickness is made small to give a thin mode shape in the vertical direction, and thus low voltage operation. Essentially, this structure is tightly confined vertically and weakly confined horizontally. Such features enable the application of Effective Index Method [1,16,24] because the small etch step and large width to height ratio are the conditions of validity of this approximate method.

The third structure gives a good coupling to an optical fiber. Insertion loss is a crucial parameter for most waveguide devices, and is determined by propagation loss and losses due to mode mismatch. Fresnel reflection loss is also important, but can be reduced to insignificant levels by using $\lambda/4$ anti-reflection coatings. Mode profiles of a circularly symmetric optical fiber and a waveguide will, in general, be different, because of the differing refractive indices of the semiconductor and the fiber, as well as the differing shapes of the modes. The effects of both factors may be alleviated using appropriate waveguide designs. In structure three, the guiding layer is relatively thick, and the stripe width and height are adjusted to give a more symmetric mode shape. In this structure, the slab mode is near cut-off. Again, it is also to be pointed out that, because the rib height is nearly twice the slab thickness and the rib width is less than the rib height, the accuracy of the effective index method is expected to be poor. Figures 13.9 through 13.12 are the

TABLE 13.1

Parameters of Rib Waveguide for Calculation Benchmark

Guide	n_g	n_s	d (μm)	h (μm)	w (μm)
1	44	34	0.2	1.1	2
2	44	36	0.9	0.1	3
3	44	435	5	2.5	4

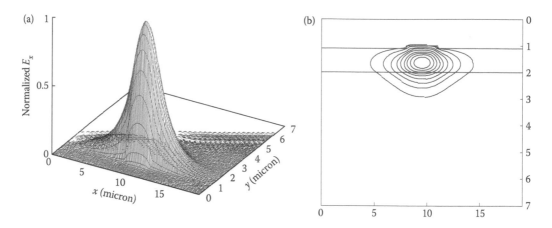

FIGURE 13.9 (a) 3D plot of TE polarized mode profile for waveguide structure with low rib. (b) Contour plot of TE polarized mode profile.

contour plot and 3D plot of the TE polarized mode of the three waveguide structure calculated by the SVMM program. Figure 13.13 shows the refractive index variation with simulation number obtained with converged solution.

The grid size h_x and h_y are 0.1. Since we assume that the field value around the computational boundary is zero, it would mean that we require a much larger computational window for both structures two and three so that the assumption would be valid. This, however, would means that we can use a coarser grid leading to reduction in computing accuracy, or maintain the grid size but face up with a huge eigenmatrix to solve. For that reason, the variable grid size comes in handy. We can avoid severe storage penalty by judiciously placing the denser mesh around the area the higher field value are assume and coarser mesh at region of a much lower field value. This would allow us to extend the boundary of the computation without incurring severe storage problem while preserving the accuracy of the computation. The choice of grid size and its influence on the accuracy of the final results would be discussed shall be illustrated in the next section.

13.4.1 CHOICE OF GRID SIZE

A judicious choice of grid size is rather to produce a plausible simulation result. To assess the effect of grid size on the accuracy of our simulation program, we compute the effective index for the TE

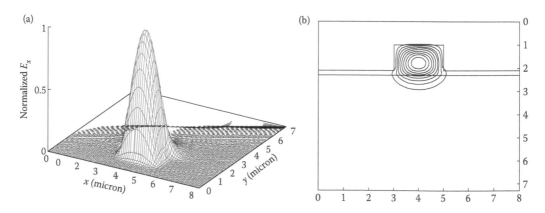

FIGURE 13.10 (a) 3D plot of TE polarized mode profile for Waveguide Structure 1. (b) Contour plot of TE polarized mode profile for Waveguide Structure 1.

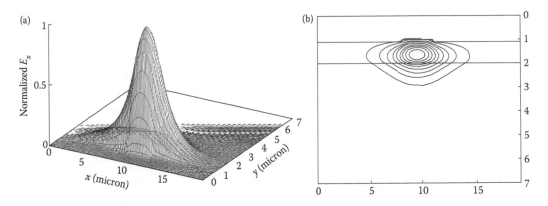

FIGURE 13.11 (a) 3D plot of TE polarized mode profile for Waveguide Structure 2. (b) Contour plot of TE polarized mode profile for Waveguide Structure 2.

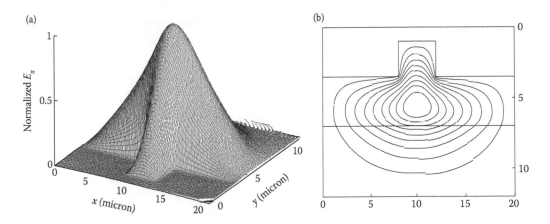

FIGURE 13.12 (a) 3D plot of TE polarized mode profile for Waveguide Structure 3. (b) Contour plot of TE polarized mode profile for Waveguide Structure 3.

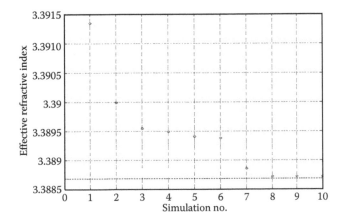

FIGURE 13.13 Refractive index variation with simulation number to obtain converged solution.

polarized mode of structure one by varying the grid size in x- and y-directions, namely h_x and h_y. We compare our result with the one simulated by P. Lusse et al. [11], which uses a dense mesh of 508×394 mesh points their Full Vectorial Finite Difference Method.

In simulations one through six, the value of $h_y = 0.1$ is kept constant while reducing h_x from 0.5 down to 0.025. As we can see, as h_x reaches 0.025, we can no longer get a significant improvement on the accuracy. Further reduction of grid size down to 0.01 would be highly impractical because we would end up with 800 grid points along the x-direction, thus paying a high penalty in terms of computer memory. In simulations seven through nine, we keep h_x at 0.025 while reducing h_y from 0.1 down to 0.025, another significant improvement in accuracy is shown and the results get very close to the one simulated by Lusse et al. [11] with both h_x and h_y equal to 0.025, a grid size of 320×292, the difference of our calculated effective index with that of Reference 11 is 2.78×10^{-5}. Simulation 10 and 11 shows how the non-uniform scheme could economize storage usage while preserving the desired accuracy. By placing denser grid mesh around the region where higher field values would be assumed and coarser mesh for region further away, we manage to reduce our mesh size from 320×292 down to 240×226 (a total reduction of 39200 points) without significant loss in accuracy, as can be seen from the graph. The non-uniform grid allocation scheme has, in this case, shown its usefulness. (It is to be repeated that each reduction of grid size need to be multiplied by 26 for that is that is the amount of workspace required by the coefficient matrix, eigenvector and the NSPCG numerical solver). Table 13.2 tabulates the effective index calculated with different grid sizes. Further Table 13.3 shows the comparisons of effective indices and normalized indices at 1550 nm wavelength.

TABLE 13.2
Calculation of Effective Index with Different Choice of Grid Size

Sim #	h_x (µm)	h_y (µm)	Xdim (µm)	Ydim (µm)	Total Grid	Effective Index
1	0.5	0.1	8.0	7.3	16×73	3913474
2	0.25	0.1	8.0	7.3	32×73	3899896
3	0.125	0.1	8.0	7.3	64×73	3895512
4	0.1	0.1	8.0	7.3	80×73	3894906
5	0.05	0.1	8.0	7.3	160×73	3894048
6	0.025	0.1	8.0	7.3	320×73	3893836
7	0.025	0.05	8.0	7.3	320×146	3888583
8	0.025	0.025	8.0	7.3	320×292	3887148
9	0.0–2.0 : 0.1 2.0–2.5 : 0.05 2.5–0 : 0.025 0–4.0 : 0.05 4.0–5.5 : 0.025 5.5–6.0 : 0.05 6.0–8.0 : 0.1	0.0025	8.0	7.3	240×292	3887162
10	0.0–2.0 : 0.1 2.0–2.5 : 0.05 2.5–0 : 0.025 0–4.0 : 0.05 4.0–5.5 : 0.025 5.5–6.0 : 0.05 6.0–8.0 : 0.1	0.0–4.0 : 0.025 4.0–7.3 : 0.05	8.0	7.3	240×226	3887165
P. Lusse	–	–	–	–	508×394	88687

TABLE 13.3

Comparisons of Effective Indices and Normalized Indices at $\lambda = 1.55$

Methods	Guide 1		Guide 2		Guide 3	
	n_{eff}	b	n_{eff}	b	n_{eff}	b
a) TE Polarised Mode						
SVMM	3887148	0.4835	3953612	0.4391	4368918	0.3782
Sv-BPM [14]	388711	0.4834	395471	0.4405	436805	0.3608
Helmholz [15]	388764	0.4839	395560	0.4416	436808	0.3614
SI [19]	38874	0.4837	39506	0.4354	43688	0.3759
SV [25]	3869266	0.4656	3954	0.4401	4368112	0.3621
FD [26]	3882623	0.4789	3952147	0.4373	436804	0.3611
b) TM Polarized Mode						
SVMM	3879173	0.4755	390647	0.3803	4368434	0.3685
Sv-BPM [14]	387924	0.4756	390693	0.3809	436772	0.3543
Helmholz [15]	387990	0.4762	390712	0.3811	346772	0.3543
SI [19]	38788	0.4752	39032	0.3763	43684	0.3669
SV [25]	3867447	0.4638	3905927	0.3796	4367719	0.3542
FD [26]	3875430	0.4718	3905701	0.3794	4367751	0.3549

13.4.2 NUMERICAL RESULTS

The following tables show the values of the propagation constants of TE and TM polarized mode for all three waveguides. The results are compared with several published results. We can see from the tables that our results compare favorably with all the other published results.

The numerical results presented so far have indicated the order of accuracy of the SVMM programs. We can observe that the exemplar results are compared well with other published results.

13.4.3 HIGHER ORDER MODES

Earlier in our discussion, we indicated that the Inverse Power Method can be used to work out the other eigenmodes of the waveguide. To illustrate that, we simulate the waveguide mode of the waveguide structure published by Rahman and Davies [22]. Table 13.4 outlines the parameters of the waveguide structure. Figures 13.14 and 13.15 show the fundamental mode and the leading asymmetric mode of the TE polarised field.

The leading asymmetric mode of Figure 13.15 can be obtained with an initial eigenvalue that is close to the eigenvalue of the leading asymmetric mode. One way to acquire a good initial guess for an independent eigenvalue is by perturbing the last few significant digits of the last calculated eigenvalue. In our case, the eigenvalue of the fundamental mode (see Figure 13.14) was calculated to be 347.78889. We proceed to the calculation of the asymmetric mode with an initial guess of 346. Other eigenmode can also be worked out in a similar fashion. However, we need to remember

TABLE 13.4

Parameters of Rib Waveguide of Reference 22 ($\lambda = 1.15\,\mu m$)

Guide	n_g	n_s	d (μm)	h (μm)	w (μm)
Reference 22	44	40	0.5	0.5	3

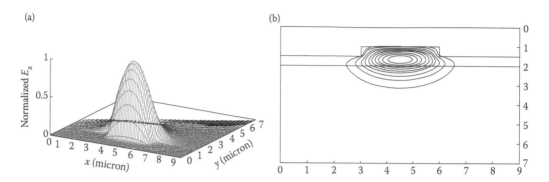

FIGURE 13.14   (a) 3D plot of fundamental mode of waveguide from Rahman and Davies, *IEE Proceedings-J*, vol. 132, pp. 349–335. (b) Contour plot of the fundamental mode of the waveguide from Rahman and Davies, *IEE Proceedings-J*, vol. 132, pp. 349–335.

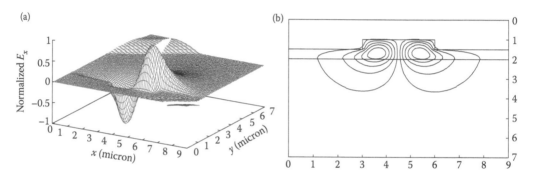

FIGURE 13.15   (a) 3D plot for the leading asymmetric mode of waveguide in from Rahman and Davies, *IEE Proceedings-J*, vol. 132, pp. 349–335. Calculated effective index = 4025302. (b) Contour plot of leading asymmetric mode of waveguide from Rahman and Davies, *IEE Proceedings-J*, vol. 132, pp. 349–335.

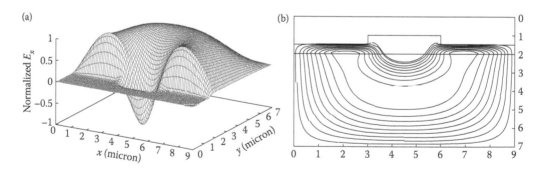

FIGURE 13.16   (a) 3D plot of the third-order mode that is not supported by the waveguide structure. Calculated effective index = 3980958, normalized index = −0.047314. (b) Contour plot of radiated mode.

that there is only a limited number of eigenmode that supported by certain waveguide structure. A good indication that the particular eigenmode is physically not feasible is an effective index which is lower than that of the refractive index of the substrate, thus giving a negative value of the normalized index. This is illustrated in Figure 13.16.

The third-order mode distribution depicted in Figure 13.16 is acquired by further reducing the initial guess of the eigenvalues from 346 to 345. As a result, we get an effective index of 398, which is lower that the refractive index of the substrate which is 40 in this case. This results in a normalized index b of -0.047. As shown in the contour plot, most of the field is radiated into the substrate of the waveguide.

This feature of SVMM that enables us to work out the higher order modes is extremely important to find out if the designed waveguide can support multi-mode operation. We will see in the next section how such a feature can be exploited in the design of single mode waveguide.

13.5 REMARKS

In this chapter, the simplified approach for analytical study of channel waveguide, the 3D version, is described using Marcatilli's method and effective index techniques. Simplified analytical dispersion relations have been obtained for these 3D waveguides. An example design of GeO_2 doped core rectangular channel waveguide is given.

Furthermore, we have successfully developed numerical techniques based on a Semivectorial Finite Difference analysis to solve the Helmholz equation. The numerical model that we have formulated can accurately and effectively model the guided modes in optical waveguides of arbitrary index profile distribution. A non-uniform mesh allocation scheme is employed in the formulation of the difference equations to free more computer memory for the computation of waveguide regions that bear greater significance. The accuracy of our computer program, SVMM is assessed by computing the propagation constants and the effective indices of several rib waveguides that have been known to be excellent benchmark waveguide structures. The results of our computation have compared favorably with other published results [14,15,19,25,26]. We then continue to simulate the optical guided modes of diffused optical waveguides in $LiNbO_3$. The computed mode size is consistent with published experimental results. Our simulations, however, have shown the inadequacies of the adopted diffusion model for its inability to model the diffused waveguide in a more robust sense. It is suggested that further research is to be conducted for a more refined and robust representation of the refractive index profile of Ti-$LiNbO_3$ diffused waveguide. Despite the shortcoming of the diffusion model that we have adopted, we have demonstrated the potential of SVMM to be used as an analytical and design tool for integrated optical waveguide.

The modulators employing ridge waveguide structures have been described in Chapter 3 and there is no need to repeat them here. The electrode design is similar to those described in Chapter 12, but because Silicon device structure is short, no travelling wave type is required. This is one of the disadvantages of weakly guided waveguide structure.

REFERENCES

1. M. Fukuma and J. Noda, Optical properties of titanium-diffused $LiNbO_3$ strip waveguides and their coupling to a fiber characteristics, *Appl. Opt.*, vol. 20, no. 4, pp. 591–597, 15 February 1980.
2. C. Bulmer et al., High efficientcy flip-chip coupling between single mode fibers and $LiNbO_3$ channel waveguides, *Appl. Phys. Lett.*, vol. 37, pp. 351–335, 1981.
3. V. Ramaswamy et al., High efficiency single mode fiber to Ti:$LiNbO_3$ waveguide coupling, *Elect. Lett.*, vol. 10, pp. 30–31, 1982.
4. R. C. Alferness et al., Efficient single mode fiber to titanium diffused lithium niobate waveguide coupling for $l = 1.32\,\mu m$, *IEEE J. Quant. Elect.*, vol. QE-18, pp. 1807–1181, 1982.
5. E. A. J. Marcatilli, Dielectric rectangular waveguide and directional couplers for integrated optics, *Bell Syst. Tech. J.*, 48, pp. 2071–2102, 1969.
6. P. G. Suchoski and R. V Ramaswamy, Minimum mode size low loss Ti:$LiNbO_3$ channel waveguides for efficient modulator operation at $1.3\,\mu m$, *IEEE J. Quant. Elect.*, vol. QE-23, no. 10, pp. 1673–1679, October 1987.
7. K. S. Chiang, Review of numerical and approximate methods for the modal analysis of general optical dielectric waveguides, *Opt. Quant. Elect.*, 26, S113–S134, 1994.

8. S. M. Saad, Review of numerical methods for the analysis of arbitrary shaped microwave and optical dielectric waveguides, *IEEE Trans. Microwave Theory Technol.*, vol. MTT-33, no. 10, pp. 894–899, October 1985.

9. S. Seki et al., Two dimensional analysis of optical waveguides with a non-uniform finite difference method, *IEE Proc. Part J. Optoelect.*, vol. 138, no. 2, pp. 123–127, April 1991.

10. M. J. Robertson et al., Semiconductor waveguides: Analysis of optical propagation in single rib structures and directional couplers, *IEE Proc. Part J. Optoelect.*, vol. 132, no. 6, pp. 336–342, December 1985.

11. P. Lusse et al., Analysis of vectorial mode fields in optical waveguides by a new finite difference method, *IEEE J. Lightwave Technol.*, vol. 12, no. 11, pp. 487–449, March 1994.

12. M. S. Stern, Semivectorial polarised *H* field solutions for dielectric waveguides with arbitrary index profiles, *IEE Proc.*, vol. 135, Pt. J, no. 5, pp. 333–338, October 1988.

13. W. Huang and H. A. Haus, A simple variational approach to optical rib waveguides, *IEEE J. Lightwave Tech.*, vol. 9, no. 1, pp. 56–61, January 1991.

14. Pao-Lo Liu and Bing-Jin Li, Semivectorial beam propagation method for analysing polarised modes of rib waveguides, *IEEE J. Quant. Electron.*, vol. 28, no. 4, pp. 778–782, April 1992.

15. Pao-Lo Liu and Bing-Jin Li, Semivectorial Helmholtz beam propagation by Lanczos reduction, *IEEE J. Quant. Electron.*, vol. 29, no. 8, pp. 2385–2389, August 1993.

16. T. M. Benson et al., Rigorous effective index method for semiconductor rib waveguides, *IEE Proc. J.*, vol. 139, no. 1, pp. 67–70, February 1992.

17. P. C. Kendall et al., Advances in rib waveguide analysis using weighted index method or the method of moments, *IEE Proc. Part J. Optoelect.*, vol. 137, no. 1, pp. 27–29, February 1990.

18. S. V. Burke, Spectral index method app. lied to rib and strip-loaded directional couplers, *IEE Proc. Part J. Optoelect.*, vol. 137, no. 1, pp. 7–10, February 1990.

19. M. S. Stern, Analysis of the spectral index method for vector modes of rib waveguides, *IEE Proc. Part J. Optoelect.*, vol. 137, no. 1, pp. 21–26, February 1990.

20. G. Ronald Hadley and R. E. Smith, Full vector waveguide modeling using an iterative finite difference method with transparent boundary conditions, *IEEE J. Lightwave Technol.*, vol. 13, no.3, pp. 465–469, March 1995.

21. T. M Benson et al., Polarisation correction applied to scalar analysis of semiconductor rib waveguides, *IEE Proc. Part J. Optoelect.*, vol. 139, no. 1, pp. 39–41, February 1992.

22. B. A. M. Rahman and J. D. Davies, Vectorial-H finite element solution of GaAs/GaAlAs rib waveguides, *IEE Proc. J.*, vol. 132, pp. 349–335.

23. Chang Min Kim and R. V. Ramaswamy, Modelling of graded index channel waveguides using non-uniform finite difference method, *IEEE J. Lightwave Technol.*, vol. 7, no. 10, pp. 1581–1589, October 1989.

24. G. B. Hocker and W. K. Burns, Mode dispersion in diffused channel waveguides by the effective index method, *Appl. Opt.*, vol. 16, no. 1, pp. 113–118.

25. M. S. Stern, Semivectorial polarised finite difference method for optical waveguides with arbitrary index profiles, *IEE Proc. Part J. Optoelect.*, vol. 135, no. 1, pp. 56–6, February 1988.

26. M. S. Stern, Rayleigh quotient solution of semivectorial field problems for optical waveguides with arbitrary index profiles, *IEE Proc. Part J. Optoelect.*, vol. 138, pp. 185–190, 1990.

27. K. Ogusu, Numerical analysis of the rectangular dielectric waveguide and its modifications, *IEEE Trans. Microwave Theory Technol.*, vol. MTT-25, no. 11, pp. 874–885, 1977.

28. J. E. Goell, A circular-harmonic computer analysis of rectangular dielectric waveguide, *Bell. Systems. Tech. J.*, Vol. 48, pp. 2133–2160, 1969.

FURTHER READING

R. C. Alferness et al., Characteristics of Ti diffused lithium niobate optical directional couplers, *Appl. Opt.*, vol. 18, no. 23, pp. 4012–4016, 1 December 1979.

R. L. Burden and J. D. Faires, *Numerical Analysis*, 4th ed., Boston: PWS-Kent Publishing Company, pp. 492–505, 1988.

W. K Burns et al., Ti diffusion in Ti:LiNbO₃ planar and channel optical waveguides, *J. Appl. Phys.*, vol. 50, no. 10, pp. 6175–6182, October 1979.

J. Ctyroky et al., 3-D analysis of LiNbO₃: Ti channel waveguides and directional couplers, *IEEE J. Quant. Electron.*, vol. QE. 20, no. 4, pp. 400–409, April 1984.

M. D. Feit et al., Comparison of calculated and measured performance of diffused channel-waveguide couplers, *J. Opt. Soc. Am.*, vol. 73, no. 10, pp. 1296–1304, October 1983.

S. Fouchet et al., Wavelength dispersion of Ti induced refractive index change in LiNbO$_3$ as a function of diffusion parameters, *IEEE J. Lightwave Tech.*, vol. LT-5, no. 5, pp. 700–708, May 1987.

G. B. Hocker and W. K Burns, Modes in diffused optical waveguides of arbitrary index profile, *IEEE J.Quan. Electron.*, vol. QE-11, no. 6, pp. 270–1975, June 1975.

K. T. Koai and Pao-Lo Liu, Modelling of Ti:LiNbO$_3$ waveguide devices: Part I-Directional Couplers, *IEEE J. Lightwave Tech.*, vol. 7. no. 3, pp. 533–539, March 1989.

K. T. Koai and Pao-Lo Liu, Modelling of Ti:LiNbO$_3$ waveguide devices: Part II-S-shaped channel waveguide bends, *IEEE J. Lightwave Tech.*, vol. 7. no. 7, pp. 1016–1022, July 1989.

S. K. Korotky et al., Mode size and method for estimating the propagation constant of single mode Ti:LiNbO$_3$ strip waveguides, *IEEE J. Quan. Electron.*, vol. QE-18, no. 10, pp. 1796–1801, October, 1982.

G. Kötitz, *Properties of Lithium Niobate*, EMIS datareviews series no. 5. London and New York: INSPEC, The Institution of Electrical Engineers, 1989.

R. K. Lagu and R. V. Ramaswamy, A variational finite-difference method for analysing channel waveguides with arbitrary index profiles, *IEEE J. Quan. Electron.*, vol. QE-22, no. 6, pp. 968–976, June, 1986.

M. Minakata, S. Shaito, and M. Shibata, Two dimensional distribution of refractive index changes in Ti diffused LiNbO$_3$ waveguides, *J. Appl. Phys.*, vol. 50, no. 5, pp. 3063–3067, May 1979.

M. Minikata et al., Precise determination of refractive index changes in Ti-diffused LiNbO$_3$ optical waveguides, *J. Appl. Phys.*, vol. 49, no. 9, pp. 4677–4682, September 1978.

D. F. Nelson and R. M. Mikulyak, Refractive indices of congruently melting lithium niobate, *J. Appl. Phys.*, vol. 45, no. 8, pp. 3688–3689, August 1974.

T. C. Oppe, W. D. Joubert, and D. R. Kincaid, *NSPCG user's guide version 1.0 — a package for solving large sparse linear systems by various iterative methods, Center for Numerical Analysis*, Austin: The University of Texas.

W. H. Press et al., *Numerical Recipes-the Art of Scientific Computing*, 3rd ed., Cambridge University Press, pp. 377–379, 2017.

Properties of Lithium Niobate, Emis Datareviews Series no.5, INSPEC publication, pp. 131–146, 1991.

N. Schulz et al., Finite difference method without spurious solutions for the hybrid-mode analysis of diffused channel waveguides, *IEEE Trans. Microwave Theory Tech.*, vol. 38, no. 6, pp. 722–729, June 1990.

A. Sharma and P. Bindal, Analysis of diffused planar and channel waveguides, *IEEE J. Quan. Electron.*, vol. 29, no. 1, pp. 150–15, January 1993.

A. Sharma and P. Bindal, An accurate variational analysis of single mode diffused channel waveguides, *Opt. Quant. Electron.*, vol. 24, pp. 1359–1371, 1992.

A. Sharma and P. Bindal, Variational analysis of diffused planar and channel waveguides and directional couplers, *J. Opt. Soc. Am. A*, vol. 11, no. 8, pp. 2244–2248, August 1994.

D. S. Smith et al., Refractive Indices of Lithium Niobate, *Opt. Comm.*, vol. 17, no. 3, pp. 332–335, June 1976.

G. Kötitz, *Properties of Lithium Niobate*, EMIS datareviews series no. 5. London and New York: INSPEC, The Institution of Electrical Engineers, 1989.

E. Strake et al., Guided Modes of Ti:LiNbO$_3$ channel waveguides: A novel quasi analytical technique in comparison with the scalar finite-element method, *IEEE J. Lightwave Tech.*, vol. 6, no. 6, pp. 1126–1135, June 1988.

M. Valli and A. Fioretti, Fabrication of good quality Ti:LiNbO$_3$ planar waveguides by diffusion in dry and wet O$_2$ atmospheres, *J. Mod. Opt.*, vol. 35, no. 6, 885–890, 1988.

14 Optical Fibers
Geometrical and Lightwaves Guiding Properties

This chapter describes the fundamental understanding of guiding lightwaves in planar and circular optical waveguides. A brief fundamental derivation of the wave equation from the Maxwell's equations is given for symmetric planar waveguides leading to the phenomenon of guiding in optical fibers and circular optical waveguides.

Only geometrical parameters of optical fibers are presented here. The attenuation and dispersion effects of modulated lightwave signals can be found in the first section (Section 14.2) of this chapter. However, non-linear effects are given here so that the distortion impacts on signals can be discussed in the next chapter.

14.1 INTRODUCTION

The term "optical fiber communications" is no longer used in research centers or university research or teaching laboratories, but is extensively deployed throughout the earth, as well as undersea. Optical fiber communications have progressed for more than 35 years, and the technology has influenced a significant revolution in the information technology-oriented society beyond the twenty-first century. It was a simple idea that was proposed in an article in the Proceedings of the Institution of Electrical Engineers of U.K. by Charles Kao and G. Hockham [1]. The idea was to guide optical waves in a wavelike lightguide, called a dielectric optical waveguide. It was thought that, as the most fundamental component of optical communications technology, guiding lightwave whose frequency is extremely high, ultra-wideband regions could be available to accommodate information channels.

This chapter is dedicated to describing a fundamental understanding of optical waveguiding and its structures, as well as the conditions under which the waveguide can be designed to guide lightwaves over extremely long distances. The basic components for optical fiber communications networks are the optical guiding media for transmission of optical signals in the form of lightwaves, called optical waveguides or optical fibers. These lightwaves transmission media must satisfy the following conditions:

First, it must be able to guide lightwaves over a long distance without significantly losing the optical energy. Second, signal transmitting through them must not be distorted, and they must be structured in such a way that they can support a number of optical electromagnetic field modes [2,8].

The main objectives of this chapter are to introduce concepts of guided modes in optical waveguides in which lightwaves are confined to one dimension only, the planar optical waveguides, and to discuss how the optical waveguides are extended to the optical fibers. The properties of the step index optical fiber that support only one mode are described and used as the fundamental elements for studying non-step index fiber types and installed throughout global networks. The non-circular or non-step optical profile optical fibers can be transformed to a circular optical fiber by using an equivalent step-index profile technique that is described later in this chapter.

Advanced design of optical fibers for dispersion-shifted, dispersion-flattening, and dispersion-compensated fibers are also briefly described. These fibers have gained much attention lately because of their potential applications in optical soliton communications systems, dispersion equalized optical fiber systems, etc.

We must note here that this chapter presents only the geometrical structures and index distribution of circular optical waveguides and guiding conditions, as well as some optical properties of the guided modes. In the next chapter, we shall describe the mechanism of distortion and attenuation of optical signals when they are transmitted over long length of optical fibers.

This chapter does not aim to give a full treatment of the theoretical development of optical waveguides, but the essentials for students of physics and engineering field that would allow them to apply to some applications, such as transmission of lightwave modulated signals and possibly fiber sensors.

This chapter is organized as follows: Section 14.2 describes the fundamental aspects of planar optical waveguides with a derivation of the wave equation from the first principle via the formation of Maxwell equation. Symmetric planar optical waveguides are given as a simple example of optical waveguiding so as to guide readers to the guided modes of planar and circular optical waveguides. The V-parameter is identical for both types of waveguides and hence easy for readers to appreciate the importance of this parameter of optical waveguides.

Section 14.3 describes the optical fibers or circular optical waveguides, including their geometrical structures and conditions for supporting one and only one mode only. An equivalent step index profile method is also presented so that different index profile fibers can be designed. Gaussian approximation of the guided single mode optical waveguide is given so that the operation parameters of optical fibers can be derived without resorting to mathematical complexity via the use of the wave equation. This is derived from practice when, measuring the mode field distribution across a single mode optical fiber, one can find that its radial distribution is very close to that of a Gaussian function. Thus, all one can do is to substitute this function to the wave equation to obtain the propagation constant, hence other guided parameters.

Section 14.4 gives a brief introduction to the manufacturing of optical fibers so that readers can appreciate the technological development of optical fibers.

14.2 DIELECTRIC SLAB OPTICAL WAVEGUIDES

Dielectric slab waveguide is the simplest optical waveguiding structure that was first investigated in the early 1960s. It is from this structure that the proposal for an optical fiber for optical communications came.

The use of optical slab waveguide is now widely adopted as the basic structure for optical integrated circuits like semiconductor lasers and optical modulators. This section gives the fundamental aspects of optical waveguiding with a symmetrical structure, that is, the optical guiding layer is sandwiched between two cladding region of smaller dielectric constant or refractive index than that of the core region.

14.2.1 STRUCTURE

A dielectric slab waveguide consists of a slab of dielectric material of refractive index n_1 embedded in a material of index n_2 as shown in Figure 14.1.

Assuming the structure is extended to infinite in the y- and z-directions, and a guiding thickness of 2a and that the materials are isotropic and lossless (i.e., the permittivities are real and scalar), this leads to a 1D wave equation for the electric field of the lightwaves and hence a simplified mathematical solution. The main reason for this approximation is that a circular optical fiber can be easily analyzed by confining the optical waveguide in the y-direction into a circular structure. The wave equation for this type of structure involves the cylindrical coordinates and is more complicated to visualize analytically.

14.2.2 NUMERICAL APERTURE

If we assume at the moment that total internal reflections at the boundaries are required for guiding, what is the acceptance angle such that lightwaves can be launched? The ray path entering the optical

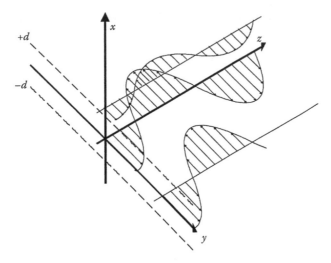

FIGURE 14.1   Real part of E-field in a slab waveguide ($t = 0$).

fiber core for total internal reflection is shown in Figure 14.2.

$$n_0 \sin \theta_0 \leq n_1 \cos \theta_c \tag{14.1}$$

By applying Snell's law at the air-core and core-cladding boundaries of the dielectric waveguide, the total internal reflection can take place only if:

$$n_1 \sin \theta_c = n_2 \sin 90° = n_2 \tag{14.2}$$

where θ_c is the critical angle such that

$$NA = (\sin \theta_0)_{max} = n_1 \cos \theta_c \tag{14.3}$$

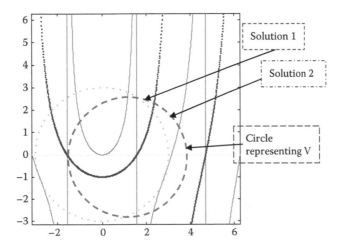

FIGURE 14.2   Graphics and solution of Eqautions 2.9, 2.10d, and 2.12. The curve indicated by is expressing the expression $v = u\tan u$, $v = u/\tan u$ and $V^2 = u^2 + u^2$.

Thus, the numerical aperture (NA) which is defined as the maximum value of $\sin\theta_o$ is given by Equation 14.1.

$$NA = \left(n_1^2 - n_2^2\right)^{1/2} \tag{14.4}$$

then by substituting Equation 14.2, we have

$$NA = \left(n_1^2 - n_2^2\right)^{1/2} \tag{14.5}$$

when the refractive indices n_1 and n_2 are different by a small index difference, then the numerical aperture is approximated by

$$NA \simeq n_2(2\Delta)^{1/2} \tag{14.6}$$

where Δ is the refractive index difference between the refractive indices between the core and the cladding regions. Thus, the refractive index difference determines the magnitude of the NA and hence the acceptance angle that an optical fiber can accept incident lightwaves. In practice, designers want to increase the numerical aperture to maximize optical power that can be coupled to an optical fiber. However, it can be seen later that the larger the NA, the larger the V-parameter and optical fibers can become, thus suffering larger delay. In designing a single mode optical fiber, we must consider several parameters that affect its performance from a communication systems point of view. Thus, we must resort to the wave equation to understand the behavior of lightwaves in a single mode optical fiber so that its transmission characteristics can be optimized.

14.2.3 Modes of Symmetric Dielectric Slab Waveguides

The number of optical guide modes of lightwaves transmitted in an optical waveguide is very important because this indicates the concentration of energy of the lightwaves. It is similar to the elastic waves propagating in a string that is vibrated with the two open or closed ends. To find the number of optical guide modes, one must find the conditions under which the lightwaves would be conformed. Normally, this condition is the eigenvalue equation derived from the wave equation that satisfies the boundary conditions.

Associated with each solution of the optical wave equation is the propagation wave number of each guided wave which are derived from the eigenvalues of the equation. Thus, the propagation constants take discrete values.

14.2.3.1 The Wave Equations

Assuming a monochromatic, single frequency or wavelength, (we mean a single one and only one frequency component of lightwaves) wave propagates in the z-direction with its electric field component given by

$$E(x, y, z) = E(x)e^{(\omega t - \beta z)} \tag{14.7}$$

that is, field dependent on x, and uniform along y-direction, β is the propagation constant along z propagating direction. Then in the absence of charges and currents, the Maxwell's equation reduces to:

$$j\beta E_y = -j\omega\mu H_x \tag{14.8}$$

$$0 = -j\omega\mu H_y \tag{14.9}$$

$$\frac{dE_x}{dx} = -j\omega\mu H_z \tag{14.10}$$

$$-j\beta H_x - \frac{dH_x}{dx} = j\omega\varepsilon E_y \tag{14.11}$$

A visualization of the E-field guided in the slab waveguide is shown in Figure 14.1. By substituting H_x from Equation 14.8 and H_z from Equations 14.10 through 14.11, we get

$$\frac{d^2 E_y}{dx^2} + (\beta^2 - \omega^2 \mu \varepsilon)E_y = 0 \tag{14.12}$$

where μ and ε are the permeability and permittivity of medium n_1 or n_2. Similarly, a wave-equation-involved H_y is given by

$$\frac{d^2 H_y}{dx^2} - +(\beta^2 - \omega^2 \mu \varepsilon)H_y = 0 \tag{14.13}$$

Equations 14.12 and 14.13 can be rewritten using $k = \omega/c$ and $c = (\mu_o \varepsilon_o)^{1/2}$, where c is the light velocity in vacuum and k is the wave number in vacuum, as

$$\frac{d^2 E_y}{dx^2} = -\left(k^2 n_j^2 - \beta^2\right)E_y \tag{14.14}$$

$$\frac{d^2 H_y}{dx^2} = -\left(k^2 n_j^2 - \beta^2\right)H_y \tag{14.15}$$

where $n_j = n_1$ or n_2 ($j = 1, 2$)

From Equations 14.14 and 14.15, we observe that the field variation along $0x$ will always exhibit:

- Sinusoidal behavior when $k^2 n_j^2 > \beta^2$ oscillating: lightwaves are guided.
- Exponential (decay) behavior when $k^2 n_j^2 < \beta^2$ evanescent lightwaves are radiating to the cladding region.

In other words, the EM field is oscillating in regions where the longitudinal propagation constant is smaller than the plane-wave propagation constant in this region and evanescent with an exponential-like behavior elsewhere.

14.2.3.2 Optical Guided-Modes

Optical waves are guided along the waveguide when their EM fields are oscillatory in the slab waveguide region and exponentially decay in the cladding region, that is,

$$kn_2 < b < kn_1 \tag{14.16}$$

We need now to define a transverse propagation constant u/a and transverse decay constant v/a as

$$\frac{u^2}{a^2} = k^2 n_1^2 - \beta^2 \tag{14.17}$$

$$\frac{v^2}{a^2} = -k^2 n_2^2 + \beta^2 \tag{14.18}$$

thus, we can observe from Equation 2.7 that

$$\frac{u^2}{a^2} + \frac{v^2}{a^2} = k^2 \left(n_1^2 - n_2^2\right) \tag{14.19}$$

or, alternatively,

$$V^2 = u^2 + v^2 = k^2 a^2 \left(n_1^2 - n_2^2\right) \tag{14.20}$$

in which the parameter V is defined as the normalized frequency (V) that is dependent only on the guide and light frequency. The V parameter and its expression is identical for both planar and optical fiber, which will be discussed in the next section. The refractive index difference term is usually much less than one. The V-parameter can be normally approximated by a simpler expression.

Equations 14.14 and 14.20 show that the field E_y for TE modes and H_y for TM mode are a linear combination of $\cos(u_x/a)$ and $\sin(u_x/a)$ inside the core layer ($|x| \leq a$) and exponentially decay form on outside ($|x| > a$) with $\exp(-v_x/a)$ for $x > a$ and $\exp(v_x/a)$ for $x < -a$. We therefore have two distinct cases: the even TE modes and odd TE modes. Their solutions are given as follows:

14.2.3.3 Even TE Modes

- For $|x| \leq a$: inside core region, oscillating waves, thus the field solution is

$$E_y \text{ or } H_x = A \cos\frac{ux}{a} \tag{14.21}$$

$$H_z = \frac{j}{\omega\mu_o}\frac{\delta E}{\delta x} = -\frac{j}{\omega\mu_o}\frac{u}{a}A\sin\frac{ux}{a} \tag{14.22}$$

- For $|x| > a$: outside core region, exponential decaying waves

$$E_y = Ce^{-\frac{V(x-a)}{a}} \tag{14.23}$$

Other H_x and H_z can also be expressed in similar expressions. Applying the boundary condition (continuity of the E—field) at the interface between the cladding and core, that is, at $x = a^+$ and $x = a^-$. Using Equations 14.21 through 14.23 have

$$-j\frac{u}{\omega\mu}\sin u = -j\frac{v}{\omega\mu}\cos u$$

$$\rightarrow v = u\tan(u) \tag{14.24}$$

This is usually called the eigenvalue equation of the wave equation which dictates the number of guided (discrete) modes and the values of the propagation constants of these guided even TE modes.

14.2.3.4 Odd TE Modes

Similarly, the solutions for guided TE odd modes can be found as follows:

$x > |a|$

$$E_y \text{ or } H_x = A\sin\frac{ux}{a} \tag{14.25}$$

$$H_z = \frac{j}{\omega\mu_o}\frac{\delta E}{\delta x} = -\frac{j}{\omega\mu_o}\frac{u}{a}A\cos\frac{ux}{a} \tag{14.26}$$

$|x| \leq a$

$$E_y = Ce^{-\frac{V(x-a)}{a}} \tag{14.27}$$

and the eigenvalue equation can be found by applying the boundary conditions:

$$v = \frac{u}{\tan u} \tag{14.28}$$

Equations 14.25 through 14.28 are obtained by using the continuity of the normal H field component at the boundaries. This is the eigenvalue equation for guided TE *odd* modes of a symmetric planar optical waveguide.

14.2.3.5 Graphical Solutions for Guided TE Modes (Even and Odd)

Combining Equations 14.20, 14.21, and 14.25 through 14.28 we observe that the waveguides can support only discrete modes and the propagation constant β related to u and v parameters can be found by solving graphically the intersection between circles of V and curves representing Equations 14.20, 14.24, and 14.25 through 14.28. These solutions are illustrated in Figure 14.2. The field distribution of the even and odd TE mode can be found in Figure 14.3.

14.2.4 Cut-Off Properties

From Figure 14.2 we observe that:
  For $V = 0$: that is, zero optical frequency or $\lambda = 0$.
  For $\lambda > 0$: we observe that we always have at least one guided mode, TE_o.

- $V < \pi/2$: one guided mode
- $V \to \pi/2$: odd TE_1 mode appears (second mode)
- $V = \pi\pi$: we have third mode (TE_2)

That is, each time V reaches a multiple integer of $\pi/2$, a new mode reaches its cut-off that is corresponding to $v = 0$ and $\beta = kn_2$. We note that for TE guided modes the optical waveguide can support at least one. However, for TM guided modes, the waveguide can support if its thickness or the optical wavelength and the refractive index difference, the v (lower case v) is at least $\pi/2$.

This is true for planar optical waveguides only. For circular optical waveguide such as in the case of circular optical fiber, the cut-off condition is different. We will deal with this type of optical waveguide in the next section.

It is noted that theoretical waveguide (planar type) always supports an optical guided mode. The cut-off is for cutting off the higher order mode to make the waveguide supports only one mode.

14.3 OPTICAL FIBER: GENERAL PROPERTIES

14.3.1 Geometrical Structures and Index Profile

An optical fiber consists of two concentric dielectric cylinders. The inner cylinder, or core, has a refractive index of $n(r)$ and radius a. The outer cylinder, or cladding, has index n_2 with $n(r) > n_2$ and a larger

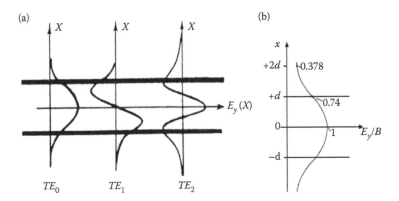

FIGURE 14.3 (a) Field distribution of TE-modes of order 0, 2 (even modes) and 1 (odd mode) and (b) field distribution of the fundamental mode.

outer radius. Core of about 4–9 μm and a cladding diameter of 125 μm are the typical values for silica-based single mode optical fiber. A schematic diagram of an optical fiber is shown in Figure 14.4.

The refractive index n of an optical waveguide is usually changed with radius r from the fiber axis $(r = 0)$ and is expressed by

$$n^2(r) = n_2^2 + NA^2 s\left(\frac{r}{a}\right) \tag{14.29}$$

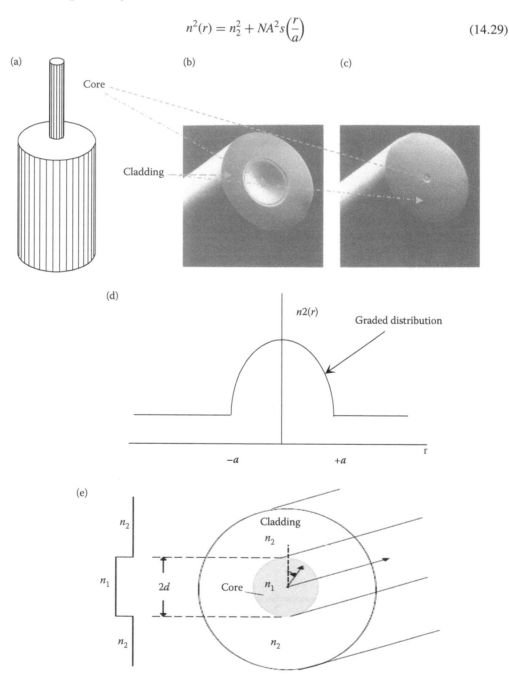

FIGURE 14.4 (a) Schematic diagram of the step-index fiber: coordinate system, structure. The refractive index of the core is uniform and slightly larger than that of the cladding. For silica glass, the refractive index of the core is about 1.478 and that of the cladding about 1.47 at 1550 nm wavelength region. (b) Cross section of an etched fiber—multi-mode type—50 micrometer diameter. (c) Single mode optical fiber etched cross section. (d) Graded index profile. (e) Fiber cross section and step index profile $d = a =$ radius of fiber.

where NA is the numerical aperture at the core axis, while $s(r/a)$ is the profile function that characterizes any profile shape ($s = 1$ at maximum) with a scaling parameter (usually the core radius).

14.3.2 INDEX PROFILES

14.3.2.1 Step-Index Profile

In a step-index profile, the refractive index remains constant in the core region, thus

$$s\left(\frac{r}{a}\right) = 1 \quad \text{for } r \leq a \tag{14.30}$$

$$s\left(\frac{r}{a}\right) = 0 \quad \text{for } r > a \tag{14.31}$$

so we have for a step index profile

$$n^2(r) = n_1^2 \quad \text{for } r < a \tag{14.32}$$

and

$$n^2(r) = n_2^2 \quad \text{for } r > a \tag{14.33}$$

14.3.2.2 Graded: Index Profile

We consider hereunder the two most common types of graded-index profiles: power-law-index and the Gaussian-profile.

14.3.2.3 Power-Law-Index Profile

The core refractive index of optical fiber is usually following a graded profile. In this case, the refractive index rises gradually from the value n_2 of the cladding glass to value n_1 at the fiber axis. Therefore, $s(r/a)$ can be expressed as

$$s\left(\frac{r}{a}\right) = \left\{1 - \left(\frac{r}{a}\right)^\alpha \quad \text{for } r \leq a \text{ and } = 0 \text{ for } r < a \tag{14.34}$$

with $\alpha = $ power exponent. Thus, the index profile distribution $n(r)$ can be expressed in the usual way by using Equations 14.34 and 14.30 and substituting $NA^2 = n_1^2 - n_2^2$.

$$n^2(r) = \begin{cases} n_1^2\left[1 - 2\Delta\left(\frac{r}{a}\right)^\alpha\right] & \text{for } r \leq a \\ n_2^2 & \text{for } r > a \end{cases} \tag{14.35}$$

with $\Delta = NA^2/n_1^2$ is the relative refractive difference. Observing the equation describing the profile shape in Equation 14.29, there are three special cases:

- $\alpha = 1$: the profile function $s(r/a)$ is linear and the profile is called a triangular profile.
- $\alpha = 2$: the profile is a quadratic function with respect to the radial distance and the profile is called the parabolic profile.
- $\alpha = \infty$; then the profile is a step type.

14.3.2.4   Gaussian: Index Profile

While in the Gaussian index profile, the refractive index changes gradually from the core center to a distance very far away from it and $s(r)$ can be expressed as

$$s\left(\frac{r}{a}\right) = e^{-\left(\frac{r}{a}\right)^2} \tag{14.36}$$

14.3.3   The Fundamental Mode of Weakly Guiding Fibers

The electric and magnetic fields $E(r, \phi, z)$ and $H(r, \phi, z)$ of the optical fibers in cylindrical coordinates can be found by solving Maxwell's equations. However, only lower-order modes of ideal step index fibers are important to the present optical fiber communications systems. The fact is that $\Delta < 1\%$, thus optical waves are weakly guided, and E and H are then approximate solutions of the scalar wave equation

$$\left[\frac{\delta^2}{\delta r^2} + \frac{1}{r}\frac{\delta}{\delta r} + k^2 n_j^2\right]\varphi(r) = \beta^2 \varphi(r) \tag{14.37}$$

where $n_j = n_1, n_2$, and $\psi(r)$ is the spatial field distribution of the nearly transverse EM waves

$$E_x = \psi(r)e^{-i\beta z}$$

$$H_y = \left(\frac{\varepsilon}{\mu}\right)^{1/2} \quad E_x = \frac{n_2}{Z_0}E_x \tag{14.38}$$

with E_y, E_z, H_x, H_z negligible, $\varepsilon = n_2^2 \varepsilon_o$ and $Z_o = (\varepsilon\mu)^{1/2}$ is the vacuum impedance. That is, the waves can be seen as a plane wave travelling down along the fiber tube. These plane waves are reflected between the dielectric interfaces; in other words, they are trapped and guided in and along the core of the optical fiber.

14.3.3.1   Solutions of the Wave Equation for Step-Index Fiber

The field spatial function $\psi(r)$ would have the form of Bessel functions (from Equation 14.37) as

$$\varphi(r) = A\frac{J_0(ur/a)}{J_0(u)} \quad \text{for } 0 < r < a \tag{14.39}$$

$$\varphi(r) = A\frac{K_0(vr/a)}{K_0(v)} \quad \text{for } r > a \tag{14.40}$$

where J_o and K_o are Bessel functions of the first kind and modified of second kind, respectively, and u and v are defined similarly, as given in Equations 14.17 and 14.18. Thus, following the Maxwell's equations relation, we can find that E_z can take two possible solutions that are orthogonal as

$$E_z = -\frac{A}{kan_2}\begin{pmatrix} \sin\phi \\ \cos\phi \end{pmatrix}\begin{cases} \dfrac{uJ_1\left(u\dfrac{r}{a}\right)}{J_0(u)} & \text{for } 0 \leq r < a \\[4mm] \dfrac{vK_1\left(\dfrac{vr}{a}\right)}{K_0(v)} & \text{for } r > a \end{cases} \tag{14.41}$$

The terms u and v must simultaneously satisfy two equations

$$u^2 + v^2 = V^2 = ka\left(n_1^2 - n_2^2\right)^{1/2} = kan_2(2\Delta)^{1/2} \tag{14.42}$$

$$u\frac{J_1(u)}{J_0(u)} = v\frac{K_1(v)}{K_0(v)} \tag{14.43}$$

where Equation 14.43 is obtained by applying the boundary conditions at the interface $r = a$ (E_z is the tangential component and must be continuous at this dielectric interface). Equation 14.43 is usually called the eigenvalue equation. The solution of this equation would give the values of β which would take discrete values and are the propagation constants of the guided lightwaves.

Equation 14.24 shows that the longitudinal field is in the order of u/kan_2 with respect to the transverse component. In practice, $\Delta \ll 1$, and by using Equation 14.42, we observe that this longitudinal component is negligible compared with the transverse component. We thus consider the mode as transversely polarized. The fundamental mode is then usually denominated as LP_{01} mode ($LP =$ Linearly Polarized) for which the field distribution is shown in Figure 14.5 and the graphical representation of the eigenvalue Equation 14.43 calculated as the variation of $\beta/k = b$ as the normalized propagation constant and the V-parameter is shown in Figure 14.6c.

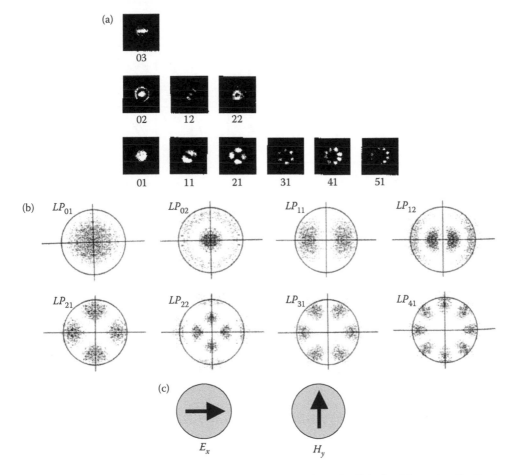

FIGURE 14.5 (a) Spectrum of guided modes in a multi-mode fiber. (b) Calculated intensity distribution of LP guided modes in a step-index optical fibers with $V = 7$. (c) Electric and magnetic field distribution of an LP_{01} mode polarized along O of the fundamental mode of a single mode fiber.

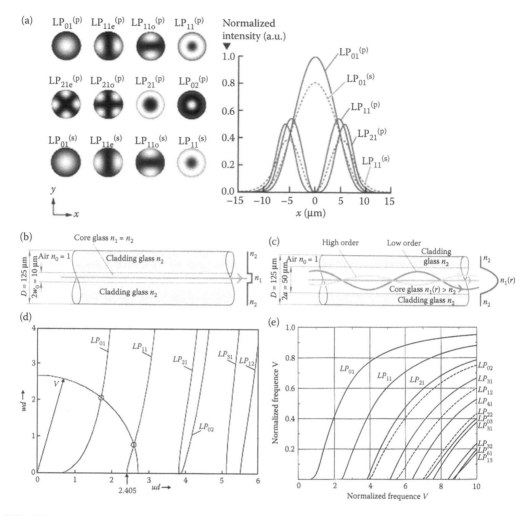

FIGURE 14.6 (a) Guided modes as seen in the tranverse plane of a circular optical fiber. (b) "Ray" model of lightwave propagating in single mode fiber. (c) Ray model for multi-mode graded-index fiber. (d) Graphical illustration of solution for eigenvalues (propagation constant–wavenumber of optical fibers). (e) b-V characteristics of guided fibers.

14.3.3.1.1 Gaussian Approximation

14.3.3.1.1.1 Fundamental Mode Revisited We note again that the **E** and **H** are approximate solutions of the scalar wave equation and the main properties of the fundamental mode of weakly guiding fibers that can be observed as follows:

- The propagation constant β (in z-direction) of the fundamental mode must lie between the core and cladding wave numbers. This means the effective refractive index of the guided mode lies with the range of the cladding and core refractive indices.
- Accordingly, the fundamental mode must be nearly a transverse electro-magnetic wave as described by Equation 14.38.

$$\frac{2\pi n_2}{\lambda} < \beta < \frac{2\pi n_1}{\lambda} \tag{14.44}$$

- The spatial dependence $\psi(r)$ is a solution of the scalar wave Equation 14.37.

14.3.3.1.2   Gaussian Approximation

The main objective is to find a good approximation for the field $\psi(r)$ and the propagation constant β. These can be found through the eigenvalue equation and the Bessel's solutions as shown in previous section. It is desirable to approximate the field accurately to obtain simple expressions to have a clearer understanding of light transmission on single mode optical fiber without going through graphical or numerical methods. Furthermore, experimental measurements and numerical solution for step and power-law profiles show that $\psi(r)$ is approximately Gaussian in appearance. We thus approximate the field of the fundamental mode as

$$\varphi(r) \cong A e^{-\frac{1}{2}\left(\frac{r}{r_0}\right)^2} \tag{14.45}$$

where is r_0 defined as the spot size that is, at which the intensity equals to e^{-1} of the maximum. If the wave Equation 14.37 is multiplied by $r\varphi(r)$ and using the identity

$$r\varphi\frac{\delta^2\varphi}{\delta r^2} + \varphi\frac{\delta\varphi}{\delta r} = \frac{\delta}{\delta r}\left(r\varphi\frac{\delta\varphi}{\delta r}\right) - r\left(\frac{\delta\varphi}{\delta r}\right)^2 \tag{14.46}$$

then, by integrating from zero to infinitive and using $[r\Psi(d\varphi/dr)]_0^\infty = 0$, we have

$$\beta^2 = \frac{\int_0^\infty \left[-\left(\frac{\delta\varphi}{\delta r}\right)^2 + k^2 n^2(r)\varphi^2\right] r\delta r}{\int_0^\infty r\varphi^2\delta r} \tag{14.47}$$

The procedure to find the spot size is then followed by substituting $\psi(r)$ (Gaussian) in Equation 14.45 into Equation 14.47, and then differentiating and setting $(\delta^2\beta)/\delta r$ evaluated at r_0 to zero, that is, the propagation constant β of the fundamental mode must give the largest value of r_0.

Knowing r_0 and β the fields E_x and H_y Equation 14.38 are fully specified.

Case 1: Step-index fiber: Substituting the step-index profile given by Equation 14.38 and $\psi(r)$ in Equation 14.45 into Equation 14.47 leads to an expression for β in term of r_0 given by

$$V = kaNA \tag{14.48}$$

The spot size is thus evaluated by setting

$$\frac{\delta^2\beta}{\delta r_0} = 0 \tag{14.49}$$

and r_0 is then given by

$$r_0^2 = \frac{a^2}{\ln V^2} \tag{14.50}$$

Substituting Equation 14.50 into Equation 14.48, we have

$$(a\beta)^2 = (akn_1)^2 - \ln V^2 - 1 \tag{14.51}$$

This expression is physically meaningful only when $V > 1$ (r_0 is positive)

Case 2: Gaussian Index Profile Fiber: Similarly for the case of a Gaussian index profile, by following the procedures for step-index profile fiber, we can obtain

$$(a\beta)^2 = (an_1 k)^2 - \left(\frac{a}{r_0}\right)^2 + \frac{V^2}{\left(\frac{a}{r_0}+1\right)} \tag{14.52}$$

and

$$r_0^2 = \frac{a^2}{V-1} \text{ by using } \frac{\delta^2\beta}{\delta r_0} = 0 \tag{14.53}$$

That is, maximizing the propagation constant of the guided waves. The propagation constant is at maximum when the "light ray" is very close to the horizontal direction. Substituting Equation 14.53 into Equation 14.52, we have

$$(a\beta)^2 = (akn_1)^2 - 2V + 1 \tag{14.54}$$

thus Equations 14.53 and 14.54 are physically meaningful only when

$$V > 1 (r_0 > 0) \tag{14.55}$$

It is obvious from Equation 14.55 that the spot size of the optical fiber with a *V*-parameter of one is extremely large. A very important point is that one must not design the optical fiber with a near unit value of the *V*-parameter. In practice, we observe that the spot size is large but finite (observable). In fact, if *V* is smaller than 1.5, the spot size becomes large, and in the next chapter this will be investigated in detail.

14.3.3.2 Cut-Off Properties

Like the case of planar dielectric waveguides, from Figure 14.6 we observe that when we have $V < 2.405$, only the fundamental LP_{01} exists. Thus we have Figure 14.7.

It is noted that for single mode operation, the *V* parameter must be less than or equal to 2.405, and in practice $V < 3$ can also be acceptable.

In fact, the value 2.405 is the first zero of the Bessel function $J_o(u)$. In practice, one cannot really distinguish between the *V* value between 2.3 and 3.0. Experimentally, observation shows that optical fiber can still support only one mode. Thus, designers do usually take the value of *V* as 3.0 or less to design a single mode optical fiber.

The *V* parameter is inversely proportional with respect to the optical wavelength. Thus, if an optical fiber is launched with lightwaves whose optical wavelength is smaller than the operating wavelength at which the optical fiber is single mode, then the optical fiber is supporting more than one mode. The optical fiber is said to be operating in a multi-mode region.

One can define the cut-off wavelength for optical fibers as follows: the wavelength (λ_c) above which only the fundamental mode is guided in the fiber, is called the cut-off wavelength λ_c. This cut-off wavelength can be found by using the *V*-parameter as $V_c = V$ (at cut-off) $= 2.405$, thus

$$\lambda_c = \frac{2\pi a NA}{V_c} \tag{14.56}$$

EXERCISE

An optical fiber has the following parameters: a core refractive index of 1.46, a relative refractive index difference of 0.3%, a cladding diameter of 125 μm, and a core diameter of 8.0 μm. (a) Find the fiber NA and hence the fiber acceptance angle. (b) What is the cut-off wavelength of

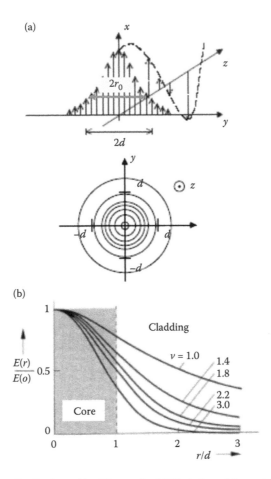

FIGURE 14.7   (a) Intensity distribution of the LP_{01} mode. (b) Variation of the spot size—field distribution with radial distance r with V as a parameter.

this fiber? (c) What is the number of optical guided modes which can be supported if the optical fiber is excited with lightwaves of a wavelength of 810 nm? (d) 1% the cladding diameter is reduced to 50 and 20 micrometers, comment on the field distribution of the guided single mode.

In practice, the fibers tend to be effectively single mode for larger values of V, say, $V < 3$ for the step profile, because the higher-order modes suffer radiation losses due to fiber imperfections. Thus, if $V = 3$, from Equation 14.42 we have $a < (3/2)\lambda NA$, in this case, that $\lambda = 1$ µm and the numerical aperture NA must be very small ($\ll 1$) for radius a to have some reasonable dimension. Usually, Δ is about 1% or less for standard single mode optical fibers in telecommunications systems.

14.3.3.3 Power Distribution

The axial power density or intensity profile $S(r)$, the z-component of the Poynting vector is given by

$$S(r) = \frac{1}{2}E_x H_y^*$$ (14.57)

Substituting Equation 14.38 into Equation 14.57, we have

$$S(r) = \frac{1}{2}\left(\frac{\varepsilon}{\mu}\right)^{1/2} e^{-\left(\frac{r}{r_0}\right)^2}$$ (14.58)

The total power is then given by

$$P = 2\pi \int_0^\infty rS(r)dr = \frac{1}{2}\left(\frac{\varepsilon}{\mu}\right)^{1/2} r_0^2 \tag{14.59}$$

and hence the fraction of power $\eta(r)$ within $0-r$ across the fiber cross section is given by

$$\eta(r) = \frac{\int_0^r rS(r)dr}{\int_0^\infty rS(r)dr} = 1 - e^{-\left(\frac{r^2}{r_0^2}\right)} \tag{14.60}$$

Table 14.1 gives the expressions for P and $\eta(r)$ of Step-index and Gaussian profile fibers (by substituting the appropriate values of r_0 into Equations 14.38 and 14.39.

As a rule of thumb and experimentally confirmed, an optical fiber is best for guided mode when the optical power contained in the core is about 80% of the total power.

EXERCISE

Using Gaussian approximation for the intensity distribution of the fundamental mode of the single mode optical fiber with $V = 2$, find the fraction of power in the core region with $a = 4\,\mu m$.

EXERCISE

Find the radius a for maximum confinement of light power, that is, maximum r_0, for step index and parabolic profile optical fibers.

14.3.3.4 Approximation of Spot-Size r_0 of a Step-index Fiber

As stated above, spot-size r_0 would play a major role in determining the performance of single mode fiber. It is useful if we can approximate the spot-size as long as the fiber is operating over a certain wavelength. When a single mode fiber is operating above its cut-off wavelength, a good

---

TABLE 14.1

Analytical Expressions for Total Optical Guided Power and Its Fractional Power Confined Inside the Core Region for Step-Index and Gaussian Index Profiles

	Step-Index	Gaussian
$S(r/a)$ for $V > 1$	$\frac{1}{2}\left(\frac{\varepsilon}{\mu}\right)^{1/2} e^{-\left(\frac{r}{a}\right)\ln V^2}$	$\frac{1}{2}\left(\frac{\varepsilon}{\mu}\right)^{1/2} e^{-\left(\frac{r}{a}\right)(V-1)}$
Power P for $V > 1$	$\frac{1}{2}\left(\frac{\varepsilon}{\mu}\right)^{1/2} \frac{a^2}{\ln V^2}$	$\frac{1}{2}\left(\frac{\varepsilon}{\mu}\right)^{1/2} \frac{a^2}{V-1}$
$\eta(r)$ for $V > 1$	$1 - e^{-\left(\frac{r}{a}\right)^2 \ln V^2}$	$1 - e^{-\left(\frac{r}{a}\right)^2 (V-1)}$
$\eta(r)$ for $V > 1$	$= 1 - \frac{1}{V^2}\,..for\,..r = a$	

approximation (greater than 96% accuracy) for r_0 is given by

$$\frac{r_0}{a} = 0.65 + 1.619\, V^{-3/2} + 2.879\, V^{-6} = 0.65 + 0.434\left(\frac{\lambda}{\lambda_c}\right)^{+3/2} + 0.0419\left(\frac{\lambda}{\lambda_c}\right)^{+6}$$

$$(14.61)$$

$$\text{for}\quad 0.8 \le \frac{\lambda}{\lambda_c} \le 2.0 \quad \text{step-index fiber}$$

EXERCISE

What is the equivalent range for the V-parameter of Equation 2.34? Inspect the b-V and $V^2(d^2(Vb)/dV^2)$ versus V and b, if possible, do a curve fitting to obtain the approximate relationship for r/a and V (MATLAB® procedure is recommended).

EXERCISE

Refer to the technical specification of Corning SMF-28 and LEAF

    a. State the core diameter of the fibers, the spot size or mode field diameters of the fibers.
    b. Thence estimate the effective areas of these fibers.
    c. What is the ration of the effective area and the physical area of the cores of the fibers.

14.3.3.5 Equivalent: Step: Index (ESI) Description

As we have seen in Section 14.2, there are two possible orthogonally polarized modes (E_x, H_y) and (E_y, H_x) that can be propagating simultaneously. These modes are usually approximated by a single linearly polarized (LP) mode. These mode properties are well known and well understood for step-index optical fibers and analytical solutions are also readily available.

Unfortunately, practical SM optical fibers never have perfect step-index profile because of the variation of the dopant diffusion and polarization. These non-step index fibers can be approximated, under some special conditions, by an equivalent-step index (ESI) profile technique.

A number of index profiles of modern single mode fibers, for example, non-zero dispersion shifted fibers, are shown in Figure 14.8. The ESI profile is determined by approximating the fundamental mode electric field spatial distribution $\psi(r)$ by a Gaussian function, as described in Section 14.2 (b). The electric field can thus be totally specified by the e^{-1} width of this function or mode spot

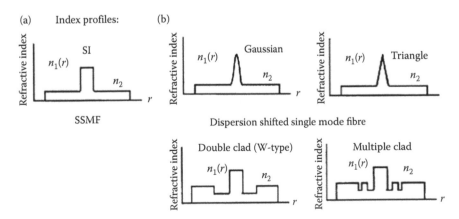

FIGURE 14.8 Index profiles of a number of modern fibers, for example, dispersion shifted single mode fibers.

size (r_0). Alternatively, the term mode field diameter (MFD) is also used and equivalent to twice the size of the mode spot size r_0.

14.3.3.5.1   Definitions of ESI Parameters

The ESI description can be used to design SM fiber with graded index, W- or segmented core profiles (under some limitations). These non-step index profiles can be described by ESI parameters denoted as follows:

V_e = effective or equivalent V-parameter
a_e = ESI core radius
λ_{ec} = ESI cut-off wavelength
Δ_e = equivalent relative index difference

These parameters are related to two moments M_0 and M_1 defined as

$$M_n = \int_0^\infty [n^2(r) - n^2(a)] r^n dr \tag{14.62}$$

For $n = 1, 2$. The effective V_e *parameter* and effective core radius r_e are given by

$$V_e^2 = 2k^2 \int_0^\infty [n^2(r) - n^2(a)] r\, dr \tag{14.63}$$

$$V_e^2 = 2k^2 M_1 \quad \text{and} \quad a_e = 2\frac{M_1}{M_0} \tag{14.64}$$

It is followed from Equations 14.63 and 14.64, the parameters λ_{ec} and Δ_e by setting

$$V_e^2 = 2k^2 a_e^2 n_1^2 \Delta_e \tag{14.65}$$

and $V_e = 2.405$ (cut-off condition for step-index). Therefore, the cut-off wavelength for an ESI profile fiber is:

$$\lambda_{ec} = \frac{2\pi\sqrt{2M_1}}{2.405} \tag{14.66}$$

It is noteworthy that V_e as given in Equation 14.42 is equivalent to the mode volume. Physically, the significance of V_e can be compared to the average density of a disk with a local density equal to $[n^2(r) - n^2(a)]$.

14.3.3.5.2   Accuracy and Limits

The ESI approximation is generally accurate to within 2% at least over the wavelength range $0.8 < \lambda/\lambda_c < 1.5$. For most practical purposes, this range is the operating wavelength to minimize the dispersion property of SM optical fibers.

14.3.3.5.3   Examples on ESI Techniques

14.3.3.5.3.1   *Graded-Index Fibers*   These index profiles of graded fibers are given by Equation 14.35. We thus have

$$n^2(r) - n^2(a) = s(r/a) = 1 - \left(\frac{r}{a}\right)^\alpha \tag{14.67}$$

Substituting Equation 14.67 into Equation 14.63 gives

$$\frac{V_e}{V} = \left(\frac{\alpha}{\alpha+2}\right)^{1/2} \tag{14.68}$$

where $V^2 = k^2 a^2 \, NA$ is the V parameter of a step-index fiber with the core index at the fiber axis of n_1. Hence we have

$$\lambda_{ec} = \frac{V}{2.405}\left(\frac{\alpha}{\alpha+2}\right)^{1/2} \tag{14.69}$$

EXERCISE

For a single mode optical fiber with a triangular profile index distribution whose equivalent V-parameter is equal to two at 1550 nm wavelength, what is the V-parameter value at the center of the core of the fiber? If the diameter of the core of the two fibers are kept identical, then what is the ration of the refractive indices at the core center of the fibers? Repeat for parabolic profile.

14.3.3.5.3.2 Graded-Index Fiber with a Central Dip The fiber index profile with central dip and graded gradually increase to the outer cladding is shown in Figure 14.9.

Similarly to Equation 14.34, for graded index fiber with maximum index at the core axis, we have

$$S(r/a) = 1 - \gamma(1-x)^\alpha \quad \text{for } 0 < r < a \tag{14.70}$$

where γ is the depth and $0 < \gamma < 1$. When $r = 0$ we have step-index profile and $r = 1$ we have the central axis refractive index is equal to the cladding index.

Using Equations 14.62 and 14.63, V_e can be easily found and given by

$$\frac{V_e^2}{V^2} = 1 - \frac{2\gamma}{(\alpha+1)(\alpha=2)} \tag{14.71}$$

14.3.3.5.4 General Method

The general technique to find the ESI parameters for optical fibers can be started by rising the stationary expression in (14.47) for expressing β of the actual fiber as compared to its equivalent propagation constant β_e as

$$\beta^2 = \beta_e^2 + k^2 \frac{\int_0^\infty \left[n^2(r) - n_e^2(r)\right] r\psi^2(r)dr}{\int_0^\infty r\psi^2(r)dr} \tag{14.72}$$

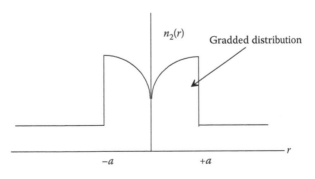

FIGURE 14.9 Refractive index profile of a graded index fiber with a central dip. This is typical profile of manufactured fiber if a good collapsing of the preform is not achieved.

where $n_e^2(r)$ is the equivalent counterpart of $n(r)$ when the fiber is expressed in its equivalent step form. The field expression $\psi(r)$ is assumed (in fact to be obtained) to be similar for the actual fiber and its step equivalence. Once the field $\psi(r)$ can be replaced by the approximate exact field shape, we can find V_e and a_e that minimize $\beta^2 - \beta_e^2$ in Equation 14.72. Generally, these parameters are functions of both V and a, thus it is impossible to get one ESI technique applicable to a wide range of wavelength and it is required to apply complicated numerical calculations.

14.4 NON-LINEAR OPTICAL EFFECTS

In this section, the non-linear effects are described. These effects play important roles in the transmission of optical pulses along single mode optical fibers. The non-linear effects can be classified into three types: the effects that change the refractive index of the guided medium due to the intensity of the pulse and the self-phase modulation; the scattering of the lightwave to other frequency-shifted optical waves when the intensity reaches over a certain threshold, the Brillouin and Raman scattering phenomena; and the mixing of optical waves to generate a fourth waves, the degenerate four wave mixing. Besides these non-linear effects there is also photorefractive effect which is due to the change of refractive index of silica due to the intensity of ultra-violet optical waves. This phenomenon is used to fabricate grating whose spacing between dark and bight region satisfying Bragg diffraction condition. These are fiber Bragg gratings and would be used as optical filters and dispersion compensator when the spacing varies or chirped.

14.4.1 Non-Linear Phase Modulation Effects

All optical transparent materials are subject to the change of the refractive index with the intensity of the optical waves, the optical Kerr effect. This physical phenomenon is originated from the anharmonic responses of electron of optical fields leading to the change of the material susceptibility. The modified refractive index $n_{1,2}^K$ of the core and cladding regions of the silica-based material can be written as

$$n_{1,2}^K = n_{1,2} + \overline{n_2} \frac{P}{A_{eff}}$$
(14.73)

where n_2 is the non-linear index coefficient of the guided medium, the average typical value of n_2 is about $2.6 \times 10^{-20}\,\mathrm{m}^2/\mathrm{W}$. P is the average optical power of the pulse and A_{eff} is the effective area of the guided mode. The non-linear index changes with the doping materials in the core. Although the non-linear index coefficient is very small, but the effective area is also very small, about 50–70 $\mu\mathrm{m}^2$, and the length of the fiber under the propagation of optical signals is very long and the accumulated phase change is quite substantial. This leads to the self-phase modulation (SPM) and cross-phase modulation (XPM) effects in the optical channels.

14.4.1.1 Self-Phase Modulation

Under a linear approximation, we can write the modified propagation constant of the guided linearly polarized mode in a single mode optical fiber as

$$\beta^K = \beta + k_0 \overline{n_2} \frac{P}{A_{eff}} = \beta + \gamma P$$
(14.74)

where $\gamma = (2\pi\overline{n_2})/(\lambda A_{eff})$.

An important non-linear parameter of the guided medium taking an effective value from one to five $(\mathrm{kmW})^{-1}$ depends on the effective area of the guided mode and operating wavelength. Thus, the smaller the mode spot size or mode field diameter, the larger the non-linear self-phase modulation effect. For dispersion compensating fiber, the effective area is about $15\,\mu m^2$, while for SSMF and NZ_DSF, the effective area ranges from 50 to $80\,\mu m^2$. Thus, the non-linear threshold power of DCF is much lower than that of SSMF and NZ-DSF. We will see later that the maximum launched power into DCF would be limited at about 0 dBm or 1 mW to avoid non-linear distortion effect, while we see about 5 dBm for SSMF.

The accumulated non-linear phase changes due to the non-linear Kerr effect over the propagation length L is given by

$$\phi_{NL} = \int_0^L (\beta^K - \beta)dz = \int_0^L \gamma P(z)dz = \gamma P_{in}L_{eff} \qquad (14.75)$$

with $P(z) = P_{in}e^{-\alpha z}$

As the representation of the attenuation of the optical signals along the propagation direction z. Notice that the non-linear SPM effect is small compared with the linear chromatic dispersion effect one can set $\phi \ll 1$ or $\phi = 0.1 rad$ and the effective length of the propagating fiber is set at $L_{eff} = 1/\alpha$ with optical losses equalized by cascaded optical amplification sub-systems. Then the maximum input power to be launched into the fiber can be set at

$$P_{in} < \frac{0.1\alpha}{\gamma N_A} \qquad (14.76)$$

For $\gamma = 2\,(W \cdot km)^{-1}$ and $N_A = 10\ \alpha = 0.2$ dB/km (or 0.0434×0.2 km^{-1}) then $P_{in} < 2.2$ mW or about 3 dBm and accordingly 1 mW for DCF. In practice, due to the randomness of the arrival "1" and "0," this non-linear threshold input power can be set at about 10 dBm as the total average power of all wavelength multiplexed optical channels launched into the propagation fiber.

14.4.1.2 Cross-Phase Modulation

The change of the refractive index of the guided medium as a function of the intensity of the optical signals can also lead to the phase of optical channels in different spectral region close to that of the original channel. This is cross-phase modulation effects (XPM). This is critical in wavelength division multiplexed (WDM) channels, and even more critical in dense WDM when the frequency spacing between channels is 50 GHz or even narrower. In such systems, the non-linear phase shift of a channel not only depends on its power, but also on that of other channels. The phase shift of the ith channel can be written as [2]:

$$\phi_N^i = \gamma L_{eff} \left(P_{in}^i + 2 \sum_{j \neq i}^M P_j \right) \qquad (14.77)$$

With M = number of multiplexed channels

The factor 2 in Equation 14.77 is due to the bipolar effects of the susceptibility of silica materials. The XPM thus depends on the bit pattern and the randomness of the synchronous arrival of the "1." It is hard to estimate, so the numerical simulation would normally be employed to obtain the cross-phase modulation distortion effects by numerical simulation using the wave propagation of the signal envelope via the non-linear Schroedinger equation. The evolution of slow varying complex envelopes $A(z, t)$ of optical pulses along a single-mode optical fiber is governed by non-linear Schroedinger equation (NLSE) [3]:

$$\frac{\partial A(z, t)}{\partial z} + \frac{\alpha}{2}A(z, t) + \beta_1 \frac{\partial A(z, t)}{\partial t} + \frac{j}{2}\beta_2 \frac{\partial^2 A(z, t)}{\partial t^2} - \frac{1}{6}\beta_3 \frac{\partial^3 A(z, t)}{\partial t^3} = -j\gamma |A(z, t)|^2 A(z, t) \qquad (14.78)$$

where z is the spatial longitudinal coordinate, α accounts for fiber attenuation, β_1 indicates DGD, β_2 and β_3 represent second- and third-order factors of fiber CD, and γ is the non-linear coefficient. This

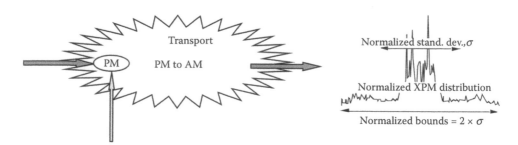

FIGURE 14.10 Illustration of XPM effects—phase modulation conversion to amplitude modulation and hence interference between adjacent channel.

equation is described in detail in Chapter 3. The phase modulation due to non-linear phase effects is then converted to amplitude modulation and thence the cross-talk to other adjacent channels. This is shown in Figure 14.10.

14.4.1.3 Stimulated Scattering Effects

Scattering of lightwave by the impurities can happen due to the absorption and vibration of the electrons and dislocation of molecules in silica-based materials. The back scattering and absorption is commonly known as Raleigh scattering losses in fiber propagation in whose phenomena the frequency of the optical carrier does not change. Other scattering processes in which the frequency of the lightwave carrier is shifted to another frequency regions is commonly known as inelastic scattering and commonly known as Raman scattering and Brillouin scattering. In both cases, the scattering of photons to a lower energy level photon with energy difference between these levels is fallen with the energy of phonons. Optical phonons are resulted from the electronic vibration for Raman scattering while acoustic phonons or mechanical vibration of the linkage between molecules lead to Brillouin scattering. At high power, when the intensity reaches over a certain threshold then the number of scattered photons is exponentially grown then the phenomena is a simulated process. Thus, the phenomena can be called as stimulated Brillouin scattering (SBS) and stimulated Raman scattering (SRS). SRS and SBS were first observed in the 1970s [3–5].

14.4.1.3.1 Stimulated Brillouin Scattering (SBS)

Brillouin scattering comes from the compression of the silica materials in the presence of an electric field called the electrostriction effect. Under the pumping of an oscillating electric field of frequency f_p, an acoustic wave of frequency F_a is generated. Spontaneous scattering is an energy transfer from the pump wave to the acoustic wave and then a phase matching to transfer a frequency shifted optical wave of frequency as a sum of the optical signal waves and the acoustic wave. This acoustic wave frequency shift is around 11 GHz with a bandwidth of around 50–100 MHz (due to the gain coefficient of the SBS) and a beating envelope would be modulating the optical signals. Thus, jittering of the received signals at the receiver would be formed, hence the closure of the eye diagram in the time domain.

   Once the acoustics wave is generated it beats with the signal waves to generate the side band components. This beating beam acts as a source and further transfer the signal beam energy into the acoustic wave energy and further amplify this wave to generate further jittering effects. The Brillouin scattering process can be expressed by the following coupled equations [6].

$$\frac{dI_p}{dz} = -g_B I_p I_s - \alpha_p I_p$$

$$-\frac{dI_s}{dz} = +g_B I_p I_s - \alpha_s I_s$$

$$(14.79)$$

I_p = intensity of pump beam
I_p = intensity of signal beam
g_B = Brillouin scattering gain coefficient
α_s, α_p = losses of signal and pump waves

The SBS gain g_B is frequency dependent with a gain bandwidth of around 50–100 MHz for pump wavelength at around 1550 nm. For silica fiber, g_B is about $5e - 11$ mW^{-1}. The threshold power for the generation of SBS can be estimated (using Equation 14.79) as

$$g_B P_{th\_SBS} \frac{L_{eff}}{A_{eff}} \approx 21 \tag{14.80}$$

with the_effective_length $L_{eff} = (1 - e^{-\alpha L}/\alpha)$
For SSMF, this SBS power threshold is about 1.0 mW. Once the launched power exceeds this power threshold level, the beam energy is reflected back. Thus, the average launched power is usually limited to a few dBm due to this low threshold power level.

14.4.1.3.2 Stimulated Raman Scattering (SRS)

Stimulated Raman scattering occurs in silica-based fiber when a pump laser source is launched into the guided medium, the scattering light from the molecules and dopants in the core region would be shifted to higher energy level and then jump down to lower energy level hence amplification of pho-tons in this level. Thus, a transfer of energy from different frequency and energy level photons occurs. The stimulated emission happens when the pump energy level reaches above the threshold level. The pump intensity and signal beam intensity are coupled via the coupled equations:

$$\frac{dI_p}{dz} = -g_R I_p I_s - \alpha_p I_p$$

$$-\frac{dI_s}{dz} = +g_R I_p I_s - \alpha_s I_s \tag{14.81}$$

I_p = intensity of pump beam
I_p = intensity of signal beam
g_B = Raman scattering gain coefficient
α_s, α_p = losses of signal and pump waves

The spectrum of the Raman gain depends on the decay lifetime of the excited electronic vibration state. The decay time is in the range of 1 ns and Raman—gain- bandwidth is about 1 GHz. In single mode optical fibers, the bandwidth of the Raman gain is about 10 THz. The pump beam wavelength is usually about 100 nm below the amplification wavelength region. Thus, to extend the gain spectra, a number of pump sources of different wavelengths are used. Polarization multiplexing of these beams is also used to reduce the effective power launched in the fiber to avoid the damage of the fiber. The threshold for stimulated Raman gain is given by

$$g_R P_{th\_SRS} \frac{L_{eff}}{A_{eff}} \approx 16 \tag{14.82}$$

with the_effective_length $L_{eff} = (1 - e^{-\alpha L}/\alpha)$ or $\approx 1/\alpha$ for long_length.
For fiber (SSMF) with an effective area of 50 μm^2, $g_R \sim 10^{-13}$ m/W, then the threshold power is about 570 mW near the C-band spectral region. This would require at least two pump laser sources that should be polarization multiplexed. The SRS is used frequently in modern optical communica-tions systems, especially when no undersea optical amplification is required, the distributed

amplification of SRS offers significant advantages as compared with lumped amplifiers such EDFA. The broadband gain and low gain ripple of SRS is also another advantage for DWDM transmission.

14.4.1.4 Four Wave Mixing

Four wave mixing (FWM) is considered as a scattering process in which three photons are mixed to generate the fourth wave. This happens when the momentum of the four waves satisfy a phase matching condition. That is, the condition of maximum power transfer. Figure 14.11 illustrates the mixing of different wavelength channels to generate interchannel cross talk. The phase matching can be represented by a relationship between the propagation constant along the z-direction in a single mode optical fiber as

$$\beta(\omega_1) + \beta(\omega_2) - \beta(\omega_3) - \beta(\omega_4) = \Delta(\omega) \qquad (14.83)$$

where $\omega_1, \omega_2, \omega_3, \omega_4$ are the frequencies of the first to fourth waves and Δ is the phase mismatching parameter. In the case that the channels are equally spaced with a frequency spacing of Ω, as in DWDM optical transmission, then we have $\omega_1 = \omega_2$; $\omega_3 = \omega_1 + \Omega$; $\omega_4 = \omega_1 - \Omega$. One can use the Taylors series expansion around the propagation constant at the center frequency of the guide carrier β_0, then we can obtain [7]:

$$\Delta(\omega) = \beta_2 \Omega^2 \qquad (14.84)$$

The phase matching is thus optimized when β_2 is zero that means that in the region where there is no dispersion FWM is maximum thus biggest interchannel crosstalk. This is the reason why dispersion-shifted fiber is not commonly used when the zero-dispersion wavelength is fallen in the spectral region of operation of channel. In modern transmission fiber, the zero-dispersion wavelength is shifted to outside the C-band, say 1510 nm, so that there is a small dispersion factor at 1550 nm and the C-band ranging from 2–6 ps/nm · km, or, for example, Corning LEAF or non-zero dispersion-shifted fibers (NZ-DSF). This small amount of dispersion is sufficient to avoid the FWM with a channel spacing of 100 or 50 GHz.

The XPM signal is proportional to instantaneous signal power. Its distribution is bounded <5 channels and otherwise effectively unbounded. Thus, the Link budgets include XPM evaluated at maximum outer bounds.

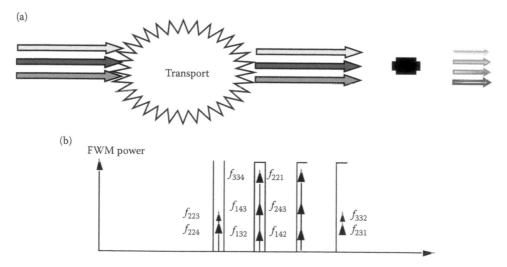

FIGURE 14.11 Illustration of four wave mixing of optical channels (a) momentum vectors of channels and (b) frequencies resulted from mixing of different channels.

14.4.2 LARGE EFFECTIVE AREA FIBERS FOR ULTRA-HIGH CAPACITY

Optical transmission technology has, in recent years, advanced rapidly with the combination of multi-level modulation formats, coherent detection, and digital signal processing (DSP) [8]. 100 Gb/s optical transport systems have been widely deployed, and 400 Gb/s and 1 Tb/s superchannels have been demonstrated in laboratories and field trials [9]. While transmission technology is necessary to transmit large capacities at higher data rates over longer distances, optical networking is now gradually evolving toward elastic/flexible optical transport network (flex OTN) or flex-grid optical network architecture [10,11] to make better use of optical network resources to accommodate the ever-increasing traffic demand. In flex OTN, the required minimum spectral resources are allocated adaptively, based on traffic demand and network conditions, and the Flex OTN may also be able to tradeoff between spectral efficiency (SE) and the all-optical reach distance. Currently, the Flex OTN can be achieved by relying on the flex OTN transponder techniques that are permissible to vary either symbol rate, code rate, or modulation formats. One example of such elastic transponders is to support several modulation formats such as polarization-division-multiplexed quadrature phase shift keying (PD-QPSK) and PD-16 quadrate amplitude modulation (PD-16QAM) [12]. A combination of transmission and networking technologies may help to make effective use of available optical networking resources. However, the most fundamental improvement to the efficiency in utilization of optical network resources is to optimize the optical fiber cabling infrastructures to provide better capacity upgrade pathways and ensure sufficient transmission margins [13]. Another effective way is to improve amplification technologies that are widely utilized in an optical network [14,15].

Historically, there is no doubt that transmission fibers have been critical to the enormous success of optical communication technology. Optical fibers, as one of the enabling technologies used for optical networks, have evolved for several decades, and their optical properties have been engineered for low system cost and high transmission performance. Recently, major optical fiber cable companies are focused on research and development (R and D) on a new class of transmission fibers [3–7,16–20] that have large effective areas (A_{eff}) and ultra-low attenuation for 400G and beyond PDM coherent transport systems. In this section, we briefly summarize the recent development of new large-area low-loss fibers and advanced amplifier technologies for next generation high-capacity terrestrial long-haul optical networks.

This section describes the key optical fiber properties of new class fibers and briefly discusses their impacts on the transmission performance for 400 Gb/s multi-level modulated coherent transport systems. Practical consideration of the large-area fibers, such as splicing and cabling for terrestrial optical networks, are briefly addressed, and the experimental results of transmission performance of new large-area low-loss fibers on a number of advanced high-order modulation format systems will also be presented.

Next-generation optical networks are under planning and 400 Gb/s and beyond would be deployed for transport systems, which employ coherent digital detection and PDM high-level modulation formats such as 16 QAM. In such systems, the chromatic dispersion and polarization-mode-dispersion (PMD) can be digitally compensated in the electrical domain; however, with wavelength division multiplexing (WDM), the cross non-linearities make neighboring channels interact depending not only on their power, but also on the state of polarization (SOP) of signals in such systems. The last is particularly problematic in polarization multiplexed coherent systems, as this is more sensitive to cross-polarization modulation (XPolM) [8], which is associated with non-linear polarization rotation in fiber transmission links. This XPolM does not have an effect on NRZ non-coherent transmission systems, but it has an impact on a coherent system because the polarization tracking in digital receivers cannot follow fast (symbol-to-symbol) polarization changes. As a result, the polarization-multiplexed coherent systems, in particular for high-order QAM, are more susceptible to fiber non-linear effects. Early research work showed at least 2–3 dB more non-linear power penalties in a QPSK coherent system compared with that in direct detection systems. This XPolX non-linear effect is stochastic in nature from fiber transmission links; hence, it impairs the effectiveness of digital compensation. In addition, the high-level modulation formats require a much higher optical signal-to noise ratio (OSNR). For example, in order to achieve the same performance at the same transmission distance, upgrading from QPSK-based 100 Gb/s to 16QAM-based 200 and 400 Gb/s requires an OSNR improvement of about 6.5 and 10 dB, respectively. Hence, it is desirable to use new fiber types that have low non-linear coefficients to retain the long-haul transmission capability when scaling up the spectral efficiency (SE) and the per-channel data rate. Recently, a new class of transmission fibers with ultra-large effective areas (A_{eff}) in a range of 120–155 μm^2 and the attenuation loss as 0.1460 dB/km has been reported for terrestrial and submarine systems [16–20]. To reduce the fiber non-linearity, which is proportional to n_2/A_{eff}, where n_2 is the non-linear refractive index, it is necessary to increase the A_{eff} of the fiber while not jeopardizing the bending performances. Step-index profiles with small index differences and large core diameters can be used to design large A_{eff} fibers. However, the light confinement in the core becomes weaker, increasing the cutoff wavelength and degrading both macro- and micro-bending performance. Placing a depressed-index region in the cladding, or placing a trench slightly apart from the core can improve the bending performance and reduce the cutoff wavelength [17]. A low Young's modulus primary coating [19] can also be used to improve the micro- and macro-bending performance and to permit a larger fiber A_{eff}.

A simple figure of merit (FOM) was proposed to compare the new class fibers with standard single-mode fiber (SSMF), which can be expressed as follows:

$$\text{Fig of merit} = 10\text{Log}_{10}\frac{A_{eff}}{A_{eff}^{SSMF}} - (\alpha - \alpha^{SSMF})L - 10\text{Log}_{10}\frac{L_{eff}}{L_{eff}^{SSMF}} \quad (14.85)$$

where the FOM (in dB) is the OSNR improvement of an optical fiber transmission system for new types of fibers compared with SMF, the first term considers the improvement from effective area (A_{eff}), and the second term is from the fiber attenuation (α in dB/km); here, L and L_{eff} is the span and the fiber effective length, respectively. It can be seen from Equation 14.1 that increasing the fiber A_{eff} is the most fundamental improvement for coherent transport, and reducing fiber attenuation can improve span loss; however, the reduction of fiber attenuation can increase the fiber effective non-linear length, hence increasing the accumulated non-linear impairments. Figure 14.12 shows an example of contour maps of relative FOM normalized to SSMF (αdB/km = 0.19 dB/km, A_{eff} = 80 μm^2, span length L = 80 km) as a function of αdB/km and A_{eff}; clearly, the OSNR performance in

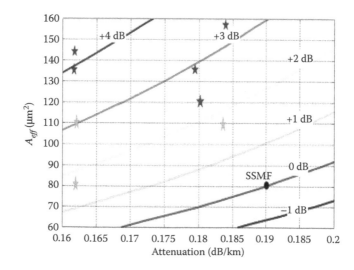

FIGURE 14.12 Relative FOM of large-area low-loss fiber normalized to SSMF for 80 km span transmission. (After B. Zhu et al. *IEEE Photonics Technol. Lett.*, vol. 22, no. 11, 2010 with permission.)

large-area low-loss fiber can be improved substantially. It should also be noted that the relative OSNR improvement in FOM depends on the span length of the transmission systems. In addition, it is also important to have high chromatic dispersion in new fibers to reduce non-linear effects in the PDM coherent systems where chromatic dispersion can be digitally compensated. We have recently developed large $A_{eof\,ff}$ TeraWave fiber for 100G and beyond terrestrial long-haul optical networks. The TeraWave fiber is a single-mode fiber with germanium-doped core and a depressed-index inner cladding region and is fabricated similarly as an All-Wave zero-water-peak fiber. Considering the features of terrestrial cabling such as craft splicing, closures, and macro- and micro-bending, the A_{eff} of TeraWave fiber is optimized to be 125 μm^2. The average fiber loss, dispersion, and dispersion slope at 1550 nm are 0.185 dB/km, 20.0 ps/nm/km, and 0.06 ps/nm^2/km, respectively. TeraWave fiber is ITU-T G.654.B [11] compliant; it uses the DLUX Ultra coating for excellent micro-bending performance, and it meets all macro-bending requirements in G.652.D and G.654.B. Volume splicing study using a commercially available splicer (e.g., FiTel S178A) with a standard splice recipe shows that the averaged splice loss between TeraWave to TeraWave is 0.04 dB/splice, and TeraWave to SMF is about 0.15 dB/splice. With optimization of splicing programs using commercially available splicers, the splicing loss between TeraWave to SSMF is below 0.10 dB/splice.

Recently, TeraWave ultra-large area single mode optical fibers have been deployed, whose core effective area of 125 μm^2 offering an ultralow-loss type of an averaged fiber loss and dispersion at 1550 nm are 0.168 dB/km and 20.0 ps/nm/km, respectively. The fiber is optimized for long-haul transmission in the C + L-bands (1530–1625 nm) at 100, 400 Gb/s, and beyond for terrestrial optical networks, and it supports greater distances between regeneration and amplification sites, helping to lower the overall cost of deploying coherent systems. We have systematically and experimentally investigated the system performance of TeraWave fiber in a 485 Gb/s coherent optical orthogonal frequency-division multiplexed (CO-OFDM) superchannel long-haul transmission system [16] and compared it with the SSMF. The CO-OFDM superchannel coherent system has advantages, including high SE, reduced guard band, and lower modulation baud rate, and it is a potential candidate for future high-capacity Tb/s-per-channel optical networks. However, the CO-OFDM superchannel systems are more susceptible to fiber non-linearity when compared with other schemes. This is because the multi-subcarrier configuration with small guard band in CO-OFDM transmission leads to the impairment of intersubcarrier non-linear interference. Such large area fibers are employed in the experiment; the 485 Gb/s CO-OFDM was generated with PDM-16QAM five-subcarrier modulation, and it was done with a dispersion uncompensated link with 80 km fiber span length using two amplification schemes, erbium doped-fiber amplifiers (EDFAs), and hybrid EDFA/Raman amplifiers. The comparison experiment showed that the optimum signal launch power into fiber spans is about 2 dB higher in TeraWave fiber than in SSMF for transmitting of 485 Gb/s CO-OFDM signals over 1600 km (20 × 80 km) spans under both EDFA-only and hybrid Raman/EDFA amplification schemes. With EDFA-only amplification, the TeraWave fiber offers an ~2 dB higher optimum Q^2 factor than the SSMF after 1600 km transmission. It is about 1 dB higher for the hybrid amplification scheme (Figure 14.13). It was also found that TeraWave fiber allows transmission more than 60% longer than for a SSMF link at a similar $Q2$ factor performance for 400G OC-OFDM systems.

We have further experimentally investigated the transmission performance of TeraWave fiber in a typical 256 Gb/s PDM-16QAM DWDM system with 37.5 and 50 GHz channel spacing. The 256 Gb/s PD-16QAM is a promising modulation format for two carrier 400 Gb/s systems due to a relatively simple scheme and maturity of opto-electronic components. The experiment was conducted with a dispersion uncompensated link with a 100 km fiber span length using three different amplification schemes, including EDFA-only, hybrid EDFA/Raman amplifiers, and all backward-pumped Raman amplifiers. It has been found that, when there is no ROADM, ten 256 Gb/s PDM-16QAM channels were transmitted over 4200, 3500, and 2000 km over TeraWave fibers within a 20% soft-decision FEC (SD-FEC) threshold, respectively, in the system with both 37.5 and 50 GHz channel spacing (see Figure 14.14). The all backward-pumped Raman amplifiers and hybrid EDFA/ Raman amplifiers can increase the transmission distances by about 100% and 70%, respectively,

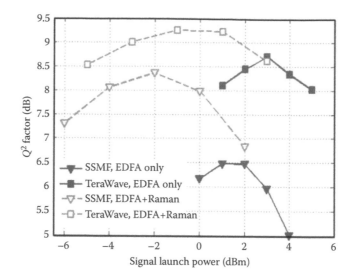

FIGURE 14.13 Measured Q^2-factor versus launch powers of 485 Gb/s CO-OFDM signal transmission over 1600 km of ULAF and SSMF. (From B. Zhu et al., *IEEE Photon. Technol. Lett.*, vol. 23, pp. 1400–1402, 2011.)

when compared with EDFA-only amplifiers. It was also found that cascaded ROADMs have a small impact on the system with the 50 GHz channel spacing. The large-area low-loss fiber has also been studied in other advanced higher-order modulation format systems (for example, time-domain hybrid 32–64 QAM, for high SE 400 Gb/s ultra-long-haul transmission systems), and it has been demonstrated that the 400 Gb/s-class DWDM signals on the standard 50 GHz ITU-T grid, which is 8.25 b/s/Hz net SE, were transmitted over 4000 km of large-area low-loss fiber with 100 km span length for a terrestrial optical network. The above results demonstrated that the large-area low-loss fibers are beneficial for system reach and margin irrespective of amplifier configuration; they increase system SE and have much better efficiency in utilization of optical network resources

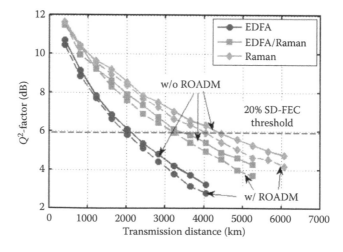

FIGURE 14.14 $Q2$ factor versus transmission distance of the 10 channel 256 Gb/s PDM 16QAM system using the three different amplification schemes without (solid curves) and with ROADMs (dashed curves) with 50 GHz channel spacing. Optimum launch powers are used.

for future Flex OTN. Further enlargement of A_{eff} (e.g., $> 200\,\mu m^2$) would inevitably deteriorate bending and cut-off behavior and cause high splicing loss. Hence, fiber design compromises must be made between improved transmission performance and limitations from careful handling and splicing of new fibers.

14.5   OPTICAL FIBER MANUFACTURING AND CABLING

This section is devoted to a brief description of the manufacturing of optical fibers and the cabling of several fibers for optical communications systems. The manufacturing techniques and cabling process affect the transmission and physical properties of the fibers. We focus on the aspects for a general understanding for optical transmission systems.

As we have described in previous sections, the standard single mode optical fiber (SSMF) structure is a cylindrical core with a refractive index slightly higher than that of the cladding region. For optical communications operating in the 1300 and 1700 nm wavelength regions, the silica material is the base material. A "pure" silica tube is the starting structure and a combination of silica, gemanimum oxide GeO_2 and P_2O_5, are then deposited inside the tube. Other dopants such as B_2O_3 and fluoride can also be used to reduce the refractive index of some small regions of the core. Once the deposition of the impurities is done (see Figure 14.17), the tube is collapsed to produce a perform as shown in Figures

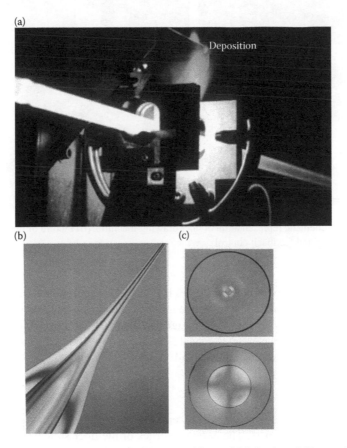

FIGURE 14.15 Schematic representation of a fiber deposition and fabrication of fiber perform, deposition of core material and collapsing, (b) fiber perform, before drawing into fiber stands, and (c) cross section of fiber perform with refractive index profile exactly the same as the fiber index profile of single mode (upper) and multi-mode (lower) types.

FIGURE 14.16   Real index profile across a SM fibre perform. Note: non-step like profile—so why modeled as step-index structure?

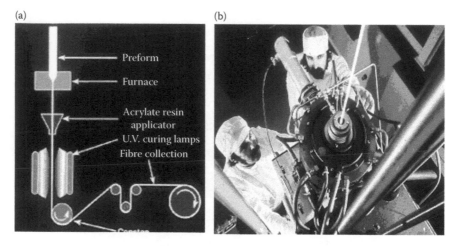

FIGURE 14.17   (a) Schematic representation of fiber drawing machine. (b) Picture of fiber microwave furnace and diameter monitoring and feedback control.

14.15 through 14.17. Also shown in these figures is a schematic representation of the fiber drawing machine and fiber drawing tower. The refractive index of the perform is also shown as noted in its caption and its details area in Figure 14.17. Figure 14.18 shows the installation of fiber cables by hanging, by ploughing, for undersea environment. This figure also shows the two optical fibers under splicing to repair a broken or damage fiber strand.

14.6   CONCLUDING REMARKS

This chapter has introduced the fundamental concepts of optical slab waveguide and then the circular optical waveguide or optical fibers. The basic properties of the fiber structures, its profile, the spot size, the cut-off wavelength, and the Gaussian approximation are described. The Gaussian approximation makes the understanding of the optical guided mode simple. It also allows us to obtain directly the optical mode distribution and thus several other approximations required to obtain the simplest form of important parameters of single mode optical fibers.

(a)

(b)

(c)

(d)

FIGURE 14.18 Installation of fiber cables (a) installation of fiber cable by hanging, (b) installation of fiber cable by ploughing, (c) installation of undersea fiber cable, and (d) splicing two optical fibers.

Once the basic properties of a single mode optical fiber are found, they form the basic set of parameters so that optical fibers whose effective index profiles are non-step can be found based on the ESI technique that would converts the parameters to an equivalent step like profile and hence other optical properties.

Only structural and waves properties of lightwave signals travelling in optical fibers are presented here. As optical communications systems engineers, we have to understand and develop techniques for analyzing and identifying the transmission of digital and analogue signals through optical fibers and understand that the attenuation and broadening of optical signals after transmitted through the transmission medium, namely attenuation and broadening via dispersion of lightwaves pulses.

REFERENCES

1. K. C. Kao and G. A. Hockham, Dielectric-fibre surface waveguides for optical frequencies, *IEE Proceedings*, vol. 133, Pt. J, no. 3, pp. 191–198, June 1986.
2. G. P. Agrawal, *Fiber Optic Communications Systems*, ed., New York, NY: J. Wiley, 2002.
3. R. H. Stolen, E. P. Ippen, and A. R. Tynes, Raman oscillation in glass optical waveguides, *Appl. Phys. Let.*, vol. 20, p. 62, 1972.
4. E. P. Ippen and R. H. Stolen, Stimulated Brillouin scattering in optical fibers, *Appl. Physics Lett.*, vol. 21, p. 539, 1972.
5. R. G. Smith, Optical power handling capacity of low loss optical fibers as determined by stimulated Raman and Brillouin scattering, *Appl. Optics*, vol. 11, pp. 2489, 1972.

6. G. P. Agrwal, *Fiber Optic Communications Systems*, ed., New York, NY: J. Wiley, 2002, p. 60.
7. G. P. Agrawal, *Fiber Optic Communications Systems*, ed., New York, NY: J. Wiley, 2002, p. 67.
8. L. N. Binh, *Advanced Digital Optical Communications Systems*, Boca Raton, FL: CRC Press, 2015.
9. L. N. Binh, *Optical Fiber Communication Systems with MATLAB® and Simulink® Models*, edn., Boca Raton, FL: CRC Press, 2016.
10. O. Gerstel, M. Jinno, A. Lord, and S. Yoo, Time elastic optical networking: A new dawn for the optical layer? *IEEE Commun. Mag.*, vol. 50, no. 2, pp. s12–s20, 2012.
11. G. Wellbrock and T. J. Xia, How will optical transport deal with future network traffic growth? *European Conf. on Optical Communication*, Cannes, France, 2014, paper Th.1.2.1.
12. A. Bononi, P. Serena, N. Rossi, and D. Sperti, Which is the dominant nonlinearity in long-haul PDM-QPSK? *European Conf. on Optical Communication*, Torino, Italy, 2010, paper Th.10.E.1.
13. G. Charlet, Impact and mitigation of non-linear effects in coherent transmission, *Optical Fiber Communication Conf.*, San Diego, California, 2009, paper NthB.4.
14. G. Charlet, Fiber characteristics for next-generation ultralong-haul transmission systems, *European Conf. on Optical Communication*, Torino, Italy, 2010, paper We.8.F.1.
15. Characteristics of a cut-off shifted single-mode optical fiber and cable, ITU-T Recommendation G.654, Oct. 2012, [Online]. Available: www.itu.int/rec/T-REC-G.654-201210-I/.
16. B. Zhu, S. Chandrasekhar, X. Liu, and D. W. Peckham, Transmission performance of a 485-Gb/s CO-OFDM superchannel with PDM-16QAM subcarriers over ULAF and SSMF-based links, *IEEE Photon. Technol. Lett.*, vol. 23, pp. 1400–1402, 2011.
17. M. Bigot-Astruc and P. Sillard, Realizing large effective area fibers, *Optical Fiber Communication Conf.*, Los Angeles, California, 2012, paper OTh4I.1.
18. S. Makovejs et al., Record-low (0.1460 dB/km) attenuation ultra-large Aeff optical fiber for submarine applications, *Optical Fiber Communication Conf.*, Los Angeles, California, 2015, paper Th5A.3.
19. Y. Yamamoto, M. Hirano, K. Kuwahara, and T. Sasaki, OSNR-enhancing pure-silica-core fibre with large effective area and low attenuation, *Optical Fiber Communication Conf.*, San Diego, California, 2010, paper OTuI2.
20. B. Zhu, Large-area low loss fibers and advanced amplifiers for high capacity long haul optical network, *European Conf. on Optical Communication*, Valencia, Spain, 2015, paper We.2.4.1.
21. B. Zhu, X. Liu, S. Chandrasekhar, D. W. Peckham, R. Lingle, Ultra-long-Haul transmission of 1.2-Tb/s multicarrier no-guard-interval CO-OFDM superchannel using ultra-large-area fiber, *IEEE Photonics Technology Letters*, vol. 22, no. 11, pp. 826–828, 2010.

FURTHER READING

G. P. Agrawal, *Fiber Optic Communications Systems*, ed., New York: Academic Press, 2002, Chapter 2.
L. B. Jeunhomme, *Single Mode Fiber Optics*, New York: Marcel Dekker, 1983.
G. Mahlkc and P. Gossing, *Fiber Optic Cables*, Chichester: Siemens A.G. J. Wiley, 1987.
D. Jones, *Optical Fiber Communications Systems*, New York: Holt, Rinhart Winston, 1988.
A. W. Snyder, Understanding monomode optical fibers, *Proc. IEEE*, vol. 69, no. 1, pp. 6–13, January 1981.
D. Keck, Single mode fibers outperform multimode cables, *IEEE Spectrum*, vol. 20, pp. 30–37, March 1983.
D. Marcuse, Loss-analysis of single-mode fiber splices, *Bell Syst. Tech. J.*, vol. 56, no. 5, pp. 703–718, 1977.
P. S. Henry, Lightwave primer, *IEEE Journal Quantum Electronics*, vol. QE-21, no. 12, pp. 1862–1879, December 1985.

15 Introduction to Design of Optical Transmission Systems

15.1 INTRODUCTION

Chapter 15 has been dedicated to the study of optical fibers and their operation parameters, and optical passive, and active components and the opto-electronic receivers, as well as optical transmitters. An optical transmission system combines all these components with appropriate optical powers at the transmitter and at the receiver to satisfy the bit-error-rate (BER) requirement or the optical signal-to-noise ratio (OSNR). Shown in Figure 15.1a is the basic structure of the transmission system of multi-span and multi-wavelength channels with mid-span optical amplifiers in bidirectional arrangement. Figure 15.1b and c shows the working fiber and protection fiber structures including add/drop mux insertion.

This chapter gives a general approach to the design of optical transmission systems, especially when optical fibers are used as the transmission medium. Non-dispersion compensating fibers and compensating fibers are employed in the transmission systems that are described.

Readers may summarize the transmission properties of various photonic and optical components that make up the transmission system, mainly their insertion losses and dispersion factors, as well as spectral properties. For example, for optical amplifiers, one needs to know the minimum input power level, the optical gain, the saturated output power level, and the noise figure. Naturally, for single mode optical fibers, the dispersion factor and the attenuation factor, as well as the non-linear threshold and factor need to be known. Thus, readers can bypass the details of the optical components and parameters essential for systems design considerations. Simulink models are based on the operational principles of each component that are formed in blocks. These blocks are interconnected together to form a transmission system. Thus, the principles of operations of each physical component are very important and must be represented faithfully in Simulink to obtain accurate simulation results.

The wavelength windows over the low-loss regions of optical fibers are given in the Appendix to show the available bands of the spectrum for uses. We can see now that the O-bands now attract interest in access and data center short haul transmission. The C-band is to be extended to the lower part of the L-band so that a total number of 100 channels can occupy matching to the gain spectral window of the modern Erbium-doped fiber amplifier.

The next section gives a generic outline of the design strategy for the transmission systems. It is followed by the detailed study of the attenuation and dispersion budget for different optical transmission structures. First, a single span transmission system is developed, then multi-span optical amplified fiber transmission systems. The techniques for generation of the dispersion and power budget are given for unidirectional and bidirectional transmission. Some typical design examples are given (see Figure 15.1a and b) for bidirectional transmission of dense wavelength division multiplexed channels in the same fiber. Bidirectional transmission means the wavelength channels are transmitted and received from both ends of the fiber link. Thus, some optical components such as optical circulators are required for routing the lightwave channels. Figure 15.1c and d shows the arrangement of bidirectional amplification and transmitter, respectively.

It is noted that only digital modulation formats are described in this chapter using intensity modulation and direct detection IM/DD. Advanced optically amplified transmission systems employing coherent detection or direct detection under external modulation formats DPSK, ASK, FSK, etc. are also given.

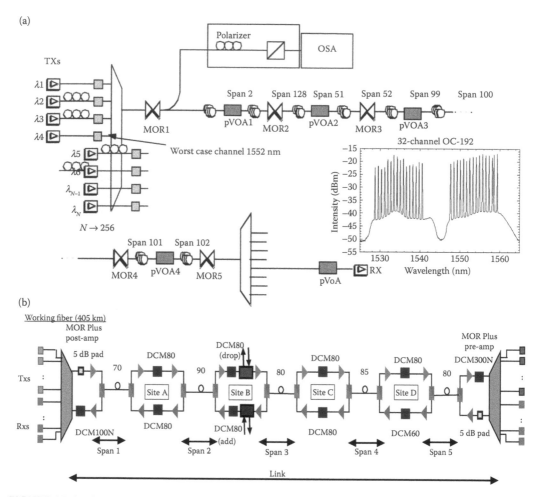

FIGURE 15.1 Optical transmission link with mid-spans and optical amplifiers—bidirection transmission in fibers and multi-wavelength channels (a) basic structure for DWDM channels—insert is the spectra of DWDM channels, (b) 405 km fiber structures. MOR = mid span optical repeater, DCM = dispersion compensating modules, DCMxxx = compensating module for xxx km of SSMF, Pad = adjustable optical attenuator. (*Continued*)

15.2 LONG HAUL SELF-COHERENT OPTICAL TRANSMISSION SYSTEMS

A long-haul transmission system, by definition, is one in which the distance between the transmitter and receiver is an important factor. That means that the maximum transmission between these sub-components of the systems must be extended without the uses of optical-electrical repeaters but optical amplifiers in mid-spans can be employed. The total distance is normally divided into a number of spans. Each span is connected through to the next by mid-span optical amplifiers/repeaters (MOR). Systems transponder will only be considered briefly as they are now very much standard sub-systems.

15.2.1 INTENSITY MODULATION DIRECT DETECTION SYSTEMS

A direct detection optical receiver requires a minimum received signal power to sustain an adequate SNR to satisfy the BER requirement. Since the fiber attenuation is dependent on the fiber length, there is a maximum path length beyond which the received signal strength would be inadequate to satisfy the SNR requirement of the receiver sub-systems.

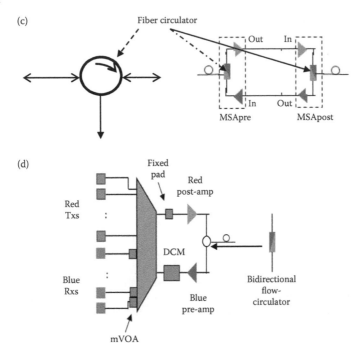

FIGURE 15.1 (Continued) Optical transmission link with mid-spans and optical amplifiers—bidirection transmission in fibers and multi-wavelength channels, (c) arrangement of optical amplifiers—mid-span amplifier (MSA) for bidirectional amplification, and (d) details of bidirectional transmitter. MSAPre and MASPost = pre- and post-optical mid-span amplification.

If, in system design, we start with specifications that include the transmitted power and the receiver sensitivity, we can estimate the maximum attenuation allowable for the transmission medium. This total attenuation includes distributed losses in fibers, coupling losses into and out of the fiber, other coupling losses at the signal monitoring site, optical add/drop mux, etc. In addition to these average losses, a system operating margin is also added to protect against aging components, occasional misalignment of fiber components or additional repairs of broken fibers by splicing.

EXAMPLE 15.1

An optical transmission system consists of (i) optical transmitter: direct intensity modulation output power = −0.5 dBm at the output of the fiber pig tail; (ii) optical receiver: receiver sensitivity = −41.5 dBm; (iii) optical fibers: average fiber attenuation and splicing losses: 0.4 dB/km; (iv) coupling losses at the receiver = 0.5 dB; (iv) operating margin = 3 dB; and (v) what is the maximum distance between the transmitter and receiver?

Solution

Optical transmitter	−0.5 dBm
Optical receiver	−41.5 dBm
Operating margin	3 dBm
Loss coupling at receiver	0.5 dB
Hence total loss allowable for fibers	41.5 − 3 − 0.5 = 38 dB
Fiber attenuation (average)	0.4 dB/km
Thus total maximum distance	38/0.4 = 38 × 2.5 km = 95 km

The receiver sensitivity is a critical parameter of the link and can be determined by the acceptable probability of error in the detection and the decision level as described in Reference 1. We recall the energy b_1 required for a bit "1" received under IM/DD as follows:

$$b_1 = \frac{q}{\Re}\left[\partial^2 G^x + \frac{2\partial i_{Neq}T_b}{qG}\right] \tag{15.1}$$

Thence, the receiver sensitivity RS is given by

$$RS = 10Log_{10}\frac{P_{av}}{P_0} \quad \text{in dBm when } P_0 = 1\,\text{mW} \tag{15.2}$$

where $P_{av} = b1/T_b$ and $P_0 = 1.0\,\text{mW}$ (the reference power level for evaluating power in dBm).

The term δ is specified by the BER curve and the statistical property of the distribution of the received "1" and "0." Hence with a specific BER, the bit period, and the photo-detector type, we could estimate the required noise equivalent current at the input of the electronic pream-plifier so that a minimum detection power should be known. This would allow us to decide whether an optical amplifier pre-amplifier is needed at the front end of the optical receiver. The decision threshold does also affect the BER and this is normally used to set the level. A typical eye diagrams at different bit rate and modulation format NRZ or RZ are shown in Figure 15.2.

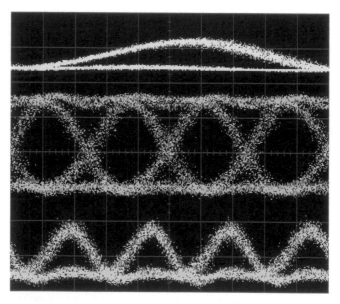

FIGURE 15.2   Typical eye diagrams after the optical receiver for different formats (NRZ or RZ) at different bit rate.

We note that the two terms within the bracket of Equation 15.1 represent the number of photons required to overcome the requirement due to the BER and the electronic noise current. This is due to the fact that the term q/R is the photon energy.

EXAMPLE 15.2

For a 10 Gb/s IM/DD transmission system. (a) What is the RS of the optical receiver if the following conditions must be met: BER = 1e-9, photodetector is PIN type with quantum efficiency of 0.8;

(b) What is the equivalent number of photons per bit that must be available at the receiver? (c) What is the number of photons required to overcome to BER requirement? (d) What is the number of photons per bit to overcome the electronic noise current? € Repeat (a)–(d) for an APD with an avalanche gain of 20.

Thus, we can see that the receiver sensitivity varies with respect to the bit rate. Naturally, the shorter the pulse width, the higher the peak power of the pulse to satisfy the average power required as estimated from the RS.

15.2.2 Loss-Limited Optical Communications Systems

The fiber loss plays an important role in the design and implementation of optical communications systems. Considering an optical transmitter generating an optical power of P_{tr}, and the optical signals are received with an optical receiver with a ability of detecting a minimum optical power of P_r at a bit rate B_r, then the maximum transmission distance would be given by

$$L = \frac{10}{\alpha_F} Log_{10} \left(\frac{P_{tr}}{P_r} \right) \tag{15.3}$$

where α is the total fiber loss per unit length (in dB/Km) including splice and connector losses. The bit rate dependence of L arises from the linear dependence of P_r on the bit rate B_r by $P_r = N_p h \nu B_r$ with N_p is the average number of photons/bit required for the receiver to detect with a certain S/N ratio.

15.2.3 Dispersion-Limited Optical Communications Systems

We have described the dispersion effect in optical fibers, which is very important for optical transmission systems when the loss is no longer a problem if optical amplifiers are employed. When the loss-limited distance is longer than that of the dispersion-limited transmission distance, then they are called dispersion-limited transmission systems.

The limit for a system with a bit rate of B_r and a total dispersion factor D_T of the fibre is given by

$$B_r L \leq \frac{1}{4|D_T|\sigma_\lambda} \tag{15.4}$$

with σ_χ is the FWHM of the source spectrum. Thus, for a standard optical fibre operating at 1550 nm, the total dispersion factor is about -15 ps/nm · km and a source FWHM of 1.0 nm the bit-rate distance product is about 166.6 (Gb/s)-Km. That is, if a 10 Gb/s operating bit rate system would be able to transmit signals for 16.6 km. If a dispersion-shifted fiber is used with a typical dispersion factor of about 1–2 ps/(nm · km) then the transmission distance can be increased about 15–7.5 times that of the SSMF or more than 100 km and up to 240 km. When these fibers are employed for extremely long distances, the polarization dispersion becomes significant and the PMD DGD parameters must be taken into account to estimate the total allowable transmission distance, especially for 10 Gb/s operating bit rate.

Several schemes are currently under investigation including the use of fiber gratings, dispersion compensators, optical signal processors/filters, and optical solitons where the non-linear effects of fibers through the self-phase modulation are use to counter the linear effects of chromatic dispersion.

15.2.4 System Preliminary Design

15.2.4.1 Single Span Optical Transmission System

In practice, the design of an optical communications system would require two principal requirements: the power budget and the rise time budget, which account for the loss limited and dispersion limited systems, respectively.

Specifications for the system would be: the system bit rate B_r and the design transmission distance. Thus, there are commonly two main cases: the point-to-point links and the ultra-long or terrestrial/ undersea links. The performance of the optical fiber systems are specified through: the bit-error-rate (BER), the operating wavelength. The cost of operating in the 810 nm window would incur lowest loss and that increases with the wavelength windows 1300 and 1550 nm. Normally, for systems operating in the 100 Mb/s with a distance less than 20 km, single mode fiber can be used, and this is the case for most LAN and Intranet applications. These systems and networks can be wired at the 810 nm window. For longer transmission systems, the two wavelength windows 1310 and 1550 nm must be used.

15.2.4.2 Power Budget

The power budget of a transmission link can be represented as shown in Figure 15.3. Depending on the availability of the power at the output of the transmitter or the receiver sensitivity of the optical receiver, one could determine the operating conditions of other sub-systems or optical components, for example, the fiber distance, the number of spans, the noise of the electronic pre-amplifiers, system operating margin, etc. Either a table of the budget of the power of the transmission link or a graph as shown in Figure 15.3 can be established.

15.2.4.3 Rise-Time/Dispersion Budget

We have previously established an approximate relationship between the dispersion, the bit rate, and the banwidth. For on–off keying (OOK), the bandwidth Δf of the detection system can be about half of the bit rate (pulse repetition rate). The relationship $B_r L \leq (1/4|D_T|\sigma_\lambda)$ can be used or the bit rate B_R can be approximately written as

$$B_R \frac{1}{4\Delta\tau} = 2\Delta f \tag{15.5}$$

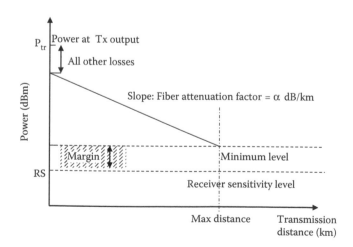

FIGURE 15.3   Power budget by graphical representation for a single transmission distance.

TABLE 15.1

RMS Pulse Width, σ, and Spectral Width, Δf of Three Typical Pulse Shapes

	$h(t)$	σ	$H(f)$	BW
Exponential pulse	$\dfrac{1}{\tau}e^{-\frac{t}{\tau}}$	τ	$\dfrac{1}{1+j\omega\tau}$	$\dfrac{0.159}{\sigma}$
Rectangular pulse	$\dfrac{1}{\tau} \quad -\dfrac{\tau}{2}\le t\le\dfrac{\tau}{2}$	$\dfrac{\tau}{2\sqrt{3}}$	$\dfrac{\sin(f\tau)}{f\tau}$	$\dfrac{0.402}{\sigma}$
Gaussian pulse	$\dfrac{1}{\sqrt{2\pi}\tau}e^{-\frac{t^2}{2\tau^2}}$	τ	$e^{-2(\pi f\tau)^2}$	$\dfrac{0.133}{\sigma}$

However, when writing the relationship between all components and sub-systems of the transmission system, we must have that the summation of all the RMS (Root mean square) of all the risetime[*] and broadening factors must be less than 70% of the bit period (NRZ or RZ bit time). This can be written as

$$\sigma = \sqrt{\sum_{i=1}^{N}\left(t_{ri}^2 + t_{fi}^2 + \Delta\tau_j^2\right)} \tag{15.6}$$

where t_{ri} and t_{fi} are is the rise and fall time, respectively, of the ith component in the transmission link and $\Delta\tau$ is the pulse broadening of the ith fiber in the link taken for both sides of a pulse after transmission through a length L optical fiber. Thus, the rise and fall time must be accounted for. Thence, we must have

$$\sigma = \sqrt{\sum_{i=1}^{N}(t_{ri}^2 + t_{fi}^2 + \Delta\tau_j^2)} \le 70\%\frac{1}{B_R} \tag{15.7}$$

That is, the total broadening of the pulse must be less than 70% of the bit period. The relationship between the RMS pulse width and the bandwidth of the system specified by the pulse shape can be found as shown in Table 15.1.

The RMS pulse width σ of a pulse shape $h(t)$ can be found by writing the normalized pulse shape and the root mean square (RMS) definition as

$$\int_{-\infty}^{+\infty} h(t)dt = 1 \tag{15.8}$$

$$\sigma = \langle t^2\rangle - \langle t\rangle^2$$

$$\text{where}\ldots\langle t\rangle = \int_{-\infty}^{+\infty} t\cdot h(t)dt\ldots\text{and}\ldots\langle t\rangle^2 = \int_{-\infty}^{+\infty} t^2\cdot h(t)dt \tag{15.9}$$

When a number of linear systems exist (e.g., optical fiber spans in cascade), then the overall impulse response can be found by convolving the impulse response of each individual fiber span.

---

[*] Rise time is measured from 10% to 90% of the rising edge of a rectangular pulse, this is usually the step function of the sub-system. Thus the rule of thumb is that the broadening factor is approximately 50% of the rise time. If the bandwidth of the sub-system is measured then the pulse broadening can be roughly $1/(4.\text{BW})$.

The resulting RMS pulse broadening can be found by taking the RMS of all pulse broadening factors. That is,

$$\sigma^2 = \sigma_1^2 + \sigma_2^2 + \sigma_3^2 + \cdots + \sigma_N^2 \tag{15.10}$$

EXAMPLE 15.4

The rise time and fall times of the transmitter and the optical detector are 1 and 0.5 ns, respectively. The bandwidth of the optical receiver is 2 GHz. The total pulse broadening of the fiber is 1.5 ns. What is the maximum bit rate of the on-off keying system?

SOLUTION

Broadening due to optical transmitter	$\frac{1}{2}$ ns
Broadening due to optical receiver	$\frac{1}{2}$ 0.5 ns
Fiber broadening	1.5 ns
Broadening due to electronic amplifiers of the optical receiver	$1/(4 \times 2 \times 1e9) = 1/8$ ns
Total RMS pulse broadening (pulse width)—impulse response	$\sigma = (0.5)^2 + (0.25)^2 + (1.5)^2 + (1/8)^2 = (0.763)^2$ ns
Thus, the maximum bit rate would be	$B_R = 1/4(\sigma) = 328.8$ MHz

15.2.4.4 Multiple-Span Optical Transmission System

For multi-span transmission systems, the power budget should be designed for the case of upgrading an existing installed optical fiber system or a to-be-installed system. For the upgrading case, normally non-amplified single mode transmission system, DCMs are to be inserted at the end of a fiber transmission spans, usually in multiple length of 40 Km section. Thus, the span length must be in order of 80 km or 100 km, depending on the quality of the single mode fibers. Two optical amplifiers are to be inserted, one at the end of the transmission fiber (normally standard single mode optical fibers) and the other at the output of the DCM. The optical power at the output of the first optical amplifier must be less than the non-linear threshold of the DCM which should be very low, about 0 dBm due to the small effective area of the DCF. The optical power at the output of the optical amplifier after the DCM must be high enough, about the launched power level into the first span. Thus, this optical amplifier must be a booster optical amplifier type. The power distribution of the optical channels along the multi-span transmission distance is shown in Figure 15.4. Ideally, the length of each span should be the same, and any excess power can be attenuated using attenuation pads. This is preferred for manufacturing of almost identical optical amplifiers in all spans and for minimizing the transmission system management. Also shown in this figure is the minimum level to ensure that this meets the minimum power level for optical amplification.

Figure 15.5 shows the dispersion budget for multi-span transmission. The minimum dispersion tolerance is to be determined depending on the modulation formats. Initially, there is no dispersion at the output of the transmitter. The dispersion is gradually increases as a function of the transmission distance in the first span. If DCM is used at the end of each span then the dispersion can be fully compensated before the transmission to the next span and so on. This dispersion map varies with respect to the change of the wavelength channel. This change can be estimated using the spectral dependence of the dispersion factor. Thus, at the end of the multi-span transmission, there is always residual dispersion at each wavelength channel. This residual dispersion can be compensated for each channel at the output of a wavelength demultiplexer before the optical signals are

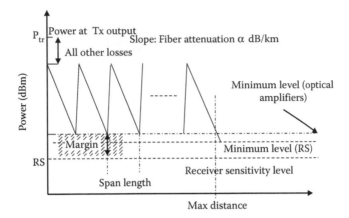

FIGURE 15.4 Power budget by graphical representation for multi-span transmission system.

received. Tuneable fiber Bragg gratings (FBG) can be used to tune to match the residual dispersion. Alternatively, this dispersion can be compensated in the electronic domain by digital signal processors.

15.2.5 Gaussian Approximation

In many cases, the optical transmitter, the impulse responses of the single mode optical fiber can be approximated as Gaussian pulse shape in the far field while it is strongly chirped in the near field region, as shown Figure 15.6. When they are in cascade, the overall impulse response of the transmission system can be represented as Gaussian. Thus, the impulse response and its Fourier transform to give the frequency response of the system are given by[*] (see Figures 15.6 and 15.7 [1] and Appendix)

$$h(t) = e^{-\frac{t^2}{2\tau^2}} \rightarrow H(\omega) = e^{-\frac{(\omega\tau)^2}{2}} \tag{15.11}$$

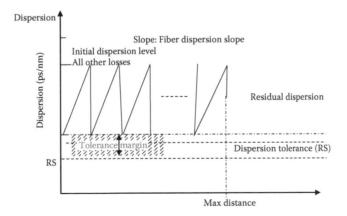

FIGURE 15.5 Dispersion budget by graphical representation for multi-span transmission system.

---

[2] See Appendix–Chapter 13 impulse responses of single mode optical fibers.

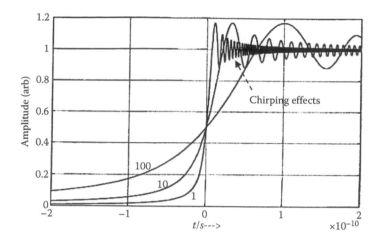

FIGURE 15.6 Fiber step response for single mode optical fiber of distance 1, 10, and 100 km and chirp effects near the edge.

The BW of this system can be found by setting the absolute value of the $H(\omega)$ to $1/(\sqrt{2})$ and solving for ω. Thence, the 3-dB bandwidth and the rise time of the system can be found with the relationship:

$$\omega_{3dB} = \frac{0.83}{\tau} \qquad (15.12)$$

If the link is made up by N fiber spans of the same dispersion factor, then each fiber length in tandem has a transfer function of

$$H_n(\omega) = e^{-\frac{N(\omega\tau)^2}{2}} \qquad (15.13)$$

The dependence of the transfer function as a square of the frequency indicates that whenever the bit rate is double the transmission distance is reduced by a factor of four. This is the "rule of thumb" for fiber transmission and for setting the dispersion tolerance as a function of bit rate.

Note that the design described in this section is given as a preliminary design to determine the selection of optical systems and electronic property of the receiver and the electrical driving unit of the transmitter.

Power budget and dispersion budget are used commly for the preliminary design. Accurate design details must be conducted by simulation and/or by laboratory demonstration prior to field installation.

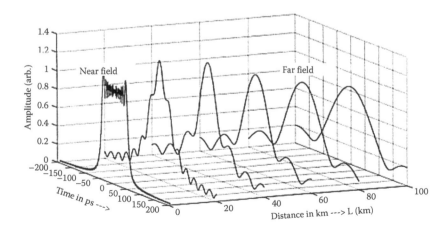

FIGURE 15.7 Gaussian impulse response of single mode optical fibers—near field and far field.

Stressed span loss (dB)	24.95
OSNR margin to $Q = 7$	3.88
Aging allowance	−4.00
w.c Rx/Tx correction	−1.90
w.c distortion dB $Q \times 2$	−0.60
Power control error	−1.50
Sum degradations (dB)	−8.00
System margin (dB)	−4.12
Guaranteed max. span loss (dB)	20.83

FIGURE 15.8   Tabulated arrangement of loss-per-span in optical transmission (in dB).

15.2.6   System Preliminary Design under Non-Linear Effects

15.2.6.1   Link Budget Measurement

Based on measuring and estimating margins on a close-to worst-case system, it is required to have a guarantee end-of-life (EOL) BER $>10^{-12}$.

Figure 15.8 shows the layout of the power budget in which an OSNR in dB of 3.88 is accounted for a Q-factor of 7 (BER = 1e-12) and allowances for aging, corrections of the transmitter and receiver, and distortion, as well as the error in power control, for example.

15.2.6.2   System Margin Measurement

The magnitude of any impairment that degrades the system to $Q = 7$ can be used as a metric for system margin. Thus, any distortion by linear and non-linear dispersion effects and/or noise contribution to the transmission of lightwave channels must be estimated and allowance of degradation of the eye diagram must be allocated in the power budget. Figure 15.9 shows various effects that would degrade the eye diagram and the aspects of noise measurement for link budget estimation.

Figure 15.9 shows the effects of the noise margin on the quality factor Q, especially the slope of the BER. The noise is loaded when cascaded optical amplifiers (MOR and MSA) are active. These noises are accumulated along the multi-span transmission line and hence superimposed on the received voltage levels. The bandwidth of the ASE noises is very wide and they are mixed with the optical signals

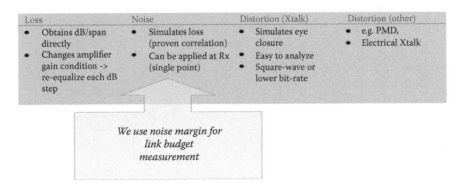

FIGURE 15.9   Budget measurement.

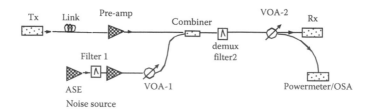

FIGURE 15.10   Arrangement for noise margin measurement—ASE = noise sources; VOA = variable optical attenuator; OSA = optical spectrum analyzer.

and noise levels in the detected eye diagram. This would decrease the degree of eye opening and some time we would measure the degradation by using the term eye opening penalty (EOP).

15.2.6.2.1   Noise Margin

A noise margin measurement can be conducted by using an ASE noise source superimposing on the optical signals, as shown in Figure 15.10. Thence, the BER and Q-factor can be determined under unloaded and loaded environments.

15.2.6.2.2   Dispersion Map

The dispersion map of the transmission system under noise margin is also important and an arrangement for such measurement can be seen in Figure 15.11a. A dispersion compensator can be inserted after the transmitter whose dispersion factor can be varied so as to determine the dispersion tolerance and the noise margin level at a specific BER. Figure 15.11b shows a typical noise margin versus dispersion

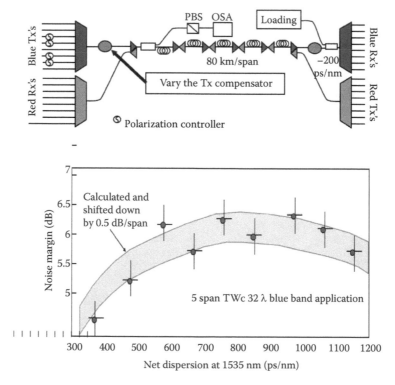

FIGURE 15.11   Dispersion map testing. TWc = total wavelength channels. Blue band = lower spectral region as compared to Red band in the C-band.

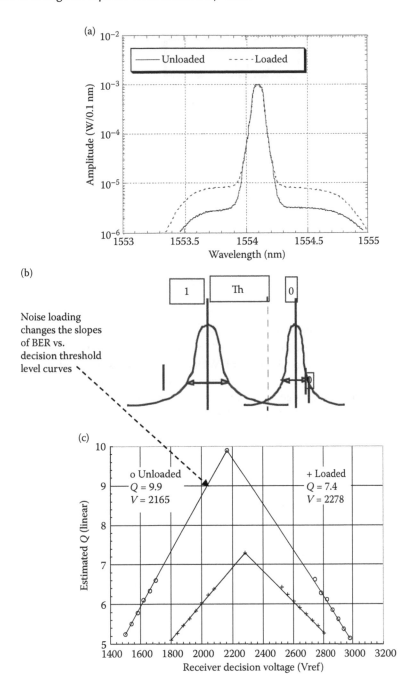

FIGURE 15.12 Noise margin measurement. (a) Noise spectra under unloading and loading condition, (b) probability distribution, and (c) BER versus decision voltage level.

for a wavelength channel. These measurements can be done by simulation. That is why a good and accurate simulation package is important for evaluation of modern optical fiber communication systems. Figure 15.12 shows the noise margin measurement of typical transmission system. Figure 15.12 (a) gives the noise spectra under unloading and loading condition, Figure 15.12 (b) the probability distribution, and Figure 15.12 (c) BER versus decision voltage level. Furthermore Figure 15.13 shows the details of Q factor as a function of sampling and decision point (voltage) of the eye diagram.

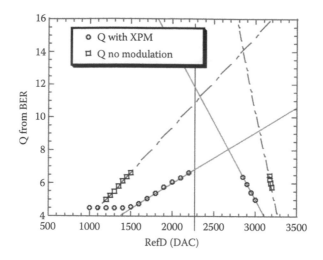

FIGURE 15.13 Details of Q factor as a function of sampling and decision point (voltage) of the eye diagram.

15.2.6.2.3 Worst Case Distortion

The worst case includes "inaccessible distortion" for worst-case model: For example, FWM, power fluctuation. This worst case is very critical for DWDM in which the wavelength channels are equally spaced, that is, when the interaction of three waves to create the fourth wave located exactly on another channel wavelength. This generates the unwanted crosstalk. Two rows of the worst case penalty are inserted in the table of Figure 15.14 to account for these circumstances. Furthermore

(a)

Stressed span loss (dB)	24.95
OSNR margin to $Q = 7$	3.88
Aging allowance	−4.00
w.c Rx/Tx correction	−1.90
w.c. distortion dB $Q \times 2$	−0.60
Power control error	−1.50
Sum degradations (dB)	−8.00
System margin (dB)	−4.12
Guaranteed max. span loss (dB)	20.83

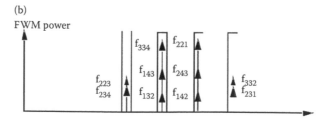

FIGURE 15.14 Worst case distortion illustrated (a) table of power budget and (b) FWM and variation of cross talk due to fluctuation of power level.

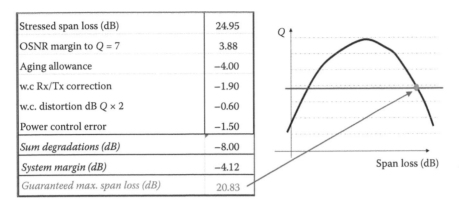

Stressed span loss (dB)	24.95
OSNR margin to $Q = 7$	3.88
Aging allowance	−4.00
w.c Rx/Tx correction	−1.90
w.c. distortion dB $Q \times 2$	−0.60
Power control error	−1.50
Sum degradations (dB)	−8.00
System margin (dB)	−4.12
Guaranteed max. span loss (dB)	20.83

FIGURE 15.15 Guarantee of span loss for quality transmission.

Figure 15.15 gives the crossing of the minimum levels and the effects of the span loss on power budget, thus gives the guarantee of span loss for quality transmission.

15.2.6.2.4 Guarantee of Maximum Span Loss
Finally, the maximum span loss can be determined to ensure that the loss of the span during installation must be met in order to achieve quality transmission. This can also be proven by (i) verification agrees with analysis so that the loss/span is within ± 0.5 dB/span (1-sigma) and the dispersion window is within ± 30 ps/nm (1-sigma) and the provisioned peak power is within ±0.25 dBm (1-sigma). The verifications are reproducible within ± 0.3 dB/span and on average verification is 0.5 dB/span better than analysis.

15.2.6.2.5 From Modeled Budget to Installation Budget
Verified modeled budgets are further adjusted to account for equipment behaviour and installation and operating procedures, for example, super-decoder penalty and equalization penalty. The verification of the modeled budget and the installed budget can be confirmed from experience of installed transmission system. The system operating margin can also be reduced from statistical data obtained from these installed transmission systems.

15.2.7 SOME NOTES ON THE DESIGN OF OPTICAL TRANSMISSION SYSTEMS

In modern optical transmission systems, the design procedures depending on specific applications need to (i) employ basic DWDM for Tb/s transmission principles and clarify the DWDM Tb/s modulation for Tx; (ii) deal with dispersion and losses in Tb/s transmission system design: advanced fibres; (iii) set engineering rules for impacts due to non-linearity effects; and (iv) examine optical amplifiers: distributed and lumped types for system design rules.

15.2.7.1 Allocations of Wavelength Channels
The DWDM channel spacing and locations of the frequencies of the channel follow the ITU standard. Such locations can be observed in Table 15.2. The frequency and frequency spacing between channels can be estimated by

$$f = \frac{c}{\lambda}$$

$$\Delta f = -\frac{c}{\lambda^2}\Delta\lambda = -\frac{f}{\lambda}\Delta\lambda \quad (15.14)$$

Thus, by selecting a reference wavelength, for example, the absorption wavelength of a chemical substance and hence the frequency of such channel, the frequencies of other wavelength channels can

TABLE 15.2
ITU Wavelength Grid for Dense Wavelength Division Multiplexing

	L-Band				C-Band				S-Band			
	100 GHz Grid		50 GHz Offset		100 GHz Grid		50 GHz Offset		100 GHz Grid		50 GHz Offset	
	THz	nm	THz	nm	THz	nm	THz	nm	THz	nm	THz	nm
1	186.00	1611.78	186.05	1611.35	191.00	1569.59	191.05	1569.18	196.00	1529.55	196.05	1529.16
2	186.10	1610.92	186.15	1610.48	191.10	1568.77	191.15	1568.36	196.10	1528.77	196.15	1528.38
3	186.20	1610.05	186.25	1609.62	191.20	1567.95	191.25	1567.54	196.20	1527.99	196.25	1527.60
4	186.30	1609.19	186.35	1608.76	191.30	1567.13	191.35	1566.72	196.30	1527.21	196.35	1526.82
5	186.40	1608.32	186.45	1607.89	191.40	1566.31	191.45	1565.90	196.40	1526.43	196.45	1526.04
6	186.50	1607.46	186.55	1607.03	191.50	1565.49	191.55	1565.08	196.50	1525.66	196.55	1525.27
7	186.60	1606.60	186.65	1606.17	191.60	1564.67	191.65	1564.27	196.60	1524.88	196.65	1524.49
8	186.70	1605.74	186.75	1605.31	191.70	1563.86	191.75	1563.45	196.70	1524.11	196.75	1523.72
9	186.80	1604.88	186.85	1604.45	191.80	1563.04	191.85	1562.63	196.80	1523.33	196.85	1522.94
10	186.90	1604.02	186.95	1603.59	191.90	1562.23	191.95	1561.82	196.90	1522.56	196.95	1522.17
11	187.00	1603.16	187.05	1602.73	192.00	1561.41	192.05	1561.01	197.00	1521.78	197.05	1521.40
12	187.10	1602.31	187.15	1601.88	192.10	1560.60	192.15	1560.20	197.10	1521.01	197.15	1520.63
13	187.20	1601.45	187.25	1601.02	192.20	1559.79	192.25	1559.38	197.20	1520.24	197.25	1519.86
14	187.30	1600.60	187.35	1600.17	192.30	1558.98	192.35	1558.57	197.30	1519.47	197.35	1519.09
15	187.40	1599.74	187.45	1599.31	192.40	1558.17	192.45	1557.76	197.40	1518.70	197.45	1518.32
16	187.50	1598.89	187.55	1598.46	192.50	1557.36	192.55	1556.95	197.50	1517.93	197.55	1517.55
17	187.60	1598.04	187.65	1597.61	192.60	1556.55	192.65	1556.15	197.60	1517.16	197.65	1516.78
18	187.70	1597.18	187.75	1596.76	192.70	1555.74	192.75	1555.34	197.70	1516.40	197.75	1516.01
19	187.80	1596.33	187.85	1595.91	192.80	1554.94	192.85	1554.53	197.80	1515.63	197.85	1515.25
20	187.90	1595.48	187.95	1595.06	192.90	1554.13	192.95	1553.73	197.90	1514.86	197.95	1514.48
21	188.00	1594.64	188.05	1594.21	193.00	1553.32	193.05	1552.92	198.00	1514.10	198.05	1513.72
22	188.10	1593.79	188.15	1593.36	193.10	1552.52	193.15	1552.12	198.10	1513.33	198.15	1512.95
23	188.20	1592.94	188.25	1592.52	193.20	1551.72	193.25	1551.31	198.20	1512.57	198.25	1512.19
24	188.30	1592.10	188.35	1591.67	193.30	1550.91	193.35	1550.51	198.30	1511.81	198.35	1511.43
25	188.40	1591.25	188.45	1590.83	193.40	1550.11	193.45	1549.71	198.40	1511.05	198.45	1510.67

(Continued)

TABLE 15.2 (Continued)
ITU Wavelength Grid for Dense Wavelength Division Multiplexing

	L-Band				C-Band				S-Band			
	100 GHz Grid		50 GHz Offset		100 GHz Grid		50 GHz Offset		100 GHz Grid		50 GHz Offset	
26	188.50	1590.41	188.55	1589.98	193.50	1549.31	193.55	1548.91	198.50	1510.28	198.55	1509.90
27	188.60	1589.56	188.65	1589.14	193.60	1548.51	193.65	1548.11	198.60	1509.52	198.65	1509.14
28	188.70	1588.72	188.75	1588.30	193.70	1547.71	193.75	1547.31	198.70	1508.76	198.75	1508.38
29	188.80	1587.88	188.85	1587.46	193.80	1546.91	193.85	1546.51	198.80	1508.01	198.85	1507.63
30	188.90	1587.04	188.95	1586.62	193.90	1546.11	193.95	1545.72	198.90	1507.25	198.95	1506.87
31	189.00	1586.20	189.05	1585.78	194.00	1545.32	194.05	1544.92	199.00	1506.49	199.05	1506.11
32	189.10	1585.36	189.15	1584.94	194.10	1544.52	194.15	1544.12	199.10	1505.73	199.15	1505.36
33	189.20	1584.52	189.25	1584.10	194.20	1543.73	194.25	1543.33	199.20	1504.98	199.25	1504.60
34	189.30	1583.68	189.35	1583.27	194.30	1542.93	194.35	1542.53	199.30	1504.22	199.35	1503.84
35	189.40	1582.85	189.45	1582.43	194.40	1542.14	194.45	1541.74	199.40	1503.47	199.45	1503.09
36	189.50	1582.01	189.55	1581.60	194.50	1541.34	194.55	1540.95	199.50	1502.71	199.55	1502.34
37	189.60	1581.18	189.65	1580.76	194.60	1540.55	194.65	1540.16	199.60	1501.96	199.65	1501.59
38	189.70	1580.35	189.75	1579.93	194.70	1539.76	194.75	1539.37	199.70	1501.21	199.75	1500.83
39	189.80	1579.51	189.85	1579.10	194.80	1538.97	194.85	1538.58	199.80	1500.46	199.85	1500.08
40	189.90	1578.68	189.95	1578.27	194.90	1538.18	194.95	1537.79	199.90	1499.71	199.95	1499.33
41	190.00	1577.85	190.05	1577.43	195.00	1537.39	195.05	1537.00	200.00	1498.96	200.05	1498.58
42	190.10	1577.02	190.15	1576.61	195.10	1536.60	195.15	1536.21	200.10	1498.21	200.15	1497.83
43	190.20	1576.19	190.25	1575.78	195.20	1535.82	195.25	1535.42	200.20	1497.46	200.25	1497.09
44	190.30	1575.36	190.35	1574.95	195.30	1535.03	195.35	1534.64	200.30	1496.71	200.35	1496.34
45	190.40	1574.54	190.45	1574.12	195.40	1534.25	195.45	1533.85	200.40	1495.97	200.45	1495.59
46	190.50	1573.71	190.55	1573.30	195.50	1533.46	195.55	1533.07	200.50	1495.22	200.55	1494.85
47	190.60	1572.88	190.65	1572.47	195.60	1532.68	195.65	1532.28	200.60	1494.47	200.65	1494.10
48	190.70	1572.06	190.75	1571.65	195.70	1531.89	195.75	1531.50	200.70	1493.73	200.75	1493.36
49	190.80	1571.23	190.85	1570.82	195.80	1531.11	195.85	1530.72	200.80	1492.99	200.85	1492.61
50	190.90	1570.41	190.95	1570.00	195.90	1530.33	195.95	1529.94	200.90	1492.24	200.95	1491.87

The International Telecommunication Union (ITU) established the following grid for Dense Wavelength Division Multiplexing systems. ITU recommends the 100 GHz (0.8 nm) spacing for DWDM operation. The 50 GHz offset (0.4 nm spacing) offers twice the bandwidth. The grid is centered at 193.10 THz (1552.52 nm). Note the grids of 50 GHz and 100 GHz spacing which are interleaved.

be estimated. Channel spacing of 50 GHz is now the normal standard for 10 and 40 Gb/s bit rate in which 44 channels can be accommodated in the C-band.

The number of WDM channels possible depends on (i) channel spacing; (ii) amplifier spectral range; (iii) source spectral width; (iv) bit rate per optical channel and modulation formats; (v) fiber dispersion; and (iv) multiplexer/demultiplexer range.

The number of channels can be estimated by $N = 1 + (\lambda_1 - \lambda_2/\Delta\lambda)$ with $\Delta\lambda$ = channel spacing, λ_1 and λ_2 are the two extreme wavelengths of the end channels.

Example: λ_1, $\lambda_2 = 1540$–1560 nm and a channel spacing of 1.6 nm (200 GHz), then N = 12.5 channels. Similarly, for 1535 and 1565 nm and spacing of 50 GHz, then N ∼ 44.

15.2.7.2   The Link Design Process

- Chose an application means: know your line rate and fiber type whether it is a currently installed fiber system required upgrading or newly installed transmission system
- Determine what amplifier configuration meets the requirements
- Check to see if the application is supported, for example, link budgets and optical ADD/DROP Multiplexer (OADM) rules are available
- Remember in span loss calculations to consider the minimum span loss: If you have to pad to reach the minimum span loss, the value after padding is used in all calculations

Choosing an application means: (i) knowing your line rate and fiber type; (ii) determining what amplifier configuration meets the requirements; and (iii) checking to see if the application is supporting whether link budgets and OADM rules are available.

15.2.7.3   Link Budget Considerations

15.2.7.3.1   Forward Error Coding

Forward error coding (FEC) is normally used at the transmitter to increase the chance of detection without error. In this process, a number of redundant bits are used and coded in the data sequence.

- FEC adds dB/span and reduces BER.
- FEC benefits are system dependent and may vary from application to applications.
- If FEC is to be used, check standards for specific information.
- FEC Reed-Solomon codes – 42.7 Gb/s for 40 Gb/s base rate; 10.7 Gb/s for 9.953 Gb/s.

15.2.7.3.2   Excess Loss (EL) – Operating Margin

- Calculating the *EL* (in dB):
  L = actual span loss; L_{total} = sum of actual losses (including fixed pads if any); n = number of spans; L_{max} = link budget max dB/span (with/without FEC)

$$EL = \sum_{L_{\text{max}} < L < L_{\text{max}}+2} L - L_{\text{max}} \tag{15.15}$$

Then the excess loss rules can be stated as

- $L < L_{\text{max}+2}$
- $L_{\text{total}} < n \times L_{\text{max}}$
- $EL > 4$ dB is not supported. FEC may be used to improve the SNR and thus increase the operating margin

$$\text{If } 2 < EL < 4 \Rightarrow L_{\text{total}} < n \times (L_{\text{max}} - 1)$$

15.2.7.3.3   Basic Steps in Design Transmission Systems

Fiber type and dispersion compensation modules:

Define the system's span losses and lengths: (i) What budgets and windows are available for this application? (ii) What extra margin (if any) must be added to the span losses? (iii) What BER is required? (iv) Is FEC an option or requirement? (v) Where the regenerator should be placed? For example to eliminate mixed fiber links where possible: (i) check the length windows to ensure there are DCM solutions available and (ii) check the losses using the excess loss rules: Attenuation Pad span losses to minimum.

If there is an OADM site, (i) check all sub-links for DCM windows and check Mid-Span-Access losses. This may place a limit on sites and number of add/drop wavelengths supported; (ii) determine the DCMs required and their locations; and (iii) take into account special OADM placement rules.

DCMs may be required at the Add Tx or Drop Rx: (i) Add Tx's may be required to have a specific chirp and (ii) use an MidSpan-Opt-Regen Plus (MOR Plus) if DCMs are required at a line amp site.

Calculate the MOR losses and add fixed pads where required.

Find the transmitter fixed pad information and note the fixed pad attenuator upgrade strategy.

Corning LEAF and Enhanced LEAF perform differently. (i) Using LEAF budgets provides a conservative but safe estimate for ELEAF (enhanced LEAF) links, however, DCMs and pads will be different. (ii) TrueWave Classic, TrueWave-RS, TrueWave+, and TrueWave− are not equivalent. (iii) Treat them as completely separate fiber types, as both link budgets and DCMs will vary between these different fiber types.

Note: Link designs for 100 GHz or 50 GHs do not apply to 200 GHz, as DWDM filter specifications are different. Use array waveguide gratings (AWG)—bidirectional filters, multiplexing, and demultiplexing.

15.2.8   LINK BUDGET CALCULATIONS UNDER LINEAR AND NON-LINEAR EFFECTS

Under the non-linear effects, the transmission performance needs to consider (i) distortion vs. noise; (ii) high optical power: distortion of the signal; and (iii) low power: low signal-to-noise ratio.

Figure 15.16 shows the variation of the Q-factor as a function of the launched power. When the launched power reaches the non-linear threshold, the distortion happens and degradation of the system performance occurs. Figure 15.17 shows the effects of the decision level on the Q factor. Gaussian distribution of the amplitude and noise levels are assumed. Optimum performance is achieved when the probability of error for detection "1" and "0" is the same.

15.2.8.1   Budget

When designing the transmission, we must keep in mind that link budgets are evaluated and verified for the system at: (i) worst case; (ii) end of life (EOL); and (iii) reference equalization (RefEQ).

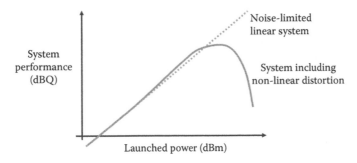

FIGURE 15.16   Variation of system performance as a function of the launched power—note linear and non-linear region.

(a)   (b)

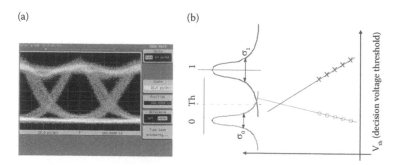

FIGURE 15.17   Eye diagram and extrapolation of Q-factor (a) received eye diagram and (b) extrapolation of Q-factor.

Budget guarantees include worst-case equalization (EOL): (a) based on worst-case implementation and power control errors over life; and (b) worst-case incorporates (i) <u>Transmitter</u>: waveform, chirp, power, wavelength, noise, and jitter; (ii) <u>Receiver</u>: complex transfer function, power, noise, jitter, and sampling; (iii) <u>Fiber</u>: dispersion, DGD, and loss; (iv) <u>Amplifiers</u>: noise figure, gain ripple and tilt, and output power; and (v) <u>System</u>: net dispersion, disp. map, channel separation, and polarization.

15.2.8.2   System Impairments

The impairments fall into two groups: distortion impairments and noise impairment:

a. Distortion impairment:
    Any bounded reduction of the inner part of the eye diagram within the sample window including (i) Rx distortions: transfer function related, (ii) Tx distortions: drive related, extinction ratio, (iii) optical path distortions: dispersion, SPM, FWM & XPM <5 channels; and (iv) PMD, cross talk, MPI

b. Noise impairment:
    Any unbounded stochastic reduction of the inner eye within the sample window including (i) receiver noise, (ii) Tx noise, (iii) optical amplifier noise, (iv) modulation instability noise; and (iv) FWM & XPM > 4 channels

c. Signal dependent noise contribution to Q is a function of distortion

    However, noise and distortion can be separated for budgeting purposes.

15.2.8.3   Power and Time Eyes

The bit error rate (BER) must include the amplitude or power eye and the time eye which may be due to jittering of the waveform on the sampling instant for evaluation of the system performance. This is illustrated in Figure 15.18. Figure 15.19 shows the eye diagram at both the transmitter and its degradation at the end of the transmission link.

15.2.8.4   Dispersion Tolerance due to Wavelength Channels and Non-Linear Effects

We must note that (a) distortion depends on net dispersion and hence on wavelength and bundled wavelength groups; (b) self-phase modulation makes distortion curves power dependent; and (c) budgets are guaranteed over a net dispersion window meaning that the dispersion windows are reduced by any errors in link dispersion values.

15.2.8.4.1   Dependence on Wavelength Channels and Launched Power

Figure 15.20 shows examples of the dispersion tolerance in terms of the variation of eye-opening penalty as a function of wavelength and launched power. It is expected that the worst performance must be targeted for the design that is the worst case scenario.

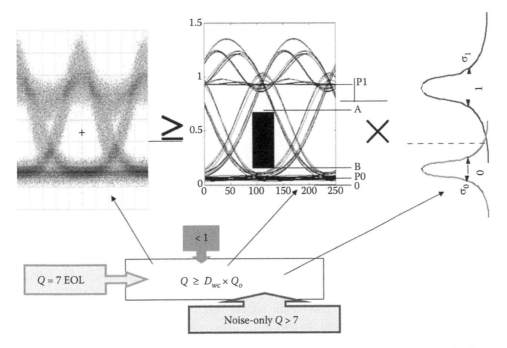

FIGURE 15.18 Illustrations of the eye diagram expected at the end of life (EOF) with the original Q factor and effects of jittering—or time Q.

The variation of the quality factor Q may also vary with respect to the total span length, that is, the optical amplification would contribute to the ASE noises and to the degradation of the quality of the system. This can be seen in Figure 15.21. Thus, the design of optical amplifiers is very important to achieving optimum performance of the transmission systems.

15.2.8.5 Example 4: Self-Coherent Six-Span Optically Amplified Link—Upgrading of Existing Transmission Systems with Standard Single Mode Optical Fibers

A transmission system requires:

- 6-span NDSF, 16 wavelength channels at 200 GHz spacing or 32 wavelength channels of 100 GHz spacing OC192 × 476 km total length
- Bidirectional transmission using one fiber only
- 1.4 dB of margin added to each calculated span loss

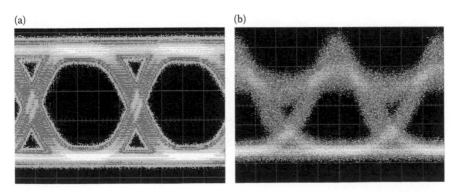

FIGURE 15.19 Eye at the transmitter output (a) and at the end (b) of the transmission link.

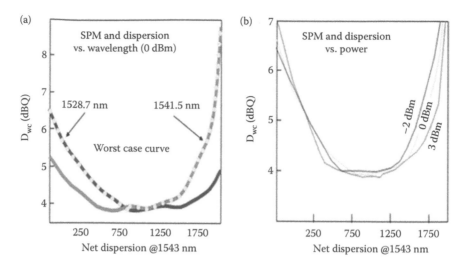

FIGURE 15.20 Eye-opening penalty versus dispersion as a function of (a) wavelength and (b) launched power.

- Fiber attenuation is 0.22 dB/km—single optical fiber
- Not using 1625 nm as reserved for online continuity monitoring

Two options, A and B, for the transmission systems are given below:

- Compare option A and option B
- Choose an application type
- Find DCM (dispersion compensation modules), fixed pad (attenuator), provisioning information for the final link design, thence select the cheapest possible solution (minimizing number of mid-span optical regenerator (MOR) as much as possible)
- Bidirectional transmission and compensation requires optical circulators that would then allow the dispersion compensation for wavelength channels of different spectral regions (reds and blues).
- Mid-span amplification: For DCM60, loss is 7.3 dB. $11 - 7.3 = 3.7$; therefore, use 4 dB attenuation pad. DCM80 $= 9.3$; therefore, use 2 dB attenuation pad

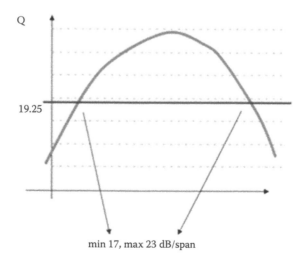

FIGURE 15.21 Variation of the quality factor Q with respect to the total span loss.

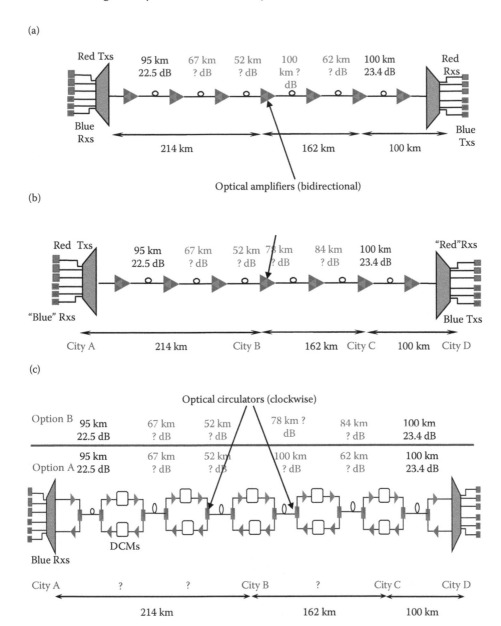

FIGURE 15.22 Design structures of multi-span link using bidirectional techniques (a) Option A, (b) Option B, and (c) detailed arrangement.

- Attenuation pads required for minimum span loss requirements as well. Example: 16.4 requires 2 dB attenuation pads etc. The design structures of multi-span link using bidirectional techniques (a) Option A, (b) Option B, and (c) detailed arrangement are given in Figure 15.22.

EXAMPLE 15.4: 16 WAVELENGTH CHANNEL 5-SPAN NDSF MOR BIDIRECTIONAL OPTICAL TRANSMISSION

We can now refer to Figures 15.22 through 15.25. With the same transmission distance as the example given above but with only five spans. Readers should inspect the design below and check to see whether they can confirm the design

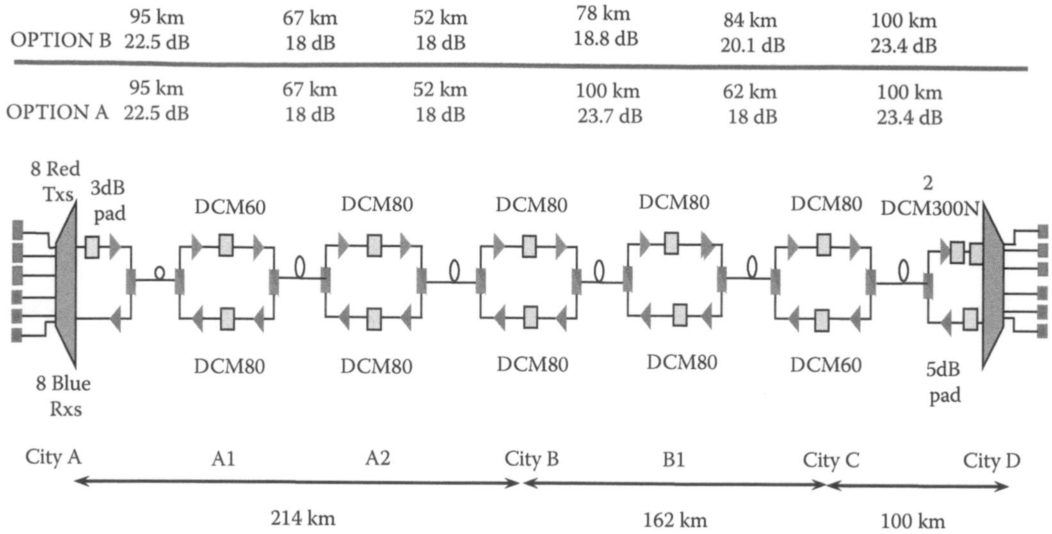

OPTION B
95 km 22.5 dB
67 km 18 dB
52 km 18 dB
78 km 18.8 dB
84 km 20.1 dB
100 km 23.4 dB
OPTION A
95 km 22.5 dB
67 km 18 dB
52 km 18 dB
100 km 23.7 dB
62 km 18 dB
100 km 23.4 dB
8 Red Txs
3dB pad
DCM60
DCM80
DCM80
DCM80
DCM80
2 DCM300N
DCM80
DCM80
DCM80
DCM80
DCM60
8 Blue Rxs
5dB pad
City A
A1
A2
City B
B1
City C
City D
214 km
162 km
100 km

FIGURE 15.23 Solution of 6-span link for 405 km distance via three cities.

2 DCM300N
8 Red Txs
3dB pad
DCM60
DCM80
DCM80
DCM80
DCM80
7/16 3/15 7/16 3/15 7/16 3/15 7/16 3/15 7/16 3/15 7/16 5/15
7/15 6/16 3/15 6/16 3/15 6/16 3/15 6/16 3/15 6/16 3/15 6/16
8 Blue Rxs
DCM80
DCM80
DCM80
DCM80
DCM60
5dB pad
City A
A1
A2
City B
B1
City C
City D
214 km
162 km
100 km

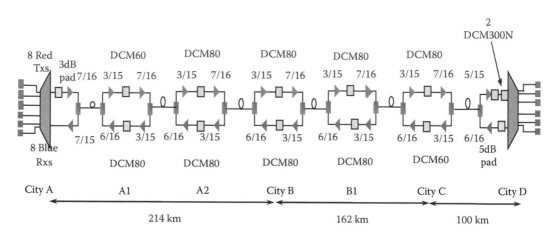

FIGURE 15.24 Transmission system solution with provisioning.

Working fiber (405 km)
Txs
Rxs
5 dB pad
DCM100N
70
DCM80
Site A
DCM80
90
DCM80 (drop)
Site B
DCM80 (add)
80
DCM80
Site C
DCM80
85
DCM80
Site D
DCM60
80
DCM300N
5 dB pad
Rxs
Txs

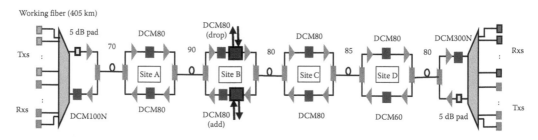

FIGURE 15.25 Transmission with provisioning for OADM insertion.

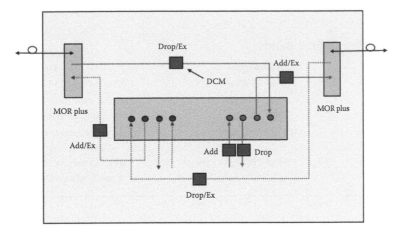

FIGURE 15.26   Structure of an optical add/drop mux (OADM). MOR = mid-span optical regenerator.

- Site A is the site closest line amplifier site to the RED transmitter (Tx)
- In each case, the chirp on the add/drop wavelengths must be positive (to compensate for the large negative DCMs within the add/drop sub-link)
- 2-span windows must be 154–166 km in length (155 km)
- 3-span windows must be 231–249 km in length (245 km)
- Rules apply for FEC with a BER of 10^{-12}

15.2.9   Engineering an OADM Transmission Link

The structure of an OADM is shown in Figure 15.26. Selected wavelength channels of the incoming multiplexed channels can be dropped or wavelength channels can be added to the transmission system, which can be observed in this diagram. The details of the optical components, such as optical filters and circulator are not shown in this section, but are observed from the system point of view.

Knowing the fiber and amplifier type and the number of wavelength channels, we can determine the express compensation rules for the link. MSA loss now includes DCMs and OADM couplers (attenuation pad up, if required). Furthermore, provisioning of powers of sites that changed amplifier type as a result of the OADM should also be made. Then determine the OADM compensation rules (in both power budget and dispersion) for the link. One should also apply these rules for both the working and protection fiber transmission paths.

APPENDIX 15A: WAVELENGTH TRANSMISSION BAND IN FIBER OPTIC NETWORK

As fiber optic networks have developed for longer distances, higher speeds and wavelength-division multiplexing (WDM) fibers have been used in new wavelength ranges, now called "bands," where fiber and transmission equipment can operate more efficiently. Single mode fiber (SMF) transmission began in the "O-band" (see Table 15.3) just above the cut-off wavelength of the SM fiber developed to take advantage of the lower loss of the glass fiber at longer wavelengths and availablility of 1310 nm diode lasers. (Originally, SMFs were developed for 850 nm lasers where the fiber core was about half what it is for today's conventional SMF (~5 microns as opposed to ~8–9 microns at 1310 nm.)

To take advantage of the lower loss at 1550 nm, fiber was developed for the C-band. As links became longer and fiber amplifiers began being used instead of optical-to-electronic-to-optical repeaters, the C-band became more important. The advent of DWDM (dense wavelength-division multiplexing) allowed multiple signals to share a single fiber by multiplexing. The use of this band was expanded and occupied by 110 channels with 50 GHz (~0.4 nm) spacing. Development

TABLE 15.3

Classifications of the Optical Bands

Band and Name	Wavelengths	Description
O-band	1260–1360 nm	Original band, PON upstream
E-band	1360–1460 nm	Water peak band
S-band	1460–1530 nm	PON downstream
C-band	1530–1565 nm	Lowest attenuation, original DWDM band, compatible with Er:doped fiber amplifiers
L-band	1565–1625 nm	Low attenuation, expanded DWDM band
U-band	1625–1675 nm	Ultra-long wavelength

of new fiber amplifiers (Raman and thullium-doped) promise to expand DWDM upward to the L-band.

Several low-cost versions of WDM are in use, and aregenerally referred to as Coarse WDM or CWDM. Most do not work over long distances so do not require amplification, broadening the wavelength choice. The most popular is FTTH PON systems, which send signals downstream to users at 1490 nm and use low-cost 1310 nm transmission upstream. Early PON systems also use 1550 downstream for TV, but that is being replaced by IPTV on the downstream digital signal at 1490 nm. Other systems use a combination of S-, C-, and L-bands to carry signals because of the lower attenuation of the fiber. Some systems even use lasers at 20 nm spacing over the complete range of 1260–1660 nm but only with low water peak fibers.

Manufacturers have been able to make fiber with low-water peaks, opening up a new transmission band (E-band), but it has not yet proven useful except for CWDM. It is probably mostly useful as an extension of the O-band, but few applications have been proposed and it is very energy-intensive for manufacturers.

The C-band can now be planned to extend to 1568 nm to match with the extension of the C-band Erbium-doped fiber amplifier to allow 110 channels with 50 GHz channel spacing.

REFERENCE

1. L.N. Binh *Digital Optical Communications*, CRC Press: Boca Raton, FL, USA, 2008.

Index